Mason's World Dictionary of Livestock Breeds, Types and Varieties

5th Edition

Revised by Valerie Porter

CABI *Publishing*

CABI *Publishing* is a division of CAB *International*

CABI Publishing
CAB International
Wallingford
Oxon OX10 8DE
UK

Tel: +44 (0)1491 832111
Fax: +44 (0)1491 833508
Email: cabi@cabi.org
Web site: www.cabi-publishing.org

CABI Publishing
10 E 40th Street
Suite 3203
New York, NY 10016
USA

Tel: +1 212 481 7018
Fax: +1 212 686 7993
Email: cabi-nao@cabi.org

A catalogue record for this book is available from the British
Library, London, UK.

Library of Congress Cataloging-in-Publication Data
Mason's world dictionary of livestock breeds, types and varieties / revised by
Valerie Porter.-- 5th ed.
      p. cm.
   Rev. ed. of: A world dictionary of livestock breeds, types, and varieties /
I.L. Mason. 4th ed. 1996.
   Includes bibliographical references (p.).
   ISBN 0-85199-430-X (alk. paper)
      1. Livestock breeds--Dictionaries. I. Porter, Valerie, 1942- II. Mason,
   I. L. (Ian Lauder), 1914- World dictionary of livestock breeds, types, and
   varieties.
SF105.M34 2002
636′003--dc21

                 2001052554

ISBN 0 85199 430 X

Typeset in Melior by Columns Design Ltd, Reading
Printed and bound in the UK by Cromwell Press, Trowbridge

Contents

Acknowledgements

Ian Mason has edited every edition of the *World Dictionary* since he conceived and compiled the first edition published in 1951. It has become the most reliable and authoritative reference work of its kind and is acknowledged and respected worldwide. After 50 years, however, he decided that he could no longer continue with what had been his life's work and I was honoured to be asked by him to edit this fifth edition on his behalf.

Above all, I want to thank him for creating the *World Dictionary* in the first place and maintaining the high standards he set for himself from the start. I am eternally grateful to him for helping me personally with endless patience, charm and hospitality in preparing to edit this 5th edition and also in answering, with the greatest courtesy and clarity, so many questions when I have been writing my own books on livestock breeds worldwide.

In addition, I should like to thank the many people in many countries who have answered my queries and those who have sent me publications about livestock in their country. Among them I should especially like to mention: John P. Boulle, Quentin Campbell and Graham Hallowell (South Africa); Wanda Feldman (Sopexa, Paris); Dr Mariano Gómez (Basque breeds); Håkan Hallander (Sweden); John Hodges (European Association for Animal Production); Richard Lutwyche (Rare Breeds Survival Trust, UK); Dr Théa Machado (Brazil); Toshinori Omi (Japan); Dr Christie Peacock (FARM-Africa, London and Addis Ababa); and Karin Wiren (Sweden).

I am also hugely indebted to Beate Scherf and her staff at FAO Rome, who have been generous in sharing information and who have worked on the *World Watch List for Domestic Animal Diversity* (*WWL-DAD*). The third edition is a massive one and has been an invaluable source, particularly concerning endangered breeds.

Finally, I must thank Rebecca Stubbs at CAB *International* for her endless patience in extending deadlines to accommodate the large amount of extra time needed for my research and pursuit of answers to queries from contacts worldwide.

Introduction

The aim of this dictionary is to list the livestock names and synonyms that may be encountered in the literature, including those not only of 'breeds' but also of other identifiable groups, types and varieties that have a common origin, or are similar in appearance or geographically linked. The main purpose is to identify each breed, type or variety and to suggest, for each of them, a name to be used in the English language.

Definition of 'Breed'

Inclusion of a name does not necessarily imply that the named group is a 'breed'. It may be no more than a local population, or a type seen in certain areas. Formally, a 'breed' is acknowledged as such when a breed society is formed to promote it or a herdbook is published. Less formally, a 'breed' (defined variously by different authors) can be described as a group of domesticates that, when bred to each other, will regularly produce offspring that are recognizably of that breed, both morphologically and in their aptitudes. Over the many centuries of livestock breeding, the criteria for identifying a breed have often been superficial and perhaps arbitrary factors such as the colour and pattern of the coat, the set of the ears or the shape of the horns. Colour, in particular, is not a reliable guide, and very often several disparate types have been lumped together simply because, for example, they are red or colour-pointed. With the rapid growth of information from gene mapping, it is becoming possible to identify breeds and their relationships with other breeds far more accurately, based on DNA rather than simply on 'looks' and known history.

Background

In order to understand the aim of this dictionary, to explain its form, and to excuse its omissions, it is necessary to consider how the idea of compiling it arose and how it has developed.

Quite early in the history of *Animal Breeding Abstracts* (*ABA*, first published in 1933 by CAB *International*), it became clear that consistency in terminology was desirable if the abstracts were to have their full value. The names of breeds of livestock formed one of the most important categories in which uniformity was essential. In preparing abstracts from many languages for a publication that was in the English language, the problem arose of how to render foreign breed names into English. At the same time even English names were liable to have several different spellings or alternative forms. A standard list was necessary to show the English forms and spellings recommended for use in *ABA*.

The next development was an attempt to avoid the confusion that can arise when the same breed has entirely different names in different parts of the world. The most notable examples of this are Shorthorn cattle (also known as Durham) and Large White pigs (also known as Yorkshire).

In order to decide which names were synonymous it was necessary to consider the characteristics and distribution of the breeds listed, and these details were included in the *World Dictionary*. To save space, and because comparable information about all breeds is hard to obtain, descriptions in this dictionary have been restricted to 'breed characters' such as colour and horns – those traits that are least affected by the environment. No attempt has been made to compile an encyclopaedia containing a complete account of each breed. The work has remained at the stage of a dictionary and, as such, it lists names that have been applied to groups of asses, buffalo, cattle, goats, horses, pigs and sheep, indicates which names are synonymous and recommends one form for English use. A brief indication is given of place of origin or present distribution, and of economic use, relationships with other groups, basic breed characters, the origin of the name and, where relevant, an indication of rarity or extinction.

Within each entry, the order in which the information is given is, where appropriate: name recommended for use in English; area or country of origin; uses; breed characters; relationship with other breeds (grouping or as variety of; origins and development; varieties or others derived from); formal recognition (naming, acceptance, breed societies and herdbooks); other names (in language of country of origin, other synonyms, proscribed names); etymology; status (if rare or extinct); and other notes and cross-references.

Names

Spelling

Many breeds are named after geographical features. For these, the spellings used are those in the *Times Atlas of the World* (1980). An exception has been made in the case of Russian names, for which the simpler system of 'anglicization' used by the National Geographic Society of America has been used (see p. xviii).

Typeface

- **Normal (roman) type** is used for existing and extinct breeds, varieties, populations, crossbreds (not true-breeding) and wild species (which may be recognized by the absence of an initial capital letter and by being followed immediately by their italicized scientific name).
- **Small capitals** (e.g. ENGLISH LONGWOOL) are used for geographical and breed groups or 'types'.
- **Italics** are used for synonyms, alternative spellings, obsolete names, misspellings, or foreign forms (and their translations).

Place of Breeding

The geographical name in parentheses after the breed name refers primarily to the place of origin of the breed. For **improved breeds**, this is clear enough; they may now have a worldwide distribution and this is indicated by the countries in which they have herdbooks, breed societies, varieties or derivatives. For **native breeds**, however, it has often been possible to give only the present distribution, whilst assuming that the place of origin was in the same area. In either case only the country, province or county may be given, with the more exact place of origin being shown by the name of the breed itself or by its synonyms or origins, e.g. 'Shorthorn (NE England) ... orig. from Holderness + Teeswater; syn. *Durham*'.

In the case of **synonyms**, the place in parentheses is the place in which this particular synonym is used.

Uses of Breeds

The abbreviations immediately after the parentheses that enclose the country name indicate the functions or 'aptitudes' of the breed, or the products obtained from it. Products in **square brackets** are past history; for example, '[dr]' indicates a former draught breed.

Horses are classified as heavy (**h**), light (**l**) or pony (**py**). In addition to recognized ponies, 'py' is used to describe any small horse, i.e. one that is, on average, less than 14 hands (140 cm) high at the withers. Horse uses are given as riding (**ri**), pack (**pa**) or draught (**dr**). **Ass** uses are similar but also include breeding for mules (**mu**).

Pigs are classified as lard (**ld**) or meat (**m**).

Cattle, buffalo and goats are described as producers of work (**dr**, **pa**, **ri**), milk (**d**), meat (**m**) and, for goats, mohair (**mo**) or cashmere (**ca**). For dual- or triple-purpose breeds the products are listed, as far as possible, in order of importance. Abbreviations do not necessarily describe conformation type. For example, the abbreviation '**d**' means only that the breed is used for milk; it does not necessarily indicate a breed of dairy type and can equally describe a low-yielding unimproved native breed.

Sheep are described as producers of milk (**d**), meat (**m**) or wool. Wool sheep are divided into:

- finewool Merino type (**fw**);
- carpet wool type (**cw**); or
- intermediates, subdivided where possible into longwool (**lw**), shortwool (**sw**), or crossbred (Merino × longwool) type (**mw**).

When information is insufficient to permit subclassification in this way, the description is simply '**w**'.

Breed Characters

Descriptions are confined to clear-cut morphological characters. Descriptions that are implicit in the breed name (e.g. Red Poll) are not always repeated. Descriptions of varieties of a breed, or breeds of a breed group, are confined to differences from the parent type.

Size

No attempt has been made to give a measure of size, however desirable this might be. Size is so much influenced by age, sex, feeding, management and environment that many figures would have to be given before it would be possible to compare different breeds. Dwarf breeds (and ponies) are noted, and sometimes there is an indication of other particularly small breeds – always bearing in mind that the latter might grow larger in a better environment.

Colour

If no colour is given, it is assumed that the colours are as follows.

- Buffalo: black, or at least dark
- Pigs: white
- Sheep: white
- Other species: no characteristic breed colour.

The colour given is the one that is typical for the breed but of course off-types occur in most breeds. The description refers to the adult female; the colour at birth may be different and, in cattle, bulls are often darker than cows.

Patterns are not always described in detail. For example, it is not stated that grey or grey-white cattle are generally darker on the shoulders than on the rest of the body, nor that many brown cattle shade to a darker colour at their extremities.

Coat

The texture of the coat is described only if it is exceptional in some way, e.g. absent (as in hairless pigs), very long, curly, or (for sheep) hair instead of wool.

Horns

Cattle

It should be assumed that cattle breeds are horned unless they are specifically described as polled (**pd**). Differences in horn size between the sexes are not mentioned but in general the cow and the castrate male have longer horns than the bull.

Goats

It is assumed that goat breeds are horned unless specifically described as polled. Between the sexes, males generally have longer horns than females.

Sheep

Owing to the wide variation among sheep, the presence or absence of horns in each sex (♂, male; ♀, female) is noted wherever the information is available. Often males are horned (**hd**) but females are polled (**pd**). As with goats, males of horned breeds generally have longer horns than females.

Tails (sheep)

Sheep tails are classified as:

- long (**lt**), i.e. down to the ground;

- short (**st**), i.e. less than half way to the hocks; or
- fat (**ft**), which may also be long (**lft**).

If the tail is not mentioned it may be assumed that it is of the thin, medium-length type so common among European breeds. No distinction is made between the docked and undocked breeds having this type of tail.

Relationship to Other Breeds

This includes origin (if known) and derivatives or varieties.

Date of origin

If two dates are given for the origin of a breed, the first is the year when selection or cross-breeding began; the second is the year when the new type was fixed or the name recognized officially. When only a single date is given it usually refers to the latter event.

Origin by cross-breeding

When a new breed has been developed by crossing two others, the founder breed names are joined in the usual way by the multiplication sign (×) and the male parental breed is put first (e.g. 'Oxford Down ... orig. ... from Cotswold × Hampshire Down'). Contrary to the usual mathematical convention, 'and' should be taken before '×' (e.g. 'orig. from Swiss Brown and Simmental, × Ukrainian Grey').

Origin by combination

The formation of a new breed by combining two existing breeds or varieties and treating them as one is indicated by the addition sign (**+**) (e.g. 'Swedish Red-and-White ... orig. from Swedish Ayrshire + Red Pied Swedish').

Formal Recognition

Breed society

If there is a society, association or club devoted to the promotion of the breed, this is indicated by '**BS**'. The date following indicates the year of its formation. Where there are breed societies in countries other than that of the breed's origin, the countries are listed, with the relevant date of formation in that country where known.

Breed standards

If there is no special society but the breed is recognized by the government and a breed standard has been established, this is indicated by '**BSd**'.

Herdbooks

If pedigrees are published in a studbook, flockbook or herdbook, this is indicated by '**HB**', followed by the date of publication of the first volume where known. Herdbooks in other countries are also listed. Sometimes early pedigrees were published in a herdbook that covered several breeds, so the date given may be earlier than the first volume of the breed society herdbook. Confusion may also have arisen here because of the use in some foreign languages of the English word 'herdbook' to mean breed society.

Synonyms

Synonyms are usually preceded by the official name of the breed in the language(s) of its country of origin. This applies particularly in the case of the common European languages, and when the recommended English name is different from the official vernacular name (e.g. with Basque breeds).

Cyrillic names (Russian, Bulgarian, Serbian) have been transliterated (see p. xviii), as well as Greek, Arabic, and Chinese and other oriental languages.

Proscribed Names

The names following 'not' at the end of an entry are not alternative names used in different times or places but are wrong forms, including not only misspellings but also definite misnomers that really refer to quite different breeds. For example, German Red cattle should not be called 'Red German', since the latter name was used until the Second World War for the Russian breed of German origin that is now called Red Steppe.

Etymology

In the majority of cases, breeds are named for the geographical region in which they originated or for the tribe that bred them, or by some descriptive feature (e.g. Shorthorn, Red Poll), or by a combination of two of these categories (e.g. Exmoor Horn, Red Bororo). It is assumed that such derivations are self-explanatory and no etymology is given.

At the same time, it must be borne in mind that the descriptions implied in a name may be inaccurate or misleading. For example, Black Persian sheep did not come from Persia, nor were Holstein-Friesian cattle originally bred in Holstein.

In all other cases the origin of the name is given, if it is known. The explanation may appear in the description of the origin of the breed or among the synonyms; otherwise it is given in square brackets, e.g. 'Criollo ... [= native]'.

A glossary of descriptive terms in foreign languages is given, to assist where no dictionary is available. Many of these terms do not appear in smaller dictionaries anyway.

Lack of Information

For the 3rd edition of the *World Watch List*, from which information has been selected for breeds not already in the 4th edition of the dictionary, representatives in each country were invited to submit details of their country's breeds to the database at FAO, Rome. Full details were requested for breeds that were endangered (either rare, or nearly extinct). *WWL* also lists, from the same sources, all the known breeds for each country but with no further information.

In some cases, especially for the latter category, all that has been gleaned is a breed name and its country, and sometimes it has not yet been possible to discover any more about the breed. Thus some entries in this edition of the dictionary simply list name and country. It is hoped that, with more detailed information becoming available as *WWL* develops and with closer contact with experts in each country, the dictionary will be able to expand on these entries. For future editions, the editor would be grateful for further details from the countries concerned.

Abbreviations

ABA	*Animal Breeding Abstracts*
adj.	adjective
Afrik.	Afrikaans
Agric.	Agricultural, Agriculture
Alb.	Albanian
BS	breed society, breed societies
BSd	breed standard
Bulg.	Bulgarian (transliterated)
C	Central
c.	century
c.	*circa* (about)
ca	cashmere
cf.	compare with
Ch.	Chinese (romanized according to the Pinyin system)
Co.	Company
Cro.	Croatian
cw	coarsewooled (i.e. mattress, mixed, or carpet)
Cz.	Czech
d	dairy (i.e. used for milk)
Dan.	Danish
Dept	Department
Dev.	Development
dist.	district
dr	draught
Du.	Dutch
E	East
e.g.	for example
esp.	especially
et al.	and others
etc.	et cetera
etym.	etymology

Exp.	Experiment(al)
fem.	feminine
Finn.	Finnish
Flem.	Flemish
Fr.	French
fr	fat-rumped
ft	fat-tailed (broad or S-shaped)
fw	finewooled (i.e. Merino type)
GB	Great Britain (i.e. England, Wales and Scotland)
geog.	geographical
Ger.	German
Gr.	Greek (transliterated)
h	heavy (horse)
HB	herdbook (or flockbook or studbook)
hd	horned
hr	long hair
Hung.	Hungarian
hy	short hair, woolless
I	Island
i.e.	that is
Imp.	Improvement
inc.	included, includes, including, inclusive
Inst.	Institute
Is	Islands
It.	Italian
Jap.	Japanese (romanized)
L	Lake
l	light (horse)
Lab.	Laboratory
Lat.	Latvian
lat.	latitude
ld	lard
lft	long fat tail
Lith.	Lithuanian
lt	long thin tail
lw	longwooled (medium fine)
m	meat
mo	mohair
mt	mountain
mts	mountains
mu	mule production
mw	medium wool (i.e. intermediate between cw and fw)
N	north
Nat.	National
Nor.	Norwegian

nr	near
NSW	New South Wales
NWFP	North West Frontier Province
obs.	obsolete
occ.	occasionally
Org.	Organization
orig.	origin, original, originally, originated
pa	pack
pd	polled, hornless
Pers.	Persian
pl.	plural
Pol.	Polish
Port.	Portuguese
prim.	primitive (i.e. native unimproved type)
Prov.	Province
py	pony (or small horse)
q.v.	which see
R	River
recog.	recognized
reorg.	reorganized
Res.	Research
resp.	respectively
ri	riding
Rom.	Romanian
Russ.	Russian (transliterated)
S	south
Serb.	Serbian (transliterated)
Serbo-cro	Serbo-Croatian
sft	short fat tail
sim.	similar
sing.	singular
Sl.	Slovak
Sn.	Slovenian
Sp.	Spanish
sp.	species
spp.	species (pl.)
St	Saint
Sta.	Station
subvar.	subvariety
sw	shortwooled (medium fine)
Swe.	Swedish
syn.	synonym(s)
Turk.	Turkish
UK	United Kingdom of Great Britain and Northern Ireland
Univ.	University

usu.	usually
var.	variety or sub-breed
vars	varieties
Vet.	Veterinary
viz.	namely
W	west
w	wooled (but not cw or fw)
WG	Chinese romanized according to Wade-Giles system
z	zebu

Pronunciation Equivalents

English	sh	ch	ts	j	zh	y	kh
French	ch	tch	ts	dj	j	y	–
Spanish	–	ch	ts	–	–	y	j or x
Turkish	ş	ç	–	c	j	y	–
Romanian	ş	c(i)	ţ	g(i)	j	i	h
Italian	sc(i)	c(i)	z	g(i)	–	i	–
German	sch	tsch	z	dj or dsch	sh	j	ch
Hungarian	s	cs	c or cz	–	zs	j	–
Polish	sz	cz	c	dż	ż	j	ch
Czech	š	č	c	dž	ž	j	ch
Croatian	š	č	c	đ	ž	j	h
Russian (transliterated)	sh	ch	ts	dzh	zh	y	kh

Transliteration of Russian

А	a	Р	r
Б	b	С	s
В	v	Т	t
Г	g	У	u
Д	d	Ф	f
ДЖ	dzh[1]	Х	kh
Е	e[2]	Ц	ts
Ж	zh	Ч	ch
З	z	Ш	sh
И	i	Щ	shch (Bulg. sht)
Й	ĭ[3]	ъ	''[4]
К	k	ы	y
Л	l	ь	'[5]
М	m	э	ė[6]
Н	n	Ю	yu
О	o	Я	ya
П	p		

For geographical and breed names:
[1] j
[2] ye initially
[3] omit after И or ы elsewhere i

[4] omit in Russ., u in Bulgarian
[5] omit
[6] e

Ass

Abbreviations used in this section:
dr = draught; ldr = light draught; m = meat; mu = mule production; pa = pack; ri = riding

Names for ass include: ane, asino, asnal, asno, baudet, burro, buruwa, Esel, jumento

Abkhasian: var. of Georgian; Russ. *Abkhazskaya*

Abyssinian: (Ethiopia)/usu. slate-grey, occ. chestnut-brown/sim. to Sudanese Pack/syn. *Ethiopian*

achdari: *see* Syrian wild ass

African wild ass: (formerly N Africa and possibly Arabia)/= *Equus africanus* Fitzinger/vars: Nubian wild ass, Somali wild ass

ainu: *see* Sardinian

akhdari: *see* Syrian wild ass

Albanian: (Albania)

Al-Barmaki: (Jordan)

Algerian: (Algeria)/dr.m/chestnut or grey

Al-Salibi: (Jordan)/syn. *Kubrusi* (= Cypriot)

American Mammoth Jackstock: (USA)/mu/usu. black, red or sorrel, occ. grey, with white underparts/orig. from Andalusian, Majorcan, Maltese, Poitou and esp. Catalan/BS 1888, HB 1911/syn. *American Jack, Jack Stock, Mammoth Jack, Mammoth Jackstock*/rare

American Spotted: *see* Spotted

Amiatina: (Mt Amiata, Grosseto, Italy)/agouti (mouse-grey) with shoulder cross, often striped legs/HB 1990/nearly extinct

Anatolian: (Turkey)/black and grey vars

Andalusian: (S Spain)/grey/HB/Sp. *Andaluza*/syn. *Andaluza-Córdoba, Campiñesa, Córdoba*/nearly extinct

Ane caraibe: *see* Caribbean

Ane gris de Provence: *see* Provence

anger: *see* Iranian onager

Apulian: *see* Martina Franca

Armenian: (Armenia)/Caucasian type/Russ. *Armyanskaya*

Asnal Catalana: *see* Catalan

Asian wild ass: *see* kiang and onager

Asinara: (Asinara I, Sardinia, Italy)/white/HB 1990/almost extinct (1 herd only)

Asnal Catalana: *see* Catalan

Atbai: *see* Etbai

Ausetan: *see* Catalan

Austro-Hungarian Albino: (Austria, Hungary and Germany)/pale yellow/Ger. *Österreichisch-Ungarischer Albinoesel*/nearly extinct (only in zoos)

Azerbaijan: Caucasian type/Russ. *Azerbaĭdzhanskaya*

Baladi: (Jordan)
Baladi: (Lower Egypt)/var. of
   Egyptian
Baluchi wild ass: *see* Indian wild ass
Benderi: (S Iran)/orig. from onager ×
   Iranian
Berga: (Catalonia, Spain)/former var.
   of Catalan
Berry Black: (France)/dark brown to
   black with pale underparts/BS
   1993/Fr. *Grand Noir du Berry*/
   nearly extinct
Berry Grey: (France)/Fr. *Petit Gris de
   Berry*/? extinct
Bokhara: *see* Uzbek
Bourbonnais: (Allier, France)/brown
   with grey belly/BS/nearly extinct
Brazilian: (São Paulo, Brazil)/syn.
   Paulista/*see also* Northeastern,
   Pêga
Bukhara: *see* Uzbek
Burkina Faso local: (Burkina Faso)/
   dr.m
burriku: *see* Sardinian
burro: (W USA)/feral or domesticated
   feral/sim. to Standard/BS
Burro Majorero: *see* Majorero

Campiñesa: *see* Andalusian
Canary Island: (Spain)/local type of
   Common Spanish/sim. to
   Andalusian
Canindé: (Ceará, Brazil)/black with
   pale belly/rarer var. of
   Northeastern
Cardão: (Ceará, Brazil)/yellow or
   bay/commoner var. of
   Northeastern
Caribbean: (Guadeloupe)/grey/Fr.
   Ane caraibe/nearly extinct
Cariovilli: (L'Aquila, Abruzzo,
   Italy)/extinct
Cassala: *see* Sudanese Riding
Castellana: *see* Castilian
Castlemorone: (Caserta, Campania,
   Italy)/nearly extinct
Castilian: (Spain)/local Common
   Spanish in Castille/Sp.
   Castellana
Catalan: (Catalonia, Spain)/ldr.mu/
   black, dark grey or brown, with
   pale underparts/ major originator
   of American Mammoth
   Jackstock/former var.: Berga/HB

1880/Sp. *Asnal Catalana*/syn.
   Ausetan, Catalonian, Vich/
   nearly extinct
CAUCASIAN: (Transcaucasia)/small
   native type/inc. Armenian,
   Azerbaijan, Dagestan, Georgian/
   Russ. *Kavkazskaya*/syn.
   Transcaucasian (Russ.
   Zakavkazskaya)
Cayor: *see* Sudano-Sahelian
CENTRAL ASIAN: (Central Asia)/small
   native type/inc. Kara-Kalpak,
   Kazakh, Kirgiz, Tajik, Turkmen,
   Uzbek/Russ. *Sredneaziatskaya*
Chad local: (Chad)/ash grey or bay-
   brown with shoulder cross
chigetai: (Mongolia and E
   Kazakhstan)/= *Equus hemionus
   hemionus* Pallas and *E. h. luteus*
   (Matschie)/vars of onager/syn.
   kulan (Kirgiz), *Mongolian wild
   ass*/not *chiggetai, djigitai,
   dzeggetai, dzhiggetai, dziggetai*/
   [Mongolian *tchikhitei* = long-
   eared, from *tchicki* = ear]/rare in
   Mongolia, extinct in Kazakhstan
Chinese: (China)/inc. large (Dezhou,
   Guanglin, Guanzhong, South
   Shanxi), intermediate (Jiama,
   Miyang, Qingyang) and small
   types (Lingxian, North China,
   South West China, Xinjiang)
Common Spanish: (Spain)/
   unimproved native type/inc.
   Canary Island local, Castilian
   local, Majorcan local, Moruna/
   Sp. *Común*
Córdoba: *see* Andalusian
Cotentin: (France)/? extinct
Cypriot: *see* Cyprus
Cyprus: (Cyprus) orig. from African
   and Asian with Catalan blood/
   not *Cypriot*/*see also* Al-Salibi

Dagestan: (Russia)/Caucasian type/
   Russ. *Dagestanskaya*
Damascus: (Syria)/usu. brown to
   black, also white/larger var. of
   Syrian
Dezhou: (N Shandong, China)/black
   with white belly/var. of Chinese
djigitai: *see* chigetai
Dongolawi: (N Sudan)/var. of
   Sudanese Riding

dzeggetai, dzhiggetai, dziggetai: *see*
chigetai

Egyptian: (Egypt)/vars: Baladi,
Hassawi, Saidi
Egyptian White: *see* Hassawi
Encartaciones: *see* Las Encartaciones
Equus africanus: *see* African wild ass
Equus africanus africanus: *see*
Nubian wild ass
Equus africanus somaliensis: *see*
Somali wild ass
Equus hemionus: *see* onager
Equus hemionus hemionus: *see*
chigetai
Equus hemionus hemippus: *see*
Syrian wild ass
Equus hemionus khur: *see* Indian
wild ass
Equus hemionus kulan: *see* kulan
Equus hemionus luteus: *see* chigetai
Equus hemionus onager: *see* Iranian
onager
Equus kiang: *see* kiang
Eroway: *see* Sudano-Sahelian
Etbai: (NE Sudan)/smaller var. of
Sudanese Riding/not *Atbai*
Ethiopian: *see* Abyssinian

Falloh, Fam: *see* Sudano-Sahelian
Four-eyebrows: *see* Jiami
French: *see* Poitou; *see also* Berry
Black, Bourbonnais, Normand,
Provence, Pyrenean

Gascon, Gascony: *see* Pyrenean
Georgian: (Georgia)/Caucasian type/
vars: Abkhasian, Kakhetian,
Meskhet-Javakhet/Russ.
Gruzinskaya
ghor-khar: *see* Indian wild ass,
Iranian onager
ghudkhur: *see* Indian wild ass
Grand Noir du Berry: *see* Berry Black
Grigio viterbese: *see* Viterbo Grey
Gruzinskaya: *see* Georgian
Guangling: (NE Shanxi, China)/dark
with pale belly/var. of Chinese
Guanzhong: (Wei R, C Shaanxi,
China)/dr/usu. black with paler
muzzle and underparts, also dark
grey with shoulder bar/large var.
of Chinese/WG *Kuan-chung*,
Russ. *Guanchzhun*/formerly
Kwanchung/rare

Had: (Senegal)/mouse-grey/var. of
Sudano-Sahelian
half-ass: *see* onager
Hamadan: (Iran, also Turkmenistan)/
usu. white/sim. to Mary/var.:
Kashan/not *Hamodan*,
Khamodan
Hassawi: (Egypt)/ri/white/var. of
Egyptian/syn. *Egyptian White*
hemione: *see* onager
hémippe de Syrie: *see* Syrian wild
ass

Indian: (India)/light grey to white
Indian wild ass: (Rann of Cutch, India;
formerly also Baluchistan)/=
Equus hemionus khur Lesson/
var. of onager/syn. *Baluchi wild
ass*, ghor-khar, ghudkhur, *Indian
onager*, *khur*, thor *char*/rare
Iranian: (Iran)/small native type/orig.
of Benderi/syn. *Persian*/*see also*
Hamadan
Iranian onager: (C Iran)/= *Equus
hemionus onager* Boddaert/pale
yellow-brown with paler
underparts, dark dorsal stripe/
var. of onager/syn. *anger*, *ghor-
khar*, *Persian onager*, *Persian
wild ass*/rare
Iraqi: (Iraq)/syn. *Shuhri*
Irpinia: (Benevento and Avellino,
Campania, Italy)
Istrian: (Croatia)/sim. to North
Adriatic
Italian: *see* Amiatina, Asinara, Irpinia,
Martina Franca, Pantelleria,
Ragusan, Sardinian Dwarf

Jackstock: *see* American Mammoth
Jackstock
Jegue: *see* Northeastern
Jiami: (Jia and Mizhi counties,
Shaanxi, China)/dark with pale
belly/var. of Chinese/syn. *Four-
eyebrows, Siumi, Swallow-
coat*/not *Jiamu*
Jinnan: *see* South Shanxi

Kakhetian: var. of Georgian/Russ.
Kakhetinskaya
Kara-Kalpak: (Uzbekistan)/Central
Asian type/Russ. *Kara-
Kalpakskaya*

Kashan: (Iran)/var. of Hamadan/not *Koshan*
Kassala: *see* Sudanese Riding
Kavkazskaya: *see* Caucasian
Kazakh: (Kazakhstan)/Central Asian type/Russ. *Kazakhskaya*
Khamodan: *see* Hamadan
khur: *see* Indian wild ass
kiang: (Tibet, Sikkim, Ladakh)/= *Equus kiang* Moorcroft/dark red-brown with white underparts and patch behind shoulder/subspp. *E. k. kiang* (Kalamaili mts, Xinjiang), *E. k. holdereri* (Jammu/Kashmir and Siddu)/not *kyang*/[Tibetan]
Kirgiz: (Kyrgyzstan)/Central Asian type/Russ. *Kirgizskaya*
Kuan-chung: *see* Guanzhong
Kubrusi: *see* Al-Salibi
kulan: (Turkmenistan and Kazakhstan)/= *Equus hemionus kulan* (Grooves and Mazák)/var. of onager/syn. *Transcaspian onager*/not *koulan*, *kulon*/[Kirgiz]/rare
Kwanchung: *see* Guanzhong
kyang: *see* kiang

Ladakhi: (Indus valley, India)/dr/ large bulging head; long body hair/sim. to Tibetan
Lagoa Dourada: *see* Pêga
Large Standard: (USA)/usu. grey-dun/HB/obs. syn. *Spanish*/*see also* Standard
Las Encartaciones: (Viscaya, Basque provs, Spain)/black, grey, white or chestnut/Sp. *Asno de las Encartaciones*/nearly extinct
Leonés-Zamorana, León-Zamora: *see* Zamorano-Leonesa
Libyan: (Libya)/smaller var. usu. bay with pale belly; larger var. (ri) dark bay or grey
Lingxian: (Ling county, Hunan, China)/var. of Chinese/rare

Majorcan: (Balearic Is, Spain)/ greyish-white/local type of Common Spanish/HB 1990/Sp. *Mallorquina*/not *Mallorcan*/ nearly extinct
Majorero: (Canary Is, Spain)/grey/

orig. from NW Africa/Sp. *Burro Majorero*/rare
Mallorcan, Mallorquina: *see* Majorcan
Mammoth Jack, Mammoth Jack Stock: *see* American Mammoth Jackstock
Martina Franca: (Apulia, Italy)/ dr.pa.ri.mu/nearly black with light underparts, reddish muzzle; colts long chestnut coat/poss. orig. from Catalan/HB 1943, BS 1948/syn. *Apulian* (It. *Pugliese*), *Martinese*/nearly extinct
Mary: (Turkmenistan)/usu. white/ sim. to Hamadan/Russ. *Maryĭskaya*/syn. *Merv*
Mauritanian: (Mauritania)/pa.dr/light grey to dark bay with black shoulder cross
Mauritian: (Senegal) *see* Sudano-Sahelian
Masai: (Kenya; Tanzania)/usu. brown
Mbaba, Mbam: *see* Sudano-Sahelian
Mediterranean: (USA) *see* Miniature
Merv: *see* Mary
Meskhet-Javakhet: (Georgia)/var. of Georgian/Russ. *Meskhet-Dzhavakhetskaya*
Mesopotamian onager: *see* Syrian wild ass
Miniature: (USA)/usu. grey or grey-dun/orig. from Sardinia (? and Sicily) 1st imported 1929/HB, BS 1990/syn. *Mediterranean*, *Sicilian*
Miyang: (Henan, China)/black with white belly/var. of Chinese
Moldavian local: (Moldova)
molente, molingianu: *see* Sardinian
Mongolian wild ass: *see* chigetai
Moroccan: (Morocco)
Moruna: (Spain)/small black type of Common Spanish/[= black]
Muscat: (Tanzania)/larger and paler than Masai/? orig. from Egyptian or Arabian

Ngounaba: *see* Sudano-Sahelian
Niger local: (Niger R, Niger)/white/ sim. to Chad local
Nordestino: *see* Northeastern
Normand: (Normandy, France)/bay-brown/BS/nearly extinct

North Adriatic: (Croatia)/sim. to Istrian

North China: (N. China)/black, grey or brown/var. of Chinese

Northeastern: (NE Brazil)/grey, agouti or purple-brown/vars Canindé, Cardão/Port. *(Jumento) Nordestino*/syn. *Jegue*/rare

Nubian wild ass: (Sudan-Eritrea border)/= *Equus africanus africanus* Fitzinger/grey with pale underparts and legs, and dark mane, dorsal stripe, tail tuft and shoulder cross-stripe/var. of African wild ass/orig. of domestic ass *Equus 'asinus'* Linnaeus/extinct 1960s (? 2 remain in captivity)

onager: = *Equus hemionus* Pallas/ pale yellowish-brown with paler underparts and darker mane and back stripe/vars: chigetai (Mongolia) = *E. h. hemionus* and *E. h. luteus*; Indian wild ass = *E. h. khur*; Iranian onager = *E. h. onager*; kulan = *E. h. kulan*; and Syrian wild ass = *E. h. hemippus*/ syn. *Asian wild ass*, *half-ass*, *hemione*/not *anger*/[Latin, from Greek = wild ass]/rare

Ossabaw Island: (Georgia, USA)/feral of Mediterranean orig.

Österreichisch-Ungarischer Albinoesel: *see* Austro-Hungarian Albino

Pantelleria: (Italy)/dark brown with pale belly/syn. *Pantesca*/not *Pantellaria*/almost extinct

Pantesca: *see* Pantelleria

Paulista: *see* Brazilian

Pêga: (Minas Gerais, Brazil)/roan or dark grey/? It. and Egyptian orig. 1810 on/HB and BS 1949/syn. *Lagoa Dourada*/[= manacles, brand of original breeder J. de Resende of Lagoa Dourada]

Pegus de Mola: *see* Sardinian

Persian: *see* Iranian

Persian wild ass: *see* Iranian onager

Petit gris de Berry: *see* Berry Grey

Poitou: (W. France)/mu/black or bay-brown; long woolly hair/HB 1864, BS 1988/Fr. *baudet de Poitou, Poitevin*/syn. *French*/ nearly extinct

poleddu: *see* Sardinian

Porro: (Senegal)/white/var. of Sudano-Sahelian

Provence: (SE France)/grey/BS/Fr. *Ane (gris) de Provence*/nearly extinct

Pugliese: *see* Martina Franca

Puttalam: (NW Sri Lanka)/blackish brown to buff with white muzzle; dwarf, feral/syn. *Puttalam Buruwa* (= Puttalam donkey)/rare

Pyrenean: (France)/black or brown with pale belly/local population; cf. Catalan/BS/Fr. *Pyrénéen, Ane des Pyrénées*/syn. *Gascon, Gascony*/nearly extinct

Qaramani: (N Yemen)/small local type

Qingyang: (E Gansu, China)/black with white belly/var. of Chinese

Ragusan: (Sicily, Italy)/pa.mu/nearly black (dark bay) with pale muzzle and belly, sometimes grey/orig. from Martina Franca and Catalan × native and Pantelleria/HB 1953/It. *Ragusana*/ syn. *Sicilian*/nearly extinct

Rifawi: (N Sudan)/white/var. of Sudanese Riding/syn. *Shindawi*/not *Riffawi*

Romagnola: (Romagna, Emilia, Italy)/nearly extinct

Romanian: (Romania)

Sahelian: *see* Sudano-Sahelian

Saidi: (Upper Egypt)/var. of Egyptian

Saloum: *see* Sudano-Sahelian

Sant' Alberto: (Ravenna and Forli, Emilia, Italy)/extinct

Sardinian: *see* Sardinian Dwarf

Sardinian Dwarf: (Sardinia, Italy)/dr/ grey or agouti/BS 1991/It. *Sarda*/ regional syn. *ainu* or *poleddu* (in Barbagia and Goceano), *burriku* (in south), *molente* (= wheat-grinding), *molingianu, Pegus de Mola*/nearly extinct (orig. type almost lost by cross-breeding with Martina Franca)

Senaar: *see* Sudanese Riding
Sëng: (Senegal)/grey to reddish-
    brown/var. of Sudano-Sahelian
Sennar: *see* Sudanese Riding
Shandong, Shantung: *see* Dezhou
Shindawi Riding: *see* Rifawi
Shuhri: *see* Iraqi
Sicilian: (Italy) *see* Ragusan
Sicilian: (USA) *see* Miniature
Sinnari: *see* Sudanese Riding
Sirou: (Senegal)/black-brown with
    white underside/var. of Sudano-
    Sahelian
Siumi: *see* Jiami
Socotra Island: (Yemen) wild ass, ?
    orig. from Nubian wild ass
Somali: (Somalia, Ogaden and N.
    Kenya)/grey/sim. to Toposa of SE
    Sudan
Somali wild ass: (N Somalia and SE
    Eritrea)/= *Equus africanus
    somaliensis* Noack/reddish-grey
    with dark mane, rarely with
    dorsal stripe and shoulder cross-
    stripe/var. of African wild
    ass/nearly extinct (from hunting)
South Shanxi: (Yuncheng basin,
    Shanxi, China)/black with white
    on nose, around eyes and on
    belly/var. of Chinese/syn. *Jinnan*
South West China: (Sichuan and
    Yunnan)/grey, black or
    brown/var. of Chinese
Spanish: (USA) *see* Large Standard
Spanish: (Spain) *see* Andalusian,
    Catalan, Common Spanish, Las
    Encartaciones, Zamorano-Leonesa
Spotted: (USA)/BS 1962/syn.
    American Spotted
Sredneaziatskaya: *see* Central Asian
Standard: (USA)/usu. grey-dun/HB/
    see also Large Standard
Subyani: (N Yemen)/Sudanese orig.
Sudanese: *see also* Toposa
Sudanese Pack: (N Sudan)/grey/sim.
    to Abyssinian
Sudanese Riding: (N Sudan)/dark
    brown or reddish-grey,
    sometimes pale grey or white;
    larger than Sudanese Pack/vars:
    Dongolawi, Etbai (NE Sudan),
    Rifawi/syn. *Kassala* (NW Eritrea),
    Sennar (Ethiopia)/not *Cassala*,
    Senaar, Sennaar, Sinnari

Sudano-Sahelian: (Senegal)/colour
    vars: Had, Porro, Sirou, Sëng/
    local names *Eroway* (Diola),
    Falloh (Mandinge), *Fam* (Serer),
    Mbaba (Peul), *Mbam* (Wolof)/
    syn. *Cayor* or *Ngounaba,
    Mauritian, Sahelian, Saloum*
Swallow-coat: *see* Jiami
Syrian: (Syria and Israel)/grey, brown
    or black/larger var.: Damascus
Syrian onager: *see* Syrian wild ass
Syrian wild ass: (Syrian and Iraq
    deserts)/= *Equus hemionus
    hemippus* I. Geoffroy/var. of
    onager/Fr. *hémippe de Syrie*/syn.
    *achdari, akhdari, Mesopotamian
    onager, Syrian onager*/extinct
    (last one shot 1927)

Tajik: (Tajikstan)/Central Asian
    type/Russ. *Tadzhikskaya*
tchikhitei: *see* chigetai
thor char: *see* Indian wild ass
Tibetan: (Tibet and Nepal)/dark brown,
    occ. dark grey or dun; small
Toposa: (SE Sudan)/sim. to Somali
Transcaspian onager: *see* kulan
Transcaucasian: *see* Caucasian
Tswana: (Botswana)
Tunisian: (Tunisia)
Turkmen: (Turkmenistan)/Central
    Asian type/Russ. *Turkmenskaya*/
    see also Mary

Uzbek: (Uzbekistan)/Central Asian
    type/Russ. *Uzbekskaya*/syn.
    Bukhara (Russ. *Bukharskaya*)/
    not *Bokhara*

Vich: *see* Catalan
Viterbo Grey: (Viterbo, Latium, Italy)/
    It. *Grigo viterbese*/extinct

wild: *see* African wild, Asian wild

Xinjiang: (Tarim basin, Xinjiang,
    China)/light grey or black/var. of
    Chinese

Zakavkazskaya: *see* Caucasian
Zamorano-Leonesa: (Zamora and
    León, Spain)/black with paler
    muzzle and underparts; long
    hair/HB 1941/syn. *Leonés-*

Zamorana, Léon-Zamora, Zamora-Leonesa/rare

Zebronkey: (USA)/name for zebra × donkey hybrid/syn. *Zeedonk*

Buffalo

Abbreviations used in this section:
d = milk; dr = draught; m = meat; ra = racing

Colour is black or dark unless otherwise indicated.

Names for buffalo include: baffarō, Büffel, carabao, karbo, karbou, karbouw, kerabou, kerbau, kerbou, krabey, kwai

African buffalo: (Africa S of Sahara)/= *Syncerus caffer* Sparrman/vars: Cape buffalo, Congo buffalo; also *S. c. aequinoctialis* (Ethiopia to Niger, red-brown to black)

Al-Ahwar: *see* Iraqi

Albanian: (Albania)/var. of European

American bison: (North America)/= *Bison bison* (Linnaeus)/not buffalo

Anatolian: (NW Turkey)/d.dr/dark grey to black, often with white on head and tail; sickle or crescentic horns/syn. *Turkish*

anoa: (Sulawesi, Indonesia)/= *Bubalus (Anoa) depressicornis* Smith/brown to black with white marks on head and feet/syn. *dwarf, pigmy* or *chamois buffalo*

arni: (NE India, SE Nepal) = *Bubalus arnee* (Kerr)/dark grey to black/ orig. of domestic buffalo (*B. 'bubalis'* Linnaeus)/♂ arna, ♀ arni/syn. *Indian wild buffalo*/not *arno*/[Hindi]/rare

Assam: (NE India)/Swamp type

Australian Swamp: (N Australia)/ feral

Azerbaijan, Azeri: *see* Caucasian

Badavan, Badavari: *see* Bhadawari

Baio: (Amazonas, Brazil)/m.dr/ brown, often with white on head and legs; long horns/Swamp type/Port. *Vermelho* (= red)/syn. *búfalo do pântano* (= swamp buffalo), *Tipo Baio*/[= bay]/rare

Baladi: (Lower Egypt)/black or grey/var. of Egyptian/subvars: Beheri, Menufi/not *Baledi*

Bangar: (N. Bangladesh)/local var.

Bangladeshi: (Bangladesh)/Swamp and River types/local vars inc. Bangar, Kachhar, Mahish, Manipuri

Beheri: (Lower Egypt)/d.m/ slate-grey/subvar. of Baladi/not *Beheiri*/[= towards the sea]

Berari: *see* Nagpuri

Bhadawari: (Agra, Uttar Pradesh, India)/d.dr/copper-coloured with pale legs/syn. *Etawah*/not *Badavan, Badavari*

Bhavanagri: *see* Jafarabadi

Białowieza bison: *see* Polish bison

Binhu: (N Hunan, China)/var. of Chinese/= *Jianghan* in Hubei/[= near the lake (Dongting)]

bison: *see* American bison, European

bison; *see also* gaur (under Cattle
– none of these are buffalo)
Bison bison: *see* American bison
Bison bonasus: *see* European bison
Bison bonasus bonasus: *see* Polish
bison
Bison bonasus caucasicus: *see*
Caucasian bison
Bombay: *see* Surti
Borneo buffalo: (N. Sarawak)/feral/
not *Bubalus arnee hosei*
(Lydekker)
Brasov: *see* Romanian
Brazilian: *see* Baio, Brazilian
Carabao, Jafarabadi, Kalaban,
Mediterranean, Murrah, Palitana,
Rosilho
Brazilian Carabao: (Marajó I, Pará,
Brazil)/m.dr/cinder grey to
roan/Swamp type (Port. *búfalo
do pântano*/var.: Rosilho/orig.
from IndoChina/Port. *Carabao
Brasileiro*/rare
Bubalus arnee: *see* arni
Bubalus arnee hosei: *see* Borneo
buffalo
Bubalus mindorensis: *see* tamarao
búfalo do pântano: *see* Brazilian
(Swamp types)
Buffalypso: (Trinidad)/m/reddish-
brown or black/orig. 1950–1970
from Murrah, Surti, Jafarabadi,
Nili and Bhadawari imported
1900–1949
Bulgarian: (Bulgaria)/d.m.(dr)/var. of
European/rare
Bulgarian Murrah: (Bulgaria)/d.m/
orig from Murrah × Bulgarian/
syn. *Mourrakh*
Burmese: (Myanmar)/Swamp type/
see also Shan
bush cow: *see* Congo buffalo

Cambodian: (plains of
Cambodia)/Swamp type/
shorthorned/Fr. *Cambodgienne*/
syn. *kravey-leu*/*see also* Moi
Cape buffalo: (E and S Africa)/=
Syncerus caffer caffer
(Sparrman)/black/var. of African
buffalo
carabao: *see* Brazilian Carabao,
Philippine carabao
caraballa: *see* Philippine carabao

Caucasian: (N Caucasus, Russia and
Transcaucasia)/d.m.dr/dark grey
to black, occ. brown, often with
white on head and tail/breed
selected 1935–1970 (recog.) in
Azerbaijan from native
type/local var.: Gilani/syn.
Azerbaijan (or *Azeri*), *Dagestan*,
Georgian, *Improved Caucasian*,
Mazandarani (Iran), *Russian*,
Transcaucasian
Caucasian bison: = *Bison bonasus
caucasicus* Grevé/var. of
European bison/syn. *mountain
bison*/extinct before 1927
Ceylon, Ceylonese: *see* Lankan
chamois buffalo: *see* anoa
Charotar: *see* Surti
Chilka: (Orissa, India)
Chinese: (SE China)/Swamp type/
recog. geog. vars: large (Haize,
Shanghai), medium (Binhu,
Dechang, Dehong), small
(Wenzhou, Xinglong, Xilin);
other named vars: Diandongnap,
Dongliu, Fuan, Fuling, Jianghan,
Poyanghu, Taiwanese, Xingyang/
Ch. *shui niu* (= water cattle)
Cluj: *see* Romanian
Congo buffalo: (W Africa and Congo
basin)/= *Syncerus caffer nanus*
(Boddaert)/red/var. of African
buffalo/syn. *bush cow*

Dagestan: *see* Caucasian
Dangyang: (Hubei, China)/inc. in
Jianghan/WG *Tang-yang*, Ger.
Dang-jiang
Danube: *see* Romanian
Deccani: *see* Surti
Dechang: (S Sichuan, China)/var. of
Chinese
Dehong: (W Yunnan, China)/var. of
Chinese
Delhi: *see* Murrah
desi: (India and Pakistan)/= native,
indigenous, hence nondescript,
breedless, unimproved/not *deshi*
Dharwari: *see* Pandharpuri
Diandongnap: (Yunnan, China)/var.
of Chinese
Dongliu: (Anhui, China)/var. of
Chinese

Durna-Thali: *see* Nagpuri
Dwab: *see* Iraqi
dwarf buffalo: *see* anoa

Egyptian: (Egypt)/d.dr/grey-black;
short curved horns/vars: Baladi,
Saidi/syn. *Gamus*
Ellichpuri: *see* Nagpuri
Etawah: *see* Bhadawari
European: (SE Europe)/d.m.(dr)/dark
grey to black, often with white
on head, tail and feet/vars:
Albanian, Bulgarian, Greek,
Italian, Macedonian, Romanian
European bison: = *Bison bonasus*
(Linnaeus)/brown; semi-wild or
in captivity/HB 1932/vars:
Caucasian bison, Polish bison/
Ger. *wisent*, Pol. and Russ.
zubr/not *aurochs*/[Latin, from
Teutonic wisand]
Fengxian Round-Barrel: *see* Jiangsu
Round-Barrel
Fuan: (coast of Fujian, China)/var. of
Chinese
Fuling: (SE Sichuan, China)/var. of
Chinese

Gamus: *see* Egyptian
Ganjam: *see* Manda
Gaulani, Gauli: *see* Nagpuri
Georgian: *see* Caucasian
Ghab: *see* Syrian
Gilani: (Iran)/local var. of Caucasian
Gir: *see* Jafarabadi
Godavari: (Godavari and Krishna
deltas, Andhra Pradesh,
India)/orig. from Murrah × desi
Gowdoo: *see* Sambalpur
Greek: (N Greece)/var. of
European/rare
Gujarati, Gujerati: *see* Surti

Haizi: (N Jiangsu, China)/var. of
Chinese/WG *Haitzü*, Russ.
Haĭtszy/not *Haidzi*/[= son of the
sea]
Hechu: (Hubei, China)/WG *Wuchü*/
no longer recog. (inc. in Jianghan)
Himalayan Mountain: *see* Nepalese
Mountain
Hungarian: (Hungary)/orig. from
Romanian (Transylvanian)/
nearly extinct

INDIAN: (India except Assam, and
Pakistan)/River type/inc. 7 recog.
breeds in India (Bhadawari,
Jafarabadi, Mehsana, Murrah,
Nagpuri, Nili-Ravi, Surti), 2 in
Pakistan (Kundi, Nili-Ravi) and
many local types (e.g. Godavari,
Jerangi, Kalahandi, Manda,
Sambalpur, Tarai)/*see also* desi
Indian wild buffalo: *see* arni
Indo-Chinese: *see* Cambodian,
Laotian, Moi, Vietnamese
Indonesian: (Indonesia)/Swamp
type/Indonesian *kerbau* (=
buffalo), also spelt *karbo*,
karbou, karbouw/syn. *Javanese*
Iranian: (Iran)/cf. Iraqi (in Khuzestan),
Caucasian (in Azerbaijan and
Gilan)/syn. *Kuhzestani, Persian*
Iranian Azara ecotype: (Iran)
Iraqi: (SE Iraq)/d./black or dark grey,
often with white on head, feet
and tail, occ. pied; sickle horns/
var.: Khuzestan/syn. *Al-Ahwar,
Dwab, Marsh*
Italian: (Caserta and Salerno, S
Italy)/d.(m)/var. of European/
orig. of Mediterranean (Brazil)

Jafarabadi: (Kathiawar, Gujarat,
India)/d/usu. black; large drooping
horns/syn. *Bhavanagri, Gir, Jaffri*/
not *Jaffarabadi, Jaffarbadi,
Jafferabadi, Jaffrabarri, Jafrabadi*/
HB in Brazil (vars in Brazil: Gir,
Palitana)
Javanese: *see* Indonesian
Jerangi: (Orissa-Andhra Pradesh border,
India)/dr/black/syn. *Zerangi*
Jianghan: (Hubei, China)/var. of
Chinese/inc. Hechu and
Dangyang (and Binhu in Hunan)
Jiangsu Round-Barrel: (China)/ WG
Chiang-su/formerly *Kiang-su*,
Fengxian Round-Barrel/no
longer recog. (inc. in Shanghai)
Jimes: (Iraq)

Kachhar: (N Bangladesh)/local var./?
Kachkar
Kalaban: (Brazil)
Kalahandi: (S Orissa, India)/d.dr/grey/
syn. *Peddakimedi* (Andhra
Pradesh)

Kanara, Kanarese: *see* South Kanara
karbo, karbou, karbouw: = buffalo
    (Indonesia)
kerabou, kerbau, kerbou: = buffalo
    (Malaysia)
kerbau-banteng: (Sumatra) *see* Murrah
kerbau-sapi: (Malaysia) *see* River
    buffalo (i.e. Indian – Murrah or
    Surti) [= cattle buffalo]
kerbau-sawah: (Malaysia) *see*
    Malaysian [= swamp buffalo]
kerbau-sungei: (Malaysia) *see* River
    buffalo (i.e. Indian – Murrah or
    Surti)
Khuzestan: (Iran)/local of Iraqi type
Khouay: (Laos)
Kiangsu: *see* Jiangsu
Kimedi: *see* Sambalpur
krabey: = buffalo (Indo-China)
krabey-beng: *see* Moi
krabey-leu: *see* Cambodian
Kuhzestani: *see* Iranian
Kujang: (Orissa, India)/d/local var.
Kundi: (N Sind, Pakistan)/d/usu.
    black, occ. brown, often with
    white on head, tail and feet/
    Murrah type/syn. *Sindhi Murrah*/
    not *Kundhi*/[= fish-hook (horns)]
kwai: = buffalo (Thailand)
Kwangtung: *see* Xinglong

Lankan: (Sri Lanka)/fawn to black/
    Swamp type/former vars:
    Mannaar, Tamankaduwa/syn.
    *Ceylon, Ceylonese, Lanka, Sri
    Lankan*
Laotian: (Laos)/Swamp type
Lime: *see* Nepalese Mountain
Linn Duong: (SE Asia)/=
    Pseudonovibos spiralis/newly
    discovered buffalo sp.
Lithuanian bison: *see* Polish bison
lowland bison: *see* Polish bison

Macedonian: (S Serbia and
    Macedonia)/var. of European
Mahish: (Bangladesh)/local var.
Mahratta, Mahratwada: *see* Nagpuri
Malabar: *see* South Kanara
Malaysian: (W Malaysia)/Swamp
    type/syn. *kerbau-sawah* (= swamp
    buffalo)/obs. syn. *Malayan*
Manda: (Orissa-Andhra Pradesh
    border, India)/d.dr/brown or
    grey/syn. *Ganjam,*

*Paralakhemundi, Parlakhemundi,
    Parlakimedi*/[= herd]
Manipuri: (Sylhet, Bangladesh)/large
    horns/local var./not *Manpuri*
Mannar: (Sri Lanka)/reddish/larger
    var. of Lankan/not *Manaar, Manar*
Marathwada: *see* Nagpuri
Marsh: *see* Iraqi
Mazandarani: (Iran) *see* Caucasian
Mediterranean: *see* Anatolian,
    Egyptian, European, Syrian
Mediterranean: (Brazil)/d/orig. from
    Italian/HB
Mediterranean: (Congo)/dr.m.d/It.
    *Bufalo Italiano, Bufalo
    Prete*/nearly extinct
Mehsana: (N Gujarat, India)/d./black,
    occ. brown or grey/orig. from
    Murray with Surti blood/not
    Mehasana, Mehsani
Menufi: (S and C Delta, Egypt)/d.m./
    dark grey; long curved horns/
    subvar. of Baladi (= local)/not
    Menoufi, Minufi
Mestizo: (Philippines)/Indian ×
    carabao
Middle Eastern: *see* Anatolian,
    Caucasian, Egyptian, Iraqi, Syrian
Mindoro: *see* tamarao
Minufi: *see* Menufi
Moi: (plateau of Cambodia)/
    longhorned/Swamp type/syn.
    krabey-beng
mountain bison: *see* Caucasian bison
Mourrakh: *see* Bulgarian Murrah
Muntenesc: *see* Romanian
Murrah group: (NW India and
    Pakistan)/d/black; short coiled
    horns/inc. Murrah (*sensu
    stricto*), Nili-Ravi, Kundi/syn.
    Delhi, kerbau banteng (Sumatra),
    kerbau sungei or *kerbau-sapi*
    (Malaysia)/not *Mura, Murra*
Murrah: (Haryana and Delhi, India)/
    see above/orig. of Bulgarian
    Murrah/HB 1940; HB also in
    Brazil/rare in Malaysia
Murrah: (Ecuador)/m.dr/rare
Myanmar: *see* Burmese

Nadiad, Nadiadi: *see* Surti
Nagpuri: (NE Maharashtra, India)/
    dr.d/usu. black, sometimes white
    on face, legs and tail; long
    horns/vars: Pandharpuri,

Purnathadi/syn. *Berari, Durna-Thali, Ellichpuri* (Berar), *Gaulani, Gauli, Marathwada, Varadi*/not *Najpuri*

Nelli: *see* Nili

Nepalese: (Nepal)/d.m/River or Swamp type/regional vars: Nepalese Hill, Nepalese Mountain, Tarai

Nepalese Hill: (C Nepal)/light grey; long horns/? Swamp type/*see also* Parkote

Nepalese Mountain: (N Nepal)/light grey or blond; medium horns; hairy/? Swamp type/syn. *Himalayan Mountain, Lime*

Netherlands Indies: *see* Indonesian

Nili: (Sutlej R. valley, Pakistan and India)/local var. of Nili-Ravi/not *Nelli*/[= blue (waters of R. Sutlej)]

Nili-Ravi: (Punjab, Pakistan and India)/d/usu. black, occ. brown, often with wall eyes and white marks on head, legs and tail/ Murrah type/vars: Nili, Ravi

Pahadi: *see* Parkote

Palestinian: *see* Syrian

Palitana: (Brazil)/var. of Jafarabadi

Pandharpuri: (SE Maharashtra, India)/var. of Nagpuri/syn. *Dharwari* (Karnataka)

pântano, búfalo do: *see* Baio

Paralakhemundi: *see* Manda

Parkote: (Nepal, hills and valleys)/d/black, no chevrons/River type/being crossed with Lime (i.e. Nepalese Mountain) and Murrah/syn. *Pahadi*

Parlakimedi: *see* Manda

Peddakimedi: *see* Kalahandi

Persian: *see* Iranian

Philippine carabao: (Philippines)/Swamp type/ [*carabao* = buffalo; ♀ is *caraballa*]

Phil-Murrah: (Philippines)/grades of Murrah (25–50%) crossed on to Philippine carabao

Phil-Ravi: (Philippines)/grades of Nili-Ravi (25–75%) crossed on to Philippine carabao

pigmy buffalo: *see* anoa

Polish bison: = *Bison bonasus bonasus* (Linnaeus)/var. of European bison/syn. *Białowieza bison, Lithuanian bison, lowland bison*

Poyanghu: (Jiangxi, China)/var. of Chinese

Prete: *see* Italian, Mediterranean

Purnathadi: (Maharashtra, India)/ d/var. of Nagpuri

Ravi: (Ravi and Chenab R valleys, Pakistan)/local var. of Nili-Ravi/ syn. *Sandal Bar*

RIVER BUFFALO: (Asia W of Assam, Egypt, SE Europe)/d.m.dr/black, occ. brown; short coiled or sickle horns/orig. from Swamp buffalo (or directly from wild arni) by selection for colour, horns and milk yield/inc. Indian, Middle Eastern European/syn. *kerbau-sungei* (Malaysia), *Riverine*

Romanian: (Romania, esp. Transylvania)/d/black/var. of European/syn. (formerly vars) *Muntenesc* (= Munenian, i.e. Walachian), *Transylvanian* (orig. of Hungarian)/not *Roumanian*/ inc. Brasov, Cluj, Danube

Rosilho: (Brazil, esp. Marajo I)/m.dr/ grey skin, cream hair/Swamp type/colour var. of Carabao/[= roan]

Russian: *see* Caucasian

Saidi: (Upper Egypt)/d.m.dr/black/ var. of Egyptian

Sambalpur: (Bilaspur, Madhya Pradesh, India)/d.dr/usu. black, occ. brown or grey/syn. *Gowdoo, Kimedi* (Andhra Pradesh)

Sandal Bar: *see* Ravi

sapi: *see* kerbau-sapi; sapi-utan (= banteng, under Cattle) [= cattle]

Shan: (Shan state, Myanmar)/heavier and darker than Burmese/syn. *Shan kywe, Shan Swamp buffalo*

Shanghai: (China)/var. of Chinese/ inc. Jiangsu Round-Barrel

shui niu: *see* Chinese

Siamese: *see* Thai

Sindhi Murrah: *see* Kundi

South Kanara: (SW Karnataka,
  India)/dr.ra/syn. *Kanara,
  Kanarese, Malabar*
Sri Lankan: *see* Lankan
Surat, Surati: *see* Surti
Surti: (N Gujarat, India)/d/black or
  brown, usu. with two white
  collars/syn. *Charotar, Deccani,
  Gujarati, Nadiadi, Surati,
  Talabda*/[from Surat]
SWAMP BUFFALO: (SE Asia)/dr/dark grey,
  occ. white; crescentic horns/orig.
  from and sim. to arni, *Bubalus
  arnee*/inc. Assam, Burmese,
  Cambodian, Chinese, Indonesian,
  Lanka, Laotian, Malaysian,
  Nepalese, Philippine, Thai,
  Vietnames/? orig. of River buffalo
Syncerus caffer: *see* African buffalo
Syncerus caffer caffer: *see* Cape
  buffalo
Syncerus caffer nanus: *see* Congo
  buffalo
Syrian: (Ghab plains, Syria, and
  Israel)/d.dr/black; sim.
  Mediterranean/syn. *Ghab*, obs.
  syn. *Palestinian*/rare in Syria,
  extinct in Israel

Taiwanese: (Taiwan)/var. of Chinese
Talabda: *see* Surti
Tamankaduwa: (Sri Lanka)/d/largest
  var. of Lankan/not *Tamamkaduwa*
tamarao: (Mindoro, Philippines)/=
  Bubalus mindorensis Heude/m./
  black, grey or brown/syn.
  Mindoro buffalo/not *tamarau,
  tamaraw, tamaron, timarau*/rare
Tang-yang: *see* Dangyang
Tarai: (N Uttar Pradesh, India, and S
  Nepal)/d.m.dr/black (occ. brown)
  with white tail/? Murrah blood/
  not *Terai*

Terai: *see* Tarai
Thai: (Thailand)/often white in N/
  Swamp type/syn. *kwai* (= buffalo),
  Siamese
timarau: *see* tamarao
Tipo Baio: *see* Baio
Toda: (Nilgiris, NW Tamil Nadu,
  India)/d/River type/rare
Transcaucasian: *see* Caucasian
Transylvanian: *see* Romanian
Trau Noi: (Vietnam)
Turkish: *see* Anatolian

Varadi: *see* Nagpuri
Vermelho: *see* Baio
Vietnamese: (Vietnam)/Swamp type

water cattle: *see* Chinese
Wenzhou: (E Zhejiang, China)/var. of
  Chinese
wisent: *see* European bison
Wuchü, Wu-tchü, Wu-tshü: *see* Hebu

Xilin: (W Guangxi, China)/var. of
  Chinese/not *Xiling*
Xinglong: (Hainan, China)/var. of
  Chinese/syn. *Kwangtung*/not
  Xinulung
Xingyang: (S. Henan, China)/var. of
  Chinese

Yanjin: (Yunnan, China)/var. of
  Chinese
Yerli Kara: (Turkey)/? = Anatolian
  Black/[= native black]
Yuandong: (Fujian, China)/var. of
  Chinese/ WG *Yuan-tung*/no
  longer recog.

Zerangi: *see* Jerangi
zubr: *see* European bison
zubro-bison: = European × American
  bison

Cattle

Abbreviations used in this section:
d = milk (not necessarily a dairy-type animal, but used for milk); dr = draught; m = meat; pd = polled

Cattle are assumed to be horned, unless otherwise indicated.

Names for cattle include: bestiama, boskap, ganado, goveče, govedo, kvæg, niu, nwar, sapi, skot, stoka, vee, Vieh

Aalstreep: *see* Witrik

abel gorriak: *see* Betizuak

Aberdeen-Angus: (NE Scotland)/ m/black; pd/orig. from Angus Doddie + Buchan Haumlie in late 18th c./var.: Red Angus/orig. of American Angus, German Angus/recog. 1835, named 1909; Polled HB (with Galloway) 1862, BS 1879; BS also Argentina 1927 (HB 1879), Canada 1906 (HB 1885), S Africa 1917 (HB 1906), New Zealand 1918, Australia (Angus) 1926, Uruguay, Zimbabwe, Ireland 1967, Greece 1970, Denmark 1973, Japan, Finland, Norway, Sweden, Mexico, Paraguay 1980; HB also Brazil 1906, Russia/syn. *Angus, Northern Scotch Polled, Polled Angus*

Aberdeen-Angus colorado: *see* Red Angus

Aberdeenshire Shorthorn: *see* Beef Shorthorn

Abgal: *see* Gasara

Abigar: (Akebo, Gambela, Ilubabor, SW Ethiopia)/d.m.dr/Sanga/ subvar. of Nuer var. of Nilotic/syn. *Anuak*

Abondance: (Haute Savoie, France)/ d.m.[dr]/darker red pied, with white confined to underside and extremities, and usu. with spectacles/French Red Pied group/recently with Red-and-White Holstein blood/HB and BS 1894/Fr. *Race d'Abondance*/syn. *Chablaisien* (from Chablais), *Pie rouge française de Montagne*/[La Chapelle d'Abondance]

Abori: *see* Red Bororo

Abruzzese: (Abruzzi, Italy)/former var. of Apulian Podolian/It. *Podolica abruzzese di montagna*/extinct

Abyssinian Highland Zebu: *see* Bale

Abyssinian Sangas: (Ethiopia)/usu. light chestnut or ash-grey; large lyre horns; small hump/true Sanga type in East Africa/inc. Danakil, Raya Azebó

ABYSSINIAN SHORTHORNED ZEBU group: (Ethiopia)/dr.m.d/often grey, fawn, reddish, also black, bay, roan, pied; horns variable, occ.

pd/Small East African Zebu group/inc. Adwa, Ambo, Arsi, Bale, Goffa, Guraghe, Hammer, Harar, JemJem, Jijjiga, Mursi, Ogaden, Smada/syn. *Ethiopian Highland Zebu* (or *Lowland* for Jijjiga Zebu, Ogaden Zebu)

ACASTINADO: (Orense, Spain)/chestnut/group inc. Limiana, Verinesa, Vianesa

Acchai: *see* Lohani

Acchami: *see* Achham

ACDC: *see* Australian Commercial Dairy Cow

Aceh: (N Sumatra, Indonesia)/z/local var. orig. from Indian Zebu × local/not *Acheen, Acheh, Achhani, Achin, Achinese*/rare

Acheen, Acheh: *see* Aceh

Achham: (W Nepal)/light brown; ♂ hd, ♀ hd or pd; dwarf/z/syn. *Sanogai* (= small cow), *Acchami*

Achhani: *see* Aceh

Achin, Achinese: *see* Aceh

Achiote: (Guatemala)/yellow/Criollo type with Shorthorn (? zebu) blood/[= anatto]

Açoreana: *see* Azores

Acreno: *see* Carreña

Adal, Adali: *see* Danakil

Adamawa: (N Cameroon and NE Nigeria)/m.d.dr/usu. red or pied/West African Zebu, Gudali group/vars: Banyo, N'Gaoundéré (typical), Yola/syn. *Adamawa Fulani, Adamawa Gudali, Cameroons Fulani, Cameroons Gudali, Goudali*/not *Adamaoua*

Adapteur: *see* Belmont Adapteur

Adar: *see* Azaouak

Aden: *see* Baherie

Adige: *see* Grey Adige

Adwa: (C zone of Tigray region, Ethiopia)/dr/usu. red, chestnut, black, roan or white/Abyssinian Shorthorned Zebu group

Afghan: (Afghanistan)/z/vars: Kandahari, Konari, Sistani, Vatani/syn. *Kabuli*

Afghan Subtropical: (Afghanistan)/d/brown or black pied (according to last cross)/orig. from Russian Brown (RB) or Dutch Black Pied (DBP) × [Jersey, DBP and RB × (Kandahari, Russian Black Pied, DBP and RB, × Kandahari, Konari, Sistani, Sahiwal, Red Sindhi and Tharparkar)]

African aurochs: (N Africa)/= *Bos primigenius opisthonomus* Pomel/var. of aurochs/? orig. of Hamitic Longhorn/syn. *B. p. hahni* Hilzheimer, *African urus*/extinct

Africander: *see* Afrikander

Africangus: (Jeanerette, Louisiana, USA)/m/orig. 1953–1963 from Afrikander (30%) × Aberdeen-Angus (70%)/not *Africagnus*

African Zebu: *see* East African Zebu, Sanga, West African Zebu

Afrikander: (S Africa)/m.dr/usu. red, also yellow strain; long lateral horns, also pd var./Sanga type/orig. from Hottentot in 18th and 19th c./orig. of Bonsmara, Drakensberger/HB 1907, BS 1912; BS also in Australia 1969/Afrik. *Afrikaner*/syn. *Africander*/not *Africaner*

AFS: *see* Australian Friesian Sahiwal

Agerolese: (Agerola, nr Sorrento, Campania, Italy)/d/chestnut to almost black, with pale dorsal stripe/local var. orig. in 19th c. from Italian Brown, Jersey, Friesian, Podolian *et al.*/recog. 1950/not *Agerose*/rare (by crossing earlier with Italian Brown and later with Friesian):

ago: *see* galiba

ah gorh: *see* galiba

AIS: *see* Illawarra

Akage Washu: *see* Japanese Brown

Akamba: *see* Kamba

Akele-Guzai: *see* Aradó

Akou: *see* White Fulani

Akshi: *see* Baladi

Alambadi: (W and NW Tamil Nadu, India)/dr/grey/z; Mysore type/syn. *Bestal, Cauvery, Kaveri, Lambadi* (Hyderabad), *Mahadeswarabetta, Malai-madu, Salem*/not *Alambady, Alambudi, Alumbadi*

Ala-Tau: (SE Kazakhstan and NE Kyrgyzstan)/d.m/usu. brown/orig. 1929–1940, esp. at Stud farm Alamedin, from Swiss Brown and Kostroma, × Kirgiz (also some Friesian, Simmental and Aulie-Ata blood)/recog. 1950; HB/Russ. *Alatauskaya*

Alb: (Württemberg, Germany)/ absorbed by German Simmental/ extinct

Albanian: (Albania)/dr.d/yellow to reddish-black/Iberian (i.e. *brachyceros*) type/sim. to Buša and Greek Shorthorn/vars inc. Albanian Dwarf, Shkodra Red/ syn. *Illyrian* (inc. Buša)

Albanian Dwarf: (Albania)/var. of Albanian/syn. *Illyrian Dwarf*

Albères: (Albères mts, E Pyrenees, France/Spain)/black, brown or blond; semi-feral/Iberian type/ vars: black, fawn (*fagina*)/Sp. *Albera*, Fr. *Massanaise* (R Massane)/rare

Albese: (Alba, Cuneo, Italy)/double-muscled var. of Piedmont

Alderney: *see* Channel Island

Alentejana: (Alentejo, Portugal)/ m.dr/golden red; long horns/sim. to Retinta/var.: Algarvia/? orig. of Southern Crioulo and Mertolenga/ HB 1973, BS/syn. *Transtagana*/ not *Alemtejo*/[= S of Tagus]

Aleppo: *see* Damascus

Algarvia: (Algarve, S Portugal)/var. of Alentejana/extinct 1970s (by crossing with Limousin, and with Charolais to give Chamusco)

Algerian: *see* Guelma type, Oran var. of Brown Atlas

Aliab Dinka: (SE of S Sudan)/d.m/ usu. white with red or black markings/Sanga type/local var. of Nilotic/*see also* Aweil Dinka, Wadei Dinka/[Dinka tribe]

Alistana-Sanabresa: (Aliste and Sanabria, NW Zamora, Spain)/ m.dr/chestnut/Morenas del Noroeste group/var.: Berciana

Allgäuer: (Allgäu, SE Bavaria, Germany)/orig. local var. of German Brown without American Brown Swiss blood/ orig. from Swiss Brown and Montafon × local since 1900/ HB/not *Algau*/rare

Allmogekor: (S Sweden)/d.m/various colours/old local type/[= peasant or village cow]/nearly extinct

Almanzoreña: (R Almanzora, Almería, Spain)/former var. of Murcian/? nearly extinct

Alpenfleckvieh: *see* Bergscheck, Simmental

Alpha 16: (France)/m/yellow/ Limousin line (with Charolais) selected for small calves/extinct

Alpine: *see* Austrian Yellow, Brown Mountain, Grey Mountain, Pinzgauer, Simmental

Alpine Hérens: (Chamonix Valley and Rhône-Alpes, France)/d.m, fighting/fawn with black belly, sometimes dark red/orig. from Hérens imported from Switzerland *c.* 1800/HB/syn. *French Hérens, Valais*/rare

Alsatian Simmental: (Alsace, France)/ d.m.dr/orig. from Swiss Simmental/ HB fused 1945 with Eastern Red Pied (Pie rouge de l'Est, now French Simmental)/Fr. *Simmenthal d'Alsace*/extinct

Altai: (Russia)/m or m.d/var. of Siberian/Russ. *Altaĭskaya*/syn. *South Siberian*/extinct

Alumbadi: *see* Alambadi

Alur: (W of L Mobutu, NE DR Congo)/d.m/red, red pied, brown or black/Sanga-Zebu intermediate/orig. from Ankole (mostly Bahima), Lugware and Nkedi/syn. *Blukwa* (name of a chief), *Nioka*

Amabowe: (Zimbabwe)/ red, red-and-white or golden brown/= Mangwato/Sanga type, Setswana group/almost extinct 1940s, remnants developed for Tuli

Amanjanja: *see* Mashona

Ambo: (Ambo, Dandi, Addis Alem and Holetta area, W Shewa, Ethiopia)/dr.m.d/Abyssinian Shorthorned Zebu group; sim. to Guraghe but larger

Ambo: (Namibia) *see* Ovambo

American Angus: (USA)/var. of Aberdeen-Angus first imported 1873/var.: Red Angus/BS 1883; HB 1886

American Beef Friesian: (Colorado,

USA)/d.m/var. of Friesian/orig.
from British Friesian imported
from Ireland 1971/BS 1972

American Brahman: *see* Brahman

American Breed: (New Mexico, USA)/
m/orig. 1948–1974 by Art Jones,
Cactus Road, Portales; ½ Brahman,
¼ Charolais, ⅛ bison, 1/16 Hereford, 1/16
Shorthorn/BS 1971

American Brown Swiss: (USA)/
d.(m)/orig. from *c.* 155 Swiss
Brown imported 1869–1906/BS
1880, HB 1889; BS also Canada
1914, UK, Argentina/syn. *Brown
Swiss*

American Dairy Cattle: (USA)/d/BS
1936 to register dairy cattle, with
performance criteria only

American Dutch Belted: *see* Dutch
Belted

American Friesian: *see* American
Beef Friesian, Holstein

American Lineback: (USA)/colour
type, inc.: (1) American "G"
(dark mahogany to black with
white back-line, tail and belly; ?
Glamorgan or Gloucester blood);
(2) coloursided (red or black
with speckled face and white
back-line, belly and lower legs; ?
Witrik or Longhorn blood)/var.:
Randall Lineback/BS 1985, HB
1987/rare

American Milking Devon: *see*
Milking Devon

American Shorthorn: (USA)/m/BS

American Wagyu: (Texas, USA)/orig.
from Hereford, American Angus,
Japanese Black and Japanese
Brown/[*Wagyu* = Japanese cattle]

American White Park: (Canada)/pd
or hd/cf. American White Park
(USA)/first imported from UK to
Riverdale Zoo in Toronto 1938,
then through NEZS; 4 estab. at
King Ranch, Texas (1941);
reintroduced from USA to
Canada 1987/rare

American White Park:
(USA)/m/white with black
points (occ. red); pd (occ. hd)/
orig. from British White
imported 1941 and 1976–1984, ?
with some White Park blood/BS

1975, supplemented by British
White BS (1987)/rare

Amerifax: (USA)/m/black or red;
pd/orig. from American Angus
(⅝) × American Beef Friesian
(⅜)/BS 1977

Amritmahal: (Karnataka, India)/dr/
grey, with lighter markings on
face and dewlap/z; Mysore type/
orig. in 17th c. from Hallikar/
former vars: (up to 1860)
Chitaldrug, Hagalvadi, Hallikar/
not *Amrat Mahal*, *Amrit mahal*,
Amrit-mahal/[= nectar department,
i.e. government dairy]

Amsterdam Island: (Indian Ocean)/
feral since 1871

AMZ: *see* Australian Milking Zebu

Anadolu Yerli Kara: *see* Anatolian
Black

Anatolian: (Syria) *see* Baladi

Anatolian Black: (C Anatolia,
Turkey)/d.dr.m/usu. black/
brachyceros type/Turk. *Anadolu
Yerli Kara* (= *Anatolian Native
Black*)/local syn. *Domestic Black*

Anatolian native: *see* Anatolian
Black, East Anatolian Red, South
Anatolian Red

Ancholi: *see* Ankole

Ancient Cattle of Wales: (Wales)/inc.
coloured vars of Welsh Black, i.e.
Belted, White, Linebacked, Blue,
Red, Smoky or Mouse-coloured/
BS 1981/Welsh *Gwartheg
Hynafol Cymru*/syn. *Coloured
Cattle of Wales*/rare

Ancient Egyptian: *see* Hamitic
Longhorn

Andalusian: (Spain)/inc. Andalusian
Black, Andalusian Blond,
Andalusian Grey, Berrendas,
Retinta/? orig. of Criollo, Texas
Longhorn

Andalusian Black: (W Andalucía,
Spain)/m.(dr)/black or black-
brown/sim. to Black Iberian/Sp.
Negra andaluza/syn. *Negra
campiñesa*, *Negra de la
Campiñas andaluzas*/rare

Andalusian Blond: (Huelva, SW
Spain)/m.(dr)/cinnamon yellow
to corn-coloured/cf. Alentejana/
now var. of and being absorbed

by Retinta/Sp. *Rubia andaluza*/
syn. *Blond Extremadura* (Sp.
Extremeña rubia)

Andalusian Grey: (mts of N Córdoba
and N Huelva, Spain)/m/blue
roan with white abdomen/Sp.
Cárdena andaluza/nearly extinct

Andalusian Pied: *see* Berrendas

Andalusian Red: (Spain)/combined
with Extremadura Red and
Andalusian Blond to form
Retinta/Sp. *Retinta andaluza*

Andaluz: (Colombia) *see* Costeño con
Cuernos

Angeln: (E Schleswig, Germany)/d/
Baltic Red type, sim. to Danish
Red/linked to German Red since
1942/BS 1879, HB 1885; HB also
Russia/Ger. *Angler*/syn. *Red
Angeln* (*Angler Rotvieh*)/*see also*
Old Red Angeln

Anglesea, Anglesey: *see* North Wales
Black

Angola: *see* Barotse, Barra do Cuanzo,
Humbi, Kwaniama, Mucubai,
Ovambo, Porto Amboim,
Tshilingue

Angola: (Sertão, NE Brazil)/
coloursided, dark red, brown or
black/? orig. from (African?) zebu
× Curraleiro in 19th c./extinct

Angone: *see* Angonia

Angoni: (W and S of L Malawi, SE
Africa)/m.d.dr/many colours/
East African Shorthorned Zebu
type/vars: Malawi Zebu,
Mozambique Angoni, Zambia
Angoni/not *Angone*/[Angoni
tribe, descended from Nguni tribe]

Angonia: (Angonia, Tete,
Mozambique)/m/usu. black, or
black with white on head, throat
or dewlap, also brown or red; hd
or pd/z/Small East African Zebu,
Angoni group/syn. *Angone,
Mozambique Angoni*/*see also*
Bovines of Tete

Angus: *see* Aberdeen-Angus,
American Angus

Angus Doddie: (NE Scotland)/orig. (+
Buchan Humlie) of Aberdeen-
Angus/[doddie = pd]/extinct

Anjou: *see* Maine-Anjou

Ankina: (USA)/m/$\frac{3}{8}$ Chianina, $\frac{5}{8}$

American Angus/superseded by
Chiangus/BS 1975–1987

Ankole: (L Mobutu to L Tanganyika,
E Africa)/often red, also fawn,
black or pied/Sanga type/vars:
Bahima, Bashi, Kigezi, Ruzizi,
Watusi/syn. *Ankole Longhorn*/
not *Ancholi, Ankoli*

Ankole-Watusi: (USA)/m/giant
horns/orig. from Ankole used to
grade Texas Longhorn/BS 1983/
syn. *Watusi*/rare

Annamese, Annamite: *see* Vietnamese

Ansbach-Triesdorfer: (Middle
Franconia, Bavaria, Germany)/
d.m.dr/yellow and red "tiger"
pattern/orig. from East Friesian
and Simmental × local (Allgäuer
and Breitenburger)/nearly extinct

Antakli, Antakya: *see* Lebanese

Antioqueña, Antioquia: *see* Blanco
Orejinegro

Anuak: *see* Abigar

Anxi: (NW Gansu, China)/var.
adapted to dry sandy regions/
Mongolian group

Aosta: (Val d'Aosta, NW Italy)/ d.(m)/
vars: Aosta Black Pied, Aosta
Chestnut, Aosta Red Pied,
Oropa/BSd 1937, HB 1958, BS/It.
Valdostana, Fr. *Valdôtaine*

Aosta Black Pied: (Val d'Aosta, NW
Italy)/var. of Aosta/It. *Valdostana
pezzata nera*, Fr. *Valdôtaine pie-
noire*/declining by spread of
Aosta Chestnut

Aosta Chestnut: (Val d'Aosta, NW
Italy)/var. of Aosta/? orig. from
Hérens × Aosta Black Pied/HB
1983/It. *Valdostana castana*, Fr.
Valdôtaine châtaigne, Châtaignée

Aosta Red Pied: (Val d'Aosta, NW
Italy)/white and red or yellowish-
red, with white head/chief var. of
Aosta/It. *Valdostana pezzata
rossa*, Fr. *Valdôtaine pie-rouge*

Apulian Podolian: (Apulia, Basilicata
and Calabria, Italy)/m.[dr]/grey/
Podolian type/former vars:
Abruzzese, Calabrian, Lucanian,
Murgese, Venetian/BS and HB
1931/It. *Podolica, Podolica
pugliese*

Aquitaine Blond: *see* Blonde
d'Aquitaine

Arab: (Eritrea) *see* Bahari
Arab: (Israel) *see* Baladi
Arab, Arab Zebu: (Sudan) *see* Large
    East African Zebu
Arab: (W Africa) *see* Azaouak, Kuri,
    Maure, Shuwa
Arab Choa, Arabe choua: *see* Shuwa
Arab Shuwa: *see* Shuwa
Arabi: (Iran) *see* Nejdi
Arabian: (Egypt) *see* Maryuti
Aracena: (Ribatejo, Portugal)/m/
    black with white back, tail and
    belly/rare
Aradó: (highlands of Eritrea and N
    Ethiopia)/dr.m/usu. shades of
    red, often red pied or black pied,
    occ. black, brown, grey or white;
    small muscular hump/Sanga-
    Zebu intermediate/orig. from
    Abyssinian zebus and
    Abyssinian Sangas/local names:
    Akele-Guzai, Asaorta (It.
    Asaortina), *Bileri* (Keren), *Tigray*
    or *Tigré* (N Ethiopia)
Aral: (Togo)/z/sim. to Shuwa/syn.
    Shuwa-Aral
Archangel: *see* Kholmogory
Ardebili: *see* Sarabi
Ardennes-Liège Red Pied: *see*
    Belgian Red Pied, Eastern Red
    Pied (Belgium)
Argentine Criollo: (NW Argentina)/
    m.dr/lyre horns/sim. to Southern
    Crioulo/vars: Chaqueño,
    Fronterizo, Patagonian Criollo,
    Serrano/BS 1984, HB 1989
Argentine Holstein: (Argentina)/
    d.(m)/Sp. *Holando-Argentino*/
    syn. *Argentine Friesian, Dutch
    Argentine*
Aria: *see* Gasara
Armorican: (N Brittany, France)/
    d.m/usu. red, also red-and-white
    or roan/sim. to Maine-Anjou but
    smaller/orig. from Durham
    Shorthorn (imported 1840–1914
    and 1951–1953) × local (Carhaix
    Red Pied and Froment du Léon)/
    named 1923; combined with
    Maine-Anjou 1962–1970 under
    name *Rouge de l'Ouest* (=
    Western Red); crossed (1966 on)
    with Meuse-Rhine-Yssel and
    German Red Pied to form

(together with the 3 pure breeds)
    Pie Rouge des Plaines (= French
    Red Pied Lowland)/HB
    1919–1966/Fr. *Amoricain*/
    purebred nearly extinct
Arouquesa: (Arouca, N Beira,
    Portugal)/m.dr.(d)/light to dark
    chestnut/? orig. from Barrosã,
    Mirandesa, Minhota *et al.*/HB
    1982/not *Aroucesa*
Arsi: (highlands of C Ethiopia)/
    dr/red, black, roan, or white or
    light grey with black spots on
    side/z/developed from
    Abyssinian Shorthorned Zebu/
    syn. *Arusi, Arussi*/not *Arssi*
Artón del Valle: *see* Hartón
Arusi, Arussi: *see* Arsi
Arussi-Galla: (Ethiopia)/Sanga
    type/extinct early 20th c.
Arvi: *see* Gaolao
Asaorta, Asaortina: *see* Aradó
Asiago: *see* Burlina
Askanian Meat: (S Ukraine)/m/new
    breed orig. from Hereford,
    Charolais and Cuban Zebu ♂♂ ×
    Red Steppe ♀♀
Assam local: (India)/local type, being
    crossed with Jersey and Holstein
    ♂♂
Astrakhan: *see* Kalmyk
Asturian: (Oviedo, N Spain)/
    m.(dr.d)/shades of red with paler
    extremities/North Spanish
    type/vars: Carreña, Casina/HB
    1933/Sp. *Asturiana*
Asturiana de las montañas: *see*
    Casina
Asturiana de los valles: *see* Carreña
Atacora, Atakora: *see* Somba
Atlas: *see* Brown Atlas
Atpadi Mahal: (India)/var. of
    Khillari/syn. *Hanam* (= south)
Aubrac: (Aveyron-Lozère, France)/
    m.[dr.d]/fawn to brown/HB 1892,
    BS 1914/syn. *Laguiole*
Augeronne: (Normandy, France)/
    nearly extinct
Aulie-Ata: (S Kazakhstan and N
    Kyrgyzstan)/d.m/usu. black with
    white markings on underside,
    occ. grey/orig. (1885–1912 and
    1926 on) from Friesian × Kazakh/
    recog. 1950/HB 1935/Russ.
    Auliéatinskaya, Auliatinskaya

Aure et Saint-Girons: (S Haute Garonne, S France)/m.d.dr/ chestnut; lyre horns/HB 1919/ syn. *Aurois, Casta, Central Pyrenean* (Fr. *Race des Pyrénées centrales), Race de St Girons et d'Aure*)/nearly extinct

aurochs: = *Bos (Bos) primigenius* Bojanus/dark brown to black; forward lyre horns/vars: *B. p. primigenius* (Europe), *B. p. opistonomus* (N Africa), *B. p. namadicus* (Asia)/orig. of domestic cattle/Ger. *Ur*, Latin *urus*, Hebrew *reem, rimu*/syn. *wild ox*/not *European bison*/extinct 1627 (Poland)

Aurochs de Heck: *see* Heck cattle

Aurochs reconsti.: *see* Heck cattle

Aurois: *see* Aure et Saint-Girons

Aussie Red: (Australia)/d/blended type inc. Illawarra and Dairy or Milking Shorthorn, with influence from Danish Red and other red dairy breeds

Australian Angus: (Australia)/BS 1926

Australian Braford: (N Queensland, Australia)/m/white face; hd or pd/cf. Braford (USA)/orig. 1946–1952 from Brahman ($\frac{1}{2}$) × Hereford ($\frac{1}{2}$)/HB 1956, BS 1962

Australian Brangus: (Queensland, Australia)/m/black; pd/cf. Brangus (USA)/orig. 1950–1960 from Brahman (3/8) × Aberdeen-Angus (5/8)/HB 1956, BS 1961

Australian Charbray: (N Australia)/ m/white to light red/cf. Charbray (USA)/orig. 1970s from Charolais ($\frac{1}{4}$–$\frac{3}{4}$) × Brahman ($\frac{3}{4}$–$\frac{1}{4}$)/BS 1977

Australian Commercial Dairy Cow: (Australia)/d/BS (? 1985) for superior animals regardless of breed/syn. *ACDC*

Australian Frieswal: (N Queensland, Australia)/d/black to red or pied, sometimes brindle/orig. 1961 on at Ayr and Kairi Research Stas by selection for tick resistance, milk let-down and milk yield from F_2 and F_3 of Sahiwal × Friesian cross/HB 1983, BS 1987/syn. *AFS, Australian Friesian Sahiwal*

Australian Grey: (Australia)/BS 1979 to register both pedigree and off-type Murray Grey and Tasmanian Grey

Australian Illawarra Shorthorn: *see* Illawarra

Australian Milking Zebu: (N of NSW, Australia)/d/Jersey colours/orig. since mid-1950s from Sahiwal and Red Sindhi × Jersey; F_3 selected for milk yield, heat tolerance and tick resistance/ recog. 1973/BS 1973/syn. *AMZ*/rare

Australian Sahiwal: (Queensland, Australia)/d/orig. from 10 Sahiwals imported 1954 used to grade Shorthorn, Devon and Jersey to at least 7/8 Sahiwal/BS 1975

Australian Shorthorn: (N Territory and NW Australia)/m/var. of Beef Shorthorn that has become naturally adapted to tropical N/syn. *Kimberley Shorthorn, North Australian Shorthorn, Northern Territory Shorthorn*

Australian White: (Australia)/m/pd/ orig. from 3 British White heifers imported 1958 in calf to White Galloway, and purebreds imported 1984/BS 1983

Austrian Blond: (Carinthia and, formerly, Styria, Austria)/d.m.dr/ nearly white/orig. from old German red and Austrian mt types/recog. 1900/with Murboden and Waldviertel = Austrian Yellow/Ger. *Blondvieh*/ syn. *Carinthian Blond* (Ger. *Kärtner Blondvieh*, Sn. *Koruška plava*), *Lavanttal* (Carinthia), *Mariahof* (Styria), *Mariahofer-Lavanttaler, Styrian Blond* (Ger. *Steierisches Blondvieh*)/nearly extinct

Austrian Brown: (W Austria)/ d.m.[dr]/Brown Mountain type/American Brown Swiss blood in 1960s/former vars: Montafon, Styrian Brown, Tyrol Brown/BS and HB/Ger. *Österreichisches Braunvieh*

Austrian Brown (Original): (Austria)/ d.m/brown with light ring (Ger. *Rehmaul*) around muzzle, black

skin/pure type of Austrian
Brown (without American
Brown Swiss blood)/Ger.
*Original Österreichisches
Braunvieh*/nearly extinct
Austrian Simmental: (Austria)/
d.m.[dr]/orig. (1830 on) from
(Swiss) Simmental and German
Simmental × local (e.g.
Bergscheck)/former vars: Danube,
East Styrian Spotted, Innviertel,
Tyrol Spotted/BS and HB/Ger.
Österreichisches Fleckvieh (=
Austrian Spotted cattle)
Austrian Yellow: (C Austria)/d.m/
white or yellow/name used 1960
for Austrian Blond + Murboden +
Waldviertel/[BS and HB]/Ger.
Österreichisches Gelbvieh/syn.
Light (or *Pale*) *Alpine* (Ger. *Lichtes
Alpenvieh*), *Light Mountain* (Ger.
Lichtes Höhenvieh), *Pale
Highland*/ extinct
Avai: *see* Ethiopian Boran
Avétonou: (Togo)/m/orig. at Avétonou
Research Centre from N'Dama ($\frac{1}{2}$),
local (Somba, Borgou and Lagune)
($\frac{1}{4}$), and Yellow Franconian or
Brown Mountain ($\frac{1}{4}$)/extinct
Avileña: (Avila, C Spain)/dr.m/black/
orig. from local black cattle
superior to and named separately
from other Serrana types (which
were later called *Negra Iberica*, i.e.
Black Iberian)/joined with Black
Iberian (1980) to form Avileña-
Black Iberian/BS 1974/syn. *Barco-
Piedrahita, Barqueña* (from Barco),
Black Carpetana, Piedrahitense
(from Piedrahita)
Avileña-Black Iberian: (C Spain)/
m.[dr]/black/orig. 1980 by union
of Avileña and Black Iberian/
BS/Sp. *Avileña-Negra Iberica*
Awankari: *see* Dhanni
Aweil Dinka: (NW of S Sudan)/d.m/
Sanga type, local var. of Nilotic;
sim. to but smaller than Aliab
Dinka
Ayrshire: (SW Scotland)/d/red-
brown and white; lyre horns/
orig. in late 18th and early 19th
c. from Teeswater *et al.* × local,
with Highland blood in 19th

c./orig. of Finnish Ayrshire,
Swedish Ayrshire; part orig. of
Red Trondheim, Norwegian Red-
and-White/recog. 1814, BS 1877,
HB 1878; BS also USA 1863 (HB
1875), Canada 1870, Australia
(HB 1892), S Africa 1916 (HB
1906), New Zealand 1909,
Kenya, Zimbabwe, Brazil,
Czechoslovakia, Colombia; HB
also Russia /obs. syn.
Cunningham, Dunlop/not *Ayr*
Azangus: (Azerbaijan) Aberdeen-
Angus crossed with zebu
Azaouak: (E Mali, WC Niger and NW
Nigeria)/m.pa.d/red, white, red-
and-white, black-and-white or
fawn with white patches/West
African (shorthorned) Zebu, sim.
to Shuwa/Hausa *Shanun Adar*,
Fulani *Azawa, Azawal* (pl.
Azawaje)/syn. *Arab,
Damerghou, Darmeghou,
Tagama* (Niger), *Tuareg* (sing.
Targui, Targhi, Targi; Fr.
Touareg)/not *Azawak*
Azawa, Azawaje, Azawak, Azawal:
see Azaouak
Azebuado: (Brazil)/intermediates
between zebu and European
cattle/inc. Angola, Brazilian
Dairy Hybrid, Canchim, China,
Dairy Indo-European, Guademar,
Ibagé, Javanês, Lavinia, Malabar,
Mantiqueira, Nilo, Pitangueiras,
Riopardense, Santa Gabriela,
Tatu/syn. *Indu-Europeu,
Zeburano*
Azerbaijan Brown: *see* Caucasian
Brown
Azerbaijan Red: (Azerbaijan)/subvar.
of Lesser Caucasus var. of
Caucasian/crossed with Swiss
Brown to form Caucasian
Brown/Russ. *Krasnaya
Azerbaĭdzhanskaya*/extinct
Azerbaijan Zebu: (SE Azerbaijan)/
usu. red, black or pied/z/cf.
Talishi/Russ. *Azerbaĭdzhanskiĭ
zebu*/syn. *Caucasian Zebu,
Talysh* (Russ. *Talyshinskiĭ*)/rare
Azores: (Portugal)/m.d/orig. from
Mirandesa and Minhota with
exotic blood/Port. *Açoreana*

Azul y Pintado: (Andes, Colombia)/ blue roan and pied/var. of Blanco Orejinegro

Babaev: (Kostroma, Russia)/orig. from Allgäuer × local/orig. (with Miskov and crossed with Swiss Brown) of Kostroma/extinct

Bachaur: (N Bihar, India)/dr/z; Grey-white Shorthorned type, sim. to Hariana/syn. *Sitamarhi*/not *Bachchaur, Bachhaur, Bachur*

Baden Spotted: *see* Messkircher

Baggara: (Darfur and Kordofan, Sudan)/m.pa.d/often white with red (or black) markings, also grey, red, black, yellow or dun/ Large East African Zebu group, North Sudan type/influenced by Fulani cattle in W and by Nilotic in S/syn. *Western Baggara*/ [*baggar* = cattle]

Baggerbont: (Netherlands)/colour var.: red or black pied with spotted legs/[= mud-pied]/rare

Bahari: (S Red Sea coast) *see* Baherie

Baharié: (W Africa) *see* Kuri [Arabic *bahar* = sea]

Baherie: (Red Sea coast of Massawa region, E Eritrea)/d.m/usu. chestnut, also pied/z/Small East African Zebu (Somali group); sim. to North Somali/orig. from Yemeni Zebu/syn. *Aden, Arab, Bahari* (= cattle from sea or ocean), *Berbera* (from Somalia)

Bahima: (Kibali-Ituri, NE DR Congo and SW Uganda)/d.m.dr/usu. dark red, also black, brown, white, grey, dun and combinations of these; large horns; small hump/ Sanga type, Ankole group/syn. *Banyoro* (incorrect), *Nsagalla* (Uganda)/not *Bahema, Wahima, Wakuma*/rare

Bahu: *see* Lugware

Baila: (Kafue flood plain, Zambia)/ var. of Barotse/orig. from Tonga by crossing with Barotse in 20th c. (hence earlier a syn. of Tonga)/ sing. *Ila*/syn. *Bashukulumpo, Mashuk, Mashukulumbwe*

Bainiu: (S China)/z/cf. Jiniu/name for small or dwarf cattle in ancient China/extinct

Bakosi: (Bangem, SW Cameroon)/ m.ri/black, brown or pied/var. of West African Savanna Shorthorn (= Muturu of Nigeria)/syn. *Bakuri, Kosi*/not *Bakossi*/rare

Bakweri: (foot of Mt Cameroon, SW Cameroon)/m.ri/var. of West African Dwarf Shorthorn/syn. *Muturu*/not *Bakuri, Bakwiri*/ nearly extinct

Baladi: (Lower Egypt)/dr/var. of Egyptian/subvar.: Menufi/syn. *Beheri*/not *Balladi, Beladi*/[= native]

Baladi: (Saudi Arabia) *see* Hassawi

Baladi: (Syria, Lebanon, Israel and Jordan)/dr/brown to black, or pied; 30% pd; small/*brachyceros* type/Jaulan is sim. but larger/ syn. *Akshi, Anatolian* (Hama), *Arab* (Israel), *Bedouin, Djebeli, Kleiti* (Homs), *Oksh* (Israel)

Bale: (high plateaux of Bale mts, Ethiopia)/dr.(m.d)/black, chestnut, white or roan; prominent hump/Abyssinian Shorthorned Zebu group/syn. *Abyssinian Highland Zebu*

Bali cattle: (Bali, also in Lombok, Timor, S Sulawesi, E Java and SE Kalimantan, Indonesia)/m.dr/ ♂ red turning black, ♀ red with white patches/orig. from banteng/orig. of Cobourg Peninsula cattle/syn. *Balinese*

Balkan: *see* Illyrian

Ballum: (SW Jutland, Denmark)/ red/part orig. of Danish Red/ syn. *Slesvig Marsh* (Ger. *Schleswigsche Marschrasse*)/ extinct

Bălţată cu negru românească: *see* Romanian Holstein

Bălţată românească: *see* Romanian Simmental

Baltic Black Pied: (Baltic States)/ d/ orig. (1830 on) from East Friesian × local/inc. Estonian Black Pied, Lithuanian Black Pied/Russ. *Pribaltiĭskaya chernopestraya*

BALTIC RED: orig. from *brachyceros* type/orig. of Bulgarian Red, Red Steppe, Romanian Red/*see* Angeln, Danish Red, Estonian Red, Latvian Brown

Bamangwato: *see* Mangwato

Bambara: (SW Mali)/m/yellow-brown to red/orig. from N'Dama × zebu (Sudanese Fulani and other)/syn. *Mandé, Méré* (= small, not zebu)

Bambawa: (Eritrea/Sudan border)/ strain of Red Desert cattle sim. to Butana/It. *Bambaua*/rare (from crossbreeding with Butana)

Bambey: (Senegal)/stabilized crossbred of Djakoré type bred at Bambey Research Centre since 1921 ($\frac{13}{16}$ N'Dama, $\frac{3}{16}$ Gobra zebu)

Bamenda: *see* N'Gaoundéré

bami: (Bhutan) *see* mithun

Bami: (Kerman province, Iran)/m/ yellow to brown, occ. black/z

Bami: (Bhutan) *see* mithun, gayal

Bamiléké: (SW Cameroon)/var. of West African Savanna Shorthorn/extinct

Bamongwato: *see* Mangwato

Bangladeshi: *see* Bengali

Banioro: *see* Nyoro

Banna: *see* Xishuangbanna

Bannai: *see* Kankrej

banteng: (SE Asia)/= *Bos (Bibos) javanicus* (d'Alton)/brown (♀) or black (adult ♂) with white stockings and rump patch/vars: tsine, Malay banteng, Borneo banteng/orig. of Bali cattle/syn. *B. banteng* Wagner, *B. sondaicus* Schlegel and Müller/not *bantin, banting*/rare

Bantu: *see* Sanga

Banyo: (Banyo and Bamenda Highlands, Cameroon and Mambila Highlands, Nigeria)/ d.m.dr/red or red pied, often white face and belly/z/Adamawa subgroup of Gudali group; var. of Adamawa with Red Bororo blood/Fr. *Foulbé de Banyo* (*Foulbé* = Fulani)

Banyoro: *see* Nyoro; used incorrectly in DR Congo to mean Bahima

Baol: (Senegal)/local var. of or syn. for Gobra zebu

Baoulé: (Côte d'Ivoire)/d.m/usu. black pied, also black, red pied with red back, occ. fawn/var. of West African Savanna Shorthorn/

[name for Savanna Shorthorn in French-speaking countries]

Baoule: (Burkina Faso) *see* Lobi, Méré

Bapedi: *see* Pedi

Baqra Maltija: *see* Maltese

Baraka, Barca: *see* Barka

Barco: *see* Avileña

Bardigiana: (Bardi, Parma, W Emilia, Italy)/dr.d.m/red/var. of Pontremolese/subvars: Cornigliese, Valtarese/extinct

Bare: *see* Kuri

Bargur: (W Coimbatore, Tamil Nadu, India)/dr/usu. red-and-white spotted/z; Mysore type/not *Barghur, Burgaur, Burghoor, Burgoor, Burgur*

Bari: (S Sudan)/very long horns/strain of Mongalla, ? with Nilotic blood

Baria: (Cochin-China, S Vietnam)/ var. of Vietnamese

Baria: (Plateau de Bemeraha, NW Madagascar)/m.d/black; small hump; feral/? orig. from zebu × humpless/nearly extinct

Barka: (lowlands of W Eritrea, also Ethiopia)/d.m/usu. white with black spots or splashes, occ. red or coloursided; occ. pd/cf. Baggara/North Sudan Zebu group/syn. *Begait*/not *Baraka, Barca*

Barkly: (N Territory, Australia)

Barotse: (W Zambia and E Angola)/ m.(d.dr)/usu. brown or black, also dark red, occ. dun or fawn/ Sanga type, sim. to Tswana/vars: Baila, Porto Amboim/syn. *Lozi, Rowzi, Rozi*/not *Barotsi*/[Barotse or Lozi tribe]

Barqueña: *see* Avileña

Barra do Cuanzo: (W Angola)/m/ light tan to red/improved Sanga var. in Cuanzo district from Porto-Amboim or Barotse with European blood, esp. Charolais

Barrosã: (Terra de Barroso, N Portugal)/dr.m/reddish-brown with black legs; wide lyre horns/BS 1985/syn. *Maiana* (from Maia)/not *Barozza*

Barroso: (Guatemala)/m.d/dun/ Criollo type/rare/[= mud-coloured]

Barundi: *see* Burundi

Barzona: (C Arizona, USA)/m/usu. red; occ. pd/orig. 1945–1968 at Bard-Kirkland Ranch from [American Angus × (Afrikander × Hereford)] × [Santa Gertrudis × (Afrikander × Hereford)]/BS 1968/[*Bar*d *Arizona*]

Basco-Béarnais: *see* Béarnais

Bashan: (Qinling mts, S Shaanxi, NE Sichuan and NW Hubei, China)/ dr.m.d/red or yellow/Changzu group/orig. 1982 by combining local cattle, inc. Chiya, Lingnan, Miaoya (= Yunba), Pingli, Qinba, Xizhen, Xuanhan/now three types: "burly", "compact" (majority) and "fine"

Bashi: (S Kivu, E DR Congo)/d.m/ usu. red, black or fawn, or mixtures of these; small hump/ smaller var. of Ankole with shorter horns; sim. to Kigezi

Bashukulumpo: *see* Baila

Basque: *see* Pyrenean, Pyrenean Blond

Bassanese: *see* Burlina

Basta: *see* Canary Island

Basuto: (Lesotho)/dr.(d.m)/usu. black/Sanga type, often with European blood; original pure type extinct, remnants now crossed with Drakensburg, Afrikaner and Friesian/rare

Batanes Black: (Batan Is, Philippines)/small-humped var. of Philippine Native

Batangas: (Luzon, Philippines)/red, yellow or black/improved humpless var. of Philippine Native/not *Batanges*

Batawana: (Ngamiland, NW Botswana)/var. of Tswana with (formerly) very long horns/syn. *Ndawana, Ngami, Tawana*/not *Batawama, Batwane, Botawana*

Batusi, Batutsi: *see* Watusi

Batwane: *see* Batawana

Bavarian Red: (Fichtelgebirge, N Bavaria, Germany)/var. of German Red/Ger. *Bayerisches Rotvieh*/syn. *Sechsämter*/ displaced by German Simmental; extinct *c.* 1940 (Red cattle in

Bavaria since 1954 are Danish Red and Angeln)

Bavarian Simmental: (Hungary)/ local Hungarian cattle upgraded by crossing with Simmental first half of 20th c., 1960–1972 with Holstein, Red-and-White Holstein and Jersey to improve milk yield; upgraded to Holstein

Bavarian Spotted: *see* Miesbacher

Bavenda: (Sibasa, NE Transvaal, S Africa)/dr.m/usu. black or black-and-white, but also many other colours and patterns; sometimes pd/Sanga type, sim. to Mashona and Nguni/syn. *Sibasa*/smaller old type nearly extinct (by crossing with larger type)

Bayaro: *see* Diali

Bayerisches Rotvieh: *see* Bavarian Red

Bazadais: (Gironde-Landes, France)/ m.[dr]/grey to grey-brown with pink mucosae/HB 1895/[from Bazas]

Bazougers Blue: (Maine-Anjou-Perche region, France)/sim. orig. to Maine; being revived/Fr. *Bleue de Bazougers*

Béarnais: (Aspe valley, Béarn, SW France)/d.m/yellow; lyre horns/remnant of Pyrenean Blond (Fr. *Blonde des Pyrénées*) not absorbed by Blonde d'Aquitaine/syn. *Basco-Béarnais*/nearly extinct

Bechuana: *see* Tswana

Bedouin: (Syria) *see* Baladi

Beefalo: (California, USA)/m/cf. cattalo/orig. 1960s by D.C. Basolo, Tracy, from $\frac{3}{8}$ bison, $\frac{3}{8}$ Charolais, $\frac{1}{4}$ Hereford/BS 1983

Beefbilde: *see* Beevbilde

Beef Brown Swiss: (USA and Canada)/= Swiss Brown (or Brown Mountain)/BS 1984

Beef Devon: *see* Devon

Beef Friesian: *see* American Beef Friesian

Beef Machine: (New Mexico, USA)/ Red Poll, Hereford, Brown Swiss, Angus, Friesian and Simmental blood

Beefmaker: (Nebraska, USA)/m/orig. 1960s by G.A. Boucher from

Charolais (*c.* 50%) × Hereford, Angus and Shorthorn, with Brown Swiss and some Brahman blood

Beefmaker: ('Wallamumbi', Armidale, NSW, Australia)/m/ white face/orig. 1973 on from (Simmental × Hereford) × Hereford

Beefmaster: (USA)/m/often dun, red-brown or pied; hd or pd/orig. 1930 on at Lasater Ranch (SW Texas to 1954, now Colorado) from Hereford (*c.* ¼), Shorthorn (*c.* ¼) and Brahman (*c.* ½)/BS 1961; BS also S Africa 1986

Beef Shorthorn: (NE Scotland)/m/ breed of Shorthorn developed in mid-19th c./var.: Australian Shorthorn/HB (Beef section) 1959; BS USA 1882 (HB 1846), New Zealand 1914 (HB 1866), Canada 1886 (HB 1867), Australia 1920 (HB 1873), Argentina 1921 (HB 1889), Uruguay, S Africa 1919 (BS 1912); HB also Russia/syn. *Korthoring* (Afrik.), *Shorthorn* (overseas), *Scotch Shorthorn, Scottish Shorthorn*/obs. syn. *Aberdeenshire Shorthorn*/rare in GB

Beef Synthetic: (Alberta Univ., Canada)/m/37% Aberdeen-Angus, 34% Charolais, 21% Galloway, 5% American Brown Swiss, 3% other breeds

Beevbilde: (England)/m/red; pd/ developed in 1960s by E. Pentecost (Notts) with 65% Lincoln Red, 30% Beef Shorthorn and 5% Aberdeen-Angus/not *Beefbilde*

Begait: *see* Barka

Beheri: (Egypt) *see* Baladi

behi auzoa: *see* Betizuak

Beijing Black Pied: (China)/var. of Chinese Black-and-White/orig. from Holstein and Japanese Friesian × Pinzhou/syn. *Peking Black-and-White*

Beiroa: (Beira, N Portugal)/dr.m.(d)/ var. of Mirandesa

Beirut: *see* Lebanese

Beja: (NE Sudan)/grey/? var. of North Sudan Zebu

Beladi, Bélédie: *see* Baladi

Belarus Black Pied: (Belarus)/syn. *Belarus Black-and-White*

Belarus Red: (Belarus)/d/Central European Red type, improved by Angeln and German Red (late 19th and early 20th c.), Polish Red and Danish Red (1920s and 1930s), Estonian Red and Latvian Brown (1950s), and Danish Red (1980s)/HB 1967/Russ. *Krasnaya belorusskaya, Krasnobelorusskaya*/obs. syn. (to 1991) *Byelorussian Red, Red White-Russian, White-Russian Red*

Belarus Synthetic: (Belarus)/m/ Maine-Anjou (¼), Limousin (¼), Salers (¼), Russian Simmental (⅛), Russian Brown (⅛)/obs. syn. *Byelorussian Synthetic*

Belgian Black Pied: (Belgium)/d/orig. 1966 from Hervé Black Pied + Polders Black Pied; crossing with Holstein since 1950/Flem. *Zwartbont ras van België*, Fr. *Pie-noire de Belgique*/syn. *Belgian Black-and-White, Pie-noire Holstein*

Belgian Blue: (C and S Belgium)/ m.(d)/white, blue pied or blue; double-muscled/cf. Bleu du Nord/now 2 lines (beef, dual-purpose)/orig. from Shorthorn (1850–1890) and Dutch Black Pied (1840–1890) × local red or red pied/orig. of Danish Blue-and-White/HB 1919; BS Ireland 1980, GB 1983, Canada 1986, USA 1988, France 1989; HB Netherlands/Flem. *Witblauw ras van België*, Fr. *Blanc-bleu belge* (= *Belgian White-Blue*), Dan. *Belgisk Blåhvidt kvæg*/syn. (to 1971) *Central and Upper Belgian* (Flem. *Ras van Midden- en Hoog-België*, Fr. *Race de la Moyenne et Haute Belgique*); syn. *Belgian Blue-White, Belgian Roan, Belgian White and Blue Pied, Belgian White Blue, Blue, Blue Belgian, Central and*

Upland, Mid-Belgian, White Meuse and Schelde

Belgian Red: (W Flanders, Belgium)/ d.m/sim. to Flemish (France)/ Shorthorn blood/HB 1919/Flem. *Rood ras van België*, Fr. *Rouge de Belgique*/syn. *Red Flemish, West Flemish Red* (Flem. *Rood ras van West-Vlaanderen* or *West Vlaamse*, Fr. *Rouge de la Flandre occidentale*)

Belgian Red Pied: (NE Belgium)/ d.m/orig. 1838 on and esp. 1878–1888 from Meuse-Rhine-Yssel × local, with some Shorthorn blood 1844–1851/inc. Eastern Red Pied (*Pie rouge de l'Est* or *Pie rouge Ardennes-Liège*)/HB 1919/Flem. *Roodbont ras van België*, Fr. *Pie-rouge de Belgique*/syn. (to 1971) *Campine Red Pied* or *Red-and-White Campine* (Flem. *Roodbont Kempisch* or *Kempen*, Fr. *Rouge-pie Campinoise* or *Pie-rouge de Campine*)

Belgian White-and-Red: (E Flanders, Belgium)/d.m/orig. in 2nd half of 19th c. from Shorthorn and Dutch × local/HB 1900/Flem. *Witrood ras van België*, Fr. *Blanc-rouge de Belgique*/syn. *East Flemish Red Pied* (Flem. *Roodbont ras van Oost-Vlaanderen* or *Oost Vlaamse*, Fr. *Pie-rouge de la Flandre orientale*), *Red-and-White East Flemish, Red Pied Flemish*

Belgian White-and-Blue Pied: *see* Belgian Blue

Belgian White-Blue: *see* Belgian Blue

Belgisk Blåhvidt Kvæg: *see* Belgian Blue

Beliy sibirskiy skot: *see* Siberian White

Belmont BX: (Queensland, Australia)/*see also* Belmont Adapteur, Belmont Red

Belmont Adapteur: (Queensland, Australia)/orig. since 1953 at National Cattle Breeding Sta., 'Belmont', Rockhampton, from Hereford × Shorthorn selected for tropical adaptation and esp.

from single ♀ with zero ticks (1980)/syn. *Adapteur*

Belmont Red: (Queensland, Australia)/m/orig. 1953–1968 at National Cattle Breeding Sta., 'Belmont', Rockhampton, from (Afrikander × Hereford) × (Afrikander × Shorthorn) selected for fertility, weight gain and tick resistance/recog. 1968, BS 1978

Belogolovokolonistskaya: *see* Ukrainian Whiteheaded

Belo slovensko govedo: *see* Slovenian White

belted: *see* Belted Galloway, belted Swiss Brown, Belted Welsh, Broadlands, Dutch Belted, Lakenvelder, Sheeted Somerset/ syn. *beltie, sheeted, white-middled*/belt due to single dominant gene

Belted Galloway: (SW Scotland)/ black or dun with dominant white belt; pd/var. of Galloway/ BS 1921 (with Dun Galloway till 1951), HB 1922; HB also registers Red Belted Galloway; BS also New Zealand 1948, USA 1951, Australia 1975, W Germany/syn. *Beltie, Sheeted* or *White-middled Galloway*/*see also* Miniature Galloway

Belted Welsh: (N Wales)/var. of Ancient Cattle of Wales/Welsh *Bolian Gwynion*/syn. *Belted Welsh Black*/rare

Bengali: (Bangladesh and Bengal, India)/small to dwarf/z/local desi vars (Bangladesh): Chittagong Red, Dacca-Faridpur, Kamdhino, Madaripur, Munshiganj, North Bangladesh Grey; *see also* Pabna

Berbera: *see* Baherie

Berciana: (El Bierzo, NW León, Spain)/var. of Alistana-Sanabresa/rare

Bergscheck: (Styria, Austria)/ d.m.dr/red with white head, forequarters, belly and legs/local breed improved since 1900 by Pinzgauer and Simmental/ absorbed by Austrian Simmental *c.* 1950 (grade was termed

Alpenfleckvieh)/syn. *Enns,*
Ennstaler Bergscheck, Ennstaler
Schecken, Helmete (= (white)
helmet), *Kampete* (= collared,
from Ger. *Kummet* = collar, i.e.
white head), *Mountain Spotted*/
extinct 1986

Berlin cattle: (Berlin zoo, Germany)/
cf. Munich cattle/orig. (1920s) by
Lutz Heck from Corsican,
Camargue and Spanish Fighting
cattle as so-called 'bred-back
aurochs'/*see also* Heck cattle

Bernese: (Switzerland)/orig. of
Simmental/syn. *Bernese*
Oberland/extinct

Bernese-Hanna, Berno-Hana: *see*
Moravian Red Pied

Bernsko-český: *see* Bohemian Red
Pied

Bernskohanacký: *see* Moravian Red
Pied

Berrendas: (mts of N and SW
Andalucía, Spain)/m.dr/used for
herding fighting bulls/vars:
Berrenda en colorado, Berrenda
en negro/orig. of Guadiana
Spotted/syn. *Berrenda andaluza,*
Berrendas españolas/[= pied]/
rare

Berrenda en colorado: (Andalucía,
Spain)/white with red head and
forequarters; red area occ. roan
(*Salinera*)/syn. *Berrenda roja*
andaluza, Capirote (= hooded),
Red Berrendo/[= red pied]/rare

Berrenda en negro: (Andalucía,
Spain)/m.dr, bullfighting/
coloursided with black head and
legs (Pinzgauer pattern)/syn.
Berrenda negra andaluza, Black
Berrendo/[= black pied]/rare

Berrenda negra andaluza: *see*
Berrenda en negro

Berrenda roja andaluza: *see*
Berrenda en colorado

Berrichonne: (Berry-Bourbonnais,
France)/dr/supplanted by
Charolais/extinct early 20th c.

Bessarabian Red: (Moldova)/name
used for Red Steppe 1918–1945
when Moldova was part of
Romania/extinct

Bestal: *see* Alambadi

Bestuzhev: (Kuibyshev, Russia)/m.d/
red, often with white marks on
head, belly and feet/orig. (1810
on) by S.P. Bestuzhev from
Friesian, Simmental and
Shorthorn × local, with
Kholmogory and Oldenburg
blood in early 20th c./HB
1928/Russ. *Bestuzhevskaya*

Betizuak: (Goizueta, N Navarre,
Spain)/m/blond to tawny red;
dwarf; feral/Basque *Betizu*/syn.
Betisoak (sing. *Bétiso*) (France),
abel gorriak, behi auzoa, etxeko
behiak, herri ganadua, herri
behiak, kata bizarrak/[Basque
behi = cattle, *izua* = wild, i.e. =
wild cattle]/nearly extinct

Better Idea: (N Dakota, USA)/
American Angus × (Brown
Swiss × Hereford) at Cedar
Ridge Ranch/rare

Bettola, Bettolese: *see* Pontremolese

Beyrouth: *see* Lebanese

Bhagnari: (Kalat, Baluchistan and N
Sind, Pakistan)/dr.d/usu. grey/z;
Grey-white Shorthorned
type/larger type in Baluchistan,
smaller in Sind/var.: Dajjal/syn.
Kachhi, Nari/not *Bhag Nari*/
[Bhag village and Nari R]

bhotea, bhotey, bhotia: *see* dzo

Białogrzbietka: *see* Polish
Whitebacked

Bianca val padana: *see* Modenese

Bielorrusian: *see* Belarus

Bierzo: *see* Berciana

Bileri: *see* Aradó

Bima: (Togo)/lyre horns/z/? syn.
Bimaji

Bimal: (coastal dunes between
Mogadishu and Meream,
Somalia)/z/cf. Gasara/small var.
of Garre/? extinct

Binda: *see* Burlina

Binga: (NW Zimbabwe)/long horns;
humpless; dwarf/extinct

Binjharpuri: (Orissa, India)/dr/z

Bionda tortonese: *see* Montana

Bisre: *see* Jaulan

Biu: (Bornu, Nigeria)/m.dr/white
with black markings; long horns/
orig. from Dwarf Muturu
(Shorthorn) × White Fulani
zebu/absorbed by zebu/? extinct

Blaarkop: *see* Groningen
    Whiteheaded
Black Afrikander: *see* Drakensberger
Black Anatolian: *see* Anatolian Black
Black-and-White: *see* Friesian
Black-and-White Friesland: *see*
    Dutch Black Pied
Black-and-White Holland: *see* Dutch
    Black Pied
Black-and-White Jutland: *see* Jutland
    Black Pied
Black-and-White Swedish: *see*
    Swedish Friesian
black baldy: = Hereford × Aberdeen-
    Angus cross
Black Berrendo: *see* Berrenda en
    negro
Black Canadian: *see* Canadienne
Black Carpetana: *see* Avileña
Black-eared White: (Colombia) *see*
    Blanco Orejinegro
Black Forest: (Germany)/
    d.m.[dr]/inc. Hinterwald and
    Vorderwald/Ger. *Wäldervieh* (=
    forest cattle)/syn. *Gelbscheck* (=
    yellow pied), *Scheckig* (= pied),
    Small Spotted Hill
Black Highland: (Ethiopia) *see*
    Abyssinian Shorthorned Zebu,
    Jem-Jem
Black Hill Zebu: (Nepal) *see*
    Nepalese Hill
Black Iberian: (mts of C Spain)/dr.m/
    orig. from local black draught
    cattle grouped as *Serrana* (=
    mountain), named after areas
    (Avileño, Piedrahitense,
    Barqueño, Pinariego,
    Guadarmeño, etc); Avileño
    became superior and separate;
    group name changed to *Negra
    iberica* to avoid confusion with
    other Serrana cattle elsewhere/
    improved by Avileña and joined
    with it 1980 to form Avileña-
    Black Iberian/Sp. *Negra
    iberica*/obs. syn. *Serrana*/rare
Black Jersey: *see* Canadienne
Black Pied: *see* Friesian
Black Pied Aosta: *see* Aosta Black
    Pied
Black Pied Breton: *see* Breton Black
    Pied

Black Pied Danish: *see* Danish Black
    Pied
Black Pied Dutch: *see* Dutch Black
    Pied
Black Pied Dutch Friesian:
    (Netherlands)/HB 1983 for orig.
    type of Dutch Black Pied free of
    Holstein blood/Du. *Zwartbont
    Fries-Hollands*/syn. *Dutch
    Friesian*/rare
Black Pied East Belgian: *see* Hervé
    Black Pied
Black Pied Hervé: *see* Hervé Black
    Pied
Black Pied Jutland: *see* Jutland Black
    Pied
Black Pied Lowland: *see* Friesian
    (esp. German Black Pied)
Black Pied Podolian: *see* Podolian
    Black Pied
Black Pied Polders: *see* Polders Black
    Pied
Black Pied Valdostana: *see* Aosta
    Black Pied
Black Pinzgau: *see* Dorna
Black Pisa milch cow: *see* Pisana
**Blacksided Trondheim and
    Nordland:** (N Norway)/d/white
    with black (occ. red) sides or
    spots on side; pd/sim. to
    Swedish Mountain and North
    Finnish/HB 1943 (on union of
    Blacksided Trondheim with
    Nordland breed), BS/Nor. *Sidet
    trønderfe og nordlandsfe* (=
    STN)/syn. *Black Trondheim,
    Coloursided Trondheim and
    Northland, Nordland, Røros*/rare
Black Spotted: *see* Swiss Black Pied
Black Spotted Jutland Milk: *see*
    Jutland Black Pied
Black Trondheim: *see* Blacksided
    Trondheim and Nordland
Black Welsh: *see* Welsh Black
Black-White: *see* Black Pied
Black Zebu: (Ethiopia) *see* Jem-Jem
    Zebu
Blanca cacareña: *see* White Cáceres
Blanca guadianese: *see* White Cáceres
Blanc-bleu: *see* Belgian Blue
Blanco Orejimono: (Andes,
    Colombia)/white with red ears/
    var. of Blanco Orejinegro/syn.
    BOM

Blanco Orejinegro: (Andes, Colombia)/m.d/white with black points/Criollo type/vars: Blanco Orejimono, Azul y Pintado/syn. *Antioquia* (Sp. *Antioqueño*), *Bon, BON*/[= white black-eared]/rare

Blanc-rouge de Belgique: *see* Belgian White-and-Red

Blässiges: *see* White-marked

Blauw: (Netherlands)/colour var.: blue roan/rare

Blauwbont: (Netherlands)/colour var.: blue roan and white/rare

Blazed: *see* White-marked

Blended Red and White Shorthorn: (UK)/d.m/Dairy Shorthorn with 50–75% blood of Red Holstein, Red Friesian, Danish Red, Meuse-Rhine-Yssel or Simmental, since 1969

Bleue de Bazougers: *see* Bazougers Blue

Bleu du Nord: (Canton of Bavai, Nord dépt, N France)/m or d.m/blue, blue pied or white/cf. Belgian Blue/HB 1982, BS 1986/syn. *Blue, Northern Blue*/rare

Blond: *see* European Blond and Yellow

Blonde d'Aquitaine: (SW France)/m.[dr]/yellow or yellow-brown/orig. 1961 from Garonnais + Quercy and then absorbed Pyrenean Blond/var.: Coopelso 93/HB; BS also Canada, GB 1971, Denmark, Australia and New Zealand, USA 1973, Ireland 1975, Belgium, Argentina; HB Netherlands/syn. *Aquitaine Blond*

Blonde des Pyrénées: *see* Béarnaise

Blonde des Pyrénées à muqueuses rosées: *see* Pyrenean Blond

Blonde du Cap Bon: *see* Cape Bon Blond

Blonde du Sud-Ouest: (SW France)/ suggested (but unused) name for Blonde d'Aquitaine + Limousin/ [= southwestern blond]

Blond Extremadura: *see* Andalusian Blond

Blond Moroccan: *see* Oulmès-Zaërs Blond

Blond Tortona: *see* Montana

Blondvieh: *see* Austrian Blond

Blond Zaërs: *see* Oulmès-Zaërs Blond

Blue: (Belgium) *see* Belgian Blue

Blue: (France) *see* Bleu du Nord

Blue: (Netherlands) *see* Blauw, Blauwbont

Blue Albion: (N Derbyshire, England)/d.m/blue roan, or blue roan and white/orig. from white Shorthorn × Welsh Black and possibly Friesian/current 'Blue Albions' probably crosses/HB 1916–1940, BS 1920–1966/syn. *Blue English, Derbyshire Blue*

Blue Grey: (Great Britain)/Whitebred Shorthorn or white Beef Shorthorn × Galloway, 1st cross

Blüem: (Switzerland)/coloursided var. of Swiss Brown

Blue Macedonian: *see* Macedonian Blue

Blue Nile: *see* Kenana

Blukwa: *see* Alur

Boccarda: *see* Burlina

Bodadi: *see* Red Bororo

Boenca: *see* N'Dama

boeuf sauvage cambodgien: *see* kouprey

Bohai Black: (NW Shandong, China)/ dr.d.m/Huanghuai type/BSd 1985/syn. *Wudi Black*

Bohemian Berne: *see* Bohemian Red Pied

Bohemian Pied: *see* Czech Pied

Bohemian Red: (Czech Republic)/ d.m.dr/Central European Red type/some Bernese and Simmental bulls used early 20th c./var: Cheb/Cz. *České červinky* (= *Czech Red*)/nearly extinct

Bohemian Red Pied: (Czech Republic)/former var. of Czech Pied/Cz. *Česky červenostrakatý*/ syn. *Bohemian Berne* (Cz. *Bernsko-český*, *Bohemian-Simmental*, *Bohemian Red Spotted*

Bohemian Spotted: *see* Czech Pied

Bohinj Cika: (Slovenia)/var. of Cika, sim. to Tolmin Cika but larger/Sn. *Bohinjska cika*/nearly extinct

Bohuskulla: (SW Sweden)/d/black, brown or white; pd/remnant of

Swedish Mountain cattle/nearly extinct

Bokoloji: *see* Sokoto Gudali

Bolian Gwynion: *see* Belted Welsh

Bolivian Criollo: (Bolivia)/ m.dr.(d)/ vars: Chaqueño, Chusco, Saavedreño, Yacumeño/BS 1988

Bolowana: (Transkei, Cape Prov., S Africa)/Sanga type/sacred herd of Chief Ngubezulu in Bomvanaland/syn. *Izankaya*/ extinct (graded to Afrikander)

BOM: *see* Blanco Orejimono

BON: *see* Blanco Orejinegro

Bonsai Zebu: (Tamun, Mexico)/Indo-Brazilian selected for small size (*c.* 100 cm high) since 1970 by Juan Manuel Berruecos Villalobos at Nat. Autonomous Univ./syn. *Bonsai Brahman*

Bonsmara: (Transvaal, S Africa)/m/ red/orig. 1936–1955 at Mara Cattle Res. Sta. by J.C. Bonsma from Afrikander ($\frac{5}{8}$), Shorthorn ($\frac{3}{16}$) and Hereford ($\frac{3}{16}$)/recog. 1956, BS 1964/[*Bonsma* and M*ara*]

Bonyhádi: (County Tolna, Hungary)/ d/former var. of Hungarian Pied/not *Bonyhád*/extinct

Boran: (N Kenya, S Ethiopia, SW Somalia)/m/usu. white or grey, also red or pied/Large East African Zebu type, shorthorned/ vars: Ethiopian Boran, Galla Boran, Orma Boran, Kenya (Improved) Boran/not *Boram, Borena*

Bordelais: (SW France)/d/black, with white-speckled body/Dutch and Breton blood (orig. larger beef type in Brittany)/nearly extinct since late 1960s (replaced by Friesian); being reconstituted since 1992 as Bordelais nouveau/ [from Bordeaux]/nearly extinct

Borero: *see* Fellata

Borgou: (N Benin)/d.m.dr/usu. white, often with black points, sometimes with black spots or black-and-white or coloursided, also black, or spotted grey or fawn/sim. to Keteku but smaller/ stabilized cross orig. from humpless (Lagune, Somba or Muturu) × zebu (usu. White Fulani); truer type in S Benin, "Borgou-zebu" in N Benin/syn. *Borgawa, Borgowa, Borgu, Kaiama, Ketaku, Keteku* (Nigeria), *Ketari*/also syn. for Keteku

Borgou: (Togo, CAR and other Francophone countries)/name used for all humpless/zebu intermediates

Bor Khalium: *see* Mongolian Yellow

Borneo banteng: (Sabah and Kalimantan)/= *Bos (bibos) javanicus lowi* (Lydekker)/var. of banteng

Borneo Zebu: (Kalimantan, Indonesia)/orig. from Ongole/vars: Kabota, Kaningan

Boro: *see* Red Bororo

Bororo: *see* Red Bororo, White Fulani

Bororodji: *see* Red Bororo

Borrié: *see* Kuri

Borroro: *see* Red Bororo

Bos (Bibos) gaurus: *see* gaur

Bos (Bibos) 'frontalis': *see* mithun

Bos (Bibos) gaurus: *see* Indian gaur

Bos (Bibos) hubbacki: *see* seladang

Bos (Bibos) gaurus readei: *see* Burmese gaur

Bos (Bibos) javanicus: *see* banteng

Bos (Bibos) sauveli: *see* kouprey

"Bos certus": (Brazil)/m/crossing of European and zebu (Nelore) started 1985 by Arno Huber to form sire line ($\frac{5}{8}$ Red Angus, $\frac{1}{4}$ Chianina, $\frac{1}{8}$ Nelore) and dam line ($\frac{5}{8}$ Nelore, $\frac{1}{4}$ Simmental, $\frac{1}{8}$ Charolais) to be crossed to form "Bos certus" (37.5% Nelore zebu, 31.25% Red Angus, 12.5% Chianina, 12.5% Simmental and 6.25% Charolais)

Bos indicus: = zebu cattle (tropical areas)

***Bos indicus* miniature:** (Columbia, Missouri, USA)/line of miniature *Bos indicus* cattle, with phenotypical similarities to laron dwarfism

Boškarin: *see* Istrian

Bosnian: *see* Buša

Bos (Poëphagus) 'grunniens': *see* yak (domestic)

Bos (Poëphagus) mutus: *see* yak (wild)
Bos primigenius: *see* aurochs
Bos primigenius opisthonomus: *see*
    African aurochs
Bos taurus: = humpless cattle
    (temperate Europe)
Botawana: *see* Batawana
Botswana beef synthetic: (Botswana)/
    m/orig. from Tswana, Tuli,
    Brahman, Bonsmara and
    Simmental (48% Sanga, 23%
    zebu, 29% European)
Boucsan: *see* Bucşan
Boudouma: *see* Kuri
Bourbonnaise: (France)/dr/
    supplanted by Charolais/extinct
    early 20th c.
Bo Vang: (Vietnam)
Bovelder: (S Africa)/in development
Bovian: (Netherlands)/m/Charolais ×
    Blonde d'Aquitaine ♂♂ for
    crossing with dairy ♀♀; not used
    since 1994
Bovines of Tete: (Tete province,
    Mozambique)/m.(d.dr)/Sanga-
    Zebu; strain of Angoni/? orig.
    from Angonia and Landim/rare
Boyenca: *see* N'Dama
Boz Step: *see* Turkish Grey Steppe
BRACHYCEROS: (Europe)/i.e. derivative
    of *Bos taurus brachyceros*
    (Adametz)/usu. red, brown or
    black; short horns; deep forehead;
    small/inc. Central Europe Red,
    Iberian *et al.*/syn. *Celtic Red*
Braford: (S of USA)/m/white face/
    Brahman × Hereford, F₁ or
    later/recog. 1954, BS 1969; BS
    also Argentina/syn. *Herebu*
    (Argentina)/not *Bradford*/*see*
    also Australian Braford,
    Brazilian Braford, Pampiano
Bragado do Sorraia: (Ribatejo,
    Portugal)/dr.m/red with spots
    inside legs (= *bragado*)/var. of
    Mertolenga/? orig. from Fighting
    bull × Alentejana/syn.
    Charnequeira (*charneca* = heath)
Bragança: (NE Portugal)/dr.m/var. of
    Mirandesa
Brahaza: *see* Red Bororo
Brah-Maine: (USA)/⅜ Brahman, ⅝
    Maine-Anjou/BS 1985
Brahman: (S of USA)/m/black skin,
    with white to black coat (usu.

silver grey); hd or pd/z/orig. from
Ongole, Kankrej, Gir and Krishna
Valley breeds imported from
India 1849–1906, and from Brazil
1924–25 and 1946, possibly with
some European blood/vars: Grey
Brahman, Red Brahman/part orig.
of Brahmousin, Brahorn, Bralers,
Brangus, Bra-Swiss, Simbrah *et*
al./BS 1924; BS also Argentina
1954, Australia 1955, S Africa
1958 (HB 1967)/syn. *American*
Brahman/not *Brama*/*see also*
Jamaica Brahman
Brahmental: *see* Simbrah
Brahmini: (India)/z/sacred free-
    roaming bull/syn. *Brahmin,*
    Sandhe (Nepal) (Nepalese *Sādhe*)
Brahmousin: (SW USA)/orig. from
    Limousin (⅝) × Brahman (⅜)/BS
    1984
Brahorn: (Texas and Louisiana,
    USA)/m/Brahman × Shorthorn,
    1st cross
Brah-Swiss: *see* Bra-Swiss
Bralers: (Texas, USA)/red/orig. from
    Salers (⅝) × Brahman (⅜)/BS
    1984/not *Braler*
Brandrood IJsselvee: (Netherlands)/
    dark red/var. of Meuse-Rhine-
    Yssel/ syn. *Brandrod*/rare (?
    extinct)
Brangus: (Texas and Oklahoma,
    USA)/m/black; pd/orig. 1932 at
    Jeanerette, Louisiana, from
    Brahman (*c.* ⅜) × Aberdeen-
    Angus (*c.* ⅝)/var.: Red Brangus/BS
    1949; BS also New Zealand,
    Argentina, S Africa 1986/*see also*
    Australian Brangus
Brangus-Ibagé: (Brazil)/synthetic
    crossbreed from Brangus and
    Ibagé
Bra-Swiss: (Texas, USA)/m/Brahman ×
    Brown Swiss cross/not *Brah-Swiss*
Braunesgebirgsvieh: *see* Brown
    Mountain
Braunscheck: *see* White-marked
Braunvieh: *see* Brown Mountain
Braunvieh alter Zuchtrichtung:
    (Germany) *see* German Brown,
    old type
bravo: *see* Fighting bull
Bravon: (Florida, USA)/m/Brahman
    × Devon cross

Brazilian Braford: (Brazil)/crossbred from Hereford ($\frac{5}{8}$) and Nelore ($\frac{3}{8}$); cf. Braford (USA)/? = Pampiano

Brazilian Gir: (Brazil)/m.d/red, white or spotted/z/orig. from Gir imported from India 1918–1962/vars: Dairy Gir, Polled Gir/orig. of Venezuelan Zebu/HB 1936, BS 1939/not *Gyr*

Brazilian Milking Crossbred: (Minas Gerais, São Paulo and Rio de Janeiro, Brazil)/d/open population, orig. from European (mainly Holstein, also Brown Swiss, Caracu, Danish Red) × zebu (mainly Guzerá, Gir) based at National Dairy Cattle Research Centre, Coronel Pacheco, Minas Gerais/Port. *Mestiço leiteiro brasileiro* (= *MLB*)/syn. *Brazilian Dairy Hybrid*

Brazilian Polled: (Goiás, Brazil)/ m/red or yellow/? orig. from Southern Crioulo/recog. 1911, [HB and BS 1939]/Port. *Mocho nacional* (= *National Polled*)/ rare/*see also* Caracu (pd var.), Polled Crioulo Pereira Camargo/rare

Brazilian zebus: *see* Dairy Zebu, Brazilian Gir, Guzerá, Indo-Brazilian, Nelore, Polled Zebu, Tabapuã/BS

bree: *see* yak ♀

Breitenburger: (NW Germany)/orig. var. of Red Pied Schleswig-Holstein

Bressane: (Ain, E France)/yellow/ absorbed in Eastern Red Pied (Pie rouge de l'Est, now French Simmental) early 20th c./not *Bressanne*/[from Bresse]/extinct

Breton Black Pied: (S Brittany, France)/d.m/lyre horns/orig. from 2 local types (small dairy in Morbihan, larger beef Bordelais around Bordeaux)/often red-and-white in Finistère/BS, HB 1919/ Fr. *Breton pie-noir*/syn. *Breton, Brittany Black-and-White, Morbihan* (Fr. *Morbihannais*)/ rare

bri: *see* yak ♀

British-Canadian Holstein-Friesian: *see* British Holstein

British Dane: (England)/d/red; hd or pd/orig. from Danish Red imported 1961 and used to grade Red Poll/named 1967; BS/not *British Red*

British Friesian: (Great Britain)/d.m/ orig. from imports (chiefly Dutch) 1860–1892 and 1914–1950/vars: Poll Friesian, Red-and-White Friesian/orig. of American Beef Friesian/heavily crossed with American Holstein/ BS 1909, HB 1912; now in Holstein UK BS and HB for British-bred black-and-white cattle, with separate classification as strain with 87.5% British Friesian blood

British Holstein: (England)/d/BS 1946 to register Holstein cattle first imported from Canada 1944–1948/recorded in Holstein UK HB as "pure imported" ("PI") strain of British-bred black-and-white cattle/obs. syn. *British-Canadian Holstein-Friesian*

British Holstein-Friesian: (UK)/ British Friesian with high proportion of North American Holstein blood, forming major of three "strains" of British-bred black-and-white cattle in Holstein UK HB (along with British Friesian and British Holstein)

British Polled Hereford: (England)/ pd var. of Hereford from Galloway cross/BS 1949/syn. *Poll Hereford*

British White: (Great Britain)/m.[d]/ white with black (occ. red) points; pd only (since 1946)/? orig. from White Park; some white Shorthorn and white Galloway blood in 1930s; Swedish Mountain blood 1949/ orig. of American White Park, Australian White/BS 1918; BS also USA 1987/syn. *Park* (1918–1946), *White Polled*

Brittany Black and White: *see* Breton Black Pied

Broadlands: (England)/belted/ imported from Netherlands to

Moor Park, Sheen, Richmond, by
Sir William Temple/extinct 1934
'brong, 'brong-'bri: *see* yak, wild
Brown Alpine: *see* French Brown,
Italian Brown
Brown Atlas: (Maghreb, N Africa)/
dr.d.m/brown or blond/Iberian
type, sim. to Libyan/formerly
divided into Guelma and
Moroccan, each with vars/Fr.
Brune de l'Atlas/being crossbred
with Holstein and Swiss Brown;
pure only in mountains/rare
Brown Finnish: *see* West Finnish
Brown Karacabey: *see* Karacabey
Brown
BROWN MOUNTAIN: (Alps)/d.m.(dr)/
grey-brown/BS 1963
(Association of Breeders of
Brown Mountain Cattle)/inc.
Swiss Brown, Austrian Brown,
German Brown and their
derivatives (American Brown
Swiss, Bulgarian Brown, French
Brown, Hungarian Brown, Italian
Brown, Romanian Brown,
Russian Brown, Sardinian
Brown, Slovenian Brown, South
African Brown Swiss, Spanish
Brown, Turkish Brown); BS also
in Brazil 1938, New Zealand/Ger.
Braunvieh (= *Brown cattle*), Fr.
Brune des Alpes, It. *Bruna alpina*
(= *Brown Alpine*)/syn. *Brown,
Brown Swiss, Grey-Brown
Mountain* (Ger. *Graubraunes
Höhenvieh*)
Brown Moroccan: *see* Moroccan
Brown
Brownsind: (Allahabad, India)/d/
orig. in 1960s at Allahabad
Agric. Inst. from American
Brown Swiss ($\frac{3}{8}$–$\frac{5}{8}$) × Red
Sindhi/not *Brown Sind*
Brown Swiss: (Europe) *see* Swiss Brown
Brown Swiss: (USA) *see* American
Brown Swiss, Beef Brown Swiss
Brună: *see* Romanian Brown
Bruna alpina: *see* Brown Mountain,
esp. Italian Brown
Bruna dels Pirineus: (Spain) *see*
Pyrenean Brown
Bruna svizzera: *see* Swiss Brown
Brune de l'Atlas: *see* Brown Atlas

Brune des Alpes: *see* Brown
Mountain, esp. French Brown
Brune suisse: *see* Swiss Brown
Bruno-sarda: *see* Sardinian Brown
Bruxo: *see* Southern Crioulo
BRWS: *see* Blended Red and White
Shorthorn
Buchan Humlie: (NE Scotland)/orig.
(+ Angus Doddie) of Aberdeen-
Angus/syn. *Polled
Aberdeenshire*/[humlie = pd]/
extinct
Bucşan: (E Romania)/sim. to
Moldavian but darker and
smaller/former var. of Romanian
Steppe/Ger. *Bukschaner*/not
Boucsan, Bukshan/extinct
Budduma: *see* Kuri
Budějovice, Budějovický: *see* Šumava
Buduma, Budumu: *see* Kuri
Budweiser: *see* Šumava
Buje: *see* Istrian
Bukedi: *see* Nkedi
Bukschaner, Bukshan: *see* Bucşan
Bulgarian Brown: (SW Bulgaria)/
d.m/orig. from native graded to
Montafon (first imported 1894)/
recog. 1951/Bulg. *B"lgarska
kafyava*/syn. *Sofia Brown* (to
1963) (Bulg. *Sofiĭska kafyava*,
Ger. *Sofioter Braunvieh*)
Bulgarian Grey: *see* Iskar
Bulgarian Red: (NE and S Bulgaria)/
d.m/orig. 1950–1960 from Swiss
Brown, Bulgarian Brown and
Bulgarian Simmental, × Iskar,
Red Sadovo and Red Steppe,
with Latvian Brown, Red Sadovo
and esp. Danish Red bulls since
1954/Bulg. *B"lgarsko cherveno
govedo*, Ger. *Bulgarisches
Rotvieh*
Bulgarian Simmental: (NW
Bulgaria)/d.m.[dr]/usu. cream
and white/orig. 1948 on from
Simmental (first imported 1894)
× Iskar/Bulg. *B"lgarska
simentalska*/syn. (to 1969) *Kula*
(Bulg. *Kulska*, Ger. *Kulaer*)
Bulgarian Steppe: *see* Iskar
Bulgarisches Grauvieh: *see* Iskar
Bunaji: *see* White Fulani
Bunyoro: (Uganda) *see* Nyoro
Buraya karpatskaya: *see* Carpathian
Brown

Buraya lavtiĭskaya: *see* Latvian Brown

Burgaur, Burghur, Burgoor, Burgur: *see* Bargur

Burlina: (Mt Grappa, Treviso-Vicenza, Venetia, Italy)/d.(m)/ black pied, also red pied/ HB 1985/syn. *Asiago, Bassanese, Binda, Boccarda, Pezzata degli altipiani* (= *Pied Highland*)/rare, by crossing with Friesian

Burmese: (C Myanmar)/dr/usu. red (*nwar shweni*) or yellow (*nwar shwewar*)/z/crossbreds with Hariana and Tharparkar are Chaubauk/*see also* Shan

Burmese banteng: *see* tsine

Burmese bison: *see* Burmese gaur

Burmese gaur: (Assam, Myanmar, Thailand, Laos, Cambodia and Vietnam)/= *Bos (Bibos) gaurus readei* (Lydekker)/var. of gaur/ Burmese *pyaung, pyoung, pyun*/ syn. *Burmese bison*/rare

Burundi: *see* Watusi

Burwash: (N Ontario, Canada)/m/ orig. 1957 on at Burwash Industrial Farm from Charolais × Hereford and other breeds; one group is $\frac{1}{2}$ Charolais, $\frac{1}{4}$ Hereford, $\frac{1}{4}$ Shorthorn

Buryat: (Siberia, Russia)

Buša: (S of R Sava, Croatia, Bosnia, Macedonia and Serbia)/d.m.dr/ red, black, grey, yellow or brown/Iberian type, sim. to Albanian, Greek Shorthorn and Rodopi/vars: Croatian Red, Lim, Macedonian Blue, Metohija Red, Pešter/orig. (with Serbian Steppe) of Kolubara and Spreča, and (with Tyrol Grey) of Gacko/ syn. *Bosnian, Illyrian* (inc. Albanian)/not *Busa, Buscha, Busha*/nearly extinct

Bushuev: (Uzbekistan)/d.m/white with black points; small hump in ♂/orig. by M.M. Bushuev from Friesian and Swiss Brown, × local zebu (1907–1923) with further Swiss Brown blood (1923–1931), and Simmental and East Friesian blood (1932–1948)/ recog. 1967/Russ. *Bushuevskaya*/ syn. *Tashkent* (Russ. *Pritashkentskaya*)

Butana: (between R Atbara and Nile, Sudan)/d.dr/usu. dark red, also light red/North Sudan Zebu group/related strains of Red Desert cattle are Bambawa, Dongola and Shendi/syn. *Red Butana, Dar El Reih*/not *Butane, Butanne*

Byelorussian: *see* Belarus

Cabannina: (Cabanne, Genoa, Italy)/ d.(m)/brown with pale dorsal stripe/HB 1985/rare, by crossing with Italian Brown

Cabella, Cabellota: *see* Montana

Cacereña, Cáceres: *see* White Cáceres

Cachena: (Entrino, S Orense, Galicia, Spain)/m.(dr.d)/chestnut or light blond with white nose ring; long lyre horns; dwarf; semi-wild/ sim. to Barrosã; Morenas del Noroeste group/HB 1990/[= very small]/rare

Cadzow: (Lanarkshire, Scotland)/ horns cf. Dynevor/ancient herd of White Park type; heterozygous for colour/syn. *Hamilton*/nearly extinct

Caiuá: (Matto Grosso, Brazil)/m/orig. in 1970s at Caiuá farm of Liquifarm from Chianina × Nelore; Caiuá 1 is $\frac{1}{2}$ Chianina, Caiuá 2 is $\frac{5}{8}$ Chianina, Caiuá 3 is $\frac{3}{4}$ Chianina

Calabrian: (Italy)/former var. of Apulian Podolian/It. *Calabrese*/ syn. *Cotrone, Crotone* (It. *Crotonese*)

Calasparreña: (Calasparra, NE Murcia, Spain)/former var. of Murcian/nearly extinct

Caldeano: (S Minas Gerais, Brazil)/d/ light red/var. of Caracu/selected (1895 on) by Dias family at Fazenda 'Recreio', Poços de Caldas/HB and BS 1960/syn. *Caracu caldeano*

Caldelana: (Castro Caldelas, N Orense, Spain)/m.dr/steely black with reddish dorsal line, born blond/Morenas del Noroeste group/rare

Calvana: (Calvana mts, Florence, Italy)/m/white with dark tail switch and mucosae/small var. of Chianina/HB 1985/nearly extinct

Camandona: (Piedmont, Italy)/local var. absorbed by Italian Brown/ extinct

Camargue: (Rhône delta, France)/ black or brown; half wild, used for bull fights/BS

Cambodian: (Cambodia)/small hump/var.: Moi/Fr. *Cambodgien*

Cambodian wild ox: *see* kou-prey

Cameroon Fulani: *see* Adamawa

Cameroon Gudali: *see* Adamawa

Campagna: (Italy)/silver grey to fawn

Campine Red Pied: *see* Belgian Red Pied

Campurriana: (Campóo, W Santander, Spain)/larger, valley var. of Santander/absorbed by Swiss Brown and Tudanca in 1940s/extinct

Canadian: *see* Canadienne

Canadian Friesian: *see* Holstein

Canadienne: (Quebec, Canada)/d/ usu. black to dark brown (born brown)/orig. from 16th and 17th c. imports from Normandy and Brittany; Brown Swiss blood (up to $\frac{1}{8}$) in 1980s/BS 1895, HB 1896/syn. *Black Canadian, Black Jersey, Canadian, French Canadian* (till 1930), *Quebec Jersey*/rare

Canary Island: (Gran Canaria, Spain)/ m.[d.dr]/blond to reddish/? orig. from Galician Blond/Sp. *Canaria*/syn. *Basta, Criolla, de la Tierra*/see also Palmera

Canchim: (São Paulo, Brazil)/m/ cream or white; small hump in ♂/orig. (1940 on) at Government Breeding Farm at São Carlos (formerly Canchim farm) from Charolais ($\frac{5}{8}$) and zebu (Indo-Brazilian) ($\frac{3}{8}$)/BS and HB 1971/ not *Canchin*

Caoyuan Red: (N and NE China)/ m.d/red/orig. from (Milking) Shorthorn (from USA) × Mongolian in 1920s and 1930s/ named 1985; BS/syn. *Chinese Red Steppe, Grassland Red*/not *Cauyen*

Cape: *see* Hatton

Cape Bon Blond: (Tunisia)/Fr. *Blonde du Cap Bon*/extinct

Capirote: *see* Berrenda en colorado

Cappe Harak: *see* Hatton

Caprivi Sanga: (Caprivi strip, NE Namibia borders)/m.dr.d/Sanga type, Ovambo Cluster

Caqueteño: (Colombia)/Criollo type

Caracu: (R Pardo valley, Minas Gerais and São Paulo, Brazil)/ m.d.dr/usu. yellow (pale to orange); long horns/orig. from Southern Crioulo/vars: Caldeano, Caracu chifrudo (= hd), Caracu mocho (= pd)/recog. 1909, BS and HB 1916/not *Caracú*/[? from Tulpian *Acarahu,* cf. R Acaracu]

Caracu caldeano: *see* Caldeano

Carazebú: (São Paulo, Brazil)/m.d/ bay to dark yellow; pd/in formation by A. Lunardelli at Valparaíso from zebu × Caracu and Brazilian Polled/not *Carezebu*

Cárdena: (Salamanca, Spain)/dark grey/var. of Morucha/[= purple]/ nearly extinct

Cárdena andaluza: *see* Andalusian Grey

Carhaix Red Pied: (Brittany, France)/ part orig. of Armorican/? extinct

Caribe: (Cuba)/d/orig. from Holstein ($\frac{5}{8}$) × Santa Gertrudis ($\frac{3}{8}$)

Carinthian Blond: *see* Austrian Blond

Carnation: (USA)/strain of Holstein with great influence on Italian Friesian

Carniella: (Carnia, Friuli, Italy)/red/ absorbed by Italian Brown in early 20th c./extinct

Carora: (Lara, Venezuela)/d/orig. 1935–1975 from American and European Brown Swiss × Criollo ($\frac{1}{4}$–$\frac{1}{2}$)/BS/Sp. *Caroreña*/syn. *Tipo carora*

Carpathian Brown: (Trans-Carpathian Ukraine)/d.m/dark brown to light grey/cf. Hungarian Brown/orig. from local graded to Brown Mountain since 1879; American Brown Swiss and Jersey blood 1980s/recog. 1972, HB 1973/Russ. *Buraya karpatskaya*

Carpetana: *see* Avileña

Carpi, Carpigiana: *see* Modenese

Carreña: (SW Oviedo, N Spain)/red to mahogany/larger, lowland var. of Asturian/being crossed with Brown Swiss and Friesian/HB, BS/syn. *Asturiana de los valles, Carreñana*/not *Acreno*

Casanareño: (R Casanare valley, NE Colombia)/tan or spotted/Criollo type/syn. *Llanero* (also W Venezuela)

Casina: (SE Oviedo, N Spain)/dr.d.m/ light to dark red/smaller, mt var. of Asturian/HB 1978, BS/syn. *Asturiana de las montañas*/[from Caso]

Casta: *see* Aure et Saint Girons

Castana: *see* Aosta Chestnut

Castilian: (Spain)/silver grey to fawn/ extinct

Castille-León: (Duero valley, Spain)/ former grouping to inc. Avileña, Black Andalusian, Morucha and Sayaguesa/Sp. *Castellano-Leonesa*

Castle Martin: (Wales)/var. of South Wales Black/extinct

Castro Caldelas: *see* Caldelana

cattalo: (Canada)/m/cf. Beefalo/orig. by C.J. Jones (Indiana, USA, 1880s), Charles Goodnight (Texas, 1880s) and Mossom M. Boyd (Ontario, 1894) from crosses of domestic cattle with American bison (initially bison ♂ × Hereford and Angus ♀♀, with F_1 and F_2 bred back to bison); Boyd stock acquired by Canadian government (project abandoned 1964 because of infertility)/named *catalo* by Jones, altered to cattalo by Goodnight *et al.*/extinct/*see also* Simmalo

Cauca: *see* Hartón

Caucasian: (Dagestan, Russia, and Transcaucasian republics)/d/inc. smaller usu. black var. in Greater Caucasus (inc. Dagestan and Georgian Mountain) and larger usu. red var. in Lesser Caucasus (Mingrelian Red) and in Transcaucasian lowlands (inc.

Azerbaijan)/Russ. *Kavkazskaya*/ syn. *Transcaucasian*

Caucasian Brown: (Azerbaijan)/d.m/ brown with dark markings/orig. *c.* 1930–1960 from Lesser Caucasus graded to Swiss Brown for 3 generations; also Kostroma and Lebedin blood/absorbed Dagestan Brown, Lorii/recog. 1960/Russ. *Kavkazskaya buraya*/syn. *Azerbaijan Brown, Caucasus Brown*/not *Caucasian Red*

Caucasian Zebu: *see* Azerbaijan Zebu

cäuri: *see* chauri

Cauvery: *see* Alambadi

CCC: *see* Costeño con Cuernos

cebú: *see* zebu

Cecina, Cecinese: *see* Chianino-Maremmana

Celtic Red: *see* brachyceros

Central American Dairy Criollo: (Costa Rica)/var. of Tropical Dairy Criollo/orig. at Turrialba in 1959 chiefly with animals from Rivas, Nicaragua, also from Honduras and local/syn. *Costa Rican Dairy Criollo, Reyna* (Nicaragua)

Central and Upland: *see* Belgian Blue

Central and Upper Belgian: *see* Belgian Blue

Central Asian: *see* Turano-Mongolian

Central Asian Zebu: (Turkmenistan, Tajikistan and Uzbekistan)/usu. with European blood (= zeboid, Russ. *zebuvidnyĭ skot*)/vars: Fergana, Khurasani, Kuramin, Turkmen/Russ. *Sredneaziatskiĭ zebu*/syn. *Turkestan Zebu* (Russ. *Turkestanskiĭ zebu*)

Central Chinese: *see* Huanghuai group

CENTRAL EUROPE RED: *brachyceros* type/inc. Bohemian Red, Belarus Red, German Red, Lišna Red, Lithuanian Red, Polish Red

Central Pyrenean: *see* Aure et Saint-Girons

Central Russian Black Pied: (C provinces of European Russia)/ d/var. of Russian Black Pied/orig. in 1920s from East Friesian ×

crossbreds by Yaroslavl, Kholmogory, Schwyz and Simmental bulls out of local cows/HB 1940/Russ. *Srednerusskaya* (or *Tsentralnaya*) *chernopestraya*

Cenubî Anadolu Kırmızı: *see* South Anatolian Red

Česká červinka: *see* Bohemian Red, Czech Red

České červenostrakatý: *see* Bohemian Red Pied

České červinky: *see* Bohemian Red

Československý červenostrakatý: *see* Czech Pied

Český strakatý: *see* Czech Pied

Ceylon: *see* Sinhala

Chablais, Chablaisien: *see* Abondance

Chaco: *see* Chaqueño

Chacuba: (Cuba)/m/orig. from Charolais ($\frac{5}{8}$) × Cuban Zebu ($\frac{3}{8}$)

Chad: *see* Kuri

Chagga: (Kilimanjaro, Tanzania)/ d.m.dr/ dwarf var. of Tanzanian Zebu/syn. *Wachagga*/ rare

Chaissi: *see* Chesi

Chakhansurri: *see* Sistani

Cham-Doc: *see* Chau-Doc

Chamusco: (S Portugal)/Charolais or Limousin × Alentajana or Algarvia/[= toasted, browned]

Chan-Doc: *see* Chau-Doc

CHANGZHU group: (basins of Changjiang (Yangtze) and Zhujiang, S China)/dr.(m)/high-humped group of Chinese yellow cattle, possibly with banteng influence (coat pattern)/inc. with small hump in ♂: Bashan, Dabieshan, Dengchuan, Ebian Spotted, Guangfeng, Guanling, Longlin, Minnan, Nanyang, Wannan, Wenling, Wenshan, Wuling, Xizhen, Zaobei, Zhoushan; and, with hump in both sexes: South Chinese Zebu (Hainan, Leizhou, Xuwen, and Yunnan Zebu)/syn. *Chang Zhu, Chinese Zebu, Southern Yellow*

Channel Island: (Channel Islands)/French orig./obs. syn. *Alderney*/see Jersey, Guernsey

Chaouia: (Constantine, Algeria)/ former subvar. of Guelma var. of Brown Atlas/? extinct

Chaqueño: (Chaco, Bolivia and Paraguay)/m/all colours/var. of Bolivian Criollo/rare

Chaqueño: (Chaco, Argentina)/m/all colours/larger var. of Argentine Criollo

Charbray: (Texas and Louisiana, USA)/m/creamy white/orig. since 1936 from Charolais ($\frac{5}{8}$–$\frac{7}{8}$) × Brahman/BS 1949–1967, HB 1949 on/not *Charbra/see also* Australian Charbray

Charford: (N Arizona, USA)/m/orig. since 1952 at Ash Fork Ranch from Charolais ($\frac{1}{2}$), Hereford ($\frac{3}{8}$) and Brahman ($\frac{1}{8}$)

Chargrey: (Victoria, Australia)/ Charolais × Murray Grey cross by Peter O'Sullivan at Cadella Park

Charnequeiro: *see* Bragado do Sorraia

Charolais: (C France)/m.[dr.d]/white or cream, with pink mucosae/ orig. in Saône and Loire in Burgundy; Shorthorn blood 1850–1880/pd var. being bred by selection in France (also USA)/part orig. of Canchim, Charbray, Charollandais, Char-Swiss, Charwiss *et al.*/BS 1864, HB 1887; BS also USA 1951, S Africa 1955, Canada 1960, GB 1962, Denmark 1963, Argentina, Brazil, Germany, Japan, Mexico and Uruguay (all by 1964), New Zealand 1972, Belgium and Paraguay 1975, Zimbabwe 1983, Finland 1986; HB also in Luxembourg 1967, Hungary 1981, Czech Republic, Netherlands, Portugal, Russia/ not *Charollais*/[from Charolles district]

Charollandais: (France)/d.m/cross (1960s) of *Charol*ais × Friesian from Hol*land*

Char-Swiss: (Nebraska, USA)/m/orig. by G.A. Boucher, Ravenna, from $\frac{3}{4}$ Charolais and $\frac{1}{4}$ American Brown Swiss/BS 1961

Char Tarlan: *see* Mongolian Black Pied

Chartley: (Wales and England)/white
with black points; long horns
often curving down/ancient herd
of White Park type, orig. in S
Wales (Lamphrey Court) and
Staffs (Chartley Park), then
Woburn, now elsewhere/?
Longhorn blood/nearly extinct

Charwiss: (California, USA)/m/
Charolais × American Brown
Swiss, 1st cross

Châtaignée: *see* Aosta Chestnut

Chaubauk: (Myanmar)/d.dr/usu.
white/orig. from Hariana and
Tharparkar × Burmese/syn.
nwar pyiase (*nwar* = cattle)

Chau-Doc: (Vietnam)/var. of
Vietnamese/not *Cham-Doc*,
Chan-Doc

chauri: (Nepal)/♀ offspring of yak ×
cattle, inc. reciprocal crosses and
backcrosses to yak/Nepali *cāuri*
(from *cāvar* = tail)/syn. *chaunri*,
chowri, *churi*, *tsauri*/see dzo

Cheb: (Czech Republic)/larger var. of
Bohemian Red/Cz. *Chebský*, Ger.
Egerländer/extinct

Cheju: (Cheju I, Korea)/yellowish-
brown; small/var. of Korean
Native/syn. *Cheju Hanwoo, Jeju,
Jeju native*

Chernigov: (NW Ukraine)/m/var. of
Ukrainian Beef/orig. 1961–1979
from Charolais ($\frac{3}{4}$), Russian
Simmental ($\frac{1}{8}$) and Ukrainian
Grey ($\frac{1}{8}$)/recog. 1979/Russ.
Chernigovskiĭ tip/not
Chernigorsk/united with Dnieper
as Ukrainian Beef 1993

Chernopestraya: *see* Russian Black
Pied

Chernopestraya litovskaya: *see*
Lithuanian Black Pied

Chernopestraya podol'skaya: *see*
Podolian Black Pied

Chernopestryĭ skot Sibiri: *see*
Siberian Black Pied

Chervena sadovska: *see* Red Sadovo

Chesi: (Syria)/d/Damascus ×
Baladi/syn. *Chaissi*/not *Chezi*

Cheurfa: (Constantine, Algeria)/light
grey/former subvar. of Guelma
var. of Brown Atlas/? extinct

Chezi: *see* Chesi

Chiangus: (USA)/m/black; pd/cf.
Ankina/Chianina × American
Angus ($\frac{1}{4}$–$\frac{3}{4}$), or their
descendants/HB 1975

Chianina: (Chiana valley, Tuscany,
Italy)/m.[dr]/white with black
points and mucosae/? orig. from
Podolian × local in early 19th
c./former vars: Calvana,
Perugina, Valdarno, Val di
Chiana/orig. (with Podolian) of
Marchigiana (improved) and
Pasturina/HB 1956; BS Brazil
1969, Canada 1971, USA 1973,
GB 1973, Australia

Chianino-Maremmana: (Tuscany,
Italy)/m.dr/grey-white/orig. from
Chianina × Maremmana/syn.
Cecinese (from Cecina),
Improved Maremmana/nearly
extinct

Chickso: (Korea)/black with
yellowish-brown "tiger
stripes"/colour var. of Korean
Native/syn. *Korean Native
Striped*

Chieftainry cattle: (W Cameroon)/
term for one of five sections for
local taurine cattle in Cameroon
(cf. Kapsiki, Namshi, Muturu
and Keteku)

Chiford: (USA)/m/cream, fawn or
red, solid or with white face; hd
or pd/Chianina (up to $\frac{3}{4}$) ×
Hereford, or their descendants/
HB

Chilean Criollo: (C Chile)/m/name
used for nondescripts of Criollo
orig. with Shorthorn and some
Friesian and Normande blood/
Sp. *Criollo chileno*/syn. *Chilean,
Costino*/pure Criollo extinct

Chillingham: (Northumberland,
England)/white with red ears
and points; horns upright, often
lyre in ♀/semi-feral herd of pure
White Park cattle, closed since
17th c. (and possibly since 13th);
reserve herd in N Scotland/rare

Chimaine: (USA)/any colour; hd or
pd/Chianina (up to $\frac{3}{4}$) × Maine-
Anjou and their descendants/HB

Chimarrão: *see* Crioulo

China: (Rio de Janeiro, Brazil)/orig.

from Southern Crioulo with
Indian Zebu blood imported by
Baron de Bom Ritiro in
1855/extinct

Chinampo: (Lower California,
Mexico)/usu. white with black or
red markings; small/Criollo type/
declining by crossing with zebu

Chinchuan, Chinchwan: *see*
Qinchuan

Chinese: *see* Changzhu group,
Huanghuai group, Mongolian,
Tibetan, yellow cattle

Chinese Black-and-White: (China)/
d.m/orig. from Holstein, Dutch
Black Pied and other Friesians
(esp. German and Japanese) ×
local Chinese/var.: Beijing Black
Pied/BS/syn. *Chinese Black
Pied, Chinese Holstein*

Chinese Holstein: *see* Chinese Black-
and-White

Chinese Red Steppe: *see* Caoyuan
Red

Chinese Yellow: *see* yellow cattle

Chinese Zebu: *see* Changzhu group

Chino Santandereano: (Lebrija
valley, Santander, Colombia)/
m.d.dr/tan to chestnut/sim. to
Costeño con Cuernos but more
milk/local Criollo var. with zebu
blood/recog. 1950/[*chino* =
hairless]/rare

Chitaldrug: (India)/former var. of
Amritmahal, now merged in
Hallikar/not *Chitraldrag,
Chittaldroog*/extinct

Chittagong Red: (Bangladesh)/z/local
var.

Chiya: *see* Bashan

Choa: *see* Shuwa

Choletais: *see* Parthenais

Cholistani: (Cholistan desert,
Bahawalpur, Pakistan)/d.m/usu.
speckled red, brown or black/z/
recent orig. ? from Sahiwal ×
desi

Cholmogor: *see* Kholmogory

Cholung: *see* Tibetan ♂

Chosen: *see* Korean Native

Chou-pei, Chowpei: *see* Zaobei

chowri, chowrie: *see* chauri

churi: *see* chauri

Chusco: (high Andes, Ecuador, Peru

and Bolivia)/Criollo type (var. of
Bolivian Criollo)/syn. *Criollo of
the Altiplano, Serrano*

Ciernostrakate: (Slovakia)/[= black
pied]

Çifteler Brown: *see* Turkish Brown

Cika: (Slovenia)/d.m/reddish with
white back stripe/sim. to
Pinzgauer but smaller/orig. after
1960 from Pinzgauer × local
blond Buša/vars: Bohinj Cika,
Tolmin Cika/Sn. *Cikasto
govedo*/nearly extinct

Çildir: (Turkey)/var. of East
Anatolian Red/not *Cildi*

Cinisara: (Cinisi, NW Palermo,
Sicily, Italy)/d/sim. to Modicana/
Podolian type but black, occ.
white tail and belly

Coastal Horned: (Colombia) *see*
Costeño con Cuernos

Coastal Polled: (Colombia) *see*
Romosinuano

Coastal Zebu: (Kenya)/inc. Duruma,
Giriama, Kamba, Taita/syn.
Lowland Zebu

Cobourg Peninsula: (Northern
Territory, Australia)/♂ red to
brown, ♀ reddish-brown with
white patches; feral/orig. from
Bali cattle imported 1827/large
herd (200) to be conserved at
Coastal Plains (Beatrice Hill) Res.
Sta./rare

Cobra: *see* Gobra

Colombian Criollo: Colombia/vars:
Caqueteño, Casanareño, Chino
Santandereano, Hartón/*see also*
Blanco Orejinegro, Costeño con
Cuernos, Romosinuano,
Sanmartinero

Colônia: (Brazil) *see* Southern Crioulo

Colonist: *see* Ukrainian Whiteheaded

Colorada: (Spain) *see* Retinta

Coloured Cattle of Wales: *see*
Ancient Cattle of Wales

coloursided: 1. White tail, back stripe
and belly; coloured head and
legs; see American Lineback,
Aracena, Gloucester, Pinzgauer,
Tux (white less extensive)./2.
White tail, back and underside;
white or parti-coloured head and
legs; see *also* American Lineback,

Black Forest (often), Blacksided
Trondheim and Nordland, East
Finnish, Irish Moiled, Istoben
(often), Kravařsko, Kłodzka,
Longhorn, Lithuanian White
Back, Polish Whitebacked,
Sudeten Spotted (often),
Telemark, Ukrainian
Whitebacked, Vostes, Witrik./3.
White with coloured points
(muzzle, inside ears, etc.); colour
on sides reduced to a few spots
only, or absent; see Blacksided
Trondheim and Nordland (often),
Blanco Orejinegro, Boran (often),
British White, Mauritius Creole
(often), North Finnish, Polish
Marsh, Swedish Mountain, White
Fulani (often), White Park/syn.
finchbacked, finched, inkone
(Zulu), *lineback, rigget,
Rückenscheckig* (Ger., = back
pied), *Sidet* (Nor.)

Coloursided Trondheim: *see*
Blacksided Trondheim and
Nordland

Colubara: *see* Kolubara

Comtois: *see* Tourache

Condroz: (Belgium)/extinct after First
World War

Coopelso 93: (SW France)/m/sire line
developed from Blonde
d'Aquitaine, Charolais and
Limousin by *Coop*érative
d'*El*evage du *S*ud-*O*uest/nearly
extinct

Cornígero de la costa: *see* Costeño
con Cuernos

Cornigliese: (Italy)/greyish-red/small
var. of Bardigiana/[from Corniglio]

Cornish: (Cornwall, England)/black/
extinct in 19th c.

Corral, Corral crioulo, Corralled: *see*
Curraleiro

Correi: *see* Gasara, Magal

Corriente: (USA)/used for rodeo/orig.
from Spanish cattle in Mexico/
BS 1982/[= running]

Corsican: (Corsica)/m /various
colours; prim./cf. Sardinian/
Iberian type/BS/Fr. *Corse*/rare

Costa: (Ecuador)/larger var. of
Ecuador Criollo, inc. El Oro and
Esmeraldas/extinct

Costa Rican Dairy Criollo: *see*
Central American Dairy Criollo

Costeño con Cuernos: (Sinú valley, N
Colombia)/m.d/usu. red or
blond, also chestnut; lyre horns/
Criollo type/orig. of
Romosinuano/syn. *Andaluz,
CCC, Cornígero de la costa,
Sinuano de cuernos* (= horned
Sinú)/[= coastal horned]/rare

Costino: *see* Chilean Criollo

Cotentin: (Normandy, France)/m.d/
brindle (yellow, red and black
combined), or spotted with
white, often with white face/?
Shorthorn blood/? orig. of
Guernsey/absorbed by
Normande/extinct

Cotrone: *see* Calabrian

Cracker: *see* Florida Cracker

Crane-coloured: *see* Hungarian Grey

Créole: (Guadeloupe) *see* Guadeloupe
Créole

Creole: (Mauritius) *see* Mauritius
Creole

Creole: (West Indies) *see* Criollo

Cretan Lowland: (Crete)/former var.
of Greek Shorthorn/syn. *Messara*

Cretan Mountain: (Crete)/former var.
of Greek Shorthorn

Crimean Red: *see* Red Steppe

Crimousin: (Cuba)/m/orig. from
Limousin ($\frac{3}{4}$) × Criollo ($\frac{1}{4}$)

Criolla: (Spain) *see* Canary Island

CRIOLLO: (Spanish America)/cf. Crioulo
(Brazil), Texas Longhorn/Sp.
orig./inc. Argentine Criollo,
Barroso, Bolivian Criollo,
Chinampo, Colombian Criollo,
Cuban Criollo, Ecuador Criollo,
Frijolillo, Tropical Dairy Criollo,
Uruguayan Criollo, Venezuelan
Criollo/syn. *Creole* (Guyana and
anglophone West Indies)/[= native]

Criollo argentino patagónico: *see*
Patagonian Criollo

Criollo Caqueteño: *see* Caqueteño

Criollo chileno: *see* Chilean Criollo

Criollo ecuatoriano: *see* Ecuador
Criollo

Criollo lechero limonero: *see* Rio
Limón Dairy Criollo

Criollo lechero tropical: *see* Tropical
Dairy Criollo

Criollo lechero venezolano: *see* Rio Limón Dairy Criollo

Criollo of the Altiplano: *see* Chusco

Crioulo: (Brazil)/inc. Brazilian Polled, Caracu, Curraleiro, Crioulo Lageano, Pantaneiro, Polled Crioulo Pereira Camargo/syn. *Chimarrão* (= cattle turned wild from living in woods)/[= native]

Crioulo de Santa Catarina: *see* Crioulo Lageano

Crioulo do Sul: *see* Southern Crioulo

Crioulo Lageano: (Lages, Santa Catarina, Brazil)/m.d/black, brown or white, or combination of these; long lyre horns sim. to Junqueiro/orig. from Southern Crioulo in late 19th c.; ? some zebu blood/vars: Franqueiro, Pedreiro/syn. *Crioulo de Santa Catarina*/not *Lajeano*/rare

Crioulo leiteiro de Irecé: *see* Irecé

Crioulo nordestino: *see* Curraleiro

Crioulo Yacumeño: *see* Yacumeño

Cristiana: (Spain)/var. of Murcian/? nearly extinct

Crno-belo: *see* Slovenian Black Pied

Crno-šaro: *see* Croatian Red

Croatian Brown: (Croatia)/Cro. *Smedje govedo*

Croatian Pied: *see* Croatian Simmental

Croatian Red: (Lika, Croatia)/usu. red, also yellow, brown or black/var. of Buša/darker var.: *crno-šaro* (= black-tinged)/Cro. *Hrvatska buša*/nearly extinct

Croatian Simmental: (Croatia)/orig. from Simmental × local/HB/Cro. *Hrvatski Simentalac*/syn. *Croatian Pied*; obs. syn. (with Serbian Pied) *Yugoslav Pied*

Croatian Steppe: (Slavonia, Croatia)/dr/Grey Steppe type sim. to Hungarian Grey but smaller/Cro. *Slavonsko Podolsko* (= *Slavonian Podolian*)/syn. *Podolsko goveče*, *Podolac* (= *Podolic*), *Slavonian Grey Steppe*; obs. syn. (with Serbian Steppe) *Yugoslav Steppe*/nearly extinct

Crotone, Crotonese: *see* Calabrian

Crvena metohijska: *see* Metohija Red

Cuanzo: *see* Barra do Cuanzo

Cuban Criollo: (E Cuba)/m.d/tan/ Criollo type with some zebu blood/var.: Tinima

Cubante: *see* Marchigiana

Cuban Zebu: (Cuba)/m/orig. from Colombian, Venezuelan and Puerto Rican imports in early 1900s

Cuiabano: *see* Pantaneiro

Çukurova: (S Turkey)/var. of South Anatolian Red/orig. from Anatolian Black × Aleppo (= Damascus)/not *Cukurova*/ extinct

Culard INRA 95: *see* INRA 95

Cumberland White: *see* Whitebred Shorthorn

Cunningham: *see* Ayrshire

Cuprem Hybrid: (Nebraska, USA)/m/ orig. 1960–1976 by A. Mulford, Kenesaw; ♀ is Red Angus ($\frac{1}{2}$), Santa Gertrudis ($\frac{1}{4}$), Limousin ($\frac{1}{4}$); ♂ is Shorthorn ($\frac{1}{2}$), Charolais ($\frac{1}{4}$), Chianina ($\frac{1}{4}$)/BS

Curamin: *see* Kuramin

Curraleiro: (NE Brazil)/m.d.dr/red, fawn or dun with pale belly and escutcheon, and darker head, shoulder and legs; short horns/ sim. to Tropical Dairy Criollo/ orig. from Beiroa type of Mirandese/var.: Irecé/syn. *Crioulo nordestino* (= *Northeastern Crioulo*), *Goias*, *Pé duro* (= hard hoof), *Sertanejo* (= inland)/[= corralled]/rare

Cutchi: (Cutch, NW Gujarat, India)/ local type with Tharparkar and Gir influence/syn. *Kachhi*

Cyprus: (Paphos and Limassol, Cyprus)/dr.m/fawn to black/sim. to Damascus, Egyptian and Lebanese/orig. from zebu × humpless/vars: Messaoria, Paphos/Greek: *Kypriaki*/rare

Czarno-biała: *see* Polish Black-and-White

Czarno biała odmiana hf: (Poland)/ d/black-and-white spotted or pied/= imported Holstein-Friesian (cf. older Friesian-based Polish Black-and-White)/syn. *cbhf*/rare

Czech Black Pied: (Czech Republic)/ d.m/orig. from Holstein and Friesian from USA, Germany, Netherlands/HB 1980/Cz. *Nižinne černostrakate*

Czech Brindled: *see* Czech Pied

Czech Mountain Spotted: (E Czech Republic)/replacement of local crossbred (Simmental × local red shorthorned cows) by red pied 1905–1935 after importation of Bernese and Moravian Red Pied (Hana) bulls; resulting types were Czech Red Pied, Hřbínecký and Czech Mountain Spotted, all fused into Czech Pied by 1955

Czechoslovakian Red and White: *see* Czech Pied

Czech Pied: (Czech Republic)/ d.m.[dr]/red-and-white/orig. from Bernese and Simmental imported 1850–1911 × Bohemian and Moravian Reds; Ayrshire and Red-and-White Holstein blood/vars: Bohemian Red Pied, Hřbínecký, Kravařský, Moravian Red Pied/till 1969 combined with Slovakian Pied as Czechoslovakian Red Pied/HB/ Cz. *Česky strakatý*/syn. *Bohemian Pied, Bohemian Spotted, Czech Brindled, Czech Red-and-White, Czech Red Spotted, Czech Simmental, Czech Spotted*

Czech Red: *see* Bohemian Red

Czech Red-and-White, Czech Red Spotted: *see* Czech Pied

Czech Simmental, Czech Spotted: *see* Czech Pied

Czechoslovakian Red Pied: (Czech Republic)/former name (till 1969) for Czech Pied and Slovakian Pied, combined/syn. *Czechoslovak Red Pied, Czechoslovak Red-and-White, Czechoslovakian Red-and-White*

Czerwono-biała: *see* Polish Red-and-White

Czerwona Polska: *see* Polish Red

Dabieshan: (E Hubei and SW Anhui, China)/dr/shades of yellow, occ. brown, rarely black, and often with white round patches along rump (? banteng influence)/high-humped Changzhu group/var.: Huangpi/named 1982/not *Dabishan*/[= Dabie Mountain]

Dacca-Faridpur: (Bangladesh)/z/ local var., sim. to Hariana/syn. *Dhaka-Faridpur*/extinct (by crossing with Sahiwal, Sindhi and Friesian)

Dagana: (Senegal)/var. of Gobra

Dagestan Brown: (Dagestan, Russia)/ d.m/orig. from Swiss Brown × Dagestan Mountain with Carpathian Brown blood/ absorbed by Caucasian Brown/ Russ. *Dagestanskaya buraya*/ extinct

Dagestan Mountain: (Russia)/dwarf subvar. of Greater Caucasus var. of Caucasian/Russ. *Dagestanskiĭ gornyĭ*/not *Daghestan, Dhagestan*/rare

Dahomey: *see* Lagune

dairy-beef synthetic: *see* Dairy Synthetic

Dairy Criollo: *see* Tropical Dairy Criollo; also Barroso, Blanco Orejinegro, Caldeano, Costeño con Cuernos, Cuban Criollo, Saavedreño

Dairy Gir: (Brazil)/d/cf. Dairy Zebu of Uberaba/var. of Brazilian Gir/BS 1967/Port. *Gir leiteiro*

Dairy Hungarian Brown: (Hungary)/ d.(m)/orig. (1950s on) from (Danish Jersey × Hungarian Brown) × (Danish Jersey × Hungarian Pied), graded to Hungarofries or Holstein/HB/ Hung. *Tejelő magyar-barna*/syn. *Hungarian Brown Dairy, New Hungarian Brown*/not *Hungarian Red Dairy*/extinct

Dairy Hungarian Pied: (Hungary) /d.m/orig. (1950s on) from Danish Jersey × Hungarian Pied/ graded to Hungarofries or Holstein/Hung. *Tejelő magyar-tarka*/syn. *Hungarian Dairy Fleckvieh, Hungarian Spotted Dairy*/not *New Hungarian Red*/ extinct

Dairy Indo-European: (Rio de Janeiro, Brazil)/d/in formation at

Instituto do Zootecnia from Friesian *et al.* × zebu/Port. *Indoeuropeu leiteiro/*? discontinued

Dairy Maine-Anjou: (Pays de la Loire, France)/d/brown/Fr. *Maine Anjou Lait*/rare

Dairy Shorthorn: (Great Britain)/ d.m/cf. Milking Shorthorn (USA)/Shorthorn group/Red Holstein and Danish Red blood since 1969 (see Blended Red-and-White Shorthorn)/BS 1905, HB (Dairy section) 1959; BS also Australia, New Zealand (Milking Shorthorn 1913)

Dairy Synthetic: (Alberta Univ., Canada)/m.d/30% Holstein, 30% American Brown Swiss, 6% Simmental, 34% beef breeds/ syn. *dairy-beef synthetic*

Dairy Zebu of Uberaba: (Minas Gerais, Brazil)/d/cf. Dairy Gir/ orig. since 1948 by selection of Brazilian Gir (*et al.*) at government farm 'Getulio Vargas'/Port. *Zebu leiteiro de Uberaba*

Dajjal: (Dera Ghazi Khan, W Punjab, Pakistan)/var. of Bhagnari (by crossing larger type with local cattle)

Dales Shorthorn: *see* Northern Dairy Shorthorn

Dali: (China)/var. of Yunnan Zebu

Dalmatian Grey: (Croatia)/m.d/orig. early 20th c. from Buša graded to Tyrol Grey with some Montafon blood/HB 1945

Damara: (S Angola and N Namibia)/ d.dr.m/bright brown, red pied or yellow pied; long horns/Sanga type, Setswana group/syn. *Herero*/rare in Namibia, nearly extinct in Angola

Damascus: (Syria)/d/brown/? zebu blood, sim. to Cyprus, Egyptian and Lebanese/syn. *Aleppo, Damascene, Halabi, Halep* (Turkey), *Shami, Shamia*

Damerghou: *see* Azaouak

Damietta: (N Delta, Egypt)/d/var. of Egyptian/syn. *Domiatta, Domiatti, Domyati, Dumiati, Dumyati*

Danakil: (Dancalia, i.e. S Eritrea, E Tigray and Welo, Ethiopia, and NE Djibouti)/d/light chestnut, blond or ash-grey; long lyre horns/Sanga type, Abyssinian group/var.: Raya-Azebó/syn. *Adal, Adali, Afar, Dancalian, Keriyu, Raya*

Danara: *see* Garre

Dancalian: *see* Danakil

Dangi: (W Ghats, NW Maharashtra, India)/dr/spotted/z/sim. to Deoni and Gir/vars: Kalakheri, Sonkheri/BSd/syn. *Ghauti, Kanada, Konkani*

Dangjiao: (China)/extinct

Danish Black Pied: (Jutland, Denmark)/d/orig. (1949) from Jutland Black Pied + Friesian, chiefly Dutch, some East Friesian and Swedish; Friesian predominates; Holstein blood since 1965 (now over 75%)/BS, HB/Dan. *Sortbroget Dansk Malkekvæg* (= *SDM*)/syn. *Black Pied Danish, Danish Black-and-White Milk*

Danish Black-and-White Milk: *see* Danish Black Pied

Danish Blue-and-White: (Denmark) /m/= Belgian Blue/HB/Dan. *Dansk Blåhvidt kvæg*/rare

Danish Jersey: (Denmark, esp. W Fünen)/var. of Jersey from imports from Sweden (1896) and Jersey (1902–1909)/BS 1902, HB 1925

Danish Red: (Denmark, islands)/d/ Baltic Red type/orig. from North Slesvig Red (+ Angeln and Ballum) × local island cattle (1841–1863); American Brown Swiss blood *et al.* since 1975/ orig. of British Dane, Bulgarian Red *et al.*/recog. 1878, HB 1885; HB also in Russia, USA (American Red Dane) 1948–1968, New Zealand/Dan. *Rødt Dansk Malkekvæg* (= *RDM*)/syn. *Fünen* (Dan. *Fynsk*), *Red Dane, Red Danish*

Danish Red (old type): *see* RDM-1970

Danish Red-and-White Shorthorn: *see* Danish Red Pied

Danish Red Pied: (Jutland,

Denmark)/d/orig. from Shorthorn with German Red Pied and Meuse-Rhine-Yssel blood in 1950s/BS/Dan. *Dansk Rødbroget Kvæg* (= *DRK*)/syn. *Danish Red-and-White, Danish Red-and-White Shorthorn*

Danish Shorthorn: (Jutland, Denmark)/d.m/usu. roan, also red, red-and-white or white/orig. from Beef Shorthorn (imported 1847) × local Danish (esp. dairy) cows, plus dairy and dual-purpose Shorthorns imported early 20th c. to form dual-purpose Danish Shorthorn/HB 1906 incl. beef, dairy and dual-purpose imported purebred Shorthorns + Danish-bred crosses; further imports up to 1950s/absorbed by Danish Red Pied 1962 (HB opened for all red-and-white crosses, inc. purebred Shorthorn) but one pure old-type Shorthorn herd maintained (1963–1970) by Lynge family and recorded as 'Old Shorthorn' 1971, = basis of present Danish Shorthorn plus imports from England, Canada and New Zealand/Dan. *Dansk Korthorn*/rare

Dansk Blåhvidt kvæg: *see* Danish Blue-and-White

Dansk Korthorn: *see* Danish Shorthorn

Dansk Rødbroget Kvæg: *see* Danish Red Pied

Danube: (Burgenland and Lower Austria)/former var. of Austrian Simmental/Ger. *Donau*/extinct

Danyang: *see* Shandong

Darbalara: (Australia)/strain of Australian Milking Shorthorn (latter now absorbed by Illawarra)/extinct

Dar El Reih: *see* Butana

Dark Andalusian: *see* Retinta

Darmeghou: *see* Azaouak

Dashtiari: (SE Baluchestan, Iran)/ m/brown, black, white or pied/z

Dauara, Daura, Dawara: *see* Garre

Dazzal: *see* Dajjal

Deep Red: (S Netherlands)/m/red,

with white spots on head/orig. since 1976 from Meuse-Rhine-Yssel/? = Brandrood IJsselvee/Du. *Brandrood Runderen*/nearly extinct

Dehong: (S Yunnan, China)/var. of Yunnan Zebu

Delami: *see* Nuba

de la Tierra: *see* Canary Island, Palmera

De Lidia: *see* Fighting bull

Demonte: (Cuneo, Italy)/d.m/red to straw-coloured/var. of Piedmont in Stura valley/extinct

Dengchuan: (NW Yunnan, China)/♂ small hump/Changzhu group/extinct

Deogir: (India)/sim. to Deoni

Deoni: (Bidar, N Karnataka, India)/ dr.d/black-and-white or red-and-white/sim. to Dangi and Gir/syn. *Dongari, Dongarpatti*

Derbyshire Blue: *see* Blue Albion

Desan: *see* Gir

desi: (India)/z/not *deshi*/[= local, indigenous]

Deutsche: *see* German

Devarakota: (Tamil Nadu, India)/ z/local var., sim. to Ongole

Devni: (Ugdir taluka, Osmanabad, Maharashtra, India)/z/local var.

Devon: (N and E Devon, SW England)/m.[dr]/cherry red/sim. to Sussex/m.d var. in E Devon, pd var. in USA/orig. of Milking Devon/HB 1851, BS 1884; BS also USA 1905 (HB 1955), Australia 1929 (HB 1873), S Africa 1917 (HB 1906); HB also Brazil 1952, New Zealand 1964/ syn. *Beef Devon, North Devon, Red Devon, Red Ruby*/not to be confused with South Devon

Dewsland: (S Wales)/var. of South Wales Black/extinct

Dexter: (Ireland)/d.m/red or black; short-legged (achondroplastic)/ orig. from Kerry in 18th c./BS and HB in Ireland 1887–1919; BS also England (imported 1886, HB 1900), USA 1911, S Africa 1958 (HB 1938), Canada 1972, Australia/syn. *Dexter-Kerry*

Dhagestan: *see* Dagestan

Dhaka-Faridpur: *see* Dacca-Faridpur
Dhanni: (Rawalpindi, N Punjab,
  Pakistan)/dr/white with black
  markings (usu. spotted or
  coloursided), black with white
  markings, red-and-white, or
  white/z/HB 1938/syn. *Awankari,
  Pahari, Pakhari, Pothwari*/not
  Dhani, Dhaniri
dhee: *see* yak ♀
dhimjo: *see* dzo
Dhofari: *see* Omani Dhofari
Diakoré: *see* Djakoré
Diali: (R Niger floodplains and
  valleys, SW Nigeria and Niger)/
  m/white or pied; variable (esp.
  horns)/West African Zebu type,
  Fulani group/sim. to Keteku; ?
  orig. from humpless West
  African Shorthorn and White
  Fulani cross/Hausa *Jalli, Shanun
  Bayaro*, Fulani *Jalliji*, Fr. *Peul
  nigérien* (= *Nigerian Fulani*)/syn.
  Djeli, Djelli, Jali, Jalli, Jeli
Didinga: (S Sudan)/var. of Mongalla
  (? with Toposa blood)
Dimdzo, dimjo, dimschu: *see* dzo
Dinka: *see* Aliab Dinka, Aweil Dinka,
  Nilotic, Wadai Dinka
Diqin: (NW Yunnan, China)/local
  var., mainly orig. from *Bos
  taurus* rather than from *B.
  indicus* types/syn. *Diqin(g)
  Yellow*
Dirbani: *see* Hassawi
Dishley: *see* Longhorn
Dishti: (N and E of Amarah, Iraq)/
  light tan to cream or golden;
  small or no hump/var. of Iraqi/[=
  living in open]
Dithmarscher: (Germany)/orig. var. of
  Red Pied Schleswig-Holstein/
  extinct
Diyarbakir: (SE Turkey)/extinct
Djafoun: *see* Red Bororo
Djakoré: (Mali) *see* Kaarta
Djakoré: (Senegal)/m/white, grey or
  yellow/Gobra zebu × N'Dama
  cross/var.: Bambey/syn. *race du
  Siné*/not *Diakoré, Djokore,
  Jakoré*
Djebeli: (Syria) *see* Baladi
Djelani: *see* Jaulan
Djeli, Djelli: *see* Diali

Djoloff: (Senegal)/largest of three vars
  of Gobra
Dnieper: (Ukraine)/m/var. of
  Ukrainian Beef/orig. 1961–1979
  from Chianina (66.7%),
  Charolais (8.3%), Russian
  Simmental (16.7%) and
  Ukrainian Grey (8.3%)/recog.
  1979; combined 1993 with
  Chernigov as Ukrainian
  Beef/Russ. *Pridneprovskiĭ tip*
Doai, Doayo: *see* Namshi
Dobrogea Red: (E Romania)/d/red,
  sometimes white on udder/
  brachyceros type/improved by
  Red Steppe and Danish Red and
  with them forms Romanian Red/
  Rom. *Roşie Dobrogeană* or
  Dobrogea/syn. *Dobruja Red*
Dobruja Red: *see* Dobrogea Red
Doğu Anadolu kırmızı: *see* East
  Anatolian Red
Døle: (SE Norway)/d.m/brown, black,
  red or fawn, with or without
  white markings; 15% pd/orig. in
  1880s from local, Ayrshire and
  Telemark; graded to Norwegian
  Red-and-White since 1950 and
  combined in Norwegian Red
  1963/recog. 1880, HB 1909,
  restarted 1922/vars:
  Gudbrandsdal, Østerdal/syn.
  Doele/not *Døla*/[= valley]/nearly
  extinct
dolinowa: *see* Polish Red (lowland
  var.)
Domaće šareno goveče: *see* Serbian
  Pied
Domiatta, Domiatti: *see* Damietta
Dominican Criollo: (Dominican
  Republic)/var. of Tropical Dairy
  Criollo
Domshino, Domshinskaya: *see*
  Yaroslavl
Domyati: *see* Damietta
Doñana: *see* Mostrenca
Donau: *see* Danube
Donayo: *see* Namshi
Dongari, Dongarpatti: *see* Deoni
Dongola: (N Sudan)/strain of Red
  Desert, sim. to Butana (and often
  crossed with it)/rare
Dongolé: *see* Kuri
Donnersberg: (Palatinate, Germany)/

d.m/red/orig. in late 18th c. from
Bernese × native; united with
Glan in 1890 to form Glan-
Dannersberg; revived/Ger.
Donnersberger Rotvieh/rare

Doppelnutzung Rotbunt: (Germany)/
= Dual-Purpose Red Pied/cf.
Dutch Red Pied Dual-Purpose,
German Red Pied

Doran: (Costa Rica)/m/white with red
speckling or patches/Criollo type
said to contain Shorthorn blood/
[corruption of Durham =
Shorthorn]

Dorna: (Suceava dist. esp. Vatra
Dornei, N Romania)/d/black with
Pinzgauer pattern/var. of
Transylvanian Pinzgau/orig. in
early 20th c. from Pinzgauer ×
Romanian Mountain (also some
Swiss Brown blood)/syn. *Black
Pinzgau*

Dörtyol: (S Turkey)/small var. of
South Anatolian Red, sim. to
Baladi of Syria/extinct

Drakensberger: (W Natal, S Africa)/
m.dr.d/shiny black/sim. to
Basuto/orig. from Sanga (inc.
Afrikander, Basuto, Nguni) with
Friesian blood; specifically
(1947) from Uys + Kemp +
Tintern Black/BS 1947/syn.
Black Afrikander

Drenthe: (Netherlands)/red-and-
white/sim. to Ayrshire
colour/extinct

dri: *see* yak ♀

dridzo: *see* dzo

Drimmon: (Ireland)/coloursided var.
of Kerry/[Gaelic *droimeann* =
white-backed]/extinct

DRK: *see* Danish Red Pied

drong, drongdri: *see* yak, wild

Drontheimer: *see* Trondheim

Droughtmaster: (N Queensland,
Australia)/m/red; hd or pd/orig.
since 1930s from Brahman ($\frac{3}{8}$–$\frac{1}{2}$)
× Shorthorn/BS 1956

Dschau-bei: *see* Zaobei

Dscholan: *see* Jaulan

Duara: *see* Garre

dulong: (NW Yunnan, China) *see*
mithun

Dumiati, Dumyati: *see* Damietta

Dun Galloway: (Scotland)/colour var.
of Galloway/BS (with Belted
Galloway) 1921–1952; HB with
Galloway Society to 1912 and
from 1920; HB also USA (with
Galloway)

Dunlop: *see* Ayrshire

Durham: *see* Shorthorn

Durham-Mancelle: *see* Maine-Anjou

Duruma: (Kenya)/m.d.dr/Small East
African zebu group, Kenya
cluster/inc. in Lowland or
Coastal Zebu along with Giriami,
Kamba and Taita

Dutch: (Netherlands) *see* Black Pied
Dutch Friesian, Dutch Black
Pied, Dutch Improved Red Pied,
Dutch Red Pied Dual-Purpose,
Groningen Whiteheaded, Meuse-
Rhine-Yssel, Red Pied Friesian

Dutch: (elsewhere) *see* Friesian

Dutch Argentine: *see* Argentine
Friesian

Dutch Belted: (USA)/d/black (occ.
red) with white belt; occ. pd/
orig. from Lakenvelder imported
from Netherlands 1838–1848
and 1906/always bred pure; reimported by Netherlands to
reconstruct Lakenvelder/BS
1868/syn. *American Dutch
Belted, Dutch Belt*/nearly extinct

Dutch Black Pied:
(Netherlands)/d.m/graded to
Holstein (70–75%) since 1970s
(see Black Pied Dutch Friesian
for orig. type free of Holstein
blood)/orig. of Friesian type/BS
and HB 1874 (NRS = Nederlands
Rundvee Stamboek)/Du.
Zwartbont (HF)/obs. syn. *Black-
and-White Holland, Black Pied
Dutch, Dutch Friesian, Dutch
Lowland, Holland, Netherlands
Black Pied, West Friesian*

Dutch Dairy: (Netherlands)/d/HB
1996 for all black pied and red
pied dairy cattle/Du. *Nederlands
Melkras*

Dutch Friesian: *see* Black Pied Dutch
Friesian

Dutch Friesian: (Gauteng, S Africa)/
m/orig. from Friesian cattle intro.
by colonial settlers/almost extinct

Dutch Improved Red Pied:
(Netherlands)/m/orig. since 1970
from Meuse-Rhine-Yssel by
selecting for double muscling,
also some foreign beef blood/HB
1988/Du. *Verbeterd Roodbont
Vleesras (VRV)* (= *Improved Red
Pied Beef breed*)
Dutch Lowland: *see* Dutch Black
Pied
**Dutch Red-and-White, Dutch Red
Pied** : *see* Dutch Improved Red
Pied, Dutch Red Pied Dual-
Purpose, Meuse-Rhine-Yssel,
Red Pied Friesian
Dutch Red Pied Dual-Purpose:
(Netherlands)/d.m/orig. since
1970s from Meuse-Rhine-Yssel
by Red Holstein crossing/HB/Du.
Roodbont Dubbeldoel (= *RDD*)
dwarf: i.e. height at withers 100 cm
or less/Europe: Albanian Dwarf,
Betizuak, Cachena, Dagestan
Mountain, Dexter, Posavina,
Rodopi, Valachian Dwarf, West
Macedonian; Asia: Achham,
Bengali, Indian Hill, Kedah-
Kelantan, Konari, Madhya
Pradesh dwarf zebus, Malnad
Gidda, Nepalese Hill, Oman
Zebu, Punganur, Sinhala,
Tibetan, Vechur; Africa: Chagga,
Kabyle, Nuba Mountain,
Socotra, West African Dwarf
Shorthorn
Dwarf Humpless: *see* West African
Shorthorn (dwarf vars)
Dwarf Shorthorn: *see* West African
Shorthorn (dwarf vars)
Dwarf Shorthorn: (coastal and forest
regions, West Africa)/= Forest
Muturu type of humpless
shorthorns (in contrast to larger
Savanna type)/syn. *Lagune* (Fr.)
Dynevor: (S Wales)/long horns
curving out, forwards and up/
ancient herd of White Park type,
orig. in Wales, now in England/
nearly extinct
dzo: (Tibet)/any yak–cattle cross/
Tibetan *mdzo*, Ch. *pian niu* (WG
p'ien niu) (= inclined cattle),
Russ. *haĭnyk* (or *hanik*) from
Mongolian *hainag* (or *khainag*)

(= yak)/syn. *bhotey* or *bhotea* (=
Tibetan) (Nepal), *dzoz, je, yakow,
zo*/yak ♂ × domestic cow, F₁, is
dridzo (C Tibet), *urang* (Nepal),
pamjommu (Assam), *maran-
hainag* (= *sun yak*) (Mongolia)/
domestic bull × yak ♀, F₁, is
pandzo (C Tibet), *dhimjo,
dimdzo, dimjo* or *dimschu*
(Nepal), *saran-hainag* (= *moon
yak*) (Mongolia)/backcross to yak
♂ is *ah gohr* or *galiba* (China),
ago (Russ.)/*see also* dzobo,
dzomo
dzobo: (Tibet)/♂ (sterile) dzo; Tibetan
mdzo-po; syn. *dzopho,
dzopkhyo, jhopke, joppa,
zebkyo, zhopkyo, zopkio* (Nepal)
dzomo: (Tibet)/♀ dzo; Tibetan *mdz-
mo*/syn. *chauri* (*q.v.*), *jommu,
jum, zhum, zomo, zum*

EAST AFRICAN SHORTHORNED ZEBU:
(Ethiopia to Malawi)/m.dr/
thoracic musculo-fatty hump/
inc. Large East African Zebu
(Boran, Karamajong, Murle,
Toposa, Turkana) and Small East
African Zebu (Abyssinian
Shorthorned Zebu group, Angoni
group, Kenya cluster, Somali
group, Southern Sudan group,
Tanzanian cluster, Teso group)/
syn. *East African Humped,
Shorthorned Zebu*
East African zebus: *see* East African
Shorthorned Zebu, Madagascar
Zebu, North Sudan Zebu, Nuba
Mountain
East Anatolian Red: (NE Turkey)/
d.m/orig. from Caucasian/vars:
Çildir, Göle/Turk. Şarkî (or *Doğu*)
Anadolu kırmızı/syn. *Eastern Red*
Eastern Black Pied: (Belgium) *see*
Hervé Black Pied
Eastern Nuer: (NE of S Sudan)/local
var. of Nilotic
Eastern Province Zebu: (Uganda) *see*
Nkedi
Eastern Red: *see* East Anatolian Red
Eastern Red and White: (France) *see*
French Red Pied group
Eastern Red Pied: (E Liège, Belgium)/
d/former name for Red Pied

(Meuse-Rhine-Yssel) in Eupen and Malmédy/inc. in Belgian Red Pied/Fr. *Pie rouge de l'Est (de la Belgique)*, *Pie rouge Ardennes-Liège*, *Pie rouge du Pays de Hervé*, *Hervé Red Pied*/extinct

Eastern Red Pied: (France) *see* French Red Pied, French Simmental

Eastern Red Polled: (Norway) *see* Red Polled Eastland

Eastern Spotted: (France) *see* French Simmental

Eastern Taurine: (E Chad)/small population of West African Shorthorns/Fr. *Taurin de l'Est*

East Finnish: (E Finland)/d.m/red-and-white (coloursided); pd/var. of Finnish/BS 1898–1950; HB formerly also Karelia, Russia/Fin. *Itäsuomen karja, Itä-Suomalainen Karja* (= *ISK*), Swed. *Ostfinsk*/syn. *East Finncattle, Red-and-White Finnish, Red Pied Karelian*/nearly extinct

East Flemish Red Pied: *see* Belgian White-and-Red

East Friesian: (Ostfriesland, Germany)/d.m/orig. var. of German Black Pied/HB 1883/Ger. *Ostfriesen, Ostfriesisches*/syn. *Black Pied East Friesian, East Friesland*/*see also* Red East Friesian, Red Pied East Friesian

East Friesian Red Pied: *see* Red Pied East Friesian

Eastland: *see* Red Polled Eastland

East Macedonian and Thrace: (Greece)/former var. of Greek Shorthorn

East Prussian: (Germany)/d.m/orig. var. of German Black Pied (till 1945)/orig. from East Friesian and Dutch Friesian, imported in mid-19th c./Ger. *Ostpreussen, Ostpreussisch Holländer*/now see Mazury

East Siberian: *see* Yakut

East Styrian Spotted: (Austria)/former var. of Austrian Simmental/Ger. *Österreichisches Fleckvieh*/extinct

Ebian Spotted: (SC Sichuan, China)/black pied or red pied; small/high-humped Changzhu group/syn. *Ebian*/rare

Ecuador Criollo: (Ecuador)/m/vars: Sierra (inc. Páramo, Hoyas), Costa (inc. El Oro, Esmeraldas)/Sp. *Criollo ecuatoriano*/rare, being graded to exotic breeds

Eesti maat~oug: *see* Estonian Native

Eesti punane: (Estonia)/?

Egerland, Egerländer: *see* Cheb

Egyptian: (Egypt)/dr.(d)/yellow-brown to black/*brachyceros* with zebu blood, sim. to Cyprus, Damascus and Lebanese/vars: Balada (inc. Menufi), Damietta, Maryuti, Saidi

Einfarbig gelbes Hohenvieh: *see* Gelbvieh

Elb- und Wilstermarsch: *see* Wilstermarsch

Eleskirt: (Turkey)/extinct

Elling: (Germany)/orig. from Grey Mountain × native Franconian red; absorbed by Yellow Franconian/extinct

El Oro: *see* Ecuador Criollo

Enderby Island: (New Zealand)/feral/Shorthorn orig. in 1890s/nearly extinct

English Longhorn: *see* Longhorn

Enns, Ennstaler: *see* Bergscheck

Enshi: (Hubei, China)/var. of Wuling

Eo: (NW Asturias, Spain)/d.m.dr/yellow, sometimes with white markings; long horns/local var. of North Spanish type/Sp. *Agrupación Eo*/extinct

Epirus: (Greece)/former var. of Greek Shorthorn

Eringer: *see* Hérens

Eskişehir Brown: (Turkey)/? var. of Turkish Brown

Esmerelda: *see* Ecuador Criollo

Estonian Black Pied: (NW and N Estonia)/d/orig. in 2nd half of 19th c. from Dutch and East Friesian × Estonian Native/inc. in Baltic Black Pied/HB 1885, named 1951/Russ. *Èstonskaya chernopestraya*/syn. *Estonian Black Spotted*/obs. syn. *Estonian Dutch-Friesian*

Estonian Brown: *see* Estonian Red

Estonian Dutch-Friesian: *see*
Estonian Black Pied
Estonian Native: (Estonia)/d.m/red or
yellow-brown with white flecks;
pd/North European Polled type/
some West Finnish and Jersey
blood (1955–1990), also Red
Holstein/orig. (with Angeln and
Danish Red) of Estonian Red/HB
1914, BS 1920/Eston. *Eesti
maatõug*, Russ. *Mestnaya
èstonskaya*/syn. *Local
Estonian*/rare
Estonian Red: (SC and E Estonia)/d/
pale to dark red/Baltic Red
type/orig. from Angeln and
Danish Red × Estonian Native in
early and mid-19th c./HB 1885/
Russ. *Krasnaya èstonskaya*/syn.
*Estonian Brown, Estonian Red-
Brown*
Èstonskaya: *see* Estonian
Ethiopian Boran: (S Ethiopia)/d.m/
Large East African (shorthorned)
zebu group/orig. (indigenous
unimproved) var. of Boran/syn.
Avai (Somalia), *Borana, Somali
Boran*
Ethiopian Highland Zebu: *see*
Abyssinian Shorthorned Zebu
Etschtaler: (Austria) *see* Grey Adige
etxeko behiak: *see* Betizuak
EUROPEAN BLACK-AND-WHITE: BS 1966
(European Confederation of
Black-and-White Breed Societies,
Fr. *Confédération Européene des
Eleveurs de la Race Pie-Noire*,
Ger. *Europäische Vereinigung
der Schwarzbuntzüchter*)/inc.
FRS, NRS and HBs of Belgian
Black Pied, British Friesian,
Danish Black Pied, French
Holstein, German Black Pied,
Italian Friesian, Swedish
Friesian/syn. *European Black
Pied, European Friesian*
EUROPEAN BLOND AND YELLOW: BS
1962 (Ger. *Europäische Blond-
und Gelbviehzüchtervereinigung*,
Fr. *Fédération Européenne des
Races Bovines Blondes et
Jaunes*)/inc. Austrian Yellow,
Blonde d'Aquitaine, Gelbvieh,
Limousin, Piedmont

EUROPEAN RED: BS 1956 (Ger.
*Europäische Vereinigung der
Rotviehzüchter*, Fr. *Fédération
Européenne des Eleveurs des
Races Bovines Rouges*)/inc.
Angeln, Belgian Red, Danish
Red, Flemish, German Red; *see
also* Baltic Red, Bulgarian Red,
Central European Red, Gorbatov
Red, Red Poll, Red Steppe,
Romanian Red, Tambov Red
European wild ox: *see* aurochs
Evolénard: (Valais, Switzerland, and
Aosta, Italy)/d.m, fighting/red-
and-white, black-and-white, dun
or with white dorsal stripe/off-
colour vars of Hérens/syn.
Patcholé/[from Evolène]/almost
extinct
Extremadura, Extremadura Red: *see*
Retinta
Extremeña: *see* Retinta
Extremeña rubia: *see* Andalusian
Blond

Faenol: *see* Vaynol
Faeroes: (Faroe Islands)/d/Nor. orig./
Dan. *Færøerne*
Fagina Albères: *see* Albères (fawn var.)
Falkland: *see* Fife Horned
Fao: *see* Jenubi
Far Eastern: *see* Russian Simmental
Faridpur: *see* Dacca-Faridpur
Felicité: (Seychelles)/black or black-
and-white; hd or pd; feral/orig.
from zebus and European
breeds/nearly extinct
Fellata: (Chad)/West African
(shorthorned) Zebu type, sim. to
Shuwa but larger hump and bred
by Fulani/syn. *Borero, Shuwa-
aral, Wadara*/not *Filata*/[Kanouri
= Fulani]
Fellata: (W Sudan and Gambela, SW
Ethiopia) *see* Red Bororo
Fémeline: (Sâone valley, E France)/
dr.d/fawn or red/sim. to
Tourache but more of lowland,
refined, dairy, 'female' type/
crossed with Shorthorn at end of
19th c. and graded to Simmental
in early 20th c. to form part of
Eastern Red Pied (now French
Simmental)/not *Feme*/[Fr.
femelle = female]/extinct

feral: *see* Amsterdam Island, Betizoak, Chillingham, Cobourg Peninsula, Enderby Island, Felicité, Hawaiian wild, Kuchinoshima, Patagonian Criollo, Swona

Fergana: (Uzbekistan)/z/var. of Central Asian Zebu (or Zeboid)/ Russ. *Ferganskiĭ zebu*/extinct

Ferrandais: (Puy-de-Dôme, France)/ d.m.[dr]/black, brick-red and white, occ. black-and-white; lyre horns/HB 1905/rare (by crossing with French Simmental)

FFPN: *see* French Holstein

FH-hitam-putin: (Indonesia)

FH-merah: (W Java, Indonesia)/d.m/ red-and-white; ♀ occ. pd/ imported from Australia and Hungary, established 1987/syn. *Red Friesian*/[FH = Holstein-Friesian]/rare

Fiemme: *see* Grey Val di Fiemme

Fife Horned: (E Scotland)/syn. *Falkland*/extinct

Fighting bull: (Spain and Portugal)/ usu. black or dark brown/bulls for fighting bred in Andalucía, Extremadura, New Castile, Salamanca, and in Ribatejo (C Portugal, = *Ribatejana*)/HB Spain 1980, Portugal 1986, BS Portugal/ Sp. *toro de lidia, toro lidiado, ganado bravo*, Port. *Touro de Lide*

Fighting bull: (Grande Camargue, France)/black/orig. imported from Spain/Fr. *Race espagnole*, Sp. *Brava*/rare

Filani: *see* Fulani

Filata: *see* Fellata

finchbacked, finched: *see* coloursided

Finnish: (Finland)/d/pd/vars: East Finnish, North Finnish, West Finnish/Finn. *Suomenkarja, Suomalainen Karja* (= *SK*), Swe. *Finsk*/syn. *Finncattle*

Finnish Ayrshire: (Finland)/d.m/var. of Ayrshire from cattle imported 1847–1923/BS 1901/Finn. *Suomen Ayrshire, Suomalainen Ayrshirekarja*, Swe. *Finsk Ayrshire*

Finsk: *see* Finnish

Fjäll: *see* Swedish Mountain

Fjällnära: (Sweden)/d/local rustic, cf. Allmoge

Fjord: *see* Vestland Fjord

Flamande: *see* Flemish

Flanders, Flandrine: *see* Flemish

Fleckvieh: *see* Simmental

Fleckvieh Fleischnutzung: *see* German Beef Simmental

Flemish: (Belgium) *see* Belgian Red, Belgian White-and-Red

Flemish (Original): (Picardie and N France)/black or red/native of N France, free from Danish Red blood (as distinct from Flemish Red)/former vars: Maroilles, Picardy/HB 1886/Fr. *Flamande originelle, Flandrine*/rare

Flemish Red: (Nord, Pas de Calais and Picardy, France)/d.m/ mahogany/Danish Red blood since 1965 (now reducing)/HB 1886, BS 1970, now inc. pure and with 20–25% Danish Red blood/Fr. *Rouge flamande*/syn. *Population rouge, Rouge du Nord*/rare

Flemish Red Pied: *see* Belgian White-and-Red

Florida Cracker: (USA)/m/all colours; small/sim. to Pineywoods/Sp. orig. in 16th c./ BS 1989, HB 1993/dwarf var.: Guinea/syn. *Florida Native, Florida Scrub*/rare

Fogera: (L Tana, Gonder and Gojam, NW Ethiopia)/dr.m.d/often white, or white with black spots, or black-and-grey; short horns, occ. pd; small hump/Zebu-Sanga intermediate/orig. from various Abyssinian Shorthorned zebu with Nilotic and Abyssinian Sanga/not *Fogara, Fogerra, Uogherá, Wagaru*

Fogha: *see* Red Bororo

Forest cattle: *see* Black Forest

Forest Muturu: (West Africa)/m.ri/ dwarf type of West African Shorthorn in coastal forests (in contrast with larger Savanna type)/syn. *Dwarf Shorthorn, Forest Shorthorn*/[Hausa, *Muturu*

= name for humpless shorthorns in English-speaking areas, cf. *Lagune* in French-speaking areas]

Formentina: *see* Reggiana

Formosa: *see* Taiwan Zebu

Fort Cross: (Ontario, Canada)/m/orig. at Fort William farm, Thunder Bay, from Charolais × (Lincoln Red × Hereford)

Foula, Foulah, Foulani, Foulawa: *see* Fulani

Foulbé: *see* Fulani

Foulbé blanc: *see* White Fulani

Foulbé de Banyo: *see* Banyo

Foulbé de Yola: *see* Yola

Fouta Djallon, Fouta Jallon: *see* N'Dama

Fouta Longhorn: *see* N'Dama

Fouta Malinke: *see* N'Dama

Franconian: *see* Yellow Franconian

Franqueiro: (N of São Paulo, Brazil)/long horns/var. of Crioulo Lageano/orig. from Southern Crioulo late 19th c./[from district of France in São Paulo]/nearly extinct

Frati: (N Italy)/d/cf. Preti/Italian Brown × Friesian, 1st cross/[It. *frate* = friar, i.e. dark brown]

Freiburg, Freiburger: *see* Fribourg

French Brown: (NE, S of Massif Central and N of Pyrenees, France)/d.m.[dr]/orig. from Swiss Brown imported 1827 on/BS and HB 1911/Fr. *Brune des Alpes* (= *Brown Alpine*)

French Canadian: *see* Canadienne

French Dairy Simmental: *see* Montbéliard

French Hérens: *see* Alpine Hérens

French Holstein: (NE and SW France)/d/orig. from Dutch Black Pied first imported in 18th c.; graded to Holstein in 1970s and 1980s/BS 1911, HB 1922/Fr. *Prim'Holstein* (named 1990, inc. Dutch Black Pied type, Holstein type, their crosses and Red Holsteins)/syn. *Française frisonne pie-noire* (*FFPN*) (1952–1990), *French Friesian*, *French Holstein-Friesian*, *Hollandaise pie-noire* (till 1952)

French Jersey: (W France)/d

FRENCH RED PIED group: (E France)/inc. (1946) Abondance, Montbéliard and Eastern Red Pied (now French Simmental)/Fr. *Race pie rouge françaises*/syn. *Eastern Red and White*, *Eastern Red Pied*, *Pied Eastern*

French Red Pied Lowland: (N Brittany, Normandy, Centre and Massif Central regions, France)/d.m/HB 1970 to combine Armorican, German Red Pied, Meuse-Rhine-Yssel and their crosses/with Red Holstein blood since 1982/Fr. *Pie rouge des plaines*

French Simmental: (EC France esp. Côte d'Or)/m.d.[dr]/French Red Pied group/orig. in early 20th c. from Simmental × local (Bressane and Fémeline) in Sâone valley; absorbed Gessien 1945, Alsatian Simmental 1947/HB 1930/Fr. *Simmental française*/obs. syn. *Tachetée de l'Est* (= *Eastern Spotted*) (1939–1959), *Pie Rouge de l'Est* (= *Eastern Red Pied*) (1959–1992), *Red Spotted*

French Western Red: (NW France)/m.d/federation 1962–1970 of Armorican and Maine-Anjou/HB 1966/Fr. *Rouge de l'Ouest*/extinct

Friauler: (Germany) *see* Friuli

Fribourg: (W Switzerland)/d.m.dr/black pied/since 1966 crossed with Holstein to give Swiss Holstein/BS and HB/Fr. *Fribourgeois*, Ger. *Freiburger*/extinct (last purebred bull died 1973

Frieiresa: (Galicia, Spain)/m.(dr.d)/light brown or chestnut shading to dark; long horns; long golden-brown fringe/part of Morenas del Noroeste group/= Mirandesa in Portugal/rare

FRIESIAN: d.(m)/black-and-white/inc. Dutch Black Pied, East Friesian and their derivatives: American Beef Friesian, Argentine Holstein, Belgian Black Pied,

British Friesian, British Holstein, Czech Black Pied, Danish Black Pied, French Holstein, Friesland, German Black Pied, Holstein (USA), Israeli Holstein, Italian Friesian, Japanese Holstein, Polish Black-and-White Lowland, Romanian Holstein, Russian Black Pied, South Korean Holstein, Spanish Friesian, Swedish Friesian, Turino *et al.*; *see also* Red-and-White Friesian/BS/HB also in Australia, Austria, New Zealand 1911, Luxembourg 1923, Brazil 1934 *et al.*/Fr. *Frisonne*, It. *Frisona*, Sp. *Frisia*/syn. *Black Pied (Lowland), Black-and-White, Dutch, Holland*/not *Frisian, Fresian*

Friesland: (S Africa)/breed of Friesian/HB 1906, BS 1912

Fries Roodbont: *see* Red Pied Friesian

Fries Rundvee Stamboek: *see* FRS

Frieswal: (N India)/d/orig. from Friesian × Sahiwal on military dairy farms (*c.* 62% Friesian); breed selection started 1984

Frijolillo: (Lower California, Mexico)/m/often speckled/Criollo type/[*fréjol* = speckled native pinto bean]

Frisia: *see* Turino

Frisia, Frisian, Frisona, Frisonne: *see* Friesian

Friuli: (Venetia, Italy)/straw-coloured/orig. (with Simmental) of Italian Red Pied/It. *Friulana*, Ger. *Friauler*/syn. *Red Friuli*/extinct

Friuli-Simmental: *see* Italian Red Pied

Froment du Léon: (St Brieuc, Brittany, France)/d/fawn, or fawn with white markings/orig. from Guernsey/HB 1907, BS/[*froment* = corn coloured]/nearly extinct

fronterizo: (N Argentina)/small var. of Argentine Criollo

FRS: (Friesland, Netherlands)/var. of Dutch Black Pied with its own BS and HB/[= *Fries Rundvee Stamboek*]

Fula: *see* N'Gabou

FULANI: (W Africa)/long horns; thoracic hump/z/subgroups: lyre-horned (Gobra, Sudanese Fulani, White Fulani); long lyre-horned (mainly Red Bororo)/*see* Adamawa, Diali, Fellata, Gobra, Red Bororo, Sudanese Fulani, White Fulani/Fr. *Peul* (*Peuhl, Peulh, Peulhé, Peul-Peul*); Hausa *Filani, Fulani, Fulawa* (sing. *Fula, Fulah*); Kanouri (Chad) *Fellata, Filata*/syn. *Foulbé, Fulbé* (sing. *Poulo, Pullo*, also *Poul-Foulo, Poul-Poulo*)/[Fulani are a cattle-owning tribe]

Fulbé: *see* Fulani, Gudali

Fünen: *see* Danish Red

Fung: *see* Kenana

Futa, Futa Jallon: *see* N'Dama

Fuzhou: (Liaodong peninsula, Liaoning, China)/m.(d.dr)/yellow-brown or red; low hump; large/Northern Yellow (Mongolian) group, influenced by Korean/syn. *Fuzhou Yellow*

Fynsk: *see* Danish Red

Gabaruyé: *see* Maure

Gabassaé: *see* Red Bororo

Gabú: *see* N'Gabou

Gacko: (SE Hercegovina, Bosnia)/grey/orig. from Tyrol Grey × Buša/Serbo-cro. *Gatačko goveče*

Gadéhé: *see* Red Bororo

Gado da Terra: (Ribatejo, Portugal)/black/sim. to Morucha and Fighting bull/[= native cattle]/rare

Gajamavu: *see* Gujamavu

Galega: *see* Minhota

galiba: (China)/backcross of dzo to yak/syn. *agoo, ah gohr*

Galician Blond: (NW Spain)/m.(dr.d)/cream to golden-red/North Spanish type; = Minhota (Portugal)/improved with Simmental (1914 on), Swiss Brown (little) and South Devon (1958)/HB 1933, BS 1973/Sp. *Rubia gallega*/orig. type (in Monteroso and Carballino) almost extinct

Galla: (Ethiopia) *see* Sanga

Galla-Azebó: *see* Raya-Azebó

Galla Boran: (E Kenya)/unimproved
   trypanosusceptible var. of Boran

Galloway: (SW Scotland)/m/
   brownish-black; pd/vars: Belted
   Galloway, Dun Galloway, Red
   Galloway, White Galloway/
   Polled HB (with Aberdeen-
   Angus) 1862, BS 1877 (Dun and
   Red now in HB, Belted and
   White have separate HBs); BS
   also USA 1882, Canada 1882 (HB
   1874), New Zealand 1949,
   Australia 1952 (HB 1963),
   Germany 1984 (HB 1982); HB
   also in S Africa, Russia, France,
   Iceland, Czech Republic/syn.
   Southern Scots Polled/*see also*
   Miniature Galloway

Galway: *see* Lord Caernarvon

Gambia Dwarf: (Gambia)/var. of West
   African Dwarf Shorthorn/
   remnant population S of Gambia
   R/syn. *West African Shorthorn
   (Gambia strain)*/nearly extinct
   (being absorbed by N'Dama)

Gambian N'Dama: (Gambia, Senegal
   and SW Mali)/fawn or white/
   larger var. of N'Dama, ? with
   zebu blood/syn. *N'Dama Grande*
   (Senegal), *N'Dama of Kaarta*
   (Mali)

ganado bravo: (Spain) *see* Fighting bull

Gangatiri: (Uttar Pradesh, India)/d/
   strain of Hariana, sim. to
   Shahabadi

Gaolao: (NE Maharashtra and SC
   Madhya Pradesh, India)/dr.d/z/
   Grey-white Shorthorned type/
   syn. *Arvi, Gaulgani*

Gaotai: (China)/extinct

Garbatov: *see* Gorbatov Red

Garfagnina: (Garfagnana, Lucca,
   Italy)/d.m.[dr]/blue-grey/HB
   1985/syn. *Grigia appeninica* (=
   Grey Apennine), *Modenese di
   Monte, Montanara* (= *mountain*),
   Nostrana (= *local*)/nearly extinct
   by crossing with Italian Brown

Garonnais: (Lot-et-Garonne, France)/
   m.dr/yellow or yellow-brown/
   HB 1898/syn. *Garonnais de
   plaine* till 1922 when *Garonnais
   de côteau* separated as Quercy;

Garonnais and Quercy rejoined
   in 1961 to form Blonde
   d'Aquitaine/extinct

Garre: (C and S Somalia)/d.m/often
   red, also red pied or black pied;
   horns small, sometimes pd/z/
   Small East African zebu, Somali
   group/var.: Bimal/syn. *Dawara,
   Gerra, Gherra*/not *Danara,
   Dauara, Daura, Duara*/[tribe in
   Wabe Shebeli area]

Gasara: (Somalia)/lead-grey or red;
   small/z/Somali group, cf.
   Bimal/name orig. applied to all
   small zebu in Somalia, now only
   those in C and N Somalia/part
   orig. of Magal/syn. *Abgal, Aria,
   Correi, razzetta delle dune* (=
   little breed of the dunes)

Gascon: (Gascony, SW France)/
   m.[dr]/grey, with or without
   black points and mucosae/vars: *à
   muqueuses noires* (= with black
   mucosae) and *aréolé*/HBs 1894,
   vars combined in single HB 1955

Gascon aréolé: (Gers, France)/
   m.[dr]/grey, white or blond (born
   red) with pink aureole round
   anus and vulva; long lyre
   horns/not to be confused with
   black-nosed Gascon à muqueuses
   noires/syn. *Mirandais*/rare

Gatačko goveče: *see* Gacko

Gâtinais, Gâtine, Gâtinelle: *see*
   Parthenais

Gaulan: *see* Jaulan

Gaulgani: *see* Gaolao

gaur: (hill forests of Indian peninsula
   and SE Asia)/= *Bos (Bibos) gaurus*
   Smith/brown to black, pale
   feet/orig. of mithun/vars: Burmese
   gaur, Indian gaur, seladang/not
   gore, gour/ [Hindi]/rare

gayal: *see* mithun

Geest: *see* Holstein Geest

Gelbes Frankenvieh: *see* Yellow
   Franconian

Gelbray: (USA)/m/orig. from Gelbvieh
   ($\frac{5}{8}$–$\frac{3}{4}$) × Brahman ($\frac{1}{4}$–$\frac{3}{8}$)/BS 1981

Gelbscheck: *see* Black Forest

Gelbvieh: (Austria) *see* Austrian
   Yellow

Gelbvieh: (C and S Germany esp.
   Bavaria)/m.d.[dr]/cream to

reddish-yellow/orig. in late 18th and early 19th c. from self-coloured Bernese and Swiss Brown × local red or red spotted/beef type (*Gelbvieh Fleischnutzung*) rare; former vars (combined *c.* 1920): Glan-Donnersberg (to 1961), Lahn, Limpurger, Yellow Franconian/BS 1899; BS also GB 1972, USA 1972, Canada 1973, S Africa 1988/syn. *Einfarbig gelbes Höhenvieh* (= *self-coloured Yellow Hill*), German Yellow/[*gelb* = yellow]

Georgian Mountain: (Georgia)/d/ black, black pied or red pied/ local Caucasian on Greater Caucasus/var.: Khevsurian/Russ. *Gruzinskii gornyi skot*/not *Georgian Alpine*

German Angus: (Germany)/m/black to dark brown or red to yellow-grey, sometimes with white marks; usu. pd/orig. (1960s) from Aberdeen-Angus × German Black Pied, German Red Pied and German Simmental/BS, HB 1956/Ger. *Deutsche Angus, Deutsches-Angus-Fleischrind*

German Beef Simmental: (Germany)/ m/var. of German Simmental/ Ger. *Fleckvieh Fleischnutzung*

German Black Pied: (N Germany)/ d.m/Friesian type/Holstein blood in 1960s and 1970s/orig. vars: East Friesian, East Prussian, Jeverland, Oldenburg-Wesermarsch/orig. of Estonian Black Pied, Lithuanian Black Pied, Polish Black-and-White Lowland *et al.*/HB 1878, BS/Ger. *Deutsche Schwarzbunte*/syn. *German Friesian, Schwarzbuntes Niederungsvieh* (= *Black-and-White Lowland* or *Black Pied Lowland*)/German Original Black Pied (without Holstein blood) remains as rare breed

German Black Pied Dairy: (E Germany)/d/orig. (1970s) from (British Friesian × Danish Jersey) × German Black Pied/HB 1970/Ger. *Schwarzbuntes Milchrind* (*SMR*)

German Brown: (SW Bavaria and E Württemberg, Germany)/ d.m.[dr]/Brown Mountain type, cf. Murnau-Werdenfels/ American Brown Swiss blood since 1966/var.: Allgäuer/HB 1893, BS/Ger. *Deutsches Braunvieh*

German Brown (old type): (Germany)/ orig. from Allgäuer and Schweizer Torfrind [= peat cow]/ old type without American blood/Germ. *Braunvieh alter Zuchtrichtung*/rare

German Friesian: *see* German Black Pied

German Red: (C Germany)/d.m.[dr]/ Central European Red type/linked with Angeln since 1942, absorbed Glan-Donnersberg 1961/former vars: Bavarian Red, Hesse-Westphalian, Silesian Red; current vars: Harze, Hesse Red, Vogelsberg, Vogtland, Westphalian Red/BS 1911/Ger. *Deutsches Rotvieh*/syn. *Middle German Red* (Ger. *Mitteldeutsches Rotvieh*), *Middle German Hill* (Ger. *Mitteldeutsches Gebirgsvieh*), *Red cattle of Central Germany, Red Hill* or *Red Mountain* (Ger. *Rotes Höhenvieh*)/rare/*see also* German Red (Highland type)

German Red (Highland type): (Germany)/red with light muzzle/var. of German Red/orig. from local red cattle with some Angeln blood/Ger. *Rotvieh Zuchtrichtung Höhenvieh*/rare

German Red Pied: (NW Germany)/ d.m/sim. to Meuse-Rhine-Yssel (interchange of blood in 19th c.)/orig. vars: Red Pied East Friesian, Red Pied Schleswig-Holstein, Red Pied Westphalian, Rhineland, united 1934/HB 1892, BS 1922/Ger. *Deutsche Rotbunte*/syn. *Red Pied Lowland* or *Red-and-White Lowland* (Ger. *Rotbuntes Niederungsvieh*)

German Shorthorn: (W Schleswig, Germany) m.d/red, white or red-and-white/orig. in mid 19th c. from Shorthorn × German Red

Pied/HB/Ger. *Land Shorthorn*/
nearly extinct

German Simmental: (S Germany)/
d.m.[dr]/cf. Austrian Simmental/
orig. in 19th c. from Bernese and
Simmental × local/vars: German
Beef Simmental, Messkircher,
Miesbacher, Württemberg
Spotted/HB 1892, BS/Ger.
Deutsches Fleckvieh/obs. syn.
*Alpenfleckvieh, Grosses
Fleckvieh, Höhenfleckvieh, Red
Spotted Highland, Scheckvieh*/
not *Brindled Highland*

German Spotted: *see* German
Simmental

German Yellow: *see* Gelbvieh

Gerra: *see* Garre

Gessien: (Gex, NE Ain, France)/orig.
from Swiss Simmental/HB fused
with Eastern Red Pied (now
French Simmental) 1945/syn.
Gex/extinct

Gex: *see* Gessien

Gezira: *see* Kenana

Ghana Dwarf Muturu: (Ada and Keta
Lagoon, SE Ghana)/black or
black pied/West African Dwarf
(Forest) Shorthorn type/syn.
Lagoon; locally known as
Muturu/nearly extinct

Ghana Sanga: (N Ghana)/usu. white,
also black; long horns/Ghana
Shorthorn (occ. N'Dama) × zebu
(usu. White Fulani, occ. Sokoto
Gudali)/inc. Ndagu, N'Dama
Sanga, White Sanga/name also
used occ. for crosses with West
African Dwarf Shorthorn

Ghana Shorthorn: (N Ghana)/m/
black-and-white pied, also black,
white, mottled black-and-white;
rarely pd/var. of West African
Savanna Shorthorn with zebu
blood/syn. *Ghanaian Shorthorn,
Gold Coast Shorthorn, West
African Shorthorn* (*WAS*)/nearly
extinct

Ghauti: *see* Dangi

Gherra: *see* Garre

Ghumsur, Ghumsuri: *see* Goomsur

Gidda: *see* Malnad Gidda

Giddu: *see* Jiddu

Gilan: *see* Mazandarani

Gimira: (Ethiopia)/humpless
longhorn/extinct

Gir: (S Kathiawar, Gujarat, India)/
d.dr.m/mottled red and white/z/
sim. to Dangi and Deoni/BSd and
HB/orig. (with Khillari) of
Nimari; orig. of Brazilian Gir/
syn. *Desan, Gujarati* (Sri Lanka),
Kathiawari, Soorthi, Surati, Surti

Giriama: (S Kenya)/local Small East
African Zebu, Kenya cluster/inc.
in Lowland or Coastal zebu along
with Duruma, Kamba and Taita

Gir leiteiro: *see* Dairy Gir

Gir mocho: *see* Polled Gir

Girolando: (Brazil)/d/orig. (1978 on)
Holstein ($\frac{5}{8}$) × Gir ($\frac{3}{8}$)/BS 1989/
[*Gir* × Ho*landés*, i.e. Turino]

Glamorgan: (Wales)/sim. to
Gloucester but light red and light
brindle with white finching/
extinct by 1900

Glan: (Palatinate, Germany)/m.d.dr/
yellow/orig. in late 18th c. from
Swiss Brown × native red cattle,
Bernese and Limpurger; united
with Donnersberg 1890 to form
Glan-Donnersberg/BS revived
1985/nearly extinct

Glan-Donnersberg: (Rhineland-
Palatinate, Germany)/orig. in
1890 from Glan + Donnersberg/
var. of Gelbvieh till 1961; since
1950 crossed with Danish Red
and inc. in German Red since
1961/Glan BS reformed 1985;
Donnersberg is again separate
breed/Ger. *Glan-Donnersberger*/
extinct

Glang, Glangmu: *see* Tibetan

Glatz Mountain: *see* Kłodzka

Gloucester: (England)/d/black-brown
with white dorsal line, tail,
escutcheon and belly/?
composite of Gloucester and
Glamorgan improved by
Shorthorn, Friesian *et al.*/BS
1919–1966 and 1973 on/syn.
*Gloucestershire, Old
Gloucester(shire)*/rare

gnag: *see* yak, domestic

Gobra: (Senegal and S Mauritania)/
m.dr/usu. white, also red pied,
yellow-brown, or white streaked

with black; very long lyre horns/
West African (lyrehorned) Zebu
type, Fulani group/orig. (with
N'Dama) of Djakoré/vars: Baol,
Dagana, Djoloff/Fr. *Zébu peul
sénégalais*/syn. *Senegal Fulani,
Senegal Zebu, Peul Zebu*/not *Cobra*

Goda: *see* Sheko

Godali: *see* Adamawa, Sokoto Gudali

Godemale, Godemar: *see* Guademar

Goffa: (around Sawla in Goffa area,
Ethiopia)/dr.(m)/usu. red; small
hump/smallest in Abyssinian
Shorthorned Zebu group/syn.
Goffa Dwarf

Goias: *see* Curraleiro

Golan: *see* Jaulan

Gold Coast Shorthorn: *see* Ghana
Shorthorn

Göle: (E Turkey)/var. of East
Anatolian Red/not *Köle*

Goleng: *see* Tibetan

Golpayegani: (Esfahan, Iran)/
d.m.(dr)/usu. black, also red/not
Golpaegani

Goodhope Red: *see* Jamaica Red

Goomsur: (Orissa, India)/dr/z, local
var./not *Ghumsur, Ghumsuri,
Gumsur*

Gorbatov Red: (Gorki, Russia)/d.m/
orig. in 19th c. from Zillertal ×
Oka/HB 1921/Russ. *Krasnaya
gorbatovskaya* or
Krasnogorbatovskaya (= *Red
Gorbatov*)/not *Garbatov*

gore: *see* gaur

Goryn: (Stolin district, Brest prov.,
Belarus)/d.m/red pied/orig. from
local crossed with old-type
Simmental and Polish Red/Russ.
Gorynskaya/nearly extinct

Goudali de l'Adamaoua: *see*
Adamawa

gour: *see* gaur

Govuvu: (Zimbabwe)/dwarf forest
cattle of Sanga type/syn. *Kavuvu,
Kwavovu*/extinct

Grade Ongole: *see* Javanese Ongole

Grassland Red: *see* Caoyuan Red

Grati: (E Java, Indonesia)/d/orig.
(1925 on) from Friesian × native
Javanese and Madura, with some
Ayrshire and Jersey blood/
Indonesian *sapi perahan Grati* (=

Grati dairy cattle)

Graubraunes Höhenvieh: *see* Brown
Mountain

Graues Gebirgsvieh: *see* Grey
Mountain

Grauvieh: *see* Grey Alpine, Tyrol
Grey

Greater Caucasus: (Russia and
Transcaucasia)/usu. black/
smaller var. of Caucasian/inc.
Dagestan Mountain, Georgian
Mountain/Russ.
Velikokavkazskaya

Greek Shorthorn: (Greece)/dr.m/
grey-blond to dark brown/Iberian
type, sim. to Albanian and Buša/
being crossed with Swiss Brown/
former vars: Cretan Lowland,
Cretan Mountain, East
Macedonian and Thrace, Epirus,
Mainland, Peloponnesus,
Thessaly, West Macedonian/rare

Greek Steppe: (NE Greece)/dr/Grey
Steppe type/vars: Katerini,
Sykia/nearly extinct

Grey: (Austria) *see* Tyrol Grey

Grey Adige: (Trentino/Alto Adige,
Italy)/dr.d.m/absorbed by Grey
Alpine/BSd 1931, HB 1934/It.
Grigia di Val d'Adige/syn.
Etschtaler (Austria), *Ulten* or
Ultinger (Ger.), *Ultimo* (It.)/
extinct

Grey Alpine: (Trentino/Alto Adige,
Italy)/d.m.[dr]/Grey Mountain
type/= Tyrol Grey (Austria)/HB
1949/It. *Grigia alpina*/syn. *South
Tyrol Grey*/not *Grey Swiss*

Grey Apennine: *see* Garfagnina

Grey Brahman: (USA)/var. of
Brahman/orig. mainly from
Ongole, Kankrej and Gir

Grey-Brown (Mountain): *see* Brown
Mountain

Grey Bulgarian: *see* Iskar

Grey Cambodian: *see* kou-prey

Grey Hungarian: *see* Hungarian Grey

Grey Iskur: *see* Iskar

Grey Jutland: *see* Jutland Grey

GREY MOUNTAIN: *see* Dalmatian Grey,
Grey Adige, Grey Alpine,
Rhaetian Grey, Tyrol Grey/Ger.
Graues Gebirgsvieh, Grauvieh

Grey Native: (Bulgaria) *see* Iskar

Grey Sindhi: *see* Thari
GREY STEPPE: (SE Europe)/dr/grey-white, grey or white; long lyre horns/*primigenius* type/inc. Croatian Steppe, Greek Steppe, Hungarian Grey, Iskar, Istrian, Mursi, Podolian, Romanian Steppe, Serbian Steppe, Turkish Grey Steppe, Ukrainian Grey/ syn. *Podolian* (Slav. *Podolska*), *Steppe*/not *Steppic Grey, Grey Stepland*
Grey Swiss: *see* Grey Alpine
Grey Transylvanian: *see* Transylvanian
Grey Ukrainian: *see* Ukrainian Grey
Grey Val d'Adige: *see* Grey Adige
Grey Val di Fiemme: (NE Italy)/ d.m.dr/graded to Italian Brown in 20th c., grade called *sorcino*/ It. *Grigia di Val di Fiemme*/ extinct
GREY-WHITE SHORTHORNED: (NW and C India, and Pakistan)/dr.(d)/z/inc. Bachaur, Bhagnari, Gaolao, Hariana, Hissar, Krishna Valley, Mewati, Nagori, Ongole, Rath, Shahabadi
Grigia appeninica: *see* Garfagnina
Grigia de Val d'Adige: *see* Grey Adige
Groningen Whiteheaded: (N Netherlands)/d.m/black (5% red) with white head and belly/HB 1906/Du. *Groninger* (or *Gronings*) *Blaarkop*, Ger. *Groninger*, Fr. *Groningue*/syn. *Zwartblaar, Zwartwitkop* (= black with white head), *Rodblaar, Rodwitkop* (= red with white head)/rare
Grossetana: (Italy)/former var. of Maremmana/extinct
Gruzian, Gruzinski: *see* Georgian
Guabiru: *see* tucura
Guadeloupe Créole: (Guadeloupe)/z
Guademar: (Bahia, Brazil)/orig. from Curraleiro with zebu blood (2 Ongole ♂♂ imported in English ship in 1868 or 1882)/not *Godemale, Godemar, Guademão, Guadiman, Gudimar*/[? from Goodman or Godman (captain of the English ship), or from *godemes* (popular Brazilian for 'Englishmen')]/extinct
Guadiana Spotted: (SE Alentejo, Portugal)/dr/red-and-white/var. of Mertolenga/orig. from Berrenda/Port. *Malhado do baixo Guadiana*
Guadiana White: *see* White Cáceres
Guadiman: *see* Guademar
Guangfeng: (NE Jiangxi, China)/ brown-black to brown-yellow; small mt type/Changzhu group/recog. 1950s/not *Guangfeng*
Guangnan: *see* Wenshan
Guanling: (S Guizhou, China)/ dr.m/♂ small hump/var. of Panjiang/syn. *Guanling Yellow, Guizhou*
Guarapuéva: *see* Igarapé
GUDALI group: (Nigeria, Cameroon, CAR)/various colours and markings/Hausa word for shorthorned shortlegged cattle, kept by Hausa and Fulani pastoralists/main subgroups: Sokoto Gudali, Adamawa (inc. Banyo, Nguandere, Yola)/syn. *Fulbé, Peuhl zebu*
Gudbrandsdal: (Norway)/dark brown or other colours/var. of Døle/ extinct
Gudimar: *see* Guademar
Guelma: (E Algeria and Tunisia)/ grey/former var. of Brown Atlas/ subvars: Chaouia, Cheurfa, Guelma (*sensu stricto*), Kabyle, Tunisian
Guernsey: (Channel Is)/d/golden brown-and-white/? French orig./BSd 1842, HB 1878; BS England 1884, USA 1877, Australia, Canada 1905, S Africa 1930, Brazil 1941, Zimbabwe, Kenya, Ireland
Guinea: (Florida, USA)/dwarf var. of Florida Cracker/nearly extinct
Guixi: *see* Longlin
Guizhou: *see* Guanling
Gujamavu: (India)/var. of Hallikar/ not *Gajamavu, Gujjamavu, Gujmavu*
Gujarat: (Brazil) *see* Guzerá
Gujarat: (India) *see* Kankrej
Gujarati: (Sri Lanka) *see* Gir
Gujerá: *see* Guzerá
Gujrati: (Brazil) *see* Guzerá
Gumsur: *see* Goomsur
Güney Anadolu Kırmızı: *see* South Anatolian Red

Güney sarisi: (Turkey)

Guraghe: (Guraghe and Hadiya areas near Ghibe tributaries, Ethiopia)/ dr.d/usu. red, chestnut or roan/ sim. to Ambo/Abyssinian Shorthorned zebu group

Gurtenvieh: (Switzerland)/belted var. of Swiss Brown

Guserá: *see* Guzerá

Gutsul'skaya: *see* Hutsul

Guzerá: (Brazil)/m.d/grey-white/z/ orig. from Kankrej imported 1875–1964/var.: Polled Guzerá/ HB 1936, BS 1939/not *Gujarat, Gujerá, Gujrati, Guserá, Guzerat, Guzerath*

Guzerá mocho: *see* Polled Guzerá

Guzerando: (Brazil)/Guzerá × Friesian (Holandés, = Turino) cross

Guzerat: (USA)/preferred name for imported Kankrej

Gwartheg Duon Cymreig: *see* Welsh Black

Gwartheg Hynafol Cymru: *see* Ancient Cattle of Wales

Gyr: *see* Brazilian Gir

Habbani: (SW Darfur, Sudan)/m/ white/cf. Nyalawi/syn. *Tetrone*

Hagalvadi: (India)/former var. of Amritmahal now merged in Hallikar/extinct

hainag, hainak, hanik, haǐnyk: *see* dzo

Hainan Humped: (Hainan I, Guandong, China)/d.m/yellow-brown, often red, occ. brown, rarely black; small/var. of Leiqiong (South China Zebu type)/syn. *Hainan, Hainan High-hump*/rare

Halabi, Halep: *see* Damascus

Halhïn Gol: (NE Mongolia)/larger var. of Mongolian/Russ. *Khalkhin-gol'skiǐ skot*/not *Khalkhingol*

Hallikar: (S Karnataka, India)/dr/ dark grey/z, Mysore type/var.: Gujamavu/orig. of Amritmahal/ not *Halikar*

Hamilton: *see* Cadzow

Hamitic Longhorn: (Ancient Egypt)/? orig. from African aurochs (*Bos primigenius opisthonomus*)/orig.

of West African longhorns, and (with zebu) of Sanga/syn. *Ancient Egyptian, Egyptian Longhorn*/extinct

Hammer: (Hammer area in S Omo, Ethiopia)/d/usu. white or grey, also chestnut or roan; prominent hump/Abyssinian Shorthorned Zebu group/var. of or descended from Boran

Hammer: (England) *see* South Devon

Hana, Haná-Berner: *see* Moravian Red Pied

Hanagamba: *see* Red Bororo

Hanam: *see* Atpadi Mahal

hanik: *see* dzo

Hanna-Berne: *see* Moravian Red Pied

Hansi, Hansi-Hissar: *see* Hissar

Hanwoo, Han Woo: *see* Korean Native

Harar: (plateaux of E and W Harerg, Ethiopia)/dr/usu. black, roan or red/Abyssinian Shorthorned Zebu group

Hard Hoof Criollo: *see* Curraleiro

Hariana: (Haryana, India)/dr.d/ z/Grey-white Shorthorned type/ var.: Hissar/BSd and HB/not *Haryana*

Hartón: (Upper Cauca valley, Colombia)/d.m/usu. tan/Criollo type/syn. *Vallecaucana, Valle de Cauca*/not *Artón del Valle, Hortón del Valle*

Harvey's cattle: *see* Tuli

Harz: (Lower Saxony, Germany)/var. of German Red/inc. Angeln and Danish Red blood/BS/Ger. *Harzer Rotvieh*/purebreds rare

Hash Cross: (Wyoming, USA)/m/orig. by E. and F. Barnes from Milking Shorthorn × Hereford 1950, with Red Angus bulls since 1956 and Highland since 1959; Beefmaker also used in separate line/ combined in Ranger breed 1970/ [*H*ighland, *A*ngus, *S*horthorn, *H*ereford]/extinct

Hassawi: (E prov., Saudi Arabia)/ d.(m)/light red; rudimentary horns; cervo-thoracic hump/orig. from *Bos indicus* and *B. taurus* (i.e. Sanga type)/syn. *Dirbani,*

Baladi/rare (by crossing with exotic breeds)

Hatton: (C Sri Lanka)/d/orig. from local crossed with European imports via Cape of Good Hope by Dutch 1765–1815, recently with Friesian, Ayrshire and Jersey/syn. *Cappe Harak* (= *Cape cattle*), *Cape*/not *Hatta*/? extinct (by crossing with European breeds)

Hawaiian wild: (SW Hawaii)/feral/ orig. from Mexican cattle first imported 1793, with British blood later/? extinct

Hays Converter: (Calgary, Alberta, Canada)/m/black with white face, occ. red with white face/ orig. 1952 on, at farm of H. Hays, from Hereford × (Holstein × Hereford) + American Brown Swiss × Hereford/BS 1975/rare

Hazake: (NW Xinjiang, China)/= Kazakh

Heck cattle: (Germany)/fawn shading to charcoal, with black extremities, light ring around muzzle; long lyre horns/orig. by Heck brothers (Lutz in Berlin and Heinz in Munich) from various breeds, inc. Highland, Corsican, Camargue, Fighting bull, Grey Steppe, White Park/ currently bred in several European countries/international BS 1995, BSd in formation/syn. *Heck's, Heck Aurochs* (Fr. *Aurochs de Heck*), *New Aurochs, Reconstituted Aurochs* (Fr. *Aurochs reconsti.*)/*see also* Berlin cattle, Munich cattle

Hedmark: (SE Norway)/d/local type of Swedish Red-and-White orig./ inc. in Norwegian Red since 1939/extinct

Heilbronn: (Württemberg, Germany)/ orig. from Bernese × German Red/part orig. of Scheinfeld and Yellow Franconian/syn. *Neckar*/extinct

Heilongjiang Dairy: *see* Pinzhou, Sanhe

Helmete: *see* Bergscheck

Herebu: *see* Braford

Hereford: (W England)/m/red with white head and underside; hd and pd vars/recog. mid-18th c., HB 1846, BS 1878; BS also in Ireland 1850, USA 1881, Australia 1885 (HB 1890), Uruguay (HB 1887), Argentina 1924 (HB 1889), Canada 1890 (HB 1899), New Zealand 1896, S Africa 1917 (HB 1906), Brazil 1906, Zimbabwe 1958, Chile, Denmark 1969, France (HB 1975), Portugal, Spain, Sweden, Zambia, Czech Republic, Estonia, Germany, Hungary 1988, Netherlands, Norway, Paraguay; HB Netherlands, Russia/syn. *Pampa* (Uruguay)/*see also* Poll(ed) Hereford

Hereford, Traditional: (UK)/m/red with white head/old type without outside influence/rare

Hereford pattern: *see* white-faced

Hereford-Shorthorn: (Australia)

Hereland: (Great Britain)/Hereford × Highland, 1st cross

Hérens: (C Valais, Switzerland)/ d.m.dr/dark red-brown/colour var.: Evolénard/orig. of Alpine Hérens/BSd 1884, BS 1917; BS USA 1980/Ger. *Eringer*/syn. *Valais*

Herero: *see* Damara

Herrgård: (Sweden)/local var. absorbed into Red Pied Swedish/[= estate]/extinct

herri behiak, herri ganadua: *see* Betizuak

Hervé Black Pied: (C Liège, Belgium)/d/orig. from Dutch Friesian × local red pied since 1860/fused with Polders Black Pied 1966 to form Belgian Black Pied/HB 1919/Fr. *Pie-noire (du Pays) de Hervé*, Flem. *Zwartbont ras van het Land van Herve*/syn. *Eastern Black Pied* (Fr. *Pie-noire de l'Est de la Belgique*, Flem. *Zwartbont ras van Oost-België*)/extinct

Hervé Red Pied: *see* Eastern Red Pied (Belgium)

Hesse Red: (Hesse, Germany)/var. of German Red/BS/Ger. *Hessisches Rotvieh*/syn. *Hessian Red*/rare

Hesse-Westphalian Red: (Germany)/
former var. of German Red inc.
Odenwald, Vogelsberg, Waldeck,
Westphalian Red/Ger. *Hessisch-
Westfälisches Rotvieh*/extinct

High-hump: *see* South China Zebu

HIGHLAND: (Germany)/inc. German
Brown, German Red, German
Simmental, German Yellow
(Gelbvieh), Mottled Hill,
Murnau-Werdenfels/Ger.
Höhenvieh/syn. *hill, mountain,
upland*

Highland: (NW Scotland)/m/usu.
red-brown, also brindle, dun or
black; long horns; long hair/part
orig. of Luing/BS 1884, HB 1885;
BS USA 1948, Canada 1964,
Sweden 1978, Germany, France/
syn. *Kiloe, Kyloe, Scotch
Highland, Scottish Highland,
West Highland*

Highland Zebu: (Kenya) *see* Kikuyu

hill: (Germany) *see* Highland group

hill: (India) *see* Indian Hill

Hinterwald: (S Black Forest, Baden,
Germany)/d.m.[dr]/local red or
yellow pied, with white head
and legs; small/orig. from Celtic
(Keltenrind)/BS 1889; BS also
Switzerland 1988/Ger.
Hinterwälder/rare/*see also*
Vorderwald

Hissar: (S Punjab, India)/dr.d/z/var.
of Hariana/orig. 1815–1898 from
Kankrej, Hariana, Gir, Nagori and
also Ongole, Tharparkar and
Krishna Valley, selected for type
and fast trotting 1899–1912,
backcrossed to Hariana (1912 on)
to improve milk yield/orig. of
Hissari/syn. *Hansi, Hansi-Hissar,
Hissar-Hansi, Hissar-Hariana*

Hissar: (N Sumatra and W Sumbawa,
Indonesia)/d/white/z/imported
from India, established 1909/syn.
Milking Zebu/rare

Hissari: (Punjab, Pakistan)/orig. from
Hissar

H'mong: (Vietnam)

Höhenfleckvieh: *see* Simmental

Höhenvieh: *see* Highland (Germany)

Holandese: *see* Turino

Holandesa vermelha (e branca): *see*
Red-and-White Friesian

Holando-Argentino: *see* Argentine
Holstein

Holderness: (E Yorkshire, England)/
sim. to Teeswater/Dutch orig./
orig. (+ Teeswater) of Shorthorn/
syn. *Yorkshire*/extinct

Holgus: (Texas, USA)/Holstein ×
Angus cross

Holland: *see* Friesian

Holland, Hollandaise, Holländisch:
see Dutch, esp. Dutch Black Pied

Hollandaise pie-noire: *see* French
Holstein

Holloway: (Ukraine)/m

Holmogor, Holmogorskaya: *see*
Kholmogory

Holmonger: (Namibia)/m.d/light dun/
orig. since 1949 by Johan *Holm* at
*Ong*anjera Farm from yellow
Afrikander × Swiss Brown

Holstein: (Germany) *see* Red Pied
Schleswig-Holstein

Holstein: (USA and Canada)/d/black-
and-white/orig. from Dutch
Black Pied imported chiefly
1857–1887/var.: Red-and-White
Holstein/orig. of British Holstein,
Danish Holstein *et al.*/USA:
Holstein BS 1871, HB 1872,
Dutch Friesian BS 1877, HB
1880, combined as Holstein-
Friesian 1885, since 1977 name
usu. simplified to Holstein;
Canada: Holstein-Friesian BS
1884, HB 1892, named changed
to Holstein 1984; BS also in
Australia/syn. *American
Friesian, Canadian Friesian,
Holstein-Friesian*

Holstein Friesian/Guzera: (Brazil)/
d/orig. (1975 on) from crossing
Red-and-White Holstein Friesian
and Guzera

Holstein Friesland: (S Africa)/d/orig.
from imports from Holland in
17th and 18th c./BS (orig. as
Friesland) 1912

Holstein Geest: (Germany)/orig. var.
of Red Pied Schleswig-Holstein/
Ger. *Holsteinische Geest*/extinct

Holstein-Jersian: (Germany)/d/being
formed from Jersey ($\frac{1}{4}$) and
Holstein-Friesian ($\frac{3}{4}$)

Holstein Marsh: (Germany) *see* Red
Pied Schleswig-Holstein
Holstein-Rbt: (Germany)/[Rbt =
Rotbunte, i.e. Red Pied]
Holstein Red Pied: (Germany) *see*
Red Pied Schleswig-Holstein
Holstein-Sbt: (Germany)/[Sbt =
Schwarzbunte, i.e. Black Pied]
Holstein Tropical: (Cuba)/d/orig.
from Holstein imported from
Canada/composite with 31/32
Holstein, 1/32 zebu (Siboney,
Mambi)
Hordaland: (W Norway)/black/orig.
(with Møre and Ramsdal) of
Vestland Fjord/extinct
Horned Lowland: (Norway) *see*
Norwegian Red-and-White
Horned Sinú: *see* Costeño con Cuernos
Hornet slettefe: *see* Norwegian Red-
and-White
Horosan: *see* Khurasani
Horro: (Horro Gudru, Welega, also
Ilubabor and Kefa, Ethiopia)/
d.m.dr/brown or red-brown;
small to medium hump/Sanga-
Zebu intermediate/orig. from
various Ethiopian highland zebu
and Nilotic Sanga, esp. Abigar/
syn. *Wallega, Wollega* (It. *Uollega*)
Hortón del valle: *see* Hartón
Hottentot: (S Africa)/Sanga type/orig.
of Afrikander/syn. *Namaqua*/
extinct
Hoyas: *see* Ecuador Criollo
Hřbinecký: (N Moravia, Czech
Republic)/d.m/red, with white
head (Hereford pattern)/orig.
from Bohemian Red ♀♀ with
Moravian Red Pied ♂♂;
gradually merged with Czech
Pied/var. of Czech Pied/syn.
Schönhengst (Ger.), *Šenhengský*
(Cz.)/nearly extinct
Hrvatska: *see* Croatian
Hrvatska buša: *see* Croatian Red
Hrvatska Simentalac: *see* Croatian
Simmental
hsaine: *see* tsine
HUANGHUAI group: (basins of
Huanghe and Huaihe, China)/
dr.m/usu. slight cervico-thoracic
hump in ♂/? orig. from
Mongolian and South China

Zebu/inc. Bohai Black, Jiaxian
Red, Jinan, Jinnan, Luxi, Pinglu
Mountain, Qinchuan/syn.
*Central Chinese, Central Plain
Yellow, cervico-thoracic humped
'yellow' cattle of Central China,
Low-humped*
Huang niu: *see* yellow cattle (China)
Huangpi: (E Hubei, China)/best var.
of Dabieshan/named 1950s/WG
Huang-p'ei, Huang-p'o/not
Hwangpei, Hwangpo
Huertana: (SE Spain)/chief var. of
Murcian/[= local]
Huguenot: (S Africa)/m/gold, occ.
red/orig. by Dennis Solomon of
Crocodile Valley Estates from
Charolais (60%) and Afrikaner
(40%)/named 1980 (to
commemorate 17th c. Huguenot
pioneers); recog. 1995
Humbe: *see* Humbi
Humbi: (SW Angola)/light to dark
fawn, often with white
underside; long lyre horns; big
dewlap/local Sanga type sim. to
Ovambo/syn. *Angolan, Humbe*
humlie: *see* Buchan Humlie
Humpless longhorns: (W Africa)/
taurines, inc. Kuri, N'Dama
Humpless shorthorns: (W, C and E
Africa)/taurines (= *Muturu* in
English-speaking areas), inc.
Savanna type (= *Baoulé* in
French-speaking areas), Dwarf or
Forest Muturu (= *Lagune* in
French-speaking areas)/see:
Bakosi, Bakweri, Baoulé, Doaya
(Namchi), Forest Muturu, Ghana
Dwarf Muturu, Ghana Shorthorn,
Kapsiki, Lagune, Liberian Dwarf
Muturu, Lobi, Logone, Savanna
Muturu Somba; also Sheko in E
Africa
Hungarian Brown: (N Hungary)/
small var. of Swiss Brown sim. to
Carpathian Brown/graded to
Holstein or Hungarofries/orig.
(with Jersey) of Dairy Hungarian
Brown/extinct
Hungarian Brown Dairy: *see* Dairy
Hungarian Brown
Hungarian Dairy Fleckvieh: *see*
Dairy Hungarian Pied

Hungarian Grey: (Hungary)/m.[dr.d]/ any shade of grey (born reddish); long horns/Grey Steppe type, cf. Transylvanian/Maremmana blood in 1930s/4 types: Primitive, Heavy Draught, Fine ('dairy') (upgraded with Montafon, Kostroma *et al.*), Large Estate/HB 1900/Hung. *Magyar szürke, Magyar alföldi*/syn. *Crane-coloured, Grey Hungarian, Hungarian Silver, Hungarian Steppe* (Ger. *Ungarisches Steppenrind*), *White Hungarian*/rare (esp. Primitive and Fine types)

Hungarian Pied: (Hungary)/m.d.[dr]/ orig. (1884 on) from Hungarian Grey graded to Simmental/ former var.: Bonyhádi/orig. (with Jersey) of Dairy Hungarian Pied/ BS and HB 1896/Hung. *Magyartarka*, Ger. *Ungarisches Fleckvieh*/syn. *Hungarian Red-and-White, Hungarian Red Pied, Hungarian (Red) Spotted, Hungarian Simmental*

Hungarian Red Dairy: *see* Dairy Hungarian Brown

Hungarian Silver: *see* Hungarian Grey

Hungarian Spotted Dairy: *see* Dairy Hungarian Pied

Hungarian Steppe: *see* Hungarian Grey

Hungarofries: (Hungary)/d.m/orig. (1963 on) from Holstein × (Danish Jersey × Hungarian Pied) or Holstein × [(Holstein × Danish Jersey) × (Holstein × Hungarian Pied)]/Hung. *Hungarofriz*/not *Hungarian Friesian, Hungaro-Fries, Hungarofriesian*

Hutsul: (Bukovaina, W Ukraine)/orig. from Moldavian × Carpathian Brown/Russ. *Gutsul'skaya*/ extinct (by crossing with Simmental)

Hwangpei, Hwangpo: *see* Huangpi

Hyogo: (Japan)/strain of Japanese Black

Ialomiţa: (Walachia, Romania)/ former var. of Romanian

Steppe/? orig. from Moldavian × Transylvanian/Rom. *Ialomiţeană*, Ger. *Jalomitzaner*/not *Ialomitza, Jalomita*/extinct

Ibagé: (Bagé, Rio Grande do Sul, Brazil)/m/black; pd/orig. since 1955 (but crossing started in 1940s) at Cinco Cruzes Res. Sta. from Nelore ($\frac{3}{8}$) × Aberdeen-Angus ($\frac{5}{8}$)/BS 1979

IBERIAN: (E and S Mediterranean)/ shades of brown or red; short horns; small/*brachyceros* type/ inc. Albanian, Brown Atlas, Buša, Corsican, Greek Shorthorn, Libyan, Rodopi, Romanian Mountain; cf. also Anatolian Black

Iberian: (W Mediterranean)/= Spanish + Portuguese

Icelandic: (Iceland)/d/red, red-and-white, brindle or brown, occ. black or grey; usu. pd/Nor. orig. since 10th c./BS 1903, HB

Igarapé: (mts and seashores of São Paulo, Brazil)/? Iberian orig./syn. *Guarapuéva, Nanico* (= dwarf)/ [= canal or sea-furrow]/extinct

IJssel: *see* Meuse-Rhine-Yssel

Ila: *see* Baila

Illawarra: (NSW Australia)/d/usu. red or roan/orig. in early 19th c. from Shorthorn with admixture of dairy breeds/inc. Australian Milking Shorthorn/BS 1930; BS also USA/syn. *AIS, Australian Illawarra Shorthorn* (till 1984), *South Coast cattle*

Illyrian: (Albania)/Iberian type, sim. to Greek Shorthorn/inc. Albanian, Buša

Ilocos: (Philippines)/usu. red, brown or pied/var. of Philippine Native/ large and small types/Sp. *Ilocano*

Iloilo: (Philippines)/usu. black, also fawn, chestnut or pied/var. of Philippine Native

Improved Boran: *see* Kenya Boran

Improved Criollo: (Venezuela) *see* Rio Limón Dairy Criollo

Improved Friuli: *see* Italian Red Pied

Improved Maremmana: *see* Chianino-Maremmana

Improved Red Pied Beef: *see* Dutch Improved Red Pied

Improved Rodopi: (Bulgaria)/d.m/ Jersey × (Bulgarian Brown × Rodopi)/Ger. *Veredeltes Rhodopenrind*

Improved Shorthorn: *see* Shorthorn

Improved Teeswater: *see* Shorthorn

Indian: (India)/inc. dairy (Gir, Sahiwal, Red Sindhi, Tharparkar), draught (Amritmahal, Bachaur, Bargur, Gaolao, Hallikar, Kenkatha, Kherigarh, Khillari, Krishna Valley, Kangayam, Malvi, Nagori, Ponwar, Siri), dual-purpose (Dangi, Deoni, Hariana, Kankrej, Mewati, Nimari, Ongole, Rath) *et al.*

Indian bison: *see* gaur

Indian gaur: (Indian peninsula)/= *Bos (Bibos) gaurus gaurus*/var. of gaur

INDIAN HILL: (Himalayas, Baluchistan, etc.)/red, black or pied; small/z/ cf. Sinhala/inc. Kumauni, Las Bela, Lohani, Morang, Nepalese Hill, Purnea, Rojhan, Siri

INDIAN ZEBU: (India and Pakistan)/ inc. Dhanni, Grey-white Shorthorned type, Gir type, Indian Hill type, Lyrehorned type, Mysore type, Red Sindhi, Sahiwal; also inc. desi/orig. of Brahman, Indo-Brazilian

indicine: = *Bos indicus* (zebu) type, in contrast to taurine (humpless)

Indo-African Zebu: *see* Mpwapwa

Indo-Brazilian: (Uberaba, Minas Gerais, Brazil)/m/white to dark grey/z/orig. 1910–1930 from Gir and Kankrej, with some Ongole, imported from India 1875–1930/ HB 1936, BS 1939/Port. *Indubrasil*/syn. *Induberaba*

Indo-Chinese: *see* Cambodian, Laotian, Moi, Vietnamese

Indo-Chinese forest ox: *see* kou-prey

Indo-europeu leiteiro: *see* Dairy Indo-European

Indonesian: *see* Bali cattle, Borneo, Javanese, Madura, Merauke

Indonesian Ongole: *see* Ongole

Induberaba: *see* Indo-Brazilian

Indú-Brasil, Indubrasil: *see* Indo-Brazilian

Indu-Europeu: *see* Azebuado

Ingessana: (S Blue Nile, Sudan)/ dwarf/z/sim. to Nuba Mountain Zebu and to Abyssinian Shorthorned Zebu but with Kenana blood/rare

inkone: colour type, i.e. coloursided/ not *nkone*/[Zulu]/*see also* Nguni, Nkone

Inkuku: (Rwanda)/common var. of Watusi Ankole as distinct from royal giant-horned Inyambo/not *Kukku, Kuku*

Innviertel: (Upper Austria)/former var. of Austrian Simmental/Ger. *Innviertler*/extinct

INRA 9: (France)/m/selected strain of double-muscled Charolais at Station Centrale de Génétique Animale, *I*nstitut *N*ational de *R*echerche *A*gronomique/extinct (replaced by INRA 95)

INRA 95: (SW France)/m/usu. white, or combinations of black, blue, red and white; double-muscled/ synthetic sire line orig. *c.* 1967 from Charolais, Blonde d'Aquitaine, Limousin and Maine-Anjou with Piedmont blood in 1973 and Belgian Blue in 1983/HB/syn. *Culard INRA 95*/rare

Inyambo: (Rwanda)/giant-lyrehorned var. of Watusi Ankole owned by king and chiefs, as distinct from common Inkuku/not *Nyambo, Nyembo*/extinct

Iranian: *see* Bami, Dashtiari, Golpayegani, Mazandarani, Nejdi, Sarabi, Sistani/obs. syn. *Persian*

Iraqi: (Iraq)/m.dr/zebu blood/vars: Dishti, Jenubi, Rustaqi, Sharabi/*see also* Kurdi

Irecé: (Bahia, NE Brazil)/d/var. of Curraleiro/Port. *Crioulo leiteiro de Irecé*/extinct

Iringa Red: (Singida region, C Tanzania)/d.m/local strain of Small East African Zebu (Tanzanian Shorthorn cluster, Tanganyikan zebu) orig. bred for colour/nearly extinct from interbreeding with Ugogo Grey

Irish: *see* Old Irish Cow

Irish Dun: (Ireland)/d/pd/sim. to Suffolk Dun/syn. *Polled Irish*/ extinct 1974

Irish Longhorn: *see* Longhorn
Irish Moiled: (NW Ireland)/d.m/usu.
red or roan coloursided, formerly
also grey, dun, black or white;
pd/Finnish blood 1950/BS
1926–1957, revived 1982, HB
1926–1966, revived 1983/syn.
Irish Polled/not *Mael, Maoile,
Maol, Maoline, Moile, Moilie,
Moyle, Moyled, Moyley, Mulline,
Mwool, Myleen*/[from Irish *maol*
= polled]/nearly extinct
Irish Polled: *see* Irish Dun, Irish
Moiled
Irish Shorthorn: = Shorthorn in
Ireland
Isigny: (Normandy, France)/dr.d/
brindle/extinct
ISK: *see* East Finnish
Iskar: (Rs Iskar, Vitt and Ossam,
Bulgaria)/dr.m.d/Grey Steppe
type with some *brachyceros*
blood/formerly larger lowland
var. of Bulgarian Grey; with
extinction of smaller mt var.
(Stara Planina), Iskar and
Bulgarian Grey now syn./HB
1929/Bulg. *Isk"rsko govedo*/syn.
*Bulgarian Grey, Grey Native
cattle* (Bulg. *Sivo mestno govedo*,
Ger. *Bulgarisches Grauvieh*),
*Bulgarian Steppe, Grey Iskar,
Iskar Grey, Vit*/not *Isker*/rare
Israeli Holstein: (Israel)/d/orig. from
local (chiefly Damascus, also
Lebanese) graded to Friesian
imported from Netherlands
1922–1933, Canada 1946, USA
1949–1955/HB, BSd 1951/syn.
Israeli Friesian
Istar-Kvarner: *see* Istrian
Istarsko govedo: *see* Istrian
Istoben: (Kirov, Russia)/d/usu. black
or black pied, occ. red or red
pied/orig. from Kholmogory (to
1912), Swiss Brown (1913 on)
and East Friesian (1930 on) ×
local/HB 1935, approved 1943/
Russ. *Istobenskaya*
Istrian: (C Istria, Croatia)/m.dr.d/var.
of Grey Steppe type/? orig. from
Italian Podolian with influence
from Romagnola/being crossed
with Swiss Brown/HB 1988/Cro.

Istarsko govečo, Istar-Kvarner/
syn. *Boškarin* (Cro.), *Istrian Grey*
(Ger. *Istrian Grauvieh* or
Kartsvieh), *Istrian Podolaz*; obs.
syn. *Buje*/nearly extinct
Italian Brown: (Italy)/d.(m)/var. of
Swiss Brown/being crossed with
American Brown Swiss/var.:
Sardinian Brown/HB 1956/It.
Bruna alpina (= Brown Alpine)/
syn. *North Italian Brown, Svitto,
Svizzera*
Italian Friesian: (Italy, esp. Po valley)/
d.(m)/orig. from imports since
1872 esp. from Netherlands,
Canada and USA/named 1951,
HB 1956/It. *Frisona italiana*/syn.
Pezzata nera (= black pied)/not
Carnation
Italian Red Pied: (NE Italy)/m.d.[dr]/
orig. from Simmental × Friuli
(1880–1900)/BSd 1931, HB
1957/It. *Pezzata rossa italiana*/
syn. *Friuli-Simmental, Improved
Friuli, Italian Simmental, Red
Pied Friuli* (It. *Pezzata rossa
friulana*), *Simmenthal Friulana*
Italian Simmental: *see* Italian Red
Pied
Itäsuomen karja: *see* East Finnish
Izankayi: *see* Bolowana

Jakoré: *see* Djakoré
Jakut, Jakutskaja: *see* Yakut
Jali, Jalli, Jalliji: *see* Diali
Jalomita, Jalomitza: *see* Ialomiţa
Jamaica Black: (Jamaica)/m/pd/
recent orig. from Aberdeen-Angus
× zebu ($\frac{1}{4}$–$\frac{3}{8}$)/BS and HB 1954
Jamaica Brahman: (Jamaica)/m/
white or grey, occ. brown; hd or
pd/z/orig. from Ongole, with
some Kankrej and Hissar blood
(all imported before 1922),
graded to American Brahman/BS
and HB 1949
Jamaica Hope: (Jamaica)/d/usu.
shades of fawn, solid or broken/
orig. 1920–1952 at Hope Stock
Farm, Kingston, from Jersey
(80%) × zebu (15%) (chiefly one
black Sahiwal ♂) with 5%
Friesian blood/BS and HB

1953/syn. *Jersey-Zebu, Montgomery-Jersey*

Jamaica Red: (Jamaica)/m/pd/orig. from Red Poll (predom.) with Devon and Indian zebu/BS 1952/syn. *Goodhope Red, Jamaica Red Poll*

Janubi: *see* Jenubi

Japanese Black: (Japan)/m/dull black with brownish hair tips/orig. 1868–1910 from Japanese Native crossed with Shorthorn, American Brown Swiss and Devon, also some Simmental, Ayrshire and (in Kagoshima only) Holstein blood/orig. types: Mishima, Tsuru/recog. 1944, HB/Jap. *Kuroge Washu* (*kuro* = black)

Japanese Brown: (Kumamoto prefecture, Kyushu, and Kochi prefecture, Shikohu, Japan)/m/ light reddish-brown/orig. 1868–1910 from Japanese Native crossed with Simmental and Korean, also some Devon blood/ vars: Kochi, Kumamoto/recog. 1944, HB 1951/Jap. *Akage Washu* (*aka* = red)

Japanese Dairy: (Japan)/? = Japanese Holstein

Japanese Holstein: (Japan)/d.(m)/ orig. from American Holstein since 1889 (chiefly since 1960)/ HB 1911/syn. *Japanese Holstein-Friesian*

Japanese Improved: (Japan)/m/orig. from Japanese Native by crossing with European breeds 1868–1910/ inc. Japanese Black, Japanese Brown, Japanese Poll, Japanese Shorthorn/recog. 1919/syn. *Kairyo-Washu* (*kairyo* = improved), *Nipponese Improved*

Japanese Native: (Japan)/m.dr/almost eliminated by crossing with European breeds 1868–1910/ existing vars: Kuchinoshima, Mishima/orig. of Japanese Improved/Jap. *Wagyu* (= Japanese cattle)

Japanese Poll: (Yamaguchi prefecture, SE Honshu, Japan)/m/black/orig. 1868–1910 from Japanese Native (Wagyu) crossed with Aberdeen-Angus; F$_1$ ♀♀ crossed with Japanese Black ♂♂ from 1975/ recog. 1944/Jap. *Mukaku Washu*/ syn. *Japanese Polled*

Japanese Shorthorn: (N Honshu and Hokkaido, Japan)/m/deep red-brown, red-and-white or roan/ orig. from Japanese Native in N Honshu I (Tohoku region) crossed with Dairy Shorthorn and Beef Shorthorn 1868–1910 (some Ayrshire and Devon blood)/BS 1957/Jap. *Nihon Tankatu Washu*

Jarlsberg: (Vestfold, Norway)/ coloursided var. of Red Polled Eastland/extinct

Jarmelista: (Jarmelo, Beira, Portugal)/ d/local var. of Mirandesa/ disappearing

Jaroslav, Jaroslavl: *see* Yaroslavl

Jathi madu: *see* Umblachery

jatsa: (Bhutan)/dr/usu. black, with brown back stripe; also pied or grey/♂ F$_1$ of mithun × Siri/cf. Selembu/not *jatsha, jesha*/*see also* MithunSiri, yanka

jatsum: (Bhutan)/d/♀ F$_1$ of mithun × Siri/cf. Selembu/not *jatshum, jescham, jesham*/*see also* yankum

Jaulan: (Jebel ed Druz, SW Syria)/ dr.m.d/black with white markings, often white head and black spectacles; 15% pd/ *brachyceros* type, sim. to Baladi but larger/syn. *Bisre, Khamissi*/ not *Djelani, Dscholan, Gaulan, Golan, Jolan, Julani*

Javanês: (Paraiba, Brazil)/orig. in mid 19th c. by Brito Bastos, Rio Formoso, from one grey zebu bull named Javanês crossed on local cows/extinct

Javanese: (Indonesia)/small hump/z

Javanese Ongole: (Java, Indonesia)/ dr/orig. from Ongole × Javanese/ Indonesian *Peranakan Ongole*/ syn. *Grade Ongole*/*see also* Sumba Ongole

Javanese Zebu: (Papua New Guinea)/ m/z/orig. from cattle imported from Java, Sumatra and Thailand in 19th c./rare

je: *see* dzo

Jeju: *see* Cheju

Jeli: *see* Diali

Jellicut: (Tamil Nadu, India)/dr/z/ var.: Kappiliyan/syn. *Kilakad, Kilakattu, Kikad, Pulikulam*/not *Jellikut*/[= bull baiting]

Jem-Jem Zebu: (high plateau of Sidamo, Ethiopia)/dr.(d.m)/usu. black, occ. with white face or patches/Abyssinian Shorthorned Zebu group/syn. *Black Highland, Black Zebu*

Jenubi: (SE and C Iraq)/d/usu. red (golden to bright bay)/z/var. of Iraqi/orig. of Rustaqi/syn. *Fao, Ma'amir, Zubairi*/not *Genubi, Janubi, Jenoubi*/[= southern]

Jerdi: (Brazil)/Jersey × Red Sindhi cross

Jersey: (Channel Islands)/d/fawn, mulberry or grey, often with black skin pigment/orig. of Danish Jersey, New Zealand Jersey; part orig. of Australian Milking Zebu, Jamaica Hope, Jersind *et al.*/BSd 1844, HB 1866; BS USA 1868, UK 1878, Australia 1900, Canada 1901, France 1903, S Africa 1920 (HB 1906), Brazil 1938, Kenya, Sweden 1955, Germany 1961, Netherlands (HB 1967), Japan, Argentina, Belgium, Colombia, Costa Rica, India, Ireland, Ecuador, Norway, Uruguay, Zimbabwe/Fr. *Jersiais*

Jersey-Zebu: *see* Jamaica Hope

Jersian: (Great Britain)/d/name for Jersey × Friesian, F_1

Jersind: (Alahabad, India)/d/orig. in 1960s at Allahabad Agric. Inst. from Jersey ($\frac{3}{8}-\frac{5}{8}$) × Red Sindhi

jescham, jesha, jesham: *see* jatsa, jatsum

Jeverländer: (Oldenburg, Germany)/ d.m/orig. var. of German Black Pied/HB 1878–1938 (joined with East Friesian/extinct

jhopke: *see* dzo ♂

Jiaxian Red: (C Henan, China)/usu. dark red/Huanghuai group/also improved var. (larger) by crossing with Danish Red *et al.*/syn. *Jiaxian*

Jiddu: (SE Ethiopia and S Somalia, also Zanzibar, Tanzania)/d.m/ white, light fawn or dark mahogany, with white around eyes, or red-and-white/Sanga-Zebu intermediate/= Tuni (Kenya)/orig. from Somali Boran, Ethiopian Boran and unidentified Sangas/It. *Giddu*/ syn. (Somalia) *Macien* (*magien* = spotted), *Sorco, Sucra, Surco, Surco Sanga, Surug, Suruq, Surgo, Surqo* (= red-and-white)/rare

Jie: *see* Karamajong

Jijjiga Zebu: (W Harer, Ethiopia)/d/ usu. chestnut, black, white, dark grey or red; short horns, sometimes pd/lowland type in Abyssinian Shorthorned Zebu group/syn. *Jijiga, Ogaden Zebu, Small Zebu*

Jinan: (Hebei, China)/d.m/Huanghuai group/orig. 1950s by merging neighbouring vars (e.g. Luxi)

Jiniu: (S China)/z/cf. Bainiu/name for small or dwarf cattle in ancient China/extinct

Jinnan: (Jinnan Basin, SW Shanxi, China)/d.m/usu. dark red/ Huanghuai group/being improved by European breeds

Jochberg Hummel: (Bezirk Kitzbühel, Tirol, Austria)/m/pd/var. of Pinzgauer/Ger. *Jochberger Hummel*/ [*hummel* = pd]/nearly extinct

Jolan: *see* Jaulan

Jolang: *see* Tibetan ♂

jommu: *see* dzo ♀

joppa: *see* dzo ♂

Jotko: (Bornu, Nigeria)/var. of Kuri, ? with zebu blood/syn. *Jotkoram*

Julani: *see* Jaulan

jum: *see* dzo ♀

Junqueiro: (Minas Gerais, Brazil)/ long horns/sim. to Crioulo Lageano/selected in 19th c. by Junqueiro family from Southern Crioulo/extinct

Jurin, Jurinskaja: *see* Yurin

Jutland: (Denmark)/m/usu. black-and-white, occ. grey-and-white/ orig. from local/HB 1881 with sections for dairy and beef types/ selected for dairy early 20th c. as

Jutland Black Pied; crossed 1950 with black-and-whites from Netherlands and Germany to form SDM (= Danish Black Pied) but some continued to breed pure/nearly extinct 1980, when BS formed to save native beef type/Dan. *Jysk Kvæg*/rare

Jutland Black Pied: (Denmark)/d/ joined (1949) with local Friesian to form Danish Black Pied; revived as Jutland Grey 1980s/ Dan. *Sortbroget Jydsk Malkekvæg* (= *SJM*)/syn. *Black Pied Jutland, Black Spotted Jutland Milk, Black-and-White Jutland*/extinct

Jutland Grey: (Denmark)/revival of Jutland Black Pied/syn. *Grey Jutland*/nearly extinct

Jysk kvæg: *see* Jutland

Kaarta: (Mali) *see* Gambian N'Dama

Kabota: (Indonesia)/var. of Borneo Zebu

Kabuli: *see* Afghan

Kabyle: (Algeria)/dwarf/former subvar. of Guelma var. of Brown Atlas/extinct

Kachcha Siri: (E Nepal)/Siri × Nepalese Hill crossbred/ [*kachcha* = inferior]

Kachhi: (India) *see* Cutchi

Kachhi: (Pakistan) *see* Bhagnari

Kaiama: *see* Keteku

Kairyo-Washu: *see* Japanese Improved

Kalakheri: (India)/black-and-white var. of Dangi

Kalmuk: (Turkey)/extinct

Kalmyk: (N of Caspian Sea, Russia)/ m/red, often white on head, belly and feet/Turano-Mongolian type/ regional vars in North Caucasus, Lower Volga, Kazakhstan and Siberia/HB/Russ. *Kalmytskaya*/ syn. *(Red) Astrakhan* (Russ. *Krasno-astrakhanskaya*)/not *Kalmuck*

Kamaduk: (India)/d/orig. from Friesian, Swiss Brown and Jersey, × Hariana, Ongole and Gir

Kamasia: (W Kenya)/local Small East African Zebu

Kamba: (SE Kenya)/local Small East

African Zebu (Kenya cluster)/ inc. in Lowland or Coastal zebu along with Duruma, Giriama, Taita/syn. *Akamba, Ukamba, Wakamba*

Kamdhino: (Bangladesh)/z/local type of Bangladeshi/? extinct

Kampeten: *see* Bergscheck

Kanada: *see* Dangi

Kandahari: (Afghanistan)/d.m/yellowish-red to black/var. of Afghan

Kandahari, Kandhari: *see* Red Kandhari

Kanem: (Chad)/Kuri × Arab (Shuwa) zebu

Kangaian: (Brazil)/imported zebu (? = Kangayam, from India)

Kanganad: *see* Kangayam

Kangayam: (SE Coimbatore, Tamil Nadu, India)/dr/white or grey (born red)/z/Mysore type/var.: Umblachery/BSd and HB/syn. *Kanganad, Kongu*/not *Kangayan*

Kaningan: (Indonesia)/var. of Borneo Zebu

Kankrej: (SE Rann of Cutch, Gujarat, India)/dr.d/grey; lyrehorned/z/ var.: Sanchori/orig. of Guzerá (Brazil)/BSd and HB/ syn. *Bannai, Guzerat* (USA), *Nagar, Talabda, Vagadia, Wagad* or *Waged* (Cutch); *Vadhiyar, Wadhiar, Wadhir* or *Wadial* (Radhanpur)/ not *Kankerej, Kankreji, Kankrij*

Kanvary: *see* Kauvery

Kaokoveld: (NW Namibia)/red, black, pied or coloursided; long horns, crescent or lyre/Sanga type/? large var. of Ovambo

Kaolib: *see* Koalib

Kao Lumpum: *see* Khao Lumpoon

Kappiliyan: (India)/var. of Jellicut

Kapsiki: (Mokolo, Mandara mts, N Cameroon)/m.ri/usu. black pied, also black, black-brown, red pied, white, fawn, wheaten grey or mottled/small var. of West African Savanna Shorthorn/ Kapsiki *Tla pseke*/syn. *Kirdi* (= pagan), *Mbuuyé*/nearly extinct

Karacabey Brown, Karacabey Montafon: *see* Turkish Brown

Karacadag: (NE Turkey)/extinct

Karachi, Red: *see* Red Sindhi

Karadi: *see* Kurdi

Karagwe Shorthorn: (Uganda)/local West African (dwarf) Shorthorn/ extinct

Karaisali: (Daglari, Turkey)/extinct

Karakalpak: (Uzbekistan)

Karamajong: (Karamoja, NE Uganda)/ d.m/usu. grey-white, fawn or tan/East African Shorthorned Zebu (large)/strains: Toposa, Turkana/inc. Jie

Karan Fries: (India)/orig. since 1971 at Nat. Dairy Res. Inst., Karnal, from Friesian × Tharparkar ($\frac{3}{8}$–$\frac{1}{2}$)/recog. 1987

Karan Swiss: (India)/d/light grey to dark brown/orig. at Nat. Dairy Res. Inst., Karnal, from American Brown Swiss × Sahiwal/recog. 1977

Karavaevo: *see* Kostroma

Karelian: (Russia) = East Finnish/ extinct

Kärtner Blondvieh: *see* Austrian Blond

Kartsvieh: *see* Istrian

Kashibi: (S Africa)

kata bizarrak: *see* Betizuak

Kataku: *see* Keteku

Katerini: (plains of NE Greece)/ m.dr.m/grey or brown; lyre horns/larger var. of Greek Steppe/nearly extinct

Kathiawari: *see* Gir

Katonta: (Karen State, Myanmar)/z

Kauvery: *see* Alambadi

Kaveri: *see* Alambadi

Kavirondo: *see* Winam

Kavkazskaya: *see* Caucasian

Kavuvu: *see* Govuvu

Kazakh: (Kyrgyzstan and Kazakhstan, and NW Xinjiang, China)/m/Turano-Mongolian type/orig. of Ala-Tau, Aulie-Ata, Kazakh White-headed, Xinjiang Brown/Russ. *Kazakhskaya*, Ch. *Hazake*/syn. *Kirgiz* (not *Kirghiz*)/ nearly extinct

Kazakh Whiteheaded: (N Kazakhstan)/ m or m.d/white head, belly and feet/orig. from Hereford (imported 1928–1932 from England and Uruguay) × Kazakh/recog. 1950, HB/Russ. *Kazakhskaya belogolovaya*

Kea: (Greece)/orig. from Swiss Brown × Greek Shorthorn/extinct 1980s by crossing with Swiss Brown and Friesian

Kedah-Kelantan: (N and E Malaysia)/ dr.m/usu. red or reddish-dun; often dwarf in E/z/orig. from Thai/syn. *Kedah-Thailand*, *Kelantan*, *Kelantan-Kedah*, *KK*, *Siam-Kedah*, *Thai-Kedah*, *Terengganu*, *Trengganu*

Keerqin: (Inner Mongolia, China)/ d.m.dr/yellow-red/orig. from Friesian × Mongolian

Kelantan, Kelantan-Kedah: *see* Kedah-Kelantan

Kelheimer: (Upper Palatinate, Bavaria, Germany)/dr.m.d/brown with white markings/sim. to Westerwald/not *Kehlheim*/ extinct *c.* 1940

Kemerovo: (Siberia, Russia)/d.m/orig. 1930–1953 from Siberian crossed first with Simmental, Kholmogory or Ukrainian Red and then with East Friesian or its grades/Russ. *Kemerovskaya*/ extinct (absorbed by Siberian Black Pied)

Kemp: (Natal, S Africa)/m.d/sim. to Afrikander but black/orig. 1911–1947 by R. Kemp from Friesian × Afrikander/early var. of Drakensberger/extinct

Kempen, Kempisch: *see* Belgian Red Pied

Kenana: (Fung, N Sudan)/d.dr/light blue-grey with black points/ North Sudan Zebu group/var.: White Nile (with Baggara blood)/ syn. *Blue Nile*, *Fung*, *Gezira*, *Northern Riveruin*, *Northern Province*, *Rufa'ai*, *Rufá'ai el Hoi* or *el Sherik*/not *Kenanna*, *Kennana*

Kenia: *see* Kinniya

Kenkatha: (R Ken, and Bundelkhand, S Uttar Pradesh and N Madhya Pradesh, India)/dr/rufous, brown or black/z; lyrehorned type sim.

to Malvi/syn. *Kenwariya*/not
Kenwaryia

Kenran: (Japan)/Mishima × Holstein
cross/[= Mishima/Dutch]

Kenwariya: *see* Kenkatha

Kenya (Improved) Boran: (C Kenya)/
m/var. of Boran improved by
selection/BS 1951/syn. *Improved
Boran*

Kenya Zebu: (CS and W Kenya)/
Small East African Zebu, inc.
Highland (Kikuyu), Lowland
(Duruma, Giriama, Kamba,
Taita), Kamasia, Kavirondo,
Kikuyu, Masai, Nandi, Samburu,
Usuk

Kerala local: (SW India)/nondescript
local as basis of Sunandini

Keriyo, Keriyu: *see* Danakil

Kerry: (SW Ireland)/d/black; lyre
horns/coloursided var.:
Drimmond/orig. of Dexter/name
first used 1776, recog. 1839, HB
1890, BS 1917; BS also UK 1892
(HB 1900), Canada 1901, USA
1911–1921, S Africa 1985/rare

Keteku: (W Nigeria)/usu. pure white,
also with black spots and
markings, occ. black, black-and-
white, light red, red-and-white,
blue/sim. to Borgou (Benin) but
taller; larger in N savanna than
in S/recent stabilized cross of
Muturu and zebu (White Fulani)/
syn. *Borgu, Kaiama*/not *Kataku,
Ketaku, Ketari, Kettije*

khainag: *see* dzo

Khairigarh: *see* Kherigarh

Khalit: (Egypt)/d.dr.m/usu. black/
Friesian × Egyptian cross

Khalkhingol: *see* Halhïn Gol

Khamala: (Betul, S Madhya Pradesh,
India)/z/local var./not *Khamla*

Khamissi: *see* Jaulan

Khandari: *see* Red Khandari

Khao Lumpoon: (N Thailand)/dr/
white/z/syn. *Kao Lumpum,
White Lumpoon*/*see also* Thai

Khargaon, Khargoni: *see* Nimari

Khariar: (Orissa, NE India)/d

Khasi: (Meghalaya, India)

Kherigarh: (N Kheri, NC Uttar
Pradesh, India)/dr/grey or white/
z; Lyrehorned type, sim. to

Ponwar/syn. *Kheri*/not
Khairigarh

Khevsurian: (Georgia)/var. of
Georgian Mountain/Russ.
Khevsurskaya gruppa/rare

Khillari: (S Maharashtra, India)/
dr/grey or white/z; Mysore type/
vars: Atpadi, Mahal, Mhaswad,
Nakali, Thillari/BSd/syn.
Mandeshi, Shikari/not *Kilhari,
Kilkaree, Killari*/[= cattle
herdsman]

Kholmogory: (Archangel, Russia)/d/
usu. black pied, also red pied,
black or red/local orig. with
Friesian blood since 1765/var.:
Pechora/HB 1927/Russ.
Kholmogorskaya/not *Cholmogor,
Holmogor, Kholmogor*

Kholmogory Hybrid: (Russia)/new
hybrid orig. from Holstein ×
Kholmogory

Kho Peun Nuang Thai E-San:
(Thailand)

Khorosanskiĭ zebu: *see* Khurasani

Khurasani: (Turkmenistan)/usu. red
or black, also brown pied or
spotted/var. of Central Asian
Zebu/Russ. *Khorosanskiĭ zebu*

Khurdi: *see* Kurdi

Khurgoni: *see* Nimari

Kigezi: (SW Uganda)/sim. to Bashi/
Sanga type/var. of Ankole,
smaller and with shorter horns
and paler coat

Kigezi Shorthorn: (Uganda)/
humpless shorthorn/extinct

Kikuyu: (C Kenya)/m/local Small
East African Zebu (Kenya
cluster)/syn. *Highland Zebu*/rare
(from upgrading with European
breeds since 1930s)

Kilakad, Kilakattu: *see* Jellicut

Kilara: (N of Chad)/colour var. of
Shuwa

Kilhari: *see* Khillari

Kilis: (S and SE Anatolia, S Turkey)/
d/var. of South Anatolian Red/
sim. to Damascus/rare

Kilkad: *see* Jellicut

Kilkaree, Killari: *see* Khillari

Kiloe: *see* Highland (Scotland)

Kimberley Shorthorn: *see* Australian
Shorthorn

Kinniya: (Mahaweli Ganga, Sri
Lanka)/grey/z; local var. sim to
Mysore type/not *Kenia*

Kirdi: *see* Kapsiki, Muturu, Namshi

Kirdi: (W Africa)/general term used
by Fulani for non-Muslim
pagans, and sometimes for
Muturu cattle

Kirgiz: *see* Kazakh

Kirko: *see* Tibetan ♂

Kisantu: (S of Kinshasa, DR Congo)/
orig. from Angola cattle with
European blood

Kistna Valley: *see* Krishna Valley

Kivu: *see* Watusi

Kiwi: (New Zealand)/usu. black/ being
developed in Waihi by Peter
Brumby *et al.* from Jersey and
Friesian (+ Ayrshire)/BS being
formed to register crossbreds

KK: *see* Kedah-Kelantan

Kladsko sudetský červený: *see*
Kłodzka

Kleiti: (Syria) *see* Baladi

Kłodzka: (Silesia, Poland)/dr.d/red or
brown with white head, back
stripe and belly/orig. from
Sudeten/Ger. *Glatzer Gebirgsrind*
(= *Glatz Mountain*), Cz. *Kladsko
sudetský červený*/syn. *Silesian
Whiteback* (Ger. *Schlesisches
Rückenscheck*)/extinct

Koalib: (Sudan) *see* Nuba, Nuba
Mountain

Kobe: (Hyogo, Japan)/m/strain for
production of specialized "Kobe
beef" or "Matsuzaka beef"

Kochi: (Kochi prefecture, Shikohu,
Japan)/yellow-brown/orig. from
crossing Japanese Brown with
Korean (from Kyushu I) and
Simmental/strain of Japanese
Brown with more Korean blood/
inc. type with orig. Korean black
skin

Kōgyû: *see* yellow cattle (China)

Köle: *see* Göle

Kolubara: (NW Serbia)/dr.m.d/grey/
var. of Serbian Steppe with Buša
blood/Serbo-cro. *Kolubarac,
Kolubarsko goveče, Colubara*/
nearly extinct

Konari: (Jalalabad, Afghanistan)/d/
red or black with white marks,

usu. white face and belly; dwarf/
var. of Afghan/not *Kunar*

Kongu: *see* Kangayam

Koni: *see* Kuri

Konkani: *see* Dangi

Konkomba: *see* Somba

Korean Black: *see* Cheju

Korean Native: (Korea)/m.dr/yellow-
brown/sim. to Mongolian and to
Japanese Native/var.: Chickso/
HB/Korean *Hanwoo*, Jap.
Chosen/syn. *Korean Brown*

Korean Native Striped: *see* Chickso

Korthoring: *see* Beef Shorthorn

Korthorn: *see* Danish Shorthorn

Kortow synthetic line: (Poland)/Pol.
Kortowska Synthetic

Koruška plava: *see* Austrian Blond

Kosi: (Cameroon) *see* Bakosi

Kosi: (Uttar Pradesh, India) *see*
Mewati

Kostroma: (Yaroslavl, Russia)/d.m/
grey/orig. 1890–1945 at
Karavaevo state farm from Swiss
Brown × improved local (Babaev
and Miskov)/recog. and HB
1944/Rus. *Kostromskaya*/syn.
Karavaevo

kou-prey: (N Cambodia, W Vietnam,
S Laos, E Thailand)/= *Bos (Bibos)
sauveli* Urbain/grey with white
stockings/syn. *Cambodian wild
ox* (Fr. *boeuf sauvage
cambodgien*), *grey Cambodian
ox, Indo-Chinese forest ox*/not
kou-proh/nearly extinct

Kourie: *see* Kuri

Krasnaya Azerbaĭdzhanskaya: *see*
Azerbaijan Red

Krasnaya belorusskaya: *see* Belarus
Red

Krasnaya ėstonskaya: *see* Estonian Red

Krasnaya gorbatovskaya: *see*
Gorbatov Red

Krasnaya kolonistskaya: *see* Red
Steppe

Krasnaya litovskaya: *see* Lithuanian
Red

Krasnaya nemetskaya: *see* Red
Steppe

Krasnaya pol'skaya: *see* Polish Red

Krasnaya stepnaya: *see* Red Steppe

Krasnaya tambovskaya: *see* Tambov
Red

Krasnaya ukrainskaya: *see* Red Steppe
Krasno-astrakhanskaya: *see* Kalmyk
Krasnobelorusskaya: *see* Belarus Red
Krasnoburaya latviĭskaya: *see* Latvian Brown
Krasnodarsk: *see* Kuban–Black Sea
Krasnogorbatovskaya: *see* Gorbatov Red
Krasnyĭ megrelskiĭ skot: *see* Mingrelian Red
Kravařský: (Silesia, Czech Republic)/d.m/coloursided/orig. from Bohemian Red, Pinzgauer and Moravian Red Pied/var. of Czech Pied/Cz. *Kravarsky skot*, Ger. *Kuhländer*/nearly extinct
Kréda: *see* Red Bororo
Krishnagiri: (Tamil Nadu, India)/z/local var/not *Krisnagiri*
Krishna Valley: (SW Maharashtra and NW Karnataka, India)/dr.d/z; Grey-white Shorthorned type/ orig. (1880 on) from Ongole, Gir and Kankrej with Mysore blood/ syn. *Kistna Valley*
Kuban–Black Sea: (N Caucasus, Russia)/m.d.dr/grey, brown or yellow/orig. in 19th c. from Swiss Brown and Simmental, × Ukrainian Grey/Russ. *Kubano-Chernomorskaya*/syn. *Krasnodarsk*/extinct
Kubano-Chernomorskayaya: *see* Kuban–Black Sea
Kuburi: *see* Kuri
Kuchinoshima: (Tokara Is, Kagoshima, Japan)/black, red, roan, grey, or black or red pied; small/feral since 1918/var. of Japanese Native/BS/nearly extinct
Kuhland, Kuhländer: *see* Kravařský
Kuku: *see* Inkuku
Kula: *see* Bulgarian Simmental
Kultak: (Turkey)/small var. of Turkish Grey Steppe/extinct
Kumamoto: (Kumamoto prefecture, Kyushu, Japan)/m/light brown or red (darker than Kochi)/strain of Japanese Brown with more Simmental blood/orig. from imported Korean cattle, later crossed with (esp.) Simmental, also Devon *et al.*

Kumauni: (N Uttar Pradesh, India)/z; hill type/not *Kumaon*
Kummet: *see* Bergscheck
Kunar: *see* Konari
Kuramin: (Uzbekistan)/z/var. of Central Asian Zebu (or zeboid)/ Russ. *Kuraminskiĭ zebu*/not *Curamin*/extinct
Kurdi: (Kurdistan)/m/black, often with light markings; small/ *brachyceros* type/not *Karadi*, *Khurdi*/rare
Kurgan: (SW Siberia, Russia)/d.m/ red, red-and-white or roan/orig. (1890 on) from Shorthorn × (Simmental, Dutch, Bestuzhev, Tagil, Red Steppe or Swiss Brown, × local Siberian)/recog. 1949, HB/Russ. *Kurganskaya*/ rare
Kuri: (Lake Chad, W Africa)/m.d/ light or white, sometimes with spots; gigantic bulbous horns/ var.: Jotko (Nigeria); also vars in Bornu and Kanem with zebu blood/Fr. *Kouri*/syn. *Baharié* (*Bare* or *Borrié*), *Buduma* (*Budduma*, *Budumu* or *Boudouma*), *Chad*, *Dongolé*, *Kuburi*, *White Lake Chad*/not *Koni*/rare (from interbreeding with Arab Shuwa, Red Fulani and zebu)
Kuroge Washu: *see* Japanese Black
Kwaniama: (Cunene, Angola)
Kwavovu: *see* Govuvu
Kyloe: *see* Highland (Scotland)
Kyoga: (C Uganda)/Small East African Zebu (Teso group) but larger than Serere, cf. Karamajong type/bred by Kuman tribe
Kypriaki: *see* Cyprus

Labedinskaya: *see* Lebedin
Ladakh Hill: (Kashmir)/black to brown with white patches; small, humpless/local hill type, sim. to Tibetan/disappearing from crossing with yak ♂♂ for hybrids (dzo), and with Jersey, Holstein *et al.*
Ladakhi: (Kashmir)/z/local var./not *Ladaki*
Lageano: *see* Crioulo Lageano

Lagone: *see* Logone
Lagoon: *see* Ghana Dwarf Muturu
Laguiole: *see* Aubrac
Lagune: (coast of Benin, also of Côte d'Ivoire and Togo)/m/usu. black, also black-and-white, fawn, grey or dark yellow, rarely red, red pied or white; occ. pd or loose hd; very small/sim. to Logone (Chad)/name for West African Dwarf (Forest) Shorthorn in French-speaking W and C Africa/syn. *Baoulé* (Gabon: Lagune imported from Benin and DR Congo late 1940s, now mixed with and indistinguishable from Baoulé), *Dahomey* or *Mayumbe* or *Mayombe* (DR Congo: Lagune introduced from 1904 on), *Lagunaire, Race des Lagunes*/rare
Lahn: (Nassau, Germany)/var. of Gelbvieh/orig. in early 19th c. from Bernese × Vogelsberg and Westerwald/syn. *Limburg*/extinct
Laisind: (Vietnam)/dr.m.d/dark brown to grey-brown/z/orig. since 1923 from Red Sindhi × Vietnamese/not *Laizind, Leizind*
Lake Chad: *see* Kuri
Lakenvelder: (Netherlands)/d/black, or red (25%), with white belt; rarely pd/Dutch Belted blood 1988/orig. of Dutch Belted (USA)/HB 1918–1931, 1979 on/[= sheeted field]/rare
Lakhalbhinda, Lakhalbunda: *see* Red Khandari
Lambadi: *see* Alambadi
Lambi Bar: *see* Sahiwal
Lancashire: *see* Longhorn
Landim: (Mozambique) *see* Nguni
Land Shorthorn: (Germany) *see* German Shorthorn
Lang: *see* Tibetan ♂
Lango: *see* Nkedi
Lanka: *see* Sinhala
Länsisuomenkarja: *see* West Finnish
Laotian: (Laos)/dr.(m)/sim. to Vietnamese
Laramane: (Albania)
Lare e Kuge: (Albania)/m.d/white with red spots/orig. from Simmental/rare

LARGE EAST AFRICAN ZEBU: (N Sudan, Eritrea, S Ethiopia, W Somalia, N Kenya, Tanzania, NE Uganda)/ inc. Boran group (Ethiopian Boran, Kenya (Improved) Boran, Orma Boran, Somali Boran, Unimproved Boran), Karamajong group (Karamajong Zebu, Toposa, Turkana), North Sudan group (Baggara, Barka, Butana, Kenana), South Sudan group (Murle, Toposa)
Las Bela: (Baluchistan, Pakistan)/z/ hill type/prim. var. of Red Sindhi/not *Lasbella*
Lateral-horned Zebu: *see* Afrikander
Latuka: (S Sudan)/var. of Mongalla/syn. *Latuke*
Latvian Black Pied: (Latvia)/d/orig. from German Black Pied/Russ. *Latviĭskaya chernopestraya*/syn. *LM*/extinct/see Baltic Black Pied
Latvian Brown: (Latvia)/d, or d.m/brown or dark red/Baltic Red type/orig. from local crossed with Angeln (mid-19th c. on) and Danish Red (late 19th c. and early 20th c.)/HB 1911/Lat. *Latvijas brūnū*, Russ. *Buraya latviĭskaya* or *Krasnoburaya latviĭskaya* (= Latvian Red-Brown, official name since 1947)/syn. *LB, Latvian Red*
Latvian Brown-and-White: *see* Latvian Red Pied
Latvian Red, Latvian Red-Brown: *see* Latvian Brown
Latvian Red Pied: (Livonia, Latvia)/ d/Russ. *Latviĭskaya krasno-pestraya*/syn. *Latvian Brown-and-White, LR*/extinct
Lavanttal: *see* Austrian Blond
La Velasquez: (Magdalena Medio valley, Caldas, Colombia)/m/red; pd/orig. (from 1955) by José Velasquez at Hazienda 'Africa', La Dorada, from Red Poll × (Red Brahman × Romosinuano)/rare
Lavinia: (W São Paulo, Brazil)/m.d/ orig. (1954 on) by Franco de Mello at Santa Maria farm, Lavinia, from Brown Swiss ($\frac{5}{8}$) × Guzerá ($\frac{3}{8}$)/BS/not *Lavina*/rare
LB: *see* Latvian Brown
Lebanese: (coast of Lebanon, Syria

and Hatay)/d.m/shades of brown/? var. of Damascus, or Damascus × Baladi intermediate/ syn. *Antakli* (from Antakya = Hatay), *Beirut, Beyrouth*

Lebaniega: (Liébana, W Santander, Spain)/smaller, highland var. of Santander/absorbed by Tudanca in 1940s/syn. *Picos de Europa*/ extinct

Lebedin: (Sumy, NE Ukraine)/d.m/ orig. 1902–1946 from Swiss Brown × local Ukrainian Grey/ recog. 1950, HB/Russ. *Lebedinskaya*/syn. *Lebedin Schwyz* (= *Lebedin Swiss*)

Lechtal: (Tyrol, Austria)/intermediate between Allgäuer and Tyrol Grey/absorbed in Tyrol Brown/ Ger. *Lechtaler*/extinct

Legítimo: *see* Southern Crioulo

Leicestershire: *see* Longhorn

Leiqiong: (Hainan I and Leizhou peninsula, S China)/d.m/yellow, brown, occ. black; high hump/ South China zebu type/orig. 1982 by combining Hainan Humped and Leizhou/[Qiong = syn. for Hainan I]

Leizhou: (N of Leizhou peninsula, Guangxi, China)/cf. Xuwen/ South China Zebu type/not *Leizhu*

Leizind: *see* Laisind

Le Mans: *see* Mancelle

Leonese: (N Spain)/dr.d.m/usu. red to chestnut/North Spanish type/ displaced by Swiss Brown/Sp. *Mantequera* (= butter producer) *leonesa*/nearly extinct since 1970s

Lepcha: *see* Tibetan

Lesser Caucasus: (Lesser Caucasus and lowlands of Transcaucasia)/ d.m.dr/usu. red, also brown or grey/larger var. of Caucasian/ subvars: Azerbaijan Red, Mingrelian Red/Russ. *Malokavkazskaya*

Levantina: *see* Murcian

lhang: *see* Tibetan ♂

Liberian Dwarf: (SE Liberia, esp. coastal areas)/m/black or pied/ var. of West African Dwarf Shorthorn/syn. *Liberian Dwarf Muturu, Muturu of Liberia*/rare

Libyan: (Libya)/m.d.dr/fawn, red or black/Iberian type, sim. to Brown Atlas/syn. *Libyan Brown Atlas, Libyan Shorthorn*

Lichtes Alpenvieh, Lichtes Höhenvieh: *see* Austrian Yellow

LID: *see* Local Indian Dairy

Lidia, lidiado: *see* Fighting bull

Liébana: *see* Lebaniega

Lietuvos baltnugariai: *see* Lithuanian White Back

Lietuvos šemieji: *see* Lithuanian Light Grey

Light Alpine: (Austria) *see* Austrian Yellow

Light Mountain: (Austria) *see* Austrian Yellow

Lim: (Serbia)/red/var. of Buša/Serbo-cro. *Polimska buša*

Limburg: *see* Lahn

Limiana: (La Limia, S Orense, Spain)/m.dr/chestnut with blackish shading, pale muzzle/ Morenas del Noroeste group/ larger type in valleys, medium in mts /HB 1990/nearly extinct

Limón, Limonero: *see* Rio Limón Criollo

Limousin: (W of Massif Central, C France)/m.[dr]/dark yellow-red/ former vars: Meymac, Meyssac, Treignac, Vendonnais/HB 1886; BS also USA and Canada 1969, GB 1970, New Zealand, Australia, Ireland 1971, S Africa 1987, Denmark, Argentina; HB also Luxembourg 1971, Netherlands

Limpurger: (Württemberg, Germany)/var. of Gelbvieh (German Yellow)/BS 1890, HB 1987/not *Limburg*/nearly extinct

Limun: (Belgium)/extinct after 1918

Lincoln Red: (E England)/m/now pd/ orig. from Shorthorn × local in early 19th c., also with Maine Anjou *et al.*/pd var. has absorbed hd var., all now pd and named Lincoln Red/HB 1822 (as var. of Shorthorn), BS 1894 (inc. in Shorthorn HB again 1935–1941); BS also Canada 1969, New Zealand, Australia 1971, S Africa/syn. *Lincoln(shire) Red Shorthorn* (to 1960)/rare

Lincolnshire Beef Poll: *see* Polled
Lincoln Red
Lineback: *see* American Lineback,
coloursided, Randall Lineback
Ling: *see* Luing
Lingnan: *see* Bashan
Liping: (Guizhou, China)
Lisasto govedo: *see* Slovenian Pied
Lišna Red: (Český Těšín, NE Moravia,
Czech Republic)/d/var. of
Moravian Red/Cz. *Lišnanský
červený*/extinct
Lithuanian Black Pied: (Lithuania)/
d/orig. from Dutch Black Pied
(also Swedish Friesian and East
Friesian) × local Lithuanian/
recog. 1951 (HB)/inc. in Baltic
Black Pied/Russ. *Chernopestraya
litovskaya*
Lithuanian Light Grey: (SW
Lithuania)/d.m/Lith. *Lietuvos
šemieji*/rare
Lithuanian Red: (Lithuania)/d/hd or
pd/Central European Red type
sim. to Polish and Belarus
Red/some Ayrshire blood in N
and Swiss Brown blood in SE,
also Danish Red and Angeln
blood/HB/Russ. *Krasnaya
litovskaya*
Lithuanian White Back: (SW
Lithuania)/d.m/cf. Polish
Whitebacked/Lith. *Lietuvos
baltnugariai*/rare
Llanero: (llanos of SW Venezuela)/
m/Criollo type/= Casanareño
(Colombia)
LM: *see* Latvian Black Pied
Lobi: (SW Burkina Faso)/m.d/dark or
pied/var. of West African
Savanna Shorthorn sim. to
Baoulé/syn. *Lobi-Gouin*, *Méré*
Local Estonian: *see* Estonian Native
Local Indian Dairy: (Malaysia)/d/
white, grey or red/z/orig. from
Kangayam, Red Sindhi,
Tharparkar, Hallikar and Ongole
imported by Indian settlers since
early 20th c./syn. *LID*
Logone: (riverine SW Chad)/var. of
West African Savanna
Shorthorn/syn. *Toupouri*/not
Lagone/nearly extinct
Lohani: (N Baluchistan to Kohat,

Pakistan)/dr.d/usu. red with
white patches/z/hill type, sim. to
Rojhan/syn. *Acchai*/not *Lohanni*
Lola: *see* Sahiwal
Longhorn: (NW and C England and
Ireland)/m/dark red brindle,
coloursided/BS, HB 1878/syn.
English Longhorn/obs. syn. (18th
c.) *Dishley, Lancashire,
Leicestershire, Warwickshire*
Longhorn: (USA) *see* Texas Longhorn
Longlin: (Baise dist., Guangxi, China)/
var. of Panjiang/syn. *Guixi*
Lord Caernarvon's breed:
(Hampshire, England)/white
with black or red spots; pd/syn.
Galway/extinct
Loriï: (Armenia)/d.m/brown with
lighter or darker patches/orig. at
Loriï State Farm by grading
Lesser Caucasus to Swiss Brown
1934–1940 and then mating *inter
se*/absorbed by Caucasian
Brown/extinct
Lorquina: (Lorca, SW Murcia,
Spain)/usu. pale, sometimes dark
red or chestnut/former var. of
Murcian with Spanish Mountain
blood/? extinct
Lourdais: (Hautes Pyrénées, France)/
d.m/white to light creamy-wheat;
lyre horns/HB 1981/nearly
extinct
Lower Guadiana Spotted: *see*
Guadiana Spotted
Lower Rhine: *see* Rhineland
Low-humped: (China) *see* Huanghuai
group
Lowland: (N Germany)/inc. Angeln,
German Black Pied, German Red
Pied, German Shorthorn/syn.
Niederungsvieh, Tieflandrind
Lowland Red Pied: (France) *see*
French Red Pied Lowland
Lowlands Black Pied: *see* Dutch
Black Pied
Lowland Zebu: (lowlands of
Ethiopia)/= Abyssinian
Shorthorned Zebu group in
lowlands, inc. Jijiga, Ogaden/*see
also* Kenya (Lowland) Zebu
Lozi: *see* Barotse
LR: *see* Latvian Red Pied
LSK: *see* West Finnish

Lucanian: (Basilicata, Italy)/former var. of Apulian Podolian/It. *Lucana*/extinct

Lucerna: (Lower Cauca valley, Colombia)/d.m./cherry red/orig. during 1937–1956 at Hacienda Lucerna from Holstein (40%), and Milking Shorthorn (30%) × Criollo (Hartón) (30%)

Lugware: (NE Kibali-Ituri, DR Congo, and W West Nile, Uganda)/often black pied, grey or red pied/cf. Mongalla/var. of Small East African Zebu, S Sudan cluster/ syn. *Bahu*/not *Lugbara, Lugwaret, Lugwari*

Luing: (Argyll, Scotland)/m/usu. red-brown, occ. yellow, roan or white/ orig. 1949–1965 by Cadzow brothers from Beef Shorthorn × Highland/BS and HB 1966; BS also Canada 1975, USA, New Zealand, Australia/ not *Ling*

Lulu: *see* Tibetan

Luso-Holandese: *see* Turino

Luxi: (W Shandong, China)/yellow or red/Huanghuai group/2 types: "tall" and "grab-ground" (for beef and dr)/being crossed with Limousin since 1985

Lyngdal: (Norway)/red; pd/joined with Vestland Red Polled, and then with Vestland Fjord (1947) to form South and West Norwegian/extinct

LYREHORNED: (W and N India, and Pakistan)/dr.(d)/z/inc. Kankrej, Kenkatha, Kherigarh, Malvi, Ponwar, Tarai, Thari, Tharparkar

lyrehorned zebu: (W Africa) *see* Gobra, Sudanese Fulani, West Fulani

Maalsalv: *see* Målselv

Ma'amir: *see* Jenubi

Maasai: *see* Masai

Maas-Rijn-IJssel: *see* Meuse-Rhine-Yssel

Macedonian Blue: (Macedonia)/usu. blue roan/cf. East Macedonian, West Macedonian/var. of Buša/ Serbo-cro. *Makedonska buša, Plava povardarska* (*plava* = blue)

Macien: *see* Jiddu

Madagascar Zebu: (Madagascar)/ m.dr.d/many colours; lyre

horns/? orig. from Indian zebus with Sanga/orig. of Rana and (with Limousin and Afrikander) of Renitelo/syn. *Malagasy, Malgache*

Madaripur: (Bangladesh)/brown/z/ local var.

Madhya Pradesh dwarf zebus: *see* Khamala, Mampati, Ramgarhi, Son Valley

Madura: (Madura, also in Flores and S Kalimantan, Indonesia)/dr.m, racing, fighting/chestnut with pale underparts; small hump/ orig. from zebu and Bali cattle/ Du. *Madoera*/syn. *Madurese*/not *Madoura*

Mael: *see* Irish Moiled

Mafia: (Tanzania) *see* Unguja Shorthorn

Mafriwal: (Malaysia)/d.(m)/in formation from Holstein and Friesian-Sahiwal × Local Indian Dairy/[*Malaysian Friesian-Sahiwal*]

Magal: (Juba, Somalia)/black/z/orig. from cross between Gasara and Jiddu/syn. *Correi*/not *Mogol*/[= black]

Maghreb: *see* Brown Atlas

Magien: *see* Jiddu

Magnum: (Iowa, USA)/commercial hybrid

Magyar: *see* Hungarian

Mahabharat Lekh: *see* Nepalese Hill

Mahadeopuri: *see* Malvi

Mahadeswarabetta: *see* Alambadi

Mahonesa: *see* Minorcan

Maia, Maiana: *see* Barrosã

Maine: *see* Mancelle

Maine-Anjou: (NW France)/m.d/red, red-and-white or roan/sim. to Armorican but larger/orig. from Shorthorn × Mancelle *c.* 1830 on/recog. 1925, HB 1917/ federated with Armorican 1962–1970 under name Rouge de l'Ouest (= Western Red)/BS USA 1969, Canada 1970, UK, New Zealand, Australia/obs. syn. *Durham-Mancelle*

Maine Anjou Lait: *see* Dairy Maine Anjou

Majorcan: (Majorca, Spain)/m/red, blond or chestnut/Sp. *Mallorquina*/nearly extinct

Makalanga, Makaranga: *see*
Mashona

Makaweli: (Kauai, Hawaii)/orig. at
Robinson Ranch from Shorthorn
and Devon

Malabar: (India) *see* Malnad Gidda

Malabar: (NE Brazil)/m/red-brown to
black/orig. in 19th c. from
Curraleiro with Indian Zebu
blood/extinct

Malagasy: *see* Madagascar Zebu

Malai-madu: *see* Alambadi

Malakan: (NE Turkey)/orig. from
Ukrainian Grey/syn. *Okranya*/
not *Malokan, Molokan*/[Molokan
was religious sect in Russia,
offshoot of Doukhobors, from
Russ. *moloko* = milk]/extinct

Malawi Zebu: (C and S Malawi)/var.
of Angoni (? with Nguni blood)/
formerly divided into South
Malawi Zebu and North Malawi
Zebu

Malayan bison: *see* seladang

Malayan gaur: *see* seladang

Malay banteng: (N Malaysia)/= *Bos
(Bibos) javanicus butleri*
(Lydekker)/var. of banteng/syn.
sapi utan (= wild cattle)/? extinct

Malgache: *see* Madagascar Zebu

Malhada do baixo Guadiana: *see*
Guadiana Spotted

Malinke: *see* N'Dama

Malir: *see* Red Sindhi

Mallorcan, Mallorquina: *see*
Majorcan

Malnad Gidda: (Karnataka, India)/d/
dark coat; small to dwarf/local
var./syn. *Malabar* (Kerala)

Malokan: *see* Malakan

Malokavkazskaya: *see* Lesser
Caucasus

Målselv: (Tromsø, N Norway)/red;
usu. pd/orig. from Dutch (to
1860) and Ayrshire (1860–1900)
× local/combined with Red
Trondheim 1951, now local var.
of Norwegian Red

Maltese: (Malta)/d/Maltese *Baqra-
Maltija*/syn. *Il-Maltija*/nearly
extinct

Malvi: (C Malwa, W Madhya
Pradesh, India)/dr/grey/z/
lyrehorned type/BSd/syn.

Mahadeopuri, Manthani/not
Malwa, Malwi

Mambí: (Cuba)/d/orig. since late
1960s from Holstein ($\frac{3}{4}$) × Cuban
Zebu ($\frac{1}{4}$)/not *Mamby*

Mampati: (Madhya Pradesh, India)/z/
local var.

Manapari: (Tamil Nadu, India)/z/
Kangayam × local

Mancelle: (Maine, France)/orig. (with
Shorthorn) of Maine-Anjou/syn.
Le Mans, Maine/extinct

Mandalong Special: (NSW, Australia)/
m/light cream to biscuit; pd or
hd/orig. in mid-1960s at
'Mandalong Park' nr Sydney from
Charolais, Chianina, Poll
Shorthorn, British White and
Brahman; now 62.5%
Continental, 18.75% British,
18.75% Brahman

Mandé: *see* Bambara

Mandeshi: *see* Khillari

Mandingo: *see* N'Dama

Mango: *see* Somba

Mangoni, Manguni: *see* Nkone

Mangwato: (E Botswana)/var. of
Tswana/orig. of Tuli/syn.
Amabowe (Zimbabwe),
Bamangwato, Ngwato/not
Mongconto

Manhartsberg: *see* Waldviertel

Manjaca: (coast of Guinea-Bissau)/
black/var. of West African Dwarf
Shorthorn/? extinct

Manjan'i Boina: (Mahajanga prov.,
Madagascar)/d/commercial
composite in formation (since
1980s) from Madagascar Zebu
and French Brown

Manthani: *see* Malvi

Mantequera leonesa: *see* Leonese

Mantiqueira: (SE São Paulo, Brazil)/
d/white and black (inverse of
Friesian)/in formation at
Instituto de Zootecnia,
Pindamonhangaba, from Friesian
($\frac{5}{8}$) × Gir ($\frac{3}{8}$)/syn. *Tribofe* (= 3
lungs, referring to physical
fitness)/[name of mts]

Maoile, Maol, Maoline: *see* Irish
Moiled

mao niu: *see* yak, domestic

Maraîchin: (W France)/d.m.dr/brown

or yellow; lyre horns /var. of
Parthenais retaining triple
purpose use, now mainly
m/BS/Fr. *Vendée maraîchin* (=
Vendée Marsh)/rare

Maramureş Brown: *see* Romanian
Brown

MARC I: (Nebraska, USA)/m/exp.
"breed" developed 1978–1991 at
Meat Animal Research Center,
Clay Center, Nebraska: ¼
Braunvieh, ¼ Charolais, ¼
Limousin, ⅛ Hereford, ⅛ Angus

MARC II: (Nebraska, USA) as for
MARC I but ¼ Gelbvieh, ¼
Simmental, ¼ Hereford, ¼ Angus

MARC III: (Nebraska, USA) as for
MARC I but ¼ Pinzgauer, ¼ Red
Poll, ¼ Hereford, ¼ Angus
(1980–1991)

Marchigiana: (Marche, Abruzzi and
Molise, EC Italy)/m.[dr]/white
with black points/Podolian orig.,
improved by Chianina from mid-
19th c. to early 20th c. to give
improved Marchigiana
(*Marchigiana gentile*) of plains,
as distinct from orig. Podolian
(*montanara*) of mountains and
intermediate *brina* (= grey) of
hills; Romagnola blood up to
1928, bred pure since 1932/HB
1957; BS Brazil 1972, Canada,
USA 1973, GB, Australia; HB
Netherlands/syn. *Del Cubante*
(Avellino), *Marky* (USA)

Marchois: (La Creuse, France)/
d.m/extinct (absorbed by
Limousin and Charolais)

Maremmana: (Maremma, Tuscany
and Latium, Italy)/m.[dr]/grey,
born red; long open-lyre (♀) or
crescent (♂) horns/Podolian
type/former vars: Grossetana,
Roman, Improved (*stabulata*)/
orig. (with Chianina) of
Chianino-Maremmana/BSd and
HB 1935

Mariahof, Mariahofer-Lavanttaler:
see Austrian Blond

Marianas: (Oceania)/Spanish orig./
syn. *Marianne*

Maridjadvorska: *see* Austrian Blond

Marine Landais: (France)/yellow;

lyre horns/orig. *c.* 1800/nearly
extinct

Marinera: (Balearic Is, Spain)/orig. of
Minorcan/extinct

Marinhoa: (NW coast of Portugal)/
dr.m/yellow/var. of Mirandesa/
HB 1986

Marismeña: *see* Mostrenca

Marky: *see* Marchigiana

Maroilles: (Flanders, N France)/
former var. of Flemish/Fr.
Maroillais, Marollais/extinct

Maronesa: (Terra Quente, NE
Portugal)/m.dr.(d)/black-brown/
orig. from Barrosã × Mirandesa/
HB 1989/[from Mt Marão]

Marwari: *see* Sanchori

Maryuti: (NW Egypt)/var. of Egyptian
with less zebu blood/syn.
Arabian/not *Marriouti*

Masai: (Kenya and Tanzania)/d.m/
local Small East African Zebu
(Kenya cluster and Tanzanian
cluster)/syn. *Maasai*

Masai Grey: (N Tanzania)/var. of
Small East African Zebu
(Tanzania cluster)/syn. *Maasai
Grey*

Mashakalumbe: *see* Baila

Mashona: (E Zimbabwe)/m.dr/
various colours, usu. black, often
red; often pd (*izuma*)/Sanga
type, ? with Angoni blood/BS
1950, HB 1954/Chishona
Ngombe dza Maswina, Sindebele
Amanjanja/syn. *Makalanga,
Makaranga, Ngombe dza
Vakaranga* (= cattle of the
Karanga), *Shona*

**Mashuk, Mashukalumbe,
Mashukulumbu,
Mashukulumbwe:** *see* Baila

Masny Simmental: (Czech Republic)

Massa: *see* Toupouri

Massanaise: *see* Albères

Masurenland, Masurian: *see* Mazury

Maswina: *see* Mashona

Matabele: (SW Zimbabwe)/mixed
Sanga type raided by Ndebele
1822–1893; inc. Barotse,
Batawana, Mashona, Nguni and
Tonga blood/orig. of (and syn.
for) Nkone

Mateba: (I in mouth of R Congo)/m/

brown/Sanga type/orig. from
Angola cattle crossed with
European breeds, and finally
with Afrikander (predominates)

Mattu: (India)

Maure: (SC Mauritania)/pa.ri.d/usu.
black or black pied, often dark
red in E/West African
(shorthorned) Zebu type/syn.
*Arab, Gabaruyé, Mauritanian,
Moor, Moorish*

Maurine: *see* Meymac

Mauritanian: *see* Maure

Mauritius Creole: (Indian Ocean)/d/
usu. white with black points or
with coloured flecks, also brown
or brown-and-white; pd/? Fr.
orig. in 18th c.

Mayne: *see* Yola

Mayombe, Mayumbe: *see* Lagune

Mazandarani: (N Iran)/m.d.(dr)/all
colours/z/syn. *Gilani*/not
Mazanderani

Mazury: (Olsztyn, Poland)/d.m/var.
of Polish Black-and-White
Lowland by revival of East
Prussian since 1946/Pol.
Mazurska/syn. *Masurenland,
Masurian*

Mbororo: *see* Red Bororo

M'Bougi, M'Bouyé: *see* Namshi

Mbulu: (Tanzania)/local smaller hill
strain of Small East African Zebu
(Tanzania cluster, along with
Iringa Red, Masai Grey, Mkalama
Dun, Singida White)/? extinct

Mbuuyé: *see* Kapsiki

mdzo-mo, mdzo-po: *see* dzo

Megrel, Megrelian: *see* Mingrelian
Red

Mehwati: *see* Mewati

Meiniu: (NW China)/dwarf cattle in
ancient China/extinct

Meknès Black Pied:
(Morocco)/d/black with white
belly/? orig. from breeds
imported in 18th c. (Friesian,
Breton or Bordelais) or by
selection from Brown Atlas/Fr.
Noire Pie de Meknès/extinct in
1960s by grading to Friesian

Menggu: *see* Mongolian

Menno-Friesian: (Russia)/East
Prussian Black Pied cattle taken

to Russia by Mennonites in late
18th and early 19th c./extinct

Menorcan, Menorquina: *see* Minorcan

menscha: *see* mithun

Menufi: (SE Delta, Egypt)/dr.m/var. of
Baladi/not *Menoufi, Minufi*

Merauke: (W Irian, Indonesia)/z/
small

Méré: (S Burkina Faso and NE Côte
d'Ivoire)/m.dr/often white/
Fulani Zebu × Baoulé Shorthorn
cross (or × Lobi or N'Dama); also
general term for stabilized
crosses between any zebu and
humpless breeds but suggestion
the term should be reserved for
zebu crosses with Savanna
Shorthorns/name also used for
purebred Baoulé or Lobi/[Fulani
= small, not zebu]

Méré: *see also* Baoulé, Lobi

Méré: (Mali) *see* Bambara (and other
crosses)

Mertolenga: (Mertola, SE Portugal)/
dr.m/red or red pied/orig. from
Alentejana/vars: Bragado do
Sorraia, Guadiana Spotted/HB
1977, BS/not *Mértolenga*

Messaoria: (Cyprus)/larger (lowland)
var. of Cyprus

Messara: *see* Cretan Lowland

Messkircher: (Baden, Germany)/orig.
var. of German Simmental from
Simmentals first imported
1843/BS 1882/syn. *Upper Baden
Spotted* (Ger. *Oberbädisches
Fleckvieh*)/extinct

Mestiço Leiteiro Brasileiro: *see*
Brazilian Milking Crossbred

Mestizo perijanero: *see* Perijanero

Mestnaya éstonskaya: *see* Estonian
Native

Metohija Red: (Serbia)/usu. red/var.
of Buša/Serbo-cro. *Crvena
metohijska, Metohijska buša*/
syn. *Red Metohian*

Meuse-Rhine-Yssel: (SE Netherlands)/
d.m/red-and-white, or white with
red spots/var.: Brandrood
IJsselvee (= dark red MRY)/orig.
of Dutch Red Pied Dual-Purpose,
Dutch Improved Red Pied/HB
1906; BS in Canada, GB, USA; HB
in Luxembourg 1923/Du. *Roodbont*

(= *Red Pied*) *Maas-Rijn-IJssel*, Ger.
Rotbunte holländische, Fr. *Mosane-
rhénane-ysseloise*, Eng. *Meuse-
Rhine-Issel*/syn. *Dutch Red-and-
White, MRI, MRY, Red Pied Dutch*
Mewati: (E Rajasthan, India)/dr/z/
Grey-white Shorthorned type;
sim. to Hariana with Gir blood/
syn. *Kosi* (Uttar Pradesh)/not
Mehwati, Mhewati, Mowati
Mexican Holstein: (Mexico)
Meymac: (C France)/orig. from
Limousin × Marchois/syn.
Maurine/extinct
Meyssac: (C France)/former var. of
Limousin/extinct
Mézenc: (Haute Loire-Ardèche, S
France)/dr/yellow-brown/sim. to
Villard-de-Lans/extinct in late
1960s
Mezzalina: (Sicily, Italy)/upland var.
of Sicilian/now absorbed by
Modicana
Mhaswad: (S Maharashtra, India)/
var. of Khillari
Mhewati: *see* Mewati
Miaoya: (Hubei, China)/former name
for Yunba
Mid-Belgian: *see* Belgian Blue
Midden- en Hoog-België: *see* Belgian
Blue
Middle German Red: *see* German Red
middle-horned: (Great Britain)/inc.
Devon, Gloucester, Hereford,
South Devon, Sussex, Welsh
Black
Miesbacher: (Bavaria, Germany)/orig.
var. of German Simmental/syn.
Upper Bavarian Spotted (Ger.
Oberbayrisches Alpenfleckvieh)/
extinct
Milking Criollo: *see* Tropical Dairy
Criollo
Milking Devon: (New England,
USA)/d.m.(dr)/orig. var. of
Devon/BS 1952, reorg. 1978; also
in Canada/syn. *Red Devon*/rare
Milking Shorthorn: (New Zealand)
see Dairy Shorthorn
Milking Shorthorn: (USA)/d.m./cf.
Dairy Shorthorn (Great Britain)/
var. of Shorthorn developed 1885
on, later with Red Holstein *et
al.*/pd var. from pd sire/BS 1912
Milking Zebu: (Indonesia) *see* Hissar

Min: *see* Minnan
Mineiro: *see* Southern Crioulo
Mingrelian Red: (W Georgia)/d/grey
or red/subvar. of Lesser Caucasus
var. of Caucasian/Russ. *Krasnyĭ
megrelskiĭ skot*/syn. *Megrel,
Megrelian, Mingrelian*/rare
Minhota: (NW Portugal)/dr.m.d/
yellow-brown/syn. *Galega*/?
extinct (by crossing with Barrosã,
Turino and esp. Gelbvieh)
Miniature Galloway: (Australia)/
black, chocolate or belted
(Miniature Belted Galloway)/
foundation herds at Glenblairie
and Wannawin
Miniature Hereford: (Texas, USA)/m/
Hereford selected for small size
by Rust Largent
Miniature Zebu: (USA)/under 42
inches (107 cm) high/orig. from
Indian Zebu/BS 1991/syn. *Mini
Zebu*/*see also* Bonsai Zebu
Mini-Brahman: (Mexico)/selected
from Brazilian zebus/syn. *Bonsai
Brahman*
Minnan: (coast of SC Fujian, China)/
usu. brown, also black or dark red,
often with "banteng" markings/
Changzhu group/syn. *Min*
Minorcan: (Menorca, Balearic Is,
Spain)/d.m/blond to red; usu.
pd/orig. from Marinera/Sp.
Menorquina/syn. *Mahonesa*
(from Mahón)/rare
Minufi: *see* Menufi
Mirandais: *see* Gascon aréolé
Mirandesa: (C and NE Portugal)/
dr.m/light to dark chestnut/vars:
Beiroa, Bragança, Jarmelista,
Marinhoa/orig. of Berciana,
Verinesa/HB 1977, BS/cf.
Frieiresa (SE Orense, Spain)/syn.
Ratinha/[from Miranda do Douro]
Mishima: (Mishima I, W of Honshu,
Japan)/m/black/pure var. (i.e.
original type before crossing) of
Japanese Native/BS/nearly
extinct
Miskov: (Kostroma, Russia)/m.d/
local type improved by
Yaroslavl, Kholmogory and
Ayrshire/orig. (with Babaev) of
Kostroma/extinct

mithun: (hill forests of Assam, Bhutan and NW Myanmar)/= *Bos (Bibos) 'frontalis'* Lambert/= domesticated gaur/dark slate with white stockings; born dark red/syn. *bami* or *menscha* (Bhutan), *dulong* (Yunnan) (nearly extinct), *gayal* (Hindi and Bengali); *jatsa* and *jatsum* = cross with Siri ♀ ♀/not *mithan, mythun*/[Assamese]

MithunSiri: (Bhutan)/usu. black with brown dorsal stripe, also pied or grey; horns sim. to mithun/mithun ♂ × Siri, i.e. = jatsa and jatsum/1st backcross hybrids = *yanka* (♂), *yankum* (♀); 2nd backcross = *doeb, doebum*; 3rd = *data, datum*; 4th = *thrapa, thrabum*

Mitteldeutsches Gebirgsvieh: *see* German Red

Mitteldeutsches Rotvieh: *see* German Red

Mitzan: *see* Sheko

Mkalama Dun: (Singida region, C Tanzania)/local strain of Small East African Zebu (Tanzanian cluster)/nearly extinct

MLB: *see* Brazilian Milking Crossbred

Mocaniţa: *see* Romanian Mountain

Mocho de Quilengues: *see* Tshilengue

Mocho nacional: *see* Brazilian Polled

Modenese: (Modena, Emilia, Italy)/ d.m.[dr]/white with dark points/ ? orig. from Reggiana × Romagnola/BSd and HB 1957/syn. *Carpigiana* (from Carpi), *White Po* (It. *Bianca val padana*)/rare

Modenese di Monte: *see* Garfagnina

Modicana: (Siciliy, Italy)/d.m.[dr]/ brown or dark red/Podolian type/orig. lowland var. of Sicilian/vars: Mezzalina, Montanara/BSd and HB 1952/syn. *Olivastra modicana* (= olive-coloured), *Sicilian*/rare

Modicano-Sarda, Modica-Sardinian: *see* Sardo-Modicana

Mogol: *see* Magal

Moi: (Cambodia)/var. of Cambodian/z orig./nearly extinct

Moile, Moiled, Moilie: *see* Irish Moiled

Molang: *see* Tibetan ♀

Moldavian: (Romania)/dr.d/white to ash-grey/former var. of Romanian Steppe/Rom. *Moldovenescă*/ extinct

Moldavian Black-and-White: (Moldova)

Moldavian Red Steppe: (Moldova)

Moldavian Simmental: (Moldova)

Moldovian Estonian Red: (Moldova)/d/orig. from local graded to Estonian Red/Eston. *Rosíe estona*/nearly extinct

Mölltal: (Austria, Italy)/former var. of Pinzgauer in SW Carinthia (to 1925) and NE Udine, as distinct from Pinzgauer and Pustertaler Sprinzen in Salzburg and NE Bolzano/Ger. *Mölltaler*, It. *Pezzata rossa norica* (= *Red Pied Noric*, in Friuli-Venezia Giulia region)/syn. *Mölltal-Pinzgau, Norica-Pinzgau*/not *Mölthal*/ extinct

Molokan: *see* Malakan

Monchina: (SW Viscaya and SE Santander, Spain)/m/chestnut red/[= mountain]/rare

Mongalla: (S Sudan)/all colours/var. of Small East African Zebu/inc. Bari, Didinga, Latuka/syn. *Southeastern Hills Zebu, Southern Sudan Hill Zebu*/not *Mongolla*

Mongconto: *see* Mangwato

MONGOLIAN group: (Mongolia and N China)/m.d.dr/yellow-brown or black-and-white; long thin horns/humpless cattle of Turano-Mongolian type/inc. Anxi, Mongolian (*sensu stricto*), Fuzhou, Yanbian/syn. *Northern group, Northern Yellow*

Mongolian: (Mongolia, and Inner Mongolia, China)/d.m.dr/usu. brindle or reddish-brown, sometimes black, yellow or pied/ vars: Halhïn Gol, Ujumqin/Ch. *Menggu*/not *Monggol*

Mongolian Black Pied: (Mongolia)/ d.m/orig. from Friesian × Mongolian/Mongolian *Char Tarlan*

Mongolian Whiteheaded: (Mongolia)/m/orig. from

Hereford × Mongolian/cf.
Selenge/Mongolian *Tsagaan
Tolgoit*

Mongolian Yellow-Brown:
(Mongolia)/d.m/orig. from Swiss
Brown × Mongolian/Mongolian
Bor Khalium

Mongolla: *see* Mongalla

Montafon: (Vorarlberg, Austria)/orig.
var. of Austrian Brown/orig. from
local graded to Swiss Brown/BS
1923/Ger. *Montafoner*/syn.
Vorarlberg Brown/not *Montavon*,
Montophone

Montana: (borders of Lombardy,
Emilia, Liguria and Piedmont,
Italy)/d.m.[dr]/pale straw colour,
brown at birth/sim. to
Pontremolese but smaller/syn.
(Bionda) Tortonese (from
Tortona, Alessandria), *Cabellota*
(from Cabella, Genoa), *Ottonese*
(from Ottone, Piacenza), *Red
Mountain* (It. *Montana rossa,
Rossa montanina*), *Varzese* (from
Varzi, Pavia), *Varzese Ottonese*/
nearly extinct (by crossing with
Reggiana and Italian Brown)

Montaña: (Spain) *see* Santander

Montanara: (Italy) *see* Garfagnina;
see also mt vars of Marchigiana
and Sicilian

Montanara: (Sicily, Italy)/mountain
var. of Sicilian, now absorbed by
Modicana

Montana Tropical: (Brazil) *see*
Tropical Mountain

Montbéliard: (Haute Saône-Doubs,
France)/d.m.[dr]/bright red-and-
white/French Red Pied group/
orig. from Bernese brought by
Mennonites in 18th c. (named
1872); recently with Red-and-
White Holstein blood/absorbed
Tourache *c.* 1900/HB and BS
1889; HB also in Netherlands,
Switzerland (2000, section in
Simmental HB)/syn. *French
Dairy Simmental*

Montgomery: *see* Sahiwal

Montgomery-Jersey: *see* Jamaica
Hope

Montgomeryshire: (Wales)/red,
shading to black on face, belly
and tail/extinct since 1919

Montophone: *see* Montafon

mooly, moolley: *see* mulley

moon yak: *see* dzo

Moor, Moorish: *see* Maure

Morang: (Nepal)/z/cf. Nepalese
Hill/hill type, sim. to Purnea/?
extinct

**Moravian Carpathian, Moravian
Land:** *see* Moravian Red

Moravian Red: (Czech Republic)/
d.m.dr/Central European Red
type/var.: Lišna Red/Cz.
Moravský červený/syn. *Moravian
Carpathian, Moravian
Land*/extinct *c.* 1960

Moravian Red Pied: (Czech
Republic)/former var. of Czech
Pied/Cz. *Moravský
červenostrakatý*/syn. *Bernese-
Hanna, Berno-Hana* (Cz.
Bernskohanácký), *Haná-Berner,
Hanna-Berne* (Cz.
Hanáckobernský), *Moravian Red
Spotted, Spotted Moravian*/
extinct

Moravian Sudeten: *see* Sudeten

Morbihan, Morbihannais: *see* Breton
Black Pied

Møre and Ramsdal: (W Norway)/
grey; pd/orig. (with Hordaland)
of Vestland Fjord/extinct

Morenas Gallega: *see* Morenas del
Noroeste

MORENAS DEL NOROESTE: (S Galicia
and NW Zamora, Spain)/
dr.m.d/black-brown/sim. to
Black Iberian/inc. Alistana-
Sanabresa, Caldelana, Limiana,
Sayaguesa, Verinesa, Vianesa/
syn. *Morenas gallegas* (*gallegas* =
Galician) (excluding Sayaguesa
and Verinese but inc. also
Cachena and Frieiresa)/[= dark
(cattle) of northwest]/rare

Moroccan: *see* Brown Atlas, Oulmès-
Zaërs Blond, Tidili

Moroccan Blond: *see* Oulmès-Zaërs
Blond

Moroccan Brown: *see* Brown Atlas

Morucha: (Salamanca, Spain)/ m.dr/
black or grey (*cárdena*) vars/?
orig. from Black Iberian/ BS 1974/
Sp. *Salmantina* (from Salamanca)/
not *Salmanquina*/[= black]

Moruno-Sinuano: *see* Romosinuano

Morvan: (Burgundy, France)/Fr. *Morvandelle*/extinct (replaced by Charolais)

Mosane-rhénane-ysseloise: *see* Meuse-Rhine-Yssel

Mostrenca: (Doñana Nat. Park, S Huelva, Spain)/solid red, or black, grey, blue, brown, blond, white or combinations; long horns/prim.; semi-feral/orig. from Andalusian Black and Fighting bull/Sp. *Agrupación bovina mostrenca*/syn. *Doñana, Marismeña, Palurda*/rare

Mottai madu: *see* Umblachery

Mottled Hill: (Germany)/dr.m.d/name used in Germany *c.* 1936–1945 to inc. Black Forest (Hinterwald, Vorderwald), White-marked (Kelheimer, Westerwald), Whitebacked (Glatz, Pinzgauer, Sudeten, Vosges) and Bergscheck/Ger. *Schecken und Blässen*, or *Scheckiges und rückenblessiges Höhenvieh*

Motu: (Orissa, India)/dr/z

Mountain Grey: *see* Grey Mountain

Mountain Spotted: *see* Bergscheck, Simmental

Mount Mahabharat: *see* Nepalese Hill

Moutourou: *see* Muturu

Mowati: *see* Mewati

Moyenne et Haute Belgique: *see* Belgian Blue

Moyle, Moyled, Moyley: *see* Irish Moiled

Mozambique Angoni: *see* Angonia

Mpwapwa: (E Tanzania)/d.m/usu. light to dark red/z/orig. from Sahiwal *et al.*; crossing started 1940s, breeding programme 1958; mainly Red Sindhi, Sahiwal and Tanzanian Zebu, also Boran, Ayrshire, Shorthorn and Ankole; later selected towards Sahiwal (75%)/syn. *Indo-African Zebu*/rare

MRI, MRY: *see* Meuse-Rhine-Yssel

Mrko-smeda rasa: *see* Yugoslav Brown

Mucca (nera) pisana: *see* Pisana

Mucubai: (Nambié, also Cunene, SC Angola)/Sanga type

Mukaku Washu: *see* Japanese Poll

mulley: (USA)/= pd/? orig. from Polled Derby/orig. of single standard Polled Shorthorn/not *mooly, moolley, muley*/[Gaelic *maol* = bald, hornless]

Mulline: *see* Irish Moiled

Multani: *see* Sahiwal

Munich cattle: (Munich zoo, Germany)/cf. Berlin cattle, Heck cattle/orig. (1926 on) by Heinz Heck from Corsican, (Scottish) Highland, Hungarian Grey, Friesian, Murnau-Werdenfels and Allgäuer/so-called "bred-back aurochs"

Munshiganj: (Bangladesh)/z; local var./? orig. from local Bengali × Red Sindhi/not *Munshigani, Munshigonj, Munshigunj*/extinct (by crossing with Sahiwal and Friesian)

Murboden: (Mur valley, Styria, Austria, and Slovenia)/d.m.dr/ pale yellow-brown with darker points/orig. mid-19th c. from Bergscheck, Mürztal and Mariahof/with Austrian Blond and Waldviertel = Austrian Yellow (*c.* 1960)/recog. 1869/Ger. *Murbodner*, Sn. *Pomurska*/syn. *Murboden-Mürztal, Svetlolisata* or *Pšenična* (= wheaten) (Slovenia)/rare

Murcian: (Valencia to Almería, SE Spain)/m.dr/dark chestnut/ former vars: Huertana (= Murcian *sensu stricto*), Almanzoreña, Calasparreña, Cristiana, Lorquina/Sp. *Murciana*/syn. *Levantina* (= eastern)/nearly extinct

Murgese: (Murge, Apulia, Italy)/ former var. of Apulian Podolian

Murle: (SE Sudan)/m/z/Large East African Zebu group, sim. to Toposa but smaller/? orig. from Abyssinian Shorthorned Zebu

Murnau-Werdenfels: (Weilheim-Garmisch-Landsberg, Bavaria, Germany)/d.m.[dr]/local breed sim. to German Brown but smaller and more red-yellow in colour/orig. from Oberinntal

Grey, Swiss Brown and
Murboden/BS and HB 1927/Ger.
Murnau-Werdenfelser/rare

Murray Grey: (Upper Murray R
valley, Victoria, Australia)/
m/silver-grey to grey-dun/cf.
Australian Grey/orig. from 12
grey calves (born 1905 on) out of
light roan (nearly white)
Shorthorn ♀ by Aberdeen-Angus
♂; graded to Aberdeen-Angus
and selected for grey (dun)/
absorbed Tasmanian Grey *c.*
1970/BS 1962; BS also USA
1969, Canada 1970, GB 1974,
New Zealand

Mursi: (S Albania)/dr/grey with black
points/Grey Steppe type

Mursi: (S Omo, Ethiopia)/d.(m)/
various combinations (solid rare)
of grey, white, black, chestnut
and roan, or pied or striped;
large horns, prominent hump/
Abyssinian Shorthorned Zebu
group

Mürztal: (Austria)/orig. (with
Bergscheck) of Murboden in 19th
c./recombined with Murboden
1913/extinct

Muturu: (Cameroon) *see* Bakosi,
Bakwiri

Muturu: (Liberia) *see* Liberian Dwarf

Muturu: (S Nigeria)/various colours,
esp. black or pied/sim. to Lagune
but well-muscled/name for West
African Shorthorn in English-
speaking areas (cf Lagune in
French-speaking)/vars: Savanna
Muturu (pied), Forest Muturu
(dwarf, black, brown-black)/
semi-feral in Antlantika mts,
Gongola, hunted by Koma/Fr.
Moutourou/syn. *Kirdi, Nigerian
Dwarf, Nigerian Shorthorn,
Pagan*/[Hausa = humpless]/
declining/*see also* Bakweri,
Ghana Dwarf Muturu, Liberian
Dwarf Muturu

Mwool: *see* Irish Moiled

Myanmar: *see* Burmese

Myleen: *see* Irish Moiled

Mysol: (Costa Rica)/d/local zebu ×
Criollo cross/extinct

Mysore: (S India)/rapid dr/usu. grey;
vertical horns/basic breed is
Hallikar; also inc. Alambadi,
Amritmahal, Bargur, Kangayam,
Khillari

mythun: *see* mithun

Nadjdi: *see* Nejdi

Nagar: *see* Kankrej

Nagori: (Nagaur, C Rajasthan, India)/
dr/z/Grey-white Shorthorned
type/not *Nagauri, Nagoni,
Nagore*

nak: *see* yak ♀

Nakali: (India)/var. of Khillari/not
Nakli/[= imitation]

Nama: (S Namibia)/Sanga type/orig.
from Hottentot, Ovambo,
Damara, Friesian and
Afrikander/? extinct

Namaqua: *see* Hottentot

Namchi: *see* Namshi

Namshi: (NW foothills of Poli mts,
NW Cameroon)/m.ri/black, black
pied, black with white spots,
occ. brown or brown spotted/var.
of West African Savanna
Shorthorn/Fr. *Namchi*/syn.
Doayo (kept by Doayo people),
Kirdi (= pagan), *M'Bougi* (sing.
M'Bouyé), *Poli*/not *Doai,
Donayo*/rare

Nanbu: (Japan)/var. of Japanese
Native/? extinct

Nandi: (W Kenya)/local Small East
African Zebu/syn. *Nandi Blue*

Nanico: *see* Igarapé

Nantais: (Loire, France)/d.m/brown
to silver-grey/paler var. of
Parthenais, lacking black
mucosae/BS/[from Nantes]/rare

Nanyang: (Henan and N Hubei,
China)/dr/usu. yellow, also
red/Changzhu group/mountain
and lowland vars/Russ. *Nan'yan*

naran-hainag: *see* dzo

Nari: *see* Bhagnari

Nari Master: (Baluchistan, Pakistan)/
m/fawn to red-brown/orig. since
1969 from Droughtmaster ($\frac{5}{8}$) ×
Bhagnari ($\frac{3}{8}$)

Nata, Ñata: *see* Niata

National Polled: (Brazil) *see*
Brazilian Polled

Native Black: (Turkey) *see* Anatolian Black

Natura: (Brazil)/in formation from Aberdeen-Angus × Nellore

Navarre: (Spain)/Sp. *Toro de Casta Navarre* (*casta* = breed)

ncb: *see* Polish Black-and-White Lowland

Ndagu: (Ghana)/N'Dama × Sokoto Gudali crossbreds on exp. stas

N'Dama: (Fouta Djallon, Guinea, and neighbouring countries)/d.m.dr/ usu. fawn, red or brown, shading to darker extremities, occ. pied or black; lyre or crescent horns/ West African small humpless (trypanotolerant) longhorn type/vars: Gambian N'Dama, N'Gabou/orig. of N'Damance, N'Damaze and being crossed with e.g. Breton Black Pied (Côte d'Ivoire), Abondance, Jersey, Simmental (Guinea)/syn. *Boenca* or *Boyenca* (Guinea-Bissau), *Fouta Jallon, Fouta Longhorn, Fouta Malinke, Futa, Malinke, Mandingo* (Liberia), *N'Dama Petite* (Senegal), *Outa Malinke*/not *Dama, Ndama*

N'Dama Grande: (Senegal) *see* Gambian N'Dama

N'Damance: (Côte d'Ivoire)/in formation since 1980 from (N'Dama × Fleckvieh) × (N'Dama × Abondance)

N'Dama of Kaarta: (Mali) *see* Gambian N'Dama

N'Dama Petite: (Senegal)/smaller var. of N'Dama in Senegal/orig. (with Red Poll) of Senepol/syn. *Red Senegal*

N'Dama Sanga: (Ghana)/N'Dama × zebu × Ghana Shorthorn

N'Damaze: (N Côte d'Ivoire)/dr/ yellow to red, usu. with black mucosae/orig. at Centre de Panya, Boundiali, from N'Dama × Gobra

Ndawana: *see* Batawana

Neckar: *see* Heilbronn

Nederlands Melkras: *see* Dutch Dairy

Negra andaluza: *see* Andalusian Black

Negra campiñesa: *see* Andalusian Black

Negra de las Campiñas andaluzas: *see* Andalusian Black

Negra iberica: *see* Black Iberian

Nejdi: (Khuzestan, Iran)/d/all colours; hd or pd; humped or not/orig. from Jersey and Sindhi × local/syn. *Arabi*/not *Nadjdi*

Nellore: (Asia) *see* Ongole

Nellore: (USA)/preferred name for Ongole

Nellorford: (USA)/Nelore × Hereford cross

Nelore: (Brazil)/m/white/z/orig. from Ongole imported 1895–1964/ vars: Polled Nelore, Red Nelore/HB 1936, BS 1939; BS also Argentina, Paraguay

Nelore mocho: *see* Polled Nelore

Nelore vermelho: *see* Red Nelore

Nelthropp: *see* Senepol

Nepalese, Nepali zebu: *see* Achham, Morang, Nepalese Hill, Tarai

Nepalese Hill: (Churia Ghati and Mahabharat range, Nepal)/black; small/z/cf. Morang/Indian Hill type/syn. *Black Hill Zebu, Nepali Hill Zebu*

Netherlands Black Pied: *see* Dutch Black Pied

Netherlands Indies: *see* Indonesian

New Aurochs: *see* Heck cattle

New Hungarian Brown: *see* Dairy Hungarian Brown

New Hungarian Red: *see* Dairy Hungarian Pied

New Zealand Jersey: (New Zealand)/ var. of Jersey from 19th c. imports/BS 1902, HB 1903

New Zealand Sahiwal × Holstein-Friesian: *see* Taurindicus

New Zealand Zebu: *see* Taurindicus

N'Gabou: (Guinea-Bissau and Casamance, Senegal)/white with black points/var. of N'Dama/syn. *Fula, Gabú*

N'Gami, Ngami: *see* Batawana

Nganda: (Buganda, Uganda)/usu. yellow-dun or light red, often red-and-white, occ. black or white/Sanga-Zebu intermediate/ orig. from Ankole (mainly Bahima, also Watusi) × Nkedi/ local syn. *Sese Island*/sim. vars: Kyoga, Nyoro, Serere, Toro

N'Gaoundéré: (Ngaoundéré and

Bamenda highlands, Cameroon)/
red, brown, red pied or white, also
brindle, roan, or red spotted with
white and brown-black pied/
typical var. of Adamawa/syn.
Bamenda (Nigeria), *N'Gaoundéré
Gudali*/not *Ngaundere*
Ngombe dza Maswina: *see* Mashona
Ngombe dze Vakaranga: *see*
Mashona
Ngoni: *see* Angoni
Ngoua: (Laos)
Nguni: (Swaziland and S Africa)/
d.m/white, black, brown, red,
dun or yellow (solid or in
combination), often coloursided
(*inkone*), or black-and-tan or
brindle, many recognized
patterns; lyre horns/Sanga type/
var.: Pedi/BS (S Africa) 1986/syn.
Landim or *Sul do Save*
(Mozambique), *Swazi*, *Zulu*/not
Ngune
NGUNI group: (Swaziland, E Zululand
to S Transkei, S Africa, and
Mozambique S of R Save)/
d.m.dr/group of Sanga types orig.
kept by Nguni tribes (inc. Swazi
and Zulu)/inc. Landim, Nguni,
Nkone, Pedi, Shangan
Ngwato: *see* Mangwato
Niata: (Uruguay and Argentina)/var.
of Criollo with bulldog snout
(partially dominant)/Sp. *Ñata*/
not *Nata*/[= snub-nosed]/ nearly
extinct (now only on islands off
S Chile and in Jujuy, NW
Argentina)
Niederrheiner, Niederrheinisch: *see*
Rhineland
Niederungsvieh: *see* Lowland
Nigerian Dwarf: *see* Muturu
Nigerian Fulani: *see* Diali
Nigerian Shorthorn: *see* Muturu
Nihon Tankatu Washu: *see* Japanese
Shorthorn
Nilo: (Rio de Janeiro, Brazil)/orig.
from African Zebu imported by
Emperor Pedro I in 1826 and
crossed with local cows/extinct
Nilotic: (Sudan, S of lat. 10°N)/m/
usu. off-white or cream, also red,
black or pied/Sanga type in S
Sudan and SW Ethiopia, bred by

Dinka, Nuer, Shilluk and Anuak
tribes; smaller and lighter in
west, larger and heavier in east/
local tribal vars and subvars:
Abigar, Aliab Dinka, Aweil Dinka,
Eastern Nuer, Nuer, Shilluk,
Wadai-Dinka/syn. *Southern
Sudanese, Sudanese Longhorn*
Nimari: (Nimar, SW Madhya
Pradesh, India)/dr/usu. red with
white markings/z/? orig. from Gir
and Khillari/BSd/syn. *Khargaon,
Khargoni, Khurgoni*
Nioka: *see* Alur
Nipponese Improved: *see* Japanese
Improved
Nivernaise: (C France)/dr/white/HB
1864/syn. *race bovine charolaise
améliorée dans la Nièvre*
(*Nivernais-Charolais in the
Nièvre*)/extinct early 20th c.
(supplanted by Charolais)
Nizhegorod: *see* Yurino
Nizinna czarno-biała: *see* Polish
Black-and-White (Lowland)
Nizinna czerwono-biała: *see* Polish
Red-and-White (Lowland)
Nižinne černostrakate: *see* Czech
Black Pied
Nkedi: (E and N Uganda)/many
colours/var. of Small East
African Zebu (Southern Sudan
group)/syn. *Bukedi, Eastern
Province Zebu, Lango*
nkone: *see* inkone
Nkone: (Zimbabwe)/m/usu.
coloursided, red-and-white, also
red, roan or black-and-white/
Sanga type/orig. from and sim. to
Nguni, also with Boer, Mangwato,
Batawana, Mashone, Barotse,
Tonga and Afrikander blood,
selected (1946 on) from *inkone* (=
coloursided) type (? orig. from
Nguni) in Matabele cattle/related
to Govuvo/BS 1967/syn. *Mangoni,
Manguni* (1961–1969)/not
Inkone/nearly extinct
Nordfinsk: *see* North Finnish
Nordland: *see* Blacksided Trondheim
and Nordland
Norfolk Horned: (England)/m/red,
with white or mottled face/orig.
(× Suffolk Polled) of Red Poll/

syn. *Old Norfolk, Red Norfolk*/
extinct
Norfolk Polled: *see* Red Poll
Noric: *see* Pinzgauer
Norica-Pinzgau: *see* Mölltal
Normande: (Normandy, France)/
d.m/red-brown and white, often
brindled, usu. red spectacles/
orig. from local (e.g. Cotentin
and Augeron) with some
Shorthorn and Jersey blood
1845–1860/BS 1883, reorg. 1926;
BS also Colombia, Uruguay, USA
1974, Argentina/syn. *Norman,
Normandy*
Normanzu: (Brazil)/Normande ×
zebu (Gir or Guzerá) cross
Norsk: *see* Norwegian
North Australian Shorthorn: *see*
Australian Shorthorn
North Bangladesh Grey: (N
Bangladesh)/z/local var./syn.
North Bengal Grey (India)
North Devon: *see* Devon
Northeastern Crioulo: *see* Curraleiro
Northern Blue: (France) *see* Bleu du
Nord
Northern Dairy Shorthorn: (N
England)/d.m/var. of Shorthorn
probably descended directly
from Teeswater/BS and HB
1944–1969 (then combined with
Dairy Shorthorn)/syn. *Dales
Shorthorn*
Northern Yellow group: (China) *see*
Mongolian group
Northern Province: *see* Kenana
Northern Riverain: (Sudan) *see*
Kenana, Butana
Northern Scotch Polled: *see*
Aberdeen-Angus
Northern Territory Shorthorn: *see*
Australian Shorthorn
NORTH EUROPEAN POLLED: inc.
Aberdeen-Angus, Blacksided
Trondheim and Nordland,
British White, Estonian Native,
Finnish, Galloway, Icelandic,
Irish Moiled, Pechora (pd var.),
Red Poll, Red Polled Eastland,
Swedish Polled, Vestland Polled,
Vychegda-Vym
North Finncattle: *see* North Finnish
North Finnish: (N Finland)/d/white,

with coloured ears; pd/sim. to
Swedish Mountain and to
Blacksided Trondheim and
Nordland/var. of Finnish/BS
1905–1950/Finn. *Pohjoissuomen
karja* (or *Pohjois-Suomolainen
Karja*) (= *PSK*), Swe. *Nordfinsk*/
syn. *North Finncattle*/nearly
extinct
North Italian Brown: *see* Italian
Brown
Northland: (Norway) *see* Blacksided
Trondheim and Nordland
North Malawi Zebu: (N Malawi)/
former var. of Malawi Zebu, ?
with Sanga blood/obs. syn.
Nyasaland Angoni/? extinct
North Manchurian Dairy: *see*
Pinzhou
North Shaanxi: (China)
North Slesvig Red: (Germany–
Denmark)/ sim. to Angeln but
with Shorthorn blood/absorbed
in Danish Red/Dan. *Nord Slesvig
Rød*, Ger. *Rotes Nordschleswiger*/
extinct
North Somali: (N Somalia)/usu.
chestnut or black, sometimes
white/z/cf. Aden/Somali group/
vars: eastern (Burao area in NE
prov., smaller), western (Borama/
Hargesia area in NW prov.,
highly variable, often roan or
spotted, small hump, lyre horns)
NORTH SPANISH: (N Spain)/dr.m.d/
brown to red/inc. Asturian,
Galician, Leonese, Pyrenean,
Tudanca
NORTH SUDAN ZEBU: (Sudan, N of lat.
10°N)/d.dr.m/medium-sized
thoracic or cervico-thoracic
muscular hump; short horns;
large/inc. Baggara, Barka, Beja
(?), Butana, Kenana, Nyalawi/
syn. *Arab Zebu, Sudanese
Shorthorned Zebu*
North Swedish: *see* Swedish Polled
North Wales Black: (Wales)/m/HB
1883–1904/orig. (+ South Wales
Black) of Welsh Black/syn.
Anglesey/extinct
Norwegian Red: (Norway)/d/red (or
red with some white); hd or
pd/orig. 1961 by union of

Norwegian Red-and-White (inc. Red Trondheim) and Red Polled Eastland/absorbed Døle 1963, South and West Norwegian 1968/BS and HB; BS also in USA/Nor. *Norsk rødt fe* (= *NRF*)

Norwegian Red-and-White: (SE Norway)/d/red with small white markings on legs and lower body/orig. from Swedish Red-and-White × local Ayrshire (first imported 1860s), Red Trondheim and Hedmark; named 1939/absorbed Red Trondheim in 1960/joined with Red Polled Eastland in 1961 to form Norwegian Red/BS 1923/Nor. *Norsk rødt og hvitt fe* (= *NRF*)/obs. syn. (1923–1939) *Hornet slettefe* (= *Horned lowland cattle*)/extinct

Norwegian Red Polled: *see* Red Polled Eastland

Nostrana: *see* Garfagnina

NRF: *see* Norwegian Red

NRS: (Netherlands)/= *Nederlands Rundvee Stamboek*, HB which registers Dutch Black Pied (except in Friesland), Meuse-Rhine-Yssel and Groningen Whiteheaded

Nsagalla: (Uganda) *see* Bahima

Nuba: (Sudan)/humpless shorthorn/syn. *Delami, Koalib, Nuba Shorthorn*/extinct

Nuba Mountain: (S Kordofan, Sudan)/m.dr/variable hump; dwarf/z/Small East African Zebu (S Sudan cluster); sim. to Ingessana hills dwarf and Mongalla/? orig. from Baggara × dwarf humpless/syn. *Koalib, Nuba Dwarf*

Nuer: (NW of S Sudan)/local var. of Nilotic/*see also* Eastern Nuer

Nuras: (Namibia)/m/$\frac{1}{2}$ Afrikander, $\frac{1}{4}$ Simmental, $\frac{1}{4}$ Hereford

nwar: = cattle (Myanmar)

nwar pyiase: *see* Chaubauk

nwar shweni, nwar shwewar: *see* Burmese

Nyalawi: *see* Habbani

Nyambo: *see* Inyambo

Nyasaland Angoni: *see* (North) Malawi Zebu

Nyasaland Zebu, Nyasa Zebu: *see* (South) Malawi Zebu

Nyoka: *see* Alur

Nyoro: (Bunyoro, NW Uganda)/enclave of Sanga-Zebu intermediates sim. to Nganda/not *Banioro, Banyoro, Bunyoro*

NZSHF: *see* Taurindicus

Oberbaden, Oberbädisches: *see* Messkircher

Oberbayrisches Alpenfleckvieh: *see* Miesbacher

Oberinntal: *see* Valbona

Oberinntaler Grauvieh: *see* Tyrol Grey

Oberinntal Grey: *see* Tyrol Grey

Ocampo: (Venezuela)/in formation 1940s and 1950s on Ocampo estate from Friesian × (Ongole × local Criollo); graded to American Brown Swiss in 1970s/extinct

Odenwald: (S Hesse, Germany)/subvar. of Hesse-Westphalian var. of German Red/HB 1899/Ger. *Odenwälder*/extinct

Ogaden: (Ogaden area of Somalia Region, Ethiopia, and bordering E Hararghe)/d.(m)/Abyssinian Shorthorned Zebu group/var. of Boran/*see also* Jijjiga Zebu

Oka: (Gorki, Russia)/orig. of Gorbatov Red/Russ. *Priokskaya*/extinct

Oka Black Pied: (Ryazan, Russia)/d/orig. from East Friesian or Kholmogory × Simmental and Swiss Brown grades, with Jersey bulls since 1952/Russ. *Priokskaya chernopestraya*/absorbed by Russian Black Pied/extinct

Okavango: (Okavayo R, Namibia)/Sanga type (Ovambo cluster)

Okranya: *see* Molokan

Oksh: *see* Baladi

Okuma: (Gabon)/orig. from Tuli × Nguni

Old Austrian Brown: *see* Austrian Brown (Original)

Old Braunvieh: *see* German Brown (old type)

Old Danish Red: *see* RDM-1970

Oldenburg-Wesermarsch: (Oldenburg, Germany)/d.m/orig.

var. of German Black Pied, orig.
with Shorthorn blood/orig. of
Podolian Black Pied/HB 1880/
Ger. *Oldenburger-Wesermarsch*/
syn. *Oldenburg*/not *Wesermarsh*/
extinct

Old Gloucester(shire): *see* Gloucester

Old Irish Cow: (Ireland)/d/ill-defined
type/extinct about mid-19th c.

Old Marlborough Red: (Devon,
England)/d/[? misprint for
'Malborough', town in Devon]/
extinct

Old Norfolk: *see* Norfolk Horned

Old Red Angeln: (Germany)/red or
brown with black muzzle/var. of
old Angeln/Ger. *Rotvieh alter
Angler Zuchtrichtung* (= Red
cattle of old Angler type)/nearly
extinct

Olivastra modicana: *see* Modicana

Omani Dhofari: (Dhufar, Oman)/
d.m.dr/dark coat; short horns;
small hump; small/syn. *Oman
Zebu*

Omby rana: *see* Rana

Omega 47: (France)/double-muscled
♂♂, usu. Charolais, for crossing
on older cows

Ondango: *see* Tshilengue

Ondongolo: (Transkei, Cape
Province, S Africa)/cf.
Bolowana/Sanga type/sacred
herd of Chief Tyelinzima in
Bomvanaland/extinct

Ongole: (Guntur-Nellore, S Andhra
Pradesh, India)/dr.d/Grey-white
Shorthorned type/orig. of
Javanese Ongole, Nelore/BSd
and HB, BS 1983/syn. *Nellore*,
Sumba Ongole (Indonesia)/rare

Oost Vlaamse: *see* Belgian White-
and-Red

Oran: (Algeria)/former subvar. of
Moroccan var. of Brown Atlas/?
extinct

Original Austrian Brown: *see*
Austrian Brown (Original)

Original Braunvieh: *see* German
Brown (old type)

Original Österreichisches Braunvieh:
see Austrian Brown (Original)

Oristanese, Oristano: *see* Sardo-
Modicana

Orkney: (Scotland)/var. of Shetland/
extinct

Orma Boran: (Tana R dist., SE
Kenya)/var. of Boran brought by
Oromo from Ethiopia in 15th c.;
more trypano-tolerant than
Kenya (improved) Boran/syn.
Tanaland Boran

Oropa: (Biella, Piedmont, Italy)/
d.m/red pied with white head/
larger var. of Aosta with
Simmental blood/HB 1964/It.
Pezzata rossa d'Oropa

Osnabrück Holstein: (Hungary)/?
imported from Germany

Ossolana: (NE Piedmont, Italy)/local
var. absorbed by Italian Brown/
extinct

Østerdal: (SE Norway)/black, grey or
reddish-grey/old type in valleys
of Gudbransdal and Østerdal/var.
of Døle/extinct

Österreichisches: *see* Austrian

Ostfinks: *see* East Finnish

Ostfriesen, Ostfriesisches: *see* East
Friesian

Østland: *see* Red Polled Eastland

Østlansk rødkolle: *see* Red Polled
Eastland

**Ostpreussen, Ostpreussisch
Holländer:** *see* East Prussian

Öststeierisches Fleckvieh: *see* East
Styrian Spotted

Ottone, Ottonese: *see* Montana

Oulmès-Zaërs Blond: (NW Morocco)/
pale fawn/Fr. *Blonde d'Oulmès
et des Zaërs*/syn. *Blond
Moroccan, Blond Zaërs,
Moroccan Blond, Oulmès Blond*

Outa Malinke: *see* N'Dama

Ovambo: (N Namibia)/usu. dun, also
black, red, black pied with white
dorsal and underline; large lyre
horns; small (? mineral
deficiency)/cf. Humbi/Sanga
type (Ovambo cluster), sim. to
Kaokoveld but smaller/sing.
Ambo

Ovambo cluster: Ovambo and
Southwestern group of Southern
Africa Sanga, inc. Caprivi,
Humbi, Kaokoveld, Okavango,
Ovambo

Overo colorado: (Chile)/= (German)
Red Pied in Chile

Ozierese, Ozieri: see Sardinian Brown

PA: see Spanish Brown

Pabli: (Upper Pendjari valley, NW Benin)/usu. red/var. of West African Savanna Shorthorn sim. to Somba/extinct (absorbed by Borgou

Pabna: (W Bangladesh)/d.dr/z/orig. from Dacca-Faridpur and Munshiganj upgraded by Sahiwal, Hariana or Red Sindhi (1915–1976)/syn. *Pabna milking cows*

Padana: see Modenese

Pagan: see Doayo, Kapsiki, Muturu

Pahari, Pakhari: see Dhanni

Pajuna: (mts of E Andalucía, Spain)/dr.m/red, black or blond with white muzzle/? orig. from Brown Atlas, Murcian or Avileña-Black Iberian, with Retinta blood/syn. *Serrana* (= mountain, highland)/[= rustic, primitive]/rare

Pale Alpine, Pale Highland, Pale Mountain: (Austria) see Austrian Yellow

Pallaresa: (Catalonia, Spain)/m/ white/nearly extinct

Palmera: (La Palma, Canary Is, Spain)/m.dr.d/yellow or dirty white/orig. from Galician Blond and other Spanish breeds/syn. *Palmeña, de la Tierra*/rare

Palurda: see Mostrenca

pamjommu: see dzo

Pampa: see Hereford

Pampiano: (Rio Grande do Sul, Brazil)/m/orig. 1970s from Hereford × Nelore/? = Brazilian Braford/syn. *Pampiano-Braford*

pandzo: see dzo

Panjiang: (SC China)/dr.m/Changzhu group/vars (or local names): Guanling, Longlin, Wenshan

Pankota Red: (Hungary)/m/80–95% Lincoln Red, 5–20% Hungarian Pied/Hung. *Pankotai vörös érteke*

Pantaneiro: (R Paraguay valley, SW Mato Grosso, Brazil)/dun or tan; long horns/Crioulo type/syn.

Cuiabano (from R Cuyabá), *Tucura* (popular name for stunted cattle)/[Port. *pantanal* = marsh]/rare

Pantelleria: (Italy)/d/mixed population, chiefly Modicana, also Italian Brown and Simmental blood

Panwar: see Ponwar

Paphos: (Cyprus)/smaller (hill) var. of Cyprus

Páramo: see Ecuador Criollo

Parda alpina, Parda suiza: see Spanish Brown

Pardo Suizo: (Argentina) see American Brown Swiss

Pare: (Tanzania)/Small East African Zebu type (Tanzania cluster)/rare

Park: see British White, White Park

Parthenais: (Deux-Sèvres, W France)/ m.[d.dr]/fawn/vars: Nantais, Maraîchin/BS 1893/syn. *Choletais, Gâtinais, Gâtine, Gâtinelle, Vendée-Parthenay*/ [from Parthenay]

Pasiega: (E Santander, Spain)/upland var. of Santander/absorbed or displaced by Swiss Brown in 1940s/[from Pas]/extinct

Pasturina: (Casentino, Arezzo, Tuscany, Italy)/local var. from Chianina × Podolian/extinct

Patagonian Criollo: (Los Glaciares National Park, SW Argentina)/ feral var. of Argentine Criollo/Sp. *Criollo argentino patagónico*

Patcholé: see Evolénard

Patuá: (NE Minas Gerais, Rio de Janeiro, Espírito Santo and SE Bahia, Brazil)/grey or yellow-red; small/probably some zebu blood/extinct

Pecanite: (Zimbabwe)/humpless/? Dexter blood/[grazed under pecan trees]/? extinct

Pechora: (Komi, N European Russia)/ d/black-and-white or red-and-white/cf. Vychegda-Vym/var. of Kholmogory (recog. 1972)/orig. 16th to 20th c. from Zyryanka cattle of Komi and from Mezen; improved by Kholmogory 1930–1947/pd var. extinct/Russ. *Pechorskiĭ tip Kholmogorskogo skota*/nearly extinct

Pedi: (Sekhukhuneland, E Transvaal, S Africa)/black with white belly, blue roan, grey, red, or white with black spots/var. of Nguni/ syn. *Bapedi*/rare

Pedreiro: (S Mato Grosso, Brazil)/var. of Crioulo Lageano/[Port. *pedra* = stone]/extinct

Pé duro, Pe-duro: *see* Curraleiro

Pee Wee: (Alberta Univ., Canada)/m/ composite of Charolais, Aberdeen-Angus, Galloway and Hereford, selected for low yearling weight

Peking Black-and-White, Peking Black-Pied: *see* Beijing Black Pied

Peloponnesus: (Greece)/former var. of Greek Shorthorn

Pemba: *see* Unguja

Pembroke: *see* South Wales Black

Peranakan Ongole: *see* Javanese Ongole

Perijanero: (Perija mts, W Venezuela)/m/Criollo type now crossed with zebu to give *Mestizo perijanero* (*mestizo* = mixed, crossbred)

Persian: *see* Iranian

Perugina: (Perugia, Umbria, Italy)/ former var. of Chianina/extinct

Pesisir: (Indonesia)

Pešter: (W Serbia and Montenegro)/ grey/local heavier var. of Buša/ Serbo-cro. *Pešterska buša*

Peuhl: *see* Fulani

Peuhl Voltaïque: (N and C Côte d'Ivoire)/various colours, often with black mucosae/z/Fulani group/? cf. Maure

Peuhl Zebu: *see* Gudali

Peul, Peulh, Peulhé, Peul-Peul: *see* Fulani

Peul nigérian: *see* Diali

Peul Zebu: *see* Gobra, Gudali, Sudanese Fulani

Pezzata degli altipiani: *see* Burlina

Pezzata nera: *see* Italian Friesian

Pezzata rossa friulana: *see* Italian Red Pied

Pezzata rossa norica: *see* Mölltal

Philamin: (Philippines)/m.dr/orig. $\frac{1}{2}$ Hereford, $\frac{3}{8}$ Ongole, $\frac{1}{8}$ Philippine Native/extinct *c*. 1945

Philippine Native: (Philippines)/ dr.m/small hump in ♂/orig. from Chinese and Mexican/vars: Batanes Black, Batanga, Ilocos, Iloilo

pian niu: *see* dzo

Picardy: (France)/d/former var. (by crossing) of Flemish/Fr. *Picarde*/ extinct

Picos de Europa: *see* Lebaniega

Pied Eastern: (France) *see* French Red Pied group

Pied Highland: (Italy) *see* Burlina

Piedmont: (NW Italy)/m.[d.dr]/white or pale grey, with black points/ *brachyceros* type/vars: Albese, Demonte/HB 1887–1891, 1958 on, BS 1934; BS in Brazil 1974, Argentina, USA 1984, Canada; HB in Netherlands/It. *Piemontese*/syn. *Piedmontese*

Piedrahita, Piedrahitense: *see* Avileña, Black Iberian

Piemontese: *see* Piedmont

p'ien niu: *see* dzo

Pie-noire de Belgique: *see* Belgian Black Pied

Pie-noire de l'Est de la Belgique: *see* Hervé Black Pied

Pie-noire Holstein: *see* Belgian Black Pied

Pie rouge Ardennes-Liège: *see* Eastern Red Pied

Pie-rouge de Belgique: *see* Belgian Red Pied

Pie Rouge de Campine: *see* Belgian Red Pied

Pie Rouge de Flandre orientale: *see* Belgian White-and-Red

Pie Rouge de l'Est: *see* Eastern Red Pied

Pie Rouge de l'Est: *see* French Simmental

Pie Rouge des Plaines: *see* French Red Pied Lowland

Pie rouge du Pays de Hervé: *see* Eastern Red Pied

Pie rouge française de Montagne: *see* Abondance

Pin-chou, Pinchow: *see* Pinzhou

Pineywoods: (Georgia, Alabama, Louisiana and Mississippi, USA)/m.[dr]/any colour; usu. hd/sim. to Florida Cracker/ strains incl. Conway (red-and-white), Holt (black coloursided), Griffin (yellow)/BS 1999/syn. *Southern Woods*, *Woods*/ [longleaf pine forests of Gulf coastal plains]/nearly extinct

Pingh: *see* Pingli

Pingli: *see* Bashan

Pinglu Mountain: (S Shanxi, China)/ yellow or red, also black with grey stripes; short legs/ Huanghuai group/syn. *Pinglu*

Pinzbrau: (USA)/Pinzgauer × Brown Swiss cross

Pinzgau de transilvania: *see* Transylvanian Pinzgau

Pinzgauer: (Salzburg, Austria; SE Bavaria, Germany; NE Italy)/ d.m.[dr]/red-brown, coloursided with coloured head/pd var.: Jochberg Hummel/former vars: Mölltal, Pustertaler Sprinzen/ orig. of Slovakian Pinzgau, Slovenian Pinzgau, Transylvanian Pinzgau/BS and HB in Austria, Germany and Italy; BS also in Namibia 1955, S Africa 1963, USA 1973, Brazil 1975, Canada, UK 1976

Pinzhou: (NW Heilongjiang, China)/ d/yellow pied, red pied or black pied/cf. Sanhe/Simmental × local Mongolian crosses/syn. *North Manchurian Dairy*, *Pinchow*

Pirenaica: *see* Pyrenean

Pisana: (Pisa, Tuscany, Italy)/ m.d.[dr]/chestnut to black/? orig. from Swiss Brown × Chianina in mid-19th c./HB 1985/It. *Mucca (nera) pisana* (= *(Black) Pisa milch cow*)/rare (by crossing with Italian Brown)

Pitangueiras: (São Paulo, Brazil)/ d/red; pd/orig. (1944 on) at farm of Frigorífico Anglo, Pitangueiras, from Red Poll ($\frac{5}{8}$) × zebu (chiefly Guzerá and Gir)/BS 1974

Plava: *see* Austrian Blond

Plava povardarska: *see* Macedonian Blue

Pleven, Plevna, Plevne: *see* Turkish Grey Steppe

Po: *see* Modena

Podgórska Red: (Gladyszow, W

Carpathians, Poland)/upland var. of Polish Red/remnants in Polish Beskides rescued 1994, last herd in Hanczowa, base herd formed at Gladyszow/*see also* Polish Red

Podolac: *see* Croatian Steppe

PODOLIAN: (C and S Italy)/m.[dr]/Grey Steppe type/It. *Podlica*/inc. Apulian Podolian, Cinisara, Marchigiana, Maremmana, Modicana, Romagnola

Podolian: (SE Europe) *see* Grey Steppe

Podolian Black Pied: (Kamenets-Podolsk, W Ukraine)/d/orig. from Oldenburg-Wesermarsch × local in late 19th c./absorbed by Ukrainian Black Pied/HB/Russ. *Chernopestraya podol'skaya*/syn. *Ukrainian Oldenburg*/extinct

Podolic, Podolska: (SE Europe) *see* Grey Steppe

Podolica: *see* Podolian

Podolica abruzzese di montagna: *see* Abruzzese

Podolica pugliese: *see* Apulian Podolian

Podolsko goveče: *see* Croatian Steppe

Poggese: *see* Venetian

Pohjoissuomen karja: *see* North Finnish

Polders Black Pied: (W Antwerp and N Flanders, Belgium)/d/= Dutch Black Pied in Belgium/joined with Hervé Black Pied 1966 to form Belgian Black Pied/Flem. *Zwartbont ras van de Polders*, Fr. *Pie-noire des Polders*/extinct

Polesian: (Pripet Marshes, Belarus/ Ukraine)/syn. *Polish Grey*/ extinct

Polessie: (Ukraine)/m/var. of Ukrainian Beef/orig. from Charolais ($\frac{3}{8}$), Russian Simmental ($\frac{3}{8}$) and Aberdeen-Angus ($\frac{1}{4}$); or Charolais ($\frac{3}{8}$), Russian Simmental ($\frac{1}{4}$), Aberdeen-Angus ($\frac{1}{4}$) and Chianina ($\frac{1}{8}$)

Poli: *see* Namshi

Polimska buša: *see* Lim

Polish Black-and-White (Lowland): (C, W and N Poland)/d.m/ Friesian type/orig. from German and Dutch imports in 19th and

20th c./var.: Mazury/Pol.
(*Nizinna*) *czarno-biała*) (= *ncb*)/
syn. *Polish Black Pied, Polish
Friesian, Polish Lowland*

Polish Grey: *see* Polesian

Polish Marsh: (Poland)/d/all-white
var. of Polish Whitebacked (with
colour reduced to few spots on
side)/Pol. *Żuławka*/extinct

Polish Red: (SE Poland)/m.d.(dr)/
Central European Red type/
Danish Red blood 1960s, Angeln
blood since 1980/former vars:
upland (Podgórska Red), lowland
(*dolinowa*) in Białystok,
Rawicka, Silesian Red/BS 1894,
HB 1913; HB also in Ukraine/
Pol. *Polska czerwona*, Russ.
Krasnaya pol'skaya/syn. *Polish
Red-Brown*/purebreds rare

Polish Red-and-White (Lowland):
(SW Poland)/d.m.(dr)/orig. from
German Red Pied and Meuse-
Rhine-Yssel/HB 1910/Pol.
(*Nizinna*) *czerwono-biała*

Polish Simmental: (SE Rzeszów, SE
Poland)/m.d.(dr)/var. of
Simmental/HB 1925/Pol.
Simentalska

Polish Whitebacked: (NE and C
Poland)/d/black or brown,
coloursided/cf. Ukrainian
Whitebacked/var.: Polish
Marsh/Pol. *Białogrzbietka*, Ger.
Rückenscheckig/extinct

polled: *see also* Brazilian Polled,
North European Polled

Polled Aberdeenshire: *see* Buchan
Humlie

Polled Albion: (USA)/orig. in late
19th c./BS/extinct

Polled Angus: *see* Aberdeen-Angus

Polled Charolais: (USA)

Polled Crioulo Pereira Camargo:
(Santa Catarina, Brazil)/d.m/pd
var. of Crioulo/rare

Polled Derby: (England)/? orig. of
mulley/extinct

Polled Durham: *see* Polled Shorthorn

Polled Gir: (Brazil)/pd var. of
Brazilian Gir/HB/Port. *Gir
mocho*

Polled Guzerá: (Brazil)/pd var. of
Guzerá/Port. *Guzerá mocho*

Polled Hereford: (GB) *see* Poll
Hereford

Polled Hereford: (USA and Canada)/
pd var. of Hereford/Single
Standard (orig. *c.* 1893 by
grading up Aberdeen-Angus and
Red Poll, BS 1900) and Double
Standard (orig. 1901 by
mutation, HB 1913)/syn. *Poll
Hereford* (Australia, BS 1920,
and Great Britain, BS 1955)

Polled Irish: *see* Irish Dun, Irish
Moiled

Polled Jersey: (USA)/pd var. of
Jersey/BS 1895

Polled Lincoln Red: (Lincolnshire,
England)/started *c.* 1940 by E.
Pentecost from red Aberdeen-
Angus × Lincoln Red/recog.
1952; now = Lincoln Red/obs.
syn. *Lincolnshire Beef Poll*

Polled Milking Shorthorn: *see*
Milking Shorthorn

Polled Nelore: (Brazil)/pd var. of
Nelore with blood of Brazilian
Polled/HB 1969/Port. *Nelore
mocho*

Polled Norfolk: *see* Red Poll

Polled Scots: *see* Aberdeen-Angus

Polled Shorthorn: (Ohio, USA)/m/
var. of Beef Shorthorn/(1) Single
Standard (now extinct) orig. *c.*
1870 by grading up mulley cows;
(2) Double Standard (purebred)
orig. by mutation in 1880s/BS
1889–1923, HB 1894–1917/syn.
Polled Durham (to 1919)/not *Poll
Shorthorn*

Polled Simmental: (USA)/m/orig. by
grading Angus to Simmental

Polled Sinú: *see* Romosinuano

Polled Suffolk: *see* Suffolk Polled

Polled Sussex: (England and S
Africa)/m/var. of Sussex (1950
on) from red Aberdeen-Angus ×
Sussex

Polled Welsh Black: (Wales)/var. of
Welsh Black recog. late 1940s

Polled Zebu: (Brazil)/inc. Polled Gir,
Polled Guzerá, Polled Nelore,
Tabapuã/Port. *Zebu mocho*/syn.
Indu mocho

Poll Friesian: (Great Britain)/var. of
British Friesian/BS 1960, HB 1961

Poll Hereford: (England)/pd Hereford, orig. from Galloway cross/BS 1949

Poll Hereford: *see also* Polled Hereford

Poll Shorthorn: (Australia)/var. of Beef Shorthorn, first recorded 1874/HB 1935/not *Polled Shorthorn*

Polska czerwona: *see* Polish Red

Pomurska: *see* Murboden

Pontremolese: (Magra valley in Massa and La Spezia, Ligurian Apennines, Italy)/m.d.[dr]/ yellowish-brown with pale dorsal stripe/sim. to Montana/ var.: Bardigiana/BS and HB 1935/syn. *Bettolese* (from Bettola, Piacenze)/[from Pontremoli]/nearly extinct (by crossing with Italian Brown)

Ponwar: (Puranpur, NC Uttar Pradesh, India)/dr/black-and-white/z; Lyrehorned type, sim. to Kherigarh/not *Panwar*

Porto Amboim: (N Angola and SW DR Congo)/beige to brown, black, or pied/Sanga type (Zambia/Angola cluster)/? var. of Barotse/[named after Angolan coastal town]

Portuguese Friesian: *see* Turino

Posavina: (Sava valley, Croatia/ Bosnia)/small var. of Serbian Steppe/Serbo-cro. *Posavska gulja*/syn. *Sava*/extinct

Pothwari: *see* Dhanni

Poul-Foulo, Poulo, Poul-Poulo: *see* Fulani

Prednieper: *see* Dnieper

Preti: (N Italy)/d/cf. Frati/Friesian × Italian Brown 1st cross/[It. *prete* = priest, i.e. black]

Préwakwa: (Cameroon)/m/red or red pied/z/Brahman × Adamawa, F$_1$/orig. of Wakwa/extinct

Pribaltiĭskaya chernopestraya: *see* Baltic Black Pied

Pridneprovskiĭ: *see* Dnieper

Prim'Holstein: *see* French Holstein

Prioka Red: *see* Oka

Priokskaya: *see* Oka

Pritashkentskaya: *see* Bushuev

Priuralskii: *see* Russian Simmental

Privolzhskiĭ: *see* Russian Simmental

Pšenična: *see* Murboden

pseudo-Sanga: (W and C Africa)/term sometimes used for recent zebu × humpless crossbreds (e.g. Ghana Sanga, Borgou, Keteku) to avoid confusion with term "Sanga" for older stabilized crosses (e.g. Abigar, Ankole)/*see also* Ghana Sanga

PSK: *see* North Finnish

Puerto Rican: (West Indies)/some dwarf/Sp. orig.

Puganu: Brazil/rare

Puglia, Pugliese: *see* Apulian Podolian

Pugliese del basso Veneto: *see* Venetian

Pulikulam: *see* Jellicut

Pullo: *see* Fulani

Pul-Mbor: (Cameroon)/Red Bororo × Adamawa, 1st cross/[*Peul* and *Mbor*oro]

Punganur: (Chittoor, S Andhra Pradesh, India)/dr.d/white or grey to red or brown, occ. black/ dwarf var. sim. to Mysore type/ not *Punganoor*/nearly extinct

Purnea: (NE Bihar, India)/dr/black or red/z; hill type, sim. to Morang

Pustertaler Schecken: *see* Pustertaler Sprinzen

Pustertaler Sprinzen: (NE Bolzano, Italy)/sides red-pied or black-pied (not plain red)/var. of Pinzgauer/replaced by Pinzgauer/It. *Pusteria*/syn. *Pusstataler, Pustertaler Schecken* (Ger. = spotted)/nearly extinct

pyaung: *see* Burmese gaur

Pya Zein: (Myanmar)/dr.d/z/orig. from Indian Zebu × Burmese

Pyrenean: (France) *see* Aure et Saint-Girons, Pyrenean Blond

Pyrenean: (Basque country and Navarre, Spain)/m.(d)/red to corn-coloured, paler extremities and underparts/cf. Pyrenean Blond/North Spanish type/ absorbed by Friesian and by Swiss Brown in Vascongadas/HB 1933/Sp. *Pirenaica*/obs. syn. *Basque*/rare

Pyrenean Blond: (Hautes- and Basses-Pyrénées, France)

dr.m.(d)/ yellow to yellow-brown, with pink mucosae/cf. Pyrenean/much crossed with Limousin and Garonnais and absorbed by Blonde d'Aquitaine in 1960s/Fr. *Blonde des Pyrénées à muqueuses roses*/syn. *Basque*/purebred remnants remain as Béarnais

Pyrenean Brown: (Catalonia, Spain)/m/orig. from Pyrenean and Swiss Brown/Sp. *Bruna dels Pirineus*/rare

pyoung, pyun: *see* Burmese gaur

Qiang: *see* Hainan

Qinba: (Shaanxi, China)/*see* Bashan

Qinchuan: (Guanzhong area, C Shaanxi, China)/dr.m/usu. dark purplish-red, also yellow; small cervical hump/Huanghuai group/var.: Zaosheng/BSd 1986/WG *Ch'in-ch'uan*, Russ. *Chinchuan'*, *Tsinchuan*, *Tsin'chun*; formerly *Chinchwan*

Quasah: (Queensland, Australia)/m/red/orig. from Sahiwal × selected Beef Shorthorn/not *Quasar*

Quebec Jersey: *see* Canadienne

Quercy: (Tarn-et-Garonne, France)/m.dr/yellow-brown/orig. late 19th and early 20th c. from Garonnais with Limousin blood; recog. 1920; reunited with Garonnais 1961 to form Blonde d'Aquitaine/syn. *Garonnais du Coteau* (till 1920)/extinct

Quinhentão: (São Paulo, Brazil)/red/orig. about 1870 on farms of Martinico Prado from one Ongole × Friesian ♂ crossed with Franqueiro ♀♀/extinct

Race de la Moyenne et Haute Belgique: *see* Belgian Blue

Race espagnole: *see* Fighting bull

Raetian Grey: *see* Rhaetian Grey

Rahadji, Rahaji, Rahaza: *see* Red Bororo

Ramgarhi: (E Mandla, Madhya Pradesh, India)/z; local var.

Ramo Grande: (Ilha Terceira, Azores, Portugal)/red-brown or white/

orig. from Alentejana, Mirandesa and Shorthorn/HB 1994/nearly extinct

Rana: (C Madagascar)/d.m/often speckled fawn, grey or black pied/orig. from Bordelais, Gascon, Garonnais, Breton Black Pied and Normande (imported 1926–1930) and Friesian (imported from 1945), × Madagascar Zebu/syn. *omby rana* (Swahili *ngombe* = cattle)/rare

Randall Lineback: (E USA)/var. of American Lineback/orig. 18th and 19th c. as American landrace; bred at Clayton Randall estate as closed herd for 80 years from early 20th c.; rescued from extinction 1985 by Cynthia Creech of Johnson City, Tennessee/BS briefly early 20th c./syn. *Randall Blue Lineback*/nearly extinct

Ranger: (W ranges, USA)/m/orig. 1970 by combination of Hash Cross with Ritchie herd (Wyoming) which used Simmental (1968), Hash Cross (1967), Beefmaster (1966), Hereford, Brahman and Highland × Shorthorn (1950) bulls on Brahman cross cows, and with Watson herd (California) using American Brown Swiss (1970), Red Angus × Red Holstein (1969), Beefmaster (1968), Hash Cross (1968) and dairy breed bulls (1967) on Hereford cows

Rasa (românească) de munte: *see* Romanian Mountain

Ras van Midden- en Hoog-België: *see* Belgian Blue

Rath: (Alwar, E Rajasthan, India)/dr.d/z/Grey-white Shorthorned type; sim. to Hariana/not *Rathi*/[Rath are nomadic cattle breeders)]

Rathi: (Bikaner and Ganganagar, NW Rajasthan, India)/dr.d/usu. dark red or tan, occ. spotted/z/orig. from Sahiwal and Tharparkar

Ratinha: *see* Mirandesa

Rätisches Grauvieh: *see* Rhaetian Grey

Raukolle, Rautt kollet Østlandsfe: *see* Red Polled Eastland

Rautt trønderfe og målselvfe: *see* Red Trondheim

Rawicka: (Rawicz, Poznań, Poland)/ former var. of Polish Red/extinct

Raya: *see* Danakil

Raya-Azebó: (E Tigray, Ethiopia)/dr/ Sanga (Abyssinian group)/larger var. of Danakil selected for dr/syn. *Galla-Azebó*

razzetta delle dune: *see* Gasara

RDD: *see* Dutch Red Pied Dual-Purpose

RDM-1970: (Denmark)/d.m/purebred line of old Danish Red/[= *R*ødt *D*ansk *M*alkekvæg + year 1970 to distinguish old type from new synthetic Danish Red breed]

Reconstituted Aurochs: *see* Heck cattle

Red: *see* European Red

Red: (Germany) *see* German Red

Red Aberdeen-Angus: *see* Red Angus

Red Afrikaner: *see* Afrikander

Red-and-White: (Netherlands) *see* Red Pied Friesian

Red-and-White Campine: *see* Belgian Red Pied

Red-and-White Dairy Cattle: (USA)/ BS 1964 to register Red-and-White Holsteins (90%) and other red-and-white dairy cattle (e.g. Ayrshire, Guernsey, Milking Shorthorn, Danish Red crosses)

Red-and-White East Flemish: *see* Belgian White-and-Red

Red-and-White Finnish: *see* East Finnish

Red-and-White Friesian: (São Paulo and Minas Gerais, Brazil)/Port. *Holandesa vermelha (e branca)*

Red-and-White Friesian: (Great Britain)/BS and HB 1951 for red var. of British Friesian

Red-and-White Friesian: *see also* Red-and-White Holstein, Red Pied Friesian

Red-and-White Holstein: (USA)/ d/colour var. of Holstein/syn. *Holstein-Friesian Red, Red Holstein*

Red-and-White Hungarian: *see* Hungarian Pied

Red-and-White Meuse-Rhine-Yssel: *see* Meuse-Rhine-Yssel

Red Angeln: *see* Old Red Angeln

Red Angus: (USA and Australia)/ colour var. of Aberdeen-Angus/ BS in USA 1954, Australia 1960s/syn. *Aberdeen-Angus colorado* (Argentina), *Red Aberdeen-Angus*

Red Astrakhan: *see* Kalmyk

Red Bavarian: *see* Bavarian Red

Red Belted Galloway: (Great Britain)/ colour var. of Belted Galloway/rare

Red Berrendo: *see* Berrenda en colorado

Red Bessarabian: *see* Bessarabian Red

Red Bororo: (E Niger, N Nigeria, W Chad and N Cameroon)/m/ mahogany/West African (long lyrehorned) Zebu type (Fulani group)/pl. *Bororodji*; Hausa *Abori, Rahadji* or *Rahaji*, Fulani *Bodadi* (Wodabe tribe)/syn. *Brahaza, Djafoun* (Cameroon), *Fellata* (Chad and Ethiopia), *Fogha, Gabassaé, Gadéhé, Hanagamba, Kréda* (Chad), *Mbororo, Rahaza, Red Fulani, Red Longhorn*/not *Boro, Borroro*

Red Brahman: (USA)/coat red or spotted, skin black or brown; long pendulous ears (cf. broad drooping in Grey Brahman); Gir-type horns/orig. from red Gir and Kankrej (Guzerat) imported from Brazil

Red Brangus: (Texas, USA)/m/pd/ colour var. of Brangus/orig. from Brahman × American Angus/BS 1956

Red-Brown East Friesian: *see* Red East Friesian

Red Butana: *see* Butana

Red Colonist: *see* Red Steppe

Red Crimean: *see* Red Steppe

Red Croatian: *see* Croatian Red

Red Dane, Red Danish: *see* Danish Red

Red Desert: (N Sudan)/inc. Bambawa, Butana, Dongola, Shendi

Red Devon: *see* Devon

Red East Friesian: (Germany)/d.m/ Ger. *Rote Ostfriesen*/syn. *Red-*

Brown East Friesian/extinct *c.* 1940

Red Finnish: *see* West Finnish

Red Flemish: *see* Belgian Red

Red Friesian (Indonesia): *see* FH-merah

Red Friuli: *see* Friuli

Red Fulani: *see* Red Bororo

Red Galloway: (Great Britain)/colour var. of Galloway/HB (with Galloway); HB (with Galloway) also in USA/rare

Red German: (Ukraine and Russia) *see* Red Steppe

Red Gorbatov: *see* Gorbatov Red

Red Hill: *see* German Red

Red Holstein: (Switzerland) *see* Simmental (section in HB)

Red Holstein: (USA) *see* Red-and-White Holstein

Red Kandhari: (EC Maharashtra, India)/dr.d/z/local var. sim. to Deoni but red/syn. *Lakhalbunda* (Berar)/not *Kandahari, Khandari*

Red Karachi: *see* Red Sindhi

Red Longhorn: *see* Red Bororo

Red Metohian: *see* Metohija Red

Red Mountain: (Germany) *see* German Red

Red Mountain: (Italy) *see* Montana

Red Nelore: (Brazil)/var. of Nelore/Port. *Nelore vermelho*

Red Norfolk: *see* Norfolk Horned

Red North Slesvig: *see* North Slesvig Red

Red Pied Aosta: *see* Aosta Red Pied

Red Pied Campine: *see* Belgian Red Pied

Red Pied Danish: *see* Danish Red Pied

Red Pied Dutch: *see* Dutch Improved Red Pied, Dutch Red Pied Dual-Purpose, Meuse-Rhine-Yssel, Red Pied Friesian

Red Pied East Friesian: (East Friesland, Germany)/d.m/var. of German Red Pied/HB 1878/Ger. *Rotbunte Ostfriesen*/syn. *Red-White East Friesian*/extinct

Red Pied Flemish: *see* Belgian White-and-Red

Red Pied Friesian: (Friesland, Netherlands)/d.(m)/orig. from Dutch Black Pied free of Holstein blood and carrying red factor/BS 1959, HB 1984/Du. *Roodbonte*

Fries, Fries Roodbonte/syn. *Red-and-White*/nearly extinct

Red Pied Friuli: *see* Italian Red Pied

Red Pied Holstein: *see* Red Pied Schleswig-Holstein

Red Pied Karelian: *see* East Finnish

RED PIED LOWLAND: (NW Europe)/inc. Belgian Red Pied, French Red Pied Lowland, German Red Pied, Meuse-Rhine-Yssel, Polish Red-and-White Lowland/Ger. *Rotbuntes Niederungsvieh*

Red Pied Lowland: (France) *see* French Red Pied Lowland

Red Pied Lowland: (Germany) *see* German Red Pied

Red Pied Noric: *see* Mölltal

Red Pied Schleswig-Holstein: (NW Germany)/d.m/var. of German Red Pied (? with Shorthorn blood)/orig. by union of Breitenburger, Dithmarscher, Holstein Geest, Wilstermarsch/BS and HB 1875, BSd 1934/Ger. *Rotbunte Schleswig-Holsteiner*/syn. *Holstein Marsh, Holstein Red Pied, Red Pied Holstein*/extinct

Red Pied Swedish: (Sweden)/d/orig. in late 19th c. from Ayrshire and Shorthorn × local (Herrgård and Småland)/joined with Swedish Ayrshire in 1928 to form Swedish Red-and-White/BS and HB 1892–1928/Swe. *Rödbrokig Svensk Boskap* (= *RSB*)/extinct

Red Pied Valdostana: *see* Aosta Red Pied

Red Pied Westphalian: (Münsterland, W Germany)/d.m/var. of German Red Pied/orig. from Rhineland and Meuse-Rhine-Yssel; also has Holstein, German Red Pied and Dutch Black Pied blood/HB 1892/Ger. *Rotbuntes Westfälisches, Rotbunte Westfalen*

Red Polish: *see* Polish Red

Red Poll: (E England)/m.d/orig. in early 19th c. from Suffolk Polled and Norfolk Horn/part orig. of Jamaica Red, La Velásquez, MARC III, Pitangueiras, Senepol/recog. 1847, HB 1873, BS 1888; BS also Australia, Brazil, Canada 1905, Colombia, New Zealand, S

Africa 1921 (HB 1906), USA 1883
(HB 1887), Zimbabwe 1934/
syn. *Rooi Poenskop* (Afrik.)/obs.
syn. *Norfolk Polled, Norfolk and
Suffolk Red Polled* (to 1882), *Red
Polled* (to 1908)/ rare in GB
Red Polled: (Norway) *see* Red Polled
Eastland
Red Polled: (Sweden) *see* Swedish
Red Polled
Red Polled Eastland: (SE Norway)/
d.m/red, occ. with white
"headscarf"/sim. to Swedish Red
Polled/Swedish Red-and-White
(1930s) and Ayrshire (1958) and
Dutch Black Pied blood/
coloursided var.: Jarlsberg/
combined with Norwegian Red-
and-White 1961 to form
Norwegian Red/recog. 1892,
HB/Nor. *Raukolle, Rødkolle,
Rautt* (or *Rødt*) *kollet Østlandsfe,
Østlansk rødkolle*/syn. *Eastland,
Norwegian Red Polled, Østland,
Red Polled, Red Polled Østland*/
not *Red Poll*/nearly extinct
Red Ruby: *see* Devon
Red Sadovo: (Plovdiv, Bulgaria)/
d/orig. (1883 on) at Sadovo
Agric. School from Angeln ×
Simmental and Friesian/
absorbed in Bulgarian Red *c.*
1960/Bulg. *Chervena sadovska*,
Ger. *Rotes Sadowo*/extinct
Red Senegal: *see* N'Dama Petite
Red Silesian: *see* Silesian Red
Red Sindhi: (W Sind, Pakistan)/
d/z/var.: Las Bela/part orig. of
Australian Milking Zebu,
Brownsind, Jersind/BSd and HB;
also HB India, BS Brazil 1961,
BS/HB Australia (from imports
in 1950s)/syn. *Malir*
(Baluchistan), *Red Karachi,
Sindhi, Sindi* (Brazil)/not
Scindhi, Scindi, Sindy Red
Red Speckled: *see* Simmental
Red Spotted: (France) *see* French
Simmental
Red Spotted (Highland): *see*
Simmental
Red Spotted Moravian: *see* Moravian
Red Pied
Red Steppe: (China) *see* Caoyuan Red
Red Steppe: (Ukraine and S European

Russia)/d/orig. by Mennonites
1789–1824 from Red East
Friesian and Angeln, ×
Ukrainian Grey with later some
blood of East Friesian, Swiss
Brown *et al.*/inc. Bessarabian Red
and Crimean Red/HB 1923/Russ.
Krasnaya stepnaya/ obs. syn. *Red
Colonist* (Russ. *Krasnaya
kolonistskaya*), Red German
(Russ. *Krasnaya nemetskaya*) (to
1941), *Red Ukrainian* (Russ.
Krasnaya ukrainskaya)
Red Tambov: *see* Tambov Red
Red Trondheim: (Norway)/d.m/orig.
1850–1891 from Ayrshire ×
local (Røros, Tronder)/absorbed
smaller Målselv in 1951/graded
to Norwegian Red-and-White
and finally absorbed 1960/HB
1951/Nor. *Rødt* (or *Rautt*)
trønderfe og målselvfe/not
Drontheimer, Trondhjem/extinct
Red Ukrainian: *see* Red Steppe
Red West Flemish: *see* Belgian Red
Red-White: *see* Red Pied, Red-and-
White
Red White-Russian: *see* Belarus Red
Red Wittgenstein: (Wittgenstein and
Nordrhein-Westphalia,
Germany)/dr.m.d/hill cattle/
nearly extinct
reem: *see* aurochs
Reggiana: (Reggio, Emilia, Italy)/
d.m.[dr]/red to yellow-brown/HB
1935/syn. *Formentina*/rare
Regus: (Wyoming, USA)/m/red; pd/
herd of Red Angus developed at
Beckton Stock Farm, Sheridan,
by grading Hereford to Angus
Rendena: (Brenta valley, Padua-
Vicenza, Venetia, Italy)/d.(m)/
dark chestnut/small local var. sim.
to Italian Brown/being crossed
with Italian Brown/HB 1982/syn.
Bruna di Val di Rendena
Renitelo: (Kianjasoa region,
Madagascar)/m/red, often with
paler underside and mucosae/
orig. from Limousin (1930–1962)
(25%) and Afrikander (1946–
1962) (48%), × Madagascar Zebu
(27%); crosses bred *inter se* since
1956/breed recog. 1962/[= 3
mothers]/rare

Restagi, Restaki: *see* Rustaqi

Retinta: (Extremadura and W
Andalucía, Spain)/m.(dr)/dark
red; open lyre horns/orig. by
union of Andalusian Red,
Extremadura Red, Andalusian
Blond/HB 1933, BS; BS also
Argentina/[= dark red]

Reyna: *see* Central American Dairy
Criollo

Rhaetian Grey: (NE Graubünden,
Switzerland)/d.m/cf. Tyrol
Grey/orig. from mt cattle
imported from Austria/BS
1990/Ger. *Rätisches
Grauvieh*/not *Raetian*/rare

Rhineland: (W Germany)/d.m/var. of
German Red Pied very close to
Meuse-Rhine-Yssel/HB 1875/
Ger. *Rheinisches, Rheinland*/syn.
Lower Rhine (Ger. *Niederrheiner,
Niederrheinisches*)/extinct

Rhodope: *see* Rodopi

Ribatejana: (Portugal) *see* Fighting bull

rigget: *see* coloursided

Rigi: *see* Swiss Brown

rimu: *see* aurochs

Ringamåla: (S Sweden)/d/sim. to old
type (1940s) of Swedish Red-
and-White/nearly extinct

Rio Limón Dairy Criollo:
(Venezuela)/red/var. of Tropical
Dairy Criollo/BS/Sp. *Criollo
lechero limonero*/syn. *Criollo
lechero venezolano* (=
Venezuelan Dairy Criollo),
Improved Criollo

Riopardense: (São José do Rio Pardo,
São Paulo, Brazil)/d/usu. black,
also black pied, or black, blue
roan and white/orig. (1953 on)
by Osmany Junqueira Dias at
Fazenda Graminha from Holstein
($\frac{5}{8}$) × Guzerá ($\frac{3}{8}$)

Ristagi: *see* Rustaqi

Rjavo govedo: *see* Slovenian Brown

Rødbrokig Svensk Boskap: *see* Red
Pied Swedish

Rød Dansk Malkekvæg: *see* Danish
Red

Rødkolle: *see* Red Polled Eastland

Rödkulla: *see* Swedish Red Polled

Röd Kullig Lantras: *see* Swedish Red
Polled

Rodopi: (SW Bulgaria)/brown-black
with white eel-stripe; lyre horns;
dwarf/prim. *brachyceros* type;
cf. West Macedonian/HB 1956/
Bulg. *Rodopska k"soroga*, (=
shorthorned), Ger. *Rhodopen*/not
Rodopa, Rodopit/nearly extinct

Rødt Dansk: *see* Danish Red

Rødt kollet Østlandsfe: *see* Red
Polled Eastland

Rødt trønderfe og målselvfe: *see* Red
Trondheim

Rojhan: (Dera Ghazi Khan, Punjab,
Pakistan)/usu. red with white
markings/z/hill type, sim. to
Lohani

Romagnola: (Romagna, Italy)/m.[dr]/
grey to white/Podolian type/
former vars: improved (It.
gentile) (with Chianina and
Reggiana blood 1850–1900),
mountain (It. *di montagna*) (with
Maremmana blood)/HB 1956; BS
in GB 1973, Canada 1974, USA
1974, Australia, New Zealand,
Argentina

Roman: (Italy)/former var. of
Maremmana/It. *Romana*/?
extinct

Romana Red: (Dominican Republic)/
m/z/orig. as dr in 1920s by
Central Romana Corporation
from Mysore, Nelore and Guzerá
zebus, × red Puerto Rican
Criollo/Sp. *Romana Roja*

Romanian Black Pied: *see* Romanian
Holstein

Romanian Brown: (S and E
Romania)/d.m/shades of
brown/orig. early 20th c. from
Grey Steppe graded to Swiss
Brown/used to grade Mocaniţa
and Grey Steppe in S/Rom.
Brună/syn. *Maramureş Brown*/
not *Grey*

Romanian Dwarf: (Romania)/
indigenous dwarf population of
Romanian Grey/syn. *Mocaniţa*/
nearly extinct

Romanian Grey: (Moldau, Donau-
Delta, Romania)/dr.m.d/grey
with black muzzle; huge horns/
Grey Steppe type/almost extinct
1950s by crossing with

Simmental, Swiss Brown and
Romanian Red (chiefly Danish
Red)/former vars: Bucşan,
Ialomiţa, Moldavian,
Transylvanian/Rom. *Surǎ de
Stepǎ* (= *Grey Steppe*)/syn.
*Romanian Grey Steppe,
Romanian Steppe*/rare
Romanian Holstein: (Romania)/
d.m/orig. from Holstein ×
Romanian Brown *et al.*/HB 1970/
Rom. *Bǎlţatǎ cu negru româneascǎ*
(= *Romanian Black Pied*)
Romanian Holstein-Friesian:
(Romania)/orig. imported 19th
century
Romanian Mountain: (Romania)/
d/light grey, red-brown or dark
grey/*brachyceros* type/almost
extinct 1950s by crossing with
Romanian Brown/Rom. *Rasa
(românescǎ) de munte*/syn.
Mocaniţa (= mountain peasant,
shepherd) (not *Mocanitza*),
Mountain Grey
Romanian Pinzgau: *see*
Transylvanian Pinzgau
Romanian Red: (SE Romania)/d/inc.
Danish Red, Dobrogea Red and
Red Steppe/Rom. *Roşie*/declining
Romanian Simmental: (Romania)/
m.d/yellow pied to red pied/orig.
from Grey Steppe graded to
Simmental (imports *c.* 1910–
1948)/HB 1914/Rom. *Bǎlţatǎ
românescǎ* (= *Romanian
Spotted*)/syn. *Romanian Yellow
Spotted*
Romanian Spotted: *see* Romanian
Simmental
Romanian Steppe: *see* Romanian Grey
Romark: (Canada)/Romagnola ×
Marchigiana cross
Romo: *see* Romosinuano
Romosinuano: (R Sinú valley, N
Colombia)/m/yellow to red-
brown; pd/Criollo type/orig. end
of 19th c. from Costeño con
Cuernos, ? by mutation or with
Red Poll or Red Angus blood/BS
1976; HB also Costa Rica/syn.
*Coastal Polled, Moruno-Sinuano,
Romo*/not *Romo-Sinuano*/[=
polled Sinú, *romo* = blunt]

Roodblaar: *see* Groningen
Whiteheaded
Roodbont Dubbeldoel: *see* Dutch Red
Pied Dual-Purpose
Roodbonte Fries: *see* Red Pied Friesian
Roodbont Kempisch: *see* Belgian Red
Pied
Roodbont Maas-Rijn-IJssel: *see*
Meuse-Rhine-Yssel
Roodbont ras van België: *see* Belgian
Red Pied
Roodbont ras van Ooost-Vlaanderen:
see Belgian White-and-Red
Rood ras van België: *see* Belgian Red
Rood ras van West-Vlaanderen: *see*
Belgian Red
Roodwitkop: *see* Groningen
Whiteheaded
Rooi Poenskop: *see* Red Poll
Røros: (S of Trondhjem, NC Norway)/
d/coloursided (white with red,
grey or black sides); pd/extinct/
see also Blacksided Trondheim
and Nordland
Roşie: *see* Romanian Red
Roşie Dobrogeanǎ: *see* Dobrogea Red
Rosíe estona: *see* Moldovian
Estonian Red
Rossa montanina: *see* Montana
Rotbunt, Rotbunte: *see* Red Pied
Rotbunte holländische: *see* Meuse-
Rhine-Yssel
Rotbunte Ostfriesen: *see* Red Pied
East Friesian
Rotbunte Schleswig-Holsteiner: *see*
Red Pied Schleswig-Holstein
Rotbuntes Niederungsvieh: *see*
German Red Pied, Red Pied
Lowland group
Rote Ostfriesen: *see* Red East Friesian
Rotes Höhenvieh: *see* German Red
Rotes Nordschleswiger: *see* North
Slesvig Red
Rotes Sadowo: *see* Red Sadovo
Rotfleckvieh: *see* Simmental
Rottal: (Bavaria, Germany)/absorbed
by German Simmental/extinct
Rotvieh alter Angler Zuchtrichtung:
see Old Red Angeln
Rotvieh Zuchtrichtung Höhenvieh:
see German Red (Highland type)
Rouge de Belgique: *see* Belgian Red
Rouge de la Flandre occidentale: *see*
Belgian Red

Rouge de l'Ouest: *see* French Western Red

Rouge du Nord: *see* Flemish Red

Rouge flamande: *see* Flemish Red

Rouge-pie Campinoise: *see* Belgian Red Pied

Rowzi, Rozi: *see* Barotse

RSB: *see* Red Pied Swedish

Rubia andaluza: *see* Andalusian Blond

Rubia gallega: *see* Galician Blond

Rückenblessen, Rückenscheck: *see* coloursided, whitebacked

Rückenscheckig: *see* Polish Whitebacked

Rufa'ai: *see* Kenanana

Ruggelde, Ruggelds, Ruggelings: *see* Witrik

Russian Black Pied: (Russia)/d/vars: Central Russian Black Pied, Siberian Black Pied, Ural Black Pied/recog. 1925, national strains united 1959, HB 1940/Russ. *Chernopestraya*/syn. *Russian Friesian*

Russian Brown: (Russia)/orig. from Swiss Brown and German Brown first imported early 19th c./orig. of Ala-Tau, Carpathian Brown, Caucasian Brown, Kostroma, Lebedin; part orig. of Schwyz-Zeboid/HB/Russ. *Shvitskaya* (= *Schwyz, Swiss*)

Russian Brown-zebu: *see* Schwyz-Zeboid

Russian Friesian: *see* Russian Black Pied

Russian Simmental: (Russia and Ukraine)/orig. from Swiss and German imports early 19th c. and Austrian late 19th c./vars: Far Eastern, Siberian, Steppe, Sychevka, Ukrainian, Ural (*Priuralskiĭ*), Volga (*Privolzhskiĭ*)/HB 1925

Russo-Siberian: (Russia)/d/var. of Siberian/extinct

Rustaqi: (Hilla, Iraq)/light tan/ improved var. of Iraqi/orig. from Jenubi, ? with Red Sindhi blood/ not *Restagi, Restaki, Ristagi, Rustagi*/[from government farm at Rustaq, nr Baghdad]

Ruzizi: (Ruzizi valley, Kivu region, Rwanda-Burundi; also DR Congo)/m/usu. brown, also red, red pied, black pied; long horns/Ankole group/taller var. of Watusi/syn. *Rwanda-Burundi*

RX3: (Iowa, USA)/½ Red Angus, ¼ Hereford, ¼ Red Holstein/HB 1974

Rwanda: *see* Watusi

Rwanda-Burundi: *see* Ruzizi

Saavedreño: (Santa Cruz, Bolivia)/ d/red to beige/var. of Bolivian Criollo/orig. 1970–1990 at Saavedra Exp. Sta. by selection from Bolivian Criollo, and also dairy Criollos from Costa Rica, Nicaragua, Cuba and Brazil/syn. *Santa Cruz*/rare

SAB: *see* Swedish Ayrshire

Sabre: (Lambert ranch, Refugio, Texas)/m/red/orig. (1950 on) from Sussex (⅝) × Brahman (⅜)/rare

Sachor: *see* Sanchori

Sādhe: *see* Brahmini

Sahabadi: *see* Shahabadi

Sahford: (Australia)/Sahiwal × Hereford cross

Sahiwal: (S Punjab, Pakistan)/d/usu. reddish dun with white markings/ z/part orig. of Australian Frieswal, Australian Milking Zebu, Australian Sahiwal, Frieswal, Jamaica Hope, Karan Swiss, Mpwapwa, Quasah, Taurindicus *et al.*/BSd and HB (India); BS Australia 1969, Kenya/syn. *Lambi Bar, Lola, Montgomery, Multani* (Bihar), *Teli*/not *Sanewal, Saniwal*

Saidi: (Upper Egypt)/dr.m/var. of Egyptian with more zebu blood/ not *Saiidi*/[= valley, i.e. of Nile]

saing: *see* tsine

St Croix: *see* Senepol

St Girons et Aure: *see* Aure et Saint-Girons

Sakalava: (Madagascar)/? Sanga/ extinct

saladang: *see* seladang

Salamanca: *see* Morucha

Salamanca Brindled: (Spain)/m/? orig. from Swiss Brown × Morucha/Sp. *Atigrado de Salamanca*/nearly extinct

Salem: *see* Alambadi

Salerford: (USA)/Salers × Hereford cross

Salers: (Cantal, Auvergne, France)/ m.[d.dr]/mahogany/recog. 1853; BS 1908; BS also in Canada 1974, USA 1974, UK

Salinera: *see* Berrenda en colorado

Salmanquina, Salmantina: *see* Morucha

Salorn: (Texas, USA)/m/orig. since 1980s from Salers × (Salers × Texas Longhorn) bred to (Salers × Texas Longhorn), i.e. ⅝ Salers, ⅜ Texas Longhorn

Samburu: (C Kenya)/local Small East African Zebu (Kenya cluster)

Sanabresa: *see* Alistana-Sanabresa

Sanchori: (Jodhpur, Rajasthan, India)/dr.d/z/var. of Kankrej/syn. *Marwari*/not *Sachor, Sanchore*

Sandhe: *see* Brahmini

Sanewal, Saniwal: *see* Sahiwal

SANGA: (E and S Africa)/dr/long horns; small cervico-thoracic hump/? orig. from zebu and Hamitic Longhorn/inc. Abyssinian group (Danakil, Raya-Azebo), Afrikander, Ankole group (Bahima, Bashi, Kigezi, Ruzizi, Watusi), Barotse, Barra do Cuanzo, Basuto, Bavenda, Mashona, Mateba, Nguni group (Landim, Nguni, Nkone, Pedi, Shangan), Nilotic (Abigar, Aliab Dinka, Aweil Dinka, Nuer, Shilluk), Ovambo cluster (Caprivi, Humbi, Kaokoveld, Okavango, Ovambo), Setswana group (Barotse, Damara, Tswana, Tuli), Zambia/Angola cluster (Porto-Amboim, Tonga)/syn. *Bantu*/[Galla, = ox]

Sanga: (W Africa) *see* Ghana Sanga

Sanganer: (S Africa)/m/various colours/dam-line breed in formation from Afrikander and Nguni/BS and HB/rare

SANGA-ZEBU: (E Africa)/intermediates between Sanga and East African Shorthorned Zebu/inc. Alur, Aradó, Bovines of Tete, Fogera, Horro, Jiddu, Nganda, Sukuma/syn. *Zenga*

Sanhe: (NE Inner Mongolia, China)/ d.m/usu. red-and white, also black-and-white/cf. Pinzhou/ orig. *c.* 1900 from Siberian, Simmental and Friesian, × Mongolian/WG *Sanho*/syn. *Three-river breed*

Sanjiang: (valley of Minshan range, NW Sichuan, China)/yellow/rare

San Martinero: (Meta and Caquita provs, C Colombia)/m.(d)/ yellow-red or chestnut/Criollo type sim. to Costeño con Cuernos but better conformation/HB 1960/syn. *San Martin, Sanmartiniana, SM*/rare (disappearing by crossing with Brahman)

Sanaogai: *see* Achham

Santa Catarina: *see* Crioulo Lageano

Santa Clara: (Rio Grande do Sul, Brazil)/m/? pd/orig. since *c.* 1960 by R.S. Vasconcelos at Santa Clara ranch, Rosário do Sul, from zebu (⅜) (first Brahman, later Nelore, Tabapuá and Polled Zebu) × Polled Hereford (⅝)

Santa Coloma: (Colombia)/strain of fighting cattle

Santa Cruz: *see* Saavedreño

Santa Gabriela: (NE São Paulo, Brazil)/d.m/usu. red, sometimes yellow; pd/orig. (1965 on) at Estação Experimental de Zootecnia de Sertãozinho (formerly Fazenda Santa Gabriela) from Red Pied Friesian × (red pd zebu × Devon-Guzerá), i.e. ⅜ zebu, ⅝ European/ discontinued

Santa Gertrudis: (Texas, USA)/m/ red; hd or pd/orig. 1910–1940 from Shorthorn (⅝) and Brahman (⅜)/BS 1951; BS also in Brazil 1961, Canada 1967, S Africa 1974, Mexico, Australia, Zimbabwe, New Zealand, Argentina; HB also in Russia/[ranch estab. 1854 in Texas by Capt Richard King]

Santander: (Spain)/m.d.dr/North Spanish type/vars: Campurriana, Lebaniega, Pasiega, Tudanca/syn. *Montaña*/extinct (except Tudanca)

Santander Hairless: *see* Chino
Santandereano
Saônaise: (Maine-Anjou-Perche,
France)/? orig. from local
Mancelle crossed with Durham/?
extinct or nearly extinct
sapi perahan Grati: *see* Grati
sapi utan: *see* Malay banteng
Sarabi: (Iranian Azerbaijan)/
d.m.(dr)/red/? zebu blood/syn.
Ardebili, Sarabian/not *Saribi*
saran-hainag: *see* dzo
Sardinian: (Sardinia, Italy)/m.d.[dr]/
many colours, self or pied, ♂
often black or dark red, ♀ often
red or yellow-brown, occ.
brindle/cf. Corsican/orig. type
now only in mts/orig. (with
Italian Brown) of Sardinian
Brown and (with Modicana) of
Sardo-Modicana/It. *Sarda*
Sardinian Brown: (C Sardinia, Italy)/
d/var. of Italian Brown by
grading Sardinian *c.* 1880 on/It.
Bruno-sarda/syn. *Ozierese* (from
Ozieri), *Sardo-Schwyz, Sardo-
Swiss, Svitto-Sarda*
Sardo-Modicana: (S and E Sardinia,
Italy)/m.d.[dr]/yellow-brown to
dark red/orig. from Modicana ×
Sardinian (1860 on)/HB 1936/
syn. *Modica-Sardinian,
Modicano-Sarda, Oristanese*
(from Oristano)
Sardo-Schwyz, Sardo-Swiss: *see*
Sardinian Brown
Sari Alaca: (Turkey)/at Kazova
(Tokat) State farm
Şarkî Anadolu kırmızı: *see* East
Anatolian Red
Sauerland: *see* Westphalian Red
Sava: *see* Posavina
Savanna Muturu: (Côte d'Ivoire to
Cameroon)/m/larger type of
Muturu (cf. smaller Forest
Muturu)
Savanna Shorthorn: *see* West African
Shorthorn (Savanna vars)
Save: *see* Nguni
Savinja Grey: (Upper Savinja, N
Slovenia)/d.m/orig. from
Austrian Blond and Istrian
improved by Slovenian Brown
and American Brown Swiss/

Serbo-cro. *Sivka iz Gorne
Savinje*
Savolarda, Savoy: *see* Tarina
Sayaguesa: (Sayago, SW Zamora,
Spain)/m.dr/black, with brown
back stripe in ♂/Morenas del
Noroeste group/syn. *Zamorana*
Scanian: *see* Skåne
Schecken und Blässen: *see* Mottled
Hill
Scheckig: *see* Black Forest
**Sheckiges und rückenblessiges
Höhenvieh:** *see* Mottled Hill
Scheckvieh: *see* German Simmental
Scheinfeld: (Germany)/orig. in mid-
19th c. from Heilbronn × native
Franconian red/extinct (absorbed
by Yellow Franconian)
Schiltern: *see* Waldviertel
Schlesisch: *see* Silesian
Schlesisches Rückenscheck: *see*
Kłodzka
Schleswigsche Marschrasse: *see*
Ballum
Schiltern: *see* Waldviertel
Schönhengst: *see* Hřbínecký
Schwarzbunt: *see* Black Pied
Schwarzbuntes Milchrind: *see*
German Black Pied Dairy
Schwarzbuntes Niederungsvieh: *see*
German Black Pied
Schwarzfleckvieh: (Switzerland) *see*
Swiss Holstein
Schweizerisches Braunvieh: *see*
Swiss Brown
Schweizer Rotfleckvieh: *see*
Simmental
Schwyz: *see* Swiss Brown
Schwyz española: *see* Spanish
Brown
Schwyz-Zeboid: (Tajikistan)/m.d/
orig. (1937 on) from Russian
Brown × Tajik zeboid/Russ.
Shvitsezebuvidnyǐ skot/syn.
Russian Brown-zebu, Swiss-zebu
Scindhi, Scindi: *see* Red Sindhi
Scotch Highland, Scottish Highland:
see Highland
**Scotch Shorthorn, Scottish
Shorthorn:** *see* Beef Shorthorn
Scotch Polled: *see* Aberdeen-Angus,
Galloway
Scrub: (USA) *see* Florida Cracker
Scutari: *see* Shkodra

SDM: *see* Danish Black Pied
SDM-1965: (Denmark)/d.m/Friesian line (Danish Black Pied) without Holstein blood/nearly extinct
Sechsämter: *see* Bavarian Red
Sechuana: *see* Tswana
Seferihisar: (Izmir, W Turkey)/d/red/ local var./? orig. from Simmental × Aleppo (i.e. Damascus)/extinct
Seistani: *see* Sistani
Sekgatla: (Bakgatla Reserves, S Botswana)/Sanga (Setswana group)/inc. in Southern Tswana/rare (from interbreeding with Tswana and crossbreeding with Afrikander)
seladang: (Malay peninsula)/= *Bos (Bibos) gaurus hubbacki* (Lydekker)/var. of gaur/♂ being crossed experimentally with Holstein-Friesian at Central Animal Res. Inst., Kluang, Johor/syn. *Malayan bison, Malayan gaur*/not *saladang, sladang*/[Malay]/rare
selembu: (Malaysia)/= seladang × cattle hybrid/cf. Jatsa, Jatsum
Selenge: (N Mongolia)/m/dark red with white head/cf. Mongolian Whiteheaded/orig. from Kazakh Whiteheaded × Mongolian/Russ. *Selenga*/[name of river]
Senegal Fulani, Senegalese Fulani: *see* Gobra
Senegal N'Dama (large type): *see* Gambian N'Dama (= N'Dama Grande)
Senegal N'Dama (small type): *see* N'Dama Petite
Senegal Zebu: *see* Gobra
Senegambia Shorthorn: (Gambia and Casamance, S Senegal)/local West African Shorthorn/extinct
Senepol: (St Croix, US Virgin Is)/ d.m/red; pd/orig. 1918–1949 from Red Poll × N'Dama Petite imported in 1860 to Nelthropp herd, St Croix/BS 1976/syn. *Nelthropp*/not *Nelthrop*
Sengologa: (SC Botswana)/var. of Tswana belonging to Bangologa/ nearly extinct (from interbreeding with Tswana)
Šenhengský: *see* Hřbínecky

Seraya ukrainskaya: *see* Ukrainian Grey
Serbian Pied: (Serbia)/orig. from Simmental × local/Serbo-cro. *Domaće šareno goveče* (= local pied cattle)/obs. syn. (with Croatian Simmental) *Yugoslav Pied*
Serbian Steppe: (SE Vojvodina, NE Serbia)/dr/Grey Steppe type/ vars: Kolubara and Spreča (with Buša blood), Posavina/obs. syn. (with Croatian Steppe) *Yugoslav Steppe*/nearly extinct
Serenge, Serenli: *see* Tuni
Serere: (Serere peninsula, Uganda)/ sim. to Nkedi but larger and with smaller hump/Small East African Zebu type (Teso group)/rare
Seroukrainskaya: *see* Ukrainian Grey
Serrana: (Spain) *see* Spanish Mountain (Black Iberian, Pajuna)
Serrano: (Andes) *see* Chusco
serrano: (SE Salta, Argentina)/ medium var. of Argentine Criollo
Sertanejo: *see* Curraleiro
Sese Island, Sesse Island: *see* Nganda
Seshaga: (SC Botswana)/Sanga (Setswana group)/var. of Tswana belonging to Bashaga/nearly extinct (from interbreeding with Tswana to consolidate Tswana)
Setswana: *see* Tswana
Setswana group: (Southern Africa)/ group of similar Sanga breeds that used to occupy arid grasslands of middle and northern parts of Southern Africa and S part of C Africa/inc. Barotse, Damara, Tswana, Tuli; sometimes also inc. Sangas of Zambia, Angola and Botswana
Shahabadi: (W Bihar and E Uttar Pradesh, India)/dr.d/grey/z/sim. to Hariana and Gangatiri/not *Sahabadi, Shahbadi*
Shahjadpur: (Bangladesh)
Shakhansurri: *see* Sistani
Shami, Shamiua: *see* Damascus
Shan: (Shan state, Myanmar)/grey, also spotted/z/syn. *Shan nwar* (= Shan cattle)/*see also* Burmese
Shandong: (China)/m.dr/yellowish-brown, occ. dun or black/Central

Chinese type, ? = Luxi/WG *Shan-tung*, Russ. *Shan'dun*/syn. *Danyang* (WG *Tanyang*) (S Jiangsu), *Syhyang* (N Jiangsu)

Shangai: (China)/extinct

Shangan: (S Africa)/m.dr.d/Sanga (Nguni group)/syn. *Shangaan*/ nearly extinct

Shanun Adar: *see* Azaouak

Shanun Bayaro: *see* Diali

Sharabi: (Tigris valley N of Mosul, Iraq)/black coloursided; small hump in ♂/not *Sarabi*

Shashi: *see* Tarime

sheeted: *see* belted

Sheeted Somerset: (England)/red, belted; hd or pd/syn. *Somerset, Somersetshire Sheeted, White-sheeted Somerset*/extinct *c.* 1890

Sheko: (SW Kefa, SW Ethiopia)/d/ brown or black-and-white; hd or pd; humpless or tiny hump/syn. *Goda, Mitzan*/rare

Shendi: (N Sudan)/var. of Butana (or related strain)/rare (from inbreeding with Red Butana)

Shetland: (Scotland)/d.m/formerly various colours, now black-and-white, occ. red-and-white/ Shorthorn, Angus and Friesian blood/BS 1910, HB 1912–1921 and 1982 on/var.: Orkney/syn. *Zetland*/nearly extinct

Shikari: *see* Khillari

Shilluk: (N of S Sudan)/d.m/local var. of Nilotic/not *Shilluck*

Shimane: (Japan)/strain of Japanese Black

Shkodra Red: (Albania)/var. of Albanian/It. *Scutari*

Shoa: *see* Shuwa

Shona: *see* Mashona

SHORTHORN: (NE England)/m.d/red, roan or white, occ. red-and-white or roan-and-white/orig. from Holderness + Teeswater in late 18th c./inc. Beef Shorthorn, Dairy Shorthorn, Milking Shorthorn (USA), Northern Dairy Shorthorn, Polled Shorthorn (USA), Poll Shorthorn (Australia), Whitebred Shorthorn/part orig. of Armorican, Belgian breeds, Danish Red Pied, German

Shorthorn, Illawarra, Japanese Shorthorn, Lincoln Red, Main-Anjou *et al.*/HB 1822, BS 1875; BS also Canada 1881, S Africa 1912 (HB 1906); HB also Brazil 1906/obs. syn. *Durham, Improved Shorthorn, Improved Teeswater*

Shorthorned Sahelian Zebu: *see* West African Zebu (shorthorned)

Shorthorned Zebu: *see* East African Shorthorned Zebu, Grey-white Shorthorned (India), West African Zebu (shorthorned)

Shuwa: (Chad, NE Nigeria and N Cameroon)/pa.d.m/usu. dark red, also black or pied/West African Zebu (shorthorned) type, sim. to Azaouak/? some West African Shorthorn (humpless) blood/ vars: Kilara, Toupouri/Fr. *Choa, Hausa Wadara* (Nigeria)/syn. *Arab* (Chad), *Arab Choa, Arabe Choua* (Cameroon), *Arab Shuwa, Shuwa-Aral* (Togo), *Tur*

Shvitsezebuvidnyĭ skot: *see* Schwyz-Zeboid

Shvitskaya: *see* Russian Brown

shweni, shwewar: *see* Burmese

Siamese: *see* Thai

Siam-Kedah: *see* Kedah-Kelantan

Sibasa: *see* Bavenda

Siberian: (Russia)/d/Turano-Mongolian type/vars: Altai, Russo-Siberian, West Siberian, Yakut/Russ. *Sibirskiĭ skot*/ extinct by 1960 (except Yakut)

Siberian Black Pied: (W Siberia, Russia)/d.m/var. of Russian Black Pied/orig. since 1929 from East Friesian × local Siberian/ Russ. *Chernopestryĭ skot Sibiri, Sibirskaya chernopestraya*

Siberian White: (Novosibirsk, Russia)/white with black ears/ Russ. *Beliy sibirskiy skot*/extinct in 1950s by crossing with Simmental and Black Pied/rare

Sibirsk: *see* Siberia

Siboney: (Cuba)/d/orig. since late 1960s from Holstein (⅝) and Cuban Zebu (⅜)/not *Syboney*

Sicilian: (Sicily, Italy)/dr.m.d/dark red/orig. Iberian type decimated 1860 and crossed with Chianina,

Reggiana, Calabrian *et al.* to
produce Mezzalina, Modicana
and Montanara vars; now
Sicilian = Modicana/extinct
Sidet trønderfe og nordlandsfe: *see*
Blacksided Trondheim and
Nordland
Siebenburgisch: *see* Transylvanian
Siegerland, Siegerländer: *see*
Westphalian Red
Sierra: *see* Ecuador Criollo
Sierra Leone: *see* N'Dama
Silesian Red: (Poland)/var. of Polish
Red (to 1945 var. of German Red)
with Danish Red blood/Pol.
Śląska czerwona, Ger.
Schlesisches Rotvieh/extinct
Silesian White-Back: *see* Kłodzka
Simbra: (S Africa)/m/in development
from crosses of Simmental ×
Brahman and vice versa
Simbrah: (USA)/cross of Simmental
($\frac{1}{2}$–$\frac{3}{4}$) × Brahman ($\frac{1}{2}$–$\frac{1}{4}$) registered
by American Simmental
Association/BS also Argentina/
syn. (to 1978) *Brahmental*/not
Simbra
Simbrangerford: (Texas, USA)/
Simmental × Brangus ×
Hereford
Simbrasil: (Brazil)
Simford: (NSW, Australia)/m/deep
honey to red, with white
face/orig. at 'Waterloo Station'
from Simmental × Hereford/rare
Simford: (Israel)/m/orig. 1970s from
Simmental × Hereford
Sim-Luing: (NW Scotland)/m/
Simmental × Luing 1st cross/HB
Simmalo: (California, USA)/m/orig.
by Mel Lauriton from $\frac{1}{2}$
Simmental, $\frac{1}{4}$ bison, $\frac{1}{4}$ Hereford
SIMMENTAL: inc. Simmental
(Switzerland) and its derivatives:
Austrian Simmental, Bulgarian
Simmental, Croatian Simmental,
Czech Pied, French Red Pied
group, German Simmental,
Hungarian Pied, Italian Red Pied,
Polish Simmental, Romanian
Simmental, Russian Simmental,
Serbian Pied, Slovakian Pied,
Slovenian Pied/BS also in
Southern Africa 1950 (HB

Namibia 1921, S Africa 1951),
Brazil 1963, Argentina 1966, USA
1968, Canada 1969, GB 1971,
Ireland 1971, Australia 1971,
Sweden 1975, China 1981, New
Zealand, Zambia, Zimbabwe,
Uruguay/Ger. *Fleckvieh* (= pied or
spotted cattle) or *Rotfleckvieh* (=
red pied cattle)
Simmental: (W Switzerland)/
d.m.dr/dun-red and white, or
leather-yellow and white, with
white face/orig. from Bernese/
Red Holstein (RH) blood in
1960s and 1970s/part orig. of
Simbra, Simbrah,
Simbrangerford, Simbrasil,
Simford, Sim-Luing, Simmalo *et
al.*/recog. 1862, BS and HB 1890;
divided since 1994 into 3
sections: Simmental (< 14% RH),
Tachetée Rouge (14–74% RH)
and Red Holstein (75–100%
RH)/Fr. *Tachetée rouge* (or *Pie
rouge*) *du Simmental*, Ger.
Simmentaler Fleckvieh/syn.
Spotted, Swiss Red Spotted (Ger.
Schweizer Rotfleckvieh), *Swiss
Simmental Spotted*
Simmental d'Alsace: *see* Alsatian
Simmental
Simentalsko: *see* Slovenian Pied
Simmenthal-Friulana: *see* Italian Red
Pied
Sinan: (Guizhou, China)/var. of
Wuling
Sindhi: *see* Red Sindhi
Sindhi, Grey: *see* Thari
Sindi: *see* Red Sindhi
Sindi: (Brazil)/orig. from Red Sindhi
imported 1930 and 1952/BS 1961
Siné, race du: *see* Djakoré
Singhi: (Somalia)/z/extinct
Singida White: (Tanzania)/local
strain of Small East African Zebu
(Tanzania cluster), shorthorned/
rare
Sinhala: (Sri Lanka)/z/usu. black or
red, sometimes pied; dwarf/sim.
to Indian hill type/syn. *Lanka*
Sinú: *see* Costeño con Cuernos
(horned), Romosinuano (polled)
Sinuano de cuernos: *see* Costeño con
Cuernos

Siri: (W Bengal and Bhutan)/dr.d/
black, red, black-and-white or
red-and-white/z/hill type, with
Tibetan (humpless) blood/orig.
(× Nepalese Hill) of Kachcha Siri
and (with mithun) of jatsa/syn.
Trahbum (Bhutanese)/rare
(declining through crossing with
Jersey)

Sistani: (E Iran and SW Afghanistan)/
m/black or pied; hd or pd or
scurs/z/syn. *Chakhansurri,
Shakhansurri* (Afghanistan)/not
Seistani

Sitamarhi: *see* Bachaur

Siva rasa: (Serbia)/d.m.dr/grey, ♂
with white back line/composite
of Wipptal and Oberinntaler
(Tyrol Grey)/[*siva* = grey]/nearly
extinct

Sivka iz Gorne Savinje: *see* Savinja
Grey

Sivo mestno govedo: *see* Iskar

Sivo-stepska: *see* Yugoslav Steppe

SJB: *see* Swedish Jersey

SJM: *see* Jutland Black Pied

Skåne: (Sweden)/orig. from Red Pied
Holstein and Dutch, × local/syn.
Scanian/extinct (absorbed by
Red Pied Swedish)

SKB: *see* Swedish Polled

Skotsky náhorni skot: (Czech
Republic) = Highland (Scotland)

sladang: *see* seladang

Śląska czerwona: *see* Silesian Red

Slavonian Grey Steppe: *see* Croatian
Steppe

Slavonian Podolian: (Slavonia,
Croatia)/grey; lyre horns/Grey
Steppe type/Cro. *Slavonski
podolac*/syn. *Slavonian
Syrmian*/nearly extinct/*see also*
Croatian Steppe

Slavonian Syrmian: *see* Slavonian
Podolian

Slavonski podolac: *see* Croatian
Steppe

Slavonsko Podolsko: *see* Croatian
Steppe

SLB: *see* Swedish Friesian

Slesvig Marsh: *see* Ballum

Slovakian Pied: (Slovakia)/d.m./orig.
from Simmental × local/HB
1925/Sl. *Slovenský strakatý*/syn.

*Slovakian Red-and-White,
Slovakian Simmental,
Slovak(ian) Spotted, Slovakian
Yellow Spotted, Slovakish
Yellow, Slovak Pied*

Slovakian Pinzgau: (Slovakia)/
d.m/orig. from Pinzgauer ×
local/HB 1925/Sl. *Slovenský
Pincgavský*

Slovakian Red: (Slovakia)/d.m.dr/
Central European Red type sim.
to Polish Red/Sl. *Slovenský
červený*/extinct 1965

Slovenaćko belo goveče: *see*
Slovenian White

Slovenian Black Pied: (Slovenia)/
d/orig. from Danish Black Pied,
German Black Pied and Holstein/
HB 1960/Sn. *Crno-belo*

Slovenian Brown: (Slovenia)/
d.m/orig. 1900–1910 from
Austrian, Swiss and Italian
Brown × Mölltal and Murboden/
HB 1909/Sn. *Rjavo govedo*

Slovenian Pied: (Slovenia)/orig. since
1885 from Austrian, German and
Swiss Simmental/syn. *Lisasto
govedo, Simentalsko, Slovenian
Simmental*

Slovenian Pinzgau: (Slovenia)/d/
local (Mölltal and Murboden)
graded to Pinzgauer/Sn.
Pinzgavska, Pinzgavac/obs. syn.
Yugoslav Pinzgau/rare

Slovenian Red Pied: (Slovenia)/
almost extinct through
absorption into Simmental and
Slovenian Brown

Slovenian Simmental: *see* Slovenian
Pied

Slovenian White: (Slovenia)/Sn. *Belo
slovensko govedo*, Serbo-cro.
Slovenaćko belo goveče/extinct
since 1950

Slovenský: *see* Slovakian

SM: *see* Sanmartinero

Smada: (NW Ethiopia)/dr.(d)/usu.
black, also red, roan or black-
and-white/Abyssinian
Shorthorned Zebu type

Småland: (Sweden)/local/extinct
(absorbed into Red Pied
Swedish)

Small East African Zebu: (E Africa)/

smaller type in wetter habitats, with greater variability in size and conformation than Large East African Zebu/groups inc. Abyssinian Shorthorned, Angoni group, Kenya cluster, Madagascar group, Southern Sudan Shorthorned, Somali Shorthorned, Tanzanian cluster (Tanganyika and Zanzibar), Teso group

Small Spotted Hill: (Germany) *see* Black Forest

Small Zebu: (Ethiopia) *see* Jijjiga Zebu

Smedje govedo: *see* Croatian Brown

SMR: *see* German Black Pied Dairy

Socotra: (Socotra I, Indian Ocean)/ short horns; dwarf; humpless/? extinct

Sofia Brown: *see* Bulgarian Brown

Sofioter Braunvieh: *see* Bulgarian Brown

Sokoto Gudali: (NW Nigeria)/d.dr/ white, grey or dun/West African Zebu (shorthorned) type/Fulani *Gudali*/syn. *Bokoloji, Godali, Sokoto*

Solognote: (Sologne, France)/d/ small/extinct 1940s

Solomon Red: (Solomon Is)/orig. 1970s–1980s from Brahman, Santa Gertrudis, Hereford, Shorthorn and Droughtmaster/ rare

SOMALI group: (Somalia)/m.d.dr/ horns short, sometimes absent/ Small East African Zebu type/ inc. Baherie, Garre, Gasara, North Somali Zebu/syn. *Small zebus of Somalia, Somaliland Zebu*

Somali Boran: *see* Ethiopian Boran

Somaliland Zebu: *see* Somali

Somba: (Atakora Highlands, N Togo and N Benin, esp. Boukombé dist.)/m/black, black pied or red pied; occ. pd/var. of West African Savanna Shorthorn, raised by Somba in Benin and by Tamberma in Togo/syn. *Atacora* (Benin), *Konkomba* or *Mango* (Togo)

Somerset: *see* Sheeted Somerset

Sonkheri: (India)/red-and-white var. of Dangi

Son Valley: (Madhya Pradesh, India)/z/local var.

sorcino: *see* Grey Val di Fiemme

Sorco: *see* Jiddu

Sør og vestlandsfe: *see* South and West Norwegian

Sorraia: *see* Bragado do Sorraia

Sortbroget Dansk Malkekvæg: *see* Danish Black Pied

Sortbroget Jydsk Malkekvæg: *see* Jutland Black Pied

Sorthi: *see* Gir

South African Brown Swiss: (S Africa)/d.m/orig. from American, Swiss and German imports since 1907/BS 1926, HB 1954; separate HB 1974 for dual-purpose type; recog. as separate breed 1996

South African Dairy Swiss: (S Africa)/d/shades of brown/orig. from South African Brown Swiss/BS, HB 1974; recog. as separate breed 1995

South Anatolian Red: (Turkey)/ d/vars: Aleppo (= Damascus), Çukurova, Dörtyol, Kilis/Turk. *Cenubî* (or *Güney*) *Anadolu Kırmızı*/syn. *Southern Yellow-Red*

South and West Norwegian: (SW Norway)/d/usu. red, also dun; hd or pd/orig. 1947 by joining Lyngdal and Vestland Red Polled with Vestland Fjord/absorbed by Norwegian Red 1968/HB 1947/ Nor. *Sør og vestlandsfe* (= *South and Westland cattle*)/? extinct

South Bravon: (USA)/Brahman × South Devon cross

South China Zebu: (Guangdon, Guangxi, Fujian and S Yunnan, China)/dr/usu. light brown, also black, brown or dun; hump in ♂ and ♀/Changzhu group/inc. Hainan Humped, Leizhou, Yunnan Zebu/syn. *High-hump, South China Draught*

South Coast: (Australia) *see* Illawarra

South Devon: (SW England)/m.[d]/ light red/BS and HB 1891; BS also S Africa 1914, USA 1974, Canada 1974, New Zealand,

Australia, Ireland/obs. syn.
Hammer, South Hams/not *Devon*
Southeastern Hills Zebu: (Sudan) *see*
Mongalla
Southern: (India) *see* Umblachery
Southern Crioulo: (Brazil, S of lat.
18–20°S)/m/brown, orange or
pied; long horns/sim. to
Argentine Criollo, Texas
Longhorn, Romosinuano/Port.
orig. (? from Alentejana)/orig. of
Caracu, Crioulo Lageano,
Franqueiro, Junqueiro/Port.
Crioulo do Sul/syn. *Bruxo,*
Colônia (from Colonia do
Sacramento, Uruguay), *Legítimo,*
Mineiro (from Minas Gerais)/not
Criôlo/[= native]/extinct
Southern Island group: (Is and
peninsulas of E and S Asia)/
collective term for native types
Southern Sudanese: *see* Nilotic
Southern Sudan Hill Zebu: *see*
Mongalla
Southern Sudan Shorthorned Zebu:
(S Sudan, Uganda, DR Congo)/
Small East African Zebu cluster
inc. Lugware, Mongalla, Nkedi,
Nuba Mountain Zebu
Southern Tswana: (S Botswana)/var.
of Tswana with Afrikander
blood/inc. Sekgatla/syn.
Dikgomo tsa Borwa (= cattle of
the south)
Southern Ukrainian: (Ukraine)/
m/var. of Ukrainian Beef/orig.
from Charolais ($\frac{1}{2}$), Hereford ($\frac{1}{4}$),
Red Steppe ($\frac{1}{4}$)/Russ.
Yuzhnoukrainskaya
Southern Woods: *see* Pineywoods
Southern Yellow: *see* Changzhu
group
Southern Yellow-Red: *see* South
Anatolian Red
South Hams: *see* South Devon
South Korean Holstein: (S Korea)/
d.(m)/orig. from Holstein (chiefly
USA) since 1963/HB 1966/syn.
South Korean Holstein-Friesian
South Malawi Zebu: (S Malawi)/
former var. of Malawi Zebu/obs.
syn. *Nyasa Zebu, Nyasaland Zebu*
South Siberian: *see* Altai
South Tyrol Grey: *see* Grey Alpine

South Wales Black: (S Wales)/m.d/
vars: Castle Martin, Dewsland/
orig. (+ North Wales Black) of
Welsh Black/HB 1874–1904/syn.
Pembroke/extinct
Southwestern Blond: *see* Blonde du
Sud-Ouest
Spanish Brown: (N Spain)/orig. from
Swiss Brown imported since
1850 but esp. 1880–1910/HB/Sp.
Parda alpina (= *PA*)/syn. *Parda*
suiza, Schwyz española/old type
now rare
Spanish Friesian: (Spain)/being
graded to Holstein since 1970
Spanish Mountain: *see* Black Iberian,
Pajuna
Spanish Pied: *see* Berrendas
Spotted: *see* Simmental
Spotted cattle of the lower
Guadiana: *see* Guadiana Spotted
Spotted Moravian: *see* Moravian Red
Pied
Spotted Mountain: *see* Simmental
Spotted Romanian: *see* Romanian
Simmental
Spotted Valdostana: *see* Aosta
Spreča: (NE Bosnia)/yellow-grey/
orig. from Serbian Steppe with
Buša blood/Serbo-cro. *Sprečko*
goveče/syn. *Tuzla*/nearly extinct
SRB: *see* Swedish Red-and-White
Sredneaziatskiĭ zebu: *see* Central
Asian zebu
Srednerusskaya chernopestraya: *see*
Central Russian Black Pied
Stara Planina: (W Bulgaria)/dr/dark
grey/smaller mountain var. of
Bulgarian Grey with more
brachyceros blood/Bulg.
Staroplaninska k"soroga (=
shorthorned)/extinct
Steierisch, Steiermärkler: *see* Styrian
Steppe: *see* Grey Steppe, Russian
Simmental
Steppe Red: *see* Caoyuan Red, Red
Steppe
Steppic Gray: *see* Grey Steppe
STN: *see* Blacksided Trondheim and
Nordland
Styrian Blond: *see* Austrian Blond
Styrian Brown: (Austria)/former var.
of Austrian Brown/Ger.
Steierisches Braunvieh/extinct

Styrian Spotted: *see* East Styrian
Spotted
Sucra: *see* Jiddu
Sudanese Fulani: (Mali)/m/usu. light
grey with dark spots, also with
white back and black speckles/
West African Zebu (lyrehorned)
type (Fulani group)/var.:
Toronké/Fr. *Zébu peul
soudanais*/syn. *Peul Zebu*
Sudanese Longhorn: *see* Nilotic
Sudanese Shorthorned Zebu: *see*
North Sudan Zebu
Sudeten: (Czech Republic)/d.m.dr/
orig. from Central European
Red/vars: Sudeten Pied (often
coloursided), Sudeten Red/orig.
of Kłodzka/Cz. *Sudetský*/syn.
Moravian Sudeten/extinct
Suffolk Polled: (England)/dun or red-
brown/orig. (× Norfolk Horned)
of Red Poll/syn. *Suffolk Dun*/
extinct
Suiá: (Matto Grosso, Brazil)/
Marchigiana × Nelore cross at
Liquifarm's Suiá Farm
Suisbú: (N Argentina)/m/orig. from
American Brown Swiss × zebu/
[Pardo *Suizo* × *zebú*]/extinct in
1980s
Suizo Pardo: (Mexico)/= American
Brown Swiss
Suk: (NW Kenya)/local Small East
African Zebu, ? with Karamajong
blood/cf. Usuk (Uganda)
Suksun: (Perm, Russia)/d/red/orig. in
late 19th c. from Danish Red ×
local, followed by Angeln in
early 20th c., with Red Steppe,
Estonian Red and Latvian Brown
blood 1933–1938/HB 1941/Russ.
Suksunskaya
Sukthi: (Albania)
Sukuma: (S of L Victoria,
Tanzania)/d.m/usu. red, light
dun, red roan or blue roan/
Sanga-Zebu intermediate/orig.
from Ankole and Tanzanian
Zebu/syn. *Tinde*
Sul do Save: *see* Nguni
Šumava: (Bohemia, Czech Republic)/
d.m.dr/yellow to grey/sim. to
Bergscheck/Cz. *Šumavský*/syn.
Budějovice (Cz. *Budějovický*, Ger.
Budweiser)/extinct *c.* 1945

Sumba Ongole: (Indonesia) *see*
Ongole
Sunandini: (Kerala, India)/d.(dr)/
shades of fawn, brown or grey-
brown/orig. 1965–1985 from
European ($\frac{1}{2}$) × local zebu;
European was Swiss Brown
(1965–1974) imported from
Switzerland, Jersey (1968–1987)
from Australia and New Zealand,
American Brown Swiss
(1981–1984) and Holstein
(1981–1987); zebu mainly local
nondescript but also Sahiwal,
Gir and Kankrej/named 1979
sun yak: *see* dzo
Suomen, Suomalainen: *see* Finnish
Sură de Stepă: *see* Romanian Grey
Surati: *see* Gir
Surco, Surco Sanga: *see* Jiddu
Surgo, Surqo: *see* Jiddu
Surti: *see* Gir
Surug, Suruq: *see* Jiddu
Sussex: (Sussex and Kent, England)/
m.[dr]/red/sim. to Devon/var.:
Polled Sussex/BS and HB 1879;
BS also USA 1884 (reformed
1966), S Africa 1920 (HB 1906),
New Zealand, Zambia
Svensk: *see* Swedish
Svensk kullig boskap: *see* Swedish
Polled
Svensk Låglands boskap: *see*
Swedish Friesian
Svensk rödbrokig boskap: *see* Red
Pied Swedish
Svetlolisata: *see* Murboden
Svitto: *see* Italian Brown
Svitto-Sarda: *see* Sardinian Brown
Svizzera: *see* Italian Brown
Swabian Spotted: *see* Württemberg
Spotted
Swazi: *see* Nguni
Swedish Ayrshire: (Sweden)/d/orig.
from Ayrshire (imported 1847–
1905) × local/joined with Red
Pied Swedish 1928 to form
Swedish Red-and-White/BS
1899–1928, 1952 on, HB 1901–
1927, 1954 on (subsection in
Swedish Red-and-White HB)/
Swe. *Svensk Ayrshire boskap* (=
SAB) (*boskap* = cattle)

Swedish Friesian: (S Sweden)/d.(m)/ orig. from Dutch imports 1860–1907 and recent, orig. with local and East Friesian blood/HB 1891, BS 1913/Swe. *Svensk Låglands boskap* (= *SLB*)/syn. *Black-and-White Swedish, Swedish Holstein, Swedish Lowland*

Swedish Highland: *see* Swedish Mountain

Swedish Holstein: *see* Swedish Friesian

Swedish Jersey: (S Sweden)/d/var. of Jersey first imported 1890s and reimported from Denmark after 1945/BS 1949, HB 1955/Swe. *Svensk Jersey boskap* (= *SJB*)

Swedish Lowland: *see* Swedish Friesian

Swedish Mountain: (N Sweden)/ d/white with red or black points and side spots/var. of Swedish Polled; sim. to North Finnish and to Blacksided Trondheim and Nordland/HB 1892, BS 1920–1938/Swe. *Fjällras*/syn. *Swedish Highland, Swedish White Polled*/rare

Swedish Polled: (N Sweden)/d/red, red-and-white or white/inc. (since 1938) Swedish Mountain, Swedish Red Polled/BS and HB 1938/Swe. *Svensk kullig boskap* (= *SKB*) (*kullig* = polled)/syn. *North Swedish*/rare

Swedish Red-and-White: (Sweden)/ d/red with small white markings/orig. 1928 from Red Pied Swedish + Swedish Ayrshire/BS and HB 1928/Swe. *Svensk röd och vit boskap* (= *SRB*)/syn. *Swedish Red Spotted*

Swedish Red Pied: *see* Red Pied Swedish

Swedish Red Polled: (C Sweden)/ d/var. of Swedish Polled/sim. to Red Polled Eastland (Norway) and West Finnish (some recent imports)/HB 1913, BS 1913–1938/Swe. *Rödkulla, Röd kullig lantras, Röd kullig boskap*/rare

Swiss Black-and-White: *see* Swiss Holstein

Swiss Black Pied, Swiss Black Spotted: *see* Swiss Holstein

Swiss Brown: (E Switzerland)/ d.m.dr/grey-brown/colour vars: Blüem, Gurtenvieh/American Brown Swiss blood in 1960s and 1970s/orig. of Brown Mountain breeds/BS 1897, HB 1878 reorg. 1911/Ger. *Schweizerisches Braunvieh*, Fr. *Brune suisse*, It. *Bruna svizzera*/syn. *Schwyz, Suíça Parda* (Brazil); obs. syn. *Rigi*/not *Schwitz, Schwiz*

Swiss Brown (Original): (Switzerland)/BS 1980 for orig. Swiss Brown breed without American blood/Ger. *Schweizer Original Braunvieh*

Swiss Holstein: (Fribourg, Switzerland)/d/orig. 1965 on from Holstein × Fribourg (97% Holstein in 1991)/BS/Ger. *Schweizerische Holstein*/syn. (to 1991) *Swiss Black Pied* or *Swiss Black Spotted* (Fr. *Tachetée noire*, Ger. *Schwarzfleckvieh*)

Swiss Red Spotted: *see* Simmental

Swiss Simmental Spotted: *see* Simmental

Swiss Tachetée: *see* Simmental

Swiss-zebu: *see* Schwyz-Zeboid

Swona: (I of Swona, S Orkney, Scotland)/mostly red-brown or black, also black-and-white or white; ♂ hd or pd, ♀ pd/ interbred group orig. from Aberdeen-Angus × Shorthorn; feral since 1974/nearly extinct

Syboney: *see* Siboney

Sychevka: (Smolensk, Russia)/ d.m/var. of Russian Simmental/ orig. (1880 on) from Simmental × local/recog. 1950; HB/Russ. *Sychëvskaya*/syn. *Sychevka Simmental, Western Simmental*/ not *Sychovsk*

Syhyang: *see* Shandong

Sykia: (Khalkidiki, NE Greece)/m.dr/ grey; lyre horns/smaller var. of Greek Steppe/nearly extinct

Tabapuã: (São Paulo, Brazil)/m/grey-white; pd/z/orig. from one pd Guzerá × Nelore ♂ born 1940

and used on Nelore ♀♀ on farm of Ortenblad family, 'Agua Milagrossa', Tabapuã/BS and HB 1959

Tachetée de l'Est: (France) *see* French Simmental

Tachetée noire: *see* Swiss Holstein

Tachetée rouge du Simmental: *see* Simmental

Tagama: *see* Azaouak

Tagil: (Sverdlovsk, S Urals, Russia)/ d/usu. black pied, also black, red or red pied/orig. in mid-19th c. from Friesian, Kholmogory *et al.* × local/HB 1931/Russ. *Tagil'skaya*

Taino: (Cuba)/d/orig. Criollo ($\frac{3}{8}$), Holstein ($\frac{5}{8}$)

Taita: (SE Kenya)/Small East African Zebu (Kenya cluster, lowland or coastal type, cf. Duruma, Giriama, Kamba)/syn. *Taveta*

Taiwan Black: (Taiwan)/dr.m/small

Taiwan Yellow: (Taiwan)/local Chinese yellow cattle of Changzhu type, with possible Philippine influence/extinct (from crossing with Indian breeds to form Taiwan Zebu) but name still used

Taiwan Zebu: (Taiwan)/dr/usu. yellow, red or brown, also black or grey/orig. from Taiwan Yellow with Red Sindhi and Kankrej blood since 1910/syn. *Formosa Draught*, *Taiwan Yellow*

Tajik: *see* Central Asian Zebu

Talabda: *see* Kankrej

Talysh: *see* Azerbaijan Zebu

Tamankaduwa: (NE Sri Lanka)/usu. white or light-coloured/z/local var./orig. from imported Indian breeds

Tambov Red: (Tambov, Russia)/ m.d/orig. early 19th c. from Zillertal × local with some Devon and Simmental blood/ recog. 1948; HB/Russ. *Krasnaya tambovskaya*, *Krasnotambovskaya*

Tanaland Boran: *see* Orma Boran

Tanganyika Zebu: *see* Tanzanian Zebu

Tanjore: *see* Umblachery

Tanyang: *see* Shandong

TANZANIAN: (Tanzania and Zanzibar)/ Small East African Zebu cluster, inc. Tanganyika Shorthorn group (Chagga, Pare, Tanzanian Zebu, Tarime, Ugogo Grey) and Zanzibar Zebu

Tanzanian Zebu: (Tanzania)/= Small East African Zebu in Tanzania/ dwarf var.: Chagga; local strains orig. bred for colour (no longer distinctive): Iringa Red, Masai Grey, Mbulu, Mkalama Dun, Singida White; other local or tribal types: Pare, Tarime, Ugogo Grey/ syn. *Tanganyika Zebu*, *Tanzanian Shorthorned Zebu*, *TSZ*/not *Tanzania Zebu*

Tapi, Tapti: *see* Thillari

Tarai: (S Nepal)/usu. white or grey/ z; Lyrehorned type/Hariana influence/not *Terai*

Tarentaise: (French Alps)/d.m/fawn to yellow/= Tarina (Italy), Tarine (Albania)/recog. 1859, HB 1880; BS in Canada 1973, USA 1973, Australia/not *Tarentais*, *Tarente*

Targhi, Targi, Targui: *see* Azaouak

Tarime: (Mara region, Tanzania)/ Small East African Zebu (Tanzania cluster), substantial taurine influence (? Ankole blood)/vulnerable/syn. *Shashi*

Tarina: (Susa and Chirone valleys, Turin, Italy)/d.m/= Tarentaise (France)/BS and HB 1888, reorg. 1949/Fr. *Tarine*/syn. *Savoy* (It. *Savoiarda*)/nearly extinct

Tarine: (Albania)/d.m/reddish/= Tarentaise (France)/nearly extinct

Taro: *see* Valtarese

Tashkent: *see* Bushuev

Tasmanian Grey: (Tasmania, Australia)/m/sim. to Murray Grey and absorbed by it c. 1979/ orig. from crossbred Aberdeen-Angus × white Shorthorn born at Parknook in 1938/BS/extinct/ *see also* Australian Grey

Tattabareji: *see* Yola

Tatu: (Rio de Janeiro, Brazil)/orig. from crosses of Red Sindhi or Sahiwal ♂♂ imported about 1850/extinct

Taurache: *see* Tourache

Tauricus: (KwaZulu/Natal, S Africa)/ d/in development from Nguni and Jersey crosses

Taurindicus: (Ruakura, New Zealand)/Sahiwal × Friesian cross with 25% (T25), 37.5% (T40) or 50% (T50) Friesian/syn. *NZSHF* (= *New Zealand Sahiwal × Holstein-Friesian*)

Taurindicus: (Tanga, Tanzania)/d/ herd of crossbred European dairy cattle × East African Zebus started 1946, disbanded 1966/extinct

Taurine: *see* West African humpless

Taveta: *see* Taita

Tawana: *see* Batawana

Taylor: (Patna, Bihar, India)/d/red, grey or black; no hump/orig. from 4 Shorthorn and Channel Island bulls imported by Commissioner Taylor in 1856, × local zebu/not *Taypor*/rare

Teck: (Württemberg, Germany)/extinct (absorbed by German Simmental)

Teeswater: (Durham, England)/sim. to Holderness/Dutch orig./orig. (+ Holderness) of Shorthorn/syn. *(unimproved) Shorthorn*/extinct

Tejelő magyar-barna: *see* Dairy Hungarian Brown

Tejelő magyar-tarka: *see* Dairy Hungarian Pied

Telemark: (C and SE Norway)/d/ coloursided, usu. red or brindle; lyre horns/var.: Valdres/recog. 1856, BS 1895, HB 1926/Nor. *Telemarkfe*/rare

Teli: *see* Sahiwal

Terai: *see* Tarai

Terengganu: *see* Kedah-Kelantan

Terreña: (Alava and Bizkaia, Spain)/ dr.m/brown/vars: Terreña gorbeana (smaller), Terreña de la Sierra (larger)/not *Terrana*/rare

Teso: *see* Nkedi

Teso group: (Uganda)/Small East African Zebu group, inc. Kyoga, Serere, Teso (= Nkedi), Usuk

Tete: *see* Bovines of Tete

Tetrone: *see* Habbani

Texas Longhorn: (USA)/all colours and patterns (red commonest base colour)/cf. Criollo, Southern Crioulo/Spanish orig. (? Andalusian) first imported in 1640/BS 1964/syn. *Longhorn*

Texon: (Texas, USA)/orig. *c.* 1989 by Frank Lloyd of Euless from Devon × Texas Longhorn

Thai: (Thailand)/dr/yellow, red, brown or black/z/orig. of Kedah-Kelantan/syn. *Siamese*/*see also* Khao Lumpoon

Thai-Kedah: *see* Kedah-Kelantan

Thailand Fighting: (S Thailand)/usu. red or brown, also dark or black/z

Thanh-Hoa: (N Vietnam)/var. of Vietnamese

Thari: (SE Sind, Pakistan)/d.dr/grey/ z; Lyrehorned type/orig. of Tharparkar/syn. *Grey Sindhi*

Tharparkar: (SW Rajasthan, India)/ d.dr/grey/z; Lyrehorned type/ orig. from Thari/BSd and HB/ syn. *White Sindhi* (Sri Lanka)/ not *Thar Parkar*

Therkuthi madu: *see* Umblachery

Thessaly: (Greece)/former var. of Greek Shorthorn

Thibar: (Tunisia)/m.d/usu. red-brown, also red, fawn or ash-grey/40–50% z/orig. from Brown Atlas crossed with Modicana (1897), Ongole (1907), Charolais (1908–1910), Red Sindhi, Tarentaise, Montbéliard (1917)

Thillari: (India)/white/var. of Khillari/syn. *Tapi, Tapti*/[= cattle breeder]

Thrace: (Greece) *see* East Macedonian and Thrace

Thracian: (Turkey) *see* Turkish Grey Steppe

Three-river breed: (China) *see* Sanhe

Tibetan: (Tibet and Qinghai, China)/ d.m.pa/black or yellow (solid or pied), also brindle and coloursided/sim. to Mongolian but dwarf/syn. *humpless dwarf cattle of Tibet, Kirko* (Khumbu, Nepal) (♂ *Jolang, Cholung*; ♀ *Molang*), *Lepcha* (Sikkim), *Lulu* (Mustang, Nepal), *Goleng* or *Glang* (♂ *Lang, Lhang*; ♀ *Glangmu*)

Tidili: (W High Atlas and Anti-Atlas, Morocco)/d/usu. brown or black

Tieflandrind: *see* Lowland

Tigray, Tigré: *see* Aradó

Tinde: *see* Sukuma

Tinima: (Cuba)/m/darker and more hump in ♂/double-muscled strain of Criollo

Tinos: (Greece)/orig. from Near East cattle/pure on Tinos; crossed on Ikaria, Samos, Rhodes/? extinct

Tintern Black: (Natal, S Africa)/ m.d/early var. of Drakensberger/ orig. from Afrikander × local in early 20th c. in Tintern herd/ extinct

Tipo Carora: *see* Carora

Tiroler Braunvieh: *see* Tyrol Brown

Tiroler Fleckvieh: *see* Tyrol Spotted

Tiroler Grauvieh: *see* Tyrol Grey

Tirolese: *see* Tux-Zillertal

Tla pseke: *see* Kapsiki

Tolmin Cika: (Slovenia)/var. of Cika, sim. to Bohinj Cika but smaller/ Sn. *Tolminska cika*/nearly extinct

Tonga: (S Zambia)/usu. red, black or pied, occ. light brown, or dun spotted; long horns/Sanga type (Zambia/Angola cluster), smaller than Barotse/orig. (by crossing with Barotse) of Baila/obs. syn. *Baila*

Toposa: (SE Sudan)/m/all colours/ East African Zebu sim. to Karamajong and Turkana; sim. to but larger than Murle

Toro: (W Uganda)/Sanga-Zebu intermediate sim. to Nganda but with more Ankole blood

toro de lidia: (Spain) *see* Fighting bull

Toronké Fulani: (Mali)/m.d/usu. white/var. of Sudanese Fulani sim. to Gobra

Tortona, Tortonese: *see* Montana

Touareg: *see* Azaouak

Toubou: (E Niger)/Red Bororo × Kuri cross

Toupouri: (Chad)/var. of Shuwa/syn. *Massa*/see also Logone

Tourache: (Franche Comté, E France)/dr.d/usu. dark red and white/sim. to Fémeline but coarser, highland, more bull-like type/absorbed by Montbéliard in late 19th and early 20th c./syn. *Comtois, Tourache*/[Fr. *taureau* = bull]/extinct

Touro de Lide: *see* Fighting bull

Traditional Hereford: *see* Hereford (Traditional)

Trahbum: *see* Siri

Transcaucasian: *see* Caucasian

Transtagana: *see* Alentejana

Transylvanian: (Romania)/huge horns/cf. Hungarian Grey/former var. of Romanian Steppe/Rom. *Transilvăneană*, Ger. *Siebenburgisch*/syn. *Grey Transylvanian, Transylvanian Steppe*/extinct

Transylvanian Pinzgau: (NW Romania)/d.m.dr/red with white back, escutcheon and belly/black var.: Dorna/orig. from Romanian Grey graded to Pinzgauer in early 20th c./HB 1959/Rom. *Pinzgau de Transilvania*/syn. *Romanian Pinzgau*/rare

Treignac: (France)/former var. of Limousin/extinct

Trengganu: *see* Kedah-Kelantan

Tribofe: *see* Mantiqueira

Tronder: (Trondheim, Norway)/d/ red/extinct

Trondheim, Trondhjem: *see* Blacksided Trondheim, Red Trondheim

Tropical: (Córdoba, Argentina)/d/ orig. 1959 on from Argentine Holstein × zebu/extinct

Tropical Dairy Criollo: (Costa Rica, Dominican Republic, Mexico, Nicaragua and Venezuela)/ d/light dun to deep red, often with black points and around eyes/vars: Central American Dairy Criollo, Dominican Criollo, Rio Limón Criollo/Sp. *Criollo lechero tropical*/syn. *Improved Criollo, Milking Criollo*

Tropical Mountain: (Brazil)/m/ composite, in formation, based on (1) zebu breeds (Gir, Guzerá, Indo-Brazilian, Nelore, Tabapũa, Boran), (2) tropically adapted European breeds (Afrikander, Belmont Red, Bonsmara, Caracu, Romosinuano, Senepol, Sangas), (3) British breeds (Aberdeen-Angus, Devon, Hereford, Red Angus, Red Poll, South Devon) and (4) continental European breeds (Charolais, Gelbvieh, Swiss Brown, Simmental) with maximum 25% Nelore inheritance, 50% of combined groups 1 and 2/syn. *Montana Tropical*

Tropicana: (Argentina)/d/orig. from Guernsey × zebu/extinct

Tsagaan Tolgoit: *see* Mongolian Whiteheaded

tsauri: *see* chauri

Tsentralnaya chernopestraya: *see* Central Russian Black Pied

Tshilengue: (Chongoroi and Quilengues, SW Angola)/ m/brownish with black spots, black hips and white muzzle/ Sanga type/syn. *Mocho de Quilengues, Ondango*

Tsinchuan, Tsin'chun: *see* Qinchuan

tsine: (Myanmar, N Thailand, Laos and Cambodia)/= *Bos (Bibos) javanicus birmanicus*/var. of banteng/syn. *Burmese banteng*/ not *hsaine, saing, tsaine, tsaing*

TSSHZ-1: (Tajikistan)/d.m/orig. from Russian Brown × Tajik zebu/ recog. 1985

Tsuru: (Japan)/dr/pre-crossbreeding strain of Japanese Black since Edo era (1600–1876)

Tswana: (Botswana)/m.(d.dr)/black, grey, brown, white or pied; long horns/Sanga type (Setswana group), sim. to Barotse and Tuli/ local vars: Batawana, Mangwato, Sengologa, Seshaga, Southern Tswana, being consolidated as Tswana/syn. *Bechuana, Sechuana, Setswana, West Sanga* (S Africa)

TSZ: *see* Tanzanian Zebu

Tuareg: *see* Azaouak

tucura: (Mato Grosso and Pará, Brazil)/popular name for stunted cattle (or horse), often applied to Pantaneiro/syn. *Guabiru* (= rat)/ [= grasshopper]

Tudanca: (SW Santander, Spain)/ m.(dr.d)/♂ black, ♀ black to grey, brown or hazel/mt (and only surviving) var. of Santander; declining by crossing with Swiss Brown/BS

Tuli: (SW Zimbabwe)/m/golden-brown, yellow or red; hd or pd/Sanga type (Setswana group)/orig. by selection (1946 on) from Mangwato (Amabowe) var. of Tswana for golden-brown colour and pd/BS 1961; BS also S Africa/syn. *Harvey's cattle* (Len Harvey = farmer who collected original Amabowe remnants in Tuli area)

Tulim: (S Africa)

Tuni: (E Kenya)/d.m.dr/usu. red with diffuse white spots, also red or pied/Sanga-Zebu intermediate/= Jiddu/syn. *Serenli, Serengi*

Tunisian: *see* Brown Atlas, Cape Bon Blond

tur: *see* aurochs

Tur: *see* Shuwa

Turano-Mongolian: (Central Asia)/ humpless, inc. Kalmyk, Kazakh, Mongolian, Siberian, Tibetan

Turino: (Portugal and Brazil)/d.(m)/ orig. from Dutch Black Pied imported in 18th c./HB/syn. *Frisia, Holandese, Luso-Holandese, Portuguese Friesian*/ not *Tourino*/[Port. (old slang) = dandy]

Turkana: (NW Kenya)/East African Zebu sim. to Karamajong and Toposa

Turkestan Zebu: *see* Central Asian Zebu

Turkish Brown: (NW Turkey)/d.m/ orig. initially at Karacabey State farm in 1925 from Montafon (and Swiss Brown) × Turkish Grey Steppe (and a few Anatolian Black)/syn. *Çifteler Brown, Karacabey Brown, Karacabey Montafon*

Turkish Grey Steppe: (NW Turkey)/ m.d.dr/Grey Steppe type/orig. from Iskar/former var.: Kultak/ Turk. *Boz Step* (= *Grey Steppe*), *Plevne*/syn. *Pleven, Plevna, Thracian, Turkish Grey*

Turkish native: *see* Anatolian Black, East Anatolian Red, South Anatolian Red

Turkmen: (Turkmenistan)/var. of Central Asian Zebu

Tux: (Tyrol, Austria)/black or reddish-black with white tail and belly/former var. of Tux-Zillertal/ Ger. *Tuxer*/nearly extinct

Tux-Zillertal: (Tyrol, Austria)/black or red with white tail and belly/? orig. from Hérens/former vars: Tux, Zillertal/Ger. *Tux-Zillertaler*/syn. *Tirolese, Tyrolean*/nearly extinct

Tuy-Hoa: (SE Vietnam)/var. of Vietnamese

Tuzla: *see* Spreča

Tyrol, Tyrolean, Tyrolese: *see* Tux-Zillertal, Tyrol Grey

Tyrol Brown: (Austria)/former var. of Austrian Brown/orig. from Montafon, Lechtal and Swiss Brown/Ger. *Tiroler Braunvieh*/ extinct

Tyrol Grey: (Tyrol, Austria)/d.m.[dr]/ silver-grey/cf. Rhaetian Grey/= Grey Alpine (Italy) but also inc. Etschtaler, Wipptal/orig. (with Buša) of Dalmatian Grey, Gacko/ BS 1924, HB 1926/Ger. *Tiroler Grauvieh*/syn. *Oberinntaler Grauvieh, Tyrolean Grey, Valbona* (Albania)

Tyrol Spotted: (Tyrol, Austria)/former var. of Austrian Simmental/name Simmental first used 1870, named *Unterinntaler Fleckvieh* 1890, changed to *Tiroler Fleckvieh* 1938/BS 1906, HB 1933/extinct

Uberaba: *see* Dairy Zebu of Uberaba

Uckermärker: (Germany)/m/beef sire for German Black Pied Dairy ♀♀/orig. from Charolais × German Simmental (beef type)/ rare

Uganda Zebu: *see* Karamajong, Nkedi

Ugogo Grey: (Dodoma area, C Tanzania)/d.m/Small East African Zebu (Tanzania cluster); most distinctive (but smaller) strain of Tanzanian Zebu/syn. *Ugogo Zebu*

Ugoi: (Tanzania)/Sanga type/extinct

Ujumqin: (Inner Mongolia, China)/ var. of Mongolian/syn. *Wuzhumuqin*

Ukamba: *see* Kamba

Ukrainian Beef: (Ukraine)/m/orig. since late 1960s from Chianina ($\frac{3}{8}$), Charolais ($\frac{3}{8}$), Russian Simmental ($\frac{1}{8}$), Ukrainian Grey ($\frac{1}{8}$)/vars: Chernigov, Dnieper, combined 1993 as Ukrainian Beef; former vars: Polessie, Southern, Volynsk, Znamensk/ syn. *Ukrainian Beef Synthetic, Ukrainian Meat*

Ukrainian Black Pied: (Ukraine)/cf. Russian Black Pied/recog. 1995/ Ukrainian *Ukraïnska cherno-ryaba*, Russ. *Ukrainskaya chernopestraya*

Ukrainian Dairy Red: (Ukraine)/ d/orig. (from 1989) from Red-and-White Holstein × Red Steppe/recog. 1992

Ukrainian Grey: (C Ukraine)/ m.dr.d/Grey Steppe type/recog. 1910, HB 1935/Russ. *Seraya ukrainskaya* or *Seroukrainskaya* (= *Grey Ukrainian*)/syn. *Ukrainian Grey Steppe*/rare

Ukrainian Oldenburg: *see* Podolian Black Pied

Ukrainian Red: *see* Red Steppe

Ukrainian Red-and-White: (Ukraine)/d/in formation (1992 on) from local Simmental by crossing with Red-and-White Holstein *et al.* for dairy type

Ukrainian Simmental: (Ukraine)/var. of Russian Simmental

Ukrainian Whitebacked: (NW Ukraine)/m/black, red, brown or grey, with white, coloursided/cf. Polish Whitebacked/extinct

Ukrainian Whiteheaded: (NW Ukraine)/d/red (or black) with white head, feet and belly and black eye rings/orig. end of 18th and early 19th c. from Dutch cattle (Groningen Whiteheaded) brought by Mennonites, × local Ukrainian Grey and Polesian/ HB 1926/Russ. *Ukrainskaya belogolovaya*/obs. syn. (to *c.* 1945) *Whiteheaded Colonist* (Russ. *Belogolovokolonistskaya*)/rare

Ukrainskaya belogolovaya: *see* Ukrainian Whiteheaded

Ulten, Ultinger: *see* Grey Adige
Ultimo: *see* Grey Adige
Umblachery: (Thanjanvur, Tamil
Nadu, India)/dr/grey with white
points and back line; born red or
brown with white markings;
steers dehorned and ears
clipped/z/local var. sim. to
Kangayam but smaller/syn. *Jathi
madu, Mottai madu, Southern,
Tanjore, Therkuthi madu*/not
Umbalachery, Umblacheri/rare
Ungarisches Fleckvieh: *see*
Hungarian Pied
Ungarisches Steppenrind: *see*
Hungarian Grey
Unguja Shorthorn: (Pemba and Mafia
Is, Tanzania)/humpless
shorthorn/syn. *Pemba, Mafia
Shorthorn*/extinct
Unterinntal: *see* Tyrol Spotted
Uogherá: *see* Fogera
Uollega: *see* Horro
Upper Baden Spotted: *see*
Messkircher
Upper Bavarian Spotted: *see*
Miesbacher
Upper Swabian Spotted: *see*
Württemberg Spotted
ur: *see* aurochs
Ural: *see* Russian Simmental
Ural Black Pied: (Sverdlovsk,
Chelyabinsk and Perm, Russia)/
d/var. of Russian Black Pied/orig.
1937 on, esp. at Istok State Farm,
from East Friesian × Tagil and
backcross to Friesian/HB
1951/Russ. *Ural'skaya
chernopestraya*
urang: *see* dzo
Urinsk: *see* Yurino
Urla: (Turkey)/extinct
Uruguayan Criollo: (NE Uruguay)/
sim. to Argentine Criollo/single
herd at Chuy
urus: *see* aurochs
Usuk: (C Uganda)/Small East African
Zebu (Teso (= Nkedi) group)/?
orig. from Nkedi and
Karamajong; larger and heavier
than Nkedi/rare (by
interbreeding with Nganda and
Ankole)/*see also* Suk (Kenya)
Uys: (Natal, S Africa)/d.m.dr/early

var. of Drakensberger/? orig. in
late 19th c. by D. Uys from black
Friesian × Afrikander, ? with
Zulu blood, or from Vaderlander
(= Groningen) × local/extinct

Vaalbonte: (Netherlands)/dun-and-
white colour var./rare
Vadhiyar: *see* Kankrej
Vagadia: *see* Kankrej
Vakaranga: *see* Mashona
Valachian Dwarf: (Slovakia)/brown
pied or red pied/Cz. *Valašský*/
not *Walachian*/extinct
Valais: *see* Hérens
Valašský: *see* Valachian Dwarf
Valbona: (Albania)/d.m/grey/= Tyrol
Grey/syn. *Grauvieh* (Ger.),
Oberinntal (Ger.)/rare
Val d'Adige: *see* Grey Adige
Val d'Aosta: *see* Aosta
Valdarno: (Florence and Pisa, Italy)/
former var. of Chianina/extinct
Val di Chiana: (Siena, Italy)/orig. var.
of Chianina/extinct
Valdostana: *see* Aosta
Valdôtaine: *see* Aosta
Valdres: (Norway)/var. of Telemark/
nearly extinct
Val d'Ultimo: *see* Grey Adige
Vale: (Netherlands)/dun colour var./
rare
Vallecaucana, Valle de Cauca: *see*
Hartón
Vallée d'Aosta: *see* Aosta
Valle del Taro: *see* Valtarese
Val Padana: *see* Modenese
Valtarese: (Italy)/dark reddish-grey
var. of Bardigiana/syn. *Valle del
Taro*/extinct
Väne: (Västergötland, SW Sweden)/
d.m/red, white or black/ remnants
of old peasant breed rediscovered
early 1990s/Swe. *Väneko*/[village
of Väne-Ryr]/rare
Varzese, Varzi: *see* Montana
Varzese Ottonese: *see* Montana
Västfinsk: *see* West Finnish
Vatani: (Afghanistan)/usu. black, also
dark brown/var. of Afghani/[=
village (cattle)]
Vaynol: (England)/more prim. var. of
White Park; horns usu. sim. to
Chillingham/orig. feral in Argyll;

at Vaynol (N Wales) 1872–1980
and 1992 on, now also in
England; owned by Rare Breeds
Survival Trust since 1984/?
Indian zebu blood in 1880s/HB
1990/syn. *Faenol*/nearly extinct
Vechur: (Kottayam, S Kerala, India)/
d/red, black or fawn and white/
dwarf var. of local Kerala
nondescript/z/syn. *Dwarf Cow*/
nearly extinct
Velásquez: *see* La Velásquez
Velikokavkazskaya: *see* Greater
Caucasus
Vendée Marsh: *see* Maraîchin
Vendéen group: (Franc)/inc.
Maraîchin, Nantais, Parthenais
Vendée-Parthenay: *see* Parthenais
Vendonnais: (France)/former var. of
Limousin/extinct
Venetian: (S Venetia, Italy)/former
var. of Apulian Podolian/It.
Pugliese del basso Veneto/syn.
Poggese/extinct
Venezuela Criollo: *see* Llanero,
Perijanero, Rio Limón Criollo
Venezuelan Dairy Criollo: *see* Rio
Limón Dairy Criollo
Venezuelan Zebu: (Venezuela)/m/
usu. red or pied/z/orig. from
Brazilian Gir with some Nelore
and Indo-Brazilian blood/BS/Sp.
Cebú venezolano
Verbeterd Roodbont Vleesras: *see*
Dutch Improved Red Pied
Verdeltes Rhodopenrind: *see*
Improved Rodopi
Verinesa: (Verin, S Orense, Spain)/
Morenas del Noroeste group/
orig. from Vianesa by grading to
Mirandesa
Vestland Fjord: (W Norway)/d.m/
dun, red, black, brown or
brindle, with white markings,
coloursided; hd or pd/orig.
1866–1895 from black hd
(Hordaland) and grey pd (Møre
and Ramsdal)/combined with
Lyngdal and Vestland Red Polled
1947 to form South and West
Norwegian/Nor. *Vestlandsk
fjordfe*/syn. *Fjord*, *West Coast
Fjord*, *Western Fjord*, *Westland
Horned*/rare

Vestland Red Polled: (Norway)/red,
sometimes with white markings/
combined with Lyngdal, then
with Vestland Fjord 1947 to form
South and West Norwegian/Nor.
Vestlandsk raudkolle/syn. *West
Coast Red Polled*, *Westland
Polled*/rare
Vianesa: (Viana del Bollo, SE Orense,
Spain)/m.dr.(d)/chestnut/Morenas
del Noroeste group/orig. of
Verinesa/HB 1990/rare
Victoria: (Texas, USA)/m/orig. 1946
from Hereford ($\frac{3}{4}$) and Brahman ($\frac{1}{4}$)
Vietnamese: (Vietnam)/z orig./vars:
Baria, Chau-Doc, Thanh-Hoa,
Tuy-Hoa/syn. *Annamese*,
Vietnam Yellow
Villard-de-Lans: (Isère, SE France)/
m.d.[dr]/yellow-brown/orig.
(from 1850) from local Fémeline
and Bressane; sim. to Mézenc/
federated with Blonde
d'Aquitaine (as var.) 1969–
1976/recog. 1863; HB 1978/not
Villars de Lans/rare
Vit: *see* Iskar
Vladimir: *see* Yaroslavl
Vogelsberg: (Hesse, Germany)/var. of
German Red/BS 1885/Ger.
Vogelsberger/nearly extinct
Vogesen: *see* Vosges
Vogtland: (Saxony, Germany)/var. of
German Red/displaced by German
Simmental in 1935/BS 1885/Ger.
Vogtländer or *Vogtländisches
Rotvieh*/nearly extinct
Volga: *see* Russian Simmental
Vollega: *see* Horro
Volynsk: (Ukraine)/m/var. of
Ukrainian Beef/orig. from
Aberdeen-Angus ($\frac{1}{4}$), Limousin
($\frac{1}{4}$), Hereford ($\frac{1}{4}$), Russian Black
Pied ($\frac{1}{4}$)/recog. 1994/Russ.
Volynskaya
Vorarlberg Brown: *see* Montafon
Vorderwald: (C Black Forest, Baden,
Germany)/d.m.[dr]/red pied with
white face and legs/HB 1896/Ger.
Vorderwälder/*see also* Hinterwald
Vosges: (Alsace, NE France)/d.m.[dr]/
black-and-white, coloursided/HB
1928/Fr. *Vosgienne*, Ger.
Vogesen/rare

VRV: *see* Dutch Improved Red Pied
Vychegda-Vym: (Komi, Russia)/d/ red,
red pied, black or black pied;
pd/absorbed by Kholmogory/Russ.
Vychegodsko-vymskaya/extinct

Wachagga: *see* Chagga
Wadai-Dinka: (Sudan) *see* Nilotic
Wadai-Dinka: (DR Congo)/Sanga/
orig. from Nilotic introduced
early 20th c./? extinct
Wadara, Wadera: *see* Shuwa
Wadhiar: *see* Kankrej
Wadial: *see* Kankrej
Wagad: *see* Kankrej
Wagara: *see* Fogera
Wagyu: *see* Japanese Native; *see* also
American Wagyu
Wahima: *see* Bahima
Wakamba: *see* Kamba
Wakuma: *see* Bahima
Wakwa: (Cameroon) /m/red or red
pied/z/orig. (1953 on) from
Préwakwa (Brahman ×
Adamawa) bred *inter se* and
backcrossed to both breeds; $\frac{1}{2}$
Brahman, $\frac{1}{2}$ N'Gaoundéré (var. of
Adamawa)/nearly extinct
(programme not sustained, small
pure herd maintained on station
only)
Walachian: *see* Valachian
Waldeck: (N Hesse, Germany)/
subvar. of Hesse-Westphalian
var. of German Red/Ger.
Waldecker/extinct
Wäldervieh: *see* Black Forest
Waldviertel: (Lower Austria)/d.m/
white, blond, red or brown/ orig.
from Celtic and German Red,
crossed with Glan-Donnersberg
1940/with Austrian Blond and
Murboden = Austrian Yellow *c.*
1960/Ger. *Waldviertler
Blondvieh*/syn. *Manhartsberg,
Schiltern* (Czech Republic)/not
Waldriert/rare
Wallega: *see* Horro
Wannan: (S Anhui, China)/Changzhu
group
Wanniu: (Gansu and Shanxi, China)/
name for humpless cattle in
ancient China/extinct
Warwickshire: *see* Longhorn

WAS: *see* Ghana Shorthorn, West
African Shorthorn
Watende: (Kenya)/usu. black, occ.
brown; short horns; small hump/
Small East African Zebu, Kenya
cluster/bred by Watende/Kuria
tribes to S of Kavirondo Gulf of L.
Victoria in Suna-Isebania area/
rare
Watusi: (Rwanda, Burundi, and
neighbouring areas of N Kivu,
DR Congo, and NW Tanzania)/
brown, red or black, often with
white patches; huge horns/var. of
Ankole bred by Tutsi/strains:
Inkuku, Inyambo/not *Barundi,
Batusi, Batutsi, Watussi*
Watusi: (USA) *see* Ankole-Watusi
Weebollabolla Shorthorn:
(Australia)/m/local strain of
Shorthorn developed by Munro
family since mid-19th c., closed
herd since 1904/syn.
Weebollabolla/also in S Africa
Weiling: (China)
Weissblaue: *see* Belgian Blue
Welsh Black: (Wales)/m.(d)/orig.
1904 from North Wales Black +
South Wales Black/vars: Ancient
Cattle of Wales, Polled Welsh
Black/HB 1874, BS 1904; BS also
Canada 1971, New Zealand
1974, USA 1975/Welsh *Gwartheg
Duon Cymreig*
Welsh runt: (English Midlands)/=
Welsh Black bullock
Wenling Humped: (EC Zhejiang,
China)/dr.m/yellow to brown,
often with black backline/z/
Changzhu group/not *Wenlin*
Wenshan: (SE Yunnan, China)/dr.m/
yellow; small/ Changzhu
group/var. of Panjiang/syn.
Guangnan
Werdenfels: *see* Murnau-Werdenfels
Wesermarsch: *see* Oldenburg-
Wesermarsch
West African Dwarf Shorthorn: *see*
West African Shorthorn (Dwarf
vars)
West African humpless: (West Africa)/
see Kuri, N'Dama, West African
Shorthorn/syn. *Taurine* (Fr.)
West African longhorns: (West

Africa)/? orig. from Hamitic
Longhorn/see Kuri, N'Dama

WEST AFRICAN SHORTHORN: (Liberia to
Cameroon)/usu. black, also pied
or brown/West African small
humpless (trypanotolerant)
type/inc. Savanna Shorthorn
vars (Bakosi, Bamiléké, Baoulé,
Ghana Shorthorn, Kapsiki, Lobi,
Logone, Namshi, Pabli, Savanna
Muturu, Somba) and Dwarf
Shorthorn vars (Bakweri, Forest
Muturu, Gambia Dwarf, Lagune,
Liberian Dwarf, Manjaca)

West African small humpless: (West
Africa)/trypanotolerant/inc.
N'Dama and West African
Shorthorn group

WEST AFRICAN ZEBU: (West Africa,
mainly dry savanna and Sahelian
belts)/inc. shorthorned
(Azaouak, Fellata, Maure,
Shuwa, Sokoto), medium-horned
(Adamawa, Diali), lyrehorned
(Gobra, Sudanese Fulani, White
Fulani) and long lyrehorned (Red
Bororo); also grouped as Fulani
(Diali, Bororo, Gobra, Red
Fulani, Sudanese Fulani, White
Fulani), Gudali (Adamawa group
and Sokoto) and others
(Azaouak, Maure, Shuwa)

West Coast: (Norway) *see* Vestland
Western Baggara: *see* Baggara
Western Fjord: *see* Vestland Fjord
Western Red: *see* French Western
Red (Rouge de l'Ouest)
Western Red Polled: *see* Vestland
Red Polled
Western Simmental: (Russia) *see*
Sychevka
Westerwald: (Rhineland-Nassau,
Germany)/dr.m.d/red-brown
with white markings/sim. to
Kelheimer/Ger. *Westerwälder/*
syn. *Wittgensteiner Blässvieh/*
extinct *c.* 1940
Westfälisches Rotvieh: *see*
Westphalian Red
West Finnish: (W Finland)/d/red/var.
of Finnish/part orig. of Estonian
Native/BS 1906–1950/Finn.
Länsisuomenkarja, Länsi-
Suomalainen Karja (= *LSK*), Swe.

*Västfinsk/*syn. *Brown Finnish,*
Red Finnish

West Flemish, West Flemish Red: *see*
Belgian Red
West Friesian: *see* Dutch Black Pied
West Highland: (Scotland) *see* Highland
Westland: (Norway) *see* Vestland
West Macedonian: (Greece)/dwarf/
var. of Greek Shorthorn/cf.
Macedonian Blue, Rodopi
West Nile: (NW Uganda) *see* Lugware
Westphalian Red: (S Westphalia,
Germany)/var. of German Red,
with Westerwald blood/BS, HB
1890/Ger. *Westfälisches Rotvieh/*
syn. *Sauerland, Siegerland* (Ger.
Siegerländer)
Westphalian Red Pied: *see* Red Pied
Westphalian
West Sanga: *see* Tswana
West Siberian: (Russia)/m/var. of
Siberian; sim. to Kalmyk/extinct
West Vlaamse: *see* Belgian Red
whitebacked: colour pattern/see
coloursided, esp. Lithuanian
White Back, Polish Whitebacked,
Ukrainian Whitebacked, Witrik;
see also Kłodzka, Pinzgauer,
Sudeten/Ger. *Rückenblessen,*
Rückenscheck
White Bororo: *see* White Fulani
Whitebred Shorthorn: (Cumbria, N
England)/m/white/var. of
Shorthorn for crossing with
Galloway to produce Blue
Grey/HB 1961, BS 1962/syn.
Cumberland White
White Cáceres: (Extremadura,
Spain)/m/white to yellow; lyre
horns/? white var. of Retinta/Sp.
*Blanca cacereña/*syn. *Blanca*
guadianese (= *White*
Guadiana)/rare
white-face: colour pattern/i.e.
Hereford pattern (white head,
belly, feet and tail tip)/see
Hereford, Kazakh Whiteheaded
and Hřbínecký (white more
extensive), Groningen
Whiteheaded, Ukrainian
Whiteheaded and Yaroslavl
(white less extensive)
White Finnish: *see* North Finnish
White Forest: *see* White Park

White Fulani: (N Nigeria and Cameroon)/dr.d.m/usu. white with black skin and points/West African Zebu (lyrehorned) type/Hausa *Bunaji*, Fulani *Yakanaji*, Fr. *Foulbé blanc*/syn. *Akou, White Bororo, White Kano*

White Galloway: (Great Britain)/ white with coloured points/ colour var. of Galloway/HB (with Belted Galloway) 1982; HB in USA (with Galloway) 1973/rare

White Guadiana: *see* White Cáceres

Whiteheaded Colonist: *see* Ukrainian Whiteheaded

Whiteheaded Kazakh: *see* Kazakh Whiteheaded

White Horned: (GB) *see* White Park

White Hungarian: *see* Hungarian Grey

White Kano: *see* White Fulani

White Lake Chad: *see* Kuri

White Lumpoon: *see* Khao Lumpoon

White-marked: (Germany)/dr.m.d/ brown with white markings/inc. Kelheimer, Westerwald/Ger. *Blässiges, Braunscheck*/syn. *Blazed*

White Meuse and Schelde: *see* Belgian Blue

white-middled: *see* belted

White Nile: (Kosti district, Sudan)/ var. of Kenana with Baggara blood

White Park: (Great Britain)/white with black points; long horns/ ancient herds or vars: Chillingham, Cadzow, Chartley, Dynevor, Vaynol; also several new herds/? orig. of British White/HB 1919 (Park Cattle Society), revived 1974, BS 1983; BS also USA 1991/syn. *Park, White Forest, White Horned, Wild White*/nearly extinct

White Park: (USA) *see* American White Park

White Po: *see* Modenese

White Polled: (GB) *see* British White

White-Russian Red: *see* Belarus Red

White Sanga: (S Ghana)/White Fulani × Ghana Shorthorn

White-sheeted Somerset: *see* Sheeted Somerset

White Siberian: *see* Siberian White

White Sindhi: *see* Tharparkar

White Slovenian: *see* Slovenian White

White Tortona: *see* Montana

White Welsh: (Wales)/var. of Ancient Cattle of Wales with colour of White Park/rare

wild: *see* aurochs, banteng, feral, gaur, kou-prey, yak

Wild White: *see* White Park

Wilstermarsch: (Germany)/orig. var. of Red Pied Schleswig-Holstein/ BS 1855/Ger. *Elb- und Wilstermarsch*/extinct

Winam: (lowlands of L Victoria Basin in Nyanza and Western provs, W Kenya)/= Kavirondo/Small East African Zebu (Kenya cluster)/cf. Kavirondo

Wipptal: (Tyrol, Austria)/Ger. *Wipptaler*/extinct (absorbed by Tyrol Grey)

Witblauw ras van Belgie: *see* Belgian Blue

Witrik: (Netherlands)/coloursided (red or black) var./syn. *Aalstreep* (= eelstripe) (Holland), *Ruggelde, Ruggelds, Ruggelings, Witrug*/[= white back]/rare

Witrood ras van Belgie: *see* Belgian White-and-Red

Wittgenstein: *see* Red Wittgenstein

Wittgensteiner Blässvieh: *see* Westerwald

Wodabe: (Cameroon)/intermediate between White Fulani and Red Bororo

Wokalup: (W Australia)/m/cross derived from Charolais, Brahman, Friesian and Aberdeen-Angus or Hereford/ syn. *Wokalup Multibreed*

Wollega: *see* Horro

Woods: (USA) *see* Pineywoods

Wudi Black: *see* Bohai Black

Wuling: (NW Hunan, SW Hubei and NE Guizhou, China)/small mt var. of Changzhu group/local names or vars: Enshi, Sinan, Xiangxi

Württemberg Brown: (Germany)/Ger. *Württembergischen braunvieh*

Württemberg Spotted: (Germany)/ orig. var. of German Simmental/

Ger. *Württemberger Fleckvieh*/
syn. *Upper Swabian Spotted*/not
Wurtemburg
Wuzhumuqin: *see* Ujumqin

Xiangxi: (Hunan, China)/var. of
Wuling
Xinjiang Brown: (NW Xinjiang,
China)/d.m/brown, also yellow
or pied/orig. from Swiss Brown
× Kazakh with German Brown
and Austrian Brown blood in
1977 and 1988/not *Xingjiang*
Xishuangbanna: (S Yunnan, China)/
var. of Yunnan Zebu/syn. *Banna*
Xizhen: (S Shaanxi, China)/usu.
yellow-brown, sometimes brown,
rarely black/Changzhu group/
see Bashan
Xuanhan: (Sichuan, China)/small/not
Xuahan/*see* Bashan
Xuwen: (peninsular Guangdong prov.
and Hainan prov., China)/
Changzhu group/small/cf.
Leizhou

Yacumeño: (Beni, E Bolivia)/m/dun
or red/var. of Bolivian Criollo
selected since 1961 at Espiritu
ranch on R Yacumá from nearly
extinct Beni Criollo/named
1964/syn. *Criollo yacumeño*/rare
yak, domestic: (Tibetan plateau,
Himalayas, Altai, Mongolia)/=
Bos (Poëphagus) 'grunniens'
Linnaeus/usu. black or brown,
also white, grey, blue roan or
pied/Tibetan *g.yag* (♂), Ch. *mao
niu* (= hairy cattle); ♀ *dri*
(Tibetan *'bri*, syn. *bree, dhee*) or
nak (Tibetan *gnag*; for yak-cattle
cross, see dzo
yak, wild: (NC Tibet)/= *Bos
(Poëphagus) mutus* Przewalski/
dark brown to black with grey
dorsal line/orig. of domestic
yak/Tibetan *'brong* (= *drong*), ♀
'brong-'bri (= *drongdri*)/rare
Yakanaji: *see* White Fulani
yakow: *see* dzo
Yakut: (Yakutia, Russia)/m.d/black,
red or spotted with white back
line/formerly var. of Siberian/

Russ. *Yakutskiĭ skot*/syn. *East
Siberian*/not *Yakutian*/rare
Yanbian: (SE Jilin, China)/dr.m/
yellow/Mongolian and Korean,
× local
Yangba: (China)/extinct
yanka: (Bhutan)/♂ backcross of
jatsum (mithun × Siri) to Siri
♂/syn. *yanku*
yankum: (Bhutan)/♀ backcross of
jatsum (mithun × Siri) to Siri ♂
Yaroslavl: (Russia)/d/usu. black (occ.
red) with white head and
feet/orig. in late 19th c., with
Kholmogory blood (*et al.*)/HB
1924/Russ. *Yaroslavskaya*/syn.
Domshino (Russ. *Domshinskaya*),
Vladimir/not *Jaroslav, Jaroslavl,
Yaroslav*
Yellow: (Europe) *see* Austrian
Yellow, Gelbvieh
Yellow and Blond Mountain: *see*
European Blond and Yellow
Yellow and Pale Highland: *see*
European Blond and Yellow
yellow cattle: (China)/= common
cattle (as distinct from buffalo
and yak), not a description of
colour: 45% dark red to dark
yellow, 11% black, others pale
yellow, 6% "tiger stripe", rarely
white head or white back/inc.
humpless (Mongolian or
Northern Yellow group), zeboid
(Huanghuai or low-humped
group) and humped (Changzhu
or high-humped group)/Ch.
Huang niu, Jap. *Kôgyû*
Yellow Dane: (USA)/extinct late
18th c.
Yellow Franconian: (Bavaria,
Germany)/only surviving var.
(and hence syn.) of Gelbvieh/
orig. in late 19th c. from
Simmental, Elling, Scheinfeld *et
al.*/Ger. *Gelbes Frankenvieh*/syn.
Franconian
Yellow Hill: (Germany) *see* Gelbvieh
Yellow Pied: (Germany) *see* Black
Forest
Yemeni Zebu: (Yemen)/dr/usu. red, also
brown, grey or fawn, occ. black
Yola: (NE Nigeria and SW
Cameroon)/red, black, brown or

dun, pied or speckled with white, or blue roan/z/var. of Adamawa with Muturu blood/ Fulani *Tattabareji* (= speckled)/ syn. *Foulbé de Yola, Mayne*/ extinct in Nigeria, rare (? extinct) in Cameroon

Yorkshire: *see* Holderness

Yssel: *see* Meuse-Rhine-Yssel

Yugoslav Brown: (Yugoslavia)/ d.m.dr/brown/Yug. *Mrko-smeda rasa* (= Dark Brown)/rare

Yugoslav Pied: *see* Croatian Simmental, Serbian Pied, Slovenian Pied

Yugoslav Pinzgau: *see* Slovenian Pinzgau

Yugoslav Podolian: *see* Yugoslav Steppe

Yugoslav Steppe: (Croatia and Serbia)/dr.m.d/grey, born reddish; long horns/obs. name for Grey Steppe type in former Yugoslavia/syn. *Podolska, Sivo-stepska, Slavonian Podolian, Yugoslav Podolian*/see Croatian Steppe, Istrian, Serbian Steppe

Yunba: *see* Bashan

Yunnan Zebu: (S and SW Yunnan, China)/South China Zebu type/vars: Dali, Dehong, Xishuangbanna, ? Zhaotong/syn. *Yunnan High-hump, Yunnan Humped*

Yurino: (Mari Republic, Russia)/ d.m/red or brown/orig. from Gorbatov Red and Zillertal (1812–1880) and Swiss Brown (1880–1908), × local, with some Simmental blood/recog. 1943, HB 1937/Russ. *Yurinskaya*/syn. *Nizhegorod*/not *Urinsk, Yurin, Yurinsky*/nearly extinct

Yuzhnoukrainskaya: *see* Southern Ukrainian

Zaërs: *see* Oulmés-Zaërs Blond

Zambia Angoni: (NE Zambia)/var. of Angoni

Zamorana: *see* Sayaguesa

Zanzibar Zebu: (Zanzibar and Pemba Is, Tanzania)/d.m.dr/usu. red (light red, dun, roan, brindle), black or grey/var. of Small East

African Zebu (Tanzanian cluster)/some Indian, Somali and Boran blood; smaller and better milkers on Pemba than on Unguja/nearly extinct

Zaobei: (W Hubei, China)/dr.m/ yellow, brown or red/Changzhu group, sim. to Nayang/WG *Chou-pei*, Ger. *Dschau-bei*/syn. *Chowpei*

Zaosheng: (W Gansu, China)/ upgraded to Qinchuan 1960s– 1980s/not *Zaocheng*/rare

Zavot: (NE Turkey)/d/orig. from Schwyz and Simmental (in Caucasus)/pl. *Zavotlar*/[Russ. *zavod* = factory]

zebkyo: *see* dzo ♂

zeboid: = intermediates between zebu and humpless cattle/inc. Changzhu and Huanghuai groups, Sanga and many new breeds (e.g. Brangus, Droughtmaster, Santa Gertrudis)/ Russ. *Zebuvidnyĭ*/syn. *Azebuado* (Brazil)

ZEBU: = humped cattle/inc. Indian, East African, West African, South China, SE Asian, Brazilian *et al.*/orig. of zeboid breeds/Sp. *cebú*, Fr. *zébu*/syn. *Brahman* (USA)/[? from Port. *gebo* = hump, or from Tibetan *zeu* or *zeba*]

Zebu leiteiro de Uberaba: *see* Dairy Zebu of Uberaba

Zebu mocho: *see* Polled Zebu

Zebu Peul: *see* Fulani

Zébu peul sénégalais: *see* Gobra

Zébu peul soudanais: *see* Sudanese Fulani

Zeburano: *see* Azebuado

Zebuvidnyĭ: *see* zeboid

Zenga: (E Africa)/term for Zebu– Sanga crosses in region where types overlap, esp. E African highlands (Sangas already present, ? mixed with large concentrations of Asian zebu)/ inc. Aradó, Alur (Nioka, Blukwa), Bovines of Tete, Fogera, Horro, Jiddu (Tuni), Nganda, Sukuma (Tinde)/syn. *Sanga-Zebu*

Zetland: *see* Shetland
Zhaotong: (NE Yunnan, China)/? var.
of Yunnan Zebu
zhopkyo: *see* dzo ♂
Zhoushan: (Zhoushan Qundao Is,
Zhejiang, China)/dr.(m)/
black/sim. to Dangjiao Black
(Shanghai)/Changzhu group/rare
zhum: *see* dzo ♀
Zillertal: (Austria)/red or red-brown/
var. of Tux-Zillertal/part orig. of
Gorbatov Red, Tambov Red,
Yurino/Ger. *Zillertaler*/extinct
Znamensk: (Ukraine)/m/var. of
Ukrainian Beef/orig. from
Aberdeen-Angus ($\frac{5}{8}$), Charolais
($\frac{1}{4}$), Russian Simmental ($\frac{1}{8}$)/Russ.
Znamenskaya
zo: *see* dzo
zomo: *see* dzo ♀

zopkio: *see* dzo ♂
Zubairi: *see* Jenubi
Żuławka: *see* Polish Marsh
Zulu: *see* Nguni
zum: *see* dzo ♀
Zwartblaar: *see* Groningen
Whiteheaded
Zwartbont Fries-Hollands: *see* Black
Pied Dutch Friesian
Zwartbont ras van België: *see*
Belgian Black Pied
Zwartbont ras van de Polders: *see*
Polders Black Pied
**Zwartbont ras van het Land van
Herve:** *see* Hervé Black Pied
Zwartbont ras van Oost-België: *see*
Hervé Black Pied
Zwartwitkop: *see* Groningen
Whiteheaded
Zwerg-Zebu: (Germany) = dwarf zebu

Goat

Abbreviations used in this section:
ca = cashmere; d = milk; dr = draught; hd = horned; m = meat; mo = mohair;
pd = polled

Goats are horned unless otherwise indicated

Names for goat include: bok, cabra, capra, chèvre, geit, get, getter, kambing,
keçi, koza, koze, kozy, yagi, ziege

Aardi, A'ardiyah: *see* Aradi
Abaza: (NE Turkey)/d/pinkish white
with coloured marks around
mouth and eyes and on legs; ♂
hd, ♀ usu. pd/syn. *Abkhasian*
(Russ. *Abkhazskaya*)/not
Abchasan
Abergelle: (S Tigray and N Wollo,
Ethiopia/Eritrea)/reddish-
brown/Coastal or Rift Valley
group/orig. from SW Asia
Abgal: (NE Somalia)/usu. pd/var. of
Somali
Abkhasian: *see* Abaza
Abyssinian ibex: (Semien, Ethiopia)/
= *Capra walie* Rüppell (or
C. ibex walie)/syn. *wali*/ nearly
extinct
Abyssinian Short-eared: (Ethiopia)/
many local vars, inc. Central
Highland, Kefa, Western
Highland and Western Lowland
Acamurçada: (Brazil)/= Chamois
Coloured, imported 1910, 1925/
syn. *Sundgau*/[camurça =
chamois]
ACC: *see* Canary Island
Adal: (Dancalia, Ethiopia)/d.(m)/
often tricoloured (white, light

brown and black), also white;
erect ears/syn. *Afar, Danakil*
Adamello Blond: (Brescia, Lombardy,
Italy)/orig. from Toggenburg;
sim. to Bormina/It. *Bionda
dell'Adamello*/rare
Adany: (Iran)/d
Adi Keçi: *see* Anatolian Black
Afar: *see* Adal
Afghan Native Black: *see* Vatani
Africander, Afrikaner: *see* Boer
African Dwarf: *see* West African Dwarf
African Pygmy: (USA)/m.d/often
agouti with dorsal face stripe;
achondroplastic dwarf/orig. from
West African Dwarf 1930–1960/
vars for milk or meat/part orig. of
Kinder and Pygora/HB and BS
1975; HB also Canada/syn.
American Pygmy/rare
Agrigentina: *see* Girgentana
agrimi: *see* Cretan wild goat
Agrupación caprina canaria: *see*
Canary Island
Ak keçi: *see* Çukurova
Akyab: (Jaffna, Sri Lanka)/d/often
cream or fawn; usu. hd/local
dwarf short-legged var./?
Burmese origin/not *Akyat*

Alashan Down: (Helan Shan, Inner Mongolia, China)/ca/white/syn. *Erlongshan Down*/not *Downy*

Albanian: (Albania)/d.m.hr/Balkan type/larger mountain vars: Capore, Dragobia, Hasi, Mati, Shyta; large lowland var.: Velipoja; smaller plains vars: Dukati, Muzhake/*see also* Kolonja Black, Liqenasi, Merturi, Moker, Murme e Zeze, Tranga, Vendi

Albas Down: (Inner Mongolia, China)/ca/white/not *Downy*

Albères: *see* Catalan

Alcaçuz: (NE Brazil)/black; short hair; dwarf/sim. to Graúna/ [Arab. *arq-as-sus* = leguminous savanna shrub]/nearly extinct

Alemã: (Brazil) *see* Toggenburg

Alentejana: (SW Alentejo, S Portugal)/m/var. of Charnequeira/syn. *Machuna*/not *Alemtejana*/name also used for Serpentina

Aleppo: (Turkey) *see* Damascus

Algarvia: (Algarve, S Portugal)/d.(m)/ usu. white with brown or black spots/? orig. from Charnequeira/ nearly extinct

Algerian: (Algeria)/local types: Arabia, Mekatia, Montagnarde des Aurés; *see also* Kabyle, Mzabite

Algerian Red: *see* Mzabite

Aljabali: *see* Syrian Mountain

Allmogegetter: (C and N Sweden)/ d.m.pelt, fur/white with black and brown markings/[= peasantry goat]/nearly extinct

Alman Alaca: *see* German Fawn

Alpen Ziege: *see* German Improved Fawn

Alpina: (Brazil)/inc. Alemã (Toggenburg), Alpina francesca (French Alpine), Branca alemã, Parda alemã, Parda alpina, Parda sertaneja

ALPINE: (Europe, S America)/inc. Brown Alpine, French Alpine, Italian Alpine, Oberhasli, Swiss Mountain

Alpine: (Albania)/d/brown; hanging ears/nearly extinct

Alpine: (USA)/d/inc. esp. French Alpine, also British Alpine, Oberhasli and Rock Alpine/ colour vars: Chamoisée, Cou Blanc, Cou Clair, Cou Noir, Pied, Sundgau

Alpine chamoisée: *see* French Alpine

Alpine ibex: (Italian Alps, reintroduced into Swiss, French, Austrian and Bavarian Alps)/= *Capra ibex ibex* Linnaeus/var. of ibex/syn. *alpine wild goat*/rare

Alpine Polychrome: (Savoie and Haute Savoie, France)/var. of French Alpine/nearly extinct

alpine wild goat: *see* Alpine ibex

Altai Mountain: (Gorno Altai, Siberia, Russia)/ca.m/black with grey undercoat/local improved since 1938 by Don (after failure with Angora and Orenburg)/ recog. as breed group 1968 and as breed 1982/Russ. *Gornoaltaĭskaya*

'American' goat: (USA)/name suggested in early 20th c. for common milking goats in USA

American Lamancha: *see* Lamancha

American Oberhasli: *see* Oberhasli

American Pygmy: *see* African Pygmy

American Tennessee Fainting goat: *see* Myotonic

Anatolian Black: (Turkey)/m.d.hr/ usu. black, also brown, grey or pied/Syrian type/Turk. *Kıl-Keçi* (= hair goat)/syn. *Adi Keçi* (= ordinary goat), *Kara Keçi* (= black goat), *Kılgoat*, *Turkish Native*

Andalusian Black Mountain: (N Jaén, S Ciudad Real and Albacete, Spain)/m.(d)/usu. black, sometimes pied or blue-roan/Sp. *Negra serrana andaluza*/syn. *Castiza*, *Negra serrana* (= Black Mountain)

Andalusian White: (mts of Andalucía, Spain)/m/twisted horns; roman nose/var.: Barros/ Sp. *Blanca andaluza*/syn. *Blanca serrana* (= White Mountain), *Cordobesa* (from Córdoba), *Serrana andaluza* (= Andalusian mountain)/with Castille Mountain = Spanish White

Anglo-Nubian: (Great Britain)/d/

various colours and patterns;
long lop ears; hd or pd/orig. in
late 19th c. from Oriental lop-
eared (Zaraibi, Chitral,
Jamnapari) × Old English/named
c. 1893; HB 1910, BS; HB also
Australia, USA, Canada/syn.
Nubian (USA)

Anglo-Nubian-Swiss: *see* British

Anglo-Swiss: (Great Britain)/term
used in late 19th and early 20th
c. for crossbreds from British and
Swiss breeds (Alpine, Saanen,
Toggenburg)

Angora: (C Turkey)/mo.m/usu.
white/orig. of Indian Mohair,
Soviet Mohair/BS USA 1900, S
Africa 1921 (HB 1906), Australia
1975, France 1982, UK 1984,
Canada; also in Argentina and
Lesotho/Turk. *Ankara*, *Tiftik-
Keçi* (= mohair goat)/syn.
Sybokke (S Africa)

Angora-Don: (Russia)/ca/orig. from
Angora × Don/Russ. *Angoro-
Pridonskaya*/extinct

Ankara: *see* Angora

Antilles: *see* Creole

Appenzell: (NE Switzerland)/d.(m)/
white; pd or hd; long hair/HB
1900/Ger. *Appenzeller*, It.
Appenzellese/syn. *Züricher
Ziege* (Ger.)/rare

Apulian: *see* Garganica, Murge

Aquila: (Abruzzo, Italy)/d.m/white,
brown, grey or black; hd or pd/
Toggenburg, Alpine, Maltese and
Girgentana blood/It. *di L'Aquila*

Aquitaine: *see* Pyrenean

Arab: (Teheran, Iran)/d/black or dark
brown; long twisted horns/? =
Syrian Mountain

Arab: (Somalia) *see* Somali Arab

Arab: (W Africa) *see* Sahelian

Arabi: (Iraq) *see* Iraqi

Arabia: (Laghouat, Algeria)/local
population of steppe goats, cf.
Makatia

Arabian ibex: *see* South Arabian ibex

Aradi: (N Saudi Arabia)/d/usu. black
with parts of nose, mouth and
ears often white; long-haired/
Arabic *A'ardiyah*/not *Aardi*,
Ardhi

Arapawa: (I in Marlborough Sounds,
New Zealand)/brown, black or
pied; long twisted horns; feral/?
orig. from Old English/nearly
extinct (small group removed to
Plimoth Plantation, Mass., USA)

Ardhi: *see* Aradi

Argentata dell'Etna: *see* Etna Silver

Argentinian Criollo: (Argentina)/
m.d/also mo (Patagonia)

Arsi-Bale: (highlands of Arsi, Bale
and S Shewa, Ethiopia)/usu.
white (♂) or brown (♀), also
black-and-grey, roan, red/Coastal
or Rift Valley group

Arusha: (N Tanzania)/var. of (or local
name for) Small East African

Askanian Mohair: (Ukraine)/mo/cf.
Soviet Mohair/extinct

Asmari: (Afghanistan)/white, pied or
coloured/larger than Vatani

Aspri mitou: *see* Machaeras

Assam Hill: (NE India)/m/usu. white
or black; long hair/small var. of
Black Bengal/syn. *Khasi*

Aswad: (Saudi Arabia)/d.m/black;
lop ears; often hd; long
hair/Syrian type/syn. *Baladi*

Attaouia: (SE Morocco)/local name

Attaq: (S Yemen)/sim. to Taiz Black
but coat less curly

Auckland Island: (New Zealand)/
usu. white or pied/first
introduced 1865/extinct on
island, now herd on mainland
only

Aulo: *see* Nepal Hill, Tarai

Aurès: (Algeria) *see* Montagnard des
Aures

Australian Cashmere: (Australia)/
orig. from Australian feral/BS

Australian feral: (Australia)/ca.mo

Auvergne: *see* Massif Central

Azerbaijan: (Armenia and
Azerbaijan)/m.d/usu. black
dappled with red, also black,
red, pied or grey/Russ.
Azerbaĭdzhanskaya/syn. *Long-
haired Caucasian*, *South
Caucasian*, *Transcaucasian*/rare

Azpi gorran: (Basque provs,
Spain)/extinct

Azul: (NE Brazil)/blue-grey
roan/being selected for colour
within SRD and Crioulo/syn.

Azulona, Azulā, Zulanha/[=
blue]/rare

Bach Thao: (Vietnam)/d.(m)/indigenous
type improved with imports from
India/not *De Back Thao*
Bae: (Laos)
Bagot: (England)/black forequarters,
white forequarters; feral/sim. to
Valais Blackneck/? orig. from
Swiss imports in 14th c. or from
British feral/formerly feral at
Blithfield Hall, Staffordshire,
scattered since 1957/ HB/[name
of family]/nearly extinct
Baguirmi: (Chad)/intermediate
between Sahelian and West
African Dwarf
Bahu: *see* Rwanda-Burundi
Baigani: (S Orissa, India)/smaller var.
of Ganjam
Baladi: (Lower Egypt)/d.hr/♂ hd, ♀
hd or pd/Syrian type/var.:
Sharkawi/syn. *Bedouin, Egyptian*
Baladi: (Saudi Arabia) *see* Aswad
Baladi: (Syria) *see* Damascus or
Syrian Mountain
BALKAN: (SE Europe)/inc. Albanian,
Bulgarian, Croatian, Greek, Red
Bosnian
Balkanska: *see* Croatian
Ball Field: *see* Spanish (USA)
Baltistani: (N Kashmir, Pakistan)/
d.m.hr/black with some white;
small
Baluchi: *see* Khurasani
Banat White: (W Romania)/d/orig.
from (Dutch) Saanen and
German Improved White ×
local/Rom. *Rasa de Banat* or
Vansaanen
Bangladesh Dwarf: *see* Black Bengal
Banjiao: (E Sichuan, China)/m/usu.
white/[= flat horn]
Bantu: (N Transvaal, S Africa)/
m/short coat, ears and horns
Barbari: (urban areas of Sind and
Punjab, Pakistan, and Punjab,
Haryana and Uttar Pradesh,
India)/d.m./often white with red-
brown spots; small; short coat,
ears and horns/pd var.: Thori-
Bari/not *Bar-Bari, Barbary, Bari,
Barri*/[? from Berbera in Somalia]

Bari, Barri: *see* Barbari
bariolée: *see* Sahelian
Barka: (lowlands of W and S Eritrea)/
white with brown patches; lop
ears/'Nubian' group
Barki: (NW Egypt)/m.d.hr/usu. black,
often white spots on head and
legs/sim. to Libyan/syn. *Sahrawi*
(= desert)
Barreña: (Sierra Morena, Andalucía
and SE Extremadura, Spain)/
m.d/pied, often chestnut
hindquarters and head with light
forequarters and legs; usu.
hd/var. of Andalusian White
with blood of goats from Barros,
Extremadura/not *Barro*
Barren Island: (Andaman Is, India)/
feral
Barrosã: *see* Bravia
Barwna uszlachetniona: *see* Polish
Improved Fawn
Bashkir: (NW Bashkiria, Russia)/
d.m/var. of North Russian/Russ.
Bashkirskaya koza reduralya (=
Cis-Ural Bashkir goat)/extinct
Bashkir: (Trans-Ural) *see* Orenburg
Basilicata: (S Brazil)/Italian type,
variable
Bastarda: (Italy) *see* Benevento
Batha: (Chad)/local Sahelian
Batina: (Oman)/m./indigenous/syn.
Batinah, Omani Batina
Batu: (Sri Lanka)/indigenous
Bayan-ulgiin khar: *see* Uuliin Bor
Béarn, Béarnais: *see* Pyrenean
Bechuanaland: *see* Tswana
beden: *see* Nubian ibex
Bedouin: *see* Syrian type
Beetal: (Punjab, Pakistan and India)/
d.m/usu. red, black or pied; long
ears; sim. to Jamnapari but
smaller and ♂ long twisted
horns/not *Betal*
Beiari: (Azad Kashmir, Pakistan)/
d.m/white and grey; long
ears/orig. from Beetal ×
Sindhi/syn. *Chamber*
Beiroa: (Beira, Portugal)/d.m/var. of
Charnequeira but smaller, less
spiral horns or pd
Beishan: *see* Guizhou White
Beladi: *see* Sudanese Nubian

Belaya dagestanskaya: *see* Dagestan White

Beldia: (Morocco)/= local, native

Belgian Fawn: (Belgium)/d.(m)/ brown; pd or hd/orig. from local graded to Chamois Coloured/HB 1931/Fr. *Chamoisé*, Flem. *Hertegeit*, Du. *Hertkleurig*/rare

Belgian Red: (W Flanders, Belgium)/ d.m/red or red-and-white/Du. *Rood van Belgie*, Fr. *Rouge de Belgique*/rare

Belgian Saanen: (N Belgium)/d/ white; hd or pd/orig. from local graded to Saanen imported from Switzerland/HB 1931/syn. *Belgian White, Campine*/rare

Benadir: (Webi Shibeli, S Somalia)/ d.m/often red-spotted or black-spotted; lop ears/vars: Bimal, Garre, Tuni/syn. *deguen* or *digwain* (= long ears)

Benevento: (Campania, Italy)/d.m/ tawny red and white; lop ears; usu. pd/? orig. from local with Maltese, Garganica and Alpine blood/It. *di Benevento*/syn. *Bastarda, Capra del Fortore*/ nearly extinct

Bengal: *see* Black Bengal

Berari: (Nagpur and Wardha, N Maharashtra, and Nimar, Madhya Pradesh, India)/ d.m/usu. black

Berber: (Maghreb, N Africa)/d.m/in E sim. to Libyan; in W variable in colour and with shorter ears, horns and hair/inc. Moroccan Berber

Berry-Touraine: *see* Cou-Clair de Berry

Betal: *see* Beetal

bezoar: (Caucasus and S Anatolia through Iran to Baluchistan and Sind, Pakistan)/= *Capra aegagrus* Erxleben/scimitar (or sabre) horns with sharp anterior edge/vars: Cretan wild, Sind wild/orig. of domestic goat *C. 'hircus'* Linnaeus/Persian *pazan* or *pasang*/syn. *Persian wild, wild goat*/[Persian *pád-zahr*, Arabic *bazahr* = antidote (stomach concretion)]

Bhotia: *see* Tibetan

Bhuj: (NE Brazil)/d.m/usu. black, with white or spotted long lop ears; roman nose/orig. from Kutchi/Port. *Bhuj Brasileira*/not *Bhuji, Buhj, Buj, Buji*/[town in N Gujarat, India]

Bhungri: *see* Gujarati

Biała uszlachetniona: *see* Polish Improved White

Biela krátkosrstá koza: *see* Slovakian White Short-haired

Big Black in Shanxi: *see* Shanxi Large Black

Bikanari, Bikaneri: *see* Nachi

Bílá bezrohá krátkosrstá koza: *see* Czech White Short-haired (pd)

Bílá krátkosrstá koza: *see* Czech White Short-haired

Bimal: (S Somalia)/d/white with small dark spots/small coastal var. of Benadir

Binbei Dairy: (China)/d/in formation

Bionda dell'Adamello: *see* Adamello Blond

Biritinga: (Bahia, Brazil)/dark with large light brown or cream spots/local Crioulo/[name of town; possibly from Tupi words *biri, tinga* (= white)]/rare

Black, Black Bedouin: *see* Syrian type

Black-Back: *see* Moxotó

Black Bengal: (Bengal, Bihar, Orissa and NE India, and Bangladesh)/ m.d/usu. black, also brown, white or grey; short ears and coat; bearded; small to dwarf; prolific/var: Assam Hill/syn. *Bangladesh Dwarf, Bengal*

Black Forest: (Germany)/part orig. (with Chamois Coloured) of German Improved Fawn/Ger. *Schwarzwald*/? extinct

Blackhead Ninglang: (Yunnan, China)

Black Moroccan: *see* Moroccan Black

Black Mountain: (Spain) *see* Castiza

Blackneck, Blackthroat: *see* Valais Blackneck

Black Pied Zeeland and South Holland: *see* Dutch Pied

Black Ticino: *see* Verzasca

Black Verzasca: *see* Verzasca

Blanca andaluza: *see* Andalusian White

Blanca celtibérica: *see* Castille Mountain

Blanca de Rasquera: *see* Rasquera White

Blanca española: *see* Spanish White

Blanca serrana: *see* Andalusian White, Spanish White

Blanc néerlandais: *see* Dutch White

Blended goat: (Tanzania)/m/orig. 50% Kamori, 30% Boer, 15% local; now also being crossed with Anglo-Nubian for milk

Bligon: *see* Peranakan Ettawah

Blue goat: *see* Azul, Jining Grey

Boer: (S Africa)/m.(d)/many colours but mainly white, with red head; usu. lop ears; usu. hd./orig. in E Cape from local goats (inc. Bantu and spotted Namaqua Hottentot) with European, Angora and Indian blood/long-haired (Jas Boer), pd (long-eared) and Improved vars selected to eliminate tassels and speckled, dappled or piebald colours; old types included roan (speckled), brindle (Briekwa), mouse-eared (with short ears)/orig. of German Meat goat, Pafuri/BS 1959, HB Germany 1980, BS in GB/Ger. *Burenziege*/syn. *Africander, Afrikaner, South African common goat*

Bohemian: *see* Czech

Bonte Geit: *see* Dutch Pied

Booted goat: *see* Stiefelgeiss

Boran: *see* Somali

Bormina: (Bormio, Sondrio, Lombardy, Italy)/sim. to Adamello Blond/orig. from Toggenburg/not *Bomina*/nearly extinct

Bornova: (Ege region, Turkey)/d.m/ orig. (1985) from Anglo-Nubian × (German Improved White × Maltese) and Anglo-Nubian × [GIW × (GIW × Maltese)]

Bornu White: (Nigeria)/var. of Savanna Brown/syn. *Budumu* (Chad), *White Borno*

Bosnian: *see* Red Bosnian

bouquetin: *see* ibex

bouquetin des Pyrénées: *see* Pyrenean ibex

bouquetin du Gerez: *see* Portuguese ibex

Branca alemã: *see* German Improved White

Branca sertaneja: (Paraíba, Brazil)/ d.m/white/Saanen × local (used as umbrella term for native goats with dairy aptitude and alpine character during breed development)/[= white of the *sertão* (= bush, outback)]/rare

Braune Harzer Ziege: *see* Harz

Bravia: (Minho and Trás-os-Montes, N Portugal)/m/usu. black or brown, also spotted/sim. to Charnequeira/syn. *Barossã, Brava, cabra da Serra*/[= wild]

Brazilian: *see* Acamurçada, Alcaçuz, Alemã, Alpina, Azul, Basilicata, Bhuj, Biritinga, Branca sertaneja, Brazilian Saanen, Brejo, Brown Alpine, Camurça de Sundgau, Canindé, Colônia, Crioulo, Curaçá, Egyptian, Gambia Dwarf, Graúna, Guariba, Gurguéia, Indienne, Mambrina, Marota, Meísta, Meridional, Mestiço, Moxotó, Nambi, Nubiana, Orelha de Onça, Parda alemã, Parda sertaneja, Repartida, SRD, Tauá, Uauá, UDB

Brazilian Saanen: (Brazil)/inc. German Improved White, Dutch White, also Saanen imported from England, USA, Canada, France, Switzerland, Netherlands and New Zealand

Brejo: (Paraíba, Brazil)/local variable Crioulo population in humid zone/[= marshland, swamp]

Brienz: *see* Oberhasli-Brienz

British: (Great Britain)/d/inc. crossbreds from British, Swiss and Anglo-Nubian pedigree parents/HB 1896/syn. *Anglo-Nubian-Swiss*

British Alpine: (Great Britain)/ d/black with light points and face stripes; hd or pd;/orig. from Swiss Mountain (imported 1903) and British/recog. 1921/HB 1925, BS; HB also S Africa 1922, Australia

British Cashmere: (Great Britain)/ orig. from British feral improved by imports from Iceland, Tasmania, New Zealand and Russia/BS

British feral: (mts of Wales, Scotland and N England)/all colours; wide horns in ♂; long coat/diverse orig.

British Saanen: (Great Britain)/d./ white; hd or pd/orig. from Saanen (imported 1903 and 1922) and British/recog. 1921; HB 1925, BS 1980

British Toggenburg: (Great Britain)/d/ brown with light points and face stripes; pd or hd/orig. from Toggenburg (imported 1890s, 1904, 1922 and 1965) and British/recog. 1921; HB 1925, BS/rare

Brown Alpine: (Brazil)/d/usu. fawn with dark face, belly, lower legs and back line/includes all Alpine breeds except Saanen and Toggenburg/Port. *Parda alpina* [*parda* = fawn (colour)]

Brown Bengal: *see* Black Bengal

Brown goat: (Ethiopia) *see* Central Highland

Brown Kano: *see* Kano Brown

Brush: *see* Spanish

Buchi: (Azad Kashmir, Pakistan)/ m.d.hr/black or grey/orig. from Kooti/crossed with Labri to reduce ear length and produce Shurri/[= short ear]

Buduma: *see* Bornu White

Bugi Tori: (Pakistan)

Bugri: (Pakistan)/m/syn. *Bujri*

Bukovica: (Dalmatia, Croatia)/d/ white with black saddle

Bulgarian: (Bulgaria)/Balkan type/Bulg. *mestna koza* (= local goat)/syn. *Bulgarian Landrace*

Bulgarian White Dairy: (Bulgaria)/d/ orig. at Kostinbrod from Saanen × Bulgarian/Bulg. *B"lgarska byala mlechna*/syn. *Bulgarian White Milk*

Bündner: *see* Grisons

Bunte deutsche Edelziege: *see* German Improved Fawn

Burenziege: *see* Boer

Burmese: (Meiktila and Myngyan dists, Myanmar)/red or fawn with white neck band, also pied/syn. *Htain San, Jade Ni*

Burundi: (Burundi)/black, brown, grey, white, often mixed colours; usu. bearded; usu. short coat/ taller than West African Dwarf, sim. to Small East African/*see also* Rwanda-Burundi

cabra da Serra: *see* Bravia

cabra do Gerez: *see* Portuguese ibex

cabra montés: *see* Spanish ibex

cabra montés portuguesa: *see* Portuguese ibex

cabra montez de Portugal: *see* Portuguese ibex

Cabri pays: *see* Réunion Créole

Cagnanese: (Gargano peninsula, Apulia, Italy)/twisted-horn var. of Garganica

Calabrian: (Italy)/fawn or brown/ local var./It. *Calabrese*

Cambrian: (Wales)/white and black; long silky hair; feral/extinct

Cameroon: (Great Britain)/grey with a little black and white; long coat; achondroplastic dwarf/var. of Pygmy/orig. from West African Dwarf/syn. *Blue*/rare

Cameroon Dwarf: (Cameroon)/local West African Dwarf

Camosciata alpina: *see* Chamois Coloured

Camosciata dei Grigioni: *see* Grisons Chamois-coloured

Camosciata delle Alpi: (Italian Alps)/ d/chamois coloured/cf. Chamois-Coloured (Switzerland) and French Alpine/HB 1986/syn. *Passiria Mountain* (It. *Capra di Montagna di Passiria*, Ger. *Passeira Gebirgsziege*)

Campine, Campinoise: *see* Belgian Saanen

Campobasso: (Molise, Italy)/m.d./usu. white, grey or brown; long hair/? orig. from local with Maltese, Garganica and Alpine blood/It. *di Campobasso*/syn. *Montefalcone*/ rare

Camurça de Sundgau: (Brazil)/= Sundgau/[*camurça* = chamois]

Canary Island: (Spain)/d.(m)/any

colour; sabre or twisted horns/inc. Majorera, Palmera, Tinerfeña/BS/Sp. *Agrupación caprina canaria* (= *ACC*)/syn. *Chèvre espagnole, Güera* (Morocco to N Mauritania)

Canindé: (Ceará and Piauí, NE Brazil)/black or dark brown, with cream or tan face stripes and belly (also vars with pale or black heads)/colour type known since 1915 and selected from SRD, ? influenced by Grisons Striped; now being crossed with British Alpine and Toggenburg/ recog. as breed 1999/[town in Ceará, also river in Piauí]

Canton Dairy: (Guangzhou, China)/d/ in formation

Capore: (Pogradec, E Albania)/pied/ larger mt var. of Albanian with long corkscrew horns/[= like a buck]

Capra aegagrus: *see* bezoar

Capra aegagrus blythi: *see* Sind wild goat

Capra aegagrus cretica: *see* Cretan wild goat

Capra caucasica: *see* west Caucasian tur

Capra cylindricornis: *see* tur

Capra falconeri: *see* markhor

Capra ibex ibex: *see* Alpine ibex

Capra ibex nubiana: *see* Nubian ibex

Capra ibex siberica: *see* Siberian ibex

Capra pyrenaica: *see* Spanish ibex

Capra pyrenaica hispanica: *see* Mediterranean ibex

Capra pyrenaica lusitanica: *see* Portuguese ibex

Capra pyrenaica pyrenaica: *see* Pyrenean ibex

Capra pyrenaica victoriae: *see* Gredos ibex

Capra walie: *see* Abyssinian ibex

Capricornis sumatraensis: *see* serow

Cara: (Scotland)/white; long hair; feral

Carpathian: (Carpathian Highlands SE Europe, esp. Romania, Poland)/d.m/many colours in Romania, usu. white in Poland;

twisted horns; long coarse hair/ Pol. *Karpacka*, Rom. *Carpatină*

Carpatina Cashgora: (Romania)/ ca.mo/syn. *Carpathian Cashgora*

Casamance: (Senegal)/local West African Dwarf

Cashgora: (Australia, New Zealand and Great Britain)/ca.mo/Angora × feral or × Cashmere, 1st cross

cashmere: fibre produced by Central Asian Cashmere goat and Russian down goat and by Australian Cashmere, British Cashmere, Chengde Polled, Chigu, Jining Grey, Kurdi, Liaoning Cashmere, Vatani and Wuan/syn. *down* (Russ. *pukh*), *pashmina*/[early spelling of Kashmir]

Castelhana: *see* Serpentina

Castellana: *see* Verata

Castille Mountain: (mts of Guadalajara, Castellón and Albacete, Spain)/m/white; twisted horns/with Andalusian White = Spanish White/Sp. *Blanca celtibérica, Serrana de Castilla y Levante*

Castiza: *see* Andalusian Black Mountain

Catalan: (SE Pyrénées Orientales, France)/d.m/usu. red, also black or pied/syn. *chèvre des Albères*/extinct since 1980

Caucasian: *see* Azerbaijan, Dagestan, Karachai, Mingrelian

Caucasian ibex, Caucasian tur: *see* tur, west Caucasian tur

Celtibérica: *see* Castille Mountain

CENTRAL ASIAN CASHMERE GOAT: ca/usu. white; erect or horizontal ears; twisted horns (occ. heteronymous); long hair/inc. Alashan Down, Albas Down, Changthangi, Hexi Down, Mongolian, Tibetan, Xinjiang/syn. *Cashmere goat, Pashmina goat*

CENTRAL ASIAN LOCAL COARSE-HAIRED GOAT: (Altai, Kazakhstan, Kyrgyzstan, Tajikistan, Turkmenistan and Uzbekistan)/ m.d.ca/usu. black, sometimes grey, occ. tan or pied (non-black commonest in E Kazakhstan and Altai)/Russ. *Mestnye grubosherstnye kozy Sredneĭ Azii*/rare

Central Highland: (highlands of N Ethiopia and S Eritrea)/usu. reddish-brown, also black, white or grey, sometimes pied or spotted/Abyssinian Short-eared group/syn. *Brown goat*

Centro de España: *see* Guadarrama

Cévennes White: (C France)/Fr. *Blanche des Cévennes*/extinct

Chad: (W Africa)/often white, also red, black or pied; long lop ears/ var. of Sahelian/syn. *Arab, Peul*

Chagga: (N Tanzania)/var. of (or local name for) Small East African

Chaidamu: (? China)

Chakarnagar Pari: *see* Jamnapari

Chamba: *see* Gaddi

Chambal Queen: *see* Jamnapari

Chamber: *see* Beiari

Chami: *see* Damascus

chamois: (mts of S and C Europe, Caucasus and E Anatolia)/= *Rupicapra rupicapra* Linnaeus/ [Fr.]

Chamois Coloured: (Switzerland)/ d.m/brown with black face stripes, back stripe, belly and legs; hd or pd/Swiss Mountain group/vars: Grisons Chamois-coloured, Gruyère, Oberhasli-Brienz/orig. of Belgian Fawn, German Improved Fawn, Camosciata delle Alpi, Oberhasli (USA)/HB 1900/Ger. *Gemsfarbige Begirgsziege*, Fr. *Chamoisé des Alpes* (or *Chamoix alpine*), It. *Camosciata alpina*/*see also* Tuxer

Chamoisé: *see* Belgian Fawn

Chamoisé des Grison: *see* Grisons Chamois-coloured

Chamoisée: (USA)/brown or bay, with black face stripes, back stripe and feet/colour var. of Alpine (i.e. French Alpine); "Two-tone Chamoisée": light forequarters, brown or grey hindquarters; "Broken Chamoisée": solid chamois-colour, banded or splashed with another colour (including white)

Changra: *see* Tibetan

Changthangi: (Ladakh, Kashmir, India)/m.ca.pa/white, also black, grey or brown; large twisted horns; small/syn. *Kashmiri, Pashmina goat*/not *Changthong, Chiangthangi*

Chappar: (Sind and Las Bela, Baluchistan, Pakistan)/m.ca.hr.d/ black, white or pied; small/syn. *Jabli, Jablu, Kohistani, Takru*/not *Chapar, Chaper, Chapper*

Charkissar (Down): *see* Uzbek Black

Charnequeira: (Portugal)/m.d/red (or pied); wide twisted lyre horns or pd/vars: Alentejana, Beiroa/? orig. of Algarvia/HB 1988/ [*charneco* = uncultivated area]

Chayangez: *see* Tibetan

Cheghu, Chegu: *see* Chigu

Chekiang: *see* Zhejiang

Chengde Polled: (N Hebei, China)/ m.ca/brown; pd/syn. *Yanshan Polled*

Chengdu Brown: (Sichuan, China)/ m.d/pale to dark brown with dark face stripes and back stripe; occ. pd; small; prolific/syn. *Chengdu Grey, Ma, Mah*/not *Chengtu*

Cheng-Kanni: (India)/var. of Kannaiadu

Cheviot: (England/Scotland borders)/blue-grey; feral/also at Lynton, Devon, and on Lundy I

Chernye pukhovye kozy Uzbekistana: *see* Uzbek Black

Chèvre bariolée: *see* Sahelian

Chèvre commune burundaise: *see* Rwanda-Burundi

Chèvre commune de l'Ouest: *see* Fossés

Chèvre commune rwandaise: *see* Rwanda-Burundi

Chèvre de Casamance: *see* Casamance

Chèvre des Albères: *see* Catalan

Chèvre des Fossés: *see* Fossés

Chèvre des Glaciers: *see* Valais Blackneck

Chèvre espagnole: *see* Canary Island

Chèvre naine de l'est: (Mauritania) *see* Djougry

Chèvre naine des Savanes: *see* West African Dwarf

Chiangthangi: *see* Changthangi
Chigu: (NE Himachal Pradesh and N of Uttar Pradesh, India)/ca.m/ usu. white; long twisted horns/ syn. *Kangra Valley*/not *Cheghu, Chegu*
Chilean Criollo: (Chile)/local Criollo
Chinese dwarf: *see* Chengdu Brown, Fuyang, Jining Grey, Tibetan
Chitral: (N Pakistan)/? = Sirli/part orig. of Anglo-Nubian/extinct
Chkalov: *see* Orenburg
Chubby: *see* Shiba
Chué: *see* Crioulo, SRD
Chungwei: *see* Zhongwei
Chyangra: *see* Tibetan
Chzhun'veiskaya: *see* Zhongwei
Cilento Black: (Salerno, Italy)/d.m/ orig. from local with Garganica blood/It. *Cilentana nera*/syn. *del Cilento, Salernitana*/rare
Cilento Fawn: (Salerno, Italy)/d.m/ red; hd or pd/orig. from local with Red Mediterranean blood/ It. *Cilentana fulva*/syn. *del Cilento, Salernitana*/rare
Cilento Grey: (Salerno, Italy)/d.m./It. *Cilentana grigia*/syn. *di Salerno*/ rare
Cis-Ural: *see* Bashkir
Co: (Ha Giang, Hau Gian and Phu Quoc, Vietnam)/m/syn. *De Co*
COASTAL group: (Ethiopia and Eritrea)/horizontal or erect ears; hd or pd; usu. bearded; some with tassels; short coats/inc. Abergelle, Afar, Arsi-Bale, Worre, Woyto-Guji/syn. *Rift Valley group*
Colônia: (NE Brazil)/dark or cream, with coloured patches or spots (pattern described as "tortoiseshell"); tall/[= colony]/ nearly extinct
Coloured Bohemian: *see* Czech Brown Short-haired
Congo Dwarf: (Congo)/var. of West African Dwarf
Congo Dwarf: (DR Congo) *see* Rwanda-Burundi
Conocchiola: (Gargano peninsula, Apulia, Italy)/shorthorned var. of Garganica
Córdobesa: *see* Andalusian White

Corsican: (Corsica)/d/all colours; long hair/Fr. *Corse*
Cosenza: (Calabria, Italy)/m.d/white, brown or black/? Maltese and Garganica blood/It. *di Cosenza*
Costeña: *see* Málaga
Côte d'Ivoire Dwarf: (Côte d'Ivoire)/ var. of West African Dwarf
cou blanc: (France)/term for pale neck (also *cou clair*) in French Alpine/[= white neck]
Cou Blanc: (USA)/colour type in (French) Alpine: forequarters white, hindquarters black, with black or grey face markings/[= white neck]
cou clair: (France)/term for pale neck (also *cou blanc*) in French Alpine/[= pale neck]
Cou Clair: (USA)/colour type in (French) Alpine: forequarters tan, saffron, off-white or shading to grey, hindquarters black/ cf. Tarentaise colour type (saffron-red head and neck, black body, black face stripes)/[= pale neck]
Cou-Clair de Berry: (C France)/d./ pale neck, shoulders and chest, with black belly, rump and feet/orig. from French Alpine × Poitou/syn. *Race mantelée de Berry Touraine*/extinct
Cou Noir: (USA)/colour type in (French) Alpine: forequarters black, hindquarters white/[= black neck]
Creole: (W Indies)/m/often black or brown, also pied; prick ears; short hair/? orig. from West African Dwarf/cf. Criollo/Fr. *Créole*/syn. *Créole antillaise* (Martinique and Guadeloupe), *West Indian*/*see also* Guadeloupe Créole, Réunion Créole
Cretan wild goat: (Crete, Greece)/= *Capra aegagrus cretica* Schinz/? var. of bezoar; probably descended from early domestic introductions/syn. *agrimi, kri-kri*
Crimean: (Crimea, Ukraine)/m.d/usu. white, white with black head, grey or pied/Russ. *Krymskaya*/ extinct
Criollo: (Spanish America, esp. Mexico, Argentina, Bolivia, Peru and Venezuela)/m.d/Sp. orig. in

16th c./? orig. of Lamancha/[= native]

Crioulo: (NE Brazil)/m.pelt/multi-coloured; short ears; short hair/ orig. from European (? Portuguese) and W African (?)/ orig. of SRD, UDB/incl. various types selected phenotypically (e.g. Azul, Graúna, Moxotó Negra, Nambi) and local designations (e.g. Biritinga, Brejo, Colônie, Meridional, Tauá, Uauá/syn. *Chué* (= ordinary, from Arab. Hisp. *xui* = little)/[= native]

Croatian: (Croatia)/m.d/grey-spotted/ Balkan type/HB/Cro. *Balkanska*

Crvena bosanska: *see* Red Bosnian

Çukurova: (SE Turkey lowlands)/ d.m/orig. 1961 on from Saanen × (Saanen × Kilis)/syn. *Ak keçi* (= white goat)/not *Cukurova*

Curaçá: (NE Brazil)/pale coat/var. of Marota/name for type of goat and also classification of its skin on export market/[Tupi *curussá*, ? = cross]/nearly extinct

Cutch-cross: (N Kerala, India)/d/usu. brown with some black or red spots, also black or white; long lop ears, roman nose; long hair/former var. of Malabari/? orig. from Gujarati and Arab

Cutchi: *see* Kutchi

Cyprus: (Cyprus)/m.d/spotted in hills, white in mts/erect ears; hd or pd; short hair/var.: Machaeras/ orig. of Peratiki/syn. *Cyprus free-range*

Cyprus tethered: *see* Peratiki

Czech Brown Short-haired: (Obvod Kilon borders, Czech Republic)/ d.m./brown with black head and back stripe; erect ears; 80% pd (*bezrohá*); short hair/HB 1928/pd var. almost extinct/composite of indigenous with Harz from Germany/ Cz. *Hnêdá (bezrohá) krátkosrstá koza*/syn. *Coloured Bohemian*/rare

Czech White Short-haired: (Czech Republic)/d.m./orig. from Saanen/pd var. (*bezrohá*) (rare)/ HB 1928/Cz. *Bílá (bezrohá) krátkosrstá koza*

Dadiangas: (S Cotabato, Mindanao, Philippines)/black, brown, white or pied; hd/larger strain of Philippine

Dagestan: (Russia)/m.d/vars: long-haired (ca) (black, white or grey); short-haired (usu. red or black, sim. to Karachai)/Russ. *Dagestanskaya*/syn. *East Caucasian*/rare

Dagestan White: (Dagestan, Russia)/ ca/orig. from Soviet Mohair × local Dagestan/being improved by Altain Mountain/Russ. *Belaya dagestanskaya*

Daira Deen Panah: *see* Dera Din Panah

Daiyunshan: (Fujian, China)/m.pelt/ black

Dalua: (S Orissa, India)/larger var. of Ganjam

Damagaran Dapple-grey: (Nigeria)/ var. of Savanna Brown

Damani: (Dera Ismail Khan, NWFP, Pakistan)/m.d/usu. black with brown head and legs/syn. *Lama*

Damara: (Namibia)/m/many colours, mostly speckled; lop ears; medium horns

Damascus: (Syria and Lebanon)/d/ usu. red or brown, also pied or grey; often tassels; hd or pd; long hair/Nubian type/syn. *Aleppo* (Turk. *Halep*), *Baladi*, *Damascene, Shami* (Fr. *Chami*)

Danakil: *see* Adal

Danish Landrace: (Denmark)/ d.(m)/white, grey, brown or black/orig. from Saanen and Harz × local/Dan. *Dansk landrace*/nearly extinct

Daqingshan: *see* Taihang

Dark Head meat goat: (Shimen county, Hunan, China)/m/large

Da Serra: (Guarda, NE Portugal)/ black; small/var. of Serrana

De Back Thao: *see* Bach Thao

Deccani: (W Andhra Pradesh, India)/d/usu. black, also pied/ sim. to Osmanabadi

De Co: *see* Co

Deer-coloured German Improved: *see* German Improved Fawn

deghier, deg-ier, deg yer: *see* Somali
deguen: *see* Benadir
Delftzijl: (S Africa)/black, brown and yellow; semi-pendulous ears/ indigenous (Pedi) goats collected at Delftzijl by Department of Development/nearly extinct
Dera Din Panah: (Punjab, Pakistan)/ d.hr.m/usu. black, also red-brown; long ears/not *Daira Deen Panah*
Derivata di Siria: *see* Red Mediterranean
Desecho Island: (Desecho and Mona Is off Puerto Rico)/feral/cf. Spanish (USA)
Desert Black: *see* Iraqi
desi: (India) = local, indigenous
Desi: (Pakistan) *see* Jattal, Sind Desi
Dhobini: (E Nepal)/white with black markings/colour var. of Nepal Hill/not *Dhboini*
Dhofari: (Dhufar, Oman)/d.m
dighi yer: *see* Somali
digwain: *see* Benadir
Diougry: *see* Djougry
Djallonké: (Togo, Burkina Faso)/one of two main types in sub-Saharan francophone Africa (*see also* Sahelien)/many traditional races (e.g. Kanem, Massakori, Batha, Lac, Mayo Kebbi); Mossi in Burkina Faso/name widespread but replaced in various countries by local terms (cf. Djallonké sheep): see West African Dwarf
Djelab: *see* Syrian Mountain
Djougry: (Mauritania)/dwarf/local type of West African Dwarf/syn. *Chèvre naine de l'Est* (= Dwarf of the East), *Diougry*
D'man: (oases of Draa and Ziz valleys, SE Morocco)/all colours, often brown or chestnut; often tassels; often pd/? syn. *Hourria*
Dodoma: (C Tanzania)/usu. white or brown, also white-and-black, white-and-brown, occ. black, black-brown or multicoloured; usu. hd
Døle: (E Norway)/d/often blue pied or brown; pd/former var. of Norwegian/Nor. *Dølageit*/syn. *Gudbrandsdal*/[= valley]
Domača bela, Domača križana: *see*

Slovenian White
Don: (R Don basin and Lower Volga, Russia)/ca.d/usu. black, occ. white/white var: Volgograd White; orig. of Altai Mountain/ HB 1934/Russ. *Pridonskaya*
Don-Kirgiz cross: (Krygyzstan)/ca/ orig. from Don × local Kirgiz/ Russ. *Pridono-kirgizskye pomes'*
down: *see* cashmere
Downy: *see* Alashan Down, Albas Down, Hexi Cashmere
Dra, Draa: *see* D'man
Dragobia: (Tropoja, N Albania)/d/ black; lop ears/larger mt var. of Albanian
Drenthe: *see* Dutch Toggenburg
Duan: (Guangxi, China)/m/white, black or pied/not *Tuan*
Dukati: (Vlora, SW Albania)/d/black/ smaller lowland var. of Albanian
Dutch Dwarf: (Netherlands)/orig. from West African Dwarf
Dutch Landrace: (Netherlands)/d/usu. white or pied; ♂ twisted horns; long hair/ BS 1982/Du. *Nederlandse Landgeit*/syn. *Veluwse* (= from Veluwe)/rare
Dutch Pied: (Netherlands)/d/usu. pied/Saanen and Toggenburg blood; orig. breed extinct/new breed with BS 1980/Ger. *Holländer Schecke*, Du. *Bonte geit*/syn. (Du.) *Zwartbonte Zeeuwse en Zuidhollandse geit* (= Black Pied Zeeland and South Holland Goat)
Dutch Toggenburg: (Drenthe, Netherlands)/d/orig. from Toggenburg × local/HB/Du. *Nederlandse Toggenburger*
Dutch White: (Netherlands, except Drenthe)/d/pd/orig. from Saanen (imported 1905–1911) × local/ HB/Du. *Nederlandse witte geit*, Fr. *Blanc néerlandais*/syn. *Dutch White Polled, Improved Dutch, Netherlands White*
dwarf: *see* Akyab, Black Bengal, Congo Dwarf, Djougry, Hejazi, Lapland Dwarf, San Clemente, Sinai, Southern Sudan, Spanish Angora, Tarai, West African Dwarf and its derivatives

Dwarf of the East: *see* Djougry
Dwarf West African: *see* West
    African Dwarf

East African Dwarf: *see* Small East
    African
East Caucasian: *see* Dagestan
east Caucasian tur: *see* tur
Eastern Dwarf: *see* Djougry
Edelziege: *see* German Improved
Egyptian: *see* Baladi, Barki, Saidi,
    Sinai, Wahati, Zaraibi
Egyptian: (Brazil)/obs. term for
    Syrian types (Baladi, Saidi and
    Zaraibi), imported *c.* 1840/obs.
    syn. *Mambrine*
Egyptian Nubian: *see* Zaraibi
English: (England)/variable but often
    light or dark fawn with dark
    stripe on front of legs and along
    back; wide sweeping horns/
    attempt to reform Old English
    from feral stock and British/BS
    1978/rare
English Guernsey: (England)/d/orig.
    1974 on from British breeds graded
    to Golden Guernsey/HB 1976
Epileptic: *see* Myotonic
Erlongshan Down: *see* Alashan Down
Erzgebirge: (Saxony, Germany)/d/
    red-brown with black back
    stripe, belly, legs and face mask;
    pd/HB 1936 (in German
    Improved Fawn HB 1928–1948)/
    not *Erzgebirgian*/nearly extinct
espagnole: *see* Canary Island
Etawah, Ettawa: *see* Jamnapari
Etna Silver: (Sicily, Italy)/d/grey; lop
    ears; occ. pd/It. *Capra argentate
    dell'Etna* (= silvered goat of
    Mount Etna)/rare

Fainting: *see* Myotonic
Faure Island: (W Australia)/mo/
    feral/closer to South African or
    Texas Angora than Australian
Fawn German Improved: *see*
    German Improved Fawn
feral: *see* Arapawa Island, Australian,
    Barren Island, British, Desecho
    Island, Faure Island, Forsyth
    Island, Galapagos, Guadalupe,
    Hawaii, Juan Fernandez, La
    Hague, Mauritius, Montecristo,

New Zealand, San Clemente,
    Santa Catalina
Fergana: *see* Uzbek Black
Fiji: (Fiji)/m/mixed colours/mixed
    origin (inc. Anglo-Nubian,
    Indian and Swiss breeds) in 19th
    and 20th c.
Finnish Landrace: (Finland)/d/usu.
    white, also grey or pied; hd or
    pd; long or short hair/HB
    1932/Finn. *Suomenvuohi,
    Suomalainen vuohi*/rare
Fleischziege: *see* German Meat goat
Flemish: (Belgium)/m.d/inc. Flemish
    White (*Witte Vlaamse*) and
    Flemish Pied or Fawn
    (*Wildkleurig-bonte Vlaamse*);
    sim. to Dutch Pied/Fr.
    Flamande/nearly extinct
Florida Native: (USA)/? = Spanish
    (USA)
Florida Sevillana: (C Seville,
    Andalucía, Spain)/d.(m)/red and
    white speckled/orig. partly from
    Málaga
Fnideq: (N Morocco)/d/black and
    brown with white patches/? orig.
    from Moroccan Black and
    Murcia-Granada
Forest goat: (Germany) *see*
    Thuringian
Forest goat: (West Africa) *see* West
    African Dwarf
Forsyth Island: (New Zealand)/feral
Fortore, Capra del: *see* Benevento
Fossés: (NW France)/Fr. *Chèvre des
    Fossés*/syn. *Chèvre commune de
    l'Ouest*/[= ditches]/nearly extinct
Fouta Djallon, Fouta Jallon: *see* West
    African Dwarf
Franconian: (S Germany)/dark brown
    with black back stripe, belly and
    lower leg/var. of German Improved
    Fawn/Ger. *Frankenziege*/rare
French Alpine: (France)/d/hd or
    pd/orig. from Swiss × local/inc.
    Alpine polychrome (Savoie and
    Haute Savoie, nearly extinct) and
    Alpine chamoisée; former vars:
    Sundgau, Tarentaise/*see also*
    Alpine (USA), also in Canada
French Alpine: (USA) *see* Alpine
    (USA)
French Saanen: (France)/d/orig. from

local graded to Saanen/HB 1939/ Fr. *Gessenay*

Frisia: (Upper Valtellina, N Lombardy, Italy)/m/black or dark brown with white face stripes/? orig. from Grisons Striped/HB/syn. *Frontalasca* (from Frontale), *Lafrisa, Valtellinese*

Frontalasca: *see* Frisia

Fuerteventura: *see* Majorero

Fugin: *see* Fuqing

Fulani: *see* Sahelian

Fuqing: (Fujian, China)/m/black-brown; usu. hd/not *Fugin, Fuqin, Fuquing*

Fuyang: (Anhui, China)/m/white; small; prolific/var. of Huanghuai

Gabaly: *see* Jabaly

Gaddi: (Himachal Pradesh and N of Uttar Pradesh, India)/hr.m.pa/ usu. white, also black or brown; long ears/sim. to Chigu but larger and live at lower altitudes/syn. *Chamba, Gadderan, Gadhairum, Kangra Valley, White Himalayan*/ [Gaddi are tribe of nomads]

Gaddi: (N Pakistan)/hr.d/usu. black, also white or grey; long ears

Gadhairun: *see* Gaddi

Galapagos: (Is off Ecuador)/usu. black, light brown, red or pied; feral/first introduced 1813

Galla: (Kenya) *see* Somali

Gambia Dwarf: (Brazil)/West African Dwarf imported 1984/rare

Ganjam: (S Orissa, India)/m.d/usu. black, occ. brown, white or pied; straight or screw upward horns/vars: Dalua (larger), Baigani (smaller)

Garganica: (Gargano peninsula, Apulia, Italy)/m.d/black to dark brown; long hair/vars: Cagnanese, Conocchiola/HB 1976

Garre: (S Somalia)/d.m/brown-and-white/var. of Benadir/not *Gerra, Gherra*

Gembrong: (E Bali, Indonesia)/white; ♂ hd, ♀ pd or hd; long forelock of ♂ used for lures for fish hooks/nearly extinct

Gemsfarbige: *see* Chamois Coloured

Gerez: *see* Portuguese ibex

Georgian: *see* Mingrelian

German Alpine: *see* German Improved Fawn

German Improved Fawn: (S Germany)/d/red-brown to fawn, with black face stripes, back stripe, belly and feet (pale var. has pale belly and brown feet); hd or pd/orig. from Chamois Coloured (first imported 1887) × native (Black Forest, Harz, Langensalza, Saxon, etc.)/vars: Franconian, Thuringian/HB 1928/Ger. *Bunte* (or *Rehfarbene* or *Rehfarbige*) *deutsche Edelziege*, Port. *Parda alemã*/syn. *Deer-coloured German Improved, German Alpine* (Ger. *Alpen Ziege*), *Alman Alaca* (Turk.)/not *Pied* (or *Spotted*) *German Improved*

German Improved White: (N Germany)/d/hd or pd/orig. from Saanen (first imported 1892) × native (Hessian, etc.)/HB 1928/Ger. *Weisse deutsche Edelziege*, Port. *Branca alemã*/not *German White Purebred* or *Thoroughbred*

German Meat goat: (Germany)/ m/orig. from Boer, imported from South Africa via Namibia and Morocco in 1979 to upgrade German breeds (e.g. German Improved White)/Ger. *Fleischziege*

German Toggenburg: *see* Thuringian

Germasia: (Malaysia)

Gerola: *see* Orobica

Gerra: *see* Garre

Gessenay: *see* Saanen

Gherra: *see* Garre

Ghana Forest: (Ghana)/var. of West African Dwarf

Girgentana: (Sicily, Italy)/d.(m)/ white with brown spots on head and neck; vertical screw horns/ HB 1976/syn. *Agrigentina* (from Agrigento)/[Girgenti is old name for Agrigento]/rare

Glaciers: *see* Valais Blackneck

goatex: = ibex × domestic goat hybrid

Gobi Wool goat: (Mongolia)/ca/dark

grey/orig. from Don × Mongolian (F$_2$ bred *inter se*)/Mongolian *Gobi gurvan saikhan, Govi gurvan saihan*

Gogo: (C Tanzania)/var. of (or local name for) Small East African

Gohilwadi: (S Kathiawar, Gujarat, India)/d.hr/Gujarati type

Golden Gessenay: *see* Golden Guernsey

Golden Guernsey: (Channel Is)/d/ cream to brown; usu. pd/orig. from local selected for colour and improved by Anglo-Nubian and improved British (Swiss) breeds 1920–1950/orig. of English Guernsey/BS (England) 1968, HB 1970/Fr. *guérnesiais*/ syn. (to 1893) *Golden Gessenay*/ rare

goral: (China, Korea, Himalayas)/= *Nemorhaedus goral* (Hardwicke)/ not *gooral, goural, Naemorhedus*

Gorane: *see* Sahelian

Gorki: (Russia)/d/white/orig. from (and sim. to) Russian White by 2 crosses of Saanen/Russ. *Gorkovskaya*/rare

Gornoaltaïiskaya: *see* Altai Mountain

goural: *see* goral

Govi gurvan saihan: *see* Gobi Wool goat

Granada: (Andalucía, Spain)/d/ black/part orig. of Murcia-Granada/HB 1933/Sp. *Granadina*

Grass goat: (Vietnam)

Grassland Dwarf: *see* West African Dwarf

Graubündner: *see* Grisons

Graúna: (NE Brazil esp. Paraíba)/ black/selected for colour from SRD/[Tupi *guaraúna* is name of black bird]/rare

Grauschwarze Gebirgsziege: *see* Peacock goat

Great Barrier Island: (New Zealand)/ feral

Gredos ibex: (C Spain)/= *Capra pyrenaica victoriae* Cabrera/var. of Spanish ibex/rare

Greek: (Greece)/d.m/usu. black, brown or pied, also grey, red or white; hd or pd/Balkan type/var: Ulokeros/*see also* Skopelos

Grey Alpine: (Italian Alps)/It. *Grigio alpina*/nearly extinct

Grey Bengal: *see* Black Bengal

Grey-Black (or Grey-Black-White) Mountain goat: *see* Peacock goat

Greyerzer: *see* Gruyère

Grey Molise: (Italy)/It. *Grigia Molisana*/rare

Grigio Alpino: *see* Grey Alpine

Grigionese, Grigioni: *see* Grisons

Grisons Chamois-coloured: (SE Switzerland)/hd/var. of Chamois Coloured/Ger. *Bündner* (or *Graubündner*) *gemsfarbige Gebirgsziege*, Fr. *Chamoisé des Grison*, It. *Camosciata dei Grigioni*

Grisons Striped: (SE Switzerland)/ d.m/black with pale face stripe and legs; hd or pd/Swiss Mountain group/HB 1935/Ger. *Bündner* (or *Graubündner*) *Strahlenziege*, It. *Grigionese strisciata*/rare

Gruyère: (Fribourg and W Berne, Switzerland)/pd/var. of Chamois Coloured; inc. *Schwarzenburg-Guggisberg*/Ger. *Greyerzer*

Gruzinskaya: *see* Mingrelian

Guadalupe Island: (Lower California, Mexico)/black, grey or white; feral/? orig. from Orenburg

Guadarrama: (Sierra de Guadarrama, C Spain)/d.m/usu. dark coloured or pied; hd or pd; long hair/Sp. *Rasa del Guadarrama*/syn. *del Moncayo y Guadarrama, del Centro de España, Guadarrameña*

Guadeloupe Creóle: (Guadeloupe)/ m.d/usu. black or chamoisée, also white, brown, fawn or grey; usu. hd; small/orig. from mixture of European, African and Indian imports

Guanzhong Dairy: (Shaanxi, China)/ d/white; usu. pd/orig. from Canadian Saanen × local since 1940s/syn. *Guanzhong Milk goat*/not *Guanzhou Dairy*

Guanzhong White: (Shaanxi, China)/m/hd or pd

Guariba: (NE Brazil)/dr/white/sim. to Marota but larger/syn. *Turino*/

[name of town in S Paulo and river in Piauá]

Gudbrandsdal: *see* Døle

Güera: (Morocco to N Mauritania)/ orig. from Canary Island/syn. *Chèvre espagnole, Spanish goat*

Guernsey: *see* English Guernsey; Golden Guernsey

Guggisberg: *see* Gruyère

Guinea, Guinean, Guinean Dwarf, Guinéenne: *see* West African Dwarf

Guizhou Black: (China)/colour var. of Guizhou White

Guizhou White: (Guizhou, China)/ m.pelt/usu. white, occ. black or pied; usu. hd/syn. *Beishan, Tongren White*

GUJARATI type: (W Gujarat and W Rajasthan, India)/d.hr/black; long lop ears (usu. white or spotted), roman nose; twisted horns; long hair/inc. Gohilwadi, Kutchi, Maraari, Mehsana, Zalawadi/syn. *Bhungri*

Gürcü: *see* Mingrelian

Gurguéia: (NE Brazil, esp. Piauí)/ straw-coloured to chamois or red, with black belly, back and legs and 'badger' face pattern/ crossed with French Alpine = Parda sertaneja/[name of river in Piauí]/nearly extinct

Gwembe goat: (Zambia)

Hailun: (Heilongjiang, China)/d/ white, also black, pied, grey, brown or yellow/orig. from Saanen and Toggenburg × local

Haimen: (Zhejiang, China)/m/white; prolific/syn. *Shanghai White, Zhejiang White*

Halep: (Turkey) *see* Damascus

Harerge Highland: (SE Ethiopia)/usu. white; hd or pd/Somali group

Harz: (Germany)/whitish-grey to reddish-brown, occ. black or black-and-brown/part orig. (with Chamois Coloured) of German Improved Fawn/Ger. *Braune Harzer Ziege*/nearly extinct

Hasi: (Kukesi, NE Albania)/m.d/ reddish; lop ears/larger mt var. of Albanian/not *Hasit*

Hasli: *see* Oberhasli-Brienz

Haute Provence: *see* Roya Vésubie

Haute Roya: *see* Roya Vésubie

Haut Valais: *see* Valais Blackneck

Hawaii ferals: (Hawaii)/usu. black or brown, also silver; feral/? orig. from introductions by Capts Cook (1778) and Vancouver (1792)/severely culled in 1970s/local syn. *kao, kunana*

Hebei Dairy: (China)/d/in formation

Hebridean: (Hebrides and Argyllshire, Scotland)/usu. white; twisted horns; long coat/ extinct

Hebsi: (Saudi Arabia)

Heilongjiang Dairy: (China)/? = Hailun

Hejazi: (Arabia)/m/usu. black; long hair/sim. to Syrian Mountain but dwarf

Hemitragus: *see* tahr

***Hemitragus jayakari*:** *see* Arabian tahr

***Hemitragus jemlahicus*:** *see* Himalayan tahr

***Hemitragus hylocrius*:** *see* Nilgiris tahr

Henan Dairy: (China)/d/in formation

Hertegeit: *see* Belgian Fawn

Hertkleurig: *see* Belgian Fawn

Hessian: (Germany)/orig. (with Saanen) of German Improved White/Ger. *Hessen*/extinct

Heuk Ymso: (Korea)/m/small

Hexi Cashmere: (N Gansu, China)/ca/ usu. white, also black, brown or pied/syn. *Hexi Down*/not *Downy*

Himalayan wild: *see* goral, serow, tahr

Hinterwald: (Germany)/local/extinct

Hnedá krátkosrstá koza: *see* Czech Brown Short-haired, Slovakian Brown Short-haired

Holländer Schecke: *see* Dutch Pied

Hongtong: (Shanxi, China)/m/white; hd or pd; prolific/var. of Huanghuai

Hottentot: (S Africa)/speckled; short-haired; small/poss. Nubian orig./ later syn. *Boer* to differentiate from Angora/*see also* Skilder

Hourria: *see* D'man

Htain San: *see* Burmese

Huai: *see* Huanghuai

Huaipi: (Henan, China)/m/white; hd
or pd; prolific/var. of Huanghuai
Huaitoutala: (Qinghai, China)/ca
Huanghuai: (China)/m/white/vars:
Fuyang, Huaipi, Xuhai/syn. *Huai*
Huertana: *see* Murcian
Hungarian Curly: (Hungary)/Hung.
Magyar tincse
Hungarian Improved: (Hungary)/d./
black, white, red or cream, with
white patches/orig. from Swiss
dairy breeds (esp. Saanen) ×
local/Hung. *Magyar kecske*

ibex: = *Capra ibex* Linnaeus/scimitar
horns with flat anterior edge/
vars: Abyssinian, Alpine,
Nubian, Siberian; *see also*
Spanish ibex/Fr. *bouquetin*, Ger.
Steinbock, It. *stambecco*
Ibicenza: (Ibiza, Spain)
Icelandic: (Iceland)/m.ca/usu. white,
black or grey, occ. pied; occ. pd;
long hair; highly inbred (closed
population for centuries)/
Icelandic: *Islenska geitin*/rare
Improved Boer: (South Africa)/m/red
head, white body; short hair;
prolific/BS 1949
Improved Dutch: *see* Dutch White
Improved German: *see* German
Improved
Improved North Russian: *see*
Russian White
Improved Polish: *see* Polish
Improved
Indian long-haired white: *see* Chigu,
Gaddi
Indian Mohair: (Maharashtra,
India)/mo/orig. from Angora ($\frac{7}{8}$)
× Sangamneri since 1973
Indienne: (Brazil)/old designation for
goats with long legs, lop ears,
roman nose, usu. imported by
zebu cattle breeders/obs. syn.
Mambrino
Indo-Chinese: (Indo-China, esp.
Tongking)/m.(d)/usu. fawn,
sometimes with white or black
extremities and back; erect ears;
short horns; short hair
Indonesian: *see* Katjang

Ingessana: (Sudan)/local var. of
Southern Sudan
Inner Mongolian Cashmere: *see*
Mongolian
Ionica: (Taranto, Apulia, Italy)/d/
white; lop ears; pd or hd/orig.
from Maltese × local/HB 1981/
not *Jonica*/[from Ionian sea]
Iraqi: (Iraq)/usu. black, usu. with
white on ears; lop ears/Syrian
type/syn. *Arabi, Desert Black*/*see
also* Kurdi
Irish: (Ireland)/d.m/grey, white or
black; long hair/HB 1918
Islenska geitin: *see* Icelandic
Israeli Saanen: (Israel)/d/orig.
1932–1950 from local Syrian
Mountain graded to Saanen
Istrian: (Gorizia, NE Italy)/d.m/
white; pd/? Slovenian orig./It.
Istriana/nearly extinct
Italian Alpine: (N Italy)/d.m/sim. to
Swiss Mountain/It. *Alpina*/*see
also* Camosciata delle Alpi, Grey
Alpine
Italian Saanen: (Italy, esp. Piedmont,
also Emilia and Apulia)/d/orig.
from Saanen

Jabal Akhdar: (Akhdar mt, Oman)/
local
Jabali, Jabel: *see* Jabaly, Syrian
Mountain
Jabaly: (mts in S and W Saudi
Arabia)/black, brown, grey, white
or mixed; prick ears; small/syn.
Gabali, Jabali, Jebeli, Jebli/[=
mountain]
Jabli, Jablu: *see* Chappar
Jade Ni: *see* Burmese
Jakharana, Jakhrana: *see* Jhakrana
Jamnapari: (Etawah, Uttar Pradesh,
India)/d.m/usu. white with tan
or fawn patches on head and
neck, often now wrongly black
with red spots etc. (from
crossbreeding); long ears, roman
nose; tall/sim. to Beetal but
larger/var.: Ramdhan/syn.
*Chambal Queen, Chakarnagar
Pari, Etawah*/not *Jamnabary,
Jamna Para, Jamunapari, Jumna
Pari, Jumnapari, Jumunapari,
Yamnapari*/rare in Etawah/

[Chakarnager region, rivers
Jamuna, Chambal and Kwari]
Jämtland: (Sweden)/var. of old
Swedish Landrace, formerly at
Thüringer Zoopark, Erfurt,
Germany/rare
Japanese Saanen: (Japan)/d
Jarakheil: (N Kashmir, Pakistan)/
hr.d.m/usu. black with white
patches, also brown with white;
long ears
Jarmelista: (Jarmelo, NE Portugal)/d/
black or brown with yellowish
streaks/ var. of Serrana/syn.
Jarmelênce
Jattal: (Azad Kashmir, Pakistan)/
m.hr/black/? cross of Pothohari
× local/syn. *Desi*
Jebel, Jebli: *see* Jabaly, Syrian
Mountain
Jhakrana: (Alwar, E Rajasthan, India)/
d/black, with white spots on ears
and muzzle/syn. *Zakhrana*/not
Jakharana, Jakhrana
Jianchang Black: (Sichuan, China)/
m/usu. black, occ. white, yellow
or roan
Jining Grey: (SW Shandong, China)/
fur pelt, ca/mixed black and
white hairs, occ. wholly black or
white; small; prolific/not *Blue
goat, Jining Gray, Jining Green*
Jonica: *see* Ionica
Jordanian: (Jordan)/often black, also
brown, fawn, white/Syrian type
Juan Fernandez: (Is off Chile)/often
reddish with black dorsal and
shoulder stripes, also black or
pied/feral; first introduced
(? from Spain) 1573–1580
Jumna Pari, Jumunapari: *see*
Jamnapari

Kabuli: *see* Vatani
Kabyle: (Kabyle and Dahra mts, N
Algeria)/hr/usu. brown, some
brown-and-black, occ. with
white markings; proportional
dwarf (dwarfism and shorter ear
length distinguish from
Montagnarde des Aurès)/Fr.
Naine de Kabylie (= Kabyle
Dwarf)
Kacang: *see* Katjang

Kacchan: (Pakistan)/m
Kafkas: (Turkey) *see* Mingrelian
Kaghani: (Hazara, N Pakistan)/
m.hr/usu. black, also grey or
white; long hair/Central Asian
cashmere group
Kail: (Pakistan)/m/? = Kooti
Kairi: *see* Khairi
Kajli: (Loralai, Baluchistan, and Dera
Ghazi Khan, Punjab, Pakistan)/
m.hr.d/usu. black, also white,
brown or grey, with face stripes;
long hair/syn. *Pahari, Pat, Pattu*
(Kashmir)
Kalahari Red: (S Africa)/m/red; lop
ears; short hair/recently bred
from Improved Boer and
unimproved indigenous goats;
slightly smaller than the former/
recog. 1998; BS
Kali: (E Nepal)/black/colour var. of
Nepal Hill
kambing: = goat (Indonesia and
Malaysia)
kambing gurum: *see* serow
Kambing Katjang: *see* Katjang
Kambing Maritja: *see* Maritja
Kamon: *see* Kamori
Kamori: (Sind, Pakistan)/d.(m)/usu.
red-brown, black or white; very
long ears/syn. *Kamorai*
(Tanzania)/not *Kamon*/[= fed on
kamo, a creeper]
Kandahari: *see* Vatani
Kandari Ka Khana: *see* Ramdhan
Kanem: (Chad)/local name for
Sahelian
Kangra Valley: *see* Chigu, Gaddi
Kannaiadu: (S Tamil Nadu, India)/m/
black, or black with white spots;
♂ hd, ♀ pd; tall/vars: Pat-Kanni,
Cheng-Kanni/syn. *Karapuadu,
Pullaiadu*/not *Kanni Adu*
Kano Brown: (Nigeria)/var. of
Savanna Brown/syn. *Kyasuwa,
Zinder Brown* (Niger)
kao: *see* Hawaiian ferals
Karachai: (N Caucasus, Russia)/
m.d.ca/pied, red, grey, black or
white/Russ. *Karachaevskaya*/
syn. *North Caucasian*/rare
Karadi: *see* Kurdi
Kara Keçi: *see* Anatolian Black
Karapuadu: *see* Kannaiadu

Karpacka: *see* Carpathian
Kashmiri, Kashmiri Pashmina: *see* Changthangi
Katchi: *see* Kutchi
Kathiawari: *see* Kutchi
Katjang: (Malaysia and Indonesia)/m/ usu. black or black-and-white, also brown or pied; erect ears; short hair; often small/var.: Maritja/? orig. of Philippine/syn. *Kacang* (Indonesia), *Kambing Katjang* (= bean goat)/not *Katjan*
Katsina Light-brown: (Nigeria)/var. of Savanna Brown
Kazakh: (Kazakhstan)/m.d.ca/Central Asian coarse-haired group
Kazimierz: (EC Poland)/d/black, with amber eyes/Pol. *Kazimierzowska*/extinct
Kefa: (Kefa, Illubabu and S Shewa, Ethiopia)/d.m/black, brown or red, sometimes pied; short ears; coarse hairy coat/Abyssinian Short-eared group
Kel: ? = Kooti
Kenya: *see* Galla, Small East African
Kenya Dual-Purpose: (Kenya)/d.m/ orig. 1990s from Small East African ($\frac{1}{4}$), Galla ($\frac{1}{4}$), Toggenburg ($\frac{1}{4}$), Anglo-Nubian ($\frac{1}{4}$)/syn. *Kenya DPG*
Khairi: (E Nepal)/brown, sometimes with white or black markings/ colour var. of Nepal Hill/not *Kairi, Khare, Khari*
Khandeshi: *see* Surti
Khare, Khari: *see* Khairi
Khasi: *see* Assam Hill
Khosa: (Ciskei and Transkei, E Cape, S Africa)/usu. white, also roan or pied/extinct (by cross-breeding with Angora for mohair)
Khurasani: (N Baluchistan, Pakistan)/d.m.(hr)/usu. black, also white or grey/syn. *Baluchi*/not *Khurassani*
Kigezi: (SW Uganda)/m/black or grey/long-haired var. of Small East African
Kigoma: (NW Tanzania)/black, also brown or black-brown, occ. pied or multicoloured, rarely white; rarely pd/indigenous in subhumid region
Kiko: (New Zealand)/m/orig. from large dairy ♂ (e.g. Anglo-Nubian) × New Zealand base stock backcrossed to dairy ♂ and then selected for twinning, growth rate and constitution/[= meat (Maori)]
Kilgoat: *see* Anatolian Black
Kilis: (SE Turkey)/d/usu. black; lop ears/orig. from Damascus × Anatolian Black
Kıl-Keçi: *see* Anatolian Black
Kil melezi: (Turkey)/= hair crossbred
Kinder: (Washington, USA)/m.d/orig. (?1980s) from African Pygmy × Anglo-Nubian/BS, HB 1988/ nearly extinct
Kirdi: (S Chad and N Cameroon)/ slightly larger var. of West African Dwarf/syn. *Kirdimi*
Kirgiz: (Kyrgyzstan)/Central Asian coarse-haired group/orig. of Don-Kirgiz cross/Russ. *Kirgizskaya*
Kodi Adu: (India)/new breed
Kohai Ghizer: (N Kashmir, Pakistan)/ d.m.hr/black with white on belly; small
Kohsitani: *see* Chappar
Kolonja Black: (Albania)/m.d
Kooti: (Azad Kashmir, Pakistan)/ d.m.hr/black-and-white/orig. of Buchi
Korean: (S Korea)/m/usu. black/vars: grey (smaller, rare), white (being graded up to Saanen)/sim. to Taiwan Black
Korean Native Black: *see* Korean
Kosi: (Cameroon)/local West African Dwarf
Kosta: (Indonesia)
Kottukachchiya: (Sri Lanka)/m/usu. black, also pied or brown; short lop ears; usu. hd; short coat/orig. at Kottukachchiya farm from S Indian imports
kri-kri: *see* Cretan wild goat
Krymskaya: *see* Crimean
Kuban ibex, Kubanski goat: *see* west Caucasian tur
kunana: *see* Hawaiian ferals
Kunyi: *see* Surti
Kurdi: (Kurdistan, Iraq and Iran)/ca/ white, black or brown/syn. *Karadi, Kurdish, Marghaz, Markhoz, Morghose,* or *Morghoz* (Iran)

Kutchi: (NW Gujarat, India)/d.m.hr/
  Gujarati type/orig. of Bhuj
  (Brazil)/syn. *Kathiawari*/not
  Cutchi, Katchi
KwaZulu Natal: (S Africa)/black,
  white, yellow or grey, whole or
  mixed; hd/Nguni herd at Bartlow
  Res. Sta./nearly extinct
Kyasuwa: *see* Kano Brown

Labi: (Sibi, Baluchistan, Pakistan)/
  short-eared var. of Lehri
Labri: (Azad Kashmir, Pakistan)/
  d.m.hr/usu. black; very long ears
Lac: (Chad)/local name for Sahelian
Ladakh Pashmina: (India)/ca/hd;
  very long straight coat/sim. to
  cashmere or Kangra (Gaddi)
  goats of Tibet
Lafrisa: *see* Frisia
La Hague wild: (Joubourg cliffs,
  Normandy, France)/feral/nearly
  extinct
Lama: *see* Damani
Lamancha: (Oregon, USA)/d/all
  colours; earless/orig. in Texas
  from short-eared Spanish goats
  from Mexico/recog. 1958; HB
  1958, BS 1969; HB also Canada/
  syn. *American Lamancha*/not
  LaMancha/[foundation inc. one
  goat from La Mancha, Spain]
Lamkana: *see* Tarai
Landim: (Tete prov. and S of
  Limpopo, Mozambique)/m/usu.
  dark brown, black or pied/orig.
  (with Boer) of Pafuri/syn.
  Mozambique/[= native]
Langensalza: (Thuringia, Germany)/
  part orig. (with Chamois
  Coloured) of German Improved
  Fawn/extinct
Lantras: *see* Swedish Landrace
Laoshan Dairy: (Shandong, China)/
  d/white/orig. from Saanen (first
  imported 1904 from Germany) ×
  local since 1919/not *Laushan,
  Loushan*
La Palma: *see* Palmero
L'Aquila: *see* Aquila
Lapland Dwarf: (N Norway)/d/usu.
  white, also yellow or pied;
  bezoar horns
Latuka-Bari: (S Sudan)/local var. of
  Southern Sudan

Laushan Dairy: *see* Laoshan Dairy
Lehri: (Sibi, Baluchistan, Pakistan)/
  m.hr/usu. black, also white or
  grey; long ears/var: Labi (short
  ears)/not *Leri, Lerri*
Leizhou: (Guangdong, China)/m/usu.
  black, also white or pied;
  prolific/South China type
Leri, Lerri: *see* Lehri
Liaoning Cashmere: (Liaoning,
  China)/ca/white; ♂ lateral
  twisted horns
Libyan: (Libya)/m.d/black, brown,
  grey, white or pied; ears usu. lop;
  long hair/Syrian type, sim. to
  Barki and Baladi (Egypt)
Liqenasi: (Albania)/d/black
Lithuanian native: (Lithuania)
Livo: *see* Val di Livo
Lohri: (Pakistan)/? = Lehri
Long-haired Caucasian: *see*
  Azerbaijan
Longlin: (Guangxi, China)/m.d/often
  white, also black-and-white,
  brown or black/not *Longling*
Longlinghuan: (mts of W Yunnan,
  China)/m/large
Lori: (S and C Iran)/d/black or brown
Loskop South: (S Africa)/various
  colours/indigenous; based on
  remnants of Khosa goats of
  Ciskei 1988–1991 at Loskop-
  South Res. Sta., Groblersdal/
  nearly extinct
Loushan Dairy: *see* Laoshan Dairy
Lunangui: (Yunnan, China)/m

Ma, Mah: *see* Chengdu Brown
Macedonian: (Greece)/hr/usu. black
  (often with white back stripe),
  sometimes grey; usu. hd; long
  hair/Balkan type/herd at
  Thessaloniki Univ.
Machaeras: (Troödos mts, Cyprus)/
  d.m/white/hd or pd/Gk *Aspri
  tou Machaera*/syn. *Aspri mitou*
  (= white short-eared)/[Machaeras
  monastery]/rare
Machuna: *see* Alentejana
Madara: (Madara I, Saga, Japan)/
  various colours; feral/possibly
  orig. from NE Asia and interbred
  with Saanen and other dairy
  breeds

Madjd: *see* Tali
Madu: *see* Matou
Magyar: *see* Hungarian
Mahrana: (NW India)/m
Majorcan: (Balearic Is, Spain)/red
with black pigmentation/Sp.
Mallorquina
Majorera: (Canary Is, esp.
Fuerteventura and Lanzarote,
Spain)/d/usu. pied; sabre horns;
short coat/var. of Canary Island
Makatia: (Laghouat, Algeria)/grey,
brown, white or blond; short
hair/var. of Sahelian
Malabari: (N Kerala, India)/d.m/
white, black, brown or pied; hd
or pd; short or long hair/orig.
Arab × Indian/formerly divided
into Tellicherry and Cutch-cross/
syn. *Tellicherry, West Coast*/not
Malbari, Tellichery
Málaga: (S Andalucía, Spain)/d/
sandy to red; pd or hd; prolific/
vars: with twisted horns (*prisca*),
with sabre horns (*clássica*) or pd
(*moderna* or *mejorada*, =
improved)/part orig. of Florida
Sevillana/HB 1977/Sp.
Malagueña/syn. *Costeña* (=
coastal)
Malawian: (Malawi)/m/var. of Small
East African
Mallorquina: *see* Majorcan
Maltese: (Malta, now mainly S Italy
and other Mediterranean
countries)/d/usu. chestnut or
brown, often cream with black
patches on head and neck, also
black, white, grey or pied;
horizontal or lop ears; usu. pd;
long or short coat/HB Italy 1976/
Maltese *Moghza Maltija*/rare in
Malta
Malwi: (Madhya Pradesh, India)/
local in Malwa region
Mamber: *see* Syrian Mountain
Mambilla: (Nigeria)/var. of Savanna
Brown
Mambrina Brasileira: (Brazil)/usu.
reddish-brown; long lop ears,
roman nose; twisted horns/orig.
from Syrian Mountain crossed with
locals/Fr. *Mambrine Brésilienne*
Mambrine: *see* Syrian Mountain

Maradi: (S Niger)/skins/red; prolific/
= Red Sokoto (Nigeria)/syn. *Red
Maradi* (Fr. *Rousse de Maradi*)
Marghaz: *see* Kurdi
Maritja: (Sulawesi, Indonesia)/m/
black, brown, white or pied/
small var. of Katjang/syn.
Kambing Maritja (= pepper goat)
markhor: (NE Afghanistan and
adjacent mt areas in S
Uzbekistan, SW Tajikstan, W
Pakistan and W Kashmir)/=
Capra falconeri Wagner/screw
horns/natural and deliberate
crossbreeding with local
domestic goats, hybrids prized as
stud animals in Pakistan/not
markhoor, markhore, markhorn/
[Persian = snake eater]
Markhoz: *see* Kurdi
Marota: (NE Brazil)/white coat, dark
skin/colour type selected from
SRD/var.: Curaçá/[soubriquet for
Portuguese in Brazil, esp. Bahía,
after independence]/rare (one
pure herd preserved)
Marungu: (DR Congo)/d.m/syn.
Mayema
Marwari: (W Rajasthan and N
Gujarat, India)/d.m.hr/black, occ.
brown with white markings; long
lop ears; usu. hd; long hair/
Gujarati type
Masai: (Kenya and N Tanzania)/m.d/
var. of Small East African
Mashona: (E and C Zimbabwe)/m/
var. of Small East African/syn.
Zimbabwe
Massakori: (Chad)/local name for
Sahelian
Massif-Central: (France)/d/brown or
black; long hair/crossed with
French Alpine in E./syn.
Auvergne/? extinct since 1980
(being revived by BS)
Matabele: *see* Ndebeli
Mati: (Mat, Albania)/d/reddish
brown/larger mt var. of Albanian
Matou: (Hubei and Hunan, China)/
m/white; pd; long or short hair/
WG *Ma-t'ou*/not *Madu*/[= horse
head]
Maure: (Mauritania and N Mali)/red
pied or fawn pied/var. of
Sahelian

Mauritanian: *see* Djougry, Güera, Maure

Mauritian: (Mauritius)/m/often black/sim. to Katjang/mixed orig.

Mawr: (Wadi Maur, N Tihama, Yemen)/d.m/almost white, with small black spots

Mayema: *see* Marungu

Mayo Kebbi: (Chad)/local name for Sahelian

Mediterranean ibex: (mts of S Spain)/= *Capra pyrenaica hispanica* Schimper/var. of Spanish ibex/nearly extinct

Megrel: *see* Mingrelian

Mehsana: (Mahesana, Gujarat, India)/d.m.hr/black, with white ears; long hair/Gujarati type

Meísta: (NE Brazil)/grey, with black back stripe and belly ("badger"), or reddish forequarters with black hindquarters (*mantelé posterieur*, cf. Repartida, *mantelé anterieur*)/[= halved]/? extinct (absorbed under name of Repartida)

Mekatia: (oasis in Ghardaqa region, Algeria)/syn. *Chèvre du Mzab*/*see also* Mzabite

Membrine: *see* Syrian Mountain

Mengrelian: *see* Mingrelian

Meridional: (S Brazil)/d.m/usu. white, sometimes straw or dark brown; small/Crioulo type/[= southern]/? extinct

Meriz: *see* Miriz

Merturi: (Albania)/d/dark brown to black

Meseta: (Castille, León and Extremadura, Spain)/m.(d.)/ yellow, grey or brown; usu. hd (twisted or sabre)/mixed population/Sp. *Raza de las Mesetas*, Fr. *Race des plateaux*/ [= tableland]

Mestiço: (Brazil)/[= of mixed blood]

mestna koza: *see* Bulgarian

Mexican Criollo: (Mexico)/d.m/orig. from 16th c. Spanish (esp. Castille Mountain, Murcian and Granada types), improved since early 20th c. with alpine dairy and Anglo-Nubian/orig. of Spanish (USA)

Mingrelian: (W Georgia)/d/brown, black, white, roan or grey (also pied or black-headed in Turkey); ♂ hd, ♀ hd or pd/larger highland and smaller lowland vars/Russ. *Megrel'skaya*/syn. Georgian (Russ. *Gruzinskaya*, Turk. *Gürcü*), *Megrel, Mengrelian, Tiflis, (West) Caucasian* (Turk. *Kafkas*)

Mini-Xiang: (China)

Miraz: *see* Miriz

Miriz: (N Iraq)/black/Angora type/not *Meriz, Miraz*/rare

Modi bakri: *see* Shekhawati

Modugh: *see* Somali

Moghza Maltija: *see* Maltese

Mohair: *see* Angora

Moker: (Albania)/d/dark red; lop ears

Moldavian: (Moldova)/local

Mona Island: *see* Desecho Island

Moncayo: *see* Guadarrama

Mongolian: (Mongolia; Inner Mongolia, Gansu and Qinghai, China)/ca.d.m/usu. white, also black, blue, grey, brown or pied/ syn. *Cashmere goat of Mongolia, Neimenggu Cashmere, Neimonggol Cashmere* (= Inner Mongolian Cashmere)

Montagnarde: (Brazil)/= serrana

Montagnarde des Aurès: (Monts des Aurès, Algeria)/fibre/longer ears and coat than Kabyle, not dwarf

Montana: (S Spain)/heavier var. of Murcia-Granada/[= mountain]

Montecristo: (Montecristo I, Tuscany, Italy)/various colours; feral

Montefalcone: *see* Campobasso

Montejaqueña: (Sierra de Ronda, Málaga, and Sierra de Grazalema, Cadiz, Andalucía, Spain)/d.m/ colours varied, often tricolour/ HB/syn. *Payoya* (Sp. *Agrupación caprina payoya*)/[from Montejaque]

Montes de Toledo: *see* Verata

Montgomery: (Java, Indonesia)/sire line (?)/rare

Monts Dahra: *see* Kabyle

Morghose, Morghoz: *see* Kurdi

Moroccan Black: (C Morocco)/m.(d)/ usu. black or brown; lop ears; long hair/Fr. *Noir marocaine*/ syn. *Moroccan berber* (mts only), local names *Attaouia* in S, *Yahyaouia* in SE

Mossi: (Burkina Faso)/various
colours; short hair; long ears;
slightly larger var. of West
African Dwarf

Mountain: *see* Syrian Mountain

Moussoro: (Chad and Mauritania)

Moxotó: (Pernambuco, NE Brazil)/
chamois pattern, white or cream
with black face stripes, back
stripe and belly/orig. from
French Alpine with white local;
colour type selected from SRD to
resemble Serpentina/vars:
Moxotó Negra, Moxoto lettera
(Dairy Moxoto)/recog. 1977/syn.
Black-Back/[from Vale do Rio
Moxotó]/rare

Moxotó Negra: (Pernambuco, NE
Brazil)/black or dark brown with
darker face stripes, back stripe
and belly/var. of Moxotó by
selection for colour since 1950s;
sometimes crossed with Graúna
or Murcian/almost extinct (one
small purebred herd)

Mozambique: *see* Landim

Mtwara: (SE coastal Tanzania)/
brown, multicoloured or black,
also pied or white; rarely pd

Mubende: (Buganda, Uganda)/
leather, m/usu. black, also pied;
sometimes pd; short hair/var. of
Small East African

Mudugh: *see* Somali

Murcia-Granada: (S Spain)/d/
mahogany or black; usu. pd/
formed in 1975 by combining
Murcian and Granada/vars:
Veguensis or de Vega (= lowland),
Montana (= mountain)/HB 1980,
BS/Sp. *Murciana-Granadina*/
suggested syn. *Orospeda*

Murcian: (Levante, Spain)/d/
mahogany/part orig. of Murcia-
Granada/HB 1933/Sp. *Murciana*

Murge: (Apulia, Italy)/d/usu. pied;
twisted horns or pd; short
hair/local var./It. *capra delle
Murge* or *Murgese*

Murme e Zeze: (Albania)/d/dark
brown to black

Muvô: *see* Nambi

Muzhake: (Saranda, S Albania)/d/

light grey to blue/lowland var. of
Albanian

Myanmar: *see* Burmese

Myotonic: (Tennessee and Texas,
USA)/m/black, white or piebald;
heavy muscling and hereditary
myotonia/imported from Asia
(possibly India) in early
1880s/BS 1988/syn. *American
Tennessee Fainting, Epileptic,
Fainting, Nervous, Original
Fainting, Stiff, Stiff-legged,
Tennessee Fainting, Texas
Wooden Leg, Wooden Leg goat*/
rare

Mzabite: (N Algeria)/d/Nubian
type/syn. *Algerian Red,
Touggourt*

Nachi: (SE Punjab, Pakistan)/
d.m.hr/usu. black, sometimes
pied/? = Marwari (India)/syn.
Bikaneri/not *Bikanari*

Nadjdi: (Khuzestan, Iran)/d.hr/grey
or brown; lop ears; tassels/not
Najdi, Nejdi

Naemorhedus: *see* goral

naine de Kabylie: *see* Kabyle

naine de l'est: *see* Djougry

naine de Savanes: *see* West African
Dwarf

Nambi: (NE Brazil)/d/no external
ears; ♂ hd, ♀ hd or pd/syn.
Muvô (= corruption of French
nouveau, new, i.e. different)/
selected for earless trait in
SRD/[= earless]/rare

Nanjiang Yellow: (Sichuan, China)/
m/orig. 1960–1990 from
Chengdu Brown and Nubian, ×
local Nanjiang

Napolitana: *see* Neapolitan

Ndebeli: (Gwanda-Tuli area, SW
Zimbabwe)/m/white, black,
brown, sometimes pied or
spotted; lop ears; hd or pd; usu.
short-haired/sim. to Tswana/syn.
Matabele

Neapolitan: (Campania, Italy)/d.m/
hd or pd/It. *Napolitana*/rare

Nederlandse: *see* Dutch

Negev: (S Israel)/Syrian type, size
intermediate between Syrian
Mountain and Hejazi

Negra serrana: *see* Andalusian Black
  Mountain
Neimenggu Cashmere: *see* Mongolian
Neimonggol Cashmere: *see* Mongolian
Nejdi: *see* Nadjdi
***Nemorhaedus goral*:** *see* goral
Nepalese: *see* Changra (Tibetan),
  Nepal Hill, Sindhal, Tarai
Nepalese Northern Hill: *see* Nepal
  Hill
Nepalese Southern Hill: *see* Nepal
  Hill
Nepal Hill: (Siwalik, Pahar and
  Mahabharat ranges, Nepal)/m.d/
  usu. brown or black (in E); short
  hair/colour vars (in E): Dhobini,
  Khairi, Kali, Singari, Seti;
  Northern Hill larger than
  Southern Hill/syn. *Aulo,*
  Nepalese Hill
Nera Verzasca: *see* Verzasca
Nervous: *see* Myotonic
Netherlands White: *see* Dutch White
Newala: (Mtwara and Newala,
  Tanzania)/m/var. of Small East
  African
New Zealand Angora: (New
  Zealand)/mo/national type
  evolving from Angora imported
  from Australia with input from
  up-graded 'bush' goats
New Zealand base stock: (New
  Zealand)/recaptured feral, bred
  in captivity
New Zealand feral: (New Zealand
  mainland and islands)/ca.m/usu.
  fawn, brown or grey, sometimes
  white or pied; short ears/orig.
  from various European, also
  African and Asian/*see also*
  Arapawa, Auckland Island,
  Forsyth Island, Great Barrier
  Island
Nguni: (Swaziland and Zululand)/
  m/various colours; twisted
  horns/intermediate between
  Small East African and southern
  African lop-eared/*see also*
  Unimproved Veld
Niafounké: (Mali) *see* Sahelian
Nigerian: *see* Nigerian Dwarf, Pygmy,
  Red Sokoto, Savanna Brown,
  West African Dwarf
Nigerian: (Great Britain)/often white

with black or brown markings or
  tricoloured; short coat;
  proportionate dwarf/var. of
  Pygmy/orig. from Southern
  Sudan
Nigerian Dwarf: (Nigeria)/local West
  African Dwarf
Nigerian Dwarf: (USA)/d/usu.
  brown, black or gold, often with
  white markings; proportionate
  dwarf/orig. from Southern Sudan
  or Small East African imported
  1950s–1960s, rather than West
  African Dwarf (see African
  Pygmy)/HB 1980, BS; HB also in
  Canada/rare
Nilotic: (S Sudan)/main var. of
  Southern Sudan
Nimari: *see* Surti
Nioro: *see* Sahelian
Niutui: (Henan, China)/local village
  goat discovered 1986
Noire marocaine: *see* Moroccan
  Black
Nongdong Black: (N. Shaanxi, China)/
  m.ca/var. of Zhiwulin Black
Nonglin: (W Yunnan, China)/m/red-
  brown; ♂ hd, ♀ pd
Nordic: (Scandinavia)/name
  suggested in 1980s to inc.
  Finnish Landrace, Norwegian
  Landrace, Swedish Landrace
Nordfjord: *see* Vestland
Nordland: (N Norway)/often blue-
  grey/former var. of Norwegian,
  sim. to Døle/Nor. *Nord-norsk geit*
  (*geit* = goat)
Norsk: *see* Norwegian
North Caucasian: *see* Karachai
North East Bosnian: (NE Bosnia)/BS
  1998
Northern Hill: *see* Nepal Hill
Northern Mountain: *see* Syrian
  Mountain
North Gujerat: *see* Gujerati
North Russian: (European Russia
  north of Smolensk to Tataria)/d/
  usu. white/var.: Bashkir/orig.
  (with Saanen) of Russian White/
  Russ. *Severorusskaya*/syn.
  Tatar/extinct
Norwegian: (Norway)/d.m/grey, blue,
  white or pied; long hair/former

vars: Døle, Nordland, Rogaland, Telemark, Vestland, all blended to form Norwegian/Nor. *Norsk*/ syn. *Norsk melkegeit, Norwegian Dairy*

Nuba Mountain: (S Sudan)/local var. of Southern Sudan

Nubian: (USA) *see* Anglo-Nubian

Nubian: (W Eritrea and NW Ethiopia)/d/black, occ. with white and red markings; long lop ears, roman nose; hairy/Nubian type

Nubian type: (NE Africa)/d/brown, red or black; long lop ears, roman nose; usu. pd; usu. short hair; tall/type inc. Barka, Damascus, Mzabite, Nubian, Shukria, Sudanese Nubian, Zaraibi and their derivatives

Nubiana: (Brazil)/d/various colours and patterns/= Anglo-Nubian, first imported in 1930s from England and USA

Nubian Dwarf: (USA)/d/miniature variety from Anglo-Nubian × West African Dwarf

Nubian ibex: (Red Sea coast, from N Eritrea via Sinai, Israel and Jordan to SW and S Arabia)/= *Capra ibex nubiana* Cuvier/var. of ibex/vars: Sinai or Syrian ibex, South Arabian ibex/syn. *beden*

Nubisk: (Denmark)/m/nearly extinct

Oberhasli: (USA)/d/chamois-coloured, occ. black; pd/orig. from Oberhasli-Brienze imported chiefly 1936/vars: Swiss Oberhasli (purebred), American Oberhasli (with other blood)/BS 1977, HB 1978/obs. syn. *Swiss Alpine*/rare

Oberhasli-Brienz: (Bernese Oberland, Switzerland)/hd or pd/var. of Chamois Coloured/Ger. *Oberhasli-Brienzer*/syn. *Brienz, Hasli, Oberhasli*

Ogaden: (Ethiopia)/var. of Somali

Okinawa: (Japan)/m/black, brown or pied

Old English: (England)/orig. native type, cf. Irish, Welsh/orig. (× oriental lop-eared) of Anglo-Nubian and (× Swiss) of British/

BS 1920–c.1930, HB 1924/ extinct 1930, being revived as English

Olierivier: *see* Savannah White

Omani Batina: *see* Batina

Oreamnus americanus: *see* Rocky Mountain goat

Orelha de Onça: (NE Brazil)/often black, brown or pied; very short rounded ears (but not a fixed characteristic)/alpine type/rare/ [= ear of lynx or ounce (sp. of wild feline) (*onça* = general term for all big cats in Brazil)]

Orenburg: (Orenburg, Chelyabinsk and SE Bashkiria, Russia)/ca.d/ usu. black, occ. tan, grey or pied (white var. in single flock)/poss. orig. of Guadalupe Island ferals/ obs. syn. *Chkalov, Trans-Ural Bashkir* (Russ. *Bashkirskaya koza Zauralya*)

Original Fainting goat: *see* Myotonic

Orobica: (Orobie Alps, Sondrio-Como-Bergamo, Lombardy, Italy)/d/grey or beige, sometimes with dark spots, occ. blue roan, rarely white; long twisted horns; long hair/HB/syn. *Valgerola* [Val Gerola, S of Morbegno]

Orospeda: *see* Murcia-Granada

Osmanabadi: (SE Maharashtra, India)/m.d/usu. black, sometimes white, brown or pied; ♂ usu. hd, ♀ hd or pd; long or short coat/sim. to Deccani/not *Oosmanabad*

Ovambo: (Namibia)/Ger. *Owamboziege*

Pafuri: (SW Mozambique)/m.d/ various colours; long ears/orig. from Boer × Landim since 1928/not *Parfuri*/nearly extinct

Pahari: *see* Kajli

Pak Angora: (Pakistan)/mo.m/orig. from Angora × local hair goats, only at Livestock Exp. Sta., Kherawala, Lineah/rare

Palestinian: *see* Syrian Mountain

Palmera: (La Palma, Canary Is, Spain)/d/usu. red; usu. spiral horns/var. of Canary Island

Palomaziege: *see* Peacock goat

Parbatsar: (Nagaur, W Rajasthan, India)/d.m/light brown to dark chocolate; lop ears

Parda alemã: (Brazil) *see* German Improved Fawn

Parda alpina: (Brazil) *see* Brown Alpine

Parda sertaneja: (Paraíba, Brazil)/ d.m./fawn with cream face stripes and belly (chamois pattern)/orig. from German Improved Fawn (*Parda alemã*) × local; not yet a fixed breed/[= brown or grey of the *sertão* (= bush, backwoods, interior)]/rare

pasan, pasang: *see* bezoar

pashmina: *see* cashmere

Pashmina goat: *see* Central Asian cashmere goat

Passeira, Passiria: *see* Camosciata delle Alpi

Pastoreño criollo: (Oaxaca, Mexico)/ m/white; short hair; wattles/ traditionally fattened for dried *chito* meat for Christmas

Pat: *see* Kajli

Pateri: (India)/var. of Surti/not *Patiri*

Pat-Kanni: (India)/var. of Kannaiadu

Patteri: (Pakistan)/m

Pattu: *see* Kajli

Payoya: *see* Montejaqueña

pazan: *see* bezoar

PE: *see* Peranakan Etawah

Peacock goat: (Graubünden, Switzerland)/d.m/white forequarters, hindlegs and tail, with black hindquarters, feet and face stripes/Swiss Mountain group, classified in 1938 as colour var. of Grisons Striped/ Ger. *Pfauenziege*/syn. *Grauschwarze Gebirgsziege* (= Grey-Black Mountain goat), *Grauschwarz-weisse Gebirgsziege* (= Grey-black-white Mountain goat), *Palomaziege*/rare

Peasant: *see* Allmogegetter, Provençal

Peranakan Etawah: (Indonesia)/m/ orig. from Jamnapari (imported from India 1918–1931) × Katjang local/syn. *Bligon*, *PE*

Peratiki: (Cyprus)/d/usu. light in colour; usu. pd; short hair/orig. from mixed population of Cyprus free-range, Maltese and Damascus/syn. *Cyprus tethered*

Persian wild: *see* bezoar

Peul: *see* Chad

Pfauenziege: *see* Peacock goat

Philippine: (Philippines)/m/vars: coarse-hair (cream, beige or light brown, with dark face stripes and back stripe; usu. pd), fine-hair (black or brown, with or without white belt; usu. hd), Dadiangas (larger strain)/? orig. from Katjang

Piamiri: (N Kashmir, Pakistan)/ hr.d.m/usu. black, occ. brown or grey-white with black-brown patches; ♀ usu. pd; small/orig. from Afghanistan

Pied German Improved: *see* German Improved Fawn

Pinzgau: (W Austria)/d.(m)/brown to red-brown with black legs, head and back stripe/fawn pd and black hd vars/selected for colour within local landraces/HB/Ger. *Pinzgauer gemsfarbige* (= chamois coloured), *Pinzgauer Ziege*/rare

Pirenaica: *see* Pyrenean

Pitiüsa: (Ibiza and Formentera, Spain)/d.m/very variable/rare

Poitou: (W France)/d.m/black-brown with pale belly and legs, white face marks; usu. pd; long hair/ HB 1952, BS 1986/Fr. *Poitevin*/ rare

Polish Improved Fawn: (Poznań and Silesia, Poland)/cf. German Improved Fawn/Pol. *Barwna uszlachetniona*

Polish Improved White: (W Poland)/ cf. German Improved White/Pol. *Biała uszlachetniona*

Polled Maguan: (China)

Pomellata: (Italy)/nearly extinct

Portuguese ibex: (Portugal)/= *Capra pyrenaica lusitanica* Schlegel/ var. of Spanish ibex/Port. *cabra montez de Portugal, cabra do Gerez*, Sp. *cabra montés portuguesa*, Fr. *bouquetin du Gerez*/extinct 1892

Portuguese Mountain: *see* Serrana
Potenza: (Basilicata, Italy)/d.m/
brown, grey or black; lop ears;
usu. hd; long hair/? Maltese,
Alpine and Garganica blood/It.
di Potenza/rare
Pothohari: (Azad Kashmir and
adjacent Punjab, Pakistan)/
d.m/black, grey or white
Pridon, Pridonskaya: *see* Don
Pridono-kirgizskye pomes': *see* Don-
Kirgiz cross
Provençal: (Haute Provence, France)/
d.m/black, brown, red or other
colours and combinations; ears
erect or semi-lop; horns scimitar
or twisted; long hair/syn.
*Commune Provencale, Payse,
Peasant*/rare
Pugliese: *see* Apulian
pukh: *see* cashmere
Pukhovye porody: *see* Russian down
goat
Pullaiadu: *see* Kannaiadu
Pygmy: (Great Britain)/any colour
except whole white; short or
long coat/vars: Cameroon (or
Blue), Nigerian, or crosses of
these two
Pygmy: (W Africa) *see* West African
Dwarf
Pygora: (Oregon, USA)/ca.mo/orig.
1980 from African Pygmy ×
Angora/BS 1987, HB/rare
Pyrenean: (French and Spanish
Pyrenees, and Cantabrian mts,
Spain)/d.m/usu. chestnut, dark
brown or black with paler belly
and feet; hd or pd; usu. long hair/
BS France/Sp. *Pirenaica*, Fr.
Pyrénéen or *Race des
Pyrénées*/syn. *Aquitaine, Béarnais*
(from Béarn)/rare in France
Pyrenean ibex: = *Capra pyrenaica
pyrenaica* Schinz/var. of Spanish
ibex/Fr. *bouquetin des Pyrénées*/
nearly extinct

Qinshan: (Jining, Shandong, China)/
pelt/black

Race des plateaux: *see* Meseta
Rahnama: (Afghanistan)/sim. to
Asmari but taller/[= leader]

Raiana: *see* Serpentina
Raini: (Raine region, Kirman, Iran)/
hr.d.m/grey, yellow, black or
(favoured) white; medium ears;
long lustrous fine hair
Ramdhan: (Uttar Pradesh, India)/
strain of Jamnapari, orig. from
cross with ♀ from Alwar,
Rajasthan/syn. *Kandari Ka
Khana*
Rasquera White: (Spain)/Sp. *Cabra
blanca de Rasquera*
Red Algerian: *see* Mzabite
Red Bosnian: (Bosnia)/d/red, grey,
black, brown or pied/Balkan
type/Serbo-cro. *Crvena bosanska*
Red Maradi: *see* Maradi
Red Mediterranean: (Sicily, Italy)/
d.m/red-brown; ♂ usu. hd, ♀
usu. pd; semi-lop ears; long
hair/? orig. from Damascus/
HB/It. *Rossa mediterranea*/syn.
Derivata di Siria (= Syrian
derivative) (until 1991), *Red
Syrian*
Red Skin: *see* Red Sokoto
Rehfarbige deutsche Edelziege: *see*
German Improved Fawn
Repartida: (Bahía, NE Brazil)/dark
forequarters with fawn, grey or
cream hindquarters (*mantelée
antérieur*), or vice versa with
dark hindquarters (*mantelée
postérieur*, i.e. Meísta)/colour
type selected from SRD/syn.
Surrão/[= divided (i.e. in
colour)]/nearly extinct
Retinta Extremeña: (Cáceres,
Extremadura, Spain)/d.m/dark
red; horns usu. twisted/syn.
Retinta Cacereña
Réunion Créole: (Réunion, Indian
Ocean)/m/semi-lop ears/mixed
orig./syn. *Cabri pays*
Rhâali: (Atlas Mts, Morocco)/black/?
related to Syrian Mountain
Ribatejana: (Ribatejo, C Portugal)/
d.m/hd/var. of Serrana
Rift Valley: *see* Coastal group
Rila Monastery: (Bulgaria)/d/Bulg.
Rilamonastirska
Roccaverano: (Le Langhe, Piedmont,
Italy)/d.m/white, chestnut or
pied; usu. pd; long or short hair/

heavily crossed with Saanen in recent years/rare

Rock Alpine: (California, USA)/d/ usu. black pied/orig. from Oberhasli × Toggenburg/recog. 1935/[bred by Mary Rock]/ extinct by 1978

Rocky Mountain goat: (N America)/= *Oreamnos americanus* (Blainville), actually a goat-like antelope

Rodamit, Rodmit: *see* Rovmit

Rogaland: (Norway)/d/former var. of Norwegian

Roman: (Italy)/m/usu. white; long hair/local var./It. *Romana*

Rood van Belgie: *see* Belgian Red

Rossa mediterranea: *see* Red Mediterranean

Rousse de Maradi: *see* Maradi

Rove: (Marseille, Provence, France)/ d.m/usu. red, also black, brown, chamois-coloured, red-and-black, black-and-tan; ears usu. semi-lop; wide twisted horns/BS 1979/[village west of Marseille]/ rare

Rovmit: (Tajikistan)/d/not *Radamit, Rodmit*/extinct

Roya Vésubie: (Alpes Maritimes, SE France)/d.m/black with white belly, or chamois-coloured with badger face, occ. tan or red; long twisted/BS 1979/syn. *Haute Provence, Haute Roya*/not *Royale Vésubie*/nearly extinct

Rupicapra rupicapra: *see* chamois

Russian down goat: (European and Asian Russia)/ca/usu. black/inc. Altai Mountain, Don, Orenburg, Uzbek Black/Russ. *Pukhovye porody*

Russian White: (N European Russia)/ d/cf. Gorki/orig. (1905 on) from Saanen × local North Russian/ Russ. *Russkaya belaya*/syn. *Improved North Russian, Russian Dairy, Russian White, Russian White Dairy*

Rwanda-Burundi: (Rwanda, Burundi and Kivu, DR Congo)/often black, also pied or tricolour/var. of Small East African/Fr. *Chèvre commune rwandaise* and *Chèvre*

commune burundaise/obs. syn. *Bahu* (DR Congo), *Congo Dwarf*

Saanen: (W and NW Switzerland)/ d/white; pd or hd/orig. of Banat White, Belgian Saanen, Brazilian Saanen, British Saanen, Bulgarian White Dairy, Czech White Short-haired, Dutch White, French Saanen, German Improved White, Gorki, Guanzhong Dairy, Israeli Saanen, Italian Saanen, Japanese Saanen, Laoshan Dairy, Polish Improved White, Russian White, Sable, Slovakian White Short-haired, Tangshan, Taurus/ HB 1890; HB also S Africa 1922, US 1954, GB 1922 (also BS), Italy 1981, Australia, Canada, Croatia/ Albanian *Sana*, Israeli *Saneen*, Serbo-cro. *Sanska*/syn. *Gessenay* (Fr.)/not *Sannen*

Saanen Kepisi: (Turkey)

Sable: (USA)/black var. of Saanen/ rare

Sahel burkinabé: *see* Voltaïque

Sahelian: (N of W Africa)/d.m/usu. white, black, reddish-brown, combined as pied or tricoloured; ears short horizontal or longer semi-lop; horns ♂ long twisted, ♀ sickle shaped; tassels common/ vars: Chad, Makatia, Maure, Tuareg, Voltaïque; *see also* Nigerian/Fr. *Sahélienne*/ syn. *Chèvre bariolée* (= variegated) (Mauritania), *Fulani, Sahel, West African Long-legged*; local names in Mali: *Gorane, Niafounké, Nioro*

Sahel Voltaïque: *see* Voltaïque

Sahely: *see* Tali

Sahrawi: *see* Barki

Saidi: (Upper Egypt)/Syrian type; sim. to Baladi but bigger/[= valley, i.e. of Nile]

St Gallen: *see* Stiefelgeiss

Sakhar: (Brannitsa mts, SE Bulgaria)/ hr/usu. black, also black-and-tan, occ. white-spotted or grey; long-haired; usu. twisted horns

Salernitana, Salerno: *see* Cilento

Saloia: *see* Serrana

Salt Range: (NW Punjab, Pakistan)/
d.hr/black-and-white or black;
lop ears; vertical screw horns

Samao: *see* Zhongwei

Samar: (Aleppo, NW Syria)/d/var. of
Syrian Mountain

Sana: (Albania) *see* Saanen

Sanbei: (Zinjiang, Inner Mongolia,
and Gansu, China)/fur/usu. black
or grey; ♂ usu. hd, ♀ pd

San Clemente: (I off California,
USA)/m/often forequarters black,
hindquarters red or tan but no
sharp division; small to dwarf/cf.
Spanish (USA)/orig. feral, ? Sp./
HB/nearly extinct (now limited
to mainland)

Sandomierz: (SE Poland)/d/pied,
usu. tricolour/Pol.
Sandomierska/extinct

Saneen: (Israel) *see* Saanen

Sangamneri: (Poona and
Ahmednagar, Maharashtra,
India)/d.m.hr/white, black,
brown or pied; coarse short
hair/part orig. of Indian Mohair

Sanluiseña Criollo: (San Luis,
Argentina)

Sanska koza: (Croatia) *see* Saanen

Sanska pasma: (Slovenia) *see* Saanen

Santa Catalina: (I off California,
USA)/brown with black
markings; feral/Sp. orig.

Sardinian: (Sardinia, Italy)/d.m/usu.
white or black pied, also grey or
brown, self or pied; twisted
horns, occ. pd; often tassels/var.:
Tavolara/HB 1981/It. *Sarda*/rare

Sardonaziege: *see* Stiefelgeiss

Savanna Brown: (N Nigeria)/m.(d)/
orig. from Red Sokoto (? by
crossing)/vars: Bornu White,
Damagaran Dapple-grey, Kano
Brown, Katsina Light-brown,
Mambilla/syn. *Nigerian*

Savannah Dwarf: *see* West African
Dwarf

Savannah White: (Olierivier, S
Africa)/m/white with black to
brown skin; lop ears; short coat/
developed from 1957 by Messrs
D.S.U. Cilliers & Sons by
selection for white coat and
black skin from mixture of

coloured indigenous goats to
white ♂, closed stud/BS 1993

Saxon: (Saxony, Germany)/part orig.
(with Chamois Coloured) of
German Improved Fawn/var:
Wiesental

Scheckige Tauernziege: *see* Tauern
Pied

Schwarzenburg-Guggisberg: *see*
Gruyère

Schwarzer Tessiner: *see* Verzasca

Schwarzhals: *see* Valais Blackneck

Schwarzwald: *see* Black Forest

Schwarzweisse Walliser Sattelziege:
see Valais Blackneck

Sciucria: *see* Shukria

Screziata: (Italy)/[= streaked,
speckled]/nearly extinct

Sempione: (Vercelli, Piedmont,
Italy)/m/white, spotted with
grey, black or chestnut/nearly
extinct

Sem Raça Definida: *see* SRD

serow: (SE Asia, Sumatra to S China
and Kashmir)/= *Capricornis
sumatraensis* (Bechstein)/Malay
kambing gurum

Serpentina: (Alentejo, Portugal)/
m.d/white with black back
stripe, belly, feet and tail/syn.
*Alentejana, Castelhana, Raiana,
Spanish*/[from Serpa]

Serrana: (C and N Portugal)/d.m/
black, brown or yellowish to tan;
hd or pd; long hair/vars: Da
Serra, Jarmelista, Ribatejana,
Transmontana/orig. probably
from Pyrenees/HB 1985/syn.
Saloia, Serra da Estrela

Serrana: (Spain) *see* Spanish White

Serrana andaluza: *see* Andalusian
White

Serrana de Castilla y Levante: *see*
Castille Mountain

Seti: (E Nepal)/white/colour var. of
Nepal Hill

Severorusskaya: *see* North Russian

Shaanan White: *see* Shannan White

Shaanbei Black: (N Shaanxi, China)/
var. of Zhiwulin Black

Shami: *see* Damascus

Shanghai White: *see* Haimen

Shannan White: (S Shaanxi, China)/
m/white; smaller horned and

larger pd vars; long hair/not
Shaanan

Shanxi Large Black: (S Shanxi,
China)/var. of Taihang/syn. *Big
Black in Shanxi*

Sharkawi: (Lower Egypt)/var. of
Baladi

Shekhawati: (W Rajasthan, India)/d/
black; pd; short coat/syn. *Modi
bakri, Modi bakkri*/[= hornless
goat]

Shiba: (Goto I, Nagasaki, Japan)/
white; miniature/orig. from small
local for research (now only at
Tokyo University and at NIAI,
Ibaraki)/syn. *Chubby*/rare

Shingari: (Kashmir, India)/m

Shukria: (W Eritrea)/d/brown/ Nubian
type but hd and long hair; sim. to
Sudanese Nubian/It. *Sciucria*

Shurri: (Azad Kashmir, Pakistan)/
d.m/white, grey, black or pied/
orig. from Buchi × Labri

Shyta: (Podgradec, E Albania)/larger
mt var. of Albanian

Siberian: *see* Altai Mountain

Siberian ibex: (C Asia, Afghanistan to
Mongolia)/= *Capra ibex siberica*
Pallas/var. of ibex

Sicilian: (Sicily, Italy)/m/long hair/
see also Etna Silver, Girgentana,
Red Mediterranean

Sikkim: (NE India)/local

Silvered goat of Mount Etna: *see* Etna
Silver

Sinai: (Egypt)/d.m/sim. to Negev but
dwarf

Sinai ibex: *see* Syrian ibex

Sinazongwe goat: (Zambia)

Sind Desi: (C Sind, Pakistan)/d.m/
usu. black, also brown, grey or
black-and-white; long lop ears,
roman nose; often corkscrew
horns/? orig. from Kamori ×
Chappar

Sindhal: (Nepal mts)/m.hr.d/white,
cream, brown, black, grey or
pied; long coarse hair/Nepali
Singhal/not *Sinhal*/[Sanskrit
sinha = horn]

Sind wild goat: (W Sind and
Baluchistan, Pakistan)/= *Capra
aegagrus blythi* Hume/var. of
bezoar/syn. *Sind ibex*

Singari: (E Nepal)/black with white
markings/colour var. of Nepal
Hill

Singhal: *see* Sindhal

Sinhal: *see* Sindhal

Sinkiang: *see* Xinjiang

Sirli: (N Pakistan)/hr.m/lop ears

Sirohi: (S Rajasthan and N Gujarat,
India)/d.m/brown, patched with
light or dark brown, occ. white;
long ears; usu. tassels

Skilder: (E Cape, S Africa)/dark or
black skin, often speckled with
red; lop ears/indigenous; slightly
larger member of Unimproved
Veld group/syn. *Speckled*/nearly
extinct (by crossing with Angora
or Namibian or grading up to
Boer)

Skopelos: (Greece)/d/reddish-brown
or black and fawn, occ. with
white patches

Slovakian Brown Short-haired:
(Slovakia)/d/hd or pd/cf. Czech
Brown Short-haired/Sl. *Hnedá
(bezrohá) krátkosrstá koza*
(*bezrohá* = pd, *koza* = goat)/not
Short-eared/rare

Slovakian Native: (Slovakia)/cf.
Carpathian/Sl. *Slovenská*/? extinct

Slovakian White Short-haired:
(Slovakia)/d/cf. Czech White
Short-haired/pd var. (*bezrohá*)
rare/orig. from Saanen/Sl. *Biela
krátkosrstá koza*/not *Short-
eared*/rare

Slovenian Alpine: (Slovenia)/d/
grey/Cro. *Srnasta Pasma* (= deer
breed)/rare

Slovenian White: (Slovenia)/d/white;
erect ears/Cro. *Domača bela
koze*/syn. *Domača križana koze*
(= crossbred)/rare

Small East African: (E Africa from
Kenya to Zimbabwe)/m/all
colours but mostly black, brown
and white combinations; short
ears; tassels common; short coat/
vars or local names: Arusha,
Chagga, Gogo, Kigezi, Malawi,
Masai, Mashona, Mubende,
Rwanda-Burundi, Zambian

Sokoto Red: *see* Red Sokoto

Somali: (Somalia, Ogaden and NE

Kenya)/m/white, occ. with reddish tinge or with coloured spots or patches; black skin; short ears; ♂ hd; ♀ hd or pd; short hair/vars or local names: Abgal, Boran, Galla, Modugh or Mudugh, Ogaden; cf. long-eared Benadir/Somali *deghier*, *deg yer* or *dighi yer* (= small ear)/syn. formerly *Galla*

Somali Arab: (coast of Somalia)/d/ usu. brown; short ears; long hair/orig. from Arabia

South African: *see* Bantu, Boer, Delftzijl, Hottentot, Kalahari Red, Khosa, KwaZulu Natal, Loskop South, Nguni, Savannah White, Skilder

South African Angora: (S Africa)/ HB/syn. *Sybokke*

South Arabian ibex: (SW Arabia)/var. of Nubian ibex/syn. *Arabian ibex*

South Caucasian: *see* Azerbaijan

South China: (Yunnan and Guangdong, S China)/m/usu. black; short ears; twisted horns/ sim. to Katjang

Southern Hill: *see* Nepal Hill

Southern Sudan: (S Sudan)/m/all colours; short ears; ♂ hd, ♀ often pd; very short hair; small to proportionate dwarf/vars: Ingessana, Latuka-Bari, Nilotic, Nuba Mountain, Toposa, Yei/syn. *Sudanese Dwarf*

Soviet Mohair: (Kazakh, Tajik, Turkmen and Uzbek republics)/ mo.m/white/orig. (1937 on) from Angora (USA) × Central Asian coarse-haired goats; recog. 1962/ Russ. *Sovetskaya sherstnaya*

Spanish: (Mauritania) *see* Canary Island, Güera

Spanish: (Portugal) *see* Serpentina

Spanish: (Texas, USA)/m/orig. from Mexican Criollo; Texas strains (inc. Kensing, Valera, Willingham) selected for meat/ syn. *Ball Field*, *Brush* or *Woods* (SE USA)/rare (through crossing with Boer)

Spanish Angora: (USA)/dwarf/cf. Pygora/orig. from crossing Angora and West African Dwarf/ rare

Spanish ibex: = *Capra pyrenaica* Schinz/Sp. *cabra montés* (= mountain goat)/vars: Gredos, Mediterranean, Portuguese, Pyrenean/rare

Spanish Mountain: *see* Andalusian Black Mountain, Spanish White

Spanish White: (Spain)/m/white, sometimes darker on head/inc. Andalusian White, Castille Mountain and their crosses/Sp. *Blanca española*/syn. *Spanish Mountain* (*Serrana española*), *White Mountain* (*Blanca serrana*)

Speckled: (S Africa) *see* Skilder

Spotted German Improved: *see* German Improved Fawn

Sri Lankan: (Sri Lanka)/m/local population with Indian blood

SRD: (NE Brazil)/m/derived from Crioulo with recent oriental lop-eared blood (chiefly Anglo-Nubian and Bhuj) and alpine (esp. Saanen, Toggenburg, Alpine)/orig. of Canindé, Marota, Moxotó, Repartida/syn. *Chué*/ [*Sem Raça Definida* = without defined breed] /*see also* UDB (= undefined breed)

Srednei Azii: *see* Central Asian

Srnasta Pasma: *see* Slovenian Alpine

stambecco, Steinbock: *see* ibex

Starkenburg: (Germany)/local improved with Saanen/extinct

Stiefelgeiss: (St Gallen, NE Switzerland)/m.d/yellow-grey with brown back stripe and feet, or brown with black back stripe and feet/Swiss Mountain group; sim. to Grisons Striped/Ger. *St Galler Stiefelgeiss* (= St Gallen booted goat)/syn. *Sardonaziege*, *St Gallen*/rare

Stiff, Stiff-legged: *see* Myotonic

Sudan Desert: (N Sudan)/m.d/white to black, often grey; ears variable; ♂ usu. maned/var.: Zaghawa/ syn. *Sudanese Desert*

Sudanese Dwarf: *see* Southern Sudan

Sudanese Nubian: (Sudan, riverain N of lat. 12°N)/d/usu. black with grey ears; long lop ears/Nubian type but usu. hd and long hair; sim to Syrian Mountain; sim. var. in Eritrea is Shukria/syn. *Beladi*

Sundgau: (S Haut Rhin, Alsace, France)/black, with white stripes on face and white markings on body/former var. of French Alpine; also (USA) colour var. of Alpine/? orig. from Toggenburg/ extinct

Suomenvuohi: *see* Finnish Landrace

Surat, Surati: *see* Surti

Surdudi: (Wadi Surdud, NE of Hodeida, N Yemen)/d.m/red-and-white

Surrão: *see* Repartida

Surti: (S Gujarat and NW Maharashtra, India)/d.m/usu. white; lop ears; short hair/? Arab blood/var.: Pateri/syn. *Khandeshi, Kunyi, Nimari,* formerly *Surat* in England/not *Surati*

Svensk Lantras: *see* Swedish Landrace

Swazi: (S Africa)/ceremonial/erect ears, large horns; medium frame/ Unimproved Veld group/*see also* Nguni

Swedish Landrace: (N Sweden)/d/ black, brown or white; hd or pd; usu. long hair/var.: Jämtland/ HB/Swed. *Svensk Lantras*

Swiss Alpine: (Switzerland) *see* Swiss Mountain

Swiss Alpine: (USA) *see* Oberhasli

Swiss Improved: *see* White Swiss Improved

Swiss Oberhasli: *see* Oberhasli

Swiss Mountain: (Switzerland)/inc. Chamois Coloured, Grisons Striped, Peacock goat, Stiefelgeiss, Verzasca/syn. *Swiss Alpine*

Sybokke: (S Africa) *see* Angora

Synthetic Angora: (Deccan plateau, India)/mo/new breed by crossing Angora with local goats

Syrian Black: *see* Syrian Mountain

Syrian derivative: *see* Red Mediterranean

Syrian ibex: (Sinai to Syria)/var. of Nubian ibex/syn. *Sinai ibex*

Syrian Mountain: (Syria, Lebanon, N Israel)/m.d.hr/usu. black, occ. with white markings, occ. grey or brown; long lop ears; long hair;

tall/Syrian type/var.: Samar/orig. of Mambrina Brasileira/syn. *Aljabali, Black Bedouin, Djelab, Jabali, Jabel, Jebel, Mamber, Mambrine, Membrine, Northern Mountain, Palestinian, Syrian Black*

Syrian type: (Middle East)/usu. black; long lop ears; usu. hd; long hair/inc. Anatolian Black, Aswad, Baladi, Iraqi, Jordanian, Negev, Saidi, Syrian Mountain/ syn. *Black, Black Bedouin*

tahr: = *Hemitragus*/inc. Arabian tahr (Oman) = *H. jayakari* (rare), Himalayan tahr = *H. jemlahicus*, and Nilgiris tahr (S India) = *H. hylocrius*/not *tehr, thar*

Taicheng: (Taicheng mts, Shanxi/ Hebei, China)/syn. *Daqingshan*

Taihang: (S Shanxi and Hebei, China)/vars: Shanxi Large Black, Taihang Black, Wuan/syn. *Taihangshan*

Taihang Black: (S Shanxi/Hebei, China)/var. of Taihang

Taiwan: (Taiwan)/m/horizontal or erect ears; occ. pd; short hair/ vars: western (usu. black, probably S China orig.), eastern (brown with black stripes, sim. to Katjang)

Taiz Black: (Ta'izz, N Yemen)/d.m/ curly coat/sim. to Attaq

Taiz Red: (Ta'izz, N Yemen)/d.m/red or brown, with white belly and black back line

Tajik: (Tajikistan)/Central Asian coarse-haired group

Tajiki: (Afghanistan) *see* Vatani

Takru: *see* Chappar

Tali: (SW Iran)/d/dirty white or reddish-brown; short hair/syn. *Madjd, Sahely*

Tan: *see* Zhongwei

Tangshan: (Hebei, China)/d/usu. pd/orig. from Saanen since 1931

Tanzania: *see* Small East African

Tanzanian: (Tanzania)/inc. Arusha, Blended, Chagga, Gogo, Kigoma, Newala

Tapri: (Pakistan)/m

Tarai: (S Nepal)/m.(d)/usu. black or

fawn, occ. white or pied; short or lop ears, roman nose; short hair; dwarf/syn. *Aulo, Lamkana*/not *Terai*

Tarentaise: (French Alps)/saffron-red head and neck (*cou clair*), black body and face bars/former var. of French Alpine

Tartar, Tatar, tatarskaya: *see* North Russian

Tauá: (NE Brazil)/? individuals with rust-coloured skin/local Crioulo type, also classification of its skin on export market/[Tupi *taguá* = yellow clay, also name of town in Bahía state]/? extinct

Tauern Pied: (Salzburg, Austria)/ d.m/black, brown and white tricoloured, or black-and-white/ Ger. *Scheckige Tauernziege* or *Tauernschecken*/rare

Taurus: (highlands of SE Turkey)/ d.m/orig. 1973 on from Saanen × (Saanen × Anatolian Black)

Tavolara: (Sardinia, Italy)/brown/var. of Sardinian at ConSDABI, Circello and Benevento/It. *Sarda di Tavolara*/nearly extinct

Teddy: (NE Punjab, Pakistan)/m/ white, brown, black or pied; hd or pd; small/probably orig. from Black Bengal

tehr: *see* tahr

Telemark: (Norway)/d/white; usu. pd/former var. of Norwegian

Tellicherry: (N Kerala, India)/short ears; short hair/syn. for and former var. of Malabari/not *Tellichery*

Tenerife: *see* Tinerfeña

Tennessee Fainting: *see* Myotonic

Terai: *see* Tarai

Teramo: (Abruzzo, Italy)/d/usu. grey, black or brown; long hair/being crossed with Garganica/It. *di Teramo*/rare

Tessin: *see* Verzasca

Texas Wooden Leg: *see* Myotonic

Thai: (Thailand)/often brown with black back stripe and shoulder stripes/sim. to Katjang

Thamud: (S Yemen) = Mawr (N Yemen) and probably its orig.

thar: *see* tahr

Theban: *see* Zaraibi

Thori Bari: (Pakistan and India)/pd var. of Barbari

Thuringer Wald: *see* Thuringian

Thuringian: (Thuringia, Germany)/ d.m/chocolate-brown with white face stripes and legs; hd or pd/orig. from Toggenburg, Harzer-Rhönziege and local Thuringian/var. of German Improved Fawn/BS, HB 1985/ Ger. *Thüringer Wald*/syn. *Forest goat, German Toggenburg*/nearly extinct

Tibetan: (Qinghai-Tibet plateau, China)/ca.(m.dr)/usu. white, also pied, black, brown or grey; small/syn. *Bhotia* or *Changra* (Mustang dist., N Nepal), *Cashmere goat of Tibet, Tibetan Cashmere*/not *Chayangez, Chyangra*

Tibetana: (Italy)/dwarf/orig. from West African Dwarf/syn. *Tibetan Dwarf*

Ticino: *see* Verzasca

Tiflis: (Turkey) *see* Mingrelian

Tiftik-Keçi: *see* Angora

Tihami: (Saudi Arabia)

Tinerfeña: (Tenerife, Canary Is, Spain)/d/usu. twisted horns/vars: usu. black or brown and long-haired in humid zone, usu. pied and short-haired in arid zone/var. of Canary Island

Toggenburg: (NE Switzerland)/d/ brown to mouse-grey with white face stripes and feet; pd or hd; long or short coat/? orig. from Appenzell and Chamoisée then selected for colour type/orig. of British Toggenburg, Dutch Toggenburg, Thuringian/HB 1890; HB also S Africa 1922, USA, GB 1905 (also BS), Australia, Canada, Belgium 1930, Austria/Ger. *Toggenburger*, Fr. *Toggenbourg*/not *Toggenborg*

Tokara: (Kagoshima, Japan)/m/brown with black back stripe or brown-and-white pied; also white in one population (possibly due to Saanen influence); prim./intro. from Okinawa/nearly extinct

Tongren White: *see* Guizhou White
Toposa: black or black-and-white/
   local larger var. of Southern
   Sudan
Touareg: *see* Tuareg
Touggourt: *see* Mzabite
Touraine: *see* Berry-Touraine
Tranga: (Mirdita, Albania)/d/light
   red; erect ears; long horns/syn.
   Tranges
Transcaucasian: *see* Azerbaijan
Transmontana: (Tras-os-Montes, NE
   Portugal)/m.(d)/yellowish/var. of
   Serrana
Tremiti: (Tremiti I, SE Italy)/It. *delle
   Tremiti*
Tswana: (Botswana)/m.d/all colours;
   lop ears; medium horns or pd/
   sim. to Ndebeli/obs. syn.
   Bechuanaland
Tuan: *see* Duan
Tuareg: (Niger and Mali)/usu. red
   pied or black pied; hd or pd/var.
   of Sahelian/Fr. *Touareg*
Tuni: (S Somali)/m/black
   forequarters/large var. of
   Benadir/not *Tunni*
Tunisian: (Tunisia)/black or reddish-
   brown; long hair/sim. to Berber
tur: (E Caucasus)/= *Capra
   cylindricornis*/syn. *(east)
   Caucasian tur*, formerly *C.
   caucasica*/not *Caucasian ibex*
turbary goat: (Europe)/scimitar
   horns/Neolothic domesticant
   (e.g. Switzerland's Neolithic
   lake-dwellings)/extinct
Turino: *see* Guariba
Turkana: (Kenya)/nomadic
Turkish Native: *see* Anatolian Black
Turkmen: (Turkmenistan)/Central
   Asian coarse-haired group
Tuxer: (Austria)/= Chamois Coloured
   in Austria

Uauá: (NE Brazil)/? skin type/local
   Crioulo/[Tupi = lampyre; also
   town in Bahía state]/? extinct
UDB: (NE Brazil)/= undefined
   breed/in Ceará state = largest
   group, incl. all that cannot be
   assigned as any other breed (odd
   colours, various degrees of
   crossbreeding with imported
   exotics, etc.)/*see also* SRD

Ugandan: *see* Kigezi, Mubende,
   Small East African
Ulokeros: (Greece)/var. of Greek with
   corkscrew horns
Unimproved Veld: (South Africa)/
   various colours; erect or lop
   ears/Nguni types, inc. Skilder,
   Swazi, Zulu/potential
   composites for d.ca.hr
Unjuul: (Bayan-Unguul dist., Central
   prov., Mongolia)/mo/orig. from
   Soviet Mohair × Mongolian
   since 1963/recog. 1982/
   Mongolian *Unjuulin tsagaan*
Upper Volta: *see* Voltaïque
Utegangsgeit: (Sogn og Fjordane, W
   Norway)/m/white with brown or
   black markings/[= ranging goat]
Uuliin Bor: (Bayan-Ölgiy prov., W
   Mongolia)/ca/brown/orig. from
   Mongolian graded to Altai
   Mountain since 1960s/recog.
   1991/Mongolian *Bayan-ulgiin
   khar*
Uzbek: (Uzbekistan)/d.m.ca/Central
   Asian coarse-haired group/orig.
   (with Angora) of Uzbek Black
   and of Soviet Mohair
Uzbek Black: (Uzbekistan)/ca.d/orig.
   as byproduct of formation of
   Soviet Mohair from Angora ×
   local Uzbek, selected for black
   colour/recog. as breed group
   1961/Russ. *Chernye pukhovye
   kozy Uzbekistana*/syn. *Fergana*;
   obs. syn. *Charkissar (Down)*

Valais Blackneck: (SW Switzerland)/
   m.d/black forequarters, white
   hindquarters; long hair/? orig. of
   Bagot/HB 1920/Fr. *Valaisan à col
   noir*, Ger. *Walliser Schwarzhals,
   Schwarzweisse Walliser
   Sattelziege*, It. *Vallesana (del
   collo nero)* (Novara), *Vallese*
   (Varese and Sondrio)/syn. *Chèvre
   des Glaciers, Haut Valais, Valais
   Blackthroat, Viège*
Val di Livo: (Como, Lombardy, Italy)/
   d/mixed population of alpine
   types/? extinct
Valdostana: (Vercelli and Turin,
   Piedmont, Italy)/[from Val
   d'Aosta]/? extinct

Valfortorina: (Puglia, Italy)/rare

Valgerola: *see* Orobica

Vallesana: *see* Valais Blackneck

Valtellinese: *see* Frisia

Vansaanen: *see* Banat White

Vatani: (Afghanistan)/ca/usu. black, also grey, white or brown; long ears; long horns; long hair and short down/syn. *Afghan Native Black, Kabuli, Kandahari, Tajiki*

Veguensi: (S Spain)/lowland var. of Murcia-Granada/syn. *de Vega* (= lowland)

Velipoja: (Shkodra, NW Albania)/ light reddish; long hair/larger lowland var. of Albanian

Veluwse: *see* Dutch Landrace

Vendi: (Albania)

Verata: (Vera, Cáceres, Spain)/m.d/ chestnut, black, blackish-brown or grey; twisted horns/syn. *Castellana* (= Castilian), *Montes de Toledo*

Verzasca: (Ticino, Switzerland)/d.m/ black or dark reddish-brown/ Swiss Mountain group/HB 1940/ It. *Nera Verzasca* (in Lombardy), Ger. *Schwarzer Tessiner* (= Black Ticino)/syn. *Black Verzasca, Ticino*

Vestland: (SW Norway)/d/usu. dark brown, ♂ long hair/former var. of Norwegian/syn. *Nordfjord, West Coast*

Vésubie: *see* Roya Vésubie

Viège: *see* Valais Blackneck

Vlaamse: *see* Flemish

Vlach: (Macedonia and Epirus, Greece)/local Greek, horns directed backwards

Vogan: (Togo)/West African Dwarf × Sahelian cross

Volgograd White: (S Russia)/white var. of Don/rare

Voltaïque: (Burkina Faso)/often white, also red pied and other colours/var. of Sahelian/syn. *Sahel voltaïque, Sahel burkinabé*

Wahati: (S Egypt, to W of Nile valley)/usu. black; long horns; long hair/[= from the oases]

wali: *see* Abyssinian ibex

Walliser Sattelziege: *see* Valais Blackneck

Walliser Schwarzhals: *see* Valais Blackneck

Way Thali: (Myanmar)/red; long lop ears/? orig. from Jamnapari in 1920s

Weisse deutsche Edelziege: *see* German Improved White

Weisse Schweizerziege: *see* White Swiss Improved

Weisse veredelte Landziege: *see* White Swiss Improved

Welsh: (Wales)/cf. Old English, Irish/extinct

West African Dwarf: (coast of W and C Africa)/m/all colours; trypanotolerant/vars and local names: Cameroon Dwarf, Casamance, Congo Dwarf, Côte d'Ivoire Dwarf, Djallonké, Djougry, Ghana Forest, Kirdi, Kosi, Mossi, Nigerian Dwarf/orig. of African Pygmy (USA), Dutch Dwarf, Nigerian Dwarf (USA), Pygmy (UK), Tibetana (Italy)/syn. *African Dwarf, Dwarf West African, Forest goat, Fouta Djallon, Grassland Dwarf* (Fr. *Chèvre naine des Savanes*), *Guinean, Guinean Dwarf, Pygmy*

West African Long-legged: *see* Sahelian

West Caucasian: *see* Mingrelian

west Caucasian tur: (W Caucasus)/= *Capra caucasica*/syn. *Caucasian ibex, Kuban ibex, Kubanski goat, west Caucasian ibex*; formerly *C. ibex severtsovi*; *see also* tur

West Coast: *see* Malabari, Vestland

Western goat: (Chad and Mauritania)/ = Sahelian

Western Highland: (Ethiopia)/white, fawn or white-and-fawn; long hair/largest in Abyssinian Short-eared group

Western Lowland: (Ethiopia)/usu. fawn (often with white patches), occ. black or grey; prick-eared; short coat; prolific/smallest in Abyssinian Short-eared group

Westphalian: (Germany)/extinct

White Bearded Bengal: *see* Black Bengal

White Borno: *see* Bornu White

White Dagestan: *see* Dagestan White

White German Improved: *see*
German Improved White
White Himalayan: *see* Chigu, Gaddi
White Mountain: *see* Spanish White
White Orenburg: *see* Orenburg
White Swiss Improved: (Aurich,
Switzerland)/pd/orig. from
Appenzell × Saanen/Ger. *Weisse
Schweizerziege, Weisse veredelte
Landziege*/syn. *Zurich* (Ger.
Zürcher)/extinct
Wiesental: (Germany)/var. of Saxon/
extinct
wild: *see* bezoar, chamois, goral, ibex,
markhor, Rocky Mountain, serow,
tahr, tur, west Caucasian tur
Windsor White: (England)/ca/royal
herd in Windsor Great Park
1828–1936/orig. from Tibetan,
later also ? Indian cashmere
types and other introductions
Wildkleurig-bonte Vlaamse: *see*
Flemish Pied
Witte Vlaamse: *see* Flemish White
Wooden Leg: *see* Myotonic
Woods: *see* Spanish
Worre: (Ethiopia)/usu. white, some
brown or black pied/smallest of
Coastal or Rift Valley group/
probably orig. from Yemen and
Saudi Arabia
Woyto-Guji: (Gamu Gofa and eastern
Sidamo lowlands, Ethiopia)/red-
brown and black patched, occ.
black or brown back stripe/
Coastal or Rift Valley group
Wuan: (S Hebei, China)/m.ca/black
head, grey body/var. of Taihang

Xinjiang: (mts of Xinjiang, China)/
ca.m.d/white, black or brown/
WG *Sinkiang*
Xuhai: (Jiangsu, China)/m/white; pd;
small/var. of Huanghuai

yaez: (Israel)/m/orig. at Kibbutz
Lahav, N Negev Desert, from
Nubian ibex × domestic goat
(Saanen, Damascus, Syrian
Mountain, Negev, Sinai); fertile
hybrids (♂ and ♀)/[*yael* (Hebrew
for ibex) + *ez* (Hebrew for goat)]
Yahyaouia: (SE Morocco)/local name
for Moroccan Black

Yamnapari: *see* Jamnapari
Yanbian Dairy: (Sichuan, China)/d/
in formation
Yangtze River Delta White:
(Changjiang delta, chiefly
Jiangsu, China)/m.hr/not
Yangtse, Yangtsi
Yanshan Polled: *see* Chengde Polled
Yashan Down: (Shandong, China)/
down
Yei: (W Equatoria, Sudan)/very small
var. of Southern Sudan, sim. to
Rwanda-Burundi
Yemeni: (Yemen)/see Attaq, Mawr or
Thamud, Surdudi, Taiz Black,
Taiz Red, Yemen Mountain
Yemen Mountain: (mts of N Yemen)/
usu. black; long hair
Yemso: (Korea)
Yichang White: (Hunan, Hubei and
Sichuan, China)/m
Yugoslav Saanen: (Yugoslavia)/d.m/
white; pd/composite of native
goats plus Saanen imported from
Switzerland and Bulgaria/Serbo-
cro. *Domaća Sanska*/rare

Zaghawa: (NW Darfur, Sudan)/black
var. of Sudan Desert
Zakhrana: *see* Jhakrana
Zalawadi: (Surendranagar and
Rajkot, Gujarat, India)/d.m.hr/
erect corkscrew horns/Gujarati
type/not *Zalawand, Zalawari*
Zambian: (Zambia)/m/var. of Small
East African
Zaraibi: (Upper Egypt)/d/often
cream, red or light brown with
dark spots, also black with spots;
roman nose; pd or hd/Nubian
type/part orig. of Anglo-Nubian/
syn. *Egyptian Nubian, Nubi,
Theban*/not *Zaraiby, Zareber*/[=
penned, barn type; cf. also
Zarabi, nr Asyut]
zebu goat: (Brazil)/name orig. given
to goats imported from India
(along with zebu cattle) late 19th
and early 20th c.; now used as
general description for lop-eared
goats in Brazil
Zhejiang White: *see* Haimen;
formerly Chekiang
Zhiwulin Black: (N Shaanxi, China)/

m.ca/vars: Nongdong Black,
Shaanbei Black

Zhongwei: (Ningxia, China)/fur,
m.ca/usu. white, occ. black; kid's
coat curly and lustrous (sim. to
Mongolian Tan lamb pelt)/sim.
to Mongolian/WG *Chung-wei*,
Russ. *Chzhun'veiskaya*/syn. *Tan*,
Samao

Zimbabwe: *see* Mashona, Ndebili

Zinder Brown: *see* Kano Brown

Zomri: (Saudi Arabia)/native goat

Zulanha: *see* Azul

Zulu: (S Africa)/erect ears, small
pointed horns; smooth hair;
small frame/Unimproved Veld
group

Zurich, Zürcher: *see* Appenzell,
White Swiss Improved

**Zwartbonte Zeeuwse en
Zuidhollandse geit:** *see* Dutch
Pied

Zwergziege: (Germany)/= dwarf goat/
nearly extinct

Horse

Abbreviations used in this section:
d = milk; dr = draught; h = heavy; l = light; m = meat; pa = pack; py = pony (or small horse, < 140 cm high); ri = riding

Names for horse or pony include: ati, caballo, calul, cavallo, cheval, häst, hest, jaca, kon, konj, myinn, paard, pferd, uma

Abtenauer: (Abtenau, Salzburg, Austria)/h; dr/often black/small local var. of Noric/nearly extinct

Abyssinian: (Ethiopia)/py-l/variable in colour, size and conformation/ syn. *Abyssinian-Galla, Galla, Yellow*

Achal-Teké: *see* Akhal-Teke

Achetta: *see* Giara Pony

Adaev: (between Caspian and Aral Seas, Kazakhstan)/l; ri.dr.m/var. of Kazakh, with Turkmen, Don, Thoroughbred and Orlov Trotter blood/Russ. *Adaevskaya*/not *Adayev*

Aegidienberger: (Germany)/nearly extinct

Aglikos Katharohaemos: (Attica, Thessaly, Macedonia, Peleponnese and C. Greece)/ri/any solid colour/ imported from UK/rare/[= English Thoroughbred]

Agricultural Riding Horse: (Belgium) *see* Belgian Warmblood

Ahal-Teke: *see* Akhal-Teke

AITPR: *see* Italian Heavy Draught

Akhal-Teke: (Turkmenistan)/l; ri/ often golden-dun/sim. to Iomud but larger/orig. from ancient Turkmenian/Russ. *Akhal-*

tekinskaya/syn. *Turk, Turki, Turkmen, Turkmenian, Turkoman*/not *Achal-Teké, Ahal-Teke, Akhal-Tekin*

Albanian: (Albania)/py; ri.dr/Balkan Pony type/vars: Albanian Mountain, Myzeqeja

Albanian Mountain: (Albania)/ smaller var. of Albanian

Albino: (USA)/l or py; ri/BS 1937 for dilute white (Creme) and dominant white heterozygotes (White), not a genetic albino/syn. *American Albino*, obs. syn. *American Creme and American White*

Algerian: *see* Barb

Alogaki: *see* Skyros Pony

Altai: (Russia)/py; ri.dr.d.m/chestnut, bay, black or grey, occ. speckled/Siberian pony group/ Russ. *Altaĭskaya*/syn. *Oirot*

Altai Productive: *see* Chara

Altér: (Portugal)/bay/state stud breeding the Lusitanian horse since 1830/BS/syn. *Altér Real*

Altmark: (C Germany)/h; dr/ chestnut, brown or grey/orig. since 1875 from Shire × Clydesdale with Belgian Draught

blood since 1920/HB 1904/Ger.
Altmärkisches Kaltblut (*Kaltblut*
= Coldblood)/rare

Alt-Oldenburger: *see* Oldenburg

Altösterreichisches Warmblut: *see*
Old Austrian Warmblood

Altwürttemberger: (Baden-
Württemberg, Germany)/l; ri/
usu. chestnut, grey, brown or
black/remains of Württemberg
not absorbed by German Riding
Horse/BS, HB 1907/syn. *Old
Wurttemberg*/nearly extinct

American Albino: *see* Albino

American Bashkir Curly: (Nevada,
USA)/l; ri/long curly winter
coat/orig. 1898 from feral horses,
crossed with curly-coated horses
from other breeds/BS 1971/syn.
American Curly/not *Curley*

American Cream Draught: (Iowa,
USA)/h; dr/cream with white
mane and tail and pink skin/orig.
in early 20th c. from cream-
coloured draught mare; recog.
1950/ HB 1935, BS 1944/syn.
American Cream/nearly extinct

**American Creme and American
White:** *see* Albino

American Curly: *see* American
Bashkir Curly

American Indian: (USA)/py-l; ri/orig.
from mustang/BS 1961 for
Western horses of Spanish and
non-Spanish orig./syn. *Indian,
Indian Pony*/*see also* Cayuse,
Cherokee, Chickasaw, Choctaw,
Florida Cracker, Lac La Croix
Indian Pony

American Miniature: (USA)/orig.
from Falabella, Shetland Pony *et
al.*/BS 1978 for horses 34 inches
or less at maturity

American Mustang: (USA)/py-l; ri/
BS, HB 1962 for Western horses
of feral ancestry, Spanish or non-
Spanish orig.

American Paint: (USA)/l; ri/cf.
Pinto/BS 1962 for pied colour
type (overo or tobiano) of Paint,
Quarter Horse or Thoroughbred
breeding

American Quarter Horse: *see* Quarter
Horse

American Quarter Running Horse:
see Quarter Horse

American Saddlebred: (Kentucky,
USA)/l; ri/bay, brown, chestnut,
grey or black/orig. in 19th c.
from Thoroughbred, Morgan,
Canadian and American Trotter/
var.: Ysabella/BS 1891, BS also
in Canada 1948/syn. *American
Saddle Horse, American
Saddler, Kentucky Saddle Horse,
Kentucky Saddler, Five-gaited
Horse, Saddlebred*

American Shetland Pony: (USA)/ri/
BS 1888/vars: Classic (=
Shetland Pony, refined), Modern
(with up to 50% Hackney or
Welsh blood)/*see also* Miniature
Shetland Pony/rare

American Spotted: *see* Appaloosa,
Colorado Ranger, Kanata Pony,
Morocco Spotted, Pony of the
Americas

American Trotter: (USA)/l/orig. in
early 19th c. from Thoroughbred,
Hackney, Morgan *et al.*/orig. of
Danish Trotter, Romanian Trotter,
Swedish Standardbred Trotter
and (with Orlov Trotter) of
Russian Trotter/BS 1871; BS also
in Canada, Finland; HB also in
Norway/syn. *American Trotting
Horse, Standardbred* (official
name)

American Walking Pony: (Georgia,
USA)/ri/orig. from Welsh Pony ×
Tennessee Walking Horse/BS
1968

American White: *see* Albino

Amur: (Siberia, Russia)/l/orig. from
Transbaikal improved by Tomsk/
Russ. *Amurskaya*/extinct (by
crossing with Orlov Trotter,
Russian Trotter, Don and
Budyonny)

Anadolu Yerli: *see* Anatolian Native

Anatolian Native: (Turkey)/py; ri.pa/
vars: Araba, Canik, Hinis/Turk.
Anadolu Yerli/syn. *Native
Turkish Pony*

Andalusian: (Guadalquivir valley,
Spain)/l; ri/usu. grey, chestnut or
bay/orig. from Arab and Barb *et
al.*/var.: Carthusian; former

strains: Córdoba, Extremadura, Marismeña, Ronda, Seville/orig. of Criollo and Colonial Spanish (America)/HB 1912; BS USA (1963), GB, France, Germany, Australia, Mexico, Costa Rica, Guatemala/Sp. *Andalusa* or *Pura raza española* (= pure Spanish breed)/syn. *Andalusian-Barb, Andalusian-Lusitanian, Andalusian-Valenzuela, Betic, Bético* (*Betica* = Andalucía), *Español-Andaluz* (= Spanish Andalusian), *Guzman, Purebred Spanish, Spanish*; obs. syn. *Castilian* (?), *Iberian Saddle Horse, Iberian War Horse, Villanos*

Andean: (Peru)/var. of Peruvian Criollo/inc. Chumbivilcas, Morochuca/Sp. *Andino*

Andravida: (Ilia, Greece)/l; ri.dr/orig. early 20th c. from Anglo-Norman × local with Nonius ♂♂ after 1920/HB 1995/syn. *Eleia, Ilia*/nearly extinct

Anglický plnokrevník: *see* Thoroughbred

ANGLO-ARAB: l/Arab × Thoroughbred 1st or later cross, or breed orig. from this cross (e.g. French Anglo-Arab, Gidran, Sardinian Anglo-Arab, Spanish Anglo-Arab)/HB USA 1930/syn. *Anglo-Arabian*

Anglo-Arab du Limousin: *see* Limousin

Anglo-Arabo-Sarda: *see* Sardinian Anglo-Arab

Anglo-Don: (Russia)/l/Thoroughbred × Don, 1st cross

Anglo-Kabarda: (N Caucasus, Russia)/l; ri/bay, black-brown or black/breed group, orig. from Thoroughbred × Kabarda/Russ. *Anglo-Kabardinskaya porodnaya gruppa*

Anglo-Karachai: (N Caucasus, Russia)/l/Thoroughbred × Karachai 1st cross/Russ. *Anglo-Karachaevskaya*

Anglo-Lusitanian: (Portugal)/l; ri/ orig. from Thoroughbred × Lusitanian/Port. *Anglo-Lusitano*/rare

Anglo-Norman: (Normandy, France)/ l/orig. in 20th c. from Norman Coach Horse with more Thoroughbred blood/orig. of Charentais, Charolais, Vendée/ BS Argentina/Fr. *Anglo-Normand*/syn. *Norman*/extinct (inc. in French Saddle Horse)

Anglo-Norman Trotter: *see* French Trotter

Anglo-Teke: (Turkmenistan)/l/ Thoroughbred × Akhal-Teke (1st) cross/Russ. *Anglo-tekinskaya*

Anlin Guoxia: (Sichua, Yunnan and Guizhou, China)/usu. bay; dwarf/orig. from Jiancheng

Annamese: *see* Vietnamese

Annamite Pony: *see* Vietnamese Pony

Appaloosa: (Oregon, USA)/l; ri/ colour type: usu. chubary (white rump usu. with small spots), also white with spots over whole body, or coloured with spots over body or on rump only, or roan/BS 1938, HB 1947; BS also in Australia, Canada, Italy (1981), Mexico, S Africa; HB in Netherlands/syn. *Palouse*/not *Apaloosa, Appolousey*/*see also* British Appaloosa, Colorado Ranger, Kanata Pony, Pony of the Americas, Toby/[from Palouse R.]

Appaloosa Pony: (USA)/sim. to Appaloosa but smaller/BS 1963/syn. *National Appaloosa Pony*

Apulian: (Italy)/l/It. *Pugliese*/*see also* Capitanata, Murgese/extinct

AraAppaloosa: (USA)/l; ri/orig. from Appalooosa with Arab blood/BS 1985

Arab: l; ri/original strains inc. Keheilan, Maneghi, Saglawi/BS in Argentina, Australia, Austria (HB 1973), Canada 1958, Czechoslovakia, England 1918 (HB 1788), Germany 1949, Hungary, India, Ireland, Israel (HB 1974), Netherlands 1935, New Zealand, Pakistan, Poland, South Africa *c.* 1961, USA 1908; HB in Belgium, Denmark, Egypt 1900, Finland 1955, France

1833, Norway, Spain, Sweden, Syria, Russia/syn. *Arabian* (USA)/*see also* Egyptian, Persian Arab, Syrian, Turkish Arab

Araba: (Turkey)/var. of Anatolian Native for pulling carriages/[= carriage]

Arab-Barb: *see* West African Barb

Arabialainen: (S Finland)/ri/grey, chestnut or bay/Arab imported from Sweden

Arabian: *see* Arab

Arabo-Haflinger: (Bavaria, Germany)/ Arab × Haflinger cross, usu. 25% Arab

Arabo-Sarda: *see* Sardinian

Arabský Plnokrvnik: (Slovakia)/ ri/white/purebred Arab in Topol'cianky (one herd only)/ nearly extinct

Araby: (Poland) *see* Czysta Krew Arabska

Aragonesa: (Aragon, Spain)/l; dr/usu. dapple grey/orig. from Percheron (+ Breton and Ardennes) × local/extinct

Arapska: (Yugoslavia)/ri/grey, bay, brown or occ. black, often with white markings/= Arab/nearly extinct

Ardahan: *see* Malakan

Ardennais du Nord: *see* Trait du Nord

Ardennes: (Belgium)/h; dr.m/lighter (mt) var. of Belgian Draught/orig. of Baltic Ardennes, Russian Heavy Draught, Swedish Ardennes/BS 1926; HB also Spain, Denmark, Luxembourg 1930/Fr. *Cheval de Trait Ardennais* or *Ardennais*/ syn. *Belgian Ardennes*

Ardennes: (France) *see* French Ardennais

Ardennes: (Poland) *see* Polish Ardens

Arenberg-Nordkirchner: (Germany)/nearly extinct

Argentine Criollo: (Argentina and Uruguay)/l; ri/Criollo type, revived 1875–1890/BS/syn. *Argentine Landrace, Criollo*

Argentine Dwarf: *see* Falabella Pony

Argentine Landrace: *see* Argentine Criollo

Argentine Polo Pony: (Argentina)/l; ri/orig. from Thoroughbred × Argentine Criollo/BS

Ariège, Ariègeois: *see* Castillonais, Mérens

Asiatic wild horse: *see* Przewalski horse

asil: (Middle East)/= purebred (Arab)

Assateague Pony: *see* Chincoteague/Assateague

Astrakhan: *see* Kalmyk

Asturian pony: (Mt Suéve, Oviedo, N Spain)/ri.dr/usu. black, also dark bay or brown/sim. to Pottok/at one time combined with Galician as *Galician-Asturian*/HB 1981, BS/Sp. *Poney asturiano, Caballo asturcón*/syn. *Suéve* /nearly extinct

Augeron: (Auge, Normandy, France)/ local HB (1913–1966) for Percheron (orig. with Norman blood)/extinct

Australian Pony: (Australia)/BS 1931 for pony of mixed English orig.

Australian Stock Horse: (Australia)/l; ri/orig. from Waler, Thoroughbred and Quarter Horse/BS 1971

Australian Waler: *see* Waler

Austrian Warmblood: (Austria)/ l/black, bay, chestnut or grey/ based on Oldenburg (heavier, Upper and Lower Austria), English Halfbred and Nonius (lighter, Burgenland and Carinthia)/old type (Old Austrian Warmblood) nearly extinct; new type based on Trakehner, Oldenburg *et al.*/Ger. *Österreichisches Warmblut*

Austria Shagya-Araber: *see* Shagya Arab

Auvergne: (France)/py/? extinct but BS formed 1998 to revive

Auxois: (Burgundy, France)/h; dr.m/ bay, roan, grey or chestnut/orig. from French heavy breeds (since 1913 only Ardennes) × local (descended from Morvan crosses)/formerly var. of French Ardennais/BS, HB 1913/Fr. *Trait Ardennais d'Auxois, Trait Auxois*/rare

Avar: (mts of Dagestan, Russia)/py; pa.ri./smallest var. of Dagestan Pony/Russ. *Avarskaya*

Avelignese: *see* Haflinger

Azerbaijan: (Azerbaijan)/l-py; ri.pa/ usu. grey or bay/improved by saddle breeds 1927–1929 and 1940s/orig. of Deliboz/Russ. *Azerbaidzhanskaya* (*kazakhskaya*)/rare

Azores Pony: (Ilha Terceira, Azores, Portugal)/ri.dr/usu. brown or bay/syn. *Garrano da Ilha Terceira*/nearly extinct

Azteca: (Mexico)/l; ri/orig. since 1972 from Andalusian × Quarter Horse and Criollo/BS in USA 1989

Bábolna Arab: *see* Shagya Arab

Baduck: *see* Batak

Bagual: (Argentina)/feral, in process of redomestication/Sp. orig., like mustang (USA) and Criollo/not *Bagnal*/[old term for wild or untamed horses abandoned in 16th and 17th c.]

Bahiano: *see* Pantaneiro

Bahr-el-Ghazal: (Chad)/var. of W African Dongola/syn. *Kréda, Ganaston*

Baiano: *see* Pantaneiro

Baicha: (SE Inner Mongolia, China)/ small strain of Chinese Mongolian/syn. *Baichayi*/not *Baichatieti*/rare

Baikal: *see* Buryat

Baise Pony: (Yunnan, Sichuan and Guangxi, SW China)/ri.pa.dr/ usu. bay; dwarf/var. of Guangxi Pony/rare

Baixo-Amazona: *see* Marajoaro

Bajau: (Malaysia)

Bakewell: *see* Vardy

Bakhtiari: (Iran)/var. of Plateau Persian

Baladi: (Egypt) *see* Egyptian

Baladi: (Jordan) *see* Jordanian

Balearic: *see* Majorcan, Minorcan

Bali: (Indonesia)/py; pa.ri/usu. dun with black back stripe, mane and tail/Indonesian pony group

Balikun: (Xinjiang, China)/py; ri.pa.dr/usu. bay or chestnut/ early orig. from Kazakh and Mongolian

BALKAN PONY: inc. Albanian, Bosnian Pony, Bulgarian Native, Greek Pony, Krk Island Pony, Macedonian Pony

Balkar: (N Caucasus, Russia)/mt var. of Kabarda

Baloch, Balouch: *see* Baluchi

Baltic Ardennes: (Baltic states)/h/ orig. from Ardennes × draught breeds and local; absorbed by Estonian Draught, Latvian Heavy Draught and Lithuanian Heavy Draught/Russ. *Baltiiskaya-Ardenskaya*/syn. *Latvian Ardennes*/extinct

Baltic Trotter: (Baltic states)/l/orig. from trotter × Konik; absorbed by Latvian, Orlov and Russian Trotters/var.: Latgale Trotter/Ger. *Panjepferd im Trabertyp*/extinct

Baluchi: (Baluchistan and Derajat, Pakistan)/l/turned-in ears/not *Baloch, Balouch*

Banamba: *see* Bélédougou

Banat: (Timiş plateau, Romania)/l; dr/orig. from Nonius, Noric, Ardennes, Oldenburg and Lipitsa/extinct

Bandiagara: (Niger bend, Mali)/local var. of Dongola-Barb orig./syn. *Gondo*

Ban-ei race horse: (Hokkaido, Japan)/ h; dr/orig. from Percheron and Breton; bred for *Ban-ei Keiba* race, in which horse pulls heavy sledge

Banker: (N Carolina, USA)/py/part Spanish orig; feral (since 16th c.), semi-feral and domestic on Outer Banks Is (inc. Shackleford)/registered in Spanish Mustang HB/syn. *Shackleford Pony*/purebreds rare

Barb: (Maghreb, N Africa)/l; ri/inc. Algerian, Moroccan, Tunisian/ orig. of West African Barb, Spanish Barb (USA)/BS France (HB 1988)/ Fr. *Barbe*/not *Barbary, Berber*/rare

Bardigiana: (Bardi, Parma, Italy)/py; m.ri.(pa.dr)/usu. bay, occ. black/improved with Freiberger since 1977/HB 1977, BS/syn. *Bardese, Cavallo montanaro* (= mountain breed)/rare

Barouéli: *see* Barwéli

Barranca: *see* Navarre Pony

Barra Pony: (Hebrides, Scotland)/var. of Hebridean Pony/extinct

Barthais: *see* Landais

Barthes Pony: *see* Landais

Barwéli: (Mali)/Fr. *cheval de Barouéli*

Bashkir: (Bashkiria, Russia)/l; d.m.ri.dr/ Mongolian type improved by various breeds/ Russ. *Bashkirskaya*

Bashkir Curly: (USA) *see* American Bashkir Curly

Basque: *see* Pottok

Basque-Navarre Pony: *see* Navarre Pony

Basseri: (Iran)/var. of Plateau Persian

Basuto Pony: (Lesotho)/ri.dr/orig. from Cape Horse 1825 on/orig. of Nooitgedacht Pony/disappearing early 20th c. by export and crossing with Arab and Thoroughbred; purebred now being revived by BS

Batak Pony: (N Sumatra)/ri/ Indonesian pony group/syn. *Deli*/not *Baduck* (Jap.), *Battak*

Bati Karadeniz Rahvan Ati: *see* Rahvan

battle horse: *see* Camargue

Bavarian Warmblood: (Bavaria, Germany)/l; ri/usu. dark chestnut/orig. 1960s from Rottal with Hanoverian, Thoroughbred and Trakehner blood; inc. in German Riding Horse/Ger. *Bayerisches Warmblut*

Bayerisches Warmblut: *see* Bavarian Warmblood

Beetewk: *see* Bityug

Beihaido: *see* Hokkaido Pony

Belarus Harness: (Belarus)/l; dr.m.d/ orig. in late 19th and early 20th c. from various draught and coach breeds (esp. Døle) × local/Russ. *Belorusskaya, Belorusskaya upryazhnaya*/obs. syn. *Byelorussian* or *White-Russian Carriage, Coach, Draught* or *Harness*

Bélédougou: (Mali)/var. of West African Barb/syn. *Banamba*

Belgian Ardennes: *see* Ardennes

Belgian Draught: (Belgium)/h; m.dr./ orig. from Flemish/var.: Ardennes/ orig. of French Ardennais, Rhenish, Soviet Heavy Draught/BS 1886, HB 1890; BS also USA 1889, Canada 1907; HB also Denmark/Flem. *Belgisch Trekpaard*, Fr. *Cheval de (gros) trait belge* (= Belgian (Heavy) Draught Horse)/syn. *Belgian*; obs. syn. (heavy type) *Brabançon* (esp. Russia)

Belgian Halfblood: *see* Belgian Sport Horse

Belgian Halfbred: *see* Belgian Sport Horse

Belgian Riding Pony: (Belgium)/Belg. *Rijpony*

Belgian Saddlebred: *see* Belgian Warmblood

Belgian Sport Horse: (Belgium)/l; ri/orig. (as *Belgian Halfblood*) from Thoroughbred and French Saddlebred × Belgian Draught/ BS 1920, HB 1921/Fr. *Cheval de Sport Belge*, Flem. *Belgisch Sportpaard*/syn. *SBS*

Belgian Trotter: (Belgium)/orig. from American, French and German Trotters/BS renamed *Fédération Belge du Trot* in 1976/Fr. *Trotteur Belge*

Belgian Warmblood: (Belgium)/l; ri/ orig. from Hanoverian, Anglo-Norman *et al.*/BS (as *Agricultural Riding Horse*) 1955, recog. 1960, renamed 1970/Flem. *Belgisch Warmbloedpaard* (= *BWP*), Fr. *Cheval de Sang Belge* (= *CSB*)/ syn. *Belgian Saddlebred*

Belgisch Trekpaard: *see* Belgian Draught

Belorusskaya: *see* Belarus

Berber: *see* Barb

Berrichon: (Berry, C France)/local HB (1923–1966) for Percheron/ extinct

Bessarabian: *see* Bulgarian Colonist, German Bessarabian

Betic, Bético: *see* Andalusian

Bético-lusitano: *see* Lusitanian

Betpakdalin: (Dzezkazgan, Kazakhstan)/subvar. of Jabe var. of Kazakh with local blood

Bhimthadi: *see* Deccani

Bhirum Pony: (N Nigeria)/dwarf/cf. Kirdi/syn. *Pagan*/not *Birom*

Bhotia Pony: (Nepal and Bhutan; Sikkim and Darjeeling, India)/ ri.pa/often white (grey) or bay/ sim. to Tibetan pony but less broad/vars: Chyanta, Tanghan, Tattu/Nepali *Bhote ghoda*/syn. *Bhutan Pony, Bhutani, Bhutia Pony*/[= Tibetan]

Bhutan, Bhutia: *see* Bhotia

Biçuk: *see* Bityug

Bigourdan: *see* Tarbes

Biłgoray: (SE Poland)/improved var. of Polish Konik/nearly extinct

Billie: *see* Quarter Horse

Bimanese: (Sumbawa, Indonesia)/var. of Sumbawa Pony/syn. *Bima*

Briom: *see* Bhirum

Bityug: (Voronezh, Russia)/h/sim. to Voronezh/orig. from heavy trotter × local in 19th c./orig. of Voronezh Heavy Draught/Russ. *Bityugskaya*, Turk. *Biçuk*/not *Beetwek, Bitjug*/extinct

Black Forest: (Germany)/h; dr/ chestnut with light mane (also one grey family)/orig. from Rhenish/HB 1896/Ger. *Schwarzwälder Füchse* (= foxy), *Schwarzwälder Kaltblut*/syn. *Black Forest Chestnut, Sankt Märgener Fuchse*/rare

Black Sea: (Krasnodar-Rostov area, Russia)/l; ri.dr./orig. from Nogai × saddle horse in 18th c., × mountain Thoroughbred, Don, Karabakh *et al.* in 19th and 20th c./absorbed by Budyonny, Don and Ukrainian Saddle Horse/Russ. *Chernomorskaya*/ nearly extinct

BLM horse: *see* mustang

blood horse: (Ger. *Vollblut*, Fr. *pur sang*)/inc. Arab, Barb, Thoroughbred

Bobo: (Burkina Faso)/py/? degenerate Barb/syn. *Bobodi*

Boer: (S Africa)/revival of old Boer (Cape Horse)/orig. of Calvinia, Cape Boer, Historic Boer/Afrik. *Boerperd*

Bohai: *see* Buohai

Bolivian Pony: (Bolivia)/almost dwarf var. of Criollo/syn. *Sunicho*/nearly extinct (by competition from donkey)

Bornu: (NE Nigeria)/var. of West African Dongola

Bosnian Pony: (Bosnia)/ri.pa/Balkan Pony type/orig. from Buša Pony in 18th c./vars: prim. Karst type in Hercegovina (brown or dun); improved by Arab in C and E Bosnia (taller, white, brown, black or chestnut)/Serbo-cro. *Bosanski brdski konj* (*brdski* = mountain)/syn. *Bosniak, Bosnian Mountain*/nearly extinct

Boulonnais: (Boulogne district, N France)/h; m.dr/usu. grey/HB 1886, BS/var.: Petit Boulonnais/rare

Bourbonnais: (Allier, C France)/local HB (1923–1966) for Percheron (but BS still extant)

Bovenlander: (Netherlands) *see* Oldenburg

Brabançon, Brabant: *see* Belgian Draught (excluding Ardennes var.)

Brandenburg: (Germany)/l; ri/inc. in Edles Warmblut/Ger. *Brandenburger Warmblut*

Brazilian: *see* Brazilian Sport Horse, Campolina, Crioulo, Mangalarga, Northeastern

Brazilian Pony: (S and SE Brazil)/HB 1971

Brazilian Sport Horse: (São Paulo, Brazil)/l; ri/orig. since 1970 from many saddle breeds × Crioulo/ BS 1977/Port. *Cavalo de Hipismo, Brasileiro de Hipismo* (*BH*) (*hipismo* = horse-racing)/ syn. *Brazilian*

Breton: (Brittany, France)/h; m.dr/ usu. chestnut or chestnut roan/ vars: Breton Draught, Breton Post-horse, Breton Small Mountain Draught/orig. of Hispano-Bretona/ HB 1909, BS; HB also Spain

Bréton Cerdà: *see* Hispano-Bretona

Bretona Ceretana: *see* Hispano-Bretona

Breton Cob: *see* Breton Post-horse

Breton Draught: (Brittany, France)/ var. of Breton/Fr. *Trait breton*

Bréton Empordanès: (Spain)/h; dr/ black or chestnut/influenced by Breton Post-horse

Breton Post-horse: (Brittany, France)/ var. of Breton/orig. *c.* 1850–1914 from Hackney × local/Fr. *Postier breton*/syn. *Breton Cob, Norfolk Breton*

Breton Saddle Horse: *see* Corlais

Breton Small Mountain Draught: (Brittany, France)/l; dr./var. of Breton

British Appaloosa: (UK)/l; ri/BS 1976 for spotted horses of various orig.

British Riding Pony: (UK)/ Thoroughbred or Arab × British pony breed/HB

British Spotted Horse and Pony: (England)/BS 1946–1976, replaced by British Appaloosa and British Spotted Pony societies

British Spotted Pony: (England)/ coloured spots on white background, or *vice versa*/BS 1946 (with Spotted Horse), split 1976/rare

British Warmblood: (England)/l; ri/ orig. from Hanoverian, Dutch Warmblood and Danish Warmblood, × Thoroughbred/BS 1977, recog. 1994

bronco: (USA)/unruly type of Western pony, rodeo bucking horse/[Sp. *bronco* = rough, coarse]

broomtail: (USA)/feral, remnants of mustang

brumby: (Australia)/l/feral/not *brumbie*

Buckskin: (USA)/l; ri/colour type: BS 1962 and 1971 for buckskin (golden with black dorsal stripe, mane, tail and on legs), dun, red dun or grulla (Sp., = crane, i.e. greyish yellow) horses of any breed

Bucovina: (Romania)/l; dr/orig. from Hutsul ($\frac{5}{8}$) × Romanian Light Draught ($\frac{3}{8}$)/nearly extinct

Budyonny: (Russia)/l; ri.dr./usu. chestnut or bay/orig. 1921–1949 by S.M. Budënnyĭ from Don, Thoroughbred and Black Sea/HB 1951/Russ. *Budënnovskaya*/not *Budjonny*

Bulgarian: *see also* Danube, East Bulgarian, Pleven

Bulgarian Colonist: (S Bessarabia)/ py/sim. to Dobrogea; orig. from Bulgarian, Moldavian and Ukrainian/syn. *Bessarabian*/ extinct

Bulgarian Heavy Draught: (Bulgaria)/h; dr/orig. (1950 on) from Ardennes, Soviet Heavy Draught and esp. Russian Heavy Draught, × East Bulgarian, Arab and crossbred draught ♀♀/Bulg. *Tezhkovozna*

Bulgarian Mountain: (Bulgaria)/inc. Karakachan, Rila Mountain, Stara Planina

Bulgarian Native: (Bulgaria)/py/ Balkan Pony type/vars: Deli-Orman, Dolny-Iskar, Karakachan, Rila Mountain, Stara Planina/ extinct except Karakachan

bull horse: *see* Camargue

Buohai: (NE Shandong, China)/l-h; ri.dr/often bay or chestnut/orig. 1952–1963 from Soviet breeds (inc. Soviet Heavy Draught and Ardennes)/HB 1974/not *Bohai*

Burgdorf: (Switzerland)/h.dr/regional var. and heavier strain of original Freiberger/orig. from Ardennes × Jura/extinct

Burguete: (Navarre, Spain)/l; m.dr/ usu. chestnut, also black or bay/ orig. in late 19th c. from Navarre Pony graded to Breton/Sp. *Caballo de Burguete* or *Burguetana*

Burmese: (Myanmar)/larger and with thinner hair coat than Shan Pony/Burmese *Myinn* (= horse)

Buryat: (Buryatia, Russia)/py; ri.dr.pa/usu. grey, bay or sorrel with zebra stripes/Siberian Pony group/Russ. *Buryatskaya*/syn. *Baikal, Buryat-Mongolian, Transbaikal* (Russ. *Zabaikal'skaya*)/not *Buriat*/rare

Buša Pony: (Croatia and Bosnia)/orig. of Bosnian Pony and Posavina/ extinct

BWP: *see* Belgian Warmblood
Byelorussian: *see* Belarus

Caballo asturcón: *see* Asturian pony
CAITPR: *see* Italian Heavy Draught
Calabrian: (Calabria, Italy)/l; ri/orig.
   from Oriental × local, improved
   by Salernitana and then by
   Thoroughbred/It. *Calabrese*/rare
Calul de Sport: *see* Romanian Sport
   Horse
Calvinia: (S Africa)/l/orig. from Boer
   crossed with Thoroughbred,
   Hackney and Cleveland Bay/
   extinct (replaced by Cape Boer)
Camargue: (Rhône delta, S France)/
   py for herding bulls/white (born
   grey) or grey (born black or
   brown); semi-feral/recog. 1968;
   HB 1978, BS/Fr. *Camarguais*/
   syn. *battle horse, bull horse*/rare
Cambodian: (Cambodia)/South-East
   Asia pony group, sim. to Thai
   and Vietnamese/Fr. *Cambodgien*
Cameroon Pony: *see* Kirdi Pony
Campeiro: (Santa Catarina, Brazil)/?
   English Thoroughbred and Arab
   blood introduced 1912/nearly
   extinct
Campolina: (Minas Gerais, Brazil)/l;
   ri.(dr)/usu. bay, sorrel or
   chestnut/orig. in mid-19th c. by
   C. Campolina from native horse
   (Crioulo) with blood of Anglo-
   Norman, Holstein, American
   Saddlebred and Mangalarga/BS
   1938, reformed 1951, 1970
   on/not *Campolino*
Canadian: (Quebec, Canada)/l; ri.dr/
   usu. black or bay/orig. in late 17th
   and early 18th c. from Arab,
   Breton, Anglo-Norman/var.:
   Canadian Pacer/HB 1885, BS 1895,
   revived 1909 on/Fr. *Canadien*/ obs.
   syn. *French Canadian*/rare
Canadian Cutting Horse: (Canada)/l;
   ri/BS for horses of any breed
   (often Quarter Horse) able to
   work with cattle esp. at rodeos
Canadian Hunter: *see* Canadian
   Sport Horse
Canadian Pacer: (Canada)/var. of
   Canadian with blood of
   Narragansett/extinct

Canadian Pinto: (Canada)
Canadian Rustic Pony: (Manitoba
   and Saskatchewan, Canada)/
   ri/grey, buckskin or bay/orig.
   from Arab (+ Caspian) ($\frac{3}{8}$), Welsh
   Mountain Pony ($\frac{3}{8}$) and bred-back
   'Tarpan' ($\frac{1}{4}$)/BS 1989; BS in USA
   1978/rare
Canadian Sport Horse: (Canada)/l;
   ri/BS 1926 as Canadian Hunter,
   Saddle and Light Horse
   Improvement Society, renamed
   1987/obs. syn. *Canadian Hunter*
   (1970–1984)
Canary Island Pony: (Spain)/Sp. *Jaca
   canaria*/extinct
Canik: (NE Turkey)/local var. of
   Anatolian Native/not *Çenik,
   Cenik*
Cantabrian Pony: *see* North Spanish
   Pony
Cape Boer: (S Africa)/ri/orig. from
   Cape Horse/nearly extinct 1948;
   since upgraded with Arab and
   esp. American Saddle Horse
   ♂♂/BS 1957, HB (inc. F1
   animals)/Afrik. *Kaapse Boerperd*
Cape Harness: (S Africa)/l; dr/orig.
   from Anglo-Arab, Friesian and
   Hackney/extinct (replaced by
   Friesian and Flemish)
Cape Horse: (S Africa)/l; ri.dr/orig.
   from Oriental imported 1652–
   1778, Thoroughbred 1782–1860
   and Hackney 1860–1891; nearly
   extinct after 1900/orig. of Basuto
   Pony, Cape Boer, Namaqua Pony/
   syn. *Boer* (old), *Hantam, South
   African*/extinct
Capitanata: (Foggia, Apulia, Italy)/l;
   dr./improved by Maremmana,
   Murgese and Anglo-Norman/*see
   also* Apulian
Carakachanski Kon: *see* Karakachan
Carpathian Pony: *see* Hutsul
Carrossier normand: *see* Norman
   Coach Horse
Carthusian: (Spain)/usu. grey/strain
   of Andalusian/Sp. *Cartuja,
   Cartujana*/syn. *Zamorano,
   Zapata*/rare
Caspian: (Gilan and Mazandaran,
   Iran)/ri.dr/usu. bay, grey or
   chestnut, occ. black; small horse

(36–44 inches)/BS; BS also GB 1976, USA 1987, Australia, New Zealand/syn. *Caspian Miniature, Caspian Pony*/rare

Castilian: (Spain)/l/? = Andalusian/ Sp. *Castellana, de Castilla*/ extinct

Castillonnais: (Ariège, Pyrenees, France)/l; dr.ri/bay or chestnut/ HB 1980, BS/Fr. *Cheval ariègeois de Castillon*/nearly extinct

Catalan, Catalonian: *see* Hispano-Bretona

Catria: (Pesaro, Marche, Italy)/ dr.ri.m/bay/orig. from Haflinger and Freiberger × Maremmana/ HB 1990/It. *Cavallo del Catria* (= of Catria mt)/rare

Cavallo Agricolo Italiano da Tiro Pesante Rapido: *see* Italian Heavy Draught

Cavallo da sella svizzera: *see* Swiss Warmblood

Cavallo del Catria: *see* Catria

Cavallo montanaro: *see* Bardigiano

Cavallo Pentro: (Molise, Italy)/orig. from native and Breton horses

Cavalo de Hipismo: *see* Brazilian Sport Horse

Cayusse: (USA)/var. of American Indian/extinct

Celebes: *see* Macassar Pony

CELTIC PONY: inc. Connemara, Faeroes, Hebridean, Iceland and Shetland Ponies; also Faco (Sp., ? = Galician), Garrano (Port.), Pottok (Basq.)

Cenik: *see* Canik

Cerbat: (Arizona, USA)/l/Sp. orig.; feral strain of Spanish Mustang/ syn. *Cerbat Mountain Spanish Mustang*/nearly extinct

Çerkes atlar: *see* Uzunyayla

Český chladnokrevník: *see* Czech Coldblood

Český klusák: *see* Czech Trotter

Český teplokrevník: *see* Czech Warmblood

Chaidamu: (China)

Chakouyi: (silk road, China)/py; ri/usu. bay, also grey or black/[= by-route transmission station]

Chapman Horse: *see* Cleveland Bay

Chara: (Altai, Russia)/l; m.ri.dr/

recent orig. from heavy draught, trotter and saddle breeds, × local/syn. *Altai Productive*

Charentais: (W France)/l/derivative of Anglo-Norman and Anglo-Arab/inc. in French Saddlebred/ extinct

Charolais: (C France)/l/orig. from Anglo-Norman × local/inc. in French Saddlebred/not *Charollais*/extinct

Charysh: (Siberia, Russia)/local var./rare

Cheju: (Korea)/py; ri.dr/South-East Asia pony group/syn. *Korean* (Jap. *Chosen*), *Saishu Island*/not *Corean*/nearly extinct

Cheju racing horse: (Korea)/native, ? with Thoroughbred blood

Chejudo LIP: (Korea)/= Chejudo Livestock Production Institute horse/closely related to Tsushima native (i.e. Taishu Pony on Tsushima I, Nagasaki, Japan) [Chejudo = Cheju Island]

Chenarani: (NE Iran)/l/Plateau Persian × Turkoman cross/not *Tcheraran*

Chernomorskaya: *see* Black Sea

Cherokee: (SE USA, removed to Oklahoma)/var. of American Indian/nearly extinct

Cheval ariègeois de Castillon: *see* Castillonnais

Cheval de (gros) trait belge: *see* Belgian Draught

Cheval de Sang Belge: *see* Belgian Warmblood

Cheval de Sport Belge: *see* Belgian Sport Horse

Cheval de Trait Ardennais: *see* Ardennes

Chickasaw: (Tennessee and N Carolina, USA)/American Indian orig./BS 1959 for horses resembling orig. Chickasaw type/ not *Chicksaw*/[Chickasaw Indians]/ nearly extinct (absorbed into Spanish Mustang population)

Chilean: (Chile)/l.ri/Criollo type, first imported 1541/HB 1893, revived 1913/Sp. *Chileno*/syn. *Chilean Criollo*/rare

Chilkov: (Buryatia, Russia)/orig. from heavy draught × Buryat in 17th c./Russ. *Chilkovskaya*/extinct

Chilote: (Chiloé I, Chile)/py; ri.dr./ various colours but mostly dark/ Spanish orig.; ? Criollo group/ rare

Chincoteague/Assateague Pony: (Assateague I, Maryland, USA)/semi-feral since 1669/BS (Chincoteague) 1989/rare

Chinese developed breeds: *see* Buohai, Guanzhong, Heihe, Heilongjiang, Jilin, Jinzhou, Keerqin, Sanhe, Shandan, Tieling Harness, Yili, Yiwu

Chinese native breeds: *see* Baise, Balikun, Chakouyi, Chinese Kazakh, Chinese Mongolian, Datong Pony, Erlunchun, Guanxi, Guizhou, Guoxia, Hequ, Jianchang, Jinhong, Lichuan, Sini, Tibetan Pony, Yanqi, Yunnan

Chinese Kazakh: (W Xinjiang, China)/py; ri.pa.d.m/usu. bay, also chestnut or black/orig. from Kazakh

Chinese Miniature: (Sichuan, Guizhou, Yunnan and Guangxi, SW China)/pa.ri/usu. bay, some chestnut/discovered 1986–1990, possibly of separate origin from other Chinese ponies/local (county) vars: smallest in Debao, Ningqiang, Sichuan; larger in Yunnan, Guizhou/BS and HB 1987; separate Guizhou BS 1988/syn. *Chinese Minihorse*/*see also* Guoxia

Chinese Mongolian: (Inner Mongolia and bordering provinces, China)/ py; ri.pa.dr.d.m/usu. bay, grey or black/same orig. as Mongolian/ vars: Baicha, Wuzhumuxin, Wushen/orig. of Erlunchun, Sanhe, Yanqi/ Ch. *Menggu*/not *Monggol*

Chistokrovnaya Arabskaya: *see* Russian Arab

Chistokrovnaya verkhovaya: *see* Thoroughbred

Choctaw: (SE USA, removed to Oklahoma)/var. of American Indian/inc. 'medicine hut', 'brown ticked' (*oscuro rabicano*) etc./nearly extinct

Chola: (Peru)/medium var. of Peruvian Criollo/syn. *Serrana* (= mountain)

Chosen: *see* Cheju

chubary: *see* Appaloosa

Chukorova: *see* Çukorova

Chumbivilcas: (Cuzco and Apurimac, Peru)/ri/usu. bay or grey/subvar. of Andean

Chummarti: (Tibet, and Kumaur dist, Himachal Pradesh, India)/l/sim. to Spiti

Chumysh: (NE Altai, Russia)/l; ri.dr/ breed group, orig. from Siberian Pony improved by Russian breeds (1770–1850) and by trotters and draught breeds (1850–1917)/Russ. *Chumyshskaya porodnaya gruppa*/rare

Chunk Morgan: (USA)/old heavy strain of Morgan, sim. to Lippitt Morgan/rare

Chuvash: (Chuvasia, Russia)/py/local horse of forest type improved by trotter, Soviet Heavy Draught and Vladimir Heavy Draught/Russ. *Chuvashskaya*/extinct

Chyanta: (Nepal)/ri/smaller var. of Bhotia Pony/Nepali *Cyāntā* (= dwarf)

cimarron: (Spanish America)/feral horse; cf. mustang/Port. *cimarrão* (Brazil)

Ciociaro: *see* Esperia Pony

Circassian horse: *see* Uzunyayla

Cirit: (Turkey)/py; ri/orig. from Anatolian Native, Turkish Arab and East Anatolian (Kurd), or from Arab × Karabair/[*cirit* or *jereed* is ancient game played on horseback]

Claybank Dun: *see* Isabella

Cleveland Bay: (E Yorkshire, England)/l; dr/bay with black points/former var.: Yorkshire Coach Horse (with Thoroughbred blood) 1886–1937/ BS 1884; BS also in USA 1879/obs. syn. *Chapman Horse*/nearly extinct in GB

Clydesdale: (Scotland)/h; dr/usu. bay, brown or black, with white on face and feet/sim. to Shire (Shire blood); orig. from Great Horse/orig. of Vladimir Heavy Draught (Russia)/named 1826, BS 1877, HB 1878; BS also Australia, Canada 1886, New Zealand 1911, S Africa, USA 1879/rare in GB

cob: (Great Britain)/l; ri/any heavyweight, short-legged riding horse (hack)/syn. *English cob, riding cob/see also* Breton Cob, Irish Cob, Norman Cob, Welsh Cob

COLDBLOOD: (Germany)/= heavy draught/inc. Black Forest, Rhenish, Saxony-Thuringian, Schleswig, South German/Ger. *Kaltblut*

Cold Deck: *see* Quarter Horse

Colombian Criollo: (Colombia)/l; ri/ Criollo type/syn. *Colombian Paso Fino, Colombian Walking Horse*

COLONIAL SPANISH: (USA)/orig. from Spanish breeds (e.g. Andalusian, Sorraia) since 1521; cf. Criollo/ inc. feral strains (e.g. Cerbat and Kiger mustang), native American strains (e.g. Cherokee, Chickasaw, Choctaw), rancher strains (e.g. Wilbur-Croce), Mexican strains, and SE strains (e.g. Florida Cracker)/for BSs, see American Indian, Chickasaw, Florida Cracker, Southwest Spanish Mustang, Spanish Barb, Spanish Mustang/syn. *Spanish American, Spanish Colonial*

Colorado Ranger: (Colorado, USA)/l; ri/colour type; spotted sim. to Appaloosa/BS 1938/syn. *Rangerbred*

Comtois: (Doubs, E France)/h-l; m.dr/ bay or chestnut/local breed sim. to Freiberger (Switzerland) improved by Ardennes/HB 1919, BS/syn. *Maiche*/[from Franche Comté]

Conestoga: (Pennsylvania, USA)/l; dr/orig. late 18th c. from Flemish and English breeds/extinct before 1900

Connemara Pony: (W Ireland)/ri.dr/ usu. grey, black, bay, brown or dun/orig. from Irish Hobby/BS 1923, HB 1926; BS also England 1947 (HB 1900), USA 1956, Sweden 1965, France (HB 1969), Germany, Netherlands; HB also Australia, Finland 1971

Copper Bottom: *see* Quarter Horse

Córdoba: (Andalucía, Spain)/former strain of Andalusian/Sp. *Cordobesa*/extinct

Corean: *see* Cheju

Corlais: (Corlay, Brittany, France)/l; ri/orig. in 19th c. from Thoroughbred (and Arab) × local/syn. *Breton Saddle Horse*/ extinct (inc. in French Saddlebred)

Corsican Pony: (Corsica, France)/ ri.pa/usu. bay/BS/Fr. *Corse*/? extinct

Cossack: (Russia)/orig. in 18th c. from Nogai × Mongolian and Kalmyk/orig. of Don/syn. *Old Don*/extinct

Costa Rica Saddle Horse: (Costa Rica)/l; ri/? Criollo type/Sp. *Caballo costaricense de paso*

Costeño: (Peru)/l; ri/larger var. of Peruvian Criollo/subvars: Costeño de Paso (or Peruvian Paso) and Costeño de Paso aclimitado a la altura (= coastal gaited horse adapted to the highlands)

Cotocoli: *see* Koto-Koli

cow pony: (USA)/type of Western pony orig. from mustang *et al.* or any horse used to work cattle

Cracker: *see* Florida Cracker

Creme and White: *see* Albino

Cremonese: (Cremona, Italy)/h; dr/ orig. from Belgian Draught with Breton and Percheron blood/ remnants absorbed by AITPR/ syn. *Padana* (from Po valley)/ extinct

Cretan: *see* Messara

CRIOLLO: (Spanish America)/l or py/ sim. to (or =) Crioulo/Sp. (Andalusian) orig. from 16th c. on/inc. Argentine Criollo, Bolivian Pony, Chilean,

Colombian Criollo, Cuban
Criollo, Cuban Trotter, Mexican
Pony, Peruvian Criollo,
Uruguayan Criollo, Venezuelan
Criollo/BS (InterAmerican
Federation for the Criollo Horse)
1972, accepts all colours except
pintado/[= native]

Criollo Trotter: *see* Cuban Trotter

Crioulo: (Rio Grande do Sul, Brazil)/
py/var. of (or =) Criollo/BS/*see
also* Northeastern/not *Criôlo*/[=
native]

Croatian Coldblood: (Croatia)/h
dr.m/usu. chestnut or brown/ orig.
from Belgian Draught and Noric
(but lighter)/vars: Mura, Posavina/
Cro. *Hrvatski Hladnokrvnjak,
Domać Hladnokrvan* (= local
coldblood)/syn. *Croatian Draught*;
obs. syn. (to 1992) *Jugoslav,
Yugoslav Draught*/nearly extinct

Cruce: *see* Wilbur-Cruce

CSB: *see* Belgian Warmblood

Cuban Criollo: (Cuba)

Cuban Paso: (Cuba)/l; ri/Sp. *Caballo
cubano de paso*/syn. *Cuban
Gaited Horse*/not *Cuban Trotter*

Cuban Pinto: (Cuba)/l; ri/tobiano or
overo/improved Criollo, orig.
since 1974 from pinto Criollo,
Quarter Horse and Thoroughbred/
Sp. *Pinto cubano*

Cuban Trotter: (Cuba)/ri/usu. black
or bay/orig. from Criollo
improved by Morgan/Sp. *Criollo
de Trote*/syn. *Criollo Trotter*

Çukurova: (S Turkey)/l; ri.dr./orig. from
Anatolian Native, Turkish Arab
and Uzunyayla/not *Chukorova,
Cukorova, Tschoukourova*

Cumberland Island: (S Georgia, USA)/
feral of Spanish orig./rare

Curley: *see* American Bashkir Curly

Curraleiro: *see* Northeastern

Cushendale: (N Ireland)/py/extinct

Cutchi: *see* Kathiawari

Cyāntā: *see* Chyanta

Czech Coldblood: (Czech Republic)/
h; dr./usu. chestnut, also bay/
Noric and Belgian Draught orig.
since 1900/Cz. *Český
chladnokrevník*

Czechoslovakian Saddle Pony: *see*
Slovakian Riding Pony

**Czechoslovakian Small Riding
Horse:** *see* Slovakian Riding
Pony

Czech Trotter: (Czech Republic)/
Hungarian, German, Austrian
and Russian orig./Cz. *Český
klusák*; obs. syn. (to 1992)
Czechoslovak Trotter

Czech Warmblood: (Czech Republic)/
l; ri/usu. bay or chestnut, also
grey or black/based on
Hanoverian with some Trakehner
and Thoroughbred blood/Cz.
Český teplokrevník

Czysta Krew Arabska: (Poland)/
racing/grey, bay or chestnut, occ.
black/Arab, first imported 1795,
main importations from Saudi
Arabia/syn. *Araby, Arab (PASP)*/
rare 1845–1930

Dąbrowa-Tarnowska: (S Poland)/
halfbred of Gidran orig./
combined with Sącz to form
Kraków-Rzeszow/Pol. *Dabrowo-
tarnowski, Tarnowsko-dabrowski*

Dagestan Pony: (N Caucasus, Russia)/
d/vars: Avar, Kumyk, Lezgian/
Russ. *Dagestanskii poni*/rare

Dahoman: (Poland)

Dales Pony: (NE England)/ri.dr./usu.
black, dark brown, grey or dark
bay/sim. to Fell Pony but larger,
owing to draught-horse blood/BS
(with Fell Pony) 1916 (split
1957), HB 1918/rare

Dali: (China)/var. of Yunnan

Danish: *see* Jutland

Danish Oldenborg: (Denmark)/l;
ri/usu. black or brown/orig. from
Oldenburg before it was crossed/
BS/Dan. *Dansk Oldenborg*

Danish Sport Pony: (Denmark)/
ri/often dapple grey, black,
chestnut or bay/orig. from British
ponies esp. New Forest and
Welsh/BS 1976/Dan. *Dansk
Sports Pony*/rare

Danish Trotter: (Denmark)/l/orig.
from American Trotter/BS/Dan.
Dansk Traver

Danish Warmblood: (Denmark)/l;
ri/mixed orig. since 1962/BS
1979/Dan. *Dansk Varmblod*

Dansk: *see* Danish

Danube: (Bulgaria)/l; dr/usu. black or bay/orig. (1924 on) from Nonius × Pleven, halfbred riding and local, with some Russian Trotter and Thoroughbred blood (1955–1965)/Bulg. *Dunavska*/syn. *Danubian, Dunav*

Darashuri: (Fars, Iran)/l/var. of Plateau Persian/syn. *Shirazi*/not *Darashoori, Dareshuri*

Darfur Pony: *see* Western Sudan Pony

Daror: *see* Somali Pony

Dartmoor Pony: (SW England)/ri/ usu. black, brown, bay or grey/ HB 1899, BS 1925; BS also USA, France (HB 1969), Netherlands; HB also Australia, Germany/rare

Datong Pony: (NE Qinghai, China)/ ri.pa.dr/WG *Ta-t'ung*/not *Tatung*

Davert: (Münster, Westphalia, Germany)/py/feral; sim. to Dülmen Pony/syn. *Davertnickel*/ extinct 1812

Dawan: ancient Chinese name for Fergana (Turkmenian) and hence for Akhal-Teke

Debao: (China)/smaller local var. of Chinese Miniature

Deccani: (Bombay, India)/py/syn. *Bhimthadi*/nearly extinct

Delft: *see* Sri Lankan Pony

Deli: *see* Batak

Deliboz: (W Azerbaijan)/l; ri.dr/grey or bay/orig. in early 20th c. from Turkish Arab × Karabakh and Azerbaijan in S Azerbaijan/Russ. *Delibozskaya*/[name of stud]/ nearly extinct

Deli-Orman: (N Bulgaria)/var. of Bulgarian Native/not *Deliorman*/ extinct

demi-sang: *see* halfbred

Demi-sang suisse: *see* Swiss Warmblood

Deutsches Reitpferd: *see* German Riding Horse

Deutsches Reitpony: *see* German Riding Pony

Devon Pack Horse: (England)/py/ extinct

Dhanni: (Punjab, India)/l/not *Dhunni*/extinct

Djerma: (middle Niger, W Africa)/l/ dark/orig. from Barb × Dongola

Dobrogea: (Romania)/py/former var. of Romanian, sim. to Moldavian but smaller/Rom. *Dobrogeană*/ not *Dobrudja, Dobruja*/extinct

Doğu Güney doğu Anadolu Ati: *see* Kurdi

Døle: (E Norway)/l-h; dr/usu. brown or bay, also black, occ. chestnut/ orig. from Frederiksborg × local in 18th c.; recent use of Norwegian Coldblood Trotter sires/orig. of Norwegian Coldblood Trotter, North Swedish/HB 1902, BS 1967/Nor. *Dølehest*/syn. *Døle Draught, Døle-Gudbrandsdal, Gudbrandsdal, Østland* (= Eastland) (till 1947)/not *Doele*/[= valley]/rare

Døle Trotter: *see* Norwegian Coldblood

Dolny-Iskar: (Bulgaria)/var. of Bulgarian Native/extinct

Domaći Hladnokrvan: *see* Croatian Coldblood, Yugoslavian Draught

Domaći brdski konj: *see* Yugoslav Mountain Pony

Don: (Russia)/l; ri.dr/golden chestnut/orig. from Cossack (= Old Don) improved in 19th c. successively with Persian and Karabakh, Arab and Russian Saddle Horse, and Thoroughbred/vars: eastern, large and riding/orig. (with Thoroughbred) of Anglo-Don/HB/Russ. *Donskaya*/syn. *Trans-Don*

Dongola: (N Sudan and W Eritrea)/l/ reddish-bay, often with white face blaze and feet/orig. of West African Dongola, Sudanese Country Bred/syn. *Dongalawi, Dongolas, Dongolaw*/rare

Dor: *see* Somali Pony

Do-san-ko: *see* Hokkaido Pony

Dülmen Pony: (Münsterland, Westphalia, Germany)/ri/all colours but usu. dun, with wild-type markings/orig. from semi-feral herd on Duke of Croy's estate at Meerfelder Bruch/sim.

vars or herds: Davert, Senne
(orig.)/orig. of Nordkirchen
Pony/HB 1989/Ger.
Dülmener/syn. *Münsterland*/
nearly extinct

Dunav: *see* Danube

Dutch Carriage Horse: (Netherlands)/
l; dr/HB/Du. *Nederlandse
Tuigpaard*

Dutch Coach: *see* Gelderland,
Groningen

Dutch Draught: (Netherlands)/h;
dr/orig. from Belgian Draught ×
Zeeland *c.* 1880, hence sim. to
Belgian Draught/BS 1914, HB
1915/Du. *Nederlandse
Trekpaard*/syn. *Holland Heavy,
Netherlands Draught*/rare

Dutch Trotter: (Netherlands)/orig.
from American and French
Trotters with some Orlov Trotter
blood/HB, BS 1879/Du.
Nederlandse Draver/syn.
Netherlands Trotter

Dutch Warmblood: (Netherlands)/l;
ri/orig. from Thoroughbred *et al.*
× local; absorbed Gelderland and
Groningen (partly)/BS 1969, HB;
BS in USA 1983/Du. *Nederlandse
Warmbloed, Warmbloed Paard
Nederlands* (= *WPN*)/syn. *Dutch
Warmblood Riding Horse,
Netherlands Riding Horse*

Dzhabe: *see* Jabe

East Anatolian: *see* Kurdi

East and South-East Anatolian: *see*
Kurdi

East Bulgarian: (E Bulgaria)/l; ri.dr/
usu. chestnut, bay or black/orig.
since *c.* 1900 from Thoroughbred
and English Halfbred × Anglo-
Arab, Arab and Bulgarian
Native/recog. 1951/Bulg.
Istochno-b"lgarska

Eastern: *see* Oriental

East Friesian: (Germany)/l; ri.[dr.]/
sim. to Oldenburg/old type
carriage horse nearly extinct;
remnants with Oldenburg,
Cleveland Bay *et al.* form Heavy
Warmblood/former BS 1893/Ger.
Ostfriesisches Warmblut/syn.
East Friesland

East German Saddle Horse: *see* Edles
Warmblut

Eastland: *see* Døle

East Prussian: *see* Trakehner

Edles Warmblut: (E Germany)/l;
ri.dr/orig. 1960s from
Brandenburg and Mecklenburg
with Hanoverian, Trakehner and
Thoroughbred blood/HB/syn.
*East German Saddle Horse,
Elegant Saddlehorse, Elegant
Warmblood, Riding Horse of the
GDR* or (since 1991) = light
warmblood as distinct from
heavy warmblood

Eesti: *see* Estonian

Ege Midillisi: *see* Mytilene

Egyptian: (Egypt)/l/Arab type/HB
1900/syn. *Baladi*

Egyptian Arab: (Egypt)/ri/orig.
mainly on one government farm
(El-Zahraa Arab Horse Stud
Farm, Cairo); surplus go to local
Baladi pool/syn. *Hamdani,
Kuhailan, Saklawi*/rare

Einsiedeln: (Switzerland)/var. of
Swiss Halfblood/orig. from
Anglo-Norman/Ger. *Einsiedler*/
syn. *Schwyz*/extinct/for New
Einsiedeln see Swiss
Warmblood

Elegant Warmblood: *see* Edles
Warmblut

Eleia: *see* Andravida

Emben: (W Kazakhstan)/subvar. of
Jabe var. of Kazakh

**Engleski polukrvnji zobnatičkog
tipa:** *see* Zobnatica Halfbred

Engleski Punokrvnjak: (Serbia) *see*
Thoroughbred

English Cart-horse: *see* Shire

English Cob: *see* cob

English Hack: *see* hack

English Halfblut: (S Africa)/syn.
Halfbred/almost extinct

English Hunter: *see* hunter

English Race-horse: *see*
Thoroughbred

English Thoroughbred: *see*
Thoroughbred

Equus "caballus": = domestic horse

Equus ferus przewalskii: *see*
Przwelski horse

Equus ferus gmelini: *see* tarpan

Eriskay Pony: (Hebrides, Scotland)/ born black, turning grey or white/only surviving var. of Hebridean Pony; orig. Celtic type with no imported blood/BS 1972/nearly extinct

Erlenbach: (Switzerland)/var. of Swiss Halfblood/orig. from Mecklenburg × Danish/syn. *Simmental*/extinct

Erlunchun: (Xingan mts, NE China)/ py; pa/usu. grey or bay/orig. from Chinese Mongolian/rare

Español-Andaluz: *see* Andalusian

Esperia Pony: (Ciociaria, Latium, Italy)/l; dr.pa.ri/dark bay/orig. in Ausoni mts, now mostly in France and a few around Latina and Aquila/It. *Pony di Esperia/* syn. *Pony ciociaro*/rare

Eston-Arden, Estonian Ardennes: *see* Estonian Draught

Estonian Draught: (Rakvere, Estonia)/h; dr/orig. from Swedish Ardennes (also Clydesdale and Shire) imported 1862–1935 × Estonian Native/Russ. *Èstonskii tyazhelovoz*/syn. *Eston-Arden, Estonian Ardennes*/rare

Estonian Heavy Draught: (Viru, Estonia)/dr.ri/composite of Estonian Native and Ardennais established 1921/Estonian *Eesti raskeveohobune*/nearly extinct

Estonian Native: (W Estonia)/py; ri.dr./chestnut, bay, light bay, dun or grey/Northern type; sim. to Žemaitukai/local with later influence from Arabian and Finnish/orig. (with Hackney) of Tori and (with Ardennes) of Estonian Draught/BS 1920, HB 1921/Estonian *Eesti hobune*, Russ. *Mestnaya èstonskaya, Èstonskaya loshad'*/syn. *Estonian Klepper, Estonian Pony*/not *Esthonian*/rare

European Warmblood: (S Africa)/l; ri/selected from European and Scandinavian Warmblood ♂ ♂ (esp. Hanoverian, Holsteiner, Dutch Warmblood, French Saddlebred, Oldenburger, Westphalian) with S African Thoroughbred ♀ ♀/BS 1990

European wild: *see* tarpan

Exmoor Pony: (NE Devon and W Somerset, SW England)/ri/bay, dun or brown, mealy nose, no white markings/BS 1921, HB 1961; BS in USA 1987/rare in GB

Extremadura: (Extremadura, Spain)/ former strain of Andalusian/Sp. *Extremeña*

Faca galizana: *see* Galician Pony

Faco pony: *see* Galician Pony

Faeroes Pony: (Faroe Islands, Denmark)/ri/sim. to Iceland Pony/BS, HB/Faerose *Føroyar*, Dan. *Færørne*/syn. *Faeroe Islands horse*/not *Faroe*/rare

Falabella: (Argentina)/under 33 inches (84 cm) high/orig. 1845–1893 from local with Shetland, Thoroughbred and Criollo blood 1879–1893/BS in GB, USA/syn. *Argentine Dwarf, Miniature Horse, Toy Horse*/not *Falabella pony*/[family owning herd since 1879]/rare

Feldmoching: (Bavaria, Germany)/ py/Ger. *Feldmochinger Moospferd*/extinct

Felin Pony: (Lublin, Poland)/ri.dr/ often dun, also grey, bay, black, chestnut, isabella/orig. since 1973 at Felin Exp. Sta. from Polish Konik, Arab Biłgoraj, Fjord, Małopolski, Welsh and Shetland Ponies

Fell Pony: (NW England)/ri.dr.[pa]/ sim. to Dales Pony but smaller/ BS 1893, HB 1899

Félvér: *see* Hungarian Halfbred

feral: *see* Bagual, Banker, broomtail, brumby, Cerbat, cimarron, Cumberland Island, Kaimanawa, Kiger Mustang, Lavradeiro Criollo, Misaki, mustang, Namib Desert, Puno pony, Sable Island Pony, Sri Lankan Pony, Suffield Mustang

Filipino: *see* Philippine Pony

Finnish: (Finland)/l-h; ri.dr/usu. chestnut, sometimes brown/ Northern type/vars: heavy (draught) (rare) and light (universal)/HB 1907, BS (*Suomen Hippos*)/Finn. *Suomen hevonen* (or *Suomalainen hevonen*), Swe. *Finsk*/syn. *Finnhorse*

Finnish Trotter: (Finland)/Finnish horse eligible for Trotter HB on basis of performance; there is also a Standardbred (American Trotter) HB in Finland

Finnish Warmblood: (Finland)/ri/ bay/imported from Sweden and other European countries/Fin. *Puoliverinen*/rare

Fiorello: (Italy)/improved var. of Maremmana

Five-gaited Horse: *see* American Saddle Horse, Iceland Pony

Fjord: (W Norway)/py; ri.dr/usu. dun with dark mane, tail and back stripe/HB 1910; BS also Canada 1980, France (HB 1969), GB, Germany, Netherlands, USA 1977; HB also Belgium, Denmark, Sweden/Nor. *Norges Fjordhest*, Fr. *Fjord de Norvege*/ syn. *Fjording, Nordbag, Nordfjord, Northern Dun, Norwegian Dun, Norwegian Fjord, Norwegian Pony, Vestland* (= *Westland*) (till 1947), *West Norway, West Norwegian*/not *Fiord*/rare

Flanders: *see* Flemish

Flemish: (Flanders)/h; dr/cf. Great Horse (Britain)/orig. of Belgian, Dutch *et al.*/Belg. *Vlaamse Paard*/syn. *Flanders*/extinct

Fleuve: (Senegal, W Africa)/l/orig. from Barb (from Hodh or Kayes) × local pony/orig. of Fouta/Fr. *Cheval du fleuve*/[Fr. = river]

Flores: (Indonesia)/Indonesian Pony group; sim. to or = Timor

Florida Cracker: (Florida, USA)/py-l; ri/American Indian type/BS 1989/syn. *Florida Scrub, Seminole*/[from the crack of the cowboy's whip]/rare

Forest horse: *see* Gotland Pony

Føroyar: *see* Faeroes Pony

Fouta: (Senegal, W Africa)/l/orig. from Fleuve × M'Bayar/syn. *Foutanké, Narougor*

Fox Trot: *see* Missouri Fox Trotting Horse

Franches Montagnes: *see* Freiberger

Franc Montagnard: *see* Freiberger

Frederiksborg: (Denmark)/l; ri.(dr)/ usu. chestnut with flaxen tail and mane/orig. in late 19th c. from stallions from Royal Frederiksborg Stud (flourished in 17th and 18th c. with Andalusian, Arab and Thoroughbred) × local Zealand mares; Oldenburg and East Friesian blood since 1939/HB 1890, BS/not *Fredericksborg, Fredriksborg*/rare

Freiberger: (Freiberg, NW Switzerland)/l; dr.ri/usu. bay, occ. chestnut/sim. to Comtois (France)/orig. from local, improved in late 19th and early 20th c. by imported coach (esp. Anglo-Norman) and draught horses (esp. Ardennes) and more recently by Swedish and French Halfbloods/HB/Fr. *Franches-Montagnes, Franc Montagnard*/ syn. *Jura*/not *French Mountain*

French Anglo-Arab: (esp. Limousin and SW France)/l; ri/inc. Limousin/HB 1941

French Ardennais: (Ardennes, NE France)/h; m.dr/orig. from Belgian Draught/orig. of (formerly vars) Auxois, Trait du Nord/HB 1908–1914, 1923 on, BS

French Canadian: *see* Canadian

French Coach: (USA)/= demi-sang (i.e. halfbred)/extinct

French Cob: *see* Norman Cob

Frencher: (Canada)/Thoroughbred × Canadian, 1st cross/extinct

French Mountain: *see* Freiberger

French Riding Pony: (France)/orig. 1970s from Arab *et al.* × local ponies/HB 1969, BS/Fr. *Poney Français de Selle*/syn. *French Saddle(bred) Pony*/rare

French Saddlebred: (France)/l; ri/ orig. 1958 from halfbreds, inc. Anglo-Norman and its derivatives (Charentais, Charolais, Vendéen) and also Corlais/HB 1963; BS USA 1990/Fr. *Cheval de selle français*/syn. *French Saddle Horse, French Warmblood*

French Trotter: (Normandy, France)/
l/bay or chestnut/orig. in 20th c.
from Norman Coach Horse with
Thoroughbred, Hackney and
American Trotter blood/HB
1906, breed recog. 1922, BS/Fr.
Trotteur français/syn. *Anglo-
Norman Trotter*, *Norman Trotter*
French Warmblood: *see* French
Saddlebred
Friesian: (Friesland, Netherlands)/l-
h; dr.ri/black/HB 1879, BS 1914;
BS also S Africa 1980, USA
1983/Du. *Friese, Friesch, Inlands
Fries* (= *Friesian native*)/not
Fresian, Frisian/nearly extinct
Friesian (Germany): *see* East Friesian
Fruit Tree Pony: *see* Guoxia
Fulani: *see* West African Dongola
Fuldblod: (Denmark)/= Thoroughbred
Furioso, Furioso-Northstar: *see*
Mezőhegyes Halfbred

Gaju: *see* Gayoe
Galiceño: (Mexico)/py; ri/orig. from
Galician Pony/BS USA 1959
Galician-Asturian: *see* Asturian
pony, Galician Pony
Galician Pony: (NW Spain)/m.ri.pa/
usu. chestnut, also sorrel, black or
dappled; semi-feral/cf. Garrano/
northern, central and
southeastern types/orig. of
Galiceño/HB 1994/Sp. *Faca
galizana, Jaca gallega, Poney* (or
Poni) *gallego*/syn. *Caballo
gallego de monte, Faco pony*/at
one time combined with
Asturian as *Galician-Asturian*
(Sp. *Poney galaico-asturiano* or
galaico-asturico)
Galla: *see* Abyssinian
Gallego: *see* Galician
Galloway Pony: (SW Scotland)/sim.
to Fell Pony and to Garron/
extinct by end 19th c.
Ganaston: *see* Bahr-el-Ghazal
Ganzhi: (Sichuan, China)/local var.
of Tibetan Pony/obs. syn.
Maiwa/not *Ganzi*
Garran: *see* Garron
Garrano: (Minho and Tras-os-
Montes, NW Portugal)/py;
ri.dr.pa/usu. bay, brown or dark

chestnut/cf. Galician Pony/HB
1994/Port. *Garrano Luso-Galeico*
(or *Luso-Galiziano*) (=
Lusitanian-Galician pony)/syn.
Minho pony/[= pony]/rare
Garrano da Ilha Terceira: *see* Azores
Pony
Garron: (NW Scotland)/larger
mainland var. of Highland Pony
(over 14 hands)/not *Garran*/
[Gaelic *gearran* = gelding, work-
horse]
Garwolin: (Warsaw, Poland)/l-h; dr/
local var. of Polish Draught sim.
to Sokolka/orig. *c.* 1922–1956
from Breton, Ardennes and
Boulonnais, × local/Pol.
Garwoliński
Gayoe Pony: (Sumatra, Indonesia)/
Indonesian pony group/Du.
Gaju/not *Kaju* (Jap.)
Gelderland: (C Netherlands)/l;
ri.(dr)/usu. chestnut or grey, also
bay or black, often white on legs
and face/orig. from Oldenburg
and Anglo-Norman × local/Du.
Gelders Paard/rare
Geogalidiko, Georgaludiko: *see* Pinia
German Bessarabian: (Bessarabia,
Ukraine)/l/black/mixed orig./
syn. *German Colonist*/extinct
German Coach: (USA)/= warmblood,
esp. Hanoverian and Oldenburg/
extinct
German Coach Horse: *see* Heavy
Warmblood
German Colonist: *see* German
Bessarabian
GERMAN RIDING HORSE: (Germany)/l/
union of warmblood breeds since
1975; inc. Bavarian Warmblood,
Hanoverian, Hessen, Holsteiner,
Westphalian, Württemberger,
Zweibrücken/Ger. *Deutsches
Reitpferd*/syn. *German Saddle
horse*
German Riding Pony: (W Germany)/
cf. Lewitzer Pony/orig. from
pony of any breed or cross/BS/
Ger. *Deutsches Reitpony*
German Trotter: (Germany)/orig.
from American, Russian and
French Trotters with later
English, Italian, Austro-

Hungarian and Scandinavian blood/BS 1896/Ger. *Traber*

Gharkawi: *see* Western Sudan Pony

Giara Pony: (Sardinia, Italy)/ ri.m.(dr)/usu. bay, occ. chestnut or black; mostly feral on high Giara plateau/recog. 1992; BS 1976/It. *Cavallino della Giara*/ syn. *Achetta, Quaddeddu, Sardinian Pony*/[Giara (Sard. *Yara*) = gravel]/rare

Giarab: (Sardinia, Italy)/cross between Giara Pony and Arab

Giawf: (N Yemen)

Gidran: (SE Europe)/l; ri.dr/often chestnut/strain of Anglo-Arab bred at Mezőhegyes, Hungary, since 1810s; usu. smaller in Romania than in Hungary/syn. *Gidran-Arabian*/rare

Gocan: (Mull, Scotland)/small Highland pony of Hebridean type/syn. *Mull*/extinct

Gondo: *see* Bandiagara

Goonhilly: (Cornwall, England)/ py/extinct

Gotland Pony: (C and S Sweden)/ri/ black, brown or bay/BS 1910, HB 1943; BS also USA 1960 (HB 1994)/Swe. *Skogsruss*/syn. *Gotlandsruss, Skogsbagge, Skogshäst* (= forest horse)/rare

Graditz: (Saxony, Germany)/Saxony Warmblood stud with some Thoroughbred blood

Great Horse: (Britain)/sim. to Flemish/orig. of Clydesdale, Shire/syn. *Old English Black Horse, Old English War Horse*/ extinct

Great Polish: *see* Wielkopolski

Greek Pony: (Greece)/Balkan Pony type/vars: Pinia, Pindos and Skyros Ponies/syn. *Greek Native*

Griffin: (Mongolia)

Groningen: (Netherlands)/l; ri.dr/ usu. black, bay or brown/sim. to Gelderland but heavier/orig. from local draught graded to Oldenburg (1870 on)/HB 1880–1890, reorg. 1985, BS 1982/Du. *Groninger, Groningse paard*/nearly extinct by absorption into Dutch Warmblood but saved since 1978

Guangxi Pony: (China)/dwarf/inc. Baise, Shishan

Guanzhong: (Wei He basin, Shaanxi, China)/l; ri.dr/usu. chestnut, also bay/orig. 1950s from Budyonny, Karabair and Russian Thoroughbred × local with Ardennes ♀♀ since 1958; by 1970 $\frac{1}{8}$ local blood, $\frac{1}{4}$ light, $\frac{5}{8}$ heavy

Gudbrandsdal: *see* Døle

Guelderland, Guelders: *see* Gelderland

Guizhou: (China)/py; ri.dr/syn. *Kweichow*/see also Chinese Miniature

Guoxia: (SW China)/ri.dr/usu. bay, roan or grey; dwarf/BS 1981/syn. *Rocky Mountain Pony*/[= under fruit tree]/rare

Gurgul: (Carpathians, Slovakia)/ strain of Hutsul since 1922

Guzman: *see* Andalusian

hack: (Great Britain)/l; ri/any horse suitable for riding, often Thoroughbred × Arab or pony/ syn. *English Hack*/[= hackney]

Hackney: (England)/l; dr./? orig. from Thoroughbred and local horse of Norse origin/BS 1883, HB 1884; BS also Argentina 1905 (HB), Australia, Canada 1892, Netherlands, S Africa 1962 (HB 1967), USA 1891 (horse and pony)/obs. syn. *Norfolk, Norfolk Hackney, Norfolk Roadster, Norfolk Trotter, Yorkshire Hackney, Yorkshire Trotter*/ [hackney = a horse for ordinary riding or driving]

Hackney Pony: (England)/dr/orig. from Hackney × Fell Pony 1872/ in Hackney BS and HB; BS also S Africa 1906; HB also Netherlands

Haflinger: (Tyrol, Austria/Italy, also S Germany)/py; ri.dr/chestnut with light mane and tail/light Noric and Arab blood/BS and HB Austria, Germany 1895, Italy 1973; BS also France (HB 1970), GB 1970, USA 1976, Canada 1982; HB also Belgium,

Netherlands, Luxembourg 1970,
Switzerland 1959/It. *Avelignese*/
syn. *Hafling (Mountain) Pony*/
not *Hafflinger*
Hailar: *see* Sanhe
Halbblut: *see* Halfbred
halfblood: *see* halfbred
HALFBRED: orig. from blood horse ×
native or draught/Ger. *Halbblut*,
Fr. *demi-sang*/syn. *halfblood*,
warmblood
Half Saddlebred: (USA)/l; ri/BS 1971
for crossbred American
Saddlebred
Hamdani: (Tunisia)/strain of Arab/
syn. *Keheilan*, *Saklawi*/rare
Hanoverian: (Hanover, Germany)/l;
ri/orig. (late 18th and early 19th
c.) from Thoroughbred,
Trakehner *et al.* × local/inc. in
German Riding Horse 1975/
derivatives: Brandenburg,
Mecklenburg, Westphalian/BS
1888; BS also USA 1973/Ger.
*Hannoveraner, Hannoversches
Warmblut*/syn. *Hanover*,
Hannoverian Sport Horse
Hantam: *see* Cape Horse
Hausa: (Niger and N Nigeria)/var. of
West African Dongola
HEAVY WARMBLOOD: (Saxony and
Thuringia, Germany)/l-h; dr.ri/
inc. Oldenburg, East Friesian,
Saxony Warmblood/Ger.
Schweres Warmblut/syn.
German Coach Horse/rare
Hebridean Pony: (Scotland)/smaller
orig. var. of Highland pony
(under 14 hands)/former vars:
Barra, Mull, Rhum, Skye, Uist/
syn. *Western Isles Pony*/extinct
(by crossing) except for Eriskay
Pony
Heihe: (Aihun, Heilongjiang, China)/
l; ri.dr/usu. bay or chestnut, occ.
grey or black/orig. early 20th c.
from Russian breeds (inc. Anglo-
Norman) × Chinese
Mongolian/recog. 1963
Heilongjiang: (Song-liao plain,
Heilongjiang, China)/l; ri.dr/
chestnut or bay/orig. 1950–1975
from Russian heavy and light
breeds × local/WG *Heilungkiang*

Hequ: (SE Qionghai, SW Gansu and
NW Sichuan, China)/l; dr.ri.pa/
black, bay or grey/vars: Jiaoke,
Kesheng, Suoke/WG *Khetshuĭ*,
Russ. *Khetsyui*/syn. (to 1954)
Nanfan/[= river bed (in Huang
He)]
Herati: (Afghanistan)/l
Hessen: (Germany)/l; ri/sim. to
Hanoverian/HB for horses bred
in Hessen with Hanoverian,
Holsteiner, Thoroughbred or
Trakehner ♂♂/inc. in German
Riding Horse/Ger. *Hessischer*
Highland Pony: (NW Scotland)/
ri.pa.dr/usu. grey, dun, black or
brown, with eel-stripe/vars (till
1932): Garron, Hebridean Pony/
HB 1889, BS 1923; BS also
Australia, France (HB 1969)/obs.
syn. *Scotch Pony*
Hinis: (NE Anatolia)/ri/local var. of
Anatolian Native with short
forelegs/Turk. *Hınısın Koly
Kısası* (= *Hinis short arm*)/rare
Hipismo, Cavalo de: *see* Brazilian
Sport Horse
Hirzai: (Baluchistan, Pakistan)/l;
ri/white or grey/Arab orig./rare
Hispano-Anglo-Arabe: *see* Spanish
Anglo-Arab
Hispano-Arabe: (Spain)/l; ri/orig.
from Arab × Andalusian/HB
1986
Hispano-Bretona: (Catalonia, Spain)/
m.(dr)/chestnut or sorrel/sim. to
Breton/orig. from Breton ×
Andalusian/syn. *Bretona
Ceretana, Breton Cerdà* (=
Hispano-Bretona de la Cerdaña),
Catalan, Catalonian
Historic Boer: (S Africa)/ri/bay,
palomino, black, grey or red-
roan/orig. from Cape Horse
(*Boerperd*); remnants selected by
Boerperd BS established 1973,
renamed Historiese Boerperd BS
1977, HB; recog. 1996 as separate
breed/Afrik. *Historiese Boer*
Hobby: *see* Irish Hobby
Hodh: (Mali, Mauritania)/var. of West
African Barb/syn. *Hohd*
Hokkaido Pony: (Japan)/ri.dr/var. of
Japanese Native/HB 1979,

BS/Ch. *Beihaido*/syn. *Do-san-ko* (term of endearment), *Hokkaido-Washu*/rare

Holland Heavy: *see* Dutch Draught

Holsteiner: (Holstein, Germany)/l; ri/ Yorkshire Coach Horse blood end of 19th c./inc. Holstein Marsh, Schleswig-Holstein/BS 1886; BS also USA 1978/inc. in German Riding Horse 1975/Ger. *Holsteiner Warmblut*

Hrvatski: *see* Croatian

Hrvatski Hladnokrvnjak: *see* Croatian Coldblood

Hrvatski posavac: *see* Posavina

Hsiangcheng: *see* Sikang Pony

Hsi-k'ang: *see* Sikang Pony

Hsi-ning: *see* Sining

Hucul: *see* Hutsul

Hungarian: (Hungary)/l/often dun/ Mongolian and Oriental orig., improved by Arab, Spanish and Thoroughbred/orig. of Hungarian Draught, Hungarian Dun/syn. *Hungarian Native*/extinct (graded to Hungarian Halfbred)

Hungarian: (USA)/l; ri/BS 1967 for Kisbér Halfbred (= Félvér) and Shagya Arab imported from Hungary since 1945, and their Thoroughbred and Arab crosses and descendants/rare

Hungarian Draught: (W Hungary)/h; dr/ bay, chestnut or grey/orig. from Noric (late 19th c.), Percheron and Ardennes, × Hungarian/ former vars: Mur Island, Pinkafeld/HB 1922/Hung. *Magyar hidegvérű* (= *Hungarian Cold Blood*/rare

Hungarian Dun: (Hungary)/l; ri/ unsuccessful attempt (since 1971) to re-establish ancient breed using Akhal-Teke ♂ ♂/rare

Hungarian Halfbred: (Hungary)/inc. Gidran, Kisbér and Mezőhegyes Halfbred; used for improvement of Hungarian/Hung. *Félvér*

Hungarian Sport Horse: (Hungary)/ inc. Hungarian Thoroughbred, Hungarian Trotter

Hungarian Trotter: (Hungary)/orig. from Hungarian Halfbred improved by Austrian Trotter *et al.*

Hunter: (Great Britain)/l; ri/any horse suitable for riding to hounds (fox-hunting), usu. heavy Thoroughbred or Thoroughbred cross/BS 1885 (National Light Horse Breeding Society)/syn. *English Hunter*/*see also* Canadian Hunter, Irish Hunter

Hutsul: (E Carpathians, esp. Poland, Romania, Slovakia and Ukraine)/ py; ri.dr.pa/usu. dun or bay, also chestnut or piebald/Konik type/ var.: Gurgul/BS Slovakia 1972, HB Czech Republic 1922, Poland 1955/Pol. *Hucuł* or *Huculska*, Cz. *Hucul*, Rom. *Huţul*, Ger. *Huzul*/ syn. *Carpathian Pony*/not *Hutzul*/rare

Huzul: *see* Hutsul

Ialomiţa: (Walachia, Romania)/var. of Transylvanian/not *Jalomitza*/ extinct

Iberian: *see* Andalusian

Iceland Pony: (Iceland)/ri.m/ chestnut, also bay, black, grey or dun; 5 gaits, including tølt/orig. from Scandinavian and North British types introduced in 11th and 12th c./HB 1923; BS France (HB 1969), Germany, GB, Netherlands, USA 1982, Canada 1982; HB also Belgium, Denmark, Finland 1960, Norway, Sweden/Ice. *Íslenzki hesturinn*, Dan. *Islandske Hest*, Finn. *Islannin Hevonen*/syn. *Icelandic Horse, Icelandic toelter horse, Iceland Tölter*, tølter (*tølt* is a 4-beat gait)

Ili: *see* Yili

Ilia, Ilias: *see* Andravida

Indian, Indian Pony: *see* American Indian

Indonesian ponies (South-East Asia pony group): *see* Bali, Batak, Gayoe, Kumingan, Lombok, Macassar, Periangan, Sandalwood, Sumbawa, Timor

Inlands Fries: *see* Friesian

International Striped Horse: *see* Striped Horse

Iomud: (Turkmenistan)/l; ri/grey or chestnut, occ. golden chestnut or

black/sim. to Akhal-Teke but
smaller/orig. from ancient
Turkmenian/Russ. *Iomudskaya*/
syn. *Yamud* (Iran)/not *Iomut,
Jomud, Yomud, Yomut,
Yomuth*/rare

Iranian: *see* Caspian, Chenarani,
Plateau Persian, Persian
Turkoman

Irish Cob: (Ireland)/l; ri.(dr)/usu.
Thoroughbred or Connemara ×
Irish Draught

Irish Draught: (Ireland)/l-h; dr.ri/usu.
grey/orig. 19th c. from Irish
Hobby and Great Horse/HB 1919,
BS 1976; BS also GB, N America
1993/syn. *Irish Cart-horse*/not
Irish Draught/rare

Irish Hobby: (Ireland)/orig. of
Connemara Pony and Irish
Draught/extinct

Irish Hunter: *see* Irish Sport Horse

Irish Pony: (Ireland)/ri/black, bay,
isabella, chestnut, palomino,
white or pied/composite of Arab,
Welsh, Connemara Pony and
Thoroughbred established *c.*
1970/rare

Irish Sport Horse: (Ireland)/l;
ri/orig.
from Irish Thoroughbred × Irish
Draught *et al.*/HB 1974/syn. *Irish
Hunter, Irish Warmblood*

Irish Thoroughbred: (Ireland)/strain
of Thoroughbred

Irish Warmblood: *see* Irish Sport
Horse

Isabella: (coat colour): yellow to
golden with white tail and
mane/Sp. *Isabelo*/syn. *Claybank
Dun* (USA), *Yellow Dun* (inc.
Palomino) (GB)/*see also* Ysabella

Islandske Hest: *see* Iceland Pony

Islannin Hevonen: *see* Iceland Pony

Íslenzki hesturinn: *see* Iceland Pony

Israeli Local Horse: (Israel)/l; ri./HB
1972 for mixture of many breeds,
excluding Arab and
Thoroughbred

Istochno-b"lgarska: *see* East
Bulgarian

Italian Heavy Draught: (N Italy)/h;
m.dr/chestnut, roan or bay/orig.
in early 20th c. from Ardennes,

Boulonnais and Hackney, graded
to Breton since 1926/It. *CAITPR
= Cavallo Agricolo Italiano da
Tiro Pesante Rapido*

Italian Saddlebred: (Italy)/ri/sauro,
bay, occ. grey, morello, roan/HB
1970s for crossbred and halfbred
with English, Arabian, Anglo-
Arab and progeny, and Sardinian
Anglo-Arab, and Italian
Saddlebred with Italian
ridinghorse mares with no
draught ancestry; 3 sections:
Arabian-ME, Anglo-Arab, Italian
Saddlebred/inc. Persano,
Salernitano

Italian Trotter: (N Italy)/orig. since
late 19th c. from American
Trotter, Swedish Trotter (since
1983) *et al.*/HB 1900/It. *Trotto*

Ivanovo Clydesdale: *see* Vladimir
Heavy Draught

Jabe: (W Kazakhstan)/l; m.d.dr/
heavier var. of Kazakh/subvars:
Betpakdalin, Emben, Kulandin/
Russ. *Dzhabe*/obs. syn. *Western
Kazakh*

jaca: (Spain)/= cob or pony

Jaf: (Kordestan, Iran)/l/var. of Plateau
Persian/*see also* Kurdi

Jakut, Jakutskaja: *see* Yakut

Jalomiţa, Jalomitza: *see* Ialomiţa

Japanese Native: (Japan)/South-East
Asia pony group/smaller island
type (Miyako, Taishu, Tokara and
Yonaguni) and larger mainland
type (Hokkaido, Kiso, Misaki and
Noma)/BS/rare

Javanese pony: *see* Kumingan Pony,
Periangan Pony

Jereed: *see* Cirit

Jianchang: (Liangshan mts, S
Sichuan, China)/py; ri.dr/usu.
bay, also black/not *Jiangchang*

Jiaoke: (S Gansu, China)/grey/var. of
Hequ

Jilin: (NW Jilin, China)/l-h; ri.dr/orig.
since 1950 from Russian breeds
(inc. Ardennes and Don) × local
Mongolian; $\frac{1}{4}$ local blood, $\frac{1}{4}$ light
and $\frac{1}{2}$ heavy/recog. late 1970s/
WG *Kirin*/syn. *Jilin Harness*/not
Jielin

Jinhong: (S Fujian, China)/py; ri.pa/
usu. chestnut

Jinjiang: (Fujian, China)/not *Jinjang*/
rare

Jinzhou: (Jin county, S Liaodong
peninsula, Liaoning, China)/
ri.dr.(d.m)/usu. bay/orig. since
1926 from foreign light and
heavy breeds × local Mongolian

Jomud, Jomut: *see* Iomud

Jordanian: (Jordan)/l/syn. *Baladi*,
Kudush

Jugoslavian: *see* Bosnian, Croatian,
Macedonian, Serbian, Slovenian

Jugoslovenski Kasac: *see* Yugoslav
Trotter

Jumla Pony: (Uttar Pradesh, India)/
extinct

Jura: *see* Freiberger

Jutland: (Denmark)/h; dr.m/usu.
chestnut, occ. sorrel or roan/
native Danish improved by Shire
stallion imported 1862/orig.
(with Suffolk) of Schleswig/HB
1881, BS 1888/Dan. *Jydsk*,
Jyder/syn. *Danish*/rare

Jyder, Jydsk: *see* Jutland

Kaapse Boer(perd): *see* Cape Boer

Kabarda: (N Caucasus, Russia)/l;
ri.(pa)/bay or bay-brown, occ.
black/mt var.: Balkar; Karachai
formerly considered a var./HB
1935/Russ. *Kabardinskaya*

Kabuli: (India)

kadish: (Middle East)/= impure
(Arab)

Kaimanawa 'wild horse': (C North
Island, New Zealand)/feral/rare

Kaju: *see* Gayoe Pony

Kalblodstravare: *see* Norwegian
Coldblood Trotter

Kalmyk: (Astrakhan and Volgograd,
Russia)/l; ri/being absorbed by
Don and other breeds/Russ.
Kalmytskaya/syn. *Astrakhan*/
rare

Kaltblut: *see* Coldblood

Kanata Pony: (Canada and USA)/BS
for py of Appaloosa colour
(spotted)/cf. Appaloosa Pony,
Pony of the Americas

Karabagh, Karabah: *see* Karabakh

Karabair: (Uzbekistan and N
Tajikistan)/l; ri/chestnut, grey or
black/Russ. *Karabairskaya*/not
Kara-Bair

Karabakh: (Azerbaijan)/l; ri.pa/usu.
chestnut or bay/improved in
18th c. by crossing with Arab
and Turkmen/HB 1981/Russ.
Karabakhskaya/not *Karabagh*,
Karabah/rare

Karacabey Halfbred Arab: (Turkey)/
l; ri/orig. Turkish Arab ×
Anatolian Native/Turk.
Karacabey yarımkan arap/syn.
Karacabey/[name of
stud]/extinct since 1979

Karacabey-Nonius: (Turkey)/l; dr/
orig. from Karacabey Halfbred
Arab, Anatolian Native and
Nonius/extinct since 1970

Karachai: (N Caucasus, Russia)/l;
ri.dr/usu. dark brown or black/
sim. to (formerly var. of)
Kabarda/HB 1935/Russ.
Karachaevskaya/not *Karachaev*,
Karachayevsk

Karakaçan: (Turkey)/l; dr/bay/orig.
from imported × Trakya/rare

Karakachan: (Shumen, NE Bulgaria)/
black or dark brown/only
surviving var. of Bulgarian
Native/Bulg. *Karakachanska*,
Carakachanski Kon/not
Karakatschan/[(Romanian)
nomad)]/nearly extinct

Karelian: (Karelia, Russia)/North
Russian Pony group/var.: Onega/
Russ. *Karel'skaya*/extinct

Karst: *see* Bosnian Pony

Katharoemo: (Cyprus)/ri/bay,
chestnut or grey/orig. from
Thoroughbred imported from
UK/[Gk, = thoroughbred]/rare

Kathiawari: (Gujarat, India)/l; ri.dr/
usu. chestnut, also bay, brown,
grey, dun, piebald or skewbald;
turned-in ears/sim. to Marwari/
BS/syn. *Cutchi, Kathi, Kutchi*/
not *Kathiwari*/rare

Kazakh: (Azerbaijan) *see* Azerbaijan

Kazakh: (Kazakhstan)/l; m.d.ri.pa/
Mongolian type/vars: Adaev, Jabe/
orig. of Chinese Kazakh and of
Kushum/former syn. *Kirgiz, Kirgiz
Mountain*/not *Kazah, Kazak*

Keerqin: (Inner Mongolia, China)/l-h; ri.dr/usu. bay or chestnut/orig. since 1950 from Don, other Russian breeds and esp. Ardennes and Sanhe, × local Mongolian/not *Ke-Er-Qin*

Keheilan: orig. strain of Arab/not *Khiolan, Khiolawi, Koheilan, Kuhailain, Kuhaylan*/[from Arabic *kohl* = antimony]

Kentucky Mountain: (USA)/smaller (less than 14.2 hands high) type of Rocky Mountain/BS 1989

Kentucky Saddle Horse, Kentucky Saddler: *see* American Saddlebred

Kentucky Whip: *see* Quarter Horse

Kerry Bog Pony: (Ireland)/very small/ nearly extinct

Kesheng: (SE Qinghai, China)/var. of Hequ with Mongolian blood

Khakasski: (Russia)/? S Siberia

Khetshuĭ, Khetsyui: *see* Hequ

Khiolan, Khiolawi: *see* Keheilan

Kielce: (Poland)/combined with Lublin to form Lublin-Kielce/Pol. *Kielecki*

Kiger Mustang: (Lake county, Oregon, USA)/ri/feral, also captive herds selected for dun (with back stripe and zebra stripes on legs), grullo or red dun/BS 1988/syn. *Kiger Mesteño*/not *Riger*/rare

Kirdi Pony: (R Logone basin, SW Chad and N Cameroon)/ri/cf. Bhirum/syn. *Cameroon Pony, Lakka, Logone, Mbai, M'baye, Mousseye, Musseye, Pagan, Sara*/[*kirdi* = pagan]

Kirgiz: (Kyrgyzstan)/py; ri.dr.d/often bay or grey/Mongolian type/orig. of New Kirgiz/formerly syn. or var. of Kazakh/Russ. *Kirgizskaya*/ syn. *Kirgiz Mountain*/not *Kirghiz*/rare

Kirin: *see* Jilin

Kisbér Halfbred: (Hungary)/l; ri/ English Halfbred from Kisbér stud which uses Thoroughbred ♂♂/HB 1860, BS 1989/Hung. *Kisbéri félvér*/not *Kis-Ber*/rare

Kiso: (C Honshu, Japan)/bay, chestnut, black or palomino/var. of Japanese Native/BS 1948, HB 1976/nearly extinct

Kladruby: (Czech Republic)/l; ri.dr/ Spanish-Neapolitan orig./vars: Old Kladruby Black, Old Kladruby White/HB 1579/Cz. *Kladrubský, Starokladrubský Kun* (= *Old Kladruby*), Ger. *Kladruber*/ not *Kladrub, Kladrup*/ [name of stud, founded 1562]/ rare

Klepper: *see* Estonian Pony

Knabstrupper: (Denmark)/l-py; ri.(dr)/sim. to Frederiksborg but spotted/name orig. from one spotted Spanish ♀ in 1804 on Knabstrup farm/BS 1971/Dan. *Dansk Knabstrupper Hest*/rare

Kobczyk: *see* Kopczyk Podlaski

Koheilan: *see* Keheilan

Kolyma: *see* Middle Kolyma

Konik: (Poland–Russia)/py/inc. Hutsul, Polesian, Polish Konik, Žemaitukai/[= small horse]

Kopczyk Podlaski: (Poland)/h/usu. bay or chestnut/var. of Polish Draught/orig. from Belgian Draught stallion 'Kopczyk' born 1921 at Podlaska stud × local/not *Kobczyk*/rare

Kordofani: *see* Western Sudan Pony

Korean: *see* Cheju, Taejung

Koto-Koli Pony: (N Benin and Togo)/syn. *Cotocoli, Togo Pony*

Kraków-Rzeszów: (Poland)/inc. Sącz and Dąbrowa-Tarnowska/ combined with Lublin-Kielce to form Małopolski/Pol. *Krakowsko-rzeszowski*

Krčki konj: *see* Krk Island Pony

Kréda: *see* Bahr-el-Ghazal

Krk Island Pony: (Croatia)/Balkan Pony type/Cro. *Krčki konj*, It. *Veglia*/extinct

kuda-Minahasa: (N Sulawesi, Indonesia)/ri.dr.m/usu. brown/ composite of Sandalwood and Thoroughbred, established 1968/rare

kudan Pacu Indonesia: (Indonesia)

Kuda Padi: (Malaysia)

Kudush: *see* Jordanian

Kuhailan: *see* Keheilan

Kulandin: (Kazakhstan)/subvar. of Jabe var. of Kazakh with Adaev blood

Kumingan Pony: (W Java)/ri.pa.dr/

Indonesian pony group/sim. to Timor but larger/syn. *Javanese pony*

Kumyk: (coastal plains of Dagestan, Russia)/l; d/largest var. of Dagestan Pony/Russ. *Kumykskaya*

Kurdi: (Kurdistan)/l-py; ri/Turk. *Doğu Güney doğu Anadolu Ati (= East and South-East Anatolian horse)*

Kushum: (W Kazakhstan)/l; ri.dr.m.d/ bay or chestnut/orig. (1930–1976) from trotter, Thoroughbred, Don and Budyonny × Kazakh/ heavy and saddle types/recog. 1976/Russ. *Kushumskaya*/syn. *West Kazakh Saddle-Draught* (Russ. *Zapadno-kazakhstanskaya verkhovo-upryazhnaya*)

Kustanai: (NW Kazakhstan)/l; ri/usu. bay or chestnut, also reddish-grey or brown/orig. (1887–1951) from Kazakh improved by Don, Russian Saddle Horse *et al.*/vars: general purpose, saddle, steppe (meat)/recog. 1951; HB/Russ. *Kustanaĭskaya*/not *Kustanair*

Kutchi: *see* Kathiawari

Kuznetsk: (Kemerovo and Novosibirsk, W Siberia, Russia)/ l-h; dr/breed group; Siberian Pony type improved in late 19th and early 20th c. by trotter, saddle and draught breeds (Clydesdale and Brabançon)/ light, basic and heavy types/ Russ. *Kuznetskaya porodnaya gruppa*/not *Kuznet*/rare

Kweichow: *see* Guizhou

La Barranca: *see* Pottok

Lac La Croix Indian Pony: (E Canada)/Indian pony of French ancestry/rare

Ladakhi pony: (Ladakh, Kashmir)/ pa.ri/sim. to Tibetan (apart from Zaniskari)

Laka, Lakka: *see* Kirdi

Lämminverinen Ravuri: (Finland)

Landais: (Landes, SW France)/py; ri.dr/black, bay, brown or chestnut/HB 1960, BS/syn. *Barthes* (Fr. *Barthais*)/rare

Latgale Trotter: (Latvia)/strain of Baltic Trotter/orig. from Orlov and American Trotters × local/ absorbed by Latvian/Russ. *Latgal'skiĭ rysak*/extinct

Latvian: (Latvia)/l-h; dr.ri/bay, dark bay or black, sometimes chestnut/orig. 1852 from native Northern type improved by various breeds at Riga State Stud (1893 on) and by Trakehner, Oldenburg, Groningen and Hanoverian (1921–1939)/light and heavy types/recog. 1952/ Russ. *Latviĭskaya, Latviĭskiĭ upryazhnyĭ*/syn. *Latvian Carriage, Latvian Coach, Latvian Draught* (Russ. *Latvijas Zirgi*)/rare

Latvian Ardennes: *see* Baltic Ardennes

Lavradeiro Criollo: (Roraima, C Brazil)/feral/syn. *wild horse of Roraima*/rare

Lebaniega: (Liébana, Santander, N Spain)/py/extinct

Lehmkuhlener Pony: (Germany)/ nearly extinct

Lewitzer: (E Germany)/py; ri/usu. piebald or skewbald/orig. 1976 from Arab, Trakehner or Thoroughbred × Shetland, Fjord and other ponies/syn. Pinto Typ Lewitzer/rare

Lezgian: (S Dagestan, Russia)/ri.pa/ var. of Dagestan Pony/Russ. *Lezginskaya*

Libyan Barb: (Libya)

Lichuan: (SW Hubei, China)/py; ri.pa/grey, bay, chestnut or black/ WG *Li-ch'uan*/not *Lichwan*

Lidzbark: (Olsztyn prov., Poland)/h; dr/dun or mousy, occ. bay/var. of Polish Draught/developed since 1945 by settlers from Oszmiana/ Pol. *Lidzbarski*

Liébana: *see* Lebaniega

Light Warmblood: *see* Edles Warmblut

Lijiang: (Yunnan, China)/py; pa.ri/ orig. since 1944 from Arab, Yili, Hequ, Kabarda and small Ardennes, × local

Limousin: (C France)/l/usu. bay or chestnut/var. of French Anglo-

Arab/Anglo-Arab orig. in late
19th c./HB 1909–1914/Fr. *Anglo-
Arabe du Limousin*

**Lipica, Lipicai, Lipicanac, Lipicanec,
Lipicky:** *see* Lipitsa

Lipitsa: (SE Europe)/l.ri/usu. white,
born grey; occ. dark bay or black/
orig. from Spanish and Arab at
Lipitsa stud near Trieste (founded
1580)/HB 1701, international BS;
BS also USA 1968 (Lipizzan), GB
(Lipizzaner); HB Italy 1984,
France 1989/Sn. *Lipicanec*, Cro.
Lipica or *Lipicanac*, Hung.
Lipicai, Cz. *Lipicky*, Rom.
Lipiţana, It. *Lipizzana*, Ger.
Lipizzaner/not *Lipitza*/[Lippiza =
small town N of Trieste, now in
Slovenia]/rare in any one country

Lipizzan, Lipizzaner: *see* Lipitsa

Lippitt Morgan: (USA)/historic strain
of Morgan from herd of Robert
Lippitt Knight of Randolph,
Vermont/sim. to Chunk Morgan/
BS 1994

Liptako: *see* Yagha

Lithuanian Landrace: *see* Žemaitukai

Lithuanian Heavy Draught:
(Lithuania)/ h; dr/chestnut, also
bay/orig. (1879 on) from
Žemaitukai with Arab,
Ardennes, Brabançon and Shire
blood, graded (since 1923–1925)
to Swedish Ardennes (3
crosses)/recog. 1963/Russ.
Litovskaya tyazhelovoznaya or
tyazheloupryazhnaya

Litovskaya tyazhelovoznaya: *see*
Lithuanian Heavy Draught

Little Poland: *see* Małopolski

Liukiu: *see* Miyako, Yonaguni

Ljutomer Trotter: (NE Slovenia)/l/
orig. from American Trotter ×
Anglo-Arab/Sn. *Ljutomerski
kasač*, Ger. *Ljutaner*/nearly extinct

Llanero: *see* Venezuelan Criollo

Lofoten: (Norway)/py/sim. to
Nordland/extinct

Logone: *see* Kirdi

Loire: (C France)/local Percheron HB
(1933–1966)/extinct

Lokai: (Tajikistan)/l; ri.pa/bay, grey,
chestnut *et al.*; often curly coat/
some Iomud, Akhal-Teke,

Karabair and Arab blood/orig. of
Tajik Riding Horse/Russ.
Lokaĭskaya

Lombok: (Indonesia)/dr/Indonesian
pony group/? = Macassar Pony

Long Mynd: (Shropshire, England)/
py/extinct

Losina: (Losa, Burgos, Spain)/py;
ri/black/cf. Galician, Garrano,
Mérens Pony, Navarre Pony,
Pottok, Sorraia *et al.*/nearly
extinct

Lovets: (Astrakhan, Russia)/l/ basically
Kalmyk or Kazakh breed used to
draw fish-carts/recently crossed
with Don or Orlov Trotter/Rus.
Lovetskaya/[= hunger; *rybolovets*
= fisher]/ extinct

Lower Amazon: *see* Marajoaro

Łowicz: (Warsaw, Poland)/h; dr/var.
of Polish Draught/orig. from
local improved by Belgian
Draught and esp. Ardennes, *c.*
1936–1956/Pol. *Łowicki* or
Łowisko-sochaczowski

Lublin: (Poland)/l; dr/orig. from pure
and halfbred Arab with some
blood of Thoroughbred, Halfbred
and Anglo-Arab/Pol. *Lubelski
kón szlachteny* (= Lublin
thoroughbred horse)

Lublin-Kielce: (Poland)/orig. from
Lublin and Kielce/combined
with Kraków-Rzeszów to form
Małopolski/Pol. *Lubelsko-
Kielecki*

Lundy Pony: (Lundy I, Great
Britain)/ri/orig. from Welsh/Arab
× New Forest (1928) with later
Welsh and Connemara blood/BS
1971/rare

Lusitanian: (S Portugal)/l; ri/usu.
grey, bay or chestnut/sim. to (?
orig. from) Andalusian/orig. of
Anglo-Lusitanian/HB 1973, BS;
BS also GB, France (HB 1987),
Belgium/Port. *Lusitano*/syn.
*Bético-lusitano, National,
Peninsular, Portuguese*

Lusitanian-Galician pony: *see* Garrano

Luso-galiziano: *see* Garrano

Lyngen: *see* Nordland

Lyngshest: *see* Nordland

Ma: (Laos)

Maaneghi: *see* Maneghi

Macassar Pony: (S Sulawesi, Indonesia)/Indonesian pony group/? = Lombok

Macedonian Pony: (Macedonia)/ Balkan Pony type, sim. to Bosnian Pony/Serbo-cro. *Makedonski brdski konj* (= Macedonian mountain pony)/ obs. syn. *Vardarska*

Madagascan Pony: (Madagascar)

Magyar hidegvérű: *see* Hungarian Draught

Maiche: *see* Comtois

Maine: (NW France)/local HB (1907–1966) for Percheron/syn. *Mayenne* (Fr. *Mayennais*)/extinct

Maiwa: *see* Ganzhi

Majorcan: (Balearic Is, Spain)/dr.ri/ black/HB 1988, BS/Sp. *Mallorquina*/nearly extinct

Makedonski brdski konj: *see* Macedonian Pony

Makra: (Sind, Pakistan)/l/colour type (dun)/not *Makara*/rare

Malakan: (NE Turkey)/l-h; dr/orig. from native Russian, Orlov, Bityug *et al.*/syn. *Ardahan*/not *Molokan*/[cf. Malakan cattle]/ rare

Mallani: *see* Marwari

Mallorquina: *see* Majorcan

Małopolski: (SE Poland)/l; ri.dr/orig. 1963 by combining Lublin-Kielce and Kraków-Rzeszów/HB/[= Little Poland]

Malý Športový Kôň: *see* Slovakian Sport Pony

Maneghi: strain of Arab/not *Maaneghi, Man'aqi, Manegi, Munighi, Muniki, Muniqi*

Mangalarga: (Minas Gerais and São Paulo, Brazil)/l; ri/orig. in mid-19th c. from Altér and Brazilian native (Crioulo)/vars: Marchador (or Mineiro), Paulisto/BS 1934/[= name of hacienda]

Manipuri Pony: (Manipur and Assam, India)/ri/South-East Asia pony type

Mannar Pony: *see* Sri Lankan Pony

Manx: (Isle of Man)/py/extinct

Marajoaro: (Marajó I, Brazil)/ri.dr/ var. of Northeastern/syn. *Baixo-Amazona* (= Lower Amazon)

Marchador: (Minas Gerais, Brazil)/ orig. var. of Mangalarga/BS 1949/syn. *Mangalarga Marchador, Mineiro*/[*marcha* = march]

Maremmana: (Maremma, Tuscany and Latium, Italy)/l; ri.(pa.dr)/ usu. black, brown, chestnut or bay/Oriental orig., esp. Thoroughbred/orig. 2 types: larger in Papal States (dark coat, carriage horse), more slender in Tuscany (bay, Andalusian and Middle East influence); both types still discernible/vars: Fiorello, Monterufoli Pony/BS, HB 1980/rare

Mareyeuse: *see* Petit Boulonnais

Marismeña: (marshes of lower R Guadalquivir, Spain)/former strain of Andalusian/not *Marismenna*/[Sp. *marisma* = marsh]/extinct

Marwari: (Rajasthan, India)/l/usu. black or brown with white patches, also chestnut/sim. to Kathiawari/BS/syn. *Mallani*/rare

Masuren: *see* Mazury

Mayennais, Mayenne: *see* Maine

Mazari: (Afghanistan)/l

Mazursko-Poznański: *see* Wielkopolski

Mazury: (NE Poland)/orig. from East Prussian (Trakehner)/combined with Poznań to form Wielkopolski/Pol. *Mazurski*/syn. *Mazurian*/not *Masuren*

Mbai: *see* Kirdi Pony

M'Bayar: (Baol, Senegal)/py; l.dr/ usu. bay, also grey, roan or chestnut/? degenerate Barb/orig. (with Fleuve) of Fouta/[M'Bayar is adj. from Baol]

M'Baye: *see* Kirdi Pony

McCurdy Plantation: (C Alabama)/l; ri/grey/orig. bred by Edward S. McCurdy from 1905/initially reg. with Tennessee Walking Horse BA but maintained separately in Alabama; BS 1993/nearly extinct

Mecklenburg: (Germany)/l/orig. from local graded to Hanoverian (and Holsteiner)/Ger. *Mecklenburger Warmblut*/absorbed by Edles Warmblut 1960s/extinct

Mecklenburg Coldblood: (Germany)/
h; dr/bay, chestnut, black or
grey/ orig. since 1845 from
Clydesdale and Suffolk × local
with Belgian Draught and
Ardennes blood after
1945/HB/Ger. *Mecklenburger
Kaltblut*/nearly extinct

**Medjimurena, Medjimurje,
Medjimurski konj:** *see* Mura

Megezhek: (SW Yakutia, Russia)/
dr.m/var. of Yakut with Kuznetsk
blood at end of 19th c./syn.
Megezh

Megrel: *see* Mingrelian

Menggu: *see* Chinese Mongolian

Menorquina: *see* Minorcan

Mérens Pony: (Ariège, S France)/
dr.pa.ri/black/HB 1947, BS/Fr.
*Race ariègeoise de Mérens,
Mérenguais*/rare

Messara: (Crete, Greece)/l; ri.dr/orig.
from Arab × local/HB 1945/
nearly extinct

mesteño: *see* mustang

Mestnaya èstonskaya: *see* Estonian
Native

Métis Trotter: *see* Russian Trotter

Mexican Pony: (Mexico)/Criollo type

Mezen: (NE Archangel and Komi,
Russia)/dr/North Russian Pony
group/Russ. *Mezenskaya*/not
Mezien/rare

Mezőhegyes Halfbred: (Hungary, also
in Romania)/l; dr/usu. bay, occ.
black/strain of English
Halfbred/orig. from
Thoroughbred ♂ ♂ 'Furioso'
(bought 1841) and 'North Star'
(bought 1852)/HB 1850/Hung.
Mezőhegyesi félvér/syn. *Furioso-
Northstar*/rare

Mezőhegyes Sport Horse:
(Hungary)/l; ri/Thoroughbred,
Holsteiner and Hanoverian
combined with Mezőhegyes
Halfbred

Middle Eastern: (Italy)/grey, bay,
brown or black/most typical var.
of Sicilian (much Arab influence
in Sicily from Middle East)/syn.
Puro sangue orientale/almost
extinct on Sicily

Middle Kolyma: (Yakutia, Russia)/
var. of Yakut/Russ.
Srednekolymskaya/syn.
Verkhoyansk

Midilli: *see* Mytilene Pony

Mijertinian: *see* Somali Pony

Mimoseano: *see* Pantaneiro

Mimoso: *see* Pantaneiro

Minahasa: *see* kuda-Minahasa

Mineiro: *see* Marchador

Mingrelian: (W Georgia)/py; ri.pa.dr/
usu. bay or near black/Russ.
Mingrel'skaya, Megrel'skaya/syn.
Megrel/rare

Minho Pony, Minhota: *see* Garrano

miniature horse: *see* American
Miniature, Chinese Miniature,
Falabella, South African Miniature/
syn. *toy horse, minihorse*

Miniature Shetland Pony: (USA)/HB
1972 for Shetland Ponies 34
inches (86 cm) or less

Minorcan: (Balearic Is, Spain)/dr.ri/
black/HB 1988, BS/Sp.
Menorquina/rare

Minusinsk: (S Krasnoyarsk, Russia)/l;
ri.dr.pa/nearly extinct by
crossing with Don,
Thoroughbred and trotter

Mira Pony: (Coimbra, Portugal)/Port.
garrano de Mira

Misaki: (Cape Toi, S Kyushu, Japan)/
usu. bay or black/feral, var. of
Japanese Native/BS/syn. *Wild
Horse*/nearly extinct/[= cape]

Missouri Fox Trotter: (Ozark hills,
Missouri, USA)/l; ri/BS 1948/
syn. *Missouri Fox Trotting
Horse*/not *Mississippi Fox Trotter*

Miyako: (Ryu-Kyu Is, Japan)/usu. bay
or dun/var. of Japanese Native/
BS/nearly extinct

Moldavian: (Romania)/l-py/former
var. of Romanian/Rom.
Moldovenescă/extinct

Moldovenescă: *see* Moldavian

Molokan: *see* Malakan

Monchina: (Guriezo, Cantabria,
Spain)/m.dr/Sp. *Monchinas*/rare

Monggol: *see* Chinese Mongolian

Mongolian: (Mongolia)/py; ri.dr.d.m/
all colours/same orig. as Chinese
Mongolian/vars: Forest, Gobi,
Mountain, Steppe/syn.
Mongolian Pony

Mongolian tarpan: *see* Przewalski horse

Mongolian wild: *see* Przewalski horse

MONGOLIAN type: sim. to Northern type/inc. Bashkir, Chinese Mongolian, Kazakh, Kirgiz, Mongolian, Siberian Pony, Yili/syn. *Tartar, Tatar*

Monterufoli Pony: (upper Cecine valley, Pisa, Tuscany, Italy)/ri.dr/usu. black, also bay/orig. from selection and use of Maremmana, Tolfetano and Arab ♂ ♂ for draught/var. of Maremmana/HB 1990/It. *Cavallino di Monterufoli, Monterufolina*/[Monterufoli Farm]/nearly extinct

Moospferd: *see* Feldmoching

Morab: (USA)/l; ri/Morgan × Arab cross named 1920s, breed selected since 1955; ♂ may be Morab, Arab or Morgan/BS 1973

Moravian Warmblooded: (Czech Republic)/ri/orig. from former Austro-Hungarian halfbred strains Furioso and Przedswit, further crossed with Anglo-Arabian strain Gidran and halfbred Arabian strain Shagya/ nearly extinct

Morgan: (Vermont, USA)/l; ri.dr/sim. to American Trotter/HB 1891, BS 1909; BS also in Canada 1968, GB 1975/historic strains: Chunk Morgan, Lippitt Morgan/ [orig. from Justin Morgan, a ♂ of ? Thoroughbred orig. born in 1789, and later named after his owner]

Moroccan Barb: *see* Barb

Morocco Spotted: (midwest, USA)/l; ri/colour type/orig. from Hackney, American Saddlebred, French Coach *et al.*/BS 1935/ [reputed orig. from Moroccan Barb]/? extinct

Morochuca: (Peru)/py/mt var. of Peruvian Criollo

Morvan: (Côte d'Or, France)/orig. of Auxois/Fr. *Morvandeaux*/extinct

Mossi: (Burkina Faso)/local var., of Dongola-Barb orig.

mostrenco: *see* mustang

Mountain Pleasure: (SE USA)/l; ri/ often chestnut, sorrel, palomino, or chocolate with flaxen/sim. to Rocky Mountain horse/HB 1989

Mousseye Pony: *see* Kirdi Pony

Moyle: (Idaho, USA)/l; ri/usu. bay or brown; often frontal bosses ('horns')/selected by Rex Moyle in mid 20th c. from mustangs in Utah/rare

M'Par: (Cayor, Senegal)/py/bay or chestnut/? degenerate Barb/syn. *Cayor*

MŠK: *see* Slovakian Sport Pony

Mulassière: *see* Poitou

Mull: *see* Gocan

Munighi, Muniki, Muniqi: *see* Maneghi

Münsterland: *see* Dülmen Pony

Mura: (Medimurje, NE Croatia)/ dr/brown, bay or grey/heavier var. of Croatian Coldblood; cf. Mur Island/orig. in 19th c. from native warmblood ♀ ♀, Noriker, Percheron, Ardennais and Brabant/Cro. *Medjimurena, Medjimurje, Medjimurski konj* (*konj* = horse)/rare

Muraközi: *see* Mur Island

Murgese: (Murge, Apulia, Italy)/l; dr.ri/usu. black, also grey or roan/orig. from Oriental × local/ HB 1926, BS 1948/not *Murghese*/ rare

Murinsulaner: *see* Mur Island

Mur Island: (SW Hungary)/l-h; dr/ var. of Hungarian Draught, sim. to Pinkafeld; cf. Mura/orig. from Percheron, Ardennes, Noric *et al.* × Hungarian/Hung. *Muraközi*, Ger. *Murinsulaner*/nearly extinct

Musseye Pony: *see* Kirdi Pony

mustang: (USA)/feral/chiefly Spanish orig./orig. of cow pony, Indian Pony/remnants = broomtail/for BSs see American Indian, American Mustang, Cerbat, Kiger Mustang, Spanish Barb, Spanish Mustang/syn. *American feral horse, BLM (= Bureau of Land Management) horse, Range horse*/[Sp. *mesteño* or *mostrenco* = wild or stray]

Muzakiya, Muzekja: *see* Myzeqeja

Myinn: *see* Burmese
Mytilene Pony: (NW Anatolia,
  Turkey)/ri.pa/local var./Turk.
  Midilli/syn. *Ege Midillisi, West
  Mytilene*/rare
Myzeqeja: (Albania)/plains var. of
  Albanian/not *Muzakiya,
  Muzekja, Mysekeja, Myzeqe,
  Myzeqea*

Nagai: *see* Nogai
Nagdi: (C Yemen)/var. of Yemeni
Namaqua Pony: (NW Cape Prov., S
  Africa)/sim. to Basuto Pony/orig.
  from Cape Horse in early 19th
  c./extinct
Nambu: *see* Nanbu
Namib Desert horse: (Namibia)/
  l/feral/rare
Nanbu: (NE Honshū, Japan)/var. of
  Japanese Native/not *Nambu*/
  extinct
Nanfan: *see* Hequ, Tibetan Pony
Narougor: *see* Fouta
Narragansett Pacer: (Rhode Island,
  USA)/orig. in 17th c. from Irish
  Hobby, Galloway Pony,
  Andalusian *et al.*/extinct by 1800
Narym: (Tomsk, Siberia, Russia)/var.
  of Siberian Pony/Russ.
  Narymskaya/rare
National: *see* Lusitanian
National Appaloosa Pony: *see*
  Appaloosa Pony
National Show Horse: (USA)/l; ri/
  orig. from Arab × American
  Saddlebred/BS 1982
National Spotted Saddle Horse:
  (Tennessee, USA)/l; ri/pinto/
  orig. from Tennessee Walking
  Horse, American Saddlebred and
  American Trotter/BS 1979
Native Turkey Pony: *see* Anatolian
  Native
Navarre Pony: (Spain)/pa.ri.m/usu.
  chestnut, bay or black/orig. (with
  Breton) of Burguete/Sp. *Jaca
  Navarra*/syn. *de la Barranca*;
  obs. syn. *Basque-Navarre Pony*
  (Sp. *Jaca Vasca-Navarra* or
  Poney Vasco-Navarro)/rare
Navarre: (France) *see* Tarbes
Nederlandse, Nederlandsche: *see*
  Dutch

Nefza Pony: (Tunisia)
Netherlands: *see* Dutch
Netherlands Riding Horse: *see* Dutch
  Warmblood
Neue Einsiedler: *see* Swiss
  Warmblood
New Caledonian: (New Caledonia, S
  Pacific)/BS 1845, HB/Fr. *Néo-
  caledonien*
New Cleveland Bay: *see* Yorkshire
  Coach Horse
New Forest Pony: (S Hampshire,
  England)/ri/BS 1891, HB 1910;
  HB also Australia, Belgium,
  Denmark, France, Germany,
  Sweden, USA 1990
Newfoundland Pony: (Canada)/
  ri.dr.(m)/usu. brown, black or
  bay/? orig. from S England in
  17th c./BS 1980, HB 1981/rare
New Kirgiz: (Kyrgyzstan)/l;
  m.d.ri.dr/chestnut or bay/orig.
  from Kirgiz improved in 19th
  and 20th c. (esp. since 1919) by
  Don and Thoroughbred/recog.
  1954/3 types according to size,
  and amount of Kirgiz blood/
  Russ. *Novokirgizskaya*
Ngua Noi: (Vietnam)
Nigerian: *see* West African Dongola
Ningqiang: (China)/var. of Chinese
  Minihorse
Nivernais: (Nièvre, C France)/black/
  local Percheron HB (1880–1996)/
  ? extinct
Nogai: (S Ukraine)/orig. of Cossack (=
  Old Don) and of Black Sea/not
  Nagai, Nogoay/extinct
Nogali: *see* Somali Pony
Noma: (Shikoku, Japan)/ri/smallest
  var. of Japanese Native/BS
  1978/nearly extinct
Nonius: (SE Europe)/l; dr.ri/brown,
  dark brown or black/orig. from
  Anglo-Norman ♂ 'Nonius Senior'
  born 1810 and used at
  Mezőhegyes, Hungary, from 1816
  on light mares/Hung. *Nóniusz*/rare
Nooitgedacht Pony: (E Transvaal, S
  Africa)/ri.dr/orig. (1952 on) from
  Basuto Pony with some Boer and
  Arab blood at Nooitgedacht Res.
  Sta. nr Ermelo/recog. 1976; BS
  1969/rare

Noram Trotter: (France, Italy and Austria)

Nordbag: *see* Fjord

Nordestino: *see* Northeastern

Nordfjord: *see* Fjord

Nordkirchen Pony: (Münsterland, Westphalia, Germany)/orig. from Dülmen Pony crossed with Polish Konik in 1920s and later with Arab and Welsh Pony/Ger. *Nordkirchner*

Nordland: (N Norway)/py; ri/usu. dark colours/Northern type sim. to Lofoten/composite of Lyngshest, Nordland and other small breeds from N Norway/HB 1969, BS/syn. *Lyngen* (Nor. *Lyngshest*), *Northlands Pony*/[= northland]/rare

Nordsvensk: *see* North Swedish

Norfolk: *see* Hackney

Norfolk-Breton: *see* Breton Post-horse

Norfolk Hackney: *see* Hackney

Norfolk Roadster: *see* Hackney

Norfolk Trotter: *see* Hackney

Norges Fjordhest: *see* Fjord

Noric: (Austria)/h; dr/usu. brown, bay or chestnut, also black, grey or spotted/var.: Abtenauer/BS; HB also Italy 1990/Ger. *Noriker* or *Norisches Kaltblut*/syn. *Pinzgauer*; pure Noric in SE Bavaria, Germany, called *Pinzgauer* (to 1914) or *heavy Noric* (1914–1934); var. in SW Bavaria with Warmblood influence called *Oberland* (to 1914) or *light Noric* (1914–1934); now combined and called *Oberland* (1934–1945) or South German Coldblood (*q.v.*)/[Roman Province of Noricum = Austria and S Bavaria]

Norman: *see* Anglo-Norman, Norman Coach Horse

Norman Coach Horse: (France)/orig. 1806–1860 from Hackney and Thoroughbred × native/orig. of Anglo-Norman, Norman Cob and French Trotter/Fr. *Carrossier normand*/syn. *Norman, Normandy Carriage Horse, Old Norman*/extinct

Norman Cob: (France)/l-h; dr.m/often bay or chestnut/orig. in 20th c. from Norman Coach Horse/HB 1955, BS/Fr. *Cob (Normand)*/ syn. *French Cob*/rare

Normand, Normandy: *see* Anglo-Norman

Norman Trotter: *see* French Trotter

Norsk Kaldblods Traver: *see* Norwegian Coldblood Trotter

North African Barb: *see* Barb

Northeastern: (NE Brazil)/py-l; ri/ dark bay, grey, isabella or light bay, or sorrel/var.: Marajoaro; also saddle (*sela*) and pack (*quartau*) types/Port. orig. from 16th c. on/BS 1974/Port. *Nordestino*/syn. *Crioulo Brasileiro, Curraleiro, Sertanejo* (= inland)

Northern Draught: *see* Trait du Nord

Northern type: (N Eurasia)/sim. to Konik type to the west and Mongolian type to the east/inc. Estonian Native, Finnish, Lofoten, Nordland, North Russian Pony/syn. *Northern Pony*

Northern Ardennais: *see* Trait du Nord

Northern Draught: *see* Trait du Nord

Northern Dun: *see* Fjord

Northland: *see* Nordland

North Russian Pony: (Russia)/ Northern type/inc. Karelian, Mezen, Ob, Pechora, Tavda, Tyatka

North Spanish Pony: (Spain)/inc. Asturian, Galician, Lebaniega, Losina, Navarre, Pottok/syn. *Cantabrian Pony* (Sp. *Cantabrica*)

North Star: *see* Mezőhegyes Halfbred

North Swedish: (Sweden)/l-h; dr.ri/ bay, brown, chestnut, black or dun with black points/orig. 1850–1900 from Døle (Norway) × improved local/HB 1903, BS 1924/Swe. *Nordsvensk häst*/rare

North Swedish Trotter: (Sweden)/ usu. chestnut or brown/ separated from North Swedish 1964/HB 1971/Swe. *Nordsvensk travare*

Northumberland Chapman: *see* Vardy

Norwegian Coldblood Trotter:
(Norway)/usu. chestnut, also bay,
black or brown/orig. from Døle
with foreign Trotter blood, esp.
North Swedish/HB 1939, BS
1994/Nor. *Norsk Kaldblods
Traver*/syn. *Norwegian Heavy
Trotter, Norwegian Trotter*; obs.
syn. *Døle Trotter*/rare

Norwegian Dun: *see* Fjord

Norwegian Fjord: *see* Fjord

Norwegian Pony: *see* Fjord

Norwegian Warmblood: (Norway)/l;
ri./orig. from foreign Warmblood
and Thoroughbred × Døle/rare

Novokirgiz: *see* New Kirgiz

Nowosądecki, Nowy Sącz: *see* Sącz

Ob: (W Siberia, Russia)/dr/North
Russian Pony group/Russ.
Priobskaya/syn. *Ostyak-Vogul*/
rare

Oberland: (Germany)/named used
1914–1934 for light var. of Noric
in SW Bavaria; syn. 1939–1945
for South German Coldblood

Obva: (W Siberia, Russia)/py/Russ.
Obvinskaya/extinct

Oirot: *see* Altai

Old Austrian Warmblood: (Austria)/
old type of Austrian
Warmblood/Ger.
*Altösterreichisches
Warmblut*/nearly extinct

Old Don: *see* Cossack

Oldenburg: (Germany)/l; ri.[dr]/bay,
bay-brown or black/sim. to East
Friesian and inc. in Heavy
Warmblood; some Thoroughbred
blood since late 19th c./orig. of
Austrian Warmblood, Danish
Oldenborg, Gelderland,
Groningen, Saxony Warmblood,
Silesian (Warmblood)/HB 1861,
BS/Ger. *Oldenburger Warmblut*/
syn. *Alt-Oldenburger* (= *Old
Oldenburg*), *Bovenlander*
(Netherlands, = upland)

Old English Black: *see* Great Horse

Old English War Horse: *see* Great
Horse

Old Kladruby Black: (Moravia, Czech
Republic)/black/var. of Kladruby
in Slatinany and Bzenec,
Modonin district/Cz.
Starokladrubsky Uranik/rare

Old Kladruby White: (Czech
Republic)/dr/grey/var. of
Kladruby in Kladrub Nord
Labem/orig. from Lipitsa,
Spanish and Neapolitan since
16th c./Cz. *Starokladrubsky
Belorus*/nearly extinct

Old Norman: *see* Norman Coach
Horse

Old Wurttemberg: *see*
Altwürttemberger

Onega: (Karelia, Russia)/py/var. of
Karelian/Russ. *Onezhskaya*/
extinct

Onezhskaya: *see* Onega

ORIENTAL: inc. Akhal-Teke, Arab,
Barb, Persian Arab, Syrian,
Turkish Arab, etc./syn. *Eastern*

Orlov-American Trotter: *see* Russian
Trotter

Orlovo-Amerikanskiĭ rysak: *see*
Russian Trotter

Orlov-Rostopchin: *see* Russian
Saddle Horse

Orlov Saddle Horse: (Russia)/Anglo-
Arab orig. in late 18th c./orig.
(with Rostopchin) of Russian
Saddle Horse/Russ. *Orlovskaya
verkhovaya*/syn. *Orlov Riding
Horse*/extinct

Orlov Trotter: (Russia)/l/grey, bay,
black or chestnut/orig. 1775–
1845 by Count Orlov at Khrenov
stud from Arab, Thoroughbred
and Persian, with Danish Saddle,
Dutch Draught, Mecklenburg and
Hackney blood/orig. (with
American Trotter) of Russian
Trotter/HB 1927/Russ. *Orlovskii
rysak*/syn. *Russian Trotter* (*c.*
1920–1949)/not *Orloff*

oscuro rabicano: *see* Choctaw

Österreichisches Warmblut: *see*
Austrian Warmblood

Ostfriesisches Warmblut: *see* East
Friesian

Østland: *see* Døle

**Ostpreussisches Warmblut
Trakehner Abstammung:** *see*
Trakehner

Ostyak-Vogul: *see* Ob

overo: pied horse with colour on

back, i.e. coloured horse with white markings; ? dominant gene/syn. *pio bajo*

Ox-Araber: (Denmark)/rare

Pacu Indonesia: *see* kudan Pacu Indonesia

Padana: *see* Cremonese

Padi: *see* Kuda Padi

Pagan: *see* Bhirum, Kirdi

Paint: *see* American Paint

Palatinate: *see* Zweibrücken

Palomino: (USA)/l; ri/colour type: golden with white mane and tail (i.e. golden isabella)/HB 1932, BS 1936, 1941; BS also Canada 1952, GB/not *Palamino*

Palouse: *see* Appaloosa

Panje: *see* Polish Konik

Panjepferd im Trabertyp: *see* Baltic Trotter

Pantaneiro: (N Mato Grosso, Brazil)/ l/usu. grey, also bay, black-and-white or brown/BS 1972/syn. *Bahiano* or *Baiano* (from Bahia), *Mimoseano* (from Mimoso), *Pantaneiro Criollo, Poconeano* (from Poconé)/[Port. *pantanais* = flooded lands]

Parsano: *see* Persano

Paso: *see* Colombian Criollo, Cuban Paso, Paso Fino, Peruvian Paso

Paso Fino: (USA)/l; ri/any colour; gait is a broken pace/orig. from Criollo (Caribbean, Colombian, Peruvian)/BS 1973; BS also Argentina, Canada 1985

Patibarcina: (Escambray, Cuba)/l; ri/ dun with dark eel-stripe and zebra stripes/orig. from Andalusian and Barb × Cuban Trotter

Paulisto: (São Paulo, Brazil)/more recent and lighter leggier var. of Mangalarga/orig. from Mangalarga Marchador with Arab and Thoroughbred blood

Pechora: (Komi, Russia)/ri.dr/North Russian Pony group/Russ. *Pechorskaya*/rare

Pegu: *see* Shan Pony

Pelna Krew Angielska: (Poland)/? = Thoroughbred

Peneia Pony: *see* Pinia

Peninsular: *see* Lusitanian

Percheron: (Le Perche, Normandy, France)/h; dr.m/grey, white or black/HB 1883, BS; local HBs (till 1966): Augeron, Berrichon, Bourbonnais, Loire, Maine, Nivernais, Saône-et-Loire; BS USA 1876, Canada 1907, Great Britain 1918, Argentina; HB S Africa 1940, Russia, Spain

Periangan Pony: (W Java)/ri.dr/ Indonesian pony group

Persano: (Salerno, Italy)/l; ri/bay, chestnut or grey/state stud 1763–1864 (or 1884); predominantly Sardinian, Salernitana, Arab and Thoroughbred blood/not *Parsano*/rare

Persian: *see* Iranian

Persian Arab: (Iran, esp. Kuzestan)/l; ri

Persian Turkoman: (NE Iran)/l; ri/ orig. (like Akhal-Teke and Iomud) from ancient Turkmenian/? = Yamud/syn. *Turk, Turkmen, Turkmene*

Peruvian Criollo: (Peru)/vars: Andean (inc. Morochuca), Cholal, Costeño/BS; BS also Canada

Peruvian Paso: (Peru)/var. of Costeño/BS USA 1967

petiço: (Rio Grande do Sul, Santa Catarina, Paraná and S of São Paulo, Brazil)/= small horse or pony

Petiso Argentino: (Argentina)/py; ri/ orig. early 19th c. from Shetland and Welsh Ponies (? × Criollo)

Petit Boulonnais: (N France)/smaller var. of Boulonnais/syn. *Mareyeuse* (= *fish-cart horse*)/ (nearly) extinct

Pfalz: *see* Zweibrücken

Pfälzisches Warmblut: *see* Zweibrücken

Pfalz Ardenner: (Palatinate and Saar, Germany)/h; dr/Norman and Ardennes blood in 17th c./HB 1890/Ger. *Pfalz Ardenner Kaltblut*/nearly extinct

Philippine Pony: (Philippines)/usu. bay, brown, grey or roan/South-

East Asia pony group with foreign blood/Sp. *Filipino*

Piebald and Skewbald: (Ireland)/ri./ black or brown, with white patches/composite of Irish Cob and Irish Sport Horse, established *c.* 1900/rare

Pindos Pony: (mts of Epirus and Thessaly, Greece)/ri.dr.pa/var. of Greek Pony/syn. *Pindhos, Pindus*/rare

Pinia: (W Peloponnese, Greece)/py; ri.dr/grey, blue roan, chestnut or bay/var. of Greek Pony with Anglo-Arab, Anglo-Norman and Nonius blood in early 20th c./HB 1995/syn. *Geogalidiko, Georgaludiko*/not *Peneia Pony, Pinias, Pineia*/[R Pinias]/rare

Pinkafeld: (Austria-Hungary)/h/sim. to Mur Island; orig. var. of Hungarian Draught/Hung. *Pinkafö*/ extinct

Pinto: (western USA)/l or py/ri/cf. Paint/BS 1947 (recog. 1956) for pied colour of any breed type; inc. overo and tobiano/stock, hunter, pleasure and saddle types/*see also* Canadian Pinto, Cuban Pinto, National Spotted Saddle Horse/[Sp. = painted]

Pinto Typ Lewitzer: *see* Lewitzer

Pinzgau, Pinzgauer: *see* Noric

pio alto: *see* tobiano

pio bajo: *see* overo

Piquira Pony: (SE and NE Brazil)/ri/ recent orig. from ponies (chiefly Shetland) × Brazilian Crioulo (? Northeastern)/HB 1970

Plantation Walking Horse: *see* Tennessee Walking Horse

Plateau Persian: (Iran)/l/inc. Bakhtiari, Basseri, Darashuri, Jaf, Qashqai, Sistani, Persian Arab

Pleven: (Bulgaria)/l; ri.dr/usu. chestnut/halfbred orig. since 1898; sim. to East Bulgarian with addition of Anglo-Arab and Gidran ♂ ♂ in its formation/ recog. 1951/Bulg. *Plevenska*/syn. *Plevna*

Po: *see* Cremonese

Poconéano: *see* Pantaneiro

Podlaski: *see* Kopczyk Podlaski

Poitou: (France)/h; dr.m/usu. dun, also grey, black or bay/orig. from Friesian/HB 1885, BS/Fr. *Trait poitevin-mulassier* or *Poitevine mulassière* (= *mule producer*)/ rare

Polesian: (Belarus)/py; dr/Konik type/Russ. *Polesskaya*/not *Polesié, Polessian, Polessye*/rare

Polish Ardens: (Poland)/h; dr/name suggested for horses at Bielin and Nowe Jankowica studs

Polish Draught: (Poland)/h; dr/usu. chestnut, roan or bay/orig. mainly from Ardennes and Belgian Draught/vars: Garwolin, Kopczyk Podlaski, Lidzbark, Lowicz, Sokólka, Sztum

Polish Konik: (W and NE Poland)/py; dr.ri/usu. mouse-grey with dorsal stripe, sometimes leg stripes/Konik type/improved var.: Biłgoraj/HB 1955/Pol. *Konik polski*, Ger. *Panjepferd*/syn. *Polish Pony, Polish Primitive*/[= small horse]/rare

Polish Noble Half-bred: (Poland)/ hunter

Poljakoff: (Mongolia)/? = Przewalski horse

Polo Pony: (England)/l; ri/Arab or Thoroughbred × pony

Pony Ciociaro: *see* Esperia Pony

Pony of the Americas: (USA)/ri/cf. Kanata Pony/BS 1954 for ponies with Appaloosa spotting/not *Pony of America*

Portuguese: *see* Lusitanian

Portuguese Galician: *see* Garrano

Portuguese ponies: *see* Azores, Garrano, Mira, Sorraia

Posavina: (R Sava, C and NE Croatia)/l; dr/bay, black, chestnut, white or wild type/ lighter var. of Croatian Coldblood, based on Buša Pony with Asian and Arabian blood and later Spanish, Italian, Nonius and Lipitsa/Cro. *(Hrvatski) Posavac, Posavlje, Posavska*/not *Posavian*/[R Sava valley]/rare

Postier breton: *see* Breton Post-horse

Pottok: (Basque country, France and

Spain)/py; ri.(m).[pa]/usu. black,
pied, chestnut or bay; semi-feral/
BS and HB France 1971, HB Spain
1982/pronounced *Potiok*;
pl. *Pottokak*; Basque *Pottoka*/ syn.
Basque (Sp. *Vasca*)/not *Pottak*,
Pottock/[= small horse]/ rare

Poznań: (Poland)/l; dr.ri/orig. from
Trakehner × Polish Konik with
Thoroughbred blood; combined
with Mazury to form
Wielkopolski/Pol. *Poznański*

Prejvalsky's: *see* Przewalski horse

Printer: *see* Quarter Horse

Priob: *see* Ob

Przedświt: (Poland)/halfbred strain
derived from Thoroughbred
stallion Przedświt in Hungary
(also in Slovakia)

Przewalski horse: (Mongolia)/=
Equus ferus przewalskii
Poliakov/py/red-brown with
light underparts, dark back
stripe, shoulder stripe and leg
bars; erect mane/? orig. of
domestic horse (*E. "caballus"*
Linnaeus)/BS in USA/syn.
*Asiatic wild horse, Mongolian
wild horse, Mongolian tarpan,
taki*/not *Prejvalsky's, Prezwalski,
Prjevalsky's, Przevalskii's,
Przevalsky, Przewalsky's,
Przhevalski's, Przrewalskii's*/
extinct in wild, rare in captivity

p.s.c.: *see* Thoroughbred

p.s.i.: *see* Thoroughbred

Pugliese: *see* Apulian

Puno Pony: (Chile)/feral; Sp. orig.
like Bagual and Criollo

Puoliverinen: *see* Finnish
Warmblood

Pura raza española: *see* Andalusian

Purebred Spanish: *see* Andalusian

Puro sangre de carrera: *see*
Thoroughbred

Puro sangre inglese: *see* Thoroughbred

pur sang: *see* blood horse

Qashqai: (Iran)/var. of Plateau
Persian/not *Quashquai*

Qatgani: (Afghanistan)/l

Quaddeddu: *see* Giara Pony

Quarter Horse: (Texas, USA)/l; ri/
orig. in 18th, revived in 20th c.,
from Chickasaw, Thoroughbred,
Criollo *et al.*/BS 1940; BS also
Canada 1956, GB, Argentina/syn.
*American Quarter horse, Billie,
Cold Deck, Copper Bottom,
Kentucky Whip, Printer, Quarter
Mile, Rondo, Shilo, Steeldust*/not
*American Quarter Running
Horse*/[from quarter-mile races of
colonial period]

Quarter Pony: (USA)/Quarter Horse
less than 14.2 hands (142 cm)
high/BS 1976

Rahvan: (Turkey)/py-l; ri/orig. from
Canik and Anatolian Native/
Turk. *Bati Karadeniz Rahvan Ati*
(= *West Black Sea Rahvan horse*/
[= a fast lateral walking gait]

Rajshahi Pony: (Bangladesh)

Range horse: *see* mustang

Ranger: *see* Colorado Ranger

Rangerbred: *see* Colorado Ranger

Rastopchin: *see* Rostopchin

Rhenish: (Germany)/h; dr/usu. grey,
bay, black or chestnut/orig. from
Belgian Draught/strains in
Saxony, Silesia and
Westphalia/orig. of Black
Forest/BS 1876/Ger. *Rheinisches
Kaltblut, Rheinisch-deutsches
Kaltblut*/syn. *Rhenish-Belgian,
Rhenish-German, Rhenish-
Westphalian* (Ger. *Rheinish-
Westfälisches*), *Rhineland Heavy
Draught*/rare

Rhineland: *see* Rhenish

Riding horse of the GDR: *see* Edles
Warmblut

Riger Mustang: *see* Kiger Mustang

Rila Mountain: (Bulgaria)/var. of
Bulgarian Native/extinct

Rio Grande do Sul: *see* Crioulo

Rocky Mountain Horse: (E Kentucky,
USA)/l; ri/usu. chocolate and
flaxen/sim. orig. in early 20th c.
to American Saddlebred and
Tennessee Walking Horse/BS
1986/[? orig. from gaited Spanish
♂ in Rocky Mountains,
Kentucky]/rare

Rocky Mountain Pony: *see* Guoxia

Rodopi: (Bulgaria)

românesc: *see* Romanian

Romanian: (Romania)/l or py/former vars: Dobrogea, Moldavian, Romanian Mountain, Transylvanian/Rom. *Rasa românească*/not *Roumanian, Rumanian*/extinct

Romanian Draught: (Romania)/nearly extinct

Romanian Light Draught: (Romania)/ orig. since 1943 from Ardennes, Romanian Trotter, Ialomiţa, Lipitsa, Mezőhegyes Halfbred and Arab/var.: Transylvanian Light Draught/Rom. *Semigreu Românesca*/rare

Romanian Mountain: (Romania)/py/ former var. of Romanian/Rom. *Calul românesc de munte*/ extinct

Romanian Sport Horse: (Romania)/l; ri/orig. since 1962 from Thoroughbred, Arab, Mezőhegyes Halfbred, Gidran, trotter and others/Rom. *Calul de Sport*/syn. *Romanian Saddle Horse, Romanian Sporting Horse*/rare

Romanian Trotter: (Romania)/orig. since 1900 from American Trotter (92%), Orlov Trotter and Romanian foundation stock/Rom. *Trapaş românesc*

Ronda: (Spain)/former strain of Andalusian/Sp. *Rondeña*/extinct

Rondo: *see* Quarter Horse

Roraima: *see* Lavradeiro

Rostopchin: (Russia)/l/Anglo-Arab orig. in early 19th c./orig. (with Orlov Saddle Horse) of Russian Saddle Horse/not *Rastopchin*/ extinct

Rottal: (Bavaria, Germany)/l; ri.[dr]/ usu. bay, also black or chestnut/ orig. from local, improved by Anglo-Norman, Thoroughbred and Cleveland Bay in 19th c. and by Oldenburg in 20th/HB 1906/ Ger. *Rottaler Warmblut*/[R Rott]/ nearly extinct

Royal Canadian Mounted Police Horse: (Canada)/l; ri/black/ Thoroughbred cross/rare

Rumelian Pony: (Balkans)/generic term for ponies in former European provinces of Turkey/ extinct

Russ: *see* Gotland Pony

Russian Arab: (Russia)/bred in former USSR from 1925, upgraded with imports from Hungary and France (1930) and Britain and Poland (1936)/ several vars/Russ. *Chistokrovnaya Arabskaya*/ rare

Russian Ardennes: *see* Russian Heavy Draught

Russian Clydesdale: *see* Vladimir Heavy Draught

Russian Heavy Draught: (Russia and Ukraine)/h; dr.d/usu. chestnut, also brown or bay/orig. (since mid-19th c.) from Ardennes × local; smaller than Ardennes/ Ural and Ukrainian vars/Russ. *Russkiĭ tyazhelovoz, Russkaya tyazhelovoznaya*/syn. *Russian Ardennes* (till 1952), *Russian Draught*

Russian Saddle Horse: (Russia)/l; ri/orig. in late 19th c. from Orlov Saddle Horse and Rostopchin/ partially absorbed by Ukrainian Saddle Horse since 1945/Russ. *Russkaya krovnaya verkhovaya*/ syn. *Orlov-Rostopchin* (Russ. *Orlovo-Rostopchinskaya*)/nearly extinct

Russian tarpan: *see* tarpan

Russian Trotter: (Russia)/l/bay, black or chestnut, occ. grey/orig. 1890–1926 from Orlov and American Trotters, with 2nd cross to American Trotter in 1960/HB 1927; recog. 1949/Russ. *Russkii rysak, Russkaya rysistaya*/syn. *Métis Trotter, Orlov-American Trotter, Russo-American Trotter (c.* 1920– 1949)/*see also* Orlov Trotter (syn. *Russian Trotter c.* 1920–1949)

Russo-American Trotter: *see* Russian Trotter

Ryukyu: *see* Miyako, Yonaguni

Rzeszów: *see* Kraków-Rzeszów

SA: *see* South African

Sable Island Pony: (Nova Scotia, Canada)/bay, brown, black or

chestnut/feral since 18th c.; ♂ ♂
of various breeds introduced
1800–1940/rare

Sächsisch: *see* Saxon

Sącz: (Nowy Sącz, Kraków, Poland)/
l; dr.ri/orig. from Hungarian
Halfbred in early 20th c./inc. in
Kraków-Rzeszów/Pol. *Sądecki,
Nowosądecki*

Saddlebred: *see* American
Saddlebred, French Saddlebred

Sądecki: *see* Sącz

Saglawi: strain of Arab/not
*Saglaviyah, Saklawi, Seglawi,
Siglavy*

Sahel: (Mali)/var. of West African
Barb with Arab blood

St Lawrence: (Canada)/h/orig. from
Canadian, Shire, Clydesdale *et
al.*/extinct

Saishu Island: *see* Cheju

Saklawi: *see* Saglawi

Salernitana: (Salerno, Campania,
Italy)/l/blackish with lighter
patches/orig. late 18th c. from
local with Andalusian and Arab
blood and with Thoroughbred
from early 1900s/HB 1990, BS/
syn. *Persano, Salernitano-
Persano*/nearly extinct

Samogitian: *see* Žemaitukai

Samólaco: (Sondrio, Lombardy,
Italy)/dr.ri/chestnut or bay/orig.
from Haflinger × Andalusian/not
Samolaca/nearly extinct

Sandalwood Pony: (Sumba,
Indonesia)/ri.dr/Indonesian Pony
group/Du. *Sandelhoutpaard*/syn.
Soemba, Sumbanese Pony/not
Sandelwood

Sandan: *see* Shandan

Sanfratellana: (San Fratello,
Messina, Sicily, Italy)/l; ri.m.dr/
usu. bay, occ. black; semi-feral/
orig. from local with Oriental,
Thoroughbred, Maremmana (for
coat colour) and Nonius blood/
HB 1990/rare

Sanhe: (NE Inner Mongolia, China)/l;
ri.dr/red-brown, black, brindle,
grey or isabella/orig. since 1904
from Russian breeds × Chinese
Mongolian/WG *Sanho*, Russ.
San'khè/obs. syn. (or sim. breed)

Hailar (WG *Hai-la-erh*),
Sanpeitze (WG *San-pei-tzü*)

Sankt Märgener Fuchse: *see* Black
Forest

Sanpaitze, Sanpeitze: *see* Sanhe

Saône-et-Loire: (C France)/local HB
(1927–1966) for Percheron and
Ardennes/extinct

Sappapaitze: *see* Sanhe

Sara: *see* Kirdi

Sarda: *see* Sardinian

Sardinian: (Sardinia, Italy) /l/usu.
bay, chestnut or grey/orig. from
Arab and Thoroughbred × local
in late 18th and early 19th c.
(native had Andalusian blood in
16th c. and various European
halfblood introductions in late
17th and early 18th c.)/since
1936 crossed with Thoroughbred
to form Sardinian Anglo-Arab/It.
Sarda/syn. *Arabo-Sarda, Sardo-
Arab*/rare

Sardinian Anglo-Arab: (Sardinia,
Italy)/l; ri/usu. bay or chestnut,
occ. grey/orig. from Sardinian
since 1936 by further crossing
with Thoroughbred and later
with French Anglo-Arab/It.
Anglo-Arabo-Sardo/rare

Sardinian Pony: *see* Giara Pony

Sardo-Arab: *see* Sardinian

Sárvár: (Hungary)/l; ri/orig. at
Mezőhegyes from Thoroughbred
and Furioso-North Star
(Mezőhegyes Halfbred)/[name of
Austrian stud]/rare

Sava: *see* Posavina

Saxony Coldblood: (Germany)/strain
of Rhenish/combined with
Thuringian to form Saxony-
Thuringian Coldblood/Ger.
Sächsisches Kaltblut

Saxony-Thuringian Coldblood:
(Germany)/h; dr/black, bay,
brown, chestnut or grey/orig.
from Saxony Coldblood +
Thuringian Coldblood/Ger.
*Sächsisch-Thüringisches
Kaltblut*/syn. *Saxon-Thuringa*/
rare

Saxony Warmblood: (Saxony and
Thuringia, Germany)/brown or
black/orig. from Hanoverian and

Oldenburg/absorbed by Heavy
Warmblood 1960s/HB 1872/Ger.
Sächsisches Warmblut/rare
SBS: *see* Belgian Sport Horse
SCB: *see* Sudan Country-Bred
Schlesisch: *see* Silesian
Schleswig Coldblood: (Germany)/h;
dr/usu. chestnut, also black or
grey/orig. from Jutland ×
Suffolk/BS 1891/Ger.
Schleswiger Kaltblut/syn.
Schleswig Heavy Draught/rare
Schleswig-Holstein: *see* Holsteiner
Schwarzwälder: *see* Black Forest
Schweike: *see* Sweyki
Schweizerwarmblut: *see* Swiss
Warmblood
Schweres Warmblut: *see* Heavy
Warmblood
Schwyz: *see* Einsiedeln
Scotch Pony: *see* Highland Pony
Seglawi: *see* Saglawi
Seistan: *see* Sistani
Selle français: *see* French Saddlebred
Seminole: *see* Florida Cracker
Senne: (Westphalia, Germany)/l; ri/
usu. bay or grey/orig. from semi-
wild (till 1870) improved by
Spanish and Oriental/Ger.
Senner/nearly extinct
Serrana: *see* Chola
Sertanejo: *see* Northeastern
Seville: (Spain)/former strain of
Andalusian/Sp. *Sevillana*/
extinct
Shackleford Pony: *see* Banker
Shagya Arab: (Hungary)/l; ri/usu.
grey/strain of Arab with
Hungarian blood/HB 1850; BS
USA (Shagya Arabian) 1986/syn.
Bábolna Arab, Slovakia *Arabsky
Kôn*, Austria *Shagya-Araber*/
[name of foundation Arab ♂
imported to Bábolna stud
1836]/rare
Shami: (N Yemen)/var. of Yemeni
Shandan: (Qilianshan mts, Qinghai,
China)/py-l; ri.pa.dr/usu. bay,
also chestnut/orig. since 1934 at
Shandan Horse Farm from many
breeds × local, inc. Don since
1950/recog. 1980/not *Sandan,
Skandan*
Shan Pony: (Shan state, Myanmar)/

ri.dr.pa/South-East Asia pony
group; smaller and with thicker
hair than Burmese/syn. *Pegu*
Shetland Pony: (Scotland)/ri.dr/
dwarf/BS 1890, HB 1891; BS also
Australia, Belgium 1954 (HB
1960), Denmark, France 1958
(HB 1966), Germany 1942,
Netherlands 1937; HB also
Sweden/orig. of American
Shetland Pony/syn. *Sheltie*
Shilo: *see* Quarter Horse
Shirazi: *see* Darashuri
Shire: (Midland shires, England)/h;
dr/black, brown, bay or grey/sim.
to Clydesdale/orig. from Great
Horse/BS 1878, HB 1882; BS also
USA 1885/syn. *English Cart-
horse*
Shirvan: (Azerbaijan)/l
Shishan Pony: (China)/var. of
Guangxi Pony
Siamese Pony: *see* Thai Pony
Siberian Pony: (Russia)/Mongolian
type/inc. Altai, Buryat, Narym,
Tuva, Yakut/orig. of Kuznetsk
Sichuan: (China)/var. of Chinese
Minihorse
Sicilian: (Sicily, Italy)/l; ri.m/often
bay/var.: Middle Eastern/orig.
from Oriental × native/HB/
rare/*see also* Sanfratellana
Siebenburgisch: *see* Transylvanian
Siglavy: *see* Saglawi
Sikang Pony: (E Tibet and W
Sichuan, China)/pa/often white
or grey/? orig. from Sining (or
Tibetan) × South China Pony/? =
Jianchang or Ganzhi/syn.
Hsiangcheng, Hsi-k'ang/not
Sikong/not now recog.
Silesian: (SW Poland)/l; dr/orig. from
Oldenburg, East Friesian and
Hanoverian/Pol. *Śląski*, Ger.
Schlesisches Warmblut (=
Silesian Warmblood)
Silesian Coldblood: (SW Poland)/
strain of Rhenish (1858 on)/ Ger.
Schlesisches Kaltblut
Silesian Noric: (Slezsko and N
Moravia, Czech Republic)/
dr/sorrel/imported from Austria
in 20th c./Cz. *Slezský Norik*/rare
Simmental: *see* Erlenbach

Sini: (E Inner Mongolia, China)/l; ri/
usu. bay, chestnut or black/orig.
early 20th c. from Buryat × Sanhe

Sining: (E Qinghai and SW Gansu,
China)/py; ri.pa/usu. black, also
chestnut, grey or dun/orig. from
Mongolian × Tibetan Pony/WG
Hsi-ning/not now recog.

Sistani: (Iran)/var. of Plateau Persian/
not *Seistan*

Skogsbagge, Skogshäst, Skogsruss:
see Gotland Pony

Skandan: *see* Shandan

Skyros Pony: (Greece)/ri.dr/white,
dun, bay or grey, occ. chestnut/
var. of Greek pony/BS/syn.
Alogaki (= little horse), *Skyrian*/
nearly extinct

Śląski: *see* Silesian

Slezský Norik: *see* Silesian Noric

Slovakia Arabski Kôň: *see* Shagya
Arab

Slovakian Mountain: (Slovakia)/orig.
from Czech Coldblood × Hutsul,
? with Noric and Haflinger
blood/Sl. *Slovenský horský*

Slovakian Sport Pony: (Vel'Ké Pole,
Nitra, Slovakia)/ri.dr/dun or bay,
occ. white/orig. since 1980 at
Agric. Univ., Nitra, from Welsh
Pony × (Welsh Pony × Arab)
with some Hanoverian,
Slovakian Warmblood and
Hutsul blood/Sl. *Malý Športový
Kôň (MŠK)* (= *small sport horse*),
Slovenský Športový Pony/syn.
*Slovak(ian) Riding Pony,
Slovak(ian) Saddle Pony, Small
Saddle Horse*; syn. (to 1992)
*Czechoslovakian Small Riding
Horse* or *Saddle Pony*/nearly
extinct

Slovakian Warmblood: (Slovakia)/l;
ri.dr/sim. to Czech Warmblood
but more Hungarian Halfbred
blood/Sl. *Slovenský teplokrvník*/
syn. *Slovak Warmblood*

Slovenian Coldblood: (N Slovenia)/
orig. in 1940s from Noric and
Mura/Sn. *Slovenski hladnokrvni
konj*/rare

Slovenian Warmblood:
(Slovenia)/Sn. *Slovenski
toplokrvni konj*/rare

Slovenski: *see* Slovenian

Slovenský: *see* Slovakian

Smudisch: *see* Žemaitukai

Soemba: *see* Sandalwood Pony

Sokólka: (Białystok, Poland)/l-h;
dr/usu. chestnut/var. of Polish
Draught/orig. (1922 on) from
Ardennes, Belgian Draught and
Breton Post-horse × local/Pol.
Sokólski

Somali Pony: (Somalia)/usu.
chestnut or grey/inc. Dor,
Mijertinian (or Daror), Nogali/
rare

Songhaq: (Niger bend, Mali)/l-py/var.
of West African Dongola with
Barb blood/syn. *Sonraq*

Soriana pony: (Spain)/dappled or
chestnut/cf. Losina/Sp. *Jaca
Soriana*

Sorraia: (R Sorraia, Spain and
Portugal)/py; ri.pa/usu. dun with
dorsal stripe and zebra stripes on
legs/HB Portugal 1937/Sp.
Sorraiana/nearly extinct

South African: *see* Cape Horse

South African Miniature: (S Africa)/
ri.dr/orig. by Wynand De Wet of
Lindley from Shetland Pony
since 1945/recog. 1989; BS
1984/syn. *SA Miniature Horse*

South African Saddlehorse: (S
Africa)/ri/orig. from indigenous
♀♀ upgraded by American
Saddlebred ♂♂, first imported
1917/inc. five-gaited, three-
gaited and harness classes/BS
1949

South China ponies: (China)/pa.dr/?
South-East Asia pony group/inc.
Guangxi, Guizhou, Guoxia,
Jianchang, Jinhong, Lichuan,
Yunnan; *see also* Chinese
Miniature

South-East Anatolian: *see* Kurdi

SOUTH-EAST ASIA PONY: (SE Asia)/?
orig. from Mongolian and Arab/
inc. Burmese, Cambodian, Cheju,
Indonesian, Japanese, Manipuri,
Philippine, Shan, South China,
Taejung, Thai, Vietnamese

Southern Barbarian: *see* Tibetan Pony

Southern Native (Nanfan): *see*
Tibetan Pony

South German Coldblood: (S Bavaria and Württemberg, Germany)/h-l; dr/often bay or chestnut/orig. from Noric with some Warmblood infusion/HB 1906/ Ger. *Süddeutsches Kaltblut*/syn. (1934–1945) *Oberland* (Ger. *Oberländer Kaltblut*)

Southwest Spanish Mustang: (SW USA)/py-l; ri/BS 1972 for mustangs of pure Spanish ancestry, offshoot of Spanish Mustang Association/rare

Soviet Heavy Draught: (C European Russia, esp. Mordovia and Pochinok)/h; dr.d/usu. chestnut, brown or bay/orig. from local mares graded to Brabançon since late 19th c./recog. 1952/Russ. *Sovetskii tyazhelovoz, Sovetskaya tyazhelovozskaya*/ syn. *Soviet Draught*

Spanish: *see* Andalusian

Spanish Anglo-Arab: (Spain)/l; ri/ crosses of Andalusian, Arab, Thoroughbred, Anglo-Arab and their derivatives/Sp. *Hispano-Anglo-Arabe*/syn. *Tres sangres* (= 3 bloods)

Spanish-American: *see* Colonial Spanish (USA), Criollo (Spanish America)

Spanish Barb: (Oshoto, Wyoming, USA)/l-py; ri/any colour except tobiano/BS 1972 for horse of Barb ancestry (mustang, Criollo and African Barb)/rare

Spanish Colonial: *see* Colonial Spanish, Wilbur-Croce

Spanish Colonist: *see* Wilbur-Croce

Spanish Mustang: (western USA)/l; ri/BS 1957 for mustangs of pure Spanish ancestry/*see also* Cerbat, Kiger Mustang, Southwest Spanish Mustang/rare

Spanish ponies: *see* Canary Island Pony, North Spanish Pony, Sorraia

Spanish Trotter: (Balearic Is, Spain)/HB/Sp. *Trotador español*

Spiti Pony: (Kangra valley, NE Punjab, India)/ri.pa/usu. grey or dun, occ. bay or black/sim. to Bhotia and Tibetan

Spotted: *see* Appaloosa, British Appaloosa, British Spotted,

British Spotted Pony, Colorado Ranger, Kanata Pony, Knabstrupper, Morocco Spotted, Pony of the Americas; *see also* National Spotted Saddle Horse

Srednekolymskaya: *see* Middle Kolyma

Sri Lankan Pony: (Delft I, Sri Lanka)/ dark chocolate brown to buff, white patch on forehead; feral; inbred therefore dwarf/orig. early 19th c. from Port. imports; formerly also on Mannar I, now in Puttalam and at Colombo Zoo/ nearly extinct

Standardbred: (USA)/official name for American Trotter

Stara Planina: (Bulgaria)/var. of Bulgarian native/extinct

Starokladrubsky Belorus: *see* Old Kladruby White

Starokladrubsky Uranik: *see* Old Kladruby Black

Steeldust: *see* Quarter Horse

Strelets: (SE Ukraine)/l/Arab and Thoroughbred orig. in late 19th c. at Strelets state stud/orig. of Tersk/extinct

Striped Horse: (Colorado, USA)/l-py; ri/colour type: brindle or roan-striped of any breed/BS (International Striped Horse Association) 1988

Stuhm: *see* Sztum

Sudanese Country Bred: (Sudan)/l/ bay, brown, chestnut, roan, grey or black (in order of frequency)/ orig. in 20th c. from Arab and Thoroughbred × Dongola/orig. of Tawleed/syn. *SCB, Sudan Country-Bred*

Süddeutsches Kaltblut: *see* Noric, South German Coldblood

Suéve: *see* Asturian pony

Suffield Mustang: (SE Alberta, Canada)/feral; sim. to old Morgan but more colour vars/BS 1994

Suffolk: (E England)/h; dr/chestnut/ BS 1877, HB 1880; BS also USA 1911/syn. *Suffolk Punch*/nearly extinct in GB

Sulawesi: *see* Macassar Pony

Sulebawa: (Nigeria)/var. of West African Barb/? orig. from Niger

Sumba Pony: *see* Sandalwood

Sumbar-Sandel-Arab: (Indonesia)
Sumbawa Pony: (Sumbawa,
Indonesia)/Indonesian Pony
group; sim. to Sandalwood/var.:
Bimanese/Du. *Soembawa*
Sunicho: *see* Bolivian Pony
Suoke: (W Sichuan, China)/var. of
Hequ
Suomen: *see* Finnish
Svensk: *see* Swedish
Svensk Fullblod: (Sweden) *see*
Thoroughbred
Svensk kallblodstravare: *see*
Swedish Trotter
Svensk Ridponny: *see* Swedish
Riding Pony
Svensk varmblodstravare: *see*
Swedish Trotter, Swedish
Warmblood
Swedish Ardennes: (Sweden)/h; dr/
orig. from Ardennes (imported
since 1873) × local/BS and HB
1901/Swe. *Svensk Ardenner*
**Swedish Halfblood, Swedish
Halfbred:** *see* Swedish
Warmblood
Swedish Riding Pony: (Sweden)/
Swe. *Svensk Ridponny*
Swedish Standardbred Trotter:
(Sweden)/orig. from American
Trotter with some French Trotter
blood
Swedish Trotter: *see* North Swedish
Trotter, Swedish Standardbred
Trotter
Swedish Warmblood: (Sweden)/l;
ri/orig. based on East Prussian
(Trakehner), Hanoverian and
Thoroughbred/HB; BS USA/Swe.
Svensk Varmblod/syn. *Swedish
Halfblood, Swedish Halfbred*
Sweyki: (East Prussia)/cf. Konik/orig.
(with Thoroughbred and
Trakehner) of East Prussian/Ger.
Schweike/extinct
Swiss Halfblood: (Switzerland)/l;
ri/orig. from Anglo-Norman,
Holstein, Württemberg, Hackney
et al. × local/vars: Einsiedeln,
Erlenbach/syn. *Swiss Anglo-
Norman, Swiss Halfbred*/extinct
Swiss Warmblood: (Switzerland)/l;
ri/orig. since 1960 from French,
German and Swedish ♂♂ × Swiss

Halfblood/Fr. *Demi-sang suisse,*
Ger. *Schweizerwarmblutpferd,* It.
Cavallo da sella svizzera (= *Swiss
Saddle horse*)/syn. *Neue Einsiedler*
(= *New Einsiedeln*)
Syrian: (Syria)/l/Arab type
Sztum: (N Poland)/h; dr/var. of
Polish Draught/orig. end of 19th
and early 20th c. from Belgian
Draught and Rhenish ×
local/Pol. *Sztumski,* Ger. *Stuhm*

Tadzhikskaya verkhovaya: *see* Tajik
Riding Horse
Taejung: (Korea)/South-East Asian
Pony group
Tägan: *see* Tanghan
Taishu Pony: (Tsushima I, Nagasaki,
Japan)/ri/var. of Japanese Native;
cf. Cheju/HB 1979, BS/not
Taishuba, Taishuh/nearly extinct
Tajik Riding Horse: (Tajikistan)/l/
breed group/orig. 1953–1983
from Arab Thoroughbred *et al.* ×
Lokai/Russ. *Tadzhikskaya
verkhovaya*
Takara: *see* Tokara
takie: *see* Przewalski horse
Tanghan: (Nepal and Bhutan)/ri/
larger var. of Bhotia Pony/Nepali
Tānan, Tägan/not *Tangan,
Tangum, Tangun*/[Tanghastan is
old name for mts of Bhutan]
Tarai Pony: (S Nepal)/usu. bay, occ.
white with dark skin/not *Terai*
Tarbes: (Pyrenees, France)/l/usu.
dark brown or bay/orig. from
Arab and Thoroughbred ×
Andalusian/Fr. *Tarbésan,
Tarbeux*/syn. *Bigourdan* (from
Bigorre), *Navarre*/not
Tarbenian/extinct by crossing
with Anglo-Arab and Breton
Tarnowsko-dabrowski: *see* Dąbrowa-
Tarnowska
tarpan: (E Europe)/= *Equus ferus
gmelini* Boddaert/mouse-grey
with black back stripe/orig. of
domestic horse, *E. "caballus"*/
syn. *European wild horse,
Russian tarpan*/[Kirgiz]/extinct
c. 1813 in Poland, 1876 in
Ukraine, 1918 in captivity
"Tarpan": (Poland, Germany)/bred-

back type orig. (Poland) from
Polish Konik in 1920s, or
(Germany) from Iceland Pony,
Gotland Pony and Przewalski
horse at Munich zoo in 1930s/BS
in USA 1971

tarpan, Mongolian: *see* Przewalski
horse

Tartar, Tatar: *see* Mongolian type

Tattu: (Nepal)/pa/smallest var. of
Bhotia Pony/Nepali *Tāttu*

Tatung, Ta-t'ung: *see* Datong Pony

Tavda: (W Siberia, Russia) py;
dr/North Russian pony group/
Russ. *Tavdinskaya*/syn.
Tavdinka/not *Tawda*/rare

Tawleed: (Khartoum, Sudan)/
ri/Sudanese Country Bred
upgraded by exotic breeds, esp.
Thoroughbred/rare

Täysverinen: (Finland)/ri/orig.
imported from Sweden/syn.
Thoroughbred/nearly extinct

TB, TBE: *see* Thoroughbred

Tcherarani: *see* Chenarani

Tenchong: (China)/var. of Yunnan

Tennessee Walking Horse: (USA)/l;
ri/orig. in late 19th and early
20th c. from Morgan, American
Saddlebred, American Trotter
and Thoroughbred/BS 1935; BS
Canada 1982/syn. *Plantation
Walking Horse, Tennessee
Walker*

Terai: *see* Tarai

Tersk: (N Caucasus, Russia)/l; ri/
silver-grey, bay or chestnut/orig.
at Tersk stud in 19th c. from
Strelets × Arab, Don and
Kabarda, and at Stavropol stud
1920–1940s/recog. 1949/Russ.
Terskaya/not *Terky*/rare

Tezhkovozna: *see* Bulgarian Heavy
Draught

Thai Pony: (Thailand)/ri/South-East
Asia Pony group/syn. *Siamese
Pony*

Thessalian: (Thessaly, Greece)/l;
ri.dr/improved since 1945 by
Arab, Anglo-Arab and
Lipitsa/nearly extinct

Thorcheron Hunter:
(USA)/Thoroughbred ×
Percheron/BS 1974

Thoroughbred: (England)/l; ri/orig.
from Oriental (chiefly Arab, also
Barb and Turkmen) × local British
♀♀ in 17th and early 18th c./HB
1791, BS 1917; BS also Argentina,
Australia, Canada, Croatia, Czech
Republic, Denmark, Greece 1954,
Ireland, Italy, Japan, Netherlands
1879, New Zealand, Poland,
Sweden, USA 1894 (HB 1868); HB
also Belgium 1854 (also BS),
France 1833, Germany 1847,
Israel 1984, Norway, Russia, S
Africa 1906, Spain/Ger.
Englisches Vollblut, Fr. *Pur-sang
anglais*, It. *Puro sangue inglese* (or
p.s.i.), Sp. *Puro sangre de carrera*
(or *p.s.c.*), Russ. (*Angliĭskaya*)
chistokrovnaya verkhovaya Cz.
Anglický plnokrevník/syn. *English
Race-horse, English
Thoroughbred, TB, TBE*

Thrace: *see* Trakya

Thuringian: (Germany)/h/orig. early
20th c. from Percheron, Rhenish
and Belgian, × local/combined
with Saxony Coldblood to form
Saxony-Thuringian Coldblood/
Ger. *Thüringer Kaltblut*

Thuringian Warmblood:
(Germany)/Ger. *Thüringer
Warmblut*/rare

Tibetan Pony: (Tibetan plateau,
China)/ri.pa/often yellow-dun
with dark dorsal stripe/sim. to
Bhotia Pony but more slender/
local types: mountain, plateau
(Naqu dist.), valley, Ganzhi
(Sichuan) (obs. syn. *Maiwa*),
Yushu (Qinghai), Zhongdian
(Yunnan)/syn. *Nanfan* (=
Southern Native)/not *Southern
Barbarian, Thibetan*

Tieling Harness: (Liaoning, China)/l-
h; dr/usu. bay or black, occ.
chestnut/orig. since 1951 from
Russian breeds (Ardennes,
Anglo-Norman, Percheron *et al.*)
× local/recog. 1980/syn.
Tieling/nearly extinct

Timor Pony:
(Indonesia)/ri.dr/Indonesian
pony group/sim. to or = Flores

Tiree: (Scotland)/var. of Hebridean
Pony/extinct

tobiano: colour type: pied horse with black on belly and white back, i.e. white horse with coloured patches; dominant gene/syn. *pio alto*/[Don Rafael Tobias bred such horses in Brazil *c.* 1846]

Toby: (USA)/BS for foundation Appaloosas without Quarter Horse or Thoroughbred blood

Togara: *see* Tokara

Togo Pony: *see* Koto-Koli

Tokara Pony: (Kagoshima, Japan)/ (dr.)/usu. dark brown with darker mane and tail/small (island) var. of Japanese Native; imported from Kikai Island in Amami-Oshima group about 1900/BS/not *Takara*, *Togara*, *Tokarauma*/almost extinct

Tolfetana: (Tolfa mts, Latium, Italy)/l; ri.m/usu. black or dark bay, occ. chestnut or grey/HB 1990/rare

tølter: *see* Iceland Pony

Tomsk: (Tomsk and Tobolsk, Siberia, Russia)/orig. from local improved by trotter or W European draught breeds/Russ. *Tomskaya*/extinct

Toqiali Pony: (China)

Tori: (Estonia)/l; dr.ri/black, chestnut, bay, palomino or piebald/orig. at Tori stud from Hackney × Estonian 1896–1926 with Breton Post-horse blood to 1970 and some Hanoverian and Trakehner blood in 1970s/HB/ Eston. *Tori hobune*, Russ. *Toriiskaya*/not *Toric*/rare

Torodi: (W Niger)/py/? degenerate Barb

Toy horse: *see* miniature horse

TPR: *see* Italian Heavy Draught

Traber: *see* German Trotter

Trait Ardennais d'Auxois: *see* Auxois

Trait Auxois: *see* Auxois

Trait breton: *see* Breton Draught

Trait du Nord: (N France)/h; m.dr/bay, roan, dun, occ. chestnut/ sim. to Belgian Draught/HB 1919, BS/syn. *Ardennais du Nord* (= *Northern Ardennais*) (to 1910, and 1965–1914)/[= Northern Draught]/ rare

Trakehner: (Germany and Russia)/l; ri/chestnut, bay or grey/orig. from Trakehnen stud (E Prussia) (chiefly Thoroughbred and Arab) founded 1732; removed to W Germany and Russia 1945/used to grade up Sweyki 1787–1914; orig. of Mazury/HB (Germany), BS 1878; BS USA 1974, Canada 1974, GB; HB Denmark/Russ. *Trakenenskaya*/obs. syn. *Ostpreussisches Warmblut Trakehner Abstammung* (= *East Prussian warmblood of Trakhenen descent*)/not *Trekehner*

Trakya: (Thrace, Turkey)/ri.pa.dr/ sim. to Anatolian Native but larger

Transbaikal: *see* Buryat

Trans-Don: *see* Don

Transylvanian: (Romania)/l/former var. of Romanian/? orig. from Bessarabian × Hungarian/ subvar.: Ialomiţa/Rom. *Transilvăneană*, Ger. *Siebenburgisch*/extinct

Transylvanian Light Draught: (Romania)/l; dr/var. of Romanian Light Draught/orig. (1961 on) from Ardennes × Mezőhegyes Halfbred (31%) and Lipitsa (40%)/Rom. *Tip de cal semigreu pentru Transilvania*/syn. *Transylvanian Lowland*

Trekehner: *see* Trakehner

Tres sangres: *see* Spanish Anglo-Arab

Trocha Pura Colombiana: (Colombia)

Trotador español: *see* Spanish Trotter

Trote en Gallope: (Colombia)/l; ri/orig. *c.* 1930 from Andalusian × Colombian Criollo

TROTTER: inc. American, Baltic, Belgian, Cuban, Czech, Finnish, French, German, Hungarian, Italian, Ljutomer, Netherlands, North Swedish, Norwegian, Orlov, Romanian, Russian, Spanish

Trotto: *see* Italian Trotter

Tschoukourova: *see* Çukurova

Tsushima: *see* Taishu

Tuigpaard: *see* Dutch Carriage Horse

Tunisian Barb: *see* Barb
Turk, Turki: *see* Akhal-Teke, Iomud, Persian Turkoman
Turkish Arab: (Turkey)/l/orig. from Arab and Anatolian Native
Turkmen, Turkmenian, Turkoman: *see* Akhal-Teke, Iomud, Persian Turkoman
Tushin: (E Georgia)/l-py; ri/Russ. *Tushinskaya*/rare
Tuva: (Siberia, Russia)/ri.dr/Siberian pony group/orig. of Upper Yenisei/Russ. *Tuvinskaya*/rare, by crossing with Don and Budyonny
Tuva Coach: (Tuva, Russia)/l; dr/orig. from Tuva improved by Kuznetsk and Chumysh (1870 on) and also by heavy Orlov Trotter and Don/Russ. *Tuvinskaya upryazhnaya*/syn. *Tuva Harness Horse*/extinct

Ujumqin: *see* Wuzhumuxin
Ukrainian Saddle Horse: (Ukraine)/l; ri/bay, chestnut or brown/orig. since 1945 from Trakehner, Hanoverian and Thoroughbred, × Hungarian Halfbred/has partly absorbed Russian Saddle Horse/recog. 1990/Russ. *Ukrainskaya verkhovaya* (= *Ukrainian Riding*)/syn. *Ukrainian*
Unmol: (NW Punjab, Pakistan)/l; ri/usu. bay or grey/[= priceless]/nearly extinct
Upper Yenisei: (Tuva, Russia)/l; m.dr/orig. 1893–1917 from local Tuva ♂♂ and Mongolian ♀♀ improved by Russian Trotter and draught breeds/Russ. *Verkhne-eniseiskaya*/rare
Uruguayan Criollo: (Uruguay)/tobiano acceptable/ HB, BS
Uzunyayla: (EC Turkey)/l; ri.pa.dr/ orig. since 1854 from "Circassian" (? Kabarda); Anatolian Native and Nonius blood 1930/syn. *Çerkes atlar* (= *Circassian horse*)/not *Uzuniala*/rare

Vardarska: *see* Macedonian

Vardy: (Northumbria, England)/ dr/sim. to Cleveland Bay but heavier/orig. late 18th c. from Shire × Cleveland Bay and Fell Pony/syn. *Bakewell, Northumberland Chapman*/[bred by George Vardy]/extinct by 1910
Vasca: *see* Pottok
Vasca-Navarra: *see* Navarre Pony
Veglia: *see* Krk
Vendéen: (Vendée, W France)/l/orig. from Anglo-Norman and Thoroughbred × local/inc. in French Saddlebred/extinct
Venezuelan Criollo: (Venezuela)/l-py; ri/often dun/syn. *Llanero* (= *prairie horse*)
Ventasso: (Emilia, Italy)/ri.dr/bay, brown, occ. grey or black/orig. from Maremmana, Thoroughbred, Lipitsa *et al.*, × local; name created 1970s to preserve tradition of horse-breeding in Val Enza (in Apennines around Reggio)/HB 1990/It. *Cavallo del Ventasso*/nearly extinct
Vercors: (France)/? extinct (last described 1947) but attempts being made to revive
Verkhne-eniseiskaya: *see* Upper Yenisei
Verkhoyansk: *see* Middle Kolyma
Vestland Pony: *see* Fjord
Viatka: *see* Vyatka
Vietnamese Pony: (Vietnam)/South-East Asia pony group/syn. *Annamese, Annamite*
Villanos: *see* Andalusian
Vlaamperd: (S Africa)/l; dr.ri/usu. black or bay, also brown-dun, often with mealy muzzle/orig. from Friesian ♂♂ (shipped in 1902 from Antwerp, therefore described as Flemish) on Hantam (Cape Horse) ♀♀, with East Friesian or Oldenburg blood, also Cleveland Bay/BS 1983/[= Flemish horse]/nearly extinct
Vlaamse Paard: *see* Flemish
Vladimir Heavy Draught: (Vladimir and Ivanovo, Russia)/h; dr/bay, also brown or black/orig. (1886 on) from Clydesdale, earlier with Percheron, Danish, Suffolk and

Dutch blood, and later (1919–1924) some Shire blood/ recog. 1946/Russ. *Vladimirskaya tyazhelovoznaya, Vladimirskii tyazhelovoz*/syn. *Ivanovo Clydesdale, Russian Clydesdale, Vladimir Clydesdale, Vladimir Draught*

Vollblut: *see* blood horse

Voronezh Coach: (Russia)/h-l; dr/ orig. in 20th c. from Clydesdale × heavy trotter or from Bityug/ Russ. *Voronezhskaya upryazhnaya*/syn. *Voronezh Draught*/nearly extinct

Vyatka: (Udmurt, Russia)/ri.dr/North Russian Pony group/often roan with black back stripe, shoulder marks and zebra stripes on legs/ Russ. *Vyatskaya*/not *Viatka*/ nearly extinct

Waler: (NSW, Australia)/l; ri/mixed orig., named in India 1946 /absorbed by Australian Stock Horse/[from New South *Wales*]/ nearly extinct (BS trying to revive)

Warmbloed Paard Nederlands: *see* Dutch Warmblood

WARMBLOOD: = halfbred/inc. Austrian, Bavarian, Belgian, British, Czech, Danish, Dutch, East Friesian, Edles Warmblut, Hanoverian, Heavy Warmblood, Holsteiner, Mecklenburg, Oldenburg, Rottal, Saxony, Silesian, Slovakian, Slovenian, Swedish, Swiss, Trakehner, Westphalian, Württemberger, Zweibrücken/Ger. *Warmblut*

Warmblut: *see* Warmblood

Warmblut des Zuchtverbandes für deutsche Pferde: (Germany)/ warmblood breed society for German horses

Warmblutschecken: (Germany)/ piebald/nearly extinct

Waziri: (NW Pakistan and Afghanistan)/l/sim. to Baluchi but smaller/rare

Welara Pony: (S California, USA)/ orig. from Arab × Welsh Pony/ BS 1981/not *Welera*/rare

Welsh Cob: (Wales)/l; ri.dr/orig. from Welsh Mountain Pony with blood of English riding and coach breeds

Welsh Pony: (Wales)/ri/inc. Welsh Mountain Pony and its derivatives: Riding Pony (with Arab or Thoroughbred blood) and Cob-Pony (with Welsh Cob blood)/BS 1902, HB 1903; BS also USA 1906, S Africa 1957, Belgium, Canada 1979, Denmark, Netherlands, New Zealand (all 1960s), France (HB 1969), Australia; HB also Sweden

Wenshan: (Yunnan, China)/cf. Wumeng

West African Barb: (W of West Africa)/l; ri.dr/usu. grey/orig. from Barb/vars: Bélédougou, Djerma (× Dongola), Hodh, Sahel, Southern, Sulebawa/syn. *Arab-Barb*

West African Dongola: (E of West Africa)/l/dark with white stockings/orig. from Dongola (Sudan)/vars: Bahr-el-Ghazal, Bornu, Hausa/orig. (with Barb) of Bandiagara, Djerma, Mossi, Songhaq, Yagha/syn. *Fulani*

West African ponies: *see* Bhirum, Bobo, Kirdi, Koto-Koli, M'Bayar, M'Par, Torodi

West Black Sea Rahvan horse: *see* Rahvan

Western Isles Pony: *see* Hebridean Pony

Western Kazakh: *see* Kushum

Western pony: (USA)/inc. bronco, cow pony, Indian Pony, mustang/for BSs see American Indian, American Mustang, Southwest Spanish Mustang, Spanish Barb, Spanish Mustang

Western Sudan Pony: (S Darfur and SW Kordofan, Sudan)/l-py/usu. light bay, chestnut or grey, with white markings/syn. *Darfur Pony, Gharkawi* (= western), *Kordofani*

Westfälisches: *see* Westphalian

West Kazakh, West Kazakh Saddle-Draught: *see* Kushum

Westland: *see* Fjord

West Norway, West Norwegian: *see* Fjord
West Mytilene: *see* Mytilene
Westphalian Coldblood: (Germany)/strain of Rhenish/Ger. *Westfälisches Kaltblut*/rare
Westphalian Pony: *see* Dülmen Pony
Westphalian Warmblood: (Germany)/l; ri/orig. from local graded to Hanoverian/inc. in German Riding Horse 1975/BS 1904/Ger. *Westfälisches Warmblut*
White: *see* Albino
White-Russian: *see* Belarus
Wielkopolski: (WC Poland)/l; ri.dr/orig. 1964 by combining Mazury and Poznań/HB/syn. *Mazursko-Poznański*/[Great Polish]
Wilbur-Cruce: (SW USA)/usu. chestnut/single herd of Spanish Mustang, orig. from Mexico, feral in Arizona 1885–1990/syn. *Spanish Colonist, Spanish Colonial*
wild: *see* feral, Przewalski, tarpan
wild horse of Roraima: *see* Lavradeiro Criollo
WPN: *see* Dutch Warmblood
Wuchuminsin, Wu-chu-mu-ch'in, Wuchumutsin: *see* Wuzhumuxin
Wumeng: (China)/var. of Yunnan
Württemberger: (Germany)/l; ri.[dr]/usu. brown, bay, chestnut or black, with white markings/inc. in German Riding Horse 1975; remnants = Altwürttemberger (Old Wurttemberg)/BS 1895/Ger. *Württemberger Warmblut*
Wushen: (CW Inner Mongolia, China)/var. of Chinese Mongolian
Wuzhumuxin: (E Inner Mongolia, China)/l; ri.dr/improved var. of Mongolian/WG *Wu-chu-mu-ch'in*/syn. *Ujumqin*/not *Wuchuminsin, Wuchumutsin*

Xilingol: (C Inner Mongolia, China)/l; ri.dr/orig. since 1960 from Russian Thoroughbred, Akhal-Teke and Sanhe, × Chinese Mongolian, later Kabarda and Don blood/[Xilin R]
Xin-Lijiang: (Yunnan, China)/cf. Tengchong/? = Lijiang

Yabu: (Iran and Afghanistan)/mixed type/not *Yaboo*/[= pack-horse]
Yagha: (N Burkina Faso, W Africa)/local var. of Dongola-Barb orig./syn. *Liptako*
Yakut: (Siberia, Russia)/m.d.ri.pa/Siberian Pony group/bay, grey-brown or grey with back stripe, often shoulder bar, and zebra stripes on legs/vars: northern, Middle Kolyma or Verkhoyansk; small southern; and large southern (with blood of improver breeds), inc. Megezh; ? also Central, Taiga and Vilyuisk vars/Russ. *Yakutskaya*
Yamud: *see* Iomud, Persian Turkoman
Yanqi: (N Xinjiang, China)/py; ri.dr/orig. from Mongolian with Don and Orlov Trotter blood
Yellow: *see* Abyssinian
Yellow Dun: *see* Isabella
Yemeni: (Yemen)/l; ri/Arab orig./vars: Shami, Nagdi, Yemeni (*sensu stricto*); ? Giawf
Yenisei: *see* Upper Yenisei
Yerli: *see* Anatolian Native
Yili: (NW Xinjiang, China)/l; ri.dr.(m.d)/usu. chestnut, brown or black/Mongolian type/orig. since 1900 from Russian breeds × Mongolian/WG *Ili*/not *Yilee*
Yiwu: (Hami steppe, Xinjiang, China)/ri.pa.dr/orig. since 1955 from Yili × Kazakh/rare
Yomud, Yomut, Yomuth: *see* Iomud
Yonaguni: (Ryukyu Is, Japan)/smaller island var. of Japanese Native/BS/nearly extinct
Yongnin: (Tibet, China)/rare
Yorkshire Coach Horse: (England)/var. of Cleveland Bay from Thoroughbred cross in early 19th c./BS 1886–1937, then reabsorbed/syn. *New Cleveland Bay*/extinct
Yorkshire Hackney: *see* Hackney
Yorkshire Trotter: *see* Hackney
Ysabella: (Indiana, USA)/golden, white or chestnut with flaxen, silver or white mane and tail/colour type of American Saddlebred

Yugoslav Draught: *see* Croatian Coldblood

Yugoslav Mountain Pony: (Bosnia, Macedonia and Serbia)/pa.dr/ usu. bay, also black, grey, chestnut or dun/composite of native and Arab/inc. Bosnian Pony, Macedonian Pony/Serbo-cro. *Domaći brdski konj*/nearly extinct

Yugoslav Trotter: (Serbia)/ri.dr/bay, often with white markings/ composite of Anglo-Arab and American Trotter/Serbo-cro. *Jugoslovenski Kasac*/rare

Yunnan: (China)/py; ri/vars: Dali, Tenchong, Wumeng/not *Yunan*

Yushu: (Qinghai, China)/local type of Tibetan Pony

Zabaikal: *see* Buryat

Zamorano: *see* Carthusian

Zaniskari Pony: (Leh, Ladakh, Kashmir)/ri.pa/usu. grey, also black, white or copper coloured/ not *Zanskari, Zaskari*/rare

Zapadno-kazakhstanskaya verkhovo-upryazhnaya: *see* Kushum

Zapata: *see* Carthusian

Zeeland: (Netherlands)/Du. *Zeeuwse Paard*/extinct (by crossing with Belgian Draught to form Dutch Draught)

Žemaitukai: (Lithuania)/py-l; ri.dr/ grey, black or bay/Konik type, sim. to Estonian Native/ improved since 1963 by North Swedish/Russ. *Zhemaichu, Zhmud, Zhmudka, Zhmudskaya,* Pol. *Żmudzki,* Ger. *Smudisch*/ syn. *Lithuanian Landrace, Samogitian*/not *Zemaituka, Zemaitukas*/nearly extinct

Žemaitukai (Modern Type): (Lithuania)/dr.m./bay/nearly extinct

Zhmud, Zhmudka: *see* Žemaitukai

Zhongdian: (NW Yunnan, China)/ strain of Tibetan/rare

Żmudka, Żmudzin, Żmudzki: *see* Žemaitukai

Zobnatica Halfbred: (Serbia)/l/orig. from Trakehner, Mezőhegyes Halfbred and Hanoverian at Zobnatica stud since 1946/Serbo-cro. *Engleski polukrvnji zobnatičkog tipa* (= *English Halfblood Zobnatica type*)

Zweibrücken: (Palatinate, Germany)/ l/orig. from Arab, Thoroughbred, Anglo-Norman and Hanoverian/ inc. in German Riding Horse 1975/HB 1906/Ger. *Zweibrücker*/ syn. *Pfälzisches Warmblut* (= *Palatinate Warmblood*)

Pig

Abbreviations used in this section:
ld = lard; m = meat

Colour is assumed to be white, unless otherwise indicated

Names for pig include: babi, buta, cerdo, cochino, gris, heo, porco, porcul, prase, Schwein, swinia, svinja, vark, wet, zhu

A-ba: var. of Tibetan

Abruzzese: (Abruzzi, Italy)/black; semi-lop ears/extinct

Acadie P22: (France)/m/orig. from Landrace and Duroc *c.* 1982 at Pen ar Lan, Maxent, Brittany/not *Arcadie*/nearly extinct

African Guinea: *see* Guinea Hog

Agrupación balear: *see* Majorcan

Agrupación levantina: *see* Levant type

Ahyb: *see* Hungahyb

Akha: (hills of N Thailand)/ ld.m/black; very small ears; straight tail/small indigenous type

Aksaĭ Black Pied: (Alma Ata, SE Kazakhstan)/m.ld/breed group/ orig. (1952 on) at Aksaĭ exp. farm from Large White × local with some Berkshire blood/Russ. *Aksaiskaya cherno pestraya*/rare

Alabuzin: (Kalinin, Russia)/ld/white, sometimes spotted/breed group/ orig. (late 19th and early 20th c.) from Large White (also Middle White and Large Black) × local lop-eared/Russ. *Alabuzinskaya porodnaya gruppa*/extinct by 1984

Álava: *see* Vitoria

Albanian: (Albania)/usu. white, also pied or coloured/var.: Shkodra

Alb de Banat: *see* Banat White

Alb de Ruşeţu: *see* Ruşeţu White

Alentejana: (Alentejo, Portugal)/ m/red, golden or black/Iberian type; sim. to Extremadura Red/ HB 1985, BS/syn. *Portuguese, Portuguese Retinta, Red Portuguese, Transtagana*/not *Alemtejana*

Alföldi: *see* Ancient Alföldi

Amélioré de l'Est: (Alsace, France)/ disappeared 1960s by crossing with French Landrace and by spread of Large White/[= improved eastern]/extinct/*see also* German (Improved) Landrace

American Berkshire: (USA)/m.[ld]/ var. of Berkshire from imports 1820–1850/BS 1875

American Essex: (USA)/m/black/ orig. from Black Essex first imported 1820/BS 1887, HB 1890/extinct 1967 but revived as exp. breed at Texas A&M Univ., College Station, renamed Greer-

Radeleff Miniature/syn. *Guinea-Essex*
American Hampshire: *see* Hampshire
American Landrace: (USA)/m/orig. from Danish Landrace imported 1934, and 38 Norwegian Landrace ♂♂ imported 1954/BS 1950
American Yorkshire: (USA)/orig. from Large White imported 1890–1900/BS 1893; HB also Romania 1969
Ancient Alföldi: (C Hungary)/ld/sim. to Mangalitsa and absorbed by it in early 19th c./extinct
Andalusian Black: *see* Black Iberian
Andalusian Blond: (W Andalucía, Spain)/whitish to golden/Iberian type/Sp. *Andaluza rubia (campiñesa)*/extinct
Andalusian Red: *see* Extremadura Red
Andalusian Spotted: *see* Jabugo Spotted
Andaluza campiñesa, Andaluza rubia: *see* Andalusian Blond
Andaman Islands: (India)/dwarf/feral/rare
Angeln Saddleback: (Schleswig, Germany)/black head and rump/orig. from local pied Land pig with Wessex Saddleback blood/orig. of German Red Pied; with Swabian-Hall = German Saddleback/HB 1928/Ger. *Angler Sattelschwein*/syn. *Angeln*/nearly extinct
Ankamali: (Kerala, Karnataka, Tamil Nadu and Maharashtra, India)/black with white patches or rusty grey
Ankang: (S Shaanxi, China)/var. of Hanjiang Black
Annamese, Annamite: *see* Vietnamese
Aodong: *see* Yuedong Black
Apulian: (Apulia and Lucania, Italy)/white, often with black markings, esp. on face and rump/former vars: Gargano, Lucanian, Murgese/It. *Pugliese*/syn. *Apulo-Lucanian* (It. *Appulo-luccana*), *Mascherina* (= masked)/extinct *c.* 1980
Arapawa Island: (New Zealand)/brown-and-black/feral

Arcadie P22: *see* Acadie P22
Ashanti Dwarf: (Ghana)/black, black pied, brown or white, occ. grey/var. of West African/syn. *Bush pig*
ASIAN: inc. Chinese, Indian, Indonesian, Philippine, Thai and Vietnamese breeds/orig. from Asian wild/syn. *Asiatic*
Asian wild: = Asian subspp. of *Sus scrofa*/orig. of Asian domestic breeds/syn. *Asiatic wild*
Askanian: (Ukraine)/meat type selected from Ukrainian Meat
Asturian: (N Spain)/m/black/local var. sim. to Vitoria/orig. from Celtic × Iberian/extinct (by crossing with Large White)
Austrian Forest pig: *see* Güssing Forest pig
Austrian Edelschwein: (Austria)/orig. from Edelschwein/syn. *Austrian Yorkshire*
Austrian Landrace: (Austria)/orig. from German Landrace with Swedish Landrace blood/Ger. *Veredeltes Landschwein* (= Improved Landrace)
Austrian Yorkshire: *see* Austrian Edelschwein
Axford: *see* Oxford Sandy-and-Black

Baasen: *see* Bazna
babi-Batak: *see* Batak
babi-Hutan: *see* Hutan
babi-Persilancan: *see* Persilancan
baboy: *see* Libtong
Baé: *see* Tatu
Baden-Württemberg hybrid: (Germany) Large White × German Landrace F_1
Badaung wet: (Myanmar)/[*wet* = pig]
Bagun: (N Croatia)/ld/whitish-yellow; short ears; curly coat/cf. Bakony/disappeared by crossing with Middle White/Ger. *Baguner*/extinct before 1982
Bahia, Baia, Baié: *see* Tatu
Bakony: (Transdanubia, W Hungary)/prim. 18th c. var./absorbed into Mangalitsa in early 19th c./extinct
Bakosi: (Cameroon)/var. of West African
Baldinger Spotted: (Donaueschingen,

Württemberg, Germany)/white
with black spots/Ger. *Baldinger
Tigerschwein*/extinct

Balearic: *see* Majorcan

Balinese: (Bali, Indonesia)/ld/black,
or black with white on belly and
legs; erect ears/orig. from South
China × local (which remains in
mts in E)

Băltăreţ: (Danube marshes,
Romania)/marsh var. of
Romanian Native/syn. *Marsh
Stocli*/[*bălta* = marsh]/nearly
extinct

Bamaxiang: (Bama and Tiandong
counties, Guangxi, China)/m/
black head and rump; dwarf/syn.
Bamm mini, Guangxi Bama/rare

Bamei: (Shaanxi, Gansu, Ningxia and
Qinghai, China)/ld.m/black;
prolific/North China type/vars:
Da Bamei, Er Bamei, Huzhu,
Xiaohuo/syn. *Jingchuan* (in
Gansu), *Xi*/nearly extinct

Bamm mini: *see* Bamaxiang

Banat: *see* Bazna

Banat White: (W and S Banat, W
Romania)/m/orig. in early 20th c.
from Middle White and
Edelschwein × Mangalitsa with
possibly some blood of Small
White, Berkshire, Large White
and German (Improved)
Landrace/Rom. *Porcul Alb de
Banat*/not *Porcul de Banat*/
nearly extinct (9 crossbred ♀♀ in
1999)

Banha: (Zona da Mata, Minas Gerais,
Brazil)/local prim. type

Banheco: (Guyana)

Bantu: (S Africa)/usu. brown, also
black, or white with black spots/
orig. from early European and
Asian imports

Basco-Béarnais: *see* Basque Black
Pied

Basilicata: *see* Lucanian

Basna, Basner: *see* Bazna

Basque Black Pied: (SW France)/
m.ld/black pied, esp. black head
and rump/Iberian type, sim. to
Limousin/vars: Basque, Béarn,
Bigourdain/BSd 1921, BS/Fr. *Pie

noir du pays basque*, Sp. *Vasco,
Basque Euzkal Txerria*/syn.
Basco-Béarnais, Combrune/rare
in France; extinct in N Spain but
to be reintroduced

Basque-Navarre: *see* Vitoria

Bastianella: (Faenza, Emilia, Italy)/
strain of inbred Large White
(imported 1875) used for
crossing with Romagnola; sim. to
San Lazzaro/extinct

Baston: *see* Lincolnshire Curly Coat

Batak: (Sumatra, Indonesia)/syn.
babi-Batak (*babi* = hog)

Ba Tri: (S Vietnam)/local var.

Bavarian Landrace: (Germany)/fore-
end white, hind-end red/Ger.
Halbrotes (= half red)
bayerisches Landschwein/extinct

Ba Xuyen: (Mekong delta, S
Vietnam)/black-and-white
spotted, white socks/orig. from
Berkshire × Boxu (1932–1956)

Bayeux: (Normandy, France)/m/black
spots; semi-lop ears/orig. from
Berkshire × Normand in 19th c.;
established 1928 as composite of
old Porc de l'Ouest (= West
French White) and Berkshire/HB
1928, BS/syn. *Bayeusain*/almost
disappeared 1960s (by crossing
with Piétrain); still nearly extinct

Bazna: (C Transylvania, Romania)/m/
black with white belt/orig. since
1872 from Berkshire ×
Mangalitsa/HB/Rom. *Porcul de
Banat*, Ger. *Basner*/syn.
Romanian Saddleback/not
*Baasen, Basna, Porcul Alb de
Banat*/rare

Baztán: (Navarre, Spain)/white, or
white with a few blue spots/local
Celtic cross/extinct

bearded pig: (Malaysia, Sumatra,
Borneo)/= *Sus barbatus*

Béarn: (SW France)/var. of Basque
Black Pied/Fr. *Béarnais*/extinct

Bedford: (E of USA)/orig. from Duke
of Bedford's Woburn herd/syn.
*Bedfordshire, Cumberland,
Woburn* (New York and
Massachusetts)/extinct
1855–1870

Beijiang: (China)/local var./rare

Beijing Black: (China)/m/black, occ. white markings/orig. 1962 from Berkshire and Large White × local (Dingxian, Shenxian, Zhouxian)/recog. 1982/syn. _Peking Black_

Beijing DeKalb: (China)/synthetic hybrid

Beijing Spotted: (China)/m/black, occ. white markings/orig. 1962 from Berkshire, Russian Large White and Yorkshire, × local/ recog. 1978/syn. _Beijing Coloured_

Beqra, Beiroa: _see_ Bisaro

Belarus Black Pied: (Minsk, Belarus)/ m.ld/orig. from Large White, Large Black, Berkshire and Middle White × local in late 19th c. and in 1920s/recog. as breed group 1957, as breed 1976/ Russ. _Belorusskaya chernopestraya_; obs. syn. _Byelorussian Black Pied, White-Russian Black Pied_ (or _Spotted_)

Belarus commercial hybrid: _see_ Belarus Meat

Belarus Large White: (Belarus)/ formerly var. of Russian (or Soviet) Large White/Russ. _Belorusskaya krupnaya belaya_ (= _BKB_)

Belarus Meat: (Belarus)/m/dev. since 1972 from Russian Large White, Estonian Bacon, Landrace and Swedish Large White/obs. syn. _Byelorussian Meat(y), White-Russian Meat type_

Belaya sakhalinskaya: _see_ Sakhalin White

Bela žláhtna: _see_ Slovenian White

Belgian Landrace: (N Belgium)/m/ orig. from local improved first by English breeds, then graded to German (Improved) Landrace (1930–1945) with Dutch Landrace blood 1945 on/var.: Belgian Negative (BN)/HB 1930; HB also Luxembourg 1955, Germany 1971, Czech Republic 1974, Italy/Fr. _Landrace belge_, Flem. _Belgisch Landvarken_ /syn. _Belgian Improved Landrace_ (Flem. _Veredeld Landvarken_),

Belgijska zwisloucha (Pol.) (= _Belgian Lop-eared_), _Improved Belgian_ (1956–1973) (Fr. _Indigène amélioré_), _Landrasse B_ (Germany), _Poppel_ (France)

Belgian Negative: (Belgium)/var. of Belgian Landrace selected for resistance to stress/syn. _BN_, _Landrace Belge Halothane Négatif_

Belgian Yorkshire: (Belgium)/var. of Large White from imports esp. since 1928/HB 1928/Fr. _Grand Yorkshire belge_, Flem. _Belgisch Groot Yorkshire_/syn. _Belgian Large White_/rare

Belgijska zwisloucha: _see_ Belgian Landrace

Beli Soj Mangulica: _see_ White Mangalitsa

Belted: _see_ Saddleback

Beltsville No. 1: (Maryland, USA)/m/ black with white spots; moderate lop ears/inbred orig. 1934–1951 from Danish Landrace ($\frac{3}{4}$), Poland China ($\frac{1}{4}$)/HB 1951/extinct

Beltsville No. 2: (Maryland, USA)/red with white underline, occ. black spots; short erect ears/inbred orig. 1940–1952, 58% Danish Large White, 32% Duroc, 5% Danish Landrace, 5% Hampshire/HB 1952/extinct

Benin local: (Benin)/local West African

Bentheim Black Pied: (S Oldenburg, Germany)/spotted; lop ears/ Piétrain blood since 1960/HB 1987/Ger. _Bunte Bentheimer_, _Schwarz-Weisses Bentheimer_/ syn. _Bentheimer Landschwein_, _Buntes Deutsches Schwein_ (= _Pied German pig_), _Emsländer Bunte_, _Wettringer Bunte, Wettringer Tigerschweine_/nearly extinct

Berkjala: (Philippines)/ld/black; short ears/orig. (1915 on) from $\frac{5}{8}$ Berkshire, $\frac{3}{8}$ Jalajala/extinct 1941–1946

Berkshire: (England)/m/black with white points/orig. from Chinese (Cantonese) (and Siamese) × local Old English before 1830 and Neapolitan (Casertana) in 1830/vars: American Berkshire,

Canadian Berkshire, Kagoshima Berkshire/orig. of Bayeux, Dermantsi Spotted, German Berkshire, Kentucky Red Berkshire, Murcian *et al.*/HB 1884/also in China and Russia/ rare in GB

Berlin Miniature: (Germany)

Bésaro: *see* Bisaro

BHZP: (W Germany)/commercial hybrid: cross of 4 lines orig. from German Landrace (MM), Large White (MF), Hampshire (FM) and Piétrain (FF) imported 1969 and kept closed/[= Bundes Hybrid Zucht Programm]

Biała ostroucha: *see* White Prick-eared

Biele ušl'achtilé: *see* Slovakian Improved White

Big China: (SW Ohio, USA)/part orig. of Poland China/syn. *Warren County hog*/extinct

Bigio tramacchiato: *see* Siena Grey

Bigourdain: (Bigorre, Hautes-Pyrénées, SW France)/var. of Basque Black Pied by crossing with Celtic type/syn. *Bigourdan*/ nearly extinct

Bihu: (China)/local var./rare

Bikovačka, Bikovo: *see* Subotica White

Bílé miniaturní prase: *see* Czech Miniature

Bílé ušlechtilé: *see* Czech Improved White

Bilsdale: *see* Yorkshire Blue and White

Birish: *see* Pelón

Bisaro: (N Portugal)/m/black, white or pied/Celtic type (with Large White blood)/HB 1994/Port. *Bésaro, Bísaro*/syn. *Beqra, Beiroa*/ not *Bizaro, Bizarra*/nearly extinct

Bispectacled White: *see* Rongchang

BKB: *see* Belarus Large White

Black Andalusian: *see* Black Iberian

Black-at-both-ends: *see* Jinhua

Black Canary: *see* Canary Black

Black Cuban criollo: (Cuba)/orig. as other criollos but possibly influenced by pre-Hispanic Canary Is pigs; more recently crossed with Duroc and Hampshire/Sp. *Negro Criollo Cubano*

Black Edelschwein: *see* German Berkshire

Black Emilian: *see* Parmense

Black Essex: (England)/var. of Small Black/orig. of American Essex/ syn. *Old Essex*/extinct

Black Hairless: (Extremadura and W Andalucía, Spain)/ld/slate black with spots around muzzle/var. of Black Iberian without bristles and skin folds/2 strains: Guadyerbas, Coronado/Sp. *Negra lampiña* (= black beardless)/syn. *Guadiana* (Sp. *Pelón guadianés* = *bald Guadiana*) in Badajoz/ nearly extinct

Black hairless: (Latin America) *see* Pelón, Pueblo

Black Hairy: (Extremadura, Andalucía, Castilla y León and Castilla La Mancha, Spain)/m/ black when mature, born red and black/var. of Black Iberian/Sp. *Negra entrepelada*/nearly extinct

Black Hamprace: *see* Montana No. 1

Black Iberian: (Extremadura and W Andalucía, Spain)/ld.m/var. of Iberian sim. to Extremadura Red/ subvars: Black Hairless, Black Hairy/Sp. *Negra ibérica*/syn. *Andalusian Black, Extremadura Black*

Black Madonie: *see* Madonie

Black Mallorcan: *see* Majorcan Black

Black Mangalitsa: (Syrmia, Serbia)/ ld/black-brown with yellow or silver belly/var. of Mangalitsa/ syn. *Lasasta* or *Lasica* (Serbo-cro., = weasel), *swallow-bellied* (Ger. *Schwalbenbauch Mangalitza, Schwalbenbäuchig Wollschwein*), *Syrmian* (Serbo-cro. *Sremica*)/extinct before 1982

Black Parma: *see* Parmense

Black Sicilian: (Nebrodi and Madonie mts, Sicily)/black, sometimes with white markings; often wattles/orig. from Iberian type, ? influenced by Casertana/incl. Nebrodi and Madonie populations/It. *Nera Siciliana*/syn. *Nebrodi and Madonie* (It. *Nebrodi e Madonie*), Sicilian (It. *Siciliana*)/rare

Black Slavonian: (E Slavonia,
Croatia)/m.(ld)/black; semi-lop
ears/orig. (by Count Pfeifer) from
Berkshire and Poland China ×
Black Mangalitsa, 1860 on/Cr.
Crna slavonska, Ger. *Schwarzes
slavonisches*/syn. *Faiferica*,
Pfeifer/not *Black Slavonic*/nearly
extinct

Black Suffolk: (England)/var. of
Small Black/cf. Black Essex/
extinct

Black Umbrian: *see* Maremmana

Black-White-Red Spotted: *see* Old
Swedish Spotted

Blanc de l'Ouest: *see* West French
White

Bleu de Boulogne: (Boulogne sur
Gesse, France)/blue; semi-lop
ears/var. of Gascony/extinct

Blond Mangalitsa: (N Serbia,
Hungary and Romania)/
yellowish-white, grey or reddish-
yellow/main var. of Mangalitsa/
orig. from curly-haired Šumadija
pigs/rare

Blond Wollschwein: *see* Mangalitsa

BN: *see* Belgian Negative

Bohemian Blue Spotted: *see* Přeštice

Bo Cake: (Myanmar)/black or grey;
lop ears/orig. from Large Black
imported by Rev. Cake 1940

**Bologna, Bologna Castagnana,
Bolognese:** *see* Romagnola

Bordeaux: *see* Bourdeaux

Borghigiana: (Borgo, Emilia, Italy)/
syn. *Fidenza* (It. *Fidentina*)/
extinct

Boulonnais: (Pas de Calais, France)/
Celtic type/Flemish orig. with
Craonnais and Large White
blood/inc. in West French White
1955/extinct

Bourdeaux: (Drôme, France)/black/
Celtic type/not *Bordeaux*/extinct

Boxu: (Vietnam)/orig. 1920 from
Craonnais × Sino-Vietnamese/
orig. (with European breeds) of
Ba Xuyen and Thuoc Nhieu/
extinct

Brahma: (USA)

Breitov: (Yaroslavl, Russia)/m.ld/lop
ears/orig. 1908–1934 from
Danish Landrace (1908 on),

Middle White (1913 on) and
Large White (1924–1926) × local
lop-eared/recog. 1948/HB/Russ.
Breĭtovskaya

Bresse: (Ain, France)/black pied/
Iberian type/absorbed by Large
White/Fr. *Bressane, Bresanne*/
extinct

Breton: (Brittany, France)/Celtic
type/crossed with Craonnais and
Large White and absorbed by
West French White 1955/extinct

Brinati: *see* Fumati

British Landrace: (Great Britain)/
orig. from Swedish Landrace
imported 1953–1955/BS 1953

British Lop: (SW England)/m/sim. to
old Welsh (joint HB 1926–1928),
also sim. to Large Black but
white/Landrace blood since
1953/BS and HB 1921/syn.
*Cornish White, Cornish White
Lop-eared, Devon Lop, Devon
Lop-eared, Devonshire, Long
White Lop-eared (1921–1928),
Lop White, National Long White
Lop-eared (1928–1960s), White
Large Black, White Lop*/nearly
extinct

British Saddleback: (England)/
formed 1967 by union of Essex
and Wessex Saddleback/BS
1996, HB; BS USA/rare

Bronze: (Germany)/orig. (1925 on)
from European wild boar ×
Bavarian Landrace and German
Pasture, improved by Berkshire/
extinct

BSI: (N Vietnam)/orig. from Berkshire
($\frac{1}{2}$) × Í ($\frac{1}{2}$), i.e. [Berkshire ×
(Berkshire × Í)] × [Í × (Berkshire
× Í)]/syn. *BSi*

BU: *see* Czech Improved White

Bug: *see* Nadbużańska

Bulgarian Landrace:
(Bulgaria)/m/orig. from Swedish,
Dutch and British Landraces

Bulgarian Native: *see* East Balkan

Bulgarian White: (Bulgaria)/m/orig.
from Bulgarian Native graded up
by Large White (first imported
early 20th c.) and by
Edelschwein/Bulg. *B"lgarska
byala*, Ger. *Bulgarisches*

weisses/obs. syn. *Bulgarian Improved White*

Bunte Bentheimer: *see* Bentheim Black Pied

Bunte Deutsches Schwein: *see* Bentheim Black Pied

Burkina Faso: (Burkina Faso)/local, being improved by Korhogo and Danish Landrace

Burmese: (Myanmar)/black, occ. spotted; medium lop ears/*see also* Mountain Dwarf

Bush pig: *see* Ashanti Dwarf

Byelorussian: *see* Belarus

Byfield: (Massachusetts, USA)/orig. *c.* 1800 from Bedford, Old English and Chinese/part orig. (in Ohio) of Poland China and supplanted by it/extinct after 1850

Cabano: *see* Canastrão

Cadiz Golden: *see* Dorado Gaditano

Caiwu: (China)/var. of Huang-Huai-Hai Black

Calabrian: (Calabria, Italy)/m/black; semi-lop ears/orig. from old Apulian/former vars: Catanzarese, Cosentina, Lagonegrese, Reggitana/It. *Calabrese*/rare

Calascibetta: (Sicily, Italy)/black, white or brindled/variable inland var. of Sicilian local with blood of Casertana (Neapolitan) and also Large White, Large Black *et al.*/cf. Nebrodi and Madonie

Camborough: (England)/prolific minimal disease Large White × British Landrace cross at Pig Imp. Co., Fyfield Wick, Berks/ [developed with advice from *Cam*bridge and Edin*burgh*]

Camborough 12: (England)/outdoor ♀/orig. from PIC HY or Large White × Camborough

Camborough Blue: (England)/outdoor ♀ line from Saddleback with Camborough blood

Campiñesa: *see* Andalusian Blond

Canabrid: (Canada)/terminal line

Canadian Berkshire: (Canada)/m/var. of Berkshire first imported 1830s/HB 1876/nearly extinct

Canadian Landrace: (Canada)/BS 1955

Canadian Yorkshire: (Canada)/var. of Large White/HB

Canary Black: (La Palma, Canary Is, Spain)/m/long ears/orig. from NW African pigs, slight influence from Iberian, Large Black and Berkshire/Sp. *Negro canario*/syn. *Black Canary, Cochino negro*/nearly extinct

Canastra: (Maranhão and Piauí, Brazil)/ld/usu. black, occ. spotted or red/Iberian type/? orig. from Alentejana, possibly with Berkshire blood/var.: Furão/syn. *Maxambomba, Meia perna* (= half leg), *Vermelho* (= red)/nearly extinct

Canastrão: (Minas Gerais and Rio de Janeiro, Brazil)/black, occ. with white spots; lop ears/Celtic type/? orig. from Bisaro (Beiroa), possibly with Large Black blood/vars: Capitão chico, Junqueiro/local syn.: *Cabano* (Paraiba and Goias), *Zabumba* (Bahia and Sergipe)/nearly extinct

Canastrinho: (Brazil)/usu. black; small; often pot-bellied/term used for group of breeds with Asiatic ancestors introduced by Portuguese colonists/syn. *Tatu* (Port. *Tatú*)

Canton, Cantonese: (Malaysia) *see* South China Black

Cantonese: (Zhujiang delta, Guangdong, China)/ld.m/black-and-white/Central China type/ vars: Dahuawu, Fa Yuen, Jinli, Kwanchow Wan, Liangcun, Meihua, Nipi, Sibao, Wai Chow/Ch. *Dahuabai* (= large pied white), WG *Ta-hua-pai*, Ger. *Kanton*/syn. *Chinese* (GB), *Gongdong Big Spotted, Guangdong Large White Spot, Large Black White, Macao* (Portugal and Brazil), *Pearl River Delta*

Capitão chico: (Brazil)/var. of Canastrão/syn. *Maranhão*/ [*capitão* = captain]

Cappuccia d'Anghiari, Cappuccio: *see* Chianina

Captain Cooker: (New Zealand)/
feral/introduced by Captain Cook
1769
Carélie: (Brittany, France)/selected
line of Finnish Landrace at Pen
ar Lan Breeding Co., Maxent/syn.
PECC/not *Karelia*/rare
Carioca: *see* Piau
Cariovilli: (Italy)/extinct
Caruncho: (São Paulo and Minas
Gerais, Brazil)/ld/white or sandy
with black spots (spots smaller
than on Piau), occ. red-and-white
or black/? orig. from Piau ×
Tatu/var.: Sorocaba/syn.
Carunchinho, Piau pequeño
(*pequeño* = small)/nearly extinct
Casco de mula: (Colombia)/ld/solid-
hoofed/Iberian type/syn.
Mulefoot/[hoof of she-mule]/rare
Casentinese, Casentino: *see* Chianina
Casertana: (Caserta, Campania,
Italy)/m/black or grey; usu.
tassels/Thai or Indo-Chinese
blood/syn. *Napolitana* (=
Neapolitan), *Pelatella* (=
plucked, i.e. hairless),
Teano/nearly extinct
Castagnona: *see* Romagnola
Catalan: *see* Vich
Catalina: (Santa Catalina I,
California, USA)/feral
Catanzarese: (Catanzaro, Italy)/
former var. of Calabrian/extinct
Cavallino: (NW Lucania, Italy)/m/
usu. black with abundant white-
tipped bristles/syn. *Cavallina
lucana* (= Lucania pony)/[=
pony, from legginess and speed]/
nearly extinct
Cazères: (Haute Garonne, SW
France)/white with a few blue
spots/sim. to Miélan/orig. (1860
on) from Large White or
Lauraguais, × Gascony, F$_1$ or
later/Fr. *de Cazères, Cazèrien*/
extinct 1970
CC21: (Cuba)/hybrid/[= *Cerdo
Cubano* (Cuban pig)]
Celta, Céltica: *see* Galician
CELTIC: (NW Europe)/m/usu. white,
also spotted or saddleback; long
lop ears/inc. lop-eared native
pigs (Landraces) of N Europe;

used esp. to distinguish Baztán,
Bisaro, Galician, Lermeño, Vich,
Vitoria and West French White
from Iberian types in
Mediterranean areas
Central China Double-Black: *see*
Huazhong Two-end Black
CENTRAL CHINA type: (C China)/black
or pied; small lop or semi-lop
ears/see Cantonese, Daweize,
Dongchen, Fuzhou,
Ganzhongnan Spotted, Giauli,
Gunxi, Hang, Huazhong Two-
End-Black, Jinhua, Leping,
Longyou Black, Minbei Spotted,
Monchuan, Nanyang Black,
Ningxiang, Putian, Qingping,
Shengxian Spotted, Wanzhe
Spotted, Wuyi Black, Xiangxi
Black, Yujiang
Central Russian KBV-1 and KB-KN:
see Russian Large White
Central Russian type: (Central
European Russia)/var. of Soviet
Meat/Russ. *Tsentral'nyi typ*
Cerdo coscate: *see* Mexican Wattled
Cerdo Cubano: *see* CC21
Cerdo negro canario: *see* Canary
Black
České bílé ušlechtilé plemeno: *see*
Czech Improved White
České másové: *see* Czech Meat
České výrazně masné: *see* Czech
Meat
Chalu Black: (China)/local var./rare
Changwei White: (China)/orig. from
Harbin White × local/recog.
1989
Chaparral feral: (S Texas, USA)/feral
in Chaparral Wildlife
Management Area
Charbin, Charbiner: *see* Harbin
Charolais: (France)/black pied/
extinct
Chato de Murcia, Chato murciano:
see Murcian
Chato de Vitoria, Chato Vitoriano:
see Vitoria
Chausy: (Belarus)/m/breed group/
Russ. *Chausskaya porodnaya
gruppa*/extinct
Chayuan: (China)/var. of
Ganzhongnan Spotted
Cheed: (Laos)/syn. *mou Cheed*

Chenghi New Strain: (Sichuan, China)/synthetic strain from Chenghua and Landrace

Chenghua: (lowland Sichuan, China)/ld.m/black/SW China type/Ger. *Tschen-Hua*/not *Chenhwa*

Cheng-hwai: *see* Huai

Chengxian Spotted: (China)/? = Chenghua (*hua* = spotted, pied, *xian* = county)

Chernaya moldavskaya: *see* Moldavian Black

Cheshire: (New York, USA)/m/orig. in 1850s from English pigs imported from counties of Cheshire and Yorkshire/BS 1884, last HB 1914/syn. *Jefferson County hog*/extinct 1930s

Chester White: (Chester county, SE Pennsylvania, USA)/m.[ld]/orig. in early 19th c. from English imports in late 18th and early 19th c., including Bedford/orig. of OIC (Ohio Improved Chester)/BS 1884; HB also Canada, GB 1981/syn. *Chester County White* in 1848

Chianina: (upper Chiana and Tiber valleys, Tuscany-Umbria, Italy)/ sim. to Siena Belted but grey with white head and feet/syn. *Cappuccia d'Anghiari, Cappuccio* (= hooded), *Casentino* (It. *Casentinese*)/ extinct by 1976

Chien-li: *see* Jianli

Chilin Black: *see* Jilin Black

Chin: *see* Mountain Dwarf

CHINESE: (China)/usu. black or pied; wrinkled face (and often body); straight tail/syn. *Chinese Mask*/ see Central China, Hong Kong, Lower Changjiang River Basin, North China, Plateau, South China, South-West China and Taiwan types

Chinese: (Great Britain, Portugal and Brazil)/in the past, "Chinese" = Cantonese

Chinese: (Malaysia) *see* South China, South China Black

Chinese lean-type: (China)/new ♀ line (D VI) by reciprocal crossing among Taihu, Landrace and commercial Segher's hybrids; D VI sow litters mated with Large White, Duroc or Piétrain

Chinese Mask: *see* Chinese (group)

Chinese miniature pigs: (China)/ derived from Xiang

Chinese new breeds: *see* Beijing Black, Beijing Spotted, Changwei White, Chinese lean-type Dingxian, Fannong Spotted, Fengjiang Fuzhou Black, Gansu Black, Gansu White, Ganzhou White, Guangdon Pied, Guangxi White, Hangu Black, Hanzhong White, Harbin White, Heilongjiang Spotted, Hubei White, Inner Mongolian Black, Inner Mongolian White, Jilin Spotted, Laoshan, Lijiang Liaoning Black, Lutai White, Lu White, Nanjing Black, New Huai, Ningxia Black, North-East China Spotted, Sanjiang White, Shanghai White, Shanxi Black, Shanxi Lean Meat, Wanbei, Wenzhou White, Wulanhada, Xiang White, Xinjiang Black, Xinjiang White, Xinjin, Yili White, Yimeng Black, Zhejiang *et al.*

Ching Ching: *see* Xinjin

Ching-Huai: *see* Jinhua

Chin-hua: *see* Jinhua

Chinoise: (France) *see* Meishan

Choctaw: (Alabama, transferred to Oklahoma, USA)/black; tassels; solid-hoofed/rare

Chunan Spotted: (China)/var. of Wanzhe Spotted

Chung-li: *see* Taoyuan

Chwanche: (hills of Nepal)/m/black; prim./cf. Ghori/syn. *Nepalese Dwarf*

Cinta Casaldianni: (Italy)/syn. *Casaldianni Belted*

Cinta Italiana: *see* Siena Belted

Cinta Senese: *see* Siena Belted

Cinto: *see* Siena Belted

Clawn: (Japan)/white; miniature/orig. from (Landrace × Large White) × (Göttingen Miniature × Ohmini) c. 1975 by Nihon Haigo Shiryo Ltd and now at Kagoshima Univ.; selected for white colour and

small size/ [Central Laboratory of
White Nipai]
Co: (high plateau of C Vietnam)/
black; dwarf/local type
Cochino Negro: see Canary Black
Colbrook, Colebrook: see Kolbroek
Coleshill: see Middle White
Colorada: see Extremadura Red
Coloured: (China) = pied/Ch. *Hua*
Combrune: see Basque Black Pied
Comune: (Albania)/m/brown/
[common]/rare
Cornish: see Large Black
Cornish White (Lop-eared): see
British Lop
Cornwall: see German Cornwall,
Large Black
Coronado: (Spain)/var. of Black
Hairless
Corrèze: (C France)/local population
of Craonnais × Limousin orig.,
later crossed with Large
White/Fr. *Corrèzien*/extinct
Corsican: (Corsica, France)/usu.
black with sparse white spots,
also grey, roan, fawn or white;
small and prim./BS/Fr. *Corse*/
purebreds rare
Cosentina: (Cosenza, Italy)/former
var. of Calabrian/syn. *Orielese*/
extinct
Côte d'Ivoire local: (Côte d'Ivoire)/
usu. black, or black pied; short
ears/Mediterranean type/part
orig. (with Large White and
Berkshire) of Korhogo
Cotswold: (England)/m/orig. 1961–
1969 by Cotswold Pig Dev. Co.
from Large White, Landrace,
Welsh and Wessex Saddleback;
used as ♀ line for crossing with
Large White strain to produce
Cotswold hybrid ♀♀/HB/also
Cotswold 16 (selected Cotswold),
Cotswold 29 (F₁ cross), Cotswold
90 (from Cotswold 16 and
Duroc), Cotswold Gold (with
Duroc), Cotswold Platinum
(white hybrid ♀)
Craonnais: (Craon, Mayenne, France)/
m/Celtic type/inc. in West French
White 1955/HB 1926/not *Crâon*/
extinct

Créole: see Guadeloupe Créole,
Haitian Créole
Criollo: (Spanish America)/Celtic
type/Sp. orig./see also Cuino,
Honduras Switch-tail, Mexican
Wattled, Pélon, Sampedreño,
Venezuelan Black, Zungo/[=
native]
Crna slavonska: see Black Slavonian
Cuban Creole: see Black Cuban
criollo
Cuino: (highlands of C Mexico)/ld/
black, spotted or yellow; dwarf/?
Chinese orig. in 16th c./syn.
Mexican Dwarf/nearly extinct
Cuino de Pachuca: (Mexico)/
Berkshire × Cuino
Cul Noir: see Limousin
Cumberland: (England)/m/lop
ears/recog. 1811; BS 1917/
extinct (last boar licensed
1955/56, last sow 1960)/see also
Bedford
Cuprem: (Nebraska, USA)/m/hybrid
developed by A. Mulford,
Kenesaw
Curtis Victoria: (Indiana, USA)/orig.
c. 1850 by F.D. Curtis from
various improved breeds/extinct
c. 1900/see also David Victoria
ČVM: see Czech Meat
Czech Improved White: (Czech
Republic)/orig. from Large White
(late 19th c.), Edelschwein and
German (Improved) Landrace
(1925 on) and local/inc.
Moravian Large Yorkshire/orig.
of Slovakian Improved White,
Slovakian White Meat/HB
1927/Cz. *České bílé ušlechtilé
prase* (= BU), Ger. *Tschechisches
weisses Edelschwein*/syn. *Czech
Large White*/not *White
Thoroughbred*
Czech J-Hyb: (Czech Republic)/?
crossed with Polish Large White
Czech Landrace: (Czech Republic)/
orig. from American, British,
Danish, French, German and
Swedish Landraces/HB 1961
Czech Meat: (Czech Republic)/orig.
from Belgian Landrace (½), Duroc
(¼) and Hampshire (¼)/recog.
1991/Cz. *České másové, České*

výrazně masné (*ČVM*) (= Czech Extremely Meaty)/syn. *Czech Lean*

Czech Miniature: (Czech Republic)/ orig. at Vet. Res. Inst., Brno, from Göttingen, Minnesota and Landrace/Cz. *Bílé miniaturní prase* (= White miniature pig)

Czechoslovakian Black Pied: *see* Slovakian Black Pied

Czechoslovakian Black Spotted: *see* Slovakian Black Pied

Czech synthetic lines: (Czech Republic)/SL 96 = Belgian Landrace and Hampshire; SL 98 = Belgian Landrace and Duroc

Da Bamei: (China)/var. of Bamei/[*da* = large]

Dahe: (Yunnan, China)/black or brown/var. of Wujin/WG *Ta-ho*

Daheping: (China)/var. of Xiangxi Black/syn. *Yuanling*

Dahuabai: *see* Cantonese

Dahuawu: (China)/var. of Cantonese/ [*da* = large, *hua* = pied]

Dai Bach Í: *see* DBI

Daltrain: (France)/♀ line bred by Dalland Meuse/var. of non-stress-sensitive Piétrain/syn. *X30*

Damin: (China)/var. of Min/[*da* = large]

Dan-Hybrid: (Denmark)/Danish Landrace × Danish Large White cross

Dan-Line: (Denmark)/Duroc–Hampshire cross

Dänisch Protestschwein: *see* Husum Red Pied

Danish Black Pied: (Denmark)/m/ often tassels/? orig. from unimproved Landrace with blood of coloured English breeds (e.g. Berkshire)/HB 1920, BS/ Dan. *Sortbroget*/syn. *Danish Black Spotted, Danish White and Black, Old Danish*/nearly extinct

Danish Landrace: (Denmark)/m/ Large White blood introduced at end of 19th c. (up to 1895) but selected against/ orig. of Swedish Landrace and other Landraces (esp. Dutch, French, Norwegian)/ recog. 1896/ HB/Dan. *Dansk Landrace*

Danish Landrace 1970: *see* DL-1970

Danish White and Black: *see* Danish Black Pied

Danois: *see* French Landrace

Danube White: (Bulgaria)/m/orig. from Bulgarian White, Large White, Landrace, Piétrain and Hampshire/syn. *Dunai White*

Dapaong: (Togo)/usu. white, also brown, black pied, grey or red/orig. from Large White × local/Fr. *Porc de Dapaong*

Dauphiné: (SE France)/black pied/Fr. *Dauphinois*/not *Dauphin*/extinct

Davis Victoria: (Indiana, USA)/ld/ orig. 1870–1886 by George F. Davis from Poland China, Berkshire, Chester White and Suffolk/BS 1886/syn. *Victoria*/ [from sow named Queen Victoria]/extinct 1930s/*see also* Curtis Victoria

Daweizi: (Hunan, China)/ld.m/black, usu. with white feet/Central China type, sim. to Ningxiang/ large and medium-sized vars/WG *Ta-wei-tzu*, Ger. *Da-We-Ze*, *Daweze*/not *Dawetze*

Daxiang: (China)/var. of Hang

DBI: (N Vietnam)/orig. from Large White ($\frac{1}{2}$) × Í ($\frac{1}{2}$), i.e. [LW × (LW × Í)] × [Í × (LW × Í)]/syn. *DBi*/[= Dai Bach Í]

Debau: (Guangxi, China)/local var./ rare

Défi: (France)/crossbred from Piétrain and Large White

Dehong Small-Ear: (China)/var. of Diannan Small-Ear

Dekalb: *see* Beijing DeKalb

Delta: *see* Vietnamese

Dengchang: *see* Tunchang

Dermantsi Pied: (Lukovit, N Bulgaria)/ld/black with white points, or white with black spots/orig. (1895 on) from Berkshire (and Mangalitsa since 1940) × native/Bulg. *P"stra dermanska*/syn. *Dermantsi Black Spotted* (Bulg. *Dermanska chernosharena*, Ger. *Schwarzbuntes Dermanzi*)

Deshi: (Uttar Pradesh, Bihar, Madya Pradesh and Punjab, India)/rusty

grey to brown or black/syn. *Desi/*
[= local or village]/*see also*
Ankamali, Ghori, Khanapara
Dwarf, Khasi, Nepalese Dwarf,
Sri Lanka Native
Deutsches Landrasse B: *see* German
Landrace B
Deutsches Sattelschwein: *see* German
Saddleback
Deutsches veredeltes Landschwein:
see German Landrace (improved)
Deutsches Weideschwein: *see*
German Pasture
**Deutsches weisses
Schlappohrschwein:** *see* German
Landrace
Deutsches weisses Stehohrschwein:
see Edelschwein
Devon: *see* Large Black, British Lop
Diani: (Batangas, Philippines)/black,
sometimes spotted/orig. from
Berkshire and Poland China ×
Philippine Native/extinct
Diannan Small-ear: (Yunnan, China)/
black/South China type/vars:
Dehong Small-Ear, Subi, Mengla,
Wenshan/syn. *South(ern)
Yunnan Small-eared*
Ding: *see* Dingxian
Dingxian: (Hebei, China)/black;
horizontal or lop ears/var. of
Huang-Huai-Hai Black/? orig.
from Poland China × local since
1929–30/recog. 1985/WG *Ting,
Tinghsien,* Russ. *Dinsyan/*syn.
Ding county/[*xian* = county]/rare
Dingyuan: (Anhui, China)/var. of
Huang-Huai-Hai Black
Din-Hsiang-Chi: *see* Tingshuanhsi
Dinsyan: *see* Dingxian
Diqing: (China)/var. of Tibetan
Djen-li: *see* Jianli
Djing-Hua: *see* Jinhua
Djumajliska: *see* Džumalia
DL-1970: (Denmark)/purebred line of
Danish Landrace before foreign
blood accepted *c.* 1970;
maintained as purebred/Dan.
*Dansk Landrace anno
1970/*almost extinct
Dlinnoukhaya belaya: *see* Russian
Long-eared White
DM-1: *see* Don
Dnepropetrovsk Hybrid: (Ukraine)/

orig. from Russian Large White
($\frac{5}{8}$) × Berkshire ($\frac{1}{8}$) × Landrace
($\frac{1}{4}$)/Russ. *Dnepropetrovskaya*
Dnieper: (Cherkassy, Ukraine)/ld/
black spotted/breed group; sim.
to Krolevets, Mirgorod and
Podolian/orig. from Mirgorod,
Berkshire (1911) and Large
White (1937–38), × local short-
eared/Russ. *Dneprovskaya* (or
*Pridneprovskaya) porodnaya
gruppa/*extinct by 1984
Dobrinka: (Lipetsk, Russia)/breed
group/orig. 1932 on from Large
White × local/Russ. *Dobrinskaya
porodnaya gruppa/*extinct
Dobrogea Black: (Romania)/m/black,
5% pied/orig. 1949–1967 from
Large Black × Russian Large
White/Rom. *Porcul negru de
Dobrogea/*extinct
Domaća mesnata: *see* Subotica White
Dome: *see* Ghori
Don: (Kamenski, Rostov, Russia)/
breed group/orig. from
Cornwall, Large Black and long-
eared white × local Large
White/Russ. *Pridonskaya/*extinct
by 1984
Dongbeihua: *see* North-East China
Spotted
Dongchen: (Guangxi, China)/black
head and rump/Central China
type, sim. to Monchuan
Dongchuan: (N Jiangsu, China)/
black/Lower Changjiang River
Basin type/syn. *Zhoujiazi*
Dongshan: (China)/var. of Huazhong
Two-end Black
Dongxiang: *see* Neijiang
Dongyang: (Zhejiang, China)/var. of
Jinhua/WG *Tung-yang,* Russ.
Dunsyan
Don Meat: (N Caucasus, Russia)/
m/orig. from Piétrain × North
Caucasus/? var. of Soviet Meat/
recog. 1978/Russ. *Donskoĭ tip/*
syn. *DM-1, Don meat type, Don
hybrid, Don type*
Doom: *see* Ghori
Dorado Gaditano: (Extremadura,
Andalucía, Castilla y León and
Castilla La Mancha, Spain)/m/
copper-red/Iberian/? var. of

Alentejana or Extremadura Red/
syn. *Cadiz Golden*/nearly extinct

Dorset Gold Tip: (Dorset, England)/
m/red with black markings and
gold tip to bristles; semi-lop
ears/? orig. from Tamworth ×
Berkshire, with Gloucestershire
Old Spots blood/extinct (last
boar licensed 1955/56)

DRB: (Bourgogne, France)/grey and
white/♀ composite line bred by
SCAPAAG of Dijon, with Duroc
dominant/rare

DRC: (Bourgogne, France)/m/
sometimes with red spots/♂
composite line bred by
SCAPAAG of Dijon, from Duroc
(from Denmark, Germany,
Switzerland) (dominant) and
French Landrace/rare

Dsien-li: *see* Jianli

Dunai White: *see* Danube White

Dunsyan: *see* Dongyang

Duroc: (USA)/m.[ld]/red/orig. 1822–
1877 from old Duroc of New York
+ Jersey Red of New Jersey,
combined in 1872/BS 1883; BS
also Canada; HB also Czech
Republic 1973, Germany 1974, GB
1977, Italy, Korea, Poland 1979,
Romania 1969/syn. (1877–1934)
Duroc-Jersey/[from racehorse
named after General Duroc]

Duroc SELPA: (France)/♂ line/var. of
Duroc bred by SELPA of Isle-et-
Bardais, St-Bonnet-Tronçais

Dutch Landrace: (E Netherlands)/
m/orig. from German (Improved)
Landrace (⅜) (imported since
1902) and Danish Landrace (⅝)
(imported 1929–1933) × local/
HB 1933 (dates from 1913–1918)/
Du. *Nederlands (veredeld)
Landvarken* (= *Netherlands
(Improved) landpig*), Fr. *Indigène
néerlandaise*, Ger.
Niederländisches Landschwein/
syn. *Netherlands Landrace, PRN*

Dutch Large White: *see* Dutch
Yorkshire

Dutch Yorkshire: (W Netherlands)/
m/orig. from Edelschwein and
Large White imported since *c.*
1870/HB since 1913–1918/Du.
*Nederlands Groot-Yorkshire
Varken*, Ger. *Niederländisches
Edelschwein*/syn. *Dutch Large
White, Edelvarken, Netherlands
Large White*

D VI: *see* Chinese lean-type

dwarf: *see* Andaman Island, Ashanti
Dwarf, Bamaxiang, Chwanche,
Co, Cuino, Khanapara Dwarf,
Hainan, Hezuo, Kangaroo Island,
Mountain Dwarf, Ras-n-lansa,
Tibetan, Vietnamese Potbelly,
Xiang

Džumalia: (Macedonia)/m.ld/white
with black spots; lop ears/sim. to
Resava/Serbo-cro. *Džumaliska,
Djumajliska*/rare

Eastern Bulgarian Mountain:
(Bulgaria)/? = East Balkan

East Balkan: (E Bulgaria)/ m.ld/usu.
black or dark grey, occ. pied;
prick ears; prim./only surviving
var. of Bulgarian Native/orig.
from Asian × European wild
pig/Bulg. *Istochnobalkanska*/
syn. *Kamchiya* (Bulg.
Kamchiska)/not *Eastbolckan,
Eastern Balkan*/rare

Eastern Terai: *see* Hurra

Ebei Black: (N Hubei, China)/ld/
nearly extinct

Edelschwein: (Germany)/m/orig.
from German Land pig graded up
by Large White and Middle
White imported 1880 on/recog.
1904/orig. of Latvian White,
Russian Short-eared White,
White Prick-eared (Poland)/Ger.
Deutsches (weisses) Edelschwein
(= *German (white thoroughbred
pig)*/syn. *German Large White,
German Short-eared, German
White Prick-eared* (= *Deutsches
weisses Stehohrschwein*),
*German Yorkshire, White
Edelschwein*/not *German
Pedigree, German Purebred*/*see
also* Austrian Edelschwein,
Swiss Edelschwein

Edelschwein: (Czech Republic) *see*
Czech Improved White

Edelvarken: *see* Dutch Yorkshire

Eesti Peekon: *see* ? Estonian Bacon

Eesti Suur Valge: *see* ? Estonian Large White
EKB1: *see* Estonian Large White
Emsländer Bunte: *see* Bentheim Black Pied
Enshi Black: (SW Hubei, China)
Entrepelado: *see* Black Hairy
Er Bamei: (China)/var. of Bamei/[= middle Bamei]
Erhualian: (Lower Changjiang basin, China)/heavy-skinned var. of Taihu
Ermin: *see* Zhongmin
Essex: (England)/m/white belt, feet and tail/orig. from Old English/BS 1918–1967/ combined with Wessex in 1967 to form British Saddleback/syn. *Essex Half-black, Sheeted Essex, White-shouldered Essex*/extinct/ *see also* American Essex, Black Essex
Estonian Bacon: (Estonia)/m/orig. from local crossed with Danish, Finnish and German (Improved) Landrace and Large White/recog. 1961/Russ. *Ėstonskaya bekonnaya*/syn. *Estonian Lop-eared* (Russ. *Ėstonskaya visloukhaya*)
Estonian Large White: (Estonia)/m/Russ. *Ėstonskaya krupnaya belaya*/syn. *EKB1*
Ėstonskaya visloukhaya: *see* Estonian Bacon
Estrela, Estrelense: *see* Moura
Estremadura, Estremeña: *see* Extremadura
Euroline: (England)/Meatline of Premier Piglink based on Large White
European wild: = European subspp. of *Sus scrofa*/orig. of European domestic breeds
Exi Black: (China)/var. of Huchuan Mountain/[= W Jiangxi Black]
Extremadura Black: *see* Black Iberian
Extremadura Red: (Extremadura and Andalucía, Spain)/m/chief var. of Iberian, sim. to Alentejana and Black Iberian/orig. var.: Oliventina/Sp. *Extremeña retinta* or *colorada*/syn. *Andalusian Red, Red Iberian*

Euzkal Txerria: *see* Basque Black Pied

Faentina: (Faenza, Italy)/reddish/ former var. of Romagnola/extinct
Faiferica, Fajferica: *see* Black Slavonian
Fannong Spotted: (Henan, China)/m /black-and-white pied/orig. from Russian Large White and Berkshire × local/ recog. 1983/ Ch. *Fannong hua*/ syn. *Fannong Coloured*/ [Huangfannong farm in Sihua county]
Fastback: (England)/m/cross developed 1963–1973 by British Oil and Cake Mills from (British) Saddleback, Landrace and Large White
Fa Yuen: (Hong Kong)/black spots on back/var. of Cantonese/WG *Hua-hsien*/not *Fayeun*/extinct
Fehér hússertés: *see* Hungarian White
Fengjiang: (China)/lean meat line from Fengjing × Jiangquhai
Fengjing: (S Shanghai, China)/var. of Taihu/long- and short-snouted types/Ger. *Fong Djing*/syn. *Fungcheng, Rice-bran pig*/not *Fengjiang*
feral: *see* Andaman Islands, Arapawa Island, Catalina, Captain Cooker, Chaparral, Florida swamp, Kangaroo Island, Ossabaw Island, Pineywoods Rooter, Razor Back
FH lines: (C France)/lines bred by France Hybrides, St Jean-en-Braye/012 = var. of French Landrace (♀ line); 016 = var. of Piétrain (♂ line); 019 = composite bloodline (♂ line), usu. white; 025 = var. of French Large White (♀ line)/rare
Fidentina, Fidenza: *see* Borghigiana
Finnish Landrace: (Finland)/may have blue spots; lop-eared/orig. from short-eared East Finnish and lop-eared West Finnish, with some Swedish Landrace blood in 1940s/BS 1908, HB 1914/Finn. *Maatiaissika, Suomalainen maatiaisrotu*, Swe. *Finsk lantras*

Flamand, Flandrin: *see* Flemish

Flemish: (French Flanders)/Celtic type/HB 1937–1955/inc. in West French White 1955/Fr. *Flamand, Flandrin*/extinct

Florida swamp hog: (USA)/feral/syn. *Gulf pig*

Fong Djing: *see* Fengjing

Fonte Bôa Pied: (Portugal)/Port. *Fonte Bôa malhada*

Förädlad Lantras: *see* Swedish Landrace

Forest Mountain: (Armenia)/m.ld/ grey, white or black; long-haired in winter; semi-lop ears/breed group/orig. in 1950s from Mangalitsa × (Large White × local)/Russ. *Lesogornaya porodnaya gruppa*/ obs. syn. *New Lesogor* (Russ. *Novaya lesogornaya*/nearly extinct

Forest pig: *see* Güssing Forest pig

Forlivese: (Forli, Italy)/slate-coloured with white-tipped black bristles/ former var. of Romagnola/extinct

Formosa: *see* Taiwan

Freixianda: (Alvaiázere dist., Coimbra, Portugal)/red with black spots/local var./not *Freixiandra*

French Danish: *see* French Landrace

French Landrace: (NW France)/m/ orig. from Danish Landrace (first imported 1930) with Swedish and Dutch Landrace/BS 1952/Fr. *Landrace Français*/obs. syn. *Porc français de type danois* (= *French Danish*)

French Large White: (France)/var. of Large White first imported at end of 19th c./BS 1923, HB 1926/syn. *Large White Yorkshire*

Friuli Black: (Venetia, Italy)/m/ nearly extinct 1951 by crossing with Edelschwein (1908–1940), Large White *et al*./It. *Friulana nera, Nera del Friuli*/syn. *San Daniele* (It. *Sandanielese*) /extinct *c.* 1980

Froxfield Pygmy: (Hampshire, England)/coloured, often spotted, being selected for white; miniature/orig. from Yucatan Miniature × Vietnamese Potbelly

Fuan Spotted: (Fujian, China)/local var./rare

Fugong: (Yunnan, China)

Fujian Small Pig: (Fujian, China)/ black or black-and-white/inc. Huai, Minbei Spotted, Putian Black/WG *Fukien*

Fumati: (Emilia, Italy)/light copper-coloured or greyish/F$_1$ of Bastianella or San Lazzaro (inbred Large White) × Romagnola/syn. *Brinati, Fumati di San Lorenzo di Faenza*/[= smoky]/extinct

Fumian: (China)/var. of Liang Guang Small Spotted

Fungcheng: *see* Fengjing

Furão: (Brazil)/degenerate var. of Canastra/syn. *Vara*

Fuzhou Black: (Fujian, China)/m.ld/ orig. from Berkshire and Middle White × local black

Galician: (NW Spain)/m/small reddish or black spots on head and rump/Celtic type/Sp. *Gallega*/syn. *Celta, Céltica, Santiaguesa* (from Santiago)/ nearly extinct

Gallega: *see* Galician

Gallia: (Brittany, France)/selected ♀ line/var. of French Large White at Pen ar Lan Breeding Co., Maxent/syn. *PEGG*/rare

Gandong Black: (China)/var. of Wuyi Black

Gansu Black: (China)/lop ears/orig. from Berkshire, Neijiang and local/recog. 1985

Gansu White: (China)/lop ears/orig. from Russian Large White and Landrace, × local/recog. 1988

Ganxi Two-end Black: (China)/var. of Huazhong Two-end Black

Ganzhongnan Spotted: (C Jiangxi, China)/black-and-white/Central China type/vars: Chayuan, Guanzhao, Ruyia, Zuoan/Ch. *Ganzhongnan hua*/syn. *Mid South Jianxi Spotted, South Central Jiangxi Spotted*/not *Gangzhongnan*

Ganzhou White: (Jiangxi, China)/orig. from Large White ×

local Ganzhou before
1949/recog. 1982

Gargano: (Italy)/white with black
markings on hindquarters/var. of
Apulian/extinct

Garlasco: (Lomellina, Pavia, Italy)/
grey with white head/extinct

Gascony: (Haute Garonne and Gers,
SW France)/m.ld/black/Iberian
type/former vars: Bleu de
Boulogne, Tournayais/orig. (with
Large White) of Miélan and
Cazères/HB, BS/Fr. *Porc Noir
Gascon*/rare (by crossing with
Large White)

Gempshir: *see* Hampshire

Georgian: *see* Imeretian, Kakhetian,
Kartolinian

Georgian Mangalitsa: (Georgia)/ld/
black skin, white to dark tan
hair/Russ. *Mangalitskaya*/rare

German Berkshire: (Germany)/
black/orig. from German Land
pig, graded to Berkshire/syn.
Black Edelschwein/? extinct

German Cornwall: (Germany)/
black/var. of Large Black

German Edelschwein: *see*
Edelschwein

German Grazing: *see* German Pasture

German Land pig: (Germany)/orig. of
German Landrace/Ger. *Deutsches
Landschwein*/extinct

German Landrace: (Germany)/m/
orig. from Large White or
Edelschwein × German Land pig;
Dutch Landrace blood (esp. in
Schleswig-Holstein) since 1953/
former local vars: Hoya, Meissen/
orig. of Belgian Landrace, Dutch
Landrace (with Danish), Russian
Long-eared White, Swiss
Improved Landrace/HB 1904/Ger.
Deutsche Landrasse, Serb-cro.
Njemački Landras/syn. (to 1969)
German Improved Landrace (Ger.
*Deutsches veredeltes
Landschwein*; obs. syn. *Améliorée
de l'Est* (France), *German Long-
eared*, *German White Lop-eared*
(Ger. *Deutsches weisses
Schlappohrschwein*), VDL

German Landrace B: (Germany)/
imported from Belgium/Ger.
Deutsche Landrasse B/rare

German Large White: *see* Edelschwein

German Long-eared: *see* German
Landrace

German Pasture: (Germany)/ld/black
head and rump; prick ears; rough
hair/sim. to Güstin Pasture but not
smooth-haired; with Limousin
and Hampshire blood/ BS 1899/
Ger. *Deutsches Weideschwein*/
syn. *Hanover-Brunswick* (Ger.
Hannover-Braunschweig),
Hildesheim/not *German Grazing*,
Hanover-Bismarck/extinct

German Pedigree, German Purebred:
see Edelschwein

German Red Pied: *see* Husum Red
Pied

German Saddleback: (East Germany)/
black with white saddle/orig.
1948 from Angeln Saddleback
and Swabian-Hall/HB 1970, BS
1991/Ger. *Deutsches
Sattelschwein*/rare

German Short-eared: *see*
Edelschwein

German (white) thoroughbred: *see*
Edelschwein

German Yorkshire: *see* Edelschwein

Ghori: (NE India, also Bhutan and
Bangladesh)/syn. *Dome, Doom,
Pygmy* (Assam and Bangladesh)/
[hill tribe]

Gianli: (Hubei, China)/black head
and rump/Central China type/
sim. to Monchuan

Glamorgan: *see* Welsh

Gloucestershire Old Spots:
(England)/m/white with one or
more black spots; lop ears/HB
1914, BS 1990/syn. *Gloucester,
Gloucester* (or *Gloucestershire*)
Old Spot (or *Old Spotted*)/not
Gloster/rare

Gołąb, Golamb, Gołębska: *see* Puławy

Goland: (Italy)/?

Gongdong Big Spotted: *see* Cantonese

Gongguan: (N Guangdong, China)/
ld.m/var. of Liang Guang Small
Spotted/syn. *North Kwangtung*
(= Guangdong) *Lop-eared*,
Yongyun

Gongxi White: *see* Guangxi White

Göttingen Miniature: (Göttingen
Univ., Germany)/orig. (1960 on)

from Minnesota Miniature ×
Vietnamese/coloured vars black,
brown, white or pied; white var.
has dominant white from
German Landrace/Ger. *Göttinger
Miniaturschwein* or *Minischwein*/
syn. *Göttingen mini-pig, Miniporc
Göttingen* (Fr.)

GP lines: (France)/bred by PIC France
of Ploufragan/♀ lines: 1010
(French Landrace), 1020 (French
Large White), 1075 (Duroc), 1210
(Sino-European composite); ♂
lines: 1104 (French Large White),
1111 (French Large White), 1116
(Duroc), 1120 (Piétrain), 1140
(Piétrain)

Grand Yorkshire: *see* Large White

Grand Yorkshire belge: *see* Belgian
Yorkshire

Greer-Radeleff: *see* American Essex

Grigia senese: *see* Siena Grey

Gruzinskaya: *see* Georgian

Guadiana: *see* Black Hairless

Guadeloupe Créole: (Guadeloupe
and Martinique)/long snout/
colour vars inc.: "silky black",
"domino" (black and red), grey
(roan), "chabin" (dirty white)/
BS/*see also* Haitian Créole

Guadyerbas: (? Guadiana valley,
Spain)/experimental strain of
Iberian (Black Hairless) pigs
maintained as closed population
since 1945/? syn. *Guadiana*
(*Pélon guadianés* in Badajoz),
Puebla/nearly extinct

Guanchao: *see* Guanzhao

Guangdong Large White Spot: *see*
Cantonese

Guangdong Pied: (China)/orig. from
Landrace × local/recog. 1990/
syn. *Guangdong Coloured*

Guangdong Small-ear Spotted:
(China)/var. of Liang Guang
Small Spotted/subvars: Guixu,
Huantang, Tangzhui, Zhongdong/
syn. *Guangdong Small-ear*

Guangfeng Black: (Jiangxi, China)/
var. of Yujiang

Guangxi Bama: *see* Bamaxiang

Guangxi White: (China)/orig. from
Landrace and Yorkshire ×
Luchuan/recog. 1985/not *Gongxi*

Guanling: (CS Guizhou, China)/black/
SW China type/not *Guanglin*

Guanzhao: (China)/var. of
Ganzhongnan Spotted/not
Guanchao

Guanzhong Black: (China)/obese
type

Guanzhuang Spotted: (China)/pied/
local var./rare

Guinea-Essex: see American Essex

Guinea Hog: (Alabama, USA)/usu.
black, formerly also red or sandy;
often hairy/? orig. from West
Africa/orig. of Minnesota/BS
1898, HB/syn. *African Guinea,
Guinea Forest hog, Gulf pig*/rare

Guixu: (China)/var. of Guangdong
Small-ear Spotted

Guizhong Spotted: (C Guangxi,
China)/black head and
rump/local var.

Guizhou Xiang: *see* Xiang

Gulf pig: (USA) *see* Florida swamp
hog, Guinea hog, Pineywoods
Rooter

Gunxi: (Jiangxi, China)/black head
and rump/Central China type,
sim. to Monchuan

Gurktal: (Slovenia)/ld/sim. to
Turopolje

Güssing Forest pig: (S Burgenland,
Austria)/m, hunting/various
colours; semi-feral/orig. *c.*
1974–1989 by K. Draskovich from
wild boar × Duroc, Hampshire,
Black Mangalitsa, black Yugoslav
pasture pig and Minnesota; F_1
♀♀ backcrossed to wild boar/Ger.
Güssinger Waldschwein/syn.
Austrian Forest pig

Güstin Pasture: (Rügen, Pomerania,
Germany)/local var. sim. to
German Pasture but smooth-
haired/Ger. *Güstiner
Weideschwein*/not *Güsting*

HaBac: (HaBac province, C N
Vietnam)/small local var. with
pot belly

Habai: *see* Harbin White

Haerbinbai: *see* Harbin White

Hailum: (S Thailand)/ld/black with
white marks/? orig. from Hainan/
syn. *Hainan*/not *Hailam*/rare

Hainan: (Hainan I, China)/ld/black head and back, white snout, blaze, flanks, belly and legs; prick ears, long head; smooth skin; small to dwarf/South China type/vars: Lingao, Tunchang, Wenchang

Hainan: (Thailand) *see* Hailum

Hainan: (S Vietnam) *see* Heo Hon

Haitian créole: (Haiti and Dominican Republic)/local/extinct (exterminated 1978–1982 to prevent spread of African swine fever); island repopulated with Guadeloupe Créole, Gascony and Taihu

Halbrotes bayerisches Landschwein: *see* Bavarian Landrace

Half-red Bavarian Landrace: *see* Bavarian Landrace

Hall, Hällische: *see* Swabian-Hall

Hamline: (Yorkshire, England)/m/ white or coloured ♂ line orig. by National Pig Dev. Co. from Piétrain, Duroc and Hampshire, × Welsh, Landrace and Large White

Hampen: (SW England)/outdoor ♀ line orig. by Peninsular Breeding Co. from Hampshire × Landrace

Hampline: (England)/Meatline ♂ of Premier Piglink based on Duroc

Hamprace: *see* Montana No. 1

Hampshire: (Kentucky, USA)/m.[ld]/ black with white belt/orig. in 19th c. ultimately from Old English (? from Hampshire, England)/BS 1893; BS also Canada, HB also GB 1967, Romania 1969, Czech Republic 1972, Germany 1974, Italy, Switzerland *et al.*/Russ. *Gempshire*, Yug. *Hempir*/obs. syn. *Belted, Mackay, Norfolk Thin Rind, Ring Middle, Ring Necked, Saddleback, Woburn*

Hanford Miniature: (Washington State, USA)/orig. since 1958 at Pacific North West Laboratory from Palouse and Pineywoods Rooter

Hang: (Jiangxi, China)/black head, back and rump/Central China type/vars: Daxiang, Lianhua Spotted, Wuning Spotted/syn. *Hangkou, Shanghang*/rare

Hangu Black: (China)/orig. from Berkshire, Neijiang and local/recog. 1988

Hanjiang Black: (S Shaanxi, China)/ North China type/vars: Ankang, Heihe, Shuiwei, Tiehe, Tielu

Hannover-Braunschweig: *see* German Pasture

Hanover-Bismarck: *see* German Pasture

Hanover-Brunswick: *see* German Pasture

Hanzhong White: (SW Shaanxi, China)/orig. from Russian Large White, Yorkshire and Berkshire × local/recog. 1982

Harbin White: (Heilongjiang, China)/ m.ld/white; erect ears; usu. kinky tail/orig. from Large White × local (Min), using imports from Russia (1896) and Canada (1926)/recog. 1975/Ch. *Haerbinbai*, Russ. *Kharbinskaya belaya*, Ger. *Weisse Charbiner*/ syn. *Habai*/not *Kharbin*

Hebao: *see* Xiaomin

Heihe: (Heilongjiang, China)/var. of Hanjiang Black

heihua: (China)/= black spotted/*see* North-East China Spotted

Heilongjiang Spotted: (China)/orig. 1962–1979 (named) from Kemerovo × Min

Hempir: (Serbia)/= Hampshire/rare

Hengjing: (China)/var. of Taihu

Heo Hon: (Mekong delta, S Vietnam)/ white with black back patches; small ears, long snout/orig. from Hainan (intro. from Hainan I in 18th c.)/part orig. of Sino-Vietnamese/[islands pig]/? extinct

Heo Moi: (S Vietnam)/black; small ears, pointed snout; small; prim./ orig. indigenous mountain pigs, later to Mekong delta, crossed with Chinese pigs end 19th c. to form Sino-Vietnamese type (itself then crossed with Craonnais to create Boxu)/? extinct

Hereford: (Missouri, USA)/[ld].m/red with white head, legs, belly and tail (colour pattern of Hereford cattle)/orig. 1902–1920 in Iowa and Nebraska from Chester White, Duroc, Poland China and Hampshire/BS 1934/syn. *White-faced*/rare

Hetao Lop-ear: (Inner Mongolia, China)/var. of Huang-Huai-Hai Black/Ch. *Hetao Daer* (= *Hetao Big-ear*)/rare

Hexi: (N Gansu, China)/ld/small/ nearly extinct

Hezuo: (SW Gansu, China)/black; small to dwarf/var. of Tibetan

Hildesheim: *see* German Pasture

Hohenheim: (Germany)/three F_2 families bred from wild boar, Meishan and Piétrain

Holstein hybrid: (Germany)/based on Belgian Landrace ♂ ♂ × (Edelschwein × German Landrace) ♀ ♀

Honduras Switch-tail: (C mts of Honduras)/long tail with hairy switch

Hong Kong: (China)/former vars of Cantonese in Hong Kong: Fa Yuen, Kwanchow Wan, Wai Chow

Hongqiao: (Zhejiang, China)/black/ Lower Changjiang River Basin type/rare

Hormel: *see* Minnesota Miniature, Sinclair Miniature

Hoya: (Bremen, Germany)/former local var. of German Improved Landrace/Ger. *Hoyaer*

Hsin-chin: *see* Xinjin

hua: *see* Coloured [= pied, spotted]

Hua-hsien: *see* Fa Yuen

Huai: (N Jiangsu and Anhui, China)/ black; large lop ears/North China type/var. of Huang-Huai-Hai Black/orig. (with Large White) of New Huai/subvars: Huaibei, Huainan/syn. *Jiangsu Huai*/not *Cheng-hwai, Hwai, Kiangsu Hwai, Wuai*

Huai: (SE Fujian, China)/black/South China type/inc. in Fujian Small Pig

Huaibei: (China)/northern var. of Huai (North China type)/[*bei* = northern]

Huainan: (China)/southern var. of Huai (North China type)/[*nan* = southern]

Huang-Huai-Hai Black: (N Jiangsu, N Anhui, Shandong, Shanxi, Hebei, Henan and Inner Mongolia, China)/m/North China type/vars: Caiwu, Dingxian, Dingyuan, Hetao Lop-ear, Huai (Huaibei and Huianan), Laiwu, Mashen, Shan, Shenxian, Shenzhou, Wanbei, Zhao

Huantang: (China)/var. of Guangdong Small-ear Spotted

Huazhong Two-end Black: (Lower Changjiang basin, China)/black head and tail/Central China type/vars: Dongshan, Gangxi Two-end Black, Jianli, Shaziling, Tongcheng/Ch. *Huazhongliangtouwu*/syn. *Central China Double-Black*/ [*hua* = pied, *zhong* = middle]

Hubei White: (Hubei, China)/m/lop or horizontal ears/orig. from Russian Large White and Landrace × local Tongcheng black/recog. 1986

Huchuan Mountain: (China)/black/ SW China type/vars: Exi Black, Penzhou Mountain/Ch. *Huchuan Shandi*

Huiyang Black: (China)/var. of Yuedong Black

Hülsenberger hybrids: (Germany)/ based on crossing (Hampshire × Belgian Landrace) ♂ ♂ × (Duroc × German Landrace) ♀ ♀

Hungahyb: (Hungary)/4-line hybrid derived from Hungarian White, Dutch Landrace, Hampshire and Piétrain or Belgian Landrace/syn. *Ahyb*

Hungarian Curly Coat: *see* Mangalitsa

Hungarian Landrace: (Hungary)

Hungarian White: (Hungary)/m/orig. from Large White with some blood of Middle White, Edelschwein and German (Improved) Landrace/HB 1923/ Hung. *Fehér hússertés*, Ger. *Ungarisches weisses Fleischschwein* (= Hungarian white meat pig)/syn. *Hungarian Yorkshire*

Hurra: (Tarai, Nepal)/m/black or

brown, with white on jaw and tail/syn. *Eastern terai*

Husum Red Pied: (Schleswig-Holstein, Germany)/red head and neck, rump and hindquarters/colour var. of Angeln Saddleback since end 19th c./[BS 1954]/Ger. *Rotbunte Husumer Schwein*/syn. *German Red Pied* (Ger. *Rotbuntes Schwein, Rotbunte Schleswig-Holsteiner*), *Protest pig* (Ger. *Dänisch Protestschwein*) by Danish minority because colours of Danish flag/nearly extinct

Hutan: (Indonesia)/syn. *babi-Hutan* (*babi* = hog)

Huzhu: (Qinghai, China)/black/var. of Bamei/[*zhu* = pig]

Hwai: *see* Huai

Hypor: (Netherlands)/m/hybrid orig. from Dutch Landrace, Saddleback, Hampshire, German Landrace and Large White

Í: (Red R delta, N Vietnam)/m.ld/black; wrinkled skin; ♂ smaller than ♀/orig. of BSI (with Berkshire), DBI (with Large White), Vietnamese Potbelly/syn. *i*

Iban: (Sarawak, Malaysia)/grey with white feet; small ears, long snout, straight tail; prim./syn. *Kayan*

Iberian: (S Europe)/ld/usu. coloured (red, black or pied); medium semi-erect ears, long snout/type inc. native breeds of Mediterranean countries, name used esp. to distinguish Alentejana, Basque, Gascony, Iberian and Limousin from Celtic types in N of France, Portugal and Spain/syn. *Mediterranean*

Iberian: (S Spain)/inc. Extremadura Red (dominant), Jabugo Spotted, Black Iberian/declining by crossing with Duroc *et al.*/BS/Sp. *Ibérica*/see also: Dorado Gaditano, Silvela, Torbiscal, Valdesequera

Ievlev: (Tula, Russia)/ld/usu. black spotted, also black, white or white-spotted; lop ears/breed group/orig. from Large White,

Large Black, "middle black" and "black spotted", × local/Russ. *Ievlevskaya porodnaya gruppa*, Ger. *Jewlewsker*/extinct by 1984

Ilocos: (Philippines)/usu. black with white spots/var. of Philippine Native/Sp. *Ilocano*

Imeretian: (Georgia)/local with Polish White Lop-eared blood/Russ. *Imeretinskaya*/not *Imeritian*/extinct

Improved Belgian: *see* Belgian Landrace

Improved Eastern: (France) *see* Amélioré de l'Est

Improved Indigenous: (Nigeria) *see* Nigerhyb

Improved Landrace: *see* Belgian, Dutch, German, Swiss (Ger. *Verdeltes Landschwein*)

Improved Ningan: *see* Ning-an

Indian native: *see* Ankamali, Deshi, Ghori, Khanapara Dwarf, Khasi

Indigène amélioré: (Belgium) *see* Belgian Landrace

Indo-Chinese: *see* Vietnamese

Indonesian: *see* Balinese, Javanese, Nias, Sumatran, Sumba

Inhata: (Brazil)/rare

Inner Mongolian Black: (China)/orig. from Berkshire, Neijiang and local/recog. 1983/Ch. *Neimenggu Hei*

Inner Mongolian White: (China)/orig. from Landrace and Russian Large White, × local/recog. 1988/Ch. *Neimenggu Bai*

Inobuta: (Japan)/black with red tinge/orig. from $\frac{1}{4}$ wild pig, $\frac{1}{4}$ Duroc, $\frac{1}{2}$ Berkshire/[*ino* = wild pig, *buta* = domestic pig]

INRA hyperprolific line: (France)/Large White line

Irish Grazier: (USA)/Irish orig. in early 19th c./part orig. of Poland China/extinct 1870–1890

Irish Greyhound: (Ireland)/heavy lop ears; wattles; harsh coat/[with narrow waist and long legs, like racing Greyhound]/extinct early 20th c.

Iron Age pig: (Gloucestershire, England)/attempt at Cotswold Farm Park to breed back from wild × Tamworth

Istochnobalkanska: *see* East Balkan
Italian Belted: *see* Siena Belted
Italian Landrace: (Italy)/HB
Ivanovo: (Ukraine)/breed group

Jabugo Spotted: (Huelva, Spain)/m/
red, blond or whitish, with
irregular black speckles/Iberian
type, ? var. of Iberian/? orig. end
19th c. (established 1920) from
Large White × Black and Red
Iberian (*Iberico Retinto*)/HB 1981/
Sp. *Manchada de Jabugo*/syn.
Andaluza manchada (=
Andalusian Spotted)/nearly
extinct
Jalajala: (Philippines)/var. of
Philippine Native/orig. (with
Berkshire) of Berkjala
Jasan: (Korea)
Javanese: (Java, Indonesia)/orig. from
European breeds (mainly Dutch
and British) × native
Javanese warty pig: (Java, Sulawesi
and the Philippines)/= *Sus
verrucosus*
Jefferson County hog: *see* Cheshire
Jersey Red: (New Jersey, USA)/dark
red with white patches, or sandy
and white/? orig. from 'African
Red Guinea'/named 1857/part
orig. of Duroc; absorbed by
Duroc-Jersey (= Duroc, 1960s)
1872/extinct
Jewlewsker: *see* Ievlev
Jiang Jiang: (China)/lean meat lines
in development (see Fengjiang,
Lijiang)
Jiang-Hai type: *see* Lower Changjiang
River Basin type
Jiangquhai: (N Jiangsu, China)/black/
Lower Changjiang River Basin
type/part orig. of lean meat lines
Fengjiang and Lijiang/syn.
Jiashazi, Shajiaben/not
Jangquhai, Jiangchuhai, Jingquhai
Jiangshan Black: (China)/var. of Yujiang
Jiangsu Huai: *see* Huai (North China
type)
Jianli: (S Hubei, China)/var. of
Huazhong Two-end Black/WG
Chien-li, Ger. *Djen-li, Dsien-
li*/syn. *Kienli*
Jiaoxi: (China)/var. of Taihu

Jiashazi: *see* Jiangquhai
Jia-Xing: (Poitou, France)/black;
prolific/= Jiaxing Black/imported
from China for research/nearly
extinct
Jiaxing Black: (NE Zhejiang, China)/
var. of Taihu/not *Jiashing, Jia
Xin, Kia Sing, Tia Sing*
Jia Xiu: (China)
Jilin Black: (Jilin, NE China)/m/var.
of Xinjin/Russ. *Tszilin chernaya*/
not *Chilin Black, Kirin
Black*/rare
Jilin Spotted: (China)/black-and-
white spotted; lop ears/var. of
North-East China Spotted/orig.
from Kemerovo × local/recog.
1978/syn. *Jilin Coloured, Ji
Spotted*
Jinfeng: (China)/commercial meat
type dev. from Fengjing,
Landrace and Piétrain
Jingchuan: *see* Bamei
Jingquhai: *see* Jiangquhai
Jinhua: (Upper Fuchun R, Zhejiang,
China)/ld/black-and-white, usu.
black head and rump/Central
China type, sim. to Monchuan/
vars: Dongyang, Yongkang/WG
Chin-hua, Russ. *Tsinkhua*, Ger.
*Ching-Huai, Djing-Hua,
Khinghua*/ syn. *Black-at-both-
ends pig, Two-end-black pig*/not
Jinghua, Kinhwa, Tsinghwa
Jinli: (China)/var. of Cantonese
Ji Spotted: *see* Jilin Spotted
Jungchang, Jungtschang: *see*
Rongchang
Jungkang, Junkan: *see* Yongkang
Junqueiro: (Brazil)/orig. from English
breeds × Canastrão/extinct

Ka Done: *see* Raad
Kagoshima Berkshire: (Japan)/
distinctive type of Berkshire
developed in Japan
Kahyb: (Hungary)/rotational cross of
5 lines inc. Large White,
Landrace and some Duroc,
Hampshire and Lacombe/not
Kakhib/[Kaposvar hybrid]
Kakhetian: (E Georgia)/ld.m/grey,
piglets striped; prim./breed
group/Russ. *Kakhetinskaya*/rare
Kalikin: (Ryazan, Russia)/ld/grey

pied/orig. from Large White and Berkshire × local lop-eared/ Russ. *Kalikinskaya*/extinct by 1984

Kama: (Solikamsk, Perm, Russia)/ breed group/orig. from Large White and Breitov × local lop-eared/Russ. *Prikamskaya porodnaya gruppa*/not *Kamsk*/ extinct

Kaman: (Batangas, Philippines)/red/ orig. from Duroc × Philippine Native/extinct

Kamchiya, Kamčija, Kamčik, Kamčiska: *see* East Balkan

Kang: (Laos)/syn. *mou Kang*

Kangaroo Island: (S Australia)/ black spotted; small to dwarf; feral/? orig. from British breeds released in 1801

Kanton: *see* Cantonese

Karelia: *see* Carélie

Kartolinian: (Georgia)/local with Large White blood/Russ. *Kartolinskaya*/extinct

Kayan: *see* Iban

Kazakh hybrid: *see* Semirechensk

KB-KN: (Central Russia)/strain of Russian Large White

KB-V-1: (Central Russia)/strain of Russian Large White

Keelung: *see* Ting-Shuang-Hsi

Kele: (Guizhou mts, Yunnan, China)/ ld/black/var. of Wujin

Kelomive: *see* Kemerovo

Kemerovo: (S Siberia, Russia)/ld/ black or black pied/orig. from Berkshire × local Siberian, with Large White blood/recog. 1961/ HB/Russ. *Kemerovskaya*/syn. *Kelomive, Kemiroff* (China)

Kemerovo meat type: *see* KM-1

Kemiroff: *see* Kemerovo

Kentucky Red Berkshire: (USA)/ld/ orig. from Berkshire in early 19th c./BS 1923/syn. *Red Berkshire*/ extinct

Keopra: *see* Raad

Khanapara Dwarf: (Asam, India)/? = Ghori

Kharbin: *see* Harbin

Kharkov: (Ukraine)/m/orig. from Russian Large White, Landrace and Welsh/syn. *Kharkov hybrid*

Khasi: (Meghalaya, NE India)/black with white on forehead and toes/ local indigenous potbelly/[Khasi tribe]

Khinghua: *see* Jinhua

Kiangsu Hwai: *see* Huai (North China type)

Kia Sing: *see* Jiaxing Black

Kienli: *see* Jianli

Kinhwa: *see* Jinhua

Kirin Black: *see* Jilin Black

Kirin Min: *see* Min

KM-1: (Siberia, Russia)/m/orig. 1968– 1978 (new type recog.) from (Landrace × Kemerovo) × [Landrace × (Landrace × Kemerovo)] i.e. $\frac{5}{8}$ Landrace, $\frac{3}{8}$ Kemerovo, selected for thin back fat and efficient food conversion/ syn. *Kemerovo meat type* (Russ. *Kemerovskiĭ myasnyĭ typ*)

Kolbroek: (South Africa)/usu. black, often with paler spots and patterns; short snout (cf. long-snouted Windsnyer); low belly; fatty/? orig. from Chinese intro. by Portuguese in 15th century/ [*kol* = spot, *broek* = ham; or possibly derived from ship named Coalbrook wrecked on Cape coast 1778]

Konstantinovo: (Russia)/used in cross with Russian Large White to create KB-KN

Korean Improved: (Korea)/orig. from Berkshire × North China

Korean Native: (Korea)/black; prim.

Korhogo: (N Côte d'Ivoire)/usu. black, also black pied or spotted; short ears/orig. (1930 on) from Berkshire and Large White × local West African, initially at Bouaké but later transferred to Korhogo

Koronadal: (General Santos, Cootabato, Philippines)/ash-red with dark spots/orig. from Berkjala, Poland China and Duroc

Korotkoukhaya belaya: *see* Russian Short-eared White

Krasnodar: *see* Kuban, Lesogor

Krasnodar type: (Russian and Ukraine)/ var. of Soviet Meat/syn. *Southern*

Krolevets: (N Ukraine)/ld/black pied or black/breed group/sim. to Dnieper, Mirgorod and Podolian/ orig. from English breeds × local Polesian/Russ. *Krolevetskaya porodnaya gruppa*/syn. *Polesian Lard* (Russ. *Polesskaya salnaya*)/ extinct by 1984

Krškopolje Saddleback: (Slovenia)/ m/black-and-white; lop ears/ local type with blood of German (Improved) Landrace and possibly also Berkshire and Large Black/Sn. *Krškopoljska črnopasasta, Pasasti prašič*/rare

Krupnaya belaya: *see* Russian Large White

Krusta: (Netherlands)/Large White sire line/used also in (Duroc × Krusta) × Danish Landrace cross for new sire line

Kuang-chou Wan: *see* Kwangchow

Kula: (NW Bulgaria)/ld/black, red-brown or dirty white/var. of Mangalitsa from local crossed with Šumadija and Mangalitsa/ Bulg. *Kulska*/syn. *Kula-Mangalitsa*

Kunekune: (North I, New Zealand)/ ld/cream, ginger, brown or black, or any combination of these in spots; tassels; coat short or long wavy; small/Asian and possibly European orig./BS; HB also UK/ syn. *Maori, Poaka, Pua'a*/[= fat]/ rare

Kwai: (Uttardit Prov., N Thailand)/ ld/black, sometimes white on shoulder and legs/[= buffalo]/ rare

Kwangchow Wan: (Hong Kong)/black back, white belly/var. of Cantonese/WG *Kuang-chou Wan*/extinct

Lacombe: (Alberta, Canada)/m/lop ears/orig. 1947–1958 at Lacombe Res. Centre from Danish Landrace (55%), Chester White (22%) and Berkshire (23%)/BS 1959, HB 1961/rare

Laconie line: (Brittany, France)/$\frac{1}{3}$ Hampshire, $\frac{1}{3}$ Large White, $\frac{1}{3}$ Piétrain; selected by Pen ar Lan Breeding Co., Maxent, since 1973

Laconie P77: (Brittany, France)/m/ black, brown or white/stress-resistant composite of Large White, Hampshire and Piétrain orig. 1972–73 by Pen ar Lan Breeding Co., Maxent, for crossing with Penshire P66/syn. *PELL*/rare

Ladt: (Laos)/syn. *mou Ladt*

Laiwu: (China)/var. of Huang-Huai-Hai Black

Lagonegrese: (Lagonegro, SW Lucania, Italy)/former var. of Calabrian/extinct

Lampiña negra: *see* Black Hairless

Lampiño: *see* Black Hairless

Landau: (Brazil)/orig. 1985–1992 by EMBRAPA/CNPSA from *Land*race and Pi*au*/project aborted

LANDRACE: (NW Europe)/inc. improved native white lop-eared (Celtic) breeds of NW Europe (Danish, Finnish, German, Norwegian and Swedish Landraces) and their derivatives (American, Austrian, Belgian, British, Bulgarian, Canadian, Czech, Dutch, French, Irish, Italian, Romanian and South African Landraces, and Polish White Lop-eared, Slovakian White Meat and Swiss Improved Landrace)/[= local, native, indigenous breed]

Landrace: (Denmark) *see* Danish Landrace

Landrace Belge Halothane Négatif: *see* Belgian Negative

Landrase: *see* Norwegian Landrace

Landrasse B: *see* Belgian Landrace

Landschwein: *see* German Land pig

Landvarken: *see* Belgian Landrace

Lang Hong: (N Vietnam)/black pied/? var. of Mong Caq (with black patches more randomly distributed)

Lantang: (Guangdong, China)/ld.m/ head and back black, belly and feet white/South China type/not *Lantan*

Lantian Spotted: *see* Wannan Spotted

Lantras: *see* Swedish Landrace

Lanxi Spotted: (China)/pied/local var./rare

Lan-Yu: (Hung t'ou I, Taiwan)/black; straight tail/only extant var. of

Taiwan Small-ear, studied at Taiwan Nat. Univ. since 1970s/ orig. of Lee-Sung/var.: Spotty Lanyu/syn. *Lan-Yu Small-Ear miniature pig, Taiwan Small-ear miniature*/rare

Laoshan: (Shandong, China)/m.ld/ orig. from Berkshire and Middle White × local black

Large Black: (England)/m/black, otherwise sim. to British Lop/ formerly vars in (1) Cornwall and Devon and (2) Suffolk and Essex (orig. from Small Black)/orig. of German Cornwall/HB 1899; HB also Canada, Russia; BS USA/ syn. *Cornwall, Devon, Lop-eared Black*/nearly extinct in GB and USA

Large Black White: *see* Cantonese

Large English: *see* Large White

Large Polish Long-eared: (Poland, except SE)/black, red or pied; prim./orig. of Złotniki/Pol. *Wielka polska długoucha*/syn. *local long-eared, native long-eared*/extinct

Large Spotted White: (China) *see* Cantonese

Large White: (England)/m/prick ears/orig. from local Yorkshire with Chinese (Cantonese) blood in late 18th c./orig. of American Yorkshire, Austrian Edelschwein, Belarus Large White, Belgian Yorkshire, Bulgarian Improved White, Canadian Yorkshire, Dutch Yorkshire, Edelschwein, Estonian Large White, French Large White, Hungarian White, Latvian White, Lithuanian White, Norwegian Yorkshire, Polish Large White, Romanian Large White, Russian Large White, Swedish Large White, Swiss Edelschwein, Ukrainian Large White, Ukrainian White Steppe/recog. 1868; HB 1884; HB also Finland 1914, Ireland 1969, China, Italy, Korea *et al.*/syn. *Grand Yorkshire, Large English, Large White English, Large White Yorkshire, Large York, Yorkshire*

Large White Lop-eared: *see* British Lop

Large White Ulster: (N Ireland)/m.ld/ lop ears/HB 1908–1933/syn. *Ulster, Ulster White*/extinct c. 1960 (last boar licensed 1956)

Large York: *see* Large White

Lasasta, Lasasti Soj Mangulica: *see* Black Mangalitsa

Latvian Landrace: (Latvia)

Latvian White: (Latvia)/m/orig. from Berkshire and (Russian) Short-eared White × long-eared and short-eared Latvian in late 19th c. with Large White and Edelschwein blood in 20th c./ recog. 1967/Russ. *Latviĭskaya belaya*

Lauragais: (SW France)/orig. with Gascony of Cazères/extinct

Lee-Sung: (Taiwan)/white with some black spots (or black); erect ears; miniature/orig. since 1974 from Landrace × Lan-Yu at Taiwan Nat. Univ., Taipei/[by Profs *Lee* and *Sung*]/rare

Leicoma: (Leipzig, Cottbus and Magdeburg, Germany)/m/occ. coloured/orig. from Dutch Landrace (chiefly), local Saddleback and Estonian Bacon (1971–1975) crossed with Duroc and local Landrace (1976–1981); named 1986/HB 1971

Leping: (Jiangxi, China)/m/black with white forehead, belly and feet/ Central China type/not *Lepin*

Lermeña: (Burgos, Spain)/local Celtic type, sim. to Vitoria/extinct

Lesogor: (Krasnodar, N Caucasus, Russia)/breed group of European short-eared type/Russ. *Lesogornaya porodnaya gruppa*/ [= forest-mountain]/extinct/*see also* Forest Mountain

Levant type: (Castellón to Almería, E Spain)/local graded to Large White, Berkshire (= Murcian) and Vitoria/Sp. *Agrupación levantina*/extinct except Murcian

Lewe minipig: (Germany)/? = Mountain Dwarf in Myanmar (Burma)

Liangcun: (China)/var. of Cantonese

Liang Guang Small Spotted: (Guangxi and Guangdong, China)/black back and head, white neck, shoulders and feet/ South China type/vars: Fumian, Gongguan, Guangdong Small-ear Spotted, Luchuan/Ch. *Liang Guang Xiaohua*

Liangshan: (mts of Sichuan, China)/ var. of Wujin

Lianhua Spotted: (China)/var. of Hang

Liaoning Black: (China)/orig. from Landrace or Duroc × Min

Libtong: (Luzon, Philippines)/black with white spots on forehead and white feet and tail tip/? orig. from Hampshire × Philippine Native/syn. *baboy, Tagudin*/ [Libtong village in Tagudin]/ nearly extinct

Lichahei: (China)

Lijiang: (China)/lean meat line from Lichahei × Jiangquhai

Limousin: (Haute Vienne, C France)/ ld.m/white with black head and rump, often small spots on body/ Iberian type/var. Périgord/HB 1935, BS/syn. *St Yrieix, Cul Noir*/nearly extinct (by crossing with Craonnais and Large White)

Lincolitsa: (Hungary)/Lincolnshire Curly Coat × Mangalitsa cross in 1920s/extinct

Lincoln Red: (New Zealand)/m/orig. 1940–1959 at Canterbury Agric. College, Lincoln, from Large White × Tamworth/extinct (programme terminated 1959)

Lincolnshire Curly Coat: (England)/ m.ld/lop ears/BS 1906–1960/ syn. *Baston, Lincoln, Lincoln Curly Coat, Lincoln Curly Coated White, Lincolnshire Curly-Coated*/extinct 1972 (last boar licensed 1963/64)

Lindrödssvin: (Skåne, S Sweden)/ light grey or brown, with black spots/rare

Lingao: (Hainan I, China)/dwarf/var. of Hainan/rare

Lingtouwou: *see* Middle China Lingtouwou

Linia 990: *see* Polish Synthetic

Lipetsk type: (Russia)/var. of Soviet Meat orig. from Poltava × (Poltava × Belarus commercial hybrid)/not *Lipets*

Lishigiao: (? Shanxi, China)

Lithuanian Native: (SW Lithuania)/ m.ld/white, black or tan; bearded/Lith. *Vietines kiaules* (= local pig)/syn. Lithuanian Aboriginal/rare

Lithuanian White: (Lithuania)/m or m.ld/orig. from Large White, Edelschwein and German Landrace × local Lithuanian/ recog. 1967/Lith. *Lietuvos baltųjų*, Russ. *Litovskaya belaya*

Liubai: (China)/white/local var.

Liujia: (? Guanxi, China)

Livny: (Orel, Russia)/ld/white or black-pied, occ. red-pied or black; semi-lop ears /orig. *c.* 1930 from Middle White, Large White and Berkshire, × local lop-eared/ recog. 1949/HB/Russ. *Livenskaya*/ not *Liven, Livensky*/rare

Loches: (Touraine, France)/cf. Montmorillon/extinct (absorbed by Large White)

Long-eared White: *see* Russian Long-eared White

Longlin: (NE Guangxi, China)/black with white feet/South China type/rare

Long White Lop-eared: *see* British Lop

Longyou Black: (Jiangxi, China)/ Central China type/not *Longyu*

Lop-eared Black: *see* Large Black

lop-eared white: (UK) *see* British Lop, Cumberland, Large White Ulster, Welsh

Lop White: *see* British Lop

LOWER CHANGJIANG RIVER BASIN TYPE: (China)/black; lop ears/see Dongchuan, Hongqiao, Jiangquhai, Taihu, Wei, Yangxin/ syn. *Jiang-Hai type*

Lucania pony: *see* Cavallino

Lucanian: (Italy)/var. of Apulian/syn. *Basilicata*/extinct

Luchuan: (SE Guanxi, China)/ld/ black-and-white (usu. black head and back), occ. black or white/ South China type/var. of Liang Guang Small Spotted/WG *Lu-ch'uan*, Russ. *Luchuan'chzhu*, Ger. *Lutschuan*/not *Luchwan*

Lung Kong: *see* Wai Chow

Lung-Tan-Po: *see* Taoyuan

Lutai White: (Lutai farm, Hebei,
China)/m/Russian Large White ×
Large White/rare
Lutschuan: *see* Luchuan
Lu White: (China)/orig. from Russian
Large White, Yorkshire and
"white hybrid"/recog. 1983

Maatiaissika: *see* Finnish Landrace
Macao, Macau: *see* Cantonese, Tatu
Macchiaiola: *see* Maremmana
Mackay: *see* Hampshire
Madonie: (Madonie mts, Sicily, Italy)/
population of Black Sicilian/syn.
Black Madonie (It. *Nera delle
Madonie*), *Madonie-Sicilian* (It.
Siciliana delle Madonie)/nearly
extinct
Magyar Lapàly Sertés: (Hungary)
Mai Schan, Maishan: *see* Meishan
Majorcan Black: (Balearic Is, Spain)/
ld.m/slate-grey or black; tassels/
Iberian type/BS, HB 1997/Sp.
Mallorquina, *Agrupación
balear*/syn. *Balearic*/ not
Mallorcan/nearly extinct
Majorcan Black: (Balearic Is, Spain)/
m/slate-grey or black; large lop
ears; tassels/Iberian type/BS
early 1980s, HB 1997/Sp. *Porc
Negre Mallorquí*/rare
Mała polska ostroucha: *see* Small
Polish Prick-eared
Mallorcan, Mallorquina: *see*
Majorcan
Mamellado: (Spain)/m/black; tassels;
almost hairless/strain of Iberian;
sim. to Black Hairy [= Wattles]/
rare
Managra: (Manitoba, Canada)/m/lop
ears/orig. (1956 on) from
Swedish Landrace (45%),
Wessex Saddleback (20%),
Welsh (12%), Berkshire (6%),
Minnesota No. 1 (6%), Tamworth
(6%) and Yorkshire (6%)
Manchada de Jabugo: *see* Jabugo
Spotted
Mandi: *see* Pirapetinga
Mangalitsa: (N Serbia, Hungary and
Romania)/ld.m/lop ears; curly
hair/orig. from Šumadija in late
18th c.; imported to Hungary
1833 and used to improve local
(Ancient Alföldi, Bakony,

Szalonta etc.)/vars: Blond
Managalitsa (chief), Black
Mangalitsa (swallow-bellied or
weasel), Georgian Mangalitsa,
Kula, Red Mangalitsa, White
Mangalitsa/orig. of Subotica
White/HB Hungary 1927,
Romania 1958; HB also
Switzerland 1985/Serbo-cro.
Mangulica (♀) or *Mangulac*, Ger.
Mangaliza, Hung. *Mangalica*,
Rom. *Mangaliţa*/syn. *Hungarian
Curly Coat, Porc laineux des
Pacages, Wollhaariges
Weideschwein, Wollschwein/
[Serbo-cro. = easily fattened, ?
from Rom. *mînca* = to eat]/rare
Mangouste: (Seychelles)/ld/black/
Asian orig.
Manor Hybrid: (Yorkshire, England)/
Large White × Landrace cross,
bred by Northern (later Nat.) Pig
Dev. Co., Manor Farm, Belford
Manor Meishan: (Yorkshire,
England)/m/prolific/orig. 1992
by Nat. Pig Dev. Co. from ¼
Meishan, ¼ prolific Large White
and ½ prolific Landrace
Manor Ranger: (Yorkshire, England)/
coloured/outdoor ♀/Landrace ×
Hamline 1st cross
Manx Purr: (Isle of Man)/spotted
sandy-grey; long hair; feral/
extinct end 18th c.
Maori: *see* Kunekune
Maranhão: *see* Canastrão
Marele alba: *see* Romanian Large
White
Maremmana: (Maremma, Tuscany,
Italy)/black; semi-wild/syn.
Macchiaiola (= forest pig), *Nera
umbra* (= *Black Umbrian*),
Roman/extinct by 1949
Markhyb: (Brandenburg, Germany)/
hybrid ♀♀ orig. 1991 on by
Berlin-Brandenburg Pig Breeding
and Production Union from
German Landrace, Edelschwein
and Leicoma
Marseilles: (France)/Iberian type
with English blood (1850 on)/Fr.
Marseillais/extinct (absorbed by
Large White)
Marsh Stocli: *see* Băltăreţ
Maryland No. 1: (Maryland, USA)/

m/black with white spots; erect or semi-erect ears/inbred orig. 1941–1951 by Maryland Agric. Exp. Sta. from Danish Landrace × Berkshire, backcrossed first to Berkshire and then to Danish Landrace and then bred *inter se*; ⅝ Danish Landrace, ⅜ Berkshire/ HB 1951/extinct

Mascherina: *see* Apulian

Mashen: (N Shanxi, China)/m.ld/var. of Huang-Huai-Hua Black

Mask, Masked: *see* Chinese

Mastergilt: (England)/white hybrid ♀ by Masterbreeders from Large White and Landrace

Maxambomba: *see* Canastra

Meatline: (Cambridgeshire, England)/ terminal ♂ lines bred since *c.* 1977 by Premier Piglink/inc. Euroline, Hampline, Norline

Meatmaster 555: (England)/terminal ♂ line by Masterbreeders from Whites and Duroc

Mediterranean: *see* Iberian

Mega: (France)/Belgian Landrace × Hampshire cross

Meia perna: *see* Canastra

Meihua: (China)/var. of Cantonese

Meihuaxing: (China)/var. of Yangxin

Meinung: (Taiwan)/ld/sim. to Taoyuan but smaller/WG *Mi-nung*, Jap. *Mino*/extinct by 1979

Meishan: (N Shanghai, China)/black with white feet/var. of Taihu/Ger. *Mai Schan*, Fr. *Mei-Shan*, *Chinoise*/not *Maishan*

Meishan Synthetic: (UK and China)/ 50% Meishan, 50% European Landrace or Large White

Meissen: (Saxony)/early local var. of German Improved Landrace/BS 1888/Ger. *Meissener*, *Meissner*/extinct

Meixin: (China) = Meishan × Xinjin cross

Mengla: (China)/var. of Diannan Small-ear

Menshanton: (Japan)

Meo: (N Vietnam)/local breed in western mts

Meshchovsk: (Kaluga, Russia)/breed group/orig. from Large White × local/Russ. *Meshchovskaya porodnaya gruppa*/extinct by

1984

Mexican Dwarf: *see* Cuino

Mexican Wattled: (Mexico)/red, black or spotted/Sp. *Cerdo coscate* (*cerdo* = pig)

Mi: (Jintan and Yangzhong, Jiangsu, China)/var. of Taihu

Miami: (SW Ohio, USA)/orig. from Big China, Byfield and Russian strains and local pigs of Miami valley in early 19th c./part orig. of Poland China/syn. *Warren County*/extinct

Middle China Lingtouwou: (China)

Middlesex: *see* Small White

Middle White: (England)/m/orig. from Large White × Small White/recog. 1852, HB 1884, BS 1991/also in China and Japan/ syn. *Coleshill*, *Middle White Yorkshire*, *Middle Yorkshire*, *Windsor*/nearly extinct in GB

Mid-South Jiangxi Spotted: *see* Ganzhongnan Spotted

Miélan: (Gers, SW France)/ sometimes small grey spots on rump/sim. to Cazères/orig. from Craonnais × Gascony *c.* 1880/ var.: Piégut/extinct

Miloš, Milosch, Miloševa: *see* Šumadija

Min: (Liaoning, Jilin and Heilongjiang, NE China)/m.ld/black; large lop ears; hy; prolific/North China type/vars: Damin, Xiaomin, Zhong Min/syn. *Minzhu* (= Min pig)/not *Kirin Min*, *Ming*

Minbei Black: (China)/var. of Wuyi Black

Minbei Spotted: (Fujian, China)/ black-and-white pied/Central China type/vars: Wangtai, Xiamao, Yangkou/syn. *North Fujian Spotted*

Ming: *see* Min

miniature: inc. American Essex, Berlin, Clawn, Czech, Froxfield Pygmy, Göttingen, Hanford, Lan-Yu, Lee-Sung, Minisib, Minnesota, Mitsai, Munich, NIH, Ohmini, Pitman-Moore, Sinclair, Vita Vet Lab, Yucatan/BS USA/syn. *mini-pig*

Minisib: (Russia)/orig. from Vietnamese (Í) (43–56%),

Swedish Landrace (24–32%) and wild boar (*Sus scrofa scrofa* and *S. s. nigripes*) (20–25%), selected for small size and usu. for white colour/[*Mini*ature *Sib*erian]

Minnesota Miniature: (Minnesota, USA)/spotted/small pig for medical research/selected (1949 on) at Hormel Inst., St Paul, from cross of 3 feral strains (Guinea Hog, 13%; Pineywoods Rooter, 49%; Catalina, 16%) and Ras-n-lansa (22%)/orig. of Göttingen and Czech Miniatures/syn. *Hormel*/extinct in USA

Minnesota No. 1: (USA)/m/red/orig. 1936–1946 at Minnesota Agric. Exp. Sta. from Tamworth (52%) and Danish Landrace (48%)/HB 1946/extinct 1994

Minnesota No. 2: (USA)/m/white with black spots/orig. 1941–1948 from Canadian Yorkshire (40%) and Poland China (60%)/HB 1948/extinct

Minnesota No. 3: (USA)/various colours/orig. from Gloucestershire Old Spots (31%), Poland China (21%), Welsh (13%), Large White (12%), Beltsville No. 2 (6%), Minnesota No. 1 (6%), Minnesota No. 2 (5%) and San Pierre (5%)/extinct

Minnesota No. 4: (USA)/exp. population at Minnesota Agric. Exp. Sta. from Minnesota Nos 1, 2 and 3; exp. ended early 1970s/extinct

Mino: *see* Meinung

Mi-nung, Mi-nung-chung: *see* Meinung

Minzhou, Minzhu: *see* Min

Mirgorod: (C Ukraine)/ld/usu. black pied, occ. black, black-and-tan or tan/sim. to Dnieper, Krolevets and Podolian/orig. since 1882 from Large White, Middle White and Berkshire (with some Large Black and Tamworth) × local short-eared Ukrainian/recog. 1940/Russ. *Mirgorodskaya*/syn. *Mirgorod Spotted* (Russ. *Mirgorodskaya ryabaya*)/rare

Mitsai: (Taiwan)/brown-and-white

striped; miniature; ornamental/ orig. 1989 from 2 Lan-Yu ♂♂ × 6 Duroc ♀♀ and vice versa, F$_1$ selected for brown-and-black stripes at 8 weeks as breeding stock, bred by half-sib matings for brown-and-white stripes; selection for brown-and-white stripes from 1992, colour fixed in maturity in F$_6$

Mkota: (Zimbabwe)/m/black, occ. pied/ syn. *Rhodesian Indigenous, Zimbabwe Indigenous*/not *Mukota*

MM-1: (Moscow, Russia)/strain of Russian Large White

Modena Red: (Emilia, Italy)/It. *Rossa modenese*/extinct 1873

Modenese: (Modena, Emilia, Italy)/ black-and-white/orig. (1873 on) from Large White × local Iberian type/? extinct

Moldavian Black: (Moldova)/ld/ breed group/orig. (1948 on) from Berkshire *et al.* × local black, F$_1$ bred *inter se*/Russ. *Chernaya moldavskaya porodnaya gruppa*/extinct by 1984

Moldavian Meat Type: (Moldova)/m/ hybrid orig. 1991 as ♂ line/Moldavian *Tipul Moldovenesc de Carne*/syn. *Moldavian Hybrid*

Monarch: *see* Sovereign

Monchuan: (Shanggao, Monshang, Jiangxi, China)/black head and rump/Central China type/sim. to Dongchen, Gianli, Gunxi, Jinhua, Shaziling/syn. *Monshang*

Mong: (Laos)/syn. *mou Mong*

Mong Cai: (NE Vietnam)/black head, back and rump, white shoulders, belly and snout

Mongolitsa, Mongulitsa: *see* Mangalitsa

Monshang: *see* Monchuan

Montagnola: *see* Siena Belted

Montana No. 1: (USA)/m/black/ inbred orig. 1936–1948 from Danish Landrace (55%) and black (unbelted) Hampshire (45%) crossed reciprocally and backcrossed to both breeds at Miles City, Montana/named *Black Hamprace* 1947, renamed 1948/HB 1948/? extinct

Monteiro: (Pantanal Mato-Grossense, Brazil)/orig. native in thickets, now a few breeders in Mato Grosso, Matto Grosso do Sul and Goiás/rare

Montmorillon: (Vienne, France)/orig. in late 19th c. from Large White × Craonnais/absorbed by Large White/Fr. *Montmorillonnais*/syn. *Poitou*/extinct

Moor: *see* Romagnola

Mora, Mora romagnola: *see* Romagnola

Morava: (Serbia)/m.ld/black; lop ears/cf. Resava/orig. from Berkshire × Mangalitsa or Šumadija; being improved by Large Black/Serbo-cro. *Moravka, Moravska svinja*/rare

Moravian Improved White: *see* Czech Improved White

Moravian Large Yorkshire: *see* Czech Improved White

Moscow MM-1: *see* Russian Large White

mou Cheed: *see* Cheed

mou Kang: *see* Kang

mou Ladt: *see* Ladt

mou Mong: *see* Mong

Mountain Dwarf: (Myanmar)/m/ black/syn. *Chin, Wet* (= pig)

Moura: (S Brazil)/blue roan, occ. red roan/orig. from Duroc, Canastra and Canastrão/sim. to (? =) Pereira/syn. *Estrela, Estrelense, Mouro*/[Port., = Moorish, i.e. dark]/rare

Muang Khong: *see* Muong Khuong

Mukota: *see* Mkota

Mulefoot: (Missouri, USA)/m/black, often white markings; often tassels/sim. to Poland China but solid-hoofed/? orig. from Berkshire × feral Razorback for ld/BS 1908–1975/syn. *Ozark*/ nearly extinct/*see also* Casco de mula, Choctaw

Mundi: (Minas Gerais, Brazil)/closed herd currently in formation at State School of Agric./nearly extinct

Munich Miniature: (Munich, Germany)/white skin, red or black hair, or spotted/orig. from Hanford

crossed with "Columbian Portions" pig/syn. *Troll*

Muong Khuong: (mts of N Vietnam)/ not *Muang Khong*

Murcian: (E Spain)/m/var. of Levant type; local Iberian type crossed with Berkshire and Large White ♂♂/ Sp. *Chato de Murcia, Chato murciano* (= short-nosed Murcian)/nearly extinct

Murgese: (Murge, Italy)/black/var. of Apulian/extinct

Murom: (Vladimir, Russia)/m.ld/orig. in 1930s from Large White and Lithuanian White × local long-eared/recog. 1957/HB/Russ. *Muromskaya*

Musclor: (Brittany, France)/m/black-and-white/composite ♂ line bred by SCAPAAG of Dijon, with Piétrain dominant/rare

Nadbużańska: (R Bug, Poland)/ prim./Small Polish Prick-eared type/nearly extinct

Nai-Djang: *see* Neijiang

Naima: (Brittany, France)/hybrid ♀ line produced 1992–1995 by Pen ar Lan Breeding Co., Maxent, from Gallia line × (Tiameslan × Finnish Landrace Carélie line) to be crossed with Pen ar Lan ♂♂ to produce Naipal hybrids

Naipal: (Brittany, France)/hybrid produced by Pen ar Lan Breeding Co., Maxent, from Naima × Pen ar Lan ♂♂

Nanchang (Improved White): (Jiangxi, China)

Nanjing Black: (Jiangsu, China)/orig. (1972 on) from Shanzhu, Jinhua, Berkshire, Large White and Landrace

Nanyang: (Hunan, China)/Central China type

Napolitana: *see* Casertana

National Long White Lop-eared: *see* British Lop

Neapolitan: *see* Casertana

Nebrodi and Madonie: *see* Black Sicilian

Nederlands (veredeld) Landvarken: *see* Dutch Landrace

Negra entrepelada: *see* Black Hairy
Negra iberica: *see* Black Iberian
Negra lampiña: *see* Black Hairless
Neijiang: (Sichuan, China)/ld.m/
black/South-West China type/
WG *Nei-chiang*, Russ. *Neĭ-
tszyan*, Ger. *Nai-Djang*/syn.
Dongxiang/not *Neikiang*
Neimenggu: *see* Inner Mongolian
Nepalese Dwarf: *see* Chwanche
Nera del Friuli: *see* Friuli Black
Nera delle Madonie: *see* Madonie
Nera Siciliana: *see* Black Sicilian
Nera umbra: *see* Maremmana
Netherlands (Improved) Landrace:
see Dutch Landrace
Netherlands Indies: *see* Indonesian
New Guinea Native: (Papua New
Guinea)/usu. black, also black
spotted, white, red or grey;
newborn often striped; prick-
eared; straight tail/orig. possibly
hybrids between unknown
domesticants and wild Sulawesi
Warty pig
New Haitian: (Haiti)/orig. since 1986
from Guadeloupe creole pigs and
Sino-Gascony, to replace Haitian
creole pigs exterminated 1980
New Huai: (N Jiangsu, China)/m.ld/
black; lop ears/orig. 1959 from
Large White (50%) × Huai
(50%)/recog. 1977/Ch. *Xin Huai*/
not *Sin Hwai*
New Lesogor: *see* Forest Mountain
New York Red: (Sweden)/
commercial Duroc cross
Nghia Binh White: (C Vietnam)/also a
black var./Fr. *Blanche de Nghia
Binh*
Nias: (Indonesia)
Niederländisches Edelschwein: *see*
Dutch Yorkshire
Niederländisches Landschwein: *see*
Dutch Landrace
Nigerhyb: (Nigeria)/orig. 1980–1989
by Prof. O.A. Adebambo from
Nigerian Native (62.5%) and
Hampshire (37.5%)/syn.
Improved Indigenous
Nigerian Native: (Nigeria)/black/var.
of West African/syn. *Nigerian
indigenous*

NIH Miniature: (USA)/orig. from
(Minnesota Miniature ×
Yucatan) × Vita Vet Lab/[NIH =
National Institutes of Health]
Nilo: (Brazil)/ld/black, occ. white
spots; small; hairless/
Canastrinho type/? orig. from
Canastra × Tatu/orig. (with
Canastra) of Nilo-Canastra/
BSd/rare
Nilo-Canastra: (Brazil)/ld.m/
black/Iberian type/orig. from
Nilo and Canastra/rare
Ning-an: (SE Heilongjiang, China)/
var. of Xinjin/syn. *Improved
Ningan*
Ningxia Black: (China)/orig. from
Berkshire, Russian Large White,
Xinjin and local/recog. 1982
Ningxiang: (Hunan, China)/ld/white
belly, black (or black-and-white)
back/Central China type/WG
Ning-hsiang, Russ. *Ninsya*,
Ninsyan, Ger. *Ning-Chang*/not
Ning hsing, Ningsiang, Ningxing
Nipi: (China)/var. of Cantonese
Nitra Hybrid: (Nitra, Slovakia)/orig.
from Czech Improved White,
Slovakian Black Pied, Landrace,
Hampshire, Slovakian White and
Large White
Njemački Landras: (Serbia) *see*
German Landrace
Norfolk Thin Rind: *see* Hampshire
Norline: (England)/Meatline ♂ of
Premier Piglink based on Large
White and Landrace
Normand: (Cherbourg peninsula,
Normandy, France)/m/Celtic
type/much crossed with
Craonnais and Large White, and
absorbed by West French White
1955/BS 1937/extinct
North Caucasus: (N Caucasus,
Russia)/m.ld/black pied/orig.
from local Kuban with 2 crosses
of Large White (1929–1938), 2 of
Berkshire (1938–1941) and 1 of
White Short-eared (1946)/var.:
Don/recog. 1955; HB/Russ.
Severokavkazskaya/syn.
North(ern) Caucasian,
Novocherkassk

NORTH CHINA type: (China)/black; lop ears/see Bamei, Hanjiang Black, Huang-Huai-Hai Black, Min

North-East China Spotted: (NE China)/m/black-and-white spotted/orig. from Berkshire × Min/vars: Black Spotted, Jilin Spotted, Shen Spotted/Ch. *Dongbeihua* (*dong* = east, *bei* = north, *hua* = spotted)

North East Min: *see* Min

North Fujian Black-and-White: *see* Minbei Spotted

North Kwangtung (Guangdong) Lop-eared: *see* Gongguan

North Siberian: (N Omsk and Novosibirsk, Russia)/ld/orig. since 1933 from Large White × Siberian/recog. 1942/HB/Russ. *Sibirskaya severnaya*

Norwegian Landrace: (Norway)/m/ orig. from Large White and Middle White (1880–1890) and Danish Landrace (1900) × local; Swedish Landrace blood since 1945/HB 1930/Nor. *Norsk Landrase*

Norwegian Yorkshire: (SE Norway)/ m/orig. from Large White, also Landrace blood/HB 1930/Nor. *Norsk Yorkshire*/rare

Novocherkassk: *see* North Caucasus

Novosibirsk: *see* Siberian

Novosibirsk Spotted: *see* Siberian Spotted

Numpard: (Uttardit Province, Thailand)/native type in remote hill areas

Ohio Improved Chester: *see* OIC

Ohmini: (Japan)/black, occ. black pied; lop ears/orig. since 1960 by H. Ohmi at Japanese Res. Lab. for Domestic Animals, Tochigi, by selection from Chinese ("Manchu", ? = Minzhu) miniatures (imported 1942, 1949 and 1960); later strains from crossing with Minnesota Miniature ♂♂ (imported 1952)/nearly extinct

ÖHYB: (Austria)/commercial hybrid orig. from Piétrain ×

(Edelschwein × Landrace)/*see also* WINNIE

OIC: (Ohio, USA)/ld/orig. in 19th c. from white pigs from Chester County, Pennsylvania, bred for greater size cf. Improved Chester White/BS 1897/syn. *Ohio Improved Chester*/extinct

Old Danish: *see* Danish Black Pied

Old English: (England)/orig. of Essex, Hampshire, Wessex Saddleback/ extinct

Old Glamorgan: *see* Welsh

Old Oxford: *see* Oxford Sandy-and-Black

Old Swedish Spotted: (Sweden)/syn. *Black-White-Red Spotted*/extinct 1978

Oliventina: (Badjoz, Spain)/orig. var. of Extremadura Red improved by (or orig. from) Alentejana/syn. *Raya* (= frontier, i.e. of Spain and Portugal)/[from Olivenza]

Omsk Grey: (Omsk, Russia)/m/breed group/orig. 1949–1963 from Kemerovo, Siberian Spotted and Large White × local Tara/Russ. *Omskaya seraya porodnaya gruppa*/extinct by 1984

Oplemenjena šarena mesnata rasa: *see* Serbian Spotted

OR Line: (France)/Large White line for high ovulation rate

Orielese: *see* Cosentina

Ossabaw Island: (Georgia, USA)/usu. black, also red, red with white spots, black-and-white spotted, red-and-black spotted or tan; small; obese; feral/? orig. from early Spanish pigs/HB 1986 for captive population maintained elsewhere/rare

Ouest Hybride: (France)

Oxford Sandy-and-Black: (Oxfordshire, England)/m/red with sandy hairs and black blotches; semi-lop ears/orig. type extinct after last boar licensed 1963/64, reconstructed with revived BS 1985; HB/syn. *Axford, Old Oxford, Oxford Forest, Plum Pudding pig, Sandy Oxford*/rare

Ozark: *see* Mulefoot

P76: *see* Pen ar Lan P77, Penshire 66

Pacages, Porc laineux des: *see* Mangalitsa

PAK: *see* Pakhribas

Pakhribas: (Nepal)/black/orig. at Pakhribas Agric. Centre, Dhankuta, from British Saddleback (63%), Fa Yuen (18%), Tamworth (9%), Hampshire (5%) and Large White (5%)/syn. *PAK*

Palatin: (Romania)/Celtic type/extinct

Palouse: (Washington, USA)/m/orig. 1945–1956 at Washington Agric. Exp. Sta. from Danish Landrace ♂♂ (50%) × Chester White ♀♀ (50%)/HB 1956/extinct

Pampa: (Uruguay)/local criollo

Panlang: (China)/local var./nearly extinct

Parmense: (Parma, Emilia, Italy)/dark grey/syn. *Black Emilian* (It. *Emiliana negra*), *Black Parma*, *Parmigiana*, *Reggio* (It. *Reggiana*)/extinct by 1976

Pasasti prašič: *see* Krškopolje Saddleback

Pasture: *see* German Pasture

Pau'a: (Hawaii)/native/rare

PBZ: *see* Polish Landrace

Pearl River Delta: *see* Cantonese

PECC: *see* Carélie

PEGG: *see* Gallia

Peking Black: *see* Beijing Black

Pelatella: *see* Casertana

PELL: *see* Laconie P77

Pelón: (Mexico to Colombia)/ld.m/black; small/Iberian type, sim. to (? orig. from) Black Hairless (Spain)/syn. *Black hairless of the tropics*, *Birish* (Yucatan, Mexico), *Pelón de cartago* (Costa Rica), *Pelón Tabasqueno* (Tabasco, Mexico)/[= hairless]

Pen ar Lan: *see* Carélie, Gallia, Laconie, Penshire, Tiameslan

Penbuk: (N Korea)/orig. from North China × local Korean

Penshire P66: (Brittany, France)/m/black, brown and white belted or spotted/stress-resistant line orig. 1977–1984 by Pen ar Lan Breeding Co., Maxent, from Hampshire, Large White and Duroc, for crossing with Laconie P77 to produce ♂♂ for terminal crossing, P76/syn. *PEPP*/rare

Penzhou Mountain: (China)/var. of Huchuan Mountain

PEPP: *see* Penshire P66

Pereira: (Jardinópolis, São Paulo, Brazil)/grey roan (or black), sometimes with red spots; semi-erect ears/orig. by Domiciliano Pereira Lima from Canastra and Duroc; later improved by selection at Santa Cabriela Exp. Sta. at Sertãozinho/sim. to (or =) Moura

Périgord: (France)/larger var. of Limousin, crossed with Craonnais and Large White/Fr. *Périgourdin*/almost extinct

Peris 345: (Romania)/terminal crossing strain orig. 1989 from Belgian Landrace (56%), Duroc (36%) and Hampshire (8%)/syn. *345 peris*, *Peris hybrid LS 345*, *Synthetic Peris*/nearly extinct

Perna curta: *see* Tatu

Persilancan: (Indonesia)/syn. *babi-Persilancan* (*babi* = hog)

Perugina: (Perugia, Umbria, Italy)/grey with white spots/sim. to Chianina/extinct

Petren: *see* Piétrain

PETT: *see* Tiameslan P44

Pfeifer, Pfeiffer: *see* Black Slavonian

Philippine Native: (Philippines)/black, or black with white belly/vars inc. Ilocos, Jalajala/orig. of Berkjala, Diani, Kaman, Koronadel, Libton/? extinct

Piau: (Paranaíba R basin, SW Brazil)/m.ld/white, cream, grey or sandy, with black spots, occ. tricolour or dark roan with black spots/Iberian type/orig. from Canastra or Canastrão crossed with Poland China and Duroc/vars: large, medium and small (medium and large selectively improved from 1939 at São Carlos Exp. Sta. by Antônio Teixeira Viana)/BSd/syn. *Carioca*

(Rio de Janeiro); regional names include Goiano, Francano, Triangula Mineiro/not *Piauí*/[*piau* suggests speckled or painted; name also given to various other pigs with dark spots]/rare

Piau pequeño: *see* Caruncho

PIC HY: (England)/Pig Imp. Co. hybrid ♂ from Large White, Landrace, Piétrain and Belgian Landrace

Pied German: *see* Bentheim Black Pied

Piégut: (Bussière-Badil, Dordogne, France)/local var. of Miélan since early 20th c./extinct

Piétrain: (Brabant, Belgium)/m/dirty white with black or reddish spots, often red hairs; semi-lop ears/? orig. (*c.* 1919–20) from Bayeux (? and Tamworth) × local/HB 1950, BS 1952, general recog. 1955; HB also in Brazil, France 1958, Germany 1961, GB, Italy, Luxembourg, Netherlands, Spain/Russ. *P'etren*/not *Petren*, *Pjetren*

Piétrain ReHal: (Belgium)/stress-negative Piétrain line/orig. by introgressing negative stress gene from Large White into Piétrain by successive back-crosses

Pilsen: *see* Přeštice

Pineywoods Rooter: (Louisiana, USA)/feral/syn. *Gulf pig, Pineywoods, Swamp hog*/not *Piney Woods*

Pingtan Black: (Fujian, China)/local var./rare

Pirapetinga: (Minas Gerais, Brazil)/ ld/black or violet/Chinese orig. from Nilo × Tatu/syn. *Mandi*/not *Pirapitinga*/[name of river]/ nearly extinct

Pitman-Moore Miniature: (Iowa, USA)/selected from Vita Vet Lab Minipig 1967–1973 at College of Medicine, Univ. of Iowa/extinct

Pjetren: *see* Piétrain

Plateau type: *see* Tibetan

Plaung: *see* Raad

Plum Pudding pig: *see* Oxford Sandy-and-Black

Plzeň: *see* Přeštice

PM-1: *see* Poltava

Poaka: *see* Kunekune

Podolian: (W Ukraine)/ld/black pied/ breed group/sim. to Dnieper, Krolevets and Mirgorod/orig. from Berkshire, Middle White and Large White × local/Russ. *Podol'skaya porodnaya gruppa*/ syn. *Podolian Black Pied* (Russ. *Podol'skaya chernopestraya*)/ extinct by 1984

Poitou: *see* Montmorillon

Poland China: (Ohio, USA)/m.[ld]/ black with white points/orig. 1835–1872 from Berkshire, Big China, Byfield, Irish Grazier, Miama *et al.*/part orig. of Spotted/HB 1878; HB also Canada/not *Poland-China*/[a settler from *Poland*, and Big *China*]/nearly extinct

Polesian: (Pripet Marshes, Ukraine)/ formerly var. of Small Polish Prick-eared/orig. of Krolevets/ Russ. *Polesskaya*/syn. *Sarny*/extinct

Polesian Lard: *see* Krolevets

Polesskaya: *see* Polesian

Polesskaya salnaya: *see* Krolevets

Polish Landrace: (Poland)/m/orig. from Polish Marsh improved since mid-19th c. by English and German breeds, since 1908 by German (Improved) Landrace and (one var. only) since 1954 by Swedish Landrace/lines 21, 23 and 24 carry Norwegian, German and Welsh blood, resp./HB 1962/ syn. *Polish White Lop-eared* (Pol. *Polska biała zwisłoucha* or *PBZ*, Ger. *Schlappohriges Landschwein*)

Polish Large White: (Poland)/m or m.ld/orig. since late 19th c. from Large White and Edelschwein × Polish Native, re-formed (new imports) 1947/vars (combined 1956): Pomeranian Large White, White Prick-eared, English Large White/recog. and HB 1956;

BSd/Pol. *Wielka biała polska*, *WBP*

Polish Marsh: (N Poland)/lop ears; prim./orig. of Polish Landrace/ Pol. *Żuławska*/extinct

Polish Native: *see* Large Polish Long-eared, Polish Marsh, Small Polish Prick-eared

Polish Synthetic Line 990: (Poland)/ Pol. *Linia 990*

Polish White Lop-eared: *see* Polish Landrace

Poltava: (Ukraine)/m/orig. since 1966 from Russian Large White, Mirgorod, Landrace, Piétrain and Wessex Saddleback at Poltava Pig Breeding Res. Inst., later with Duroc and Hampshire blood/ Russ. *Poltavskaya*/syn. *Poltava commercial hybrid, Poltava Meat, Poltava Meat line (PM-1)*

Pomeranian Large White: (NW Poland)/m.ld/former var. of Polish Large White/orig. in 20th c. from Large White × White Prick-eared/recog. 1936; graded to Edelschwein 1940–1945/Pol. *Wielka biała pomorska*/syn. *Pomorze*/extinct

Pomorze: *see* Pomeranian

Poppel: *see* Belgian Landrace

Porc amélioré du pays: *see* Swiss Improved Landrace

Porc français de type danois: *see* French Landrace

Porc laineux des Pacages: *see* Mangalitsa

Porcul Alb de Banat: *see* Bazna

Porcul de Banat: *see* Banat

Portek: (Mexico)/hybrid

Portuguese: *see* Alentejana, Bisaro

Portuguese Retinta: *see* Alentejana

Pot Belly: *see* Vietnamese Potbelly

Premier: (Cambridgeshire, England)/ hybrid ♀ from Premier Piglink breeding co.

Přestice: (Plzeň, Bohemia, Czech Republic)/m.ld/black pied/local lop-eared saddleback orig. crossed with Middle White, Berkshire, Large Black and Bavarian Landrace at end of 19th c., improved (1957) by Wessex Saddleback, Essex and German Saddleback/HB 1964/Cz. *Přeštické, Přeštice cernostrakato*/ obs. syn. *Bohemian Blue Spotted, Pilsen, Plzeň, Saaz*/rare

Pridneprovskaya porodnaya gruppa: *see* Dnieper

Pridonskaya: *see* Don

Prikamskaya: *see* Kama

Prisheksninsk: (Vologda, Russia)/ m.ld/var. of Large White with local blood/Russ. *Prisheksninskaya*

PRN: *see* Dutch Landrace

Proligène: (Bourgogne, France)/ ♀ lines bred by SCAPAAG of Dijon/121 (French Large White), 321 (French Landrace)/321 rare

Protest pig: *see* Husum Red Pied

PS Line: (France)/Large White line selected for prenatal survival rate

Pua'a: *see* Kunekune

Puang: *see* Raad

Pudong White: (E Shanghai, China)/ sim. to Fengjing/rare

Puebla: *see* Torbiscal

Pugliese: *see* Apulian

Puławy: (Lublin, E Poland)/m.ld/ black-and-white pied (more black than white), sometimes with small red patches; erect ears/orig. since 1926 at Borowina res. sta. (for State Res. Inst. of Farming in Puławy) from local pied pigs (orig. from Berkshire × Small Polish Prick-eared in early 20th c.) upgraded by Large White/HB 1935/Pol. *Puławska*/ syn. (1926–1943) *Gołębska* (from Gołąb)/rare

Pushi Black: (China)/var. of Xiangxi Black/syn. *Tiegu*

Putian: (Fujian, China)/black/Central China type/inc. in Fujian Small Pig/syn. *Fujian Black*

Pygmy: *see* Froxfield Pygmy, Ghori

pygmy hog: (Bhutan/Assam, India) = *Sus salvanius*/nearly extinct

Qianshao Spotted: (China)/pied/local var./rare

Qingping: (C Hubei, China)/black/ Central China type

QM Hamline: (Canada)/terminal sire

line bred by National Pig Development (also in S Africa)

Raad: (N Thailand)/black/syn. *Plaung* or *Puang* (NE), *Ka Done* or *Keopra* (S)/not *Rad*/almost extinct

Rad: *see* Raad

Ran-Yu: *see* Lan-Yu

Raoping: (China)/var. of Yuedong Black

Ras-n-lansa: (Guam, Mariana Is)/ dwarf

Rat pig: *see* Wuzhishan

Raya: *see* Oliventina

Razor Back: (S of USA)/feral/term now usu. refers to feral hogs in S Arkansas and Ozark National Forest

Red Andalusian: *see* Extremadura Red

Red Berkshire: *see* Kentucky Red Berkshire

Red Hamprace: (USA)/red with black spots/red line occurring during formation of Montana No. 1 (Black Hamprace)/extinct

Red Iberian: *see* Alentejana, Extremadura Red

Red Mangalitsa: (Crişana, Romania)/ dark to light brownish-red with black points/var. of Mangalitsa/ orig. from crossing Mangalitsa and local Salontai/Ger. *Rot Wollschwein*/rare

Red Pied Husum: *see* Husum Red Pied

Red Portuguese: *see* Alentajana

Red Wattle: (E Texas, USA)/tassels/ orig. obscure, present type derived from those found early 1970s in wooded area of E Texas/ BS, HB; BS also Canada 1984/ syn. *Red Waddle*/not *Wattler*/ rare

Reggiana, Reggio: *see* Parmense

Reggitana: (Reggio, Calabria)/former var. of Calabrian/extinct

Reichenau: *see* Rychnov

Rena: (France)/Large White × British Landrace cross

Resava: (Serbia)/m.ld/white with black spots; lop ears/cf. Morava/?

orig. from Berkshire × Mangalitsa and Šumadija/Serbo-cro. *Resavska, Resavska svinja*/ syn. *Vezičevka* (from Vezičevo)

Retinta: *see* Alentejana, Extremadura Red

Rhodesian Indigenous: *see* Mkota

Rice-bran pig: *see* Fengjing

Riminese: (Rimini, Italy)/dark red, with star on forehead/former var. of Romagnola/extinct

Ring Middle, Ring Necked: *see* Hampshire

RM: *see* Rostov Meat

Robuster: (South Africa)

Romagnola: (Emilia, Italy)/m/dark brown or copper coloured, with grey skin/former vars: Faetina, Forlivese, Riminese/F_1 from cross with Middle White and Yorkshire (Bastianella or San Lazzaro) = Fumati or Brinati/It. *Mora, Mora romagnola*/syn. *Bologna* (It. *Bolognese*), *Castagnona* (= chestnut), *Moor*/ nearly extinct by 1976 (remnants acquired by Mario Lazzari 1982 and Univ. of Torino 1998 to conserve)

Roman: *see* Maremmana

Romanian Landrace: (Romania)/m/ orig. from British and Swedish Landraces/HB 1958

Romanian Large White: (Romania)/ m/orig. from Large White/HB 1958/Rom. *Marele alba*

Romanian Meat Pig: (Bărăgon Steppe, S Romania)/m/orig. at Ruşeţu Exp. Sta. since 1950 from Russian Large White × local Stocli sows backcrossed to Large White/Rom. *Porcul românesc de carne*, Ger. *Rumänisches weisses Fleischschwein*/nearly extinct

Romanian Native: (Danube valley, S Romania)/grey; prick ears; prim./ vars: Bălţăreţ, Stocli, Transylvanian/disappearing

Romanian Saddleback: *see* Bazna

Rongchang: (uplands of Sichuan, China)/ld.m.bristles/white with black spectacles, occ. with black spots elsewhere or all white/SW

China type/WG *Jung-ch'ang*,
Russ. *Zhunchan, Zhungan*, Ger.
Jungtschang/syn. *Bispectacled
White, Jungchang White*/not
Young-tschang, Yungchang

Rossosh Black Pied: (Voronezh,
Russia)/ld/breed group/orig. 1943
from Berkshire × (Large White ×
local), with Berkshire × Mirgorod
blood in 1949/Russ.
*Rossoshanskaya chernopestraya
porodnaya gruppa*/extinct by 1984

Rostov Meat: (Russia)/m/breed
group/orig. since 1973 from
Russian Large White, Russian
Short-eared White, Piétrain and
Welsh/syn. *RM, Rostov*

Rotbunte Husumer Schwein: *see*
Husum Red Pied

Rotbunte Schleswig-Holsteiner: *see*
Husum Red Pied

Rot Wollschwein: *see* Red Mangalitsa

Rubia, Rubia campiñesa: *see*
Andalusian Blond

Ruijin: (China)/var. of Ganzhongnan
Spotted

Ruşeţu White: (Romania)/white
breed at Ruşeţu Exp. Sta./Rom.
rasa Alb de Ruşeţu

Russian Large White: (European
Russia)/m/orig. from Large White
imported since 1880 and esp.
since 1922/strains: Central Russian
KBV-1 and KB-KN, Moscow MM-
1/HB 1932/Russ. *Krupnaya
belaya*/obs. syn. *Soviet Large
White* (Ch. *Subai*)/also in China;
see also Belarus Large White,
Ukrainian Large White, Vitebsk

Russian Long-eared White: (Russia)/
m/= German (Improved)
Landrace imported to Kuibyshev
1931/Russ. *Dlinnoukhaya belaya*

Russian Short-eared White: (Russia)/
= German Edelschwein imported
to Krasnodar (N Caucasus) 1927
on/Russ. *Korotkoukhaya belaya*

Rychnov: (Czech Republic)/orig.
1865 from Large White, Middle
White and Poland China ×
local/Cz. *Rychnovské*/extinct
1933 by crossing with
Edelschwein

Saaz: *see* Přeštice

Saba: (Yunnan, China)/var. of Wujin

Saddleback: *see* Angeln, Basque,
Bazna, British, German, German
Pasture, Hampshire, Huazhong,
Husum Red Pied, Jinhua,
Limousin, Monchuan, Přeštice,
Siena, Swabian-Hall/syn. *Belted,
Sheeted*

St Yrieix: *see* Limousin

Sakhalin White: (E Siberia, Russia)/
ld/orig. from Large White
(imported 1932) × local/Russ.
Belaya sakhalinskaya

Samoan: (S Pacific)

Samólaco: (Sondrio, Lombardy,
Italy)/pied/extinct

Sampedreño: (Colombia)/Criollo
type/rare

San Daniele, Sandanielese: *see* Friuli
Black

Sandy Oxford: *see* Oxford Sandy-
and-Black

Sanjiang White: (Heilongjiang,
China)/m/lop ears/orig. from
Landrace × (Landrace × Min)
1973–1983 (recog.)/[= 3 rivers]

San Lazzaro: (Faenza, Italy)/strain of
inbred Large White (imported
1875) used for crossing with
Romagnola/sim. to Bastianella/
extinct

San Pierre: (Indiana, USA)/black-
and-white/orig. 1953 from
Chester White and Berkshire/
extinct

Santiago, Santiaguese: *see* Galician

Sardinian: (Italy)/black, grey, fulvous
or pied; long bristles/local with
Craonnais, Large White,
Berkshire and Casertana blood/It.
Sarda

Sarny: *see* Polesian

Satzeling, Saziling: *see* Shaziling

Schaumann hybrid: (Germany)/
Duroc × German Landrace F_1 at
Schaumann Res. Sta.,
Hülsenberg, *c.* 1984

Schischka: *see* Šiška

Schlappohriges Landschwein: *see*
Polish Landrace

Schumadija: *see* Šumadija

Schwäbisch-Hällisches: *see* Swabian-
Hall

Schwalbenbäuchig Wollschwein: *see*
Black Mangalitsa
Schwalbenbauch Mangalitza: *see*
Black Mangalitsa
Schwarzbuntes Dermanzi: *see*
Dermantsi Pied
Schwarzes slavonisches: *see* Black
Slavonian
Schwarz-Weisses Bentheimer: *see*
Bentheim Black Pied
Schwerfurt Meat: (Schwerin and
Erfurt, E Germany)/m/orig.
1970–1975 from Belgian
Landrace × (Piétrain ×
Lacombe), named 1986/HB
1970/Ger. *Schwerfurter
Fleischrasse* (or *Fleischschwein*)
Scutari: *see* Shkodra
Semirechensk: (SE Kazakhstan)/
m.ld/white, occ. black-pied, dark
brown or tan; short ears; hy/orig.
1947–1966 from Asiatic wild
($\frac{1}{16}$–$\frac{1}{8}$) × Large White ($\frac{3}{4}$–$\frac{7}{8}$) and
Kemerovo ($\frac{1}{16}$–$\frac{1}{8}$)/recog. 1978/
Russ. *Semirechenskaya*/syn. (till
1978) *Kazakh hybrid breed
group* (Russ. *Kazakhskaya
gibridnaya porodnaya gruppa*)
Sénone: (France)/Large White × Pen
ar Lan P88
Serbian: *see* Šumadija
Serbian Spotted: (Serbia)/m/white
with black spots; lop ears/orig.
from Mangalitsa, Large White,
Swedish Landrace, Dutch
Landrace and Piétrain/Serbo-cro.
Oplemenjena šarena mesnata rasa
(= *Improved Spotted Meat breed*)/
syn. *Yugoslav Spotted*/rare
Severokavkazskaya: *see* North
Caucasus
Shahutou: *see* Shawutou
Shajiaben: *see* Jiangquhai
Shan: (China)/var. of Huang-Huai-Hai
Black/? = Shanzhu/[mountain]
Shanggao: (Jiangxi, China)/local
breed
Shanghai White: (Shanghai, China)/
m/semi-erect ears/orig. from
Large White × Fengjing and
Meishan/recog. 1979
Shanghang: *see* Hang
Shangxian Spotted: *see* Shengxian
Spotted

Shanxi Black: (Shanxi, China)/m.ld/
orig. from Yorkshire, Berkshire,
Russian Large White, Neijiang
and Mashen/recog. 1983
Shanxi Lean Meat type: (China)/
pied/orig. from Berkshire,
Russian Large White, North-East
China Spotted and local/recog.
1990
Shanzhu: (Jiangsu, China)/black; lop
ears/orig. of Nanjing Black/not
Shantu/[*shan* = mountain, *zhu* =
pig]
Shawutou: (China)/var. of Taihu/not
Shahutou
Shaziling: (Hunan, China)/small lop
ears/var. of Huazhong Two-end
Black, sim. to Monchuan/WG *Sa-
tze-ling*, Ger. *Sa-ze-ling*/not
Saziling
Sheeted: *see* Saddleback
Sheeted Essex: *see* Essex
Shengxian Spotted: (Zhejiang,
China)/black, with white feet/
Central China type/syn. *Sheng
County Spotted*/not *Shangxian
Spotted*
Shen Spotted: (China)/var. of North-
East China Spotted/syn.
Shenyang
Shenxian: (Hebei, China)/black, hy/
var. of Huang-Huai-Hai Black/
extinct
Shenzhou: (China)/var. of Huang-
Huai-Hai Black
Shiang: *see* Xiang
Shishka: *see* Šiška
Shkodra: (Albania)/often black spots/
var. of Albanian/It. *Scutari*/not
Skutari
Short-eared White: *see* Russian
Short-eared White
Shuiwei: (China)/var. of Hanjiang
Black
Shumadija, Shumadinka: *see*
Šumadija
Siamese: *see* Akha, Hailum, Kwai,
Raad
Siamese: (Thailand)/black, with
white feet, copper-coloured
skin/name for type first imported
to UK *c.* 1732 to improve British
pigs; also influenced old
Neapolitan/syn. *Tonkey* or

Tunkey (= Tonquin district in Indo-China from which it was imported)/extinct

Sibao: (China)/var. of Cantonese

Siberian: (Omsk and Novosibirsk, Russia)/coloured; prim./orig. of Kemerovo, North Siberian, Omsk Grey, Siberian Black Pied/syn. *local Siberian, local Novosibirsk, Tara* (Russ. *Tarskaya*)/extinct

Siberian Black Pied: (N Omsk and Novosibirsk, Russia)/breed group/orig. from coloured culls in breeding North Siberian/part orig. of Omsk Grey/Russ. *Sibirskaya chernopestraya*/syn. *Novosibirsk Spotted, Siberian Spotted*/rare

Siberian type: (Russia)/var. of Soviet Meat

Sichuan White line I: (China)/line bred from Landrace, Meishan and Chenghua in closed herd

Sicilian: *see* Black Sicilian, Calascibetta, Madonie

Siena Belted: (Tuscany, Italy)/m/ black with white belt; lop ears/ part orig. of Siena Grey/It. *Cinta senese*/syn. *Cinta, Cinta italiana, Montagnola*/not *Sienna*/rare

Siena Grey: (Tuscany, Italy)/m.ld/ white with grey patches on head and rump/1st cross (F_1) from Large White × Siena Belted/It. *Grigia senese*/syn. *Bigio trimacchiato*

Silvela: (Huelva, Spain)/strain of Black Iberian

Sinchin: *see* Xinjin

Sinclair Miniature: (USA)/orig. at Sinclair Farm, Missouri, by Mike Tumbelson from Minnesota Miniature and Pitman-Moore Miniature

Sin Hwai: *see* New Huai

Sino-European: (Île-de-France, France)/line orig. 1985 on by France Hybrides, Evry, from Meishan, Jiaxing Black, Landrace and Large White/Fr. *Ligne Sino-Européene*/*see also* Tiameslan

Sino-Gascony: (SW France)/1st cross (1986) of Gascony with Meishan or Jiaxing Black/Fr. *Sino-Gascon*

Sino-Vietnamese: (S Vietnam)/orig. early 20th c. from Heo Hon crossed with Hainan and other Chinese breeds/orig. (with Craonnais) of Boxu/? extinct

Sintrão: *see* Torrejano

Šiška: (Croatia, Bosnia and Serbia)/ m/grey; prick-eared; prim./orig. of Šumadija, Turopolje/Ger. *Schischka*/not *Shishka, Siska*/extinct before 1982

Sitsin, Sitsiner, Sitszin: *see* Xinjin

Skutari: *see* Shkodra

SL96, SL98: *see* Czech synthetic lines

Slavic Meat: (N Caucasus, Russia)/ Southern and Steppe types/Russ. *Slavyanskaya Myasnaya*/syn. *Slavonic Meat, SM-1*

Slavonian: *see* Black Slavonian

Slovakian Black Pied: (Nitra, Slovakia)/ld/orig. (1952 on) from Mangalitsa, Berkshire, Large Black, Czech Improved White, Piétrain *et al.*/Sl. *Slovenské čiernostrakaté*/syn. *Czechoslovakian Black Pied*

Slovakian Improved White: (Slovakia)/= Czech Improved White in Slovakia/Sl. *Slovenské biele zošl'achtené, Biele ušl'achtilé*

Slovakian White Meat: (Slovakia)/ orig. (1962–1979, recog. 1980) from Czech Improved White (37.5%) and Canadian, Swedish and Polish Landraces (62.5%) starting with reciprocal crosses/ Sl. *Slovenské biele mäsové*, Ger. *Weisses Fleischschwein in der Slowakei*/syn. *Slovakian Improved Landschwein, Slovakian Landrace, Slovakian White*

Slovenian White: (Pomurje, Slovenia)/m/orig. from Edelschwein (*c.* 1900) with some Swedish Large White blood (1983)/Sn. *Bela žláhtna* (= improved white), Cro. *Slovenačka bela*/syn. (Sn.) *Yorkšir* /rare

Slovenské: *see* Slovakian

Slovhyb-1: (Slovakia)/orig. from Czech Improved White, Landrace and Slovakian White

Slutsk Black Pied: (W Belarus)/ld/ breed group/orig. from imported × native before 1919/Russ. *Chernopestraya slutskaya porodnaya gruppa*, Ger. *Slusker*/ extinct by 1984

SM-1: *see* Slavic Meat, Steppe Meat

small Bamei: *see* Xiaohua

Small Black: (E England)/Neapolitan blood 1830, ? Chinese blood/ vars: Black Essex, Black Suffolk/ orig. of var. of Large Black in Suffolk and Essex and of American Essex/extinct

Small Kansu: *see* Hezuo

small long-snout: (Taiwan) *see* Taiwan Small-ear

Small Polish Prick-eared: (SE Poland)/black, red, pied or white; prim./formerly inc. Polesian/orig. (with Berkshire) of Puławy/Pol. *Mała polska ostroucha*/syn. *local short-eared, native prick-eared*/extinct

small short-eared: (Taiwan) *see* Taiwan Small-ear

Small White: (England)/Chinese (Cantonese) blood 1780 on/HB 1884/syn. *Middlesex, Small Yorkshire*/extinct

SML: (Ukraine)/orig. from Poltava, Landrace, Hampshire and Duroc/[Specialized Meat Line]

Solomons native: (Solomon Is)

Somo: (Mali)/black, grey or black pied

Sopot: (Croatia)/commercial hybrid lines

Sorocaba: (São Paulo, Brazil)/m/ red/orig. 1950–1960 from red Caruncho ($\frac{3}{8}$), Tamworth ($\frac{3}{8}$) and Duroc ($\frac{1}{4}$)

Sortbroget: *see* Danish Black Pied

South African Landrace: (S Africa)/ m/orig. from Danish, Dutch and Swedish Landraces/HB 1959

South Central Jiangxi Spotted: *see* Ganzhongnan Spotted

South China: (W Malaysia)/m/head and back black, belly and legs white/orig. from Cantonese imported in 19th c./syn. *Cantonese, Local Chinese*

South China Black: (W Malaysia)/m/ black

SOUTH CHINA type: (subtropical S China)/usu. black, black-and-white; sway-backed, pot-bellied, precocious/inc. Diannan Small-ear, Hainan, Huai, Lantang, Liang Guang Small Spotted, Longlin, Wuzhishan, Xiang, Yuedong Black

South Type: (Moldova)/m/dam line, established 1990/Moldovan *Tipul Sudic*/rare

SOUTH-WEST CHINA type: (Suchuan basin and Yunnan-Guizhou plateau, China)/usu. black, also black-and-white or red; semi-lop ears/inc. Chenghua, Guanling, Kele, Huchuan Mountain, Neijiang, Rongchang, Wujin, Yanan

South Yunnan Short-eared: *see* Diannan Small-ear

Sovereign: (England)/blue/hardy ♀ line by Masterbreeders from Westrain and Landrace; earlier line of sim. orig. was Monarch

Soviet Large White: *see* Russian Large White

Soviet Meat: (Russia and Ukraine)/ orig. 1981–1991 at 19 centres from Poltava with some blood of Kharkov and Belarus hybrids/ vars: Central Russian, Don, Lipetsk, Southern (Krasnodar), Steppe Meat

SPM: (SW England)/tan and white, often belted/orig. from Duroc × Hampen/[(*S*cotbeef, *P*eninsular and *M*arks & Spencer collaboration]

spotted: (China) = pied (inc. spotted)/ Ch. *Hua*

Spotted: (Indiana, USA)/ld/black with white spots/orig. from local spotted (with Miami blood), Poland China, Gloucestershire Old Spots (1 ♂ and 1 ♀, imported 1914)/BS 1914; HB also Canada, Italy/syn. *Spot, Spotted Poland, Spotted Poland China* (to 1961)

Spotted Andalusian: *see* Jabugo Spotted

Spotted Jabugo: *see* Jabugo Spotted
Spotty Lanyu: (Hung t'ou I, Taiwan)/
inbred population selected from
Lan-Yu for spotty white coat
colour
Srem, Sremica: *see* Black Mangalitsa
Sri Lanka Native: (Sri Lanka)/usu.
black, also dark grey, tan, tan-
and-grey, tan-and-black; prim.
Staffordshire: *see* Tamworth
Steirisches Weisses Edelschwein: *see*
Styrian White Edelschwein
Steppe Meat: (Stavropol, Rostov and
Volgograd, Russia)/var. of Soviet
Meat/orig. from Poltava ($\frac{1}{2}$),
Rostov Meat ($\frac{1}{4}$) and Belarus Meat
($\frac{1}{4}$)/5 zonal types/syn. *SM-1*
Steppe Ukrainian: *see* Ukrainian
Steppe
Sterling: (England)/white/hybrid ♀
by Masterbreeders from Duroc
Stocli: (Romania)/mt var. of
Romanian Native/sing. *Stoclu*/
syn. *Mountain Stocli*/not *Stokli*/?
extinct/*see also* Băltăreţ
Stora vita Engelska: *see* Swedish
Large White
Strei: (Hunedoara, Transylvania,
Romania)/ld.m/black/orig. (1877
on) from Large Black and
Mangalitsa × local/not *Stroi*/[R
Strei]/extinct
Styrian White Edelschwein: (Styria,
Austria)/orig. 1891 on from Large
White and Edelschwein × local/
Ger. *Steirisches Weisses
Edelschwein*
Subai: *see* Russian Large White
Subi: (China)/var. of Diannan Small-
Ear
Subotica White: (N Vojvodina,
Serbia)/m/white; lop ears; curly
coat/orig. from Large White ×
Mangalitsa, with Lincolnshire
Curly Coat blood/Serbo-cro.
Domaća mesnata (= local
meat)/syn. *Bikovačka* (from
Bikovo), *Subotička Mangulica*/
nearly extinct
Suffolk: (England) *see* Black Suffolk
Suffolk: (New York, USA)/m/orig.
from Small White in 1850s/
extinct by 1940s
Sulawesi warty pig: (Sulawesi,

Indonesia)/= *Sus celebensis*/syn.
Celebes wild pig
Šumadija: (N Serbia)/ld/orig. from
Šiška/orig. of Mangalitsa/Serbo-
cro. *Šumadinka, Šumadinska
svinja*, Ger. *Schumadija*/syn.
Miloš (Serbo-cro. *Miloševa*)/not
*Shumadinka, Shumadija,
Sumadia*/extinct before 1982
Sumatran: (Indonesia)/wild type
Sumba: (Indonesia)
Surány: (E Hungary)/extinct
Sus barbatus: *see* bearded pig
Sus bucculentus: (Vietnam)/
rediscovery of wild species from
Annamite Range in Laos in 1995
Sus celebensis: *see* Sulawesi warty
hog
Sus salvanius: *see* pygmy hog
Sus scrofa: = wild boar
Sus verrucosus: *see* Javanese warty
pig
Sussex: (England)/black, blue or
slate-coloured; large lop ears/?
absorbed by Large Black/extinct
Sussex: (USA)/black (earlier belted);
lop ears/extinct soon after 1907
Suzhong: (China)/hybrid lines 1 and
2 (line 1 is lean-type maternal)
Svensk: *see* Swedish
Swabian-Hall: (Württemberg,
Germany)/white with black head
and rump; lop ears/orig. from
local Land pig with Berkshire
and Essex blood in late 19th c.
and Wessex in 1927/+ Angeln =
German Saddleback/BSd 1925,
BS till 1970, revived 1986, new
HB 1983/Ger. *Schwäbisch-
Hällisches*/not *Swabian-Halle*/
rare
swamp hog: *see* Florida swamp,
Pineywoods Ranger
Swedish Landrace: (Sweden)/m/orig.
from local (0.2%) graded to
Danish Landrace (98.5%)
imported chiefly 1914–15 and
1935–39, with some German
blood (1.3%)/BS 1907, HB
1911/orig. of British Landrace/
Swe. *Svensk Lantras, Förädlad
Lantras* (= improved Landrace)
Swedish Large White: (Sweden)/var.
of Large White from British and

German imports/Swe. *Stora vita Engelska, Svensk Yorkshire*

Swedish Spotted: *see* Old Swedish Spotted

Swiss Edelschwein: (W Switzerland)/m/orig. from Large White and Middle White/HB 1911/Ger. *Schweizerisches Edelschwein*, Fr. *Grand porc blanc*/syn. *Swiss Large White, Swiss Yorkshire*

Swiss Hampshire: (Switzerland)/Hampshire (HB) in Switzerland; recog. 1994

Swiss Improved Landrace: (C and E Switzerland)/m/orig. from Large White (imported 1880–1890) and German (Improved) Landrace (imported 1910–1953) × native, improved by Dutch Landrace and Danish Landrace (1981)/HB 1911/Ger. *Schweizerisches veredeltes Landschwein*, Fr. *Porc amélioré du pays*

Swiss Large White: *see* Swiss Edelschwein

Swiss Yorkshire: *see* Swiss Edelschwein

Synthetic Peris: *see* Peris

Syrmian: *see* Black Mangalitsa

Szalonta: (NE Hungary)/red; prim./18th c. var., absorbed into Mangalitsa in early 19th c./extinct

Tabasqueno: *see* Pelón

Tagudin: *see* Libton

Ta-ho: *see* Dahe

Ta-hua-pai: *see* Cantonese

Taichung: (C Taiwan)/ld/black/orig. 1960s at Taichung Dist. Agric. Imp. Sta., Peitou, from Taoyua × Tingshuanghsi/extinct by 1979

Taihu: (Taihu area of Jiangsu and Zhejiang, China)/m/black; wrinkled face; long lop ears; prolific/Lower Changjiang River Basin type/vars: Erhualian, Fengjing, Hengjing, Jiaoxi, Jiaxing Black, Meishan, Mi, Shawutou/not *Taiwu*/[great lake]

Taiwan Ebony: (Taiwan)/synthetic line from native/syn. *Taiwan Black*

Taiwan Miniature: *see* Lan-Yu, Lee-Sung

Taiwan Native: *see* Meinung, Taichung, Taoyuan, Tingshuanghsi; syn. *Formosa*

Taiwan Small Black: *see* Taiwan Small-ear

Taiwan Small-ear: (SW Taiwan)/black; long straight face, very small ears; prim./? orig. from local Taiwan wild (*Sus scrofa taivanus*) with S China blood/syn. *aboriginal, small short-eared, small long-snout, Taiwan Small Black*/extinct except on Lan-Yu I

Taiwan Small Red: (Wushe, Taiwan)/red; slightly dished wrinkled face; short ears; straight tail; stiff mane in ♂; prim./? orig. from S China and local Taiwan wild (*Sus scrofa taivanus*)/extinct

Taiwu: *see* Taihu

Tamworth: (England)/m/golden red/recog. 1850–1860; BS and HB 1906; BS also USA 1923, Australia; HB also Canada/syn. *Staffordshire*/nearly extinct in GB

Tangzhui: (China)/var. of Guangdong Small-ear Spotted

Tanyang: *see* Danyang

Taoyuan: (NW Taiwan)/ld/black or dark grey; lop ears; wrinkled skin; straight tail/orig. from S China/WG *T'ao-yüan*, Jap. *Toyen*/syn. *Chung-li, Lung-tan-po*/not *Taoyung, Tauyuen*/rare

Taoyuan Black: (China)/var. of Xiangxi Black/syn. *Yanquan*

Tara, Tarskaya: *see* Siberian

Tatu: (S Brazil)/ld/usu. black; hairless/Chinese orig. (? Cantonese)/orig. of Caruncho, Nilo, Pirapetinga/Port. *Tatú*/syn. *Bahia* (not *Baé, Baia, Baié*) (NE), *Macao, Macau* (Macau = Chinese port held by Portuguese from 1557), *Perna curta* (= short leg), *Tatuzinho*/[= armadillo (from shape of head)]/nearly extinct

Taung: (Katha, Myanmar)/prick ears; hy; lt/? orig. from Large White/syn. *Wetaung*

Tauyuen: *see* Taoyuan

Teano: *see* Casertana

Ta-wei-tzu: *see* Daweizi

Tetra-S: (Hungary)/3-line hybrid derived from Duroc, Dutch Landrace and Large White or Belgian Landrace

Thai: (Thailand)/usu. black/ indigenous/vars (now very sim. through crossbreeding): Akha, Hailum, Kwai, Raad

Thai Binh Black Pied: (Red R delta, NE Vietnam)/white with large black patches/Fr. *Pie-noire de Thai Binh*

Thailand: *see* Akha, Hailum, Kwai, Numpard, Raad, Siamese, Thai

Thin Rind: *see* Hampshire

Thuoc Nhieu: (Mekong delta, Vietnam)/white with pied bristles/orig. from Large White (1936–1956) and Middle White (1957) × Boxu

Tiameslan: (Brittany, France)/Sino-European composite line orig. 1983–1985 by Pen ar Lan Breeding Co., Maxent, from (Meishan × Jiaxing) ♂♂ × European ♀♀ (Laconie composite line)/not *Tia Meslan*/ nearly extinct

Tiameslan P44: (Brittany, France)/ white composite of Laconie P77, Meishan and Jiaxing; closed line since 1984/syn. *PETT*/rare

Tia Sing: *see* Jiaxing Black

Tibetan: (Qinghai-Tibet plateau, China)/black, occ. brown; hy; dwarf/Plateau type/vars: A-ba, Diqing, Hezuo/Ch. *Zhangzhu* (= Tibet pig), *Xizang*

Tiegu: *see* Pushi Black

Tiehe: (China)/var. of Hangjiang Black

Tielu: (China)/var. of Hangjiang Black

Tigerschwein: *see* Baldinger Spotted

Ting, Tinger: *see* Ding

Tinghsien: *see* Ding

Tingshuanghsi: (NE Taiwan)/ld/sim. to Taoyuan but larger frame/not *Din-Hsiang-Chi*/extinct by 1979

Tipul Sudic: *see* South Type

Tongcheng: (SE Hubei, China)/var. of Huazhong Two-end Black/WG *T'ung-ch'eng*, Ger. *Tung-tschen*

Tong Con: (Chinese border, Vietnam)/local var.

Tonkey: *see* Siamese

Torbiscal: (Extremadura, Andalucía, Castilla y León and Castilla La Mancha, Spain)/m/red, occ. white, sometimes grey hair on coronary groove/var. of Iberian; line produced *c.* 1950 (closed ever since) by blending four old strains (Ervideira, Campanario, Caldeira, Puebla) of Iberian pigs from Alentejo and Extremadura (Ervideira = golden pig from Alentejo; Puebla = hairless black from Guadiana valley; both nearly extinct)/rare

Torrejano: (C Portugal)/black and white/crossbred orig. from Bisaro/syn. *Sintrão*/[from Torres Vedras]/disappearing

Tournayais: (France)/short-eared var. of Gascony/extinct

Toyen: *see* Taoyuan

Trang Phu Kanh: (C Vietnam)/ developed from local and Yorkshire, selected for 10–20 years

Transtagana: *see* Alentejana

Transylvanian: (Romania)/var. of Romanian Native/extinct

Tribred: (England)/blue/outdoor hybrid ♀ by Premier Piglink, with Welsh blood

Troll: *see* Munich Miniature

Tschechisches weisses Edelschwein: *see* Czech Improved White

Tschen-Hua: *see* Chenghua

Tsentral'nyi typ: *see* Central Russian type

Tsinghwa: *see* Jinhua

Tsivilsk: (Chuvash Republic, Russia)/ m.ld/breed group/orig. *c.* 1956 from local Large White × local Chuvash/Russ. *Tsivilskaya porodnaya gruppa*/Ger. *Zivilsker*

Tswana: (Botswana)

Tszilin: *see* Jilin

Tszinkhua: *see* Jinhua

Tunchang: (Hainan I, China)/ld/black back, white body/var. of Hainan/ syn. *Denchang*

Tuncheng, T'ung-ch'eng, Tung-tschen: *see* Tongcheng

Tungyang: *see* Dongyang

Tunkey: *see* Siamese

Turmezei: (Croatia)/ancient var./
Hung. *Túrmezö*/extinct
Turopolje: (Croatia)/m.ld/grey, yellow
or white, with black spots; semi-
lop ears; curly hair/ orig. from
Šiška and Krškopolje with
Berkshire blood at end of 19th
c./Cro. *Turopoljska, Turopoljska
Svinja*/syn. *Zagreb*/not
Turopolier/nearly extinct (only in
Lonjsko Polje Nature Park)
Tuy Hoa Hairless: (Song Bar R delta,
Vietnam)/white/Fr. *Race sans
poils de Tuy Hoa*
Two-end-black pig: *see* Jinhua

UG hybrid: (India)/developed from
base population of indigenous
(Khasa) crossed with Hampshire/
[UG = upgraded]
UKB: (Ukraine)/type of Ukrainian
Large White (Russ. *Ukrainskaya
krupnaya belaya, = UKB*)
developed by selection; UKB-1,
then UKB-2 (recog. 1994)
Ukrainian: (Ukraine)/spotted/local
population of European short-
eared type/orig. (with English
breeds) of Dnieper, Mirgorod,
Podolian, Ukrainian Spotted
Steppe, Ukrainian White Steppe/
extinct
Ukrainian Large White: (Ukraine)/
formerly var. of Russian (or
Soviet) Large White/Russ.
Ukrainskaya krupnaya belaya
Ukrainian Meat: (Ukraine)/orig. from
Poltava by crossing with Welsh,
Ukrainian White Steppe,
Estonian Bacon and Ukrainian
Spotted Steppe
Ukrainian Spotted Steppe: (S
Ukraine)/ld.m/occ. black, also
black-and-white or black-and-tan
spotted; semi-lop ears; with
bristles, otherwise sim. to
Ukrainian White Steppe/orig.
from Ukrainian White Steppe ×
local spotted, with some
Berkshire and Mangalitsa blood/
recog. 1961/Russ. *Ukrainskaya
stepnaya ryabaya*/rare
Ukrainian White Steppe: (S Ukraine)/
m.ld/orig. since 1925 (recog.

1934) from Ukrainian with 2 or 3
crosses of Large White/Russ.
Ukrainskaya stepnaya belaya/
syn. *White Ukrainian*
Ulster, Ulster White: *see* Large White
Ulster
Umbrian: *see* Maremmana
Ungarisches weisses Fleischschwein:
see Hungarian White
Unhudo de Goias: (Goias, Brazil)/
large hoof
Upton-Meishan: (Canada)
Urzhum: (Kirov, Russia)/m./orig.
(crossing since 1893, breed dev.
since 1945) from Large White ×
local lop-eared/recog. 1957/
HB/Russ. *Urzhumskaya*

Valdesequera: (Spain)/strain of
Iberian at Valdesequera farm,
Extremadura/representative of
Extremadura Red colour
phenotype
Valtellina: (N Lombardy, Italy)/m/
black/extinct
Vara: *see* Furão
Vasco: *see* Basque
VDL: *see* German (Improved)
Landrace
Veliki Jorkir: (Croatia)/m/orig. from
Large White imported from UK
and Germany/syn. *Large White*/
[*velik* = large, *Jorkir* = Yorkshire]/
rare
Vénète: (France)/French Landrace ×
Pen ar Lan P66
Venezuelan Black: (llanos of S
Venezuela and Colombia)/semi-
feral
Veredeld Landvarken: *see* Belgian
Landrace (Improved)
Veredeltes Landschwein: *see*
Improved Landrace
Vermelho: *see* Canastra
Vezičevka, Vezičevo: *see* Resava
Vich: (Catalonia, Spain)/Celtic type,
sim. to Vitoria/syn. *Catalan*/
extinct
Victoria: *see* Curtis Victoria, Davis
Victoria
Vietines kiaules: *see* Lithuanian
Native
Vietnamese: (Vietnam)/major breeds:
Ba Xuyen, Í, Mong Cai, Thuoc
Nhieu; minor breeds: BaTri, BSI,

Co, DBI, HaBac, Lang Hong, Meo,
Muong Khuong, Nghia Binh
White, Thai Binh Black Pied,
Tong Con, Trang Phu Kanh, Tuy
Hoa Hairless; *see also* Boxu, Heo
Hon, Heo Moi, Sino-Vietnamese

Vietnamese Potbelly: (USA, GB *et
al.*)/dwarf/orig. 1960s from Í
breed of Vietnam via Sweden
and Canada/BS USA, GB/syn.
Pot belly, Vietnamese Potbellied

Vietnamese Yorkshire: (Vietnam)/
orig. from Large White imported
from Russia and Germany

Vita Vet Lab Minipig: (Marion,
Indiana, USA)/light grey/orig.
(1948 on) from Florida Swamp
pigs/extinct

Vitebsk: (Russia)/strain of Russian
Large White at Vitebsk stud/orig.
1970–1990 by selection of
Belarus Large White

Vitoria: (Álava, N Spain)/m/hairless;
lop ears/orig. from local Celtic
type improved by Large White *et
al.*/Sp. *Vitoriana, Chato de
Vitoria*/syn. *Álava, Basque-
Navarre, Chato* (= short-nosed)
vitoriano/extinct

Waddle: *see* Red Wattle

Wai Chow: (Hong Kong)/black/var. of
Cantonese/WG *Wei-chou*/syn.
Lung Kong/extinct

Wanbei: (China)/black; lop ears/orig.
from Berkshire, Yorkshire,
Rongchang and Neijiang/recog.
1988

Wanbei: (N Anhui, China)/var. of
Huang-Huai-Hai Black

Wangtai: (China)/var. of Minbei
Spotted

Wannan Spotted: (China)/pied/var. of
Wanzhe Spotted/Ch.
Wannanhua/syn. *Lantian
Spotted*

Wanshan: *see* Wenchang

Wanzhe Spotted: (Anhui/Zhejiang,
China)/m/black-and-white
pied/Central China type/vars:
Chunan Spotted, Wannan
Spotted

Warren County: *see* Big China, Miami

Wattle, Wattler: *see* Red Wattle

Wattles: *see* Mamellado

WBP: *see* Polish Large White

Weasel: *see* Black Mangalitsa

Wei: (SE Anhui, China)/black/Lower
Changjiang River Basin type/syn.
Xu, Yu/rare

Wei-chou: *see* Wai Chow

Weining: (China)/var. of Wujin

Weisse Charbiner: *see* Harbin White

**Weisses Fleischschwein in der
Slowakei:** *see* Slovakian White

Welsh: (Wales)/m/lop ears/crossed
with Swedish Landrace since
1953/BS and HB 1918/syn.
(before Landrace cross) *Old
Glamorgan*/rare

Wenchang: (Hainan I, China)/black
back and head, white body and
snout/var. of Hainan, sim. to
Tunchang but smaller/Ger.
Wenschan/not *Wanshan*

Wenshan: (China)/var. of Diannan
Small-ear

Wenzhou White: (? Zhejiang, China)/
lop ears/orig. from Landrace and
Russian Large White × local/
recog. 1980

Wessex Saddleback: (England)/m/
black with white belt/orig. from
Old English/combined with
Essex 1967 to form British
Saddleback/HB and BS 1918–
1967/syn. *Belted, Sheeted
Wessex*/extinct as separate breed

West African: (W Africa, S of lat.
10°–14° N)/black, white or pied;
dwarf; prim./Iberian type/? Port.
orig./inc. Ashanti Dwarf, Bakosi,
Nigerian Native/syn. *West
African Dwarf*/*see also* Somo

West French White: (NW France)/
m/Celtic type/formed 1955 by
joining Boulonnais, Breton,
Craonnais, Flemish, Normand/
HB 1958/Fr. *Blanc de l'Ouest*/
syn. *Porc de l'Ouest, Western
White*/nearly extinct

West Jiangxi Black: *see* Exi Black

Westrain: (England)/m/orig. by Walls
from *Wessex* × Piétrain

Wet: (Myanmar) *see* Mountain Dwarf

Wetaung: *see* Taung

Wettringer: *see* Bentheim Black Pied

White: *see* Large White, Middle
White, Small White

White Edelschwein: *see* Edelschwein
White-faced: *see* Hereford
White Large Black: *see* British Lop
White Long-eared: *see* Russian Long-eared White
White Lop: *see* British Lop
White Mangalitsa: (Serbia)/var. of Mangalitsa/Serbo-cro. *Beli Soj Mangulica*/rare
White Prick-eared: (W Poland)/ m.ld/former var. of Polish Large White/orig. since late 19th c. from Edelschwein × Polish Native; named 1918/combined with large White in 1956 to form Polish Large White/Pol. *Biała ostroucha*/syn. *Polish White Short-eared*/extinct
White Russian: *see* Belarus
White Short-eared: *see* Russian Short-eared White
White-shouldered Essex: *see* Essex
White Thoroughbred: *see* Czech Improved White
White Ukrainian: *see* Ukrainian White Steppe
Wielka biała polska: *see* Polish Large White
Wielka biała pomorska: *see* Pomeranian
Wielka polska długoucha: *see* Large Polish Long-eared
wild: (Europe and Asia)/= *Sus scrofa* Linnaeus; divided into about 20 subspp./*see also* bearded pig, Javanese warty pig, pygmy hog, Sulawesi warty pig
Willebrand: (Pouitou-Charentes, France)/composite of Hampshire and Yorkshire (from USA, 1978)/nearly extinct
Windsnyer: (South Africa)/longer-snouted indigenous type than Kolbroek
Windsor: *see* Middle White
WINNIE: (Austria)/commercial hybrid from Duroc × (Edelschwein × Landrace)
Woburn: *see* Bedford, Hampshire
Wollhaariges Weideschwein: *see* Mangalitsa
Wollschwein: *see* Mangalitsa
WPB: *see* Polish Large White

Wuanzhehua: (? Hebei, China)/ spotted
Wujin: (border of Yunnan, Guizhou and Sichuan, China)/usu. black, occ. brown; small/SW China type/vars: Dahe, Kele, Liangshan, Saba, Weining/not *Wugin, Wujing*
Wulanhada: (China)/black/orig. from Berkshire, Jilin Black, Xinjin and local/recog. 1983
Wuning Spotted: (China)/var. of Hang
Wuyi Black: (Jiangxi/Fujian, China)/ Central China type/vars: Gandong Black, Minbei Black/[Wuyi mts]
Wuzhishan: (Hainan and C mts of Guangdong)/black, often with white belly and feet; dwarf/ South China type/syn. *"Rat pig"*/not *Wuzheshan*/nearly extinct

X20, X30, X80: (France)/lines bred by Dalland Meuse of Stenay: X20 = ♀ line (French Large White); X30 = ♀ line (var. of Piétrain but not stress-sensitive; syn. *Daltrain*); 80 = ♂ line (composite)
Xi: *see* Bamei
Xiamao: (China)/var. of Minbei Spotted
Xiang: (S Guizhou and N Guangxi, China)/black, often with white belly and feet; miniature/South China type/not *Shiang, Xing*
Xiangcheng: (China)/extinct
Xiang White: (China)/orig. from Russian Large White, Landrace and Large White, × local/recog. 1989
Xiangxi Black: (Hunan, China)/ Central China type/vars: Daheping, Pushi Black, Taoyuan Black
Xiaoer: (China) = small-eared
Xiaohua: *see* Liang Guang Small Spotted
Xiaohuo: (China)/small var. of Bamei
Xiaomin: (China)/small var. of Min/syn. *Hebao*/[*xiao* = small]
Xing: *see* Xiang
Xingzi Black: (China)
Xinhuai: *see* New Huai
Xinjiang Black: (China)/orig. from

Berkshire and Middle White ×
local/recog. 1989

Xinjiang White: (C and N Xinjiang,
China)/orig. from Russian Large
White × local, with Berkshire
and Yorkshire blood/recog.
1982/not *Xinjing*

Xinjin: (NE China)/m/black with
white points/orig. in 1920s from
Berkshire × local (Min)/vars: Jilin
Black, Ning-an, Xinjin (Liaoning)/
recog. 1980/WG *Hsin-chin*, Ger.
Ching Ching/syn. *Sitsin, Sitsiner,
Sitszin*/not *Sinchin*

Xinjing White: *see* Xinjiang White

Xizang: *see* Tibetan

Xu: *see* Wei

Yanan: (Sichuan, China)/ld.m/
black/South-West China type

Yangkou: (China)/var. of Minbei
Spotted

Yangxin: (SE Hubei, China)/black/
Lower Changjiang River Basin
type/vars: Meihuaxing, Yanxin
Black/syn. *Yingxin*/rare

Yanquan: *see* Taoyuan Black

Yanxin Black: (China)/var. of
Yangxin

Yayang: (China)/local var./rare

Yili White: (NW Xinjiang, China)/m/
orig. from Russian Large White ×
local white crossbreds/recog.
1982

Yimeng Black: (Shandong, China)/
m.ld/orig. from Berkshire × local
black/recog. 1989/Ch.
Yimenghei/[Lingyi and
*Meng*yang counties]

Yingxin: *see* Yangxin

Yongkang: (Zhejiang, China)/var. of
Jinhua/WG *Yung-k'ang*, Russ.
Yunkan, Ger. *Jungkang*

Yongyun: *see* Gongguan

York, Yorkshire: *see* Large White,
Middle White, Small White

Yorkshire Blue and White: (N
England)/? orig. from Large
White × Large Black/syn.
Bilsdale Blue/extinct (last boar
licensed 1963–64)

Young-tshang: *see* Rongchang

Yu: *see* Wei

Yuanling: *see* Daheping

Yucatan Miniature: (Colorado, USA)/
grey; hairless/local Yucatan
selected for small size at
Colorado State Univ. since 1972/
syn. *Yucatan Micropig*

Yuedong Black: (E Guangdong,
China)/South China type/vars:
Huiyang Black, Raoping/syn.
Aodong

Yugoslav Meat Breed: (Serbia)

Yugoslav Spotted: *see* Serbian
Spotted

Yujiang: (Yushan, Jiangxi, and
Jiangshan, Zhejiang,
China)/black/Central China
type/vars: Guangfeng Black,
Jiangshan Black, Yushan
Black/[*Yu*shan and *Jiang*shan]

Yungchang: *see* Rongchang

Yungkang: *see* Yongkang

Yunnan Small-eared: *see* South
Yunnan Short-eared

Yushan Black: (China)/var. of
Yujiang

Zabumba: *see* Canastrão

Zagreb: *see* Turopolje

Zang, Zangzhu: *see* Tibetan

Zhang: (China)/= Tibetan

Zhao: (China)/var. of Huang-Huai-Hai
Black/syn. *Zhaozhu*

Zhejiang Medium White: (China)/?
blue spots; semi-lop ears/orig.
from Landrace and Middle White
× local/recog. 1980

Zhewan Spotted: (China)

Zhongdong: (China)/var. of
Guangdong Small-ear Spotted

Zhongmin: (China)/var. of Min/syn.
Ermin/[= middle Min]

Zhoujiazi: *see* Dongchuan

Zhunchan, Zhungan: *see* Rongchang

Zimbabwe Indigenous: *see* Mkota

Zivilsker: *see* Tsivilsk

Zixi Black: (Jiangxi, China)

Zlahtna White: *see* Slovenian White

Złotniki: (Poznań, Poland)/m.(ld)/ orig.
(1946–1962) by selection from
Large Polish Long-eared with
some Large White blood/vars:
Złotniki Spotted, Złotniki
White/HB 1962/Pol. *Złotnicka*/
[name of res. sta. run by College of
Agriculture in Poznań]/rare

Zlotniki Spotted: (NW and C
Poland)/m.ld/white with black
spots; lop ears/var. of Złotniki/
orig. from primitive (erect- and
lop-eared, rescued from
extinction by Prof.
Alexandrowicz at Złotniki Res.
Sta.) introduced from Vilnius
region in 1950s, selected for meat
and lard/Pol. *Złotnicka Pstra*/rare
Zlotniki White: (C Poland)/m/lop
ears/dam line orig. 1946–1952
by Prof. Alexandrowicz from
primitives (erect- and lop-eared)
introduced from Vilnius region
selected for meat, upgraded with
Swedish Landrace/Pol. *Złotnicka
Biała*/nearly extinct

Żuławska: *see* Polish Marsh
Zungo: (N Colombia)/ld/black or
grey; hairless/Criollo type/syn.
Zungo costeño
Zuoan: (China)/var. of Ganzhongnan
Spotted

Sheep

Abbreviations used in this section:
cw = coarsewooled; d = milk; fr = fat-rumped; ft = fat-tailed (broad or S-shaped); fw = finewooled (Merino type); hd = horned; hr = long hair; hy = hairy (i.e. short-haired, woolless); lft = long fat tail; lt = long thin tail (to ground); lw = longwooled (medium fine); m = meat; mw = medium wool (crossbred Merino/longwool, intermediate between coarse and fine wool); pd = polled; sft = short fat tail; st = short tail (halfway to hocks); sw = shortwooled (medium fine); w = wooled (if inadequate information for classification of wool type)

Colour is assumed to be white unless otherwise indicated. *Fleece* is only described if exceptional (e.g. very long, curly, or hair instead of wool). *Tails* are thin and of medium length, as in many European breeds, unless otherwise indicated. *Horn status* is noted where known, in both male and female.

Names for sheep include: birka, carneiro, defaid, får, hitsuji, juh, ovca, ovce, oveja, ovelha, ovets, ovtsi, owca, pecora, schaap, schaf, skaap, yang

Aansi: (S of N Yemen)/m.cw/brown (*bunni* [Arabic, = brown], not *bowni*), black and white vars; pd; ft/not *Ainsi, Ansi*

Abda: *see* Doukkala

Abidi: *see* Ibeidi

aborigenna kotlenska: *see* Kotel

ABRO Damline: (Scotland)/m/pd/ line to replace Border Leicester as sire of lowland ewes, developed since 1967 at Animal Breeding Res. Organisation, Edinburgh, from Finnish Landrace (47%), East Friesian (24%), Border Leicester (17%), Dorset Horn (12%), selected for 8-week litter weight

Abyssinian: *see* Ethiopian

Acchelé Guzai: *see* Akele Guzai

Acipayam: (Ege region, Turkey)/ d.m.w/orig. from Assaf × (Awassi × Dağliç), i.e. $\frac{1}{2}$ Awassi, $\frac{1}{4}$ Dağliç, $\frac{1}{4}$ East Friesian

Adal: (Dancalia, NE Ethiopia)/m/ blond (white to light brown), occ. pied or dark brown; hy; pd; often earless/syn. *Adali, Afar*

Adikarasial: *see* Kilakarsal

Adromalicha: *see* Sfakia

Afar: *see* Adal

Afghan: (Afghanistan)/inc. Afghan Arabi, Baluchi, Gadik, Ghiljai, Hazaragie, Kandahari, Karakul, Turki

Afghan Arabi: (N Afghanistan)/ cw.m/usu. black or grey with white face blaze, also white; pd; long ears; fr/vars in Herat (usu.

white with black nose, probably
Kandahari blood) and in Kataghan
and Bamyan (usu. brown,
probably crosses with Turki)

Africana: *see* Red African

Africander, Africaner: *see* Afrikander

African Hair sheep: *see* Virgin
Islands White

AFRICAN LONG-FAT-TAILED: (E and S
Africa)/hy; tail usu. long and fat:
broad with hanging tip, twisted,
cylindrical, strap-shaped, funnel-
shaped or carrot-shaped/inc.
Afrikander, Madagascar, Malawi,
Mondombes, Nguni, Rwanda-
Burundi, Sabi, Tanzania Long-
tailed, Tswana/syn. *East African
Long-fat-tailed*

AFRICAN LONG-LEGGED: hy/♂ hd, ♀
pd; lop ears; thin tail/orig. from
Ancient Egyptian/inc. Angola
Long-legged, Congo Long-legged,
Baluba, Sahelian, Sudan Desert

African Red: *see* Red African

Afridi: *see* Tirahi

Afrikander: (S Africa)/various
colours; hy; pd or hd; lft/orig.
from Hottentot/vars: Damara,
Namaqua, Ronderib/Afrik.
Afrikaner/syn. *Africander, Cape
Fat-tailed*/not *Africaner*

Afrino: (S Africa)/m.fw/orig. 1969–
1976 (named) at Carnarvon Exp.
Sta., Cape prov., from South
African Mutton Merino ×
(Ronderib Afrikander × South
African Merino)/BS 1980

Afshari: (Iranian Kurdistan)/m/? var.
of Red Karaman

Agnis: (Indiana, USA)/mahogany red
with white poll and tail tip

Agra, Agrah: *see* White Karaman

agrinon: *see* Cyprus mouflon

Agul: (Azerbaijan)/var. of Lezgian

Ainsi: *see* Aansi

Ait Barka: (SE of Marrakech,
Morocco)/black/small var. of
Berber

Ait Haddidou: (between Rich and
Moulouya valley, Morocco)/
white, often with black marks on
head/var. of Berber

Ait Mohad: (High Atlas, Morocco)/
white; pd/large var. of Berber

Akele Guzai: (Eritrea)/cw/usu.

black/var. of Ethiopian
Highland/It. *Acchelé Guzai*/syn.
Shimenzana (It. *Scimenzana*)

Akhangran Mutton-Wool:
(Uzbekistan)/m.w/orig. from
Caucasian and Lincoln ×
Jaidara/Russ. *Akhangaranskaya
myaso-sherstnaya*

Ak-Karaman: *see* White Karaman

Aknoul: (Morocco)/black/small
earless var. of Berber

Akrah: *see* White Karaman

Aktyubinsk Semicoarsewool: (NE
Kazakhstan)/m.mw/orig. from
Sary-Ja × Kazakh Fat-rumped/
Russ. *Aktyubinskaya
polugrubosherstnaya*

Alai: (S Kyrgyzstan)/m.w/occ.
coloured spots on head and legs;
♂ pd or small horns, ♀ pd;
fr/orig. 1934–1981 from Kirgiz
Fat-rumped with Précoce blood
1938–1940 and 1945, and Sary-Ja
blood since 1953/Russ.
*Alaiskaya, Alaiskaya
kurdyuchnaya* (= Alai Fat-
rumped)

Alaskan white: *see* Dall sheep

Ala-Tau argali: *see* arkhar

Al-Awass: *see* Awassi

Albanian: *see* Common Albanian

Albanian Vlach: *see* Arvanitovlach

Albanian Zackel: (Albania)/inc.
Common Albanian, Mati, Pied
Polisi, Shkodra; *see also*
Bregdetit, Golemi, Havasi, Kuçi,
Zagoria

Alcarreña: (La Alcarria, Guadalajara
and Cuenca, New Castille,
Spain)/m.mw/sometimes light
brown marks on head and legs,
occ. all black; pd/Entrefino
type/syn. *Manchega pequeña*
(i.e. small var. of Manchega)/not
Paloma

Alfort: *see* Dishley Merino

Algarve Churro: (S Faro, Portugal)/
m.cw/white with black spots on
face and feet, or black (10%);
hd/orig. from Andalusian Churro
imported *c.* 1870–1890/Port.
Churra algarvia

Algerian: *see* Algerian Arab, Beni
Guil, Raimbi

Algerian Arab: (Algeria)/m.cw/♂ hd,

♀ pd/orig. of Tadmit/syn. *Ouled Jellal* (Fr. *Ouled Djellal*), *Western Thin-tailed* (Fr. *Queue fine de l'Ouest*) (Tunisia)

Alieskie: (Tajikistan)

Aljaf: (N Iraq)/migrant var. of Iraqi Kurdi/not *Gaf*

Allmogefår: (C and N Sweden)/m/ various colours; ♂ hd, ♀ pd; st/remnants of old landrace/[= peasantry sheep]/rare

Alpagota: (Alpago, Belluno, Venetia, Italy)/d.m.cw/white with black or chestnut spots on face; pd/ Lop-eared Alpine group but small ears/composite of Lamon, Vicentina and Istrian Milk/ HB/syn. *Pagota*/rare

Alpine: *see* French Alpine, Lop-eared Alpine, Savoiarda, Swiss White Mountain, Thônes-Marthod

Alpine Lop-eared: *see* Lop-eared Alpine, Savoiarda

Alpine meat breeds: *see* Lop-Eared Alpine

Alpine Piedmont: *see* Biellese

Altai: (SW Siberia, Russia)/fw.m/orig. 1934–1949 at Rubtsovsk state farm (now Ovtsevod breeding centre) from American Rambouillet (1928–1936), Caucasian and Australian Merino (1936 on), × Siberian Merino/recog. as breed group 1940 and named *Siberian Rambouillet* (Russ. *Sibirskiĭ Rambulye*) or *Soviet Rambouillet of Siberian type* (Russ. *Sovetskiĭ Rambulye sibirskogo tipa*); recog. as breed 1968; HB/Russ. *Altaĭskaya, Altaĭskaya poroda tonkorunnykh ovets* (= Altai breed of finewooled sheep)/syn. *Altai Merino*/not *Altay, Altayan*

Altai, Altaĭskaya: *see* also Telengit

Altai Merino: *see* Altai

Altai Mountain: (Russia)/mw/breed group/orig. from Tsigai 1945 × [Merino × (Merino × local coarsewooled) 1940]/Russ. *Gorno-altaĭskaya porodnaya gruppa*

Altamurana: (Altamura, Bari, Apulia, Italy)/d.cw.(m)/occ. dark spots on face; ♂ usu pd, ♀ pd/BSd 1958, HB 1972/syn. *delle Murge*/rare

Altay Fat-rumped: (N Xinjiang, China)/var. of Kazakh Fat-rumped/not *Altai Fat-rumped*

Amalé: *see* Hamari

Amasya Herik: *see* Herik

American bighorn: *see* bighorn

American Four-horned: *see* Jacob, Navajo-Churro

AMERICAN HAIR SHEEP: (tropical America)/m/usu. tan or white, also pied; hy, ♂ usu. with mane and throat ruff (except Brazil); pd/orig. from West African Dwarf (? from C Africa, e.g. Angola)/inc. Bahama Native, Barbados Blackbelly, Morada Nova, Pelibüey, Red African, Santa Inês, Virgin Islands White; *see also* Boricua

American Karakul: (USA)/orig. from Karakul end 19th c./HB 1929/ rare

American Merino: (USA)/fw/orig. from Spanish Merino imported 1793–1811/BS 1906/vars: A type (or Vermont), B type, C type (or Delaine)

American Rambouillet: (USA)/fw.m/ ♂ hd or pd, ♀ pd/orig. from Rambouillet imported 1840– 1860 from France and 1882– c.1900 from Germany/var.: Polled Rambouillet/BS 1889; HB also Canada

American Tunis: (USA)/m.d/red face and legs, born red; pd; [ft]/orig. from Tunisian Barbary imported in 1799; almost extinct late 19th c., modern type orig. from a Tunisian ram × improved Leicester Longwool ewes, with some Southdown blood to correct big ft/BS 1896/rare

ammon: *see* argali

Ammotragus lervia: *see* aoudad

Amran Black: (N and NW of San'a, Yemen)/m.cw/pd; ft

Amran Grey: (NW and W of San'a, Yemen)/m.cw/light to dark grey; pd; ft

AMS Merino: (W Australia)/fw/hd or

pd/orig. since 1967 by Jim Shepherd from many Merino strains/BS is Australian Merino Society (as distinct from Australian Stud Merino Breeders' Association)

Anatolian fat-tailed breeds: *see* Awassi, Dağliç, Herik, Red Karaman, White Karaman

Anatolian Merino: *see* Central Anatolian Merino

Anatolian red: (S Anatolia, Turkey)/= *Ovis orientalis anatolica* Val./ var. of red sheep/syn. *Anatolian wild, Cilician*

Ancient Egyptian: (NE Africa)/hy; ♂ hd; lop ears; thin tail/orig. of African Long-legged/syn. *Hamitic Long-tailed, Ovis longipes palaeoaegypticus*/ extinct

Ancon: (Massachusetts, USA)/ achondroplastic mutant occurring in 1791 (similar mutants in Norway 1919 and Texas 1960)/syn. *Otter*/[Gr. *agkon* = elbow]/extinct *c.* 1876

Andalusian Churro: (coast of Huelva and Cadiz, Andalucía, Spain)/ m/white with black spots on face, ears brown or black; ♂ hd, ♀ hd or pd/largest var. of Spanish Churro/orig. of Algarve Churro/syn. *Atlántica, Churra Lebrijana, Lebrijana* (from Lebrija), *Marismeña* (from marshes, *marismas*, of Guadalquivir delta)/rare

Andalusian Merino: (W Andalucía, Spain)/m.mw-fw/larger var. of Spanish Merino, with coarser wool/subvar.: Grazalema/Sp. *Merina andaluza entrefina*/syn. *Campiñesa, Córdoba de la Campaña*/rare

Andi: (Dagestan, Russia)/black (sometimes with white patches on head and tail tip), or white (often with coloured markings on head, legs and tail); hd/ Caucasian Fat-tailed type/Russ. *Andiĭskaya*/declining

Andorra, Andorran: *see* Spanish Churro

Angara Merino: (Irkutsk, Russia)/var. of Krasnoyarsk/orig. from Altai (1955) × (Précoce × Buryat)/ recog. 1974/Russ. *Priangarskiĭ Merinos*

Anglesey: (NW Wales)/extinct 1929

Anglo-Merino: (England)/Negretti and Paular strains of Spanish Merino × various English breeds (esp. Southdown and Ryeland) at beginning of 18th c./syn. *English Merino*/extinct

Anglo-Merino: (Poland) *see* Polish Corriedale

Anglomerynos: *see* Polish Corriedale

Angola Long-legged: (Malanje and Moxico, Angola)/m/white, occ. pied; hy; ♂ hd, ♀ pd/African Long-legged group

Angola Maned: (Huambo, C Angola)/ m/white, black or pied; hy with mane and throat ruff in ♂; pd or hd/syn. *Coquo*/cf. West African Dwarf

Anogia: *see* Psiloris

Ansi: *see* Aansi

Ansotana: (Ansó valley, NW Huesca, Aragon, Spain)/var. of Aragonese with longer wool and semi-open fleece/syn. *Calva*, also (with Roncalesa) *Entrefina pirenaica* (= *Pyrenean Semifinewool*)/not *Paloma*

Aohan Finewool: (S Inner Mongolia, China)/fw.m/♂ hd, ♀ pd/orig. in early 1950s from Russian Merinos × Mongolian, with Australian blood in 1970s/syn. *Aohan Merino*

Aouasse: *see* Awassi

aoudad: (N Africa)/= *Ammotragus lervia* (Pallas), actually a maned beardless goat/type sp. in Maghreb, subspp. in Egypt (rare), Libya (rare), S Sahara, Sudan/Fr. *mouflon à manchettes*, Ger. *Mähnerschaf*, It. *muflone berbero*, Sp. *árrui*/syn. *arui, barbary wild sheep, bearded argali*/not *audad, udad*

APENNINE group: (Apennines from Tuscany/Emilia to Abruzzo, Italy)/m.d.mw/usu. white, sometimes black; hd or pd/mixed type/orig. from Merino × local with Bergamasca and Sopravissana blood/inc. Apennine, Cornella White, Garfagnina White, La Spezia local, Massese, Nostrana, Pagliarola, Pomarancina, Vissana, Zucca Modenese/It. *Appenninica*

Apennine: (Apennines, Italy)/ m.(d.mw)/pd; semi-lop ears/ Apennine group/local vars formerly recog. inc. Barisciano, Casciana, Casentinese, Chianina, Chietina, Corngliese, Perugian Lowland, Senese, Tarina, Varzese/HB 1981/It. *Appenninica*

Appennino-Modenese: *see* Pavullese

Appenzell: (Switzerland)/part orig. of Swiss White Mountain/Ger. *Appenzeller*/extinct

Apulian Merino: *see* Gentile di Puglia

Aquila, Aquilana: *see* Barisciano

ara-ara: (SC and C Niger)/m/Fulani group of Sahel type/syn. *are-are*

Arab: (Algeria) *see* Algerian Arab

Arab: (Egypt) *see* Barki

Arab: (Somalia) *see* Somali Arab

Arab: (Turkey) *see* Awassi

Arab: (W Africa) *see* Maure

Arabi: (Afghanistan) *see* Afghan Arabi

Arabi: (SW Iran, S Iraq, NE Arabia)/ m.cw/black, also pied or white with black head; ♂ hd, ♀ pd/cf. Shafali (Syria)/Near East Fat-tailed type/orig. of Wooled Persian (S Africa)/not *Arbi* (fem. *Arbiyah*), *Erbi*

Arabi: (Russia)/black var. of Karakul

Arabian Long-tailed: *see* Najdi

Aragats: (Armenia)/var. of Armenian Semicoarsewool/orig. at Aragats state farm from Balbas × [Rambouillet or Lincoln × (Rambouillet or Lincoln × Balbas)]/Russ. *Aragatsskaya*; obs. syn. *Armenian Semifinewooled Fat-tailed breed group* (Russ. *Armyanskaya polutonkorunnaya zhirnokhvostaya porodnaya gruppa*)

Araghi: (Iran) *see* Baluchi

Aragonese: (Aragon and W Catalonia, Spain)/m.mw/pd/Entrefino type/ vars: Ansotana, Improved (*Mejorado*) or de Valle, Monegrina, Roncalese, Turolense/BS/Sp. *Rasa aragonese*/[Sp. *rasa* = smooth (fleece)]

Aral, Aralsk: *see* South Kazakh Merino

Aranesa: (Arán valley, NW Lérida, Catalonia, Spain)/= Tarasconnais

Arapawa Island: (Marlborough Sounds, New Zealand)/fw/usu. black with white blaze, occ. white; hd; feral/orig. from Australian Merino in 1860s/syn. *Arapawa Merino*/rare

Arbi, Arbiyah: *see* Arabi

Archa: *see* Harcha

Arcott: *see* Canadian Arcott, Outaouais Arcott, Rideau Arcott/ [*Animal Research Centre Ottawa*]

ARC (dam) Strain 2: *see* Outaouais Arcott

ARC (dam) Strain 3: *see* Rideau Arcott

ARC (sire) Strain 1: *see* Canadian Arcott

Ardèche, Ardèchois: *see* Blanc du Massif Central

Ardennais tacheté: *see* Houtland

Ardennes: *see* Brabant Foxhead

Ardes: (S Puy-de-Dôme, C France)/ d.m.cw/pd/Central plateau group/orig. from Lacaune × local/not *Ardres*/extinct

are-are: *see* ara-ara

argali: (C Asia, from Himalayas to Mongolia)/= *Ovis ammon* Linnaeus/shades of brown; large horns of ammon type; st/many vars (subspp.) inc. arkhar, Marco Polo's sheep/syn. *O. polii* Blyth, *ammon*/[Mongolian = wild sheep]

Argentine Cormo: (Patagonia, Argentina)/mw/sim. to Cormo (Australia)/orig. by John Blake on Condor Estate, Rio Gallegas, from Cormo, Peppin Merino ($\frac{3}{16}$) and Corriedale; Polled Merino and Beddale blood added

later/recog. 1979, BS/Sp. *Cormo
argentino*
Argentine Criollo: (Argentina)/var. of
Criollo
Argentine Merino: (Argentina)/fw.m/
orig. from Criollo improved by
Spanish and Saxony Merinos
1813–1874, graded to
Rambouillet 1875–1890/BS
Argos: (Greece)/d.m.cw/lft/orig. from
Chios or Anatolian fat-tailed ×
Greek Zackel/syn. *Karamaniko*/
rare
arhar: *see* arkhar
Ariano, Improved: *see* Quadrella
Ariano crossbred: *see* Campanian
Barbary
Ariège, Ariègeois: *see* Tarasconnais
arkal: (Ust-Urt, between Caspian and
Aral Seas, Kazakhstan/
Turkmenistan)/= *Ovis vignei
arkal* Eversmann/var. of urial/
syn. *O. orientalis arkal,
Transcaspian urial*/not *arkar*
arkhar: (Ala-Tau, E Kyrgyzstan)/=
Ovis ammon karelini Severtzov/
var. of argali/orig. (with Merino)
of Kazakh Arkhar-Merino/syn.
*Ala-Tau argali, Tien Shan
argali*/not *arhar, arkar*
Arkhar-Merino: *see* Kazakh Arkhar-
Merino
Arles Merino: (Provence, S France)/
fw.m.(d)/♂ usu. hd, occ. pd (*motti*),
♀ usu. pd, occ. hd (*banetto*)/orig.
from local graded to Spanish
Merino in late 18th and early 19th
c. and improved by Châtillonais
var. of Précoce since 1921/former
vars: Camargue, Crau/HB 1946,
BS/Fr. *Mérinos d'Arles*/syn.
Provence Merino, Var Merino
Armenian red: (Armenia)/= *Ovis
orientalis armeniana* Nasonov/
var. of red sheep; ? = Armenian
wild (*Ovis orientalis gmelini*)/
Armenian *krasnyi samukh*/syn.
Armenian mouflon
Armenian Semicoarsewool:
(Armenia)/mw.m.d/ft/orig. from
Rambouillet and Lincoln ×
Balbas/vars: Aragats, Martunin/
Russ. *Armyanskaya
polugrubosherstnaya*

**Armenian Semifinewooled Fat-
tailed:** *see* Aragats
Armyanskaya polugrubosherstnaya:
see Armenian Semicoarsewool
**Armyanskaya polutonkorunnaya
zhirnokhvostaya:** *see* Aragats
Arrit: (Keren, Eritrea)/d.m/usu.
white, also blond, red or pied;
hy-cw; pd, ♂ occ. hd; fat at base
of tail
arrui: *see* aoudad
Arsi: (Ethiopia)/brown, grey, roan or
dun; ♂ hd, ♀ pd or hd/var. of
Ethiopian Highland with woolly
undercoat/syn. *Arusi, Arusi-Bale*
Artésien: *see* Artois
Artois: (N France)/var. of Flemish
Marsh/orig. (with Leicester) of
Boulonnais/Fr. *Artésien*/extinct
arui: *see* aoudad
Arusi, Arusi-Bale: *see* Arsi
Arvanitovlach: (Greece)/usu. white
with black around eyes/mt var.
of Greek Zackel/Gr.
Arvanitovlakhiko (= *Albanian
Vlach*)
Asaf: *see* Assaf
Asali: (Saudi Arabia)/var. of Najdi
Asblack: (Cieneguilla, Peru)/m.d.cw-
hy/prolific/orig. from Assaf
(1990) × [Barbados Blackbelly ×
Corriedale (1985)] graded to
Assaf ($\frac{3}{4}$ and $\frac{7}{8}$, and a few $\frac{15}{16}$)
Ascanian: *see* Askanian
Ashfal: *see* Shafali
Ashgur: *see* Shugor
Ashkhabad: (Turkmenistan)/superior
var. of Sary-Ja/Russ.
Ashkhabadskaya
Asiatic bighorn: see snow sheep
Askanian: (S Ukraine)/fw.m/orig.
1925–1934 from Merinos at
Askania Nova improved by
American Rambouillet/HB; HB
also in Bulgaria/Russ.
*Askaniĭskaya, Askaniĭskaya
poroda tonkorunnykh ovets* (=
Askanian breed of finewool
sheep); obs. syn. *Soviet
Rambouillet of Askanian type* (to
1934), *Askanian Rambouillet*
(Russ. *Askaniĭskiĭ
Rambulye*)/not *Ascanian*

Askanian Blackheaded: (Askania Nova, Ukraine)/w.d.m/orig. from Askanian, Suffolk and Oxford Down, × Tsigai

Askanian Corriedale: (Askania Nova, Ukraine)/m.mw/orig. from English Longwool × Askanian

Askanian Crossbred: (Askania Nova, Ukraine)/w.m/orig. from Lincoln Longwool and Tsigai, × Askanian

Assaf: (Israel)/d/orig. since 1955 from East Friesian × Israeli Improved Awassi (32.5–62.5% East Friesian)/HB/syn. *Asaf* (Turkey, 50% East Friesian)

AstKarakul': *see* Karakul

Astor urial: *see* shapo

Astrakhan: *see* Karakul

Atlantic Coast: (Morocco)/inc. Beni Ahsen, Doukkala, Zemmour/Fr. *Race côtière atlantique*

Atlántica: *see* Andalusian Churro

Auda: *see* Uda

audad: *see* aoudad

Aure-Campan: (SE Hautes-Pyrénées, France)/m.mw/pale brownish grey (*bis*), occ. black; ♂ usu. hd, ♀ usu. pd/Central Pyrenean group/former var: Campan/HB 1975/Fr. *Race d'Aure et de Campan, Aurois*/syn. *Lannemezan*

Aurina: *see* Pusterese

Aurois: *see* Aure-Campan

Ausemy, Ausimi: *see* Ossimi

Ausi, Aussi: *see* Awassi

Austral: (Valdivia, Chile)/[= southern]

"Australian": (Saudi Arabia)/d/ Border Leicester × Merino, F$_1$ ♀♀ crossed with Najdi for milk quality and yield

Australian Merino: (Australia and New Zealand)/fw/♂ hd (or pd), ♀ pd (or hd)/orig. 1797–1804 from Spanish Merino (Negretti strain) via England and Cape Colony, to give Camden Park Merino (1797–*c*.1856), improved in turn by Saxony Merino, Rambouillet, American Rambouillet, American Merino (Vermont); Anglo-Merino and English Longwool blood (1840–1870) in strong and medium strains; bred *inter se* since 1907/strains: Fine (Tasmanian), Medium Peppin, Medium Non-Peppin, Strong (South Australian); vars: AMS Merino, Booroola Merino, Fonthill Merino, Poll Merino, Trangie Fertility Merino/HB 1923, BS 1959; BS also in Uruguay

Australian Zenith: *see* Zenith

Austrian Negretti: (Austria)/fw/orig. from Negretti strain of Spanish Merino × local breeds/extinct

Auvergnat, Auvergne: *see* Central Plateau type

Avar: (Dagestan, Russia)/cw.m/white, with black or red on head and feet; ♂ usu. hd, ♀ hd or pd/ Caucasian Fat-tailed type/var.: Tlyarota/Russ. *Avarskaya*/extinct

Avikalin: (Rajasthan, India)/cw/ strain developed in 1970s at Central Sheep and Wool Res. Inst., *Avika*nagar, from Rambouillet × Malpura

Avimaans: (Rajasthan, India)/being developed at Central Sheep and Wool Res. Inst., *Avika*nagar, from Malpura, Sonadi, Dorset and Suffolk

Avivastra: (Rajasthan, India)/fw/ strain developed at Central Sheep and Wool Res. Inst., *Avika*nagar, from Rambouillet × Chokla/not *Avivastar*

Avranchin: (S Manche, N France)/ m.lw/brownish face and feet; pd/sim. to Cotentin but smaller/ orig. from Leicester Longwool (+ Southdown and Romney) × local (1830–1900)/BS 1928/syn. *Race du littoral sud de la Manche*/[= Avranches district]/rare

Awassi: (Syria, Israel, Lebanon, Jordan, C and W Iraq, and SE Turkey)/m.d.cw/white with brown head and legs, sometimes black, white, grey or spotted face, occ. all brown or black; ♂ hd, ♀ usu. pd/Near East Fat-tailed type/vars: Israeli Improved

Awassi, Ne'imi, Shafali/orig. of
Pak Awassi/also being improved
for milk production in Hungary
(imported from Israel) with
Hungarian Merino, Tsigai or
Racka/Turk. İvesi/syn. Arab
(Turkey); Baladi, Deiri (= from
Deir ez Zor), Shami (= from
Damascus) (Syria); Gezirieh
(Iraq); Syrian/not Al-Awass,
Aouasse, Ausi, Aussi, Awas,
Awasi, Awass, El Awas, Iwessi,
Ousi, Oussi, Ussy

Awsemy: see Ossimi

Azerbaijan Mountain Merino:
(Kedabek, Azerbaijan)/fw/orig.
from Kedabek Merino × Bozakh;
crossed with Askanian 1935/
recog. 1947/Russ.
Azerbaǐdzhanskiǐ gornyǐ
Merinos/not Azerbaidjan,
Azerbaidzhan

Azerbaijan Mutton–Wool:
(Azerbaijan)/breed group/Russ.
Azerbaǐdzhanskaya myaso-
sherstnaya porodnaya gruppa

Azov Tsigai: (Ukraine)/m.mw/var. of
Tsigai/orig. from Romney ×
Tsigai, F$_1$ bred inter se/recog.
1963/Russ. Priazovskaya
tsigaiskaya

Azrou: see Timahdit

Bábolna: see Prolific Bábolna

Badana: (Bragança, NE Portugal)/
m.d.cw/often brown face and
feet; ♂ usu. hd, ♀ usu. pd/HB
1991/syn. Marialveira (from
Marialva)

Badger Face Welsh Mountain:
(Wales)/white with black belly
and face stripes (reverse of
Torwen)/colour var. of Welsh
Mountain/BS 1976/Welsh
Torddu (= blackbelly)/syn.
Defaid Idloes (defaid = sheep),
Welsh Badger-faced, Welsh
Mountain Badger Faced/rare

Badghisian: (Iran)/ft

Badia, Badiota: see Pusterese

Baduvāl: see Baruwal

Baggara: (S Darfur and S Kordofan,
Sudan)/type of Sudan Desert
with Nilotic blood/not Baqqara/

[bred by cattle Arabs; bagar =
cattle]

Baghdale: (Punjab, Pakistan)/fw/
brown around eyes; ♂ occ. hd, ♀
pd/orig. at Kalabagh farm,
Mianwali, from Hissardale ($\frac{1}{4}$),
Damani ($\frac{1}{4}$) and Rambouillet
($\frac{1}{2}$)/not Bhagdale/rare

Bagnerschaf: see Roux-de-Bagnes

Bagnes: see Roux-de-Bagnes

Bagnolese: (Bagnoli Irpino, Avellino,
Campania, Italy)/m.mw/light
chestnut forequarters; pd/sim. to
Comisana (by crossing)/
purebreds rare

Bagri: (Bagar tract, around Delhi,
India)/m.cw/usu. brown or black
head/local var. of Bikaneri
group/not Bagir

Bahama Native: (Bahamas)/m/usu.
white with coloured spots; ♂ hd,
♀ pd/American Hair Sheep
group but with European
influence/syn. Long Island

Bahawalpuri: see Buchi

Bahia Ox-haired: see Santa Inês

Bahu: (Kibali-Ituri, NE DR Congo)/
brown or pied; hd; dwarf/syn.
Congo Dwarf

Baidrag: see Baydrag

Bains: see Velay Black

Bait: see Bayad

Bajad: see Bayad

Bakarwal: see Bhakarwal

Bakewell Leicester: see Leicester
Longwool

Bakhtiari: (Iran)/var. of Bakhtiari-
Luri/not Bakhtyari

Bakhtiari-Luri: (Lorestan, Iran)/
m.d.cw/white, occ. coloured
marks on head and legs, occ. all
black or brown; pd; lft/vars:
Bakhtiari, Luri

Bakkarwal: see Bhakarwal

Bakur: (Saratov, Russia)/cw/orig.
from Voloshian/nearly extinct

Bal: see Balwen

Baladi: (Egypt) see Fellahi

Baladi: (Syria) see Awassi

Balami: (N Nigeria, also N Cameroon,
and Chad)/m/sim. to Uda but
white; ♂ hd, ♀ pd/Fulani group
of Sahel type/syn. Balandji,
Balani, Balemi, Balonndi,

*Bellani, Bornou Fulani, Bornu,
Bouli* (Niger), *Fellata* (W Sudan),
Waqla, Weiladjo or *Woila*
(Cameroon), *White Bororo, White
Fulani, White Uda*

Balangir: *see* Bolangir

Balani: *see* Balami

Balbas: (SE Armenia, and
Nakhichevan, Azerbaijan)/black
or brown spots on face and feet;
♂ occ. hd, ♀ pd/Caucasian Fat-
tailed type/Russ. *Balbas,
Balbasskaya*

Baldebukov: *see* German Merino

Balearic: *see* Ibiza, Majorcan,
Minorcan, Red Majorcan

Balemi: *see* Balami

Balestra: *see* Pavullese

Bali-Bali: *see* Uda

Baljuša: (Metohija, Serbia)/black
face/Pramenka type

Balkai: *see* Balkhi

Balkar: *see* Karachai

Balkhi: (extreme NW Pakistan)/
m.cw/usu. black or grey; hd
(short); fr/? = Hissar/not *Balkai*

Balochi: *see* Baluchi

Balonndi: *see* Balami

Balouchi, Baloutche: *see* Baluchi

Baltistani: (N Kashmir, Pakistan)/
m.cw.d/often brown or tan or
pied; pd; st

Baluba: (Katanga, DR Congo)/hy,
maned; lop ears; st/African Long-
legged group

Baluchi: (SW Pakistan, E Iran and S
Afghanistan)/cw.d.m/black
marks on head and legs; ♂ hd
(Pakistan) or pd (Iran,
Afghanistan), ♀ pd; ft/syn.
(Baluchistan) *Baluchi dumba,
Mengali, Taraki, Shinwari;* (Iran)
Araghi, Farahani (Arak, WC
Iran), *Kermani* (SE Iran),
*Khorasani, Khorassan,
Khurasani, Naeini, Neini,
Yazdi*/not *Balochi, Balouchi,
Baloutche, Beluj*

Baluska: (Montenegro)/syn. for
Walachian, used for screwhorn
sheep

Balwen: (Powys, Wales)/black, dark
brown or dark grey body, with
white face, socks and tail tip/var.

of Welsh Mountain/syn. *Bal* or
*Balwen Welsh Mountain, Defaid
Bal* (*defaid* = sheep)/[= white
face]/rare

Bama Thoe: (Myanmar)

Bampton Nott: (Devon, England)/
pd/orig. (+ Southam Nott and ×
Leicester Longwool) of Devon
Longwooled/[nott = pd]/ extinct

Banamba: (Mali)/red/var. of Toronké/
syn. *Sambouru, Samburu*

Bandur: *see* Mandya

banetto: = hd ♀ Arles Merino

Bangladeshi: (Bangladesh)/cw/dwarf;
prolific/cf. Garole/syn. *Bengal,
Bera, Bhera* or *Wera* (= sheep)

Bannur: *see* Mandya

Banpala: *see* Bonpala

Bantu: (S Africa) *see* Damara, Nguni,
Tswana

Bapedi: (Sekhukhuneland, NE
Transvaal, S Africa)/m./usu.
white with reddish-brown head,
or brown-and-white, brown, occ.
black-and-white; usu. pd; lft/var.
of Nguni, ? with Namaqua blood/
BS/syn. *Pedi, Transvaal Kaffir*/
rare

Baqqara: *see* Baggara

Baraka: (SW Keren, Eritrea)/♂ hd or
pd, ♀ pd/sim. to Sudan Desert
but smaller and with long hair, ?
with Arrit blood/It. *Barca*/syn.
Begghié Korboraca, Shukria (It.
Sciucria)

Barasi: *see* White Karaman

Barawal, Barawul: *see* Baruwal

Barazi: *see* White Karaman

Barbado: (Texas, USA)/m, hunted/ tan,
tan with pale or black belly, or
pied; hy-cw; ♂ hd, ♀ pd/orig.
from Barbados Black Belly in early
20th c. with blood of American
Rambouillet, mouflon *et al.*

Barbados Black Belly: (Barbados)/
m/tan with black belly (badger
face); pd; mane in ♂; prolific/
American Hair Sheep group/BS/
orig. of Barbado

Barbaresca della Campania: *see*
Campanian Barbary

Barbaresca della Sicilia: *see* Sicilian
Barbary

Barbarin: *see* Tunisian Barbary
BARBARY: (N Africa)/cw.m/usu. white,
  also pied, black or brown; ♂ hd,
  ♀ pd/Near East Fat-tailed type/
  orig. of Barbary Halfbred,
  Campanian Barbary, Ghimi,
  Sicilian Barbary/inc. Barki,
  Libyan Barbary, Tunisian
  Barbary/Fr. *Barbarin*, It.
  Barbaresca
Barbary Halfbred: (Libya)/m.cw/♂
  usu. pd, ♀ pd; ft/orig. from
  White Karaman × Libyan
  Barbary
barbary wild sheep: *see* aoudad
Barca: *see* Baraka
Barcelonnette: *see* French Alpine
Bardoka: (Metohija, SW Serbia, also
  Montenegro and Albania)/
  d.m.cw/white, black or grey with
  black spots on face and legs; ♂
  usu. pd, ♀ pd/Pramenka type/
  syn. *Bhardok, Beloglave metohijska*
  (= *Metohija Whitehead*), *White
  Methonian*/not *Barloka*/[Alb.
  bardhë = white]/rare
Barègeois: (Barèges, S Hautes-
  Pyrénées, France)/smaller var. of
  Lourdais/HB 1975
Barga: (Huleng Boir dist., Dornod
  prov., Mongolia)/cw.m/head usu.
  tan, brown-spotted or black-
  spotted/var. of Mongolian/not
  Bargad
Barisciano: (Aquila, Abruzzo, Italy)/
  local var. of Apennine/syn.
  Aquilana
Barki: (NW Egypt)/cw.m.d/usu.
  brown or black head; ♂ hd, ♀
  pd; ft/Barbary type/syn. *Arab,
  Barqi* (correct), *Bedouin,
  Dernawi* (from Derna), *Libyan,
  Mariouti, Maryuti*/[Barka =
  Cyrenaica]
Barloka: *see* Bardoka
Barnstaple: *see* Devon Closewool
Barqi: *see* Barki
Barros Merino: (Tierra de Barros,
  Badajoz, Spain)/var. of non-
  migratory Spanish Merino/Sp.
  Merino de Barros
Baruun mongolin uutsan suult: *see*
  Kazakh Fat-rumped
Baruwal: (Nepal mts and S Sikkim,

India)/m.cw/white, usu. with
  black head, also black or pied;
  hd; very short ears/= Bonpala
  (India)/Nepali *Baduvāl,
  Baduwāl*/not *Barawal, Barawul,
  Barwal, Bharwal*
Barwal: *see* Baruwal
Basco-Béarnais: (SE Basses-Pyrénées,
  SW France)/d.m.cw/♂ hd, ♀ hd
  or pd/= Vasca Carranzana
  (Spain)/Pyrenean dairy group/
  orig. 1965 from Basque +
  Béarnais/HB 1975
Basilicata: *see* Improved Lucanian
Basque: (E Pays Basque, SW France)/
  orig. (and now var.) of Basco-
  Béarnais/Fr. *Basquais*, Basque
  muthur churria (= white
  face)/rare
Basque: (France) see also Manech
Basque: (Spain) *see* Lacho, Vasca
  Carranzana
Bastarda arianese: *see* Campanian
  Barbary
Bastarda maremmana: *see*
  Maremmana
Bastarda spagnola: *see* Maremmana
Baumshire: *see* Scottish Halfbred
Baunur: *see* Mandya
Bavarian Forest: (Bavaria, Germany)/
  m.mw/occ. brown; pd/cw var. in
  Austria (nearly extinct)/orig.
  from Zaupel, Bergshaf (= German
  Mountain) and Merinolandschaf/
  HB 1988/Ger. *(Bayern)
  Waldschaf*/rare
Bayad: (Malchin and Bugat dists, Uvs
  prov., Mongolia)/white, head and
  neck may be white or fawn with
  black or brown rings around
  eyes; hd or pd/var. of Mongolian/
  syn. *Bait, Bajad*
Baydrag: (beyond R Baydrag,
  Bayanhongor prov., Mongolia)/
  m.cw.(d)/white with (usu.) black
  or brown head; sft/var. of
  Mongolian/not *Baidrag*
Bayern Waldschaf: *see* Bavarian
  Forest
Bayinbuluke: (Xinjiang, China)/
  cw.m/black head and neck; ♂ hd
  or pd, ♀ usu. pd; ft/syn. *Chateng*
Bazakh: *see* Bozakh
Bazougers: *see* Bleu du Maine

bearded argali: *see* aoudad

Béarnais: (Béarn, SW France)/usu. red spots on face/orig. (and now var.) of Basco-Béarnais

Beddale: (W Australia)/fw.m/orig. by J.H. Shepherd from AMS Merino ($\frac{5}{8}$) and Border Leicester, Leicester Longwool, Poll Dorset and Dorset Horn

Bedouin: (Arabia) *see* Najdi

Bedouin: (Egypt) *see* Barki

Begghié abesce: *see* Ethiopian Highland

Begghié Korboraca: *see* Baraka

Beglika: (Bulgaria)/d

Beira Baixa: (Castelo Branco, E Portugal)/var. of Portuguese Merino derived from migratory Spanish Merino/Port. *Merina da Beira Baixa*

Beja: *see* Bija

Bekrit: *see* Timahdit

Bela Klementina: *see* White Klementina

Belakranjina: (Kranj, Slovenia)/ Pramenka type/Sn. *Belokranjska Pramenka*/rare

Belaslatina: (NW Bulgaria)/local improved var. of Bulgarian Native/Bulg. *Beloslatinska*/not *Bjela Slatinska*/extinct

Bela vlaška: *see* Karakachan (white)

Belclare: (Ireland)/m.lw/pd; prolific/ orig. 1976–1979 at Agric. Inst. Res. Centre, Belclare, Co. Galway, from prolific Lleyn × (Fingalway × High Fertility) with additional Finnish Landrace and Galway blood (composition 1985–7: 45% Lleyn, 32% High Fertility, 18% Fingalway, 5% Prolific Galway)/ named *Belclare Improver* 1981, BS 1985/rare

Belgian Milk Sheep: (Belgium)/d/pd; prolific/Marsh type/mixed orig. since 1945/HB/Fr. *Mouton laitier belge* (or *indigène*), Flem. *Belgisch Melkschaap*/rare/*see also* Flemish

Belgian Texel: *see* Beltex

Belice: *see* Val del Belice

Bellani: *see* Ballami

Bellary: (EC Karnataka, India) m.cw/black, grey, white with black face, or pied; ♂ hd or pd, ♀ pd; st/sim. to Deccani/not *Bellari*

Belle-Ile: (Brittany, France)/m/usu. white with brown spots on head and legs; pd; prolific/sim. to Landes de Bretagne, ? with Flemish blood from 1760/syn. *race de Deux*, = two (lambs)/ nearly extinct

Bell Multinippled: (Nova Scotia, Canada)/selected from local sheep by Graham Bell in 1890 and bred till 1922/orig. of Multinipple/extinct

Bellunese: (Mansué, Treviso, Venetia, Italy)/m.mw.d/black spots on face and legs; pd/Lop-eared Alpine group/? orig. from Alpagota and Lamon/[from Belluno]/nearly extinct

Beloglava metohijska: *see* Bardoka

Belogolovaya éstonskaya: *see* Estonian Whiteheaded

Belogradchik: *see* Replyan

Belokranjka: *see* Belakranjka

Beloslatinska: *see* Belaslatina

Beltex: (GB)/m.mw/pd/smaller but heavier type of Texel bred for double-muscling in Belgium (Ardennes) orig. imported to England as *Bel*gian *Tex*el, or Culard Texel/BS 1989

Beluj: *see* Baluchin

Bembur: *see* Vembur

Benasquinai: *see* Pallaresa

Beneventana, Benevento: *see* Campanian Barbary

Bengal: *see* Bangladeshi, Garle

Beni Ahsen: (NW Morocco)/mw.m/ white with coloured face; ♂ usu. hd, ♀ pd/Atlantic Coast type/ BSd/not *Beni Hassen*, *Bnihsen*/ declining by crossing

Beni Guil: (plateaux of E Morocco and W Algeria)/m.cw/white with tan head and legs/vars: Fartass, Harcha, Tounsint/orig. (with Berber) of Zoulay/HB (Morocco)/ Fr. *Race des plateaux de l'Est*/ syn. *Daghma*, *Petit oranais* (mutton exported from Oran, Algeria, to France), *Hamra* or *Hamyan* (Algeria)/not *Beni Ghil*, *Béni-Guill*

Beni Hassen: *see* Beni Ahsen

Beni Meskine: (Morocco)/smaller var. of Tadla with coarser wool

Bentheimer: (Bentheim, Osnabrück, Germany)/m.cw/black around eyes and on ears; pd/orig. from Drenthe var. of Dutch Heath × German Heath/HB 1934/syn. *Bentheimer Landschaf*/rare

Bera: *see* Bangladeshi

Berari: (N Maharashtra, India)/cw/? pd/local var./syn. *Black Colonial*

Berber: (Atlas mts, Morocco)/m.cw/ usu. white, also black, or white with black head; hd; small/vars: Ait Barka, Ait Haddidou, Ait Mohad, Aknoul, Guigou, Imi-n-Tanout, Marmoucha, Ouaouizart, Siroua, Tounfite/orig. (with Tadla) of South Moroccan and Zaqan, and (with Tousint) of Zoulay/syn. *Chleuh, Kabyle, Mountain Sheep*/not *Barbary*

Berbera Blackhead: *see* Somali

Berca: *see* Roussin de la Hague

Bergamasca: (Bergamo, Lombardy, Italy)/m.cw/pd/basic breed of Lop-eared Alpine group/orig. of other Lop-eared Alpine breeds and of Fabrianese, Pavullese, Perugian Lowland, Zakynthos/ HB 1942/Ger. *Bergamasker*/syn. *Gigante di Bergamo, Bergamacia* (Brazil)

Bergerá: *see* Ripollesa

Bergschaf: (Austria) *see* Tyrol Mountain

Bergschaf: (Germany) *see* German Mountain

Bergueda: *see* Ripollesa

Berkshire Knot: (England)/orig. (+ Wiltshire Horn and × Southdown) of Hampshire Down/syn. *Old Berkshire*/[knot = pd]/extinct

Berne: (Switzerland)/black var. of Jura/extinct

Berrichon: (Berry, C France)/m.sw/ pd/former vars: Boischaut, Brenne, Champagne, Crevant; modern breeds: Berrichon du Cher, Berrichon de l'Indre/not *Berri*

Berrichon de l'Indre: (Berry, C France)/m.sw/pd/orig. from Champagne and Crevant vars of Berrichon/HB 1895, BS 1946/ rare

Berrichon du Cher: (Berry, C France)/ m.mw/pd/orig. from Merino (18th c.), Southdown, Leicester Longwool and Dishley Merino (19th c.) × Champagne and Boischaut vars of Berrichon/BS 1936; BS also GB

Bertuna: *see* Tacòla

Besch da Pader: *see* Engadine Red

Beskaragai Merino: (Semipalatinsk, NE Kazakhstan)/fw/orig. from Mazaev and Novocaucasian Merinos × Kazakh Fat-rumped (early 20th c.) crossed with Rambouillet (1932), Askanian (1934 on) and Altai (1947)/ combined with Sulokol Merino to form North Kazakh Merino/ Russ. *Beskaragaĭskiĭ Merinos*/extinct

Beulah Speckled Face: (N Breconshire, Wales)/sw.m/grey-mottled face; ♂ usu. pd, ♀ pd/BS 1958/syn. *Eppynt Hill and Beulah Speckled Face*/not *Beulah Speckled-face*

Bezuidenhout Afrikander: (Orange Free State, S Africa)/m.cw/pd; ft/ orig. by W.F. Bezuidenhout of Rietfontein (1918 on) from white-wooled ♂ (? Ronderib Afrikander) × [Wooled Persian (Arabi) × Blackhead Persian]/ Afrik. *Bezuidenhout Afrikaner*/ syn. *Wooled Afrikander*/nearly extinct

Bhadarwah: *see* Gaddi

Bhagdale: *see* Baghdale

Bhakarwal: (SW Kashmir, India)/ cw.m/white with colour on face, occ. coloured; ♂ hd, ♀ pd; long lop ears/not *Bakarwal, Bakkarwal, Bhakarwala, Bharkarwad* /[= class of nomads]/declining

Bhanglung: *see* Tibetan

bharal: (China and Himalayas)/= *Pseudois nayaur* Hodgson, actually a beardless goat/syn. *blue sheep, napo*/not *bhurrel, burhal, burrhel*

Bharat Merino: (Rajasthan, India)/
fw/orig. 1980s at Central Sheep
and Wool Res. Inst., Avikanagar,
from Rambouillet and Soviet
Merino ($\frac{3}{4}$) × Chookla and Nali
($\frac{1}{4}$), with some Malpura

Bhardok: *see* Bardoka

Bharwal: *see* Baruwal

Bhera: *see* Bangladeshi

Bhote, Bhotee, Bhotia: *see* Tibetan

bhurrel: *see* bharal

Bhyanglung: *see* Tibetan

Białogłowa owca mięsna: *see* Polish
Whiteheaded Mutton

Białystok: (NE Poland)/local
population of Polish Lowland
group/orig. 1963 on from Polish
Merino × local/Pol. *Polska owca
nizinna w okręgu białystockim* (=
*Polish Lowland sheep of the
Białystok region*)

Biangi: (Tibet and Himachal
Pradesh)/cw.pa/syn. *Kangra
Valley*/not *Biangir, Birangi*

Bibrik: (NE Baluchistan, Pakistan)/
m.cw.d/black or brown on face;
hd; ft/var.: Khetrani/syn. *Bugti,
Marri*/not *Bibrick*

Biellese: (Piedmont, Italy)/m.cw/
pd/Lop-eared Alpine group/inc.
true Biellese of Vercelli and type
from rest of Piedmont that is
Piemontese alpina (= *Alpine
Piedmont*) or *nostrale* (= local)
graded to Biellese/var.: Tacòla/
BSd 1959, HB 1986/syn. *Biellese-
Bergamasca* (obs.), *Ivrea* (Aosta)/
[from Biella]

Biérois: *see* Landes de Bretagne

Bierzo White: (Léon, Spain)/
m.cw/pd/Sp. *Blanca del
Bierzo*/syn. *Churra berciana,
Churra del Bierzo*/rare

bighorn: (Rocky Mountains, N
America)/= *Ovis canadensis*
Shaw/many local vars/syn.
*American bighorn, Canadian
bighorn, Rocky Mountain*/*see
also* snow sheep (Asiatic
bighorn), Dall sheep

Bija: (Red Sea Hills, NE Sudan)/var.
of Sudan Desert/not *Beja*/rare

BIKANERI: (Rajasthan, India)/cw.m/
name formerly used to inc. all or
some of Rajasthan breeds, i.e.
Bagri, Buchi (Pakistan), Chokla,
Jaisalmeri, Magra, Malpura,
Marwari, Nali, Pubal/not
*Bikaneer, Bikanier, Bikanir,
Vicanere*

Bikaneri, Bikaneri Chokla: *see*
Magra

Bilbilitana Red: (SW Zaragoza,
Aragon, Spain)/mw.m/black
fading to red-brown, often white
face-blaze; ♂ usu. hd, ♀ pd/var.
of Castilian/Sp. *Roya bilbilitana*/
[from ancient Bilbitis]

Birangi: *see* Biangi

Birka: (Vojvodina, Syrmia, and
Slavonia, Serbia and Croatia)/
mw/black, yellow or spotted
head and legs/name used for
Merino × Tsigai or Merino ×
Pramenka crosses/syn. *Sremska*
(= *Syrmian*)/[Hung. = sheep]

Bîrsa: (Romania)/fw.m.(d)/orig. since
1953 from Île de France × (Palas
Merino × Tsigai)

Biscay: *see* Lacho

Bityug: (Voronezh, Russia)/var. of
Russian Long-tailed/Russ.
Bityugskaya/extinct

Bizet: (N Haute Loire and SE Puy-de-
Dôme, C France)/m.mw/grey-
brown with black on sides of
face and on legs; ♂ usu. hd, ♀
usu. pd; lt/Central Plateau
group/BS 1905, HB 1946/syn.
Chilhac, des Bizets/[? from *bise* =
beige]

Bjela Slatinska: *see* Belaslatina

Black Bains: *see* Velay Black

Black Belly: *see* Badger Face Welsh
Mountain, Barbados Blackbelly

Black Blaze: *see* Dutch Black Blaze

Black-Brown Mountain: *see* Swiss
Black-Brown Mountain

Black Castilian: (Old Castille, Spain)/
black, with white spot on head
and tail tip/var. of Castilian

Black Colonial: *see* Berari

Black Cotswold: (USA)/black var. of
Cotswold/BS 1990

Black Crioulo: (S Brazil)/var. of
Crioulo/Port. *Crioulo preto*, Sp.
Criollo negro/nearly extinct

Blackface: (E England) *see* Suffolk

Blackface: (Scotland) *see* Scottish Blackface

BLACKFACED MOUNTAIN: (N England and Scotland)/cw/black or mottled face and legs; hd/inc. Dalesbred, Derbyshire Gritstone, Lonk, Rough Fell, Scottish Blackface, Swaledale/syn. *Blackface Hill, Blackfaced Heath*

Black-face Manech: (SW Basque country, France)/d.m.cw/black face and legs; ♂ hd, ♀ hd/var. of Manech/Basque *muthur belza*

Black Manchega: (New Castille, Spain)/black var. of Manchega/experimental flock estab. at Valdepeñas/BS/rare

Blackface Mutton: (Germany) *see* German Blackheaded Mutton

Blackface Norfolk Horned: *see* Norfolk Horn

Black Hawaiian: *see* Hawaiian Black Buck

Blackheaded: (France) *see* French Blackheaded

Blackheaded Mutton: (Germany) *see* German Blackheaded Mutton

Blackheaded Mutton: (Poland) *see* Polish Blackheaded Mutton

Blackheaded Mutton: (Switzerland) *see* Swiss Brownheaded Mutton

Blackheaded Pleven: *see* Pleven Blackhead

Blackheaded Somali: *see* Somali

Blackheaded Synthetic: *see* Polish Blackheaded Mutton

Blackhead Persian: (S Africa)/m/hy; pd; fr/var. of Persian/orig. from Somali imported 1868/part orig. of Bezuidenhout Afrikander, Brazilian Somali, Dorper, Van Rooy, Wiltiper/HB 1906, BS 1958/Afrik. *Swartkoppersie* /not *Black Head Persian, Black-headed Persian*/name also used for Somali in E Africa

Black Leicester Longwool: (Great Britain)/var. of Leicester Longwool (? with Black Welsh Mountain blood)/HB 1986/rare

Black Maure: (SE Mauritania, NE Mali and N Senegal)/m.hr.(d)/black; long-haired; ♂ hd, ♀ usu. pd; lop ears/= Zaghawa (Sudan)/

Sahel type/Fr. *Mouton maure à poil long*/syn. *Nar*

Black Merino: (Italy) *see* Carapellese

Black Merino: (Portugal)/fw.d.m/brown to black (usu. recessive; dominant var. is Pialdo)/cf. black var. of Spanish Merino/Port. *Merino preto*

Black Merino: (Tunisia) *see* Thibar

Black Milksheep: (Germany)/var. selected from East Friesian/Ger. *Schwarzes Milchschaf*

Black Mountain: *see* Montagne noire

Black Ningsia: *see* Ningsia Black

Blacknosed Swiss: *see* Valais Blacknose

black sheep: *see* Stone sheep

Black Thibar: *see* Thibar

Black Velay: *see* Velay Black

Black Vlach: *see* Karakachan

Black Welsh Mountain: (Great Britain)/sw.m/♂ hd, ♀ pd/ornamental dominant black derivative of Welsh Mountain/orig. in mid 19th c./HB and BS 1920; BS also USA 1991

Blanc de la Suisse Occidentale: *see* Swiss White Alpine

Blanc de Lozère: *see* Blanc du Massif Central

Blanc des Alpes: *see* Swiss White Alpine

Blanc des Montagnes: *see* Swiss White Mountain

Blanc du Massif Central: (Lozère, Ardèche, Gard and Hérault, S France)/m.mw.(d); pd/orig. from Caussenard by crossing with Lacaune in early 20th c./named 1965; HB 1967, BS/syn. *Blanc de Lozère, Lozèrien, Massif Central White, Race des Montagnes de la Lozère* (= *Lozère Mountain*)

Blauköpfiges Fleischschaf: *see* Bleu du Maine

Blauwe Texelaar: *see* Blue Texel

Bleiburger: (Austria)/former var. of Carinthian/extinct

Bleu du Maine: (Mayenne, NW France)/m.mw/dark grey head and legs; pd/cf. Maine à tête blanche (absorbed), Rouge de l'Ouest/orig. from Leicester Longwool (and Wensleydale)

imported 1855–1880 × local (Choletais)/BS 1938, HB 1948; BS also in GB 1982, HB Germany 1975/Ger. *Blauköpfiges Fleischschaf*/syn. *Bazougers, Bluefaced Maine, Blue-headed Maine, Maine-Anjou, Maine à tête* (or *face*) *bleue, Mayenne Blue*/not *Bleu de Maine*

BLI: *see* Improved Border Leicester

Blinkhaar: (S Africa)/shiny coat/only recog. var. of Ronderib Afrikander/[= smooth hair]

Blinkhaar: (Namibia) *see* Damara

BLM: (NSW, Australia)/mw.m/pd/ orig. (1955 on) at Condobolin Res. Sta. from Border Leicster × Merino, selected for fertility/syn. *Border Merino*

Bluefaced Leicester: (N England)/ m.lw/English Longwool type/ orig. from Border Leicester with blue face (i.e. white hairs on black skin) and finer fleece, selected 1914–1939/BS and HB 1963; BS also USA/syn. *Hexham Leicester*

Bluefaced Maine, Blue-headed Maine: *see* Bleu du Maine

blue sheep: *see* bharal

Blue Texel: (Netherlands)/var. of Texel (1977)/BS 1983/Du. *Blauwe Texelaar*/rare

Bnihsen: *see* Beni Ahsen

Bochnia: (Kraków, Poland)/Polish Longwool group/orig. from East Friesian × local/Pol. *Owca bocheńska*

Bohemian: *see* Šumava

Böhmerwald: *see* Šumava

Boischaut: (C France)/former var. of Berrichon/syn. *Bourges*/not *Bois Chaud*/extinct

Bokara, Bokhara: *see* Bukhara

Bokino: (Tambov, Russia)/var. of Russian Long-tailed/extinct

Bolangir: (NW Orissa, India)/ cw.(?hy)/white, light brown or pied; ♂ hd, ♀ pd/not *Balangir*

Bond: (NSW, Australia)/mw.m/pd/ var. of Corriedale/orig. 1909 by Thomas Bond of "Yarren", Lockhart, from Lincoln × Tasmanian and Peppin Merinos/syn. *Commercial Corriedale* (to 1979), *Bond Corriedale*

Bonga: (Kaffa, Ethiopia)/var. of Ethiopian Highland

Bonpala: (S Sikkim, India)/cw.m/ white, black or intermediate; hd; small/= Baruwal (Nepal)/local syn. *Banpala*/not *Bonapala*/[= jungle bred, as opposed to Gharpala = home bred]/nearly extinct

Booroola Leicester: (Australia)/ m/prolific/orig. from Booroola Merino ($\frac{1}{8}$) × Border Leicester ($\frac{7}{8}$)/nearly extinct

Booroola Merino: (Australia)/fw/♂ hd, ♀ usu. pd/strain of medium non-Peppin Australian Merino carrying gene for prolificacy, selected (on ♀♀ only) since *c.* 1945 by Seears brothers at Booroola, Cooma, NSW, and by CSIRO (on ♂♂ and ♀♀) since 1958/BS in New Zealand/syn. *High-Fertility Merino*

Bordaleiro: (Portugal)/d.m.mw/ white, black or pied/cf. Entrefino (Spain)/? orig. from Merino × Churro/inc. Campaniça, Entre Douro e Minho, Saloia, Serra da Estrela/[? from Port. *bordo* = curly (wool) or from Sp. *burdo* = coarse]

Bordaleiro Churro: *see* Portuguese Churro

Borde: *see* Red Majorcan

Border Cheviot: *see* Cheviot

Borderdale: (New Zealand)/m.mw/ orig. 1930s from Border Leicester × Corriedale/HB

Border Leicester: (border of Scotland and England)/lw.m/pd/English Longwool type/orig. in late 18th and early 19th c. from Leicester Longwool with Cheviot blood/var.: Glendale (Australia)/ orig. of Bluefaced Leicester, Improved Border Leicester/recog. 1869, BS 1897, HB 1898; BS also USA 1888; HB also Canada, New Zealand

Border Merino: (Australia) *see* BLM

Border-Merino: (New Zealand)/ Border Leicester × Merino, F_1

Border-Romney: (New Zealand)/
Border Leicester × New Zealand
Romney, F$_1$
Boreray: (St Kilda, Scotland)/cw.m/
white or light tan with black, tan or
pied head and legs; hd; feral since
1930/var. of Scottish Blackface/
orig. from Scottish Blackface
(introduced 1870) × Hebridean
type of Tanface (Old Scottish
Shortwool)/syn. _Boreray Blackface_,
Hebridean Blackface/ rare
Borgotarese: (Borgotaro, Parma,
Italy)/Apennine group/extinct
1979–1983
Boricua: (Puerto Rico, USA)/red-
brown or black; hy; roman nose;
lop ears
Borino: (Australia)/orig. from Border
Leicester × Merino
Bornou Fulani: _see_ Balami
Bornu: _see_ Balami
Bororo: _see_ Uda
Bosach: _see_ Bozakh
Bosnian Mountain: (C and W Bosnia
and Hercegovina)/d.m.cw/white,
often with spots on face and legs,
occ. coloured; ♂ usu. hd, ♀ pd/
Pramenka type/vars: Kupres,
Privor, Vlašić/Serbo-cro.
Bosansko-Hercegovačka
Boujaâd: (Khouribga and Beni
Mellal, NW Morocco)/sw/
yellowish head; ♂ hd, ♀ pd/?
orig. from Tadla
Boukhara: _see_ Bukhara
Bouli: _see_ Balami
Boulonnais: (N France)/m.lw/pd/
orig. in 19th c. from Leicester
Longwool, Dishley Merino _et al._,
× Artois/HB to 1963, and 1987
on, BS revived 1984/[= Boulogne
district]/rare
Bourbonnais: _see_ Velay Black
Bourges: _see_ Boischaut
Boutsiko: (Ioánnina, Epirus, Greece)/
var. of Vlakhiko/not _Butsiko_
Bovec: (Upper Soča valley, Slovenia)/
d.m.cw/white (70%) or black;
usu. pd/orig. from Zackel type/
HB 1983/Sn. _Bovška ovca_, It.
Plezzana (from Plezzo), Ger.
Flitsch/syn. _Krainer Steinschaf_
(Austria and Germany),
Trentarka/rare

bowni: _see_ bunni
Bozakh: (Armenia and Azerbaijan)/
dirty-white or yellow-white, also
tan, grey or light red; usu. pd/
Caucasian Fat-tailed type/Russ.
Bozakh, Bozakhskaya/not
Bazakh, Bosach, Buzakh/rare
Brabant Foxhead: (Belgium)/sw/light
brown with red-brown head; ♂
occ. hd, ♀ pd/Flem. _Klein-
Brabantse Voskop_/syn. (Fr.)
Ardennes, Ardennais Roux/
nearly extinct
Brač: _see_ Island Pramenka
Bragança Galician: (Terra Fria, NE
Portugal)/m.cw/white with
coloured spots on face, occ.
black; ♂ hd, ♀ usu. pd/
Portuguese Churro type/cf.
Miranda Galician/HB 1989/Port.
Galega bragançana
Braunes Bergschaf: _see_ Brown
Mountain
Braunköpfiges Fleischschaf: _see_
Swiss Brownheaded Mutton
Brazi: _see_ White Karaman
Brazilian Somali: (NE Brazil)/m/
white with black head; hy; pd;
small fr/orig. from Blackhead
Persian (imported from W Indies
1939) × local/HB/Port. _Somali
Brasileiro_/syn. _Rabo gordo_ (= fat
rump)
Brazilian Woolless: (NE Brazil)/m/
white, pied or coloured; hy
(some cw); ♂ hd or pd, ♀
pd/mixed population/orig. from
West African (? × Crioulo)/
selected breeds: Morada Nova,
Santa Inês (with Bergamasca)/
Port. _Deslanado vermelho_ (=
red), _Deslanado branco_ (=
white), _Pelo de Boi_ (= ox-haired)/
syn. _Deslanada do Nordeste_
Brecknock Hill Cheviot:
(Breconshire, Wales)/♂ usu. pd,
♀ pd/var. of Cheviot from 1850s
with Welsh Mountain blood/BS/
syn. _Brecon Cheviot_,
Sennybridge Cheviot
Bregdetit: (Albania)/d.m.cw/white with
coloured spots on face and legs
Brenne: (C France)/former var. of
Berrichon/extinct

Brentegana: (Affi and Caprino, Verona, Italy)/m.cw/pd/Lop-eared Alpine group/? orig. from Lamon/syn. *Brentegana veronese*/[? from Brenta]/rare

Breton: *see* Belle Ile, Landes de Bretagne

Breton Dwarf: *see* Ushant

Brézi: *see* White Karaman

Breznik: (SW Bulgaria)/m.cw.d/black spots on face and sometimes legs; ♂ usu. pd, ♀ pd/local improved var. of Bulgarian Native/inc. Radomir, Sofia White/Bulg. *Breznishka*/syn. *Sofia-Breznik*/rare

Brianzola: (Brianza, Como, Italy)/m.(mw)/pd/Lop-eared Alpine group, orig. from Bergamasca or Varesina/extinct

Brière, Briéron: *see* Landes de Bretagne

Brigasca: (La Brigue, border of SE France and NW Italy)/d.m.cw/ dark red spots on head and legs (Italy); usu. hd; semi-lop ears, roman nose/sim. to Frabosana/? orig. from local with Langhe and Frabosana/It. *Brigasca*, Fr. *Brigasque*/syn. *Nostrale*, *Tchabale*, *Tchapera*, *Tendasque* (from Tende)/rare

Brigasque: *see* Brigasca

Brillenschaf: *see* Carinthian

British Friesland: (Great Britain)/BS for East Friesian imported 1953 and 1964

British Longwool: *see* English Longwool

British Milksheep: (England)/ m.d.mw/prolific/orig. during 1970s by G.L.H. Alderson from Friesian (East and West) (42%), Bluefaced Leicester (16%), Prolific (15%), Lleyn (13.5%), Polled Dorset (11.5%) and others (2%)/BS/rare

British Oldenburg: (England)/m.lw/ name given to German Whiteheaded Mutton imported in 1964 and 1969/BS

Brogne: (Breonio, Verona, Italy)/ m.mw/occ. brown spots on head; pd/Lop-eared Alpine group but nearly straight profile/orig. from Bergamasca and Lamon/rare

Broomfield Corriedale: (New Zealand)/strain selected for resistance to foot rot

Brownheaded Mutton: *see* Swiss Brownheaded Mutton

Brown Mountain: (German Alps)/ pd/smaller brown var. selected from German Mountain/? orig. from Zaupel, Steinschaf and Bergamasca/HB 1977/Ger. *Braunes Bergschaf*/rare

Brown Mountain: (Salzburg, Steiermark, Tirol and Niederösterreich, Austria)/ m.pelt.cw/pd; long lop ears; roman nose/distinct from German breed, orig. by selection from brown landraces/Ger. *Braunes Bergschaf*/rare

brumării: (Romania) *see* Ţurcana (grey var.)

Bschlabser: (Tyrol, Austria)/local var. sim. to Steinschaf but woolly face/extinct

Buchi: (Bahawalpur, Pakistan)/cw/ usu. tan spots on head and legs; hd (short); st/Bikaneri group/syn. *Bahawalpuri*/not *Buchni*/[*bucha* = stubby ear]

Budiani: (Georgia)/var. of Tushin with shorter denser wool

Bugti: *see* Bibrik

Bukhara: *see* Karakul

Bukusu: (Kenya)/local var. of Masai

Bulandshari: *see* Muzaffarnagri

Bulgarian Dairy: (Bulgaria)/d/in formation (1972) from Stara Zagora, Pleven Blackhead and East Friesian/Bulg. *B"lgarska mlechna*

Bulgarian Merino Longwool: (Bulgaria)/orig. from South Bulgarian Corriedale and East Friesian

Bulgarian Merinos: *see* Danube, Karnobat, North-East Bulgarian and Thrace Finewools

Bulgarian Native: (Bulgaria)/cw.d.m/ coloured in E, usu. white or white with coloured face in W; ♂ hd, ♀ hd or pd; lt/Zackel type/ local prim. vars inc. Central

Rodopi, Duben, Koprivshten, Karakachan, Kotel, Maritsa, Replyan, Sakar, Sredna Gora, Stara Planina, Stranja, Teteven; local improved vars inc. Breznik/ for improved breeds see Pleven Blackhead, Svishtov; for breeds improved by Tsigai see Karnobat, Shumen, Stara Zagora, White Klementina

Bulgarian Synthetic Milksheep: (Bulgaria)/orig. from Pleven Blackhead, East Friesian and Awassi

Bumfdale: (NSW, Australia)/m.mw/ orig. 1979–1992 by NSW Dept Agric., Wollongbar, from Wiltshire, Suffolk × Border Leicester, and Booroola Merino × Border Leicester (♂♂), × Dorset Horn × (Border Leicester × Merino) (♀♀) as gene pool for combined crosses, then selected for weaning weight and reproductive performance/[joke name]/nearly extinct

Bundelkhandi: *see* Jalauni

Bündner Oberland: (Graubünden, Switzerland)/cw.m/usu. white, also black, grey, brown or spotted; ♂ hd, ♀ usu. pd/prim. type with small horns, sim. to Steinschaf/orig. from Peat sheep/ last found in Lugnez, Luzein, Nalps, Tavetsch, Vrin, etc./Ger. *Bündner Oberländer*/syn. *Graubünden, Grisons, Nalps, Tavetsch*/extinct 1954 but now conservation programme to breed back from Tavetsch-like Medels sheep (*see* Tavetsch-Medel)

Bundoran Comeback: (Victoria, Australia)/mw/usu. pd/orig. (since 1971) on "Bundoran", Glenthompson, from Australian Merino strains and Polwarth (5–10%)

Bungaree Merino: (South Australia)/ strain of South Australian Merino (strong wool) with English Longwool blood

bunni: *see* Aansi (brown var.)

Burgundy Merino: *see* Châtillonais

burhal: *see* bharal

Buriat-Mongol: *see* Buryat

Burkina Faso Peul: *see* Peul Voltaïque

Burmese: (Myanmar)/m.cw/lop ears

burrhel: *see* bharal

Burri: *see* Dhamari

Burundi: *see* Rwanda-Burundi

Burundi: (Burundi)/white, black, brown and combinations; ♂ hd, ♀ pd; lop ears (often reduced); ft/African Long-fat-tailed group/ Fr. *Race ovine burundaise*/*see also* Rwanda-Burundi

Buryat: (Buryatia and Chita prov., Siberia, Russia)/m.cw/white with black or red on head; ♂ hd, ♀ pd; sft/sim. to Mongolian and Siberian/orig. (with Merino) of Chita, Transbaikal Finewool/ Russ. *Buryatskaya*/ syn. *Buryat-Mongolian*/not *Buriat-Mongol*

Busia: (Uganda border of W Kenya)/ local var. of Masai

Butana: *see* Dubasi

Buti: (India) *see* Patanwadi

Buti: (Pakistan) *see* Thalli

Butjadingen: (Oldenburg, Germany)/ Marsh type/orig. of German Whiteheaded Mutton/Ger. *Butjadinger*/extinct

Butsiko: *see* Boutsiko

Byala Klementinska: *see* White Klementina

Byala yuzhnob"lgarska: *see* White South Bulgarian

byang-lug: *see* Tibetan

Cabreña: *see* Segureña

Cadorina: (Cadore, N Belluno, Venetia, Italy)/cw.m/Lop-eared Alpine group/extinct 1970s by crossing with Lamon *et al.*

Cadzow Improver: (W Lothian, Scotland)/m.sw/♂ hd, ♀ pd/orig. 1960s from Dorset Horn and Finnish Landrace as ♂ line for crossing on hill ewes/extinct

Cagi: *see* Kage

Cagliari: *see* Sardinian (lowland var.)

Cago: *see* Kage

Caithness: *see* Keerie, North Country Cheviot

Caithness Cheviot: *see* North Country Cheviot

Cakiel: *see* Zackel

Cakiel-Fryz-Siedmiogrodzka: *see* Polish Mountain

çakrak: *see* Karayaka

Calabrese, Calabrian: *see* Sciara

Calhoor: (Iran)/var. of Sanjabi

California Red: (USA)/cw/apricot tan, with red mane in ♂; pd/orig. 1971 by Glen Spurlock from Barbado × American Tunis, recrossed to Tunis by Aimé and Paulette Soulier

California Variegated Mutant: (USA)/mw/usu. badger-face (white, grey, black or brown with black belly)/orig. in 1970 by Glen Eidman, Willows, California, from Romeldale/HB and BS/syn. *CVM*

Calikui, Calokui: *see* Kallakui

Calva: *see* Ansotana

Camargue Merino: (France)/grey/ former var. of Arles Merino/Fr. *Mérinos de la Camargue*/extinct

Cambar: *see* Kambar

Cambrai: (France)/var. of Flemish Marsh/extinct

Cambridge: (England)/m.sw/dark face; pd; prolific/orig. 1966–1976 by J.B. Owen from prolific ewes (*c.* 65% Clun Forest; 15% other British breeds, chiefly Llanwenog, Lleyn and Kerry Hill, but also Bluefaced Leicester, Border Leicester, Radnor and Ryeland; with addition of 20% Finnish Landrace) selected for high litter size/HB 1969, BS/[work started at Cambridge Univ.]

Camden Park: (NSW, Australia)/early strain (1797–*c.*1856) of Australian Merino/orig. 1797–1804 from Spanish Merino (mainly Negretti) through Gordon flock (Cape Colony) and Royal flocks (at Windsor and Kew, England), crossed with African and Asian cw and hy breeds and some English lw and sw, and backcrossed to Spanish Merino/one flock only, in

Macarthur Park (at Camden and at Trangie)/syn. *Macarthur Merino*/rare

Camerès: *see* Lacaune

Cameroon Dwarf: *see* West African Dwarf

Campan: (France)/pd/smaller former var. of Aure-Campan/extinct

Campanian Barbary: (Campania, Italy)/d.m.(cw-mw)/often dark spots on face and legs; ♂ hd or pd, ♀ pd; sft/orig. from Tunisian Barbary × local thin-tailed breed/HB 1971/It. *Barbaresca della Campania*/syn. *bastarda arianese* (= *Ariano crossbred*), *beneventana* (= from Benevento), *casalinga* or *casareccia* (= household), *coda chiatta* (= flat tail), *laticauda* (= broad tail), *nostrana* (= local), *turchessa* (= Turkish)

Campaniça: (Campo Branco, S Beja, S Portugal)/d.m.mw/white, usu. with small brown spots on head and legs; ♂ hd, ♀ pd/Bordaleiro type/HB 1987

Campbell Island: (New Zealand)/ feral/orig. from Merino and longwool crosses/nearly extinct

Campera: *see* Castilian Churro

Campidano: *see* Sardinian (lowland var.)

Campine, Campinoise: *see* Kempen Heath

Campiñesa: *see* Andalusian Merino

Campo de Calatrava: (Ciudad Real, Spain)/var. of Mestizo Entrefino-fino

Camura: *see* Red African

Canadian Arcott: (Canada)/m/face white or mottled; pd/orig. 1968–1988 (recog.) at Animal Res. Centre, Ottawa, from Suffolk (37%), Île de France (28%), Leicester Longwool (14%), North Country Cheviot (7%), Romnelet (6%) and 7 other breeds (8%), crossed 1968–1974, selected for weight and carcass 1977–1988; for crossing with Ouatouais and Rideau Arcotts/HB 1988/obs. syn. *ARC (sire) Strain 1*/rare

Canadian bighorn: *see* bighorn

Canadian Corriedale: (Alberta, Canada)/mw/orig. 1919–1934 from Corriedale × (Lincoln × Rambouillet)/HB/rare

Canaltaler: *see* Kanaltaler

Canaria: *see* Canary Island

Canaria De Pelo: (Canary Is, Spain)/ m/red; pd/hair sheep reintroduced to the islands/syn. *Canaria Hair Breed*/rare/*see also* Canary Woolless

Canary Island: (chiefly Gran Canaria, Spain)/d.cw/usu. parti-coloured; ♂ usu. hd, ♀ usu. pd; ears often vestigial/Sp. *Canaria*/*see also* Palmera

Canary Woolless: (Canary Is, Spain)/ hair sheep/orig. type (white, ♂ hd, from ? NW Africa, heavily crossed with Churro types) probably extinct; now includes Pelibüey and West African Dwarf imported from Venezuela

Çandır: (Turkey)/= Dağliç × White Karaman cross (W Anatolia), or Red Karaman × Tuj cross (NE Turkey)/syn. *Kesbır, Kesme, Mazik*/not *Candyr, Kesber, Zandir*/[= crossbred]

Cannock Chase: (England)/orig. (with Southdown *et al.*) of Shropshire/ extinct

Cantabrian Churro: *see* Spanish Churro

Cape: (S Africa)/many colours; sft/var. of Hottentot/extinct

Cape Fat-tailed: *see* Afrikander

Carakachanska: *see* Karakachan

Caralpina: *see* Ripollesa

Caraman: *see* Red Karaman, White Karaman

Caramanitico: *see* Rhodes

Cara Negra: *see* Lacho

Carapellese: (Carapelle, Foggia, Apulia, Italy)/mw.d.m/black; pd/syn. *Gentile moretta, Gentile a vello nero* (= *Improved Black*), *Merinos a vello nero* (= *Black Merino*), *Moretta*/not *Carapella*/ extinct

Cara Rubia: *see* Lacho

Cardy Welsh Mountain: (Cardiganshire, Wales)/strain of Welsh Mountain with whiter face and finer fleece/syn. *Cardigan*

Carinthian: (Austria)/cw.m.(d)/ yellowish-white with black ear tips and spectacles, occ. black; pd/Lop-eared Alpine group; sim. to Tyrol Mountain (which has been displacing it since 1938) but poorer wool and smaller ears/ orig. since early 18th c. from Paduan and Bergamasca × local (Steinschaf *et al.*)/former vars (pre-1918): Seeländer (principal, also syn.), Bleiburger, Gurktaler, Kanaltaler, Petzen, Spiegel (Tyrol), Steiner/orig. of Pusterese, Solčava, Tyrol Mountain/HB 1988/Ger. *Kärntner* or *Kärtner Brillenschaf* (= spectacled sheep)/ syn. *Brillenschaf, Seeländer*/not *German Mountain*/rare

Carinthian Mountain: *see* Tyrol Mountain

Carnabat: *see* Karnobat

Carnero de pelo de buey: *see* Pelibüey

Carnica: (Friuli, Italy)/extinct 1976–1979

Carpathian Mountain: (SW Ukraine)/ m.cw.d/♂ usu. hd, ♀ usu. pd; lt/Zackel type/var. of Voloshian/ inc. Ruthenian Zackel or Racka in Transcarpathian Ukraine, Zackel in Galicia, Turkana in N Bucovina/orig. (with Tsigai) of Ukrainian Mountain/Russ. *Gornokarpatskie ovtsi*

Carpetmaster: (New Zealand)/cw/♂ usu. hd, ♀ usu. pd/orig. from one cw Border Leicester–Romney ♂ carrying gene N[J] × cw Perendales/? extinct in New Zealand, nearly extinct in Australia

Carranzana: *see* Vasca Carranzana

Carsolina: *see* Dalmatian-Karst, Istrian Milk

Cartera: (SE Teruel, Aragon, Spain)/ m.w/pd/syn. *Cartera paloma*

casalinga, casareccia: *see* Campanian Barbary

Casciana: (Cascia, Umbria, Italy)/ local var. of Apennine

Casentinese: (Casentino, Arezzo, Tuscany, Italy)/Apennine type with much Sopravissana blood

Cassubian: *see* Pomeranian

Castilian: (Old Castille, Spain)/ d.m.(mw) (Valladolid and Zamora) or m.(cw) (Salamanca)/ white; pd/Entrefino type/vars: Black (with white spot on head and tail tip), Bilbilitana Red, Tudelana/BS/Sp. *Castellana*

Castilian Churro: (Tierra de Campos, Palencia, Spain)/m/orig. and principal var. of Spanish Churro/Sp. *Churra castellana*/ syn. *Churra campera*

Castillonnais: (Castillon, SW Ariège, France)/m.cw-mw/white with red head and legs; ♂ hd, ♀ usu. pd/Central Pyrenean group/HB 1982/syn. *Pyrénéen central à extrémités rousses* (= with red extremities), *St Gironnais* (from St Girons)/rare (by crossing with Tarasconnais)

Castlemilk Moorit: (Scotland)/w/pale red-brown (blond); hd; st/orig. by Buchanan-Jardine on Castlemilk estate, Dumfries, from Shetland crossed with mouflon (? or Soay) (1930) and with Manx Loaghton blood (1936)/HB 1974, BS 1983/ syn. *Castlemilk Shetland, Moorit Shetland* (to 1977)/nearly extinct

Casubian: *see* Kashubian

Catalan: (E Pyrenees, France)/m.mw/ white, with black or red marks on face/mixed type with blood of Lacaune, Berrichon, Île-de-France and Roussillon Red/syn. *Cerdagnole*/nearly extinct

Catalan: *see* Pallaresa

Caucasian: (N Caucasus, Russia)/ fw.m/orig. 1921–1936 in N Caucasus from Novocaucasian Merino improved by American Rambouillet and by Askanian/ HB/Russ. *Kavkazskaya, Kavkazskaya tonkorunnaya* (= *Caucasian finewool*); obs. syn. *Soviet Rambouillet of Caucasian type, Caucasian Rambouillet* (Russ. *Kavkazskiĭ Rambulye*), *Caucasian Merino*/also in Bulgaria

Caucasian Fat-tailed: (Caucasus)/ m.d.cw/sim. to Karaman/inc. Andi, Balbas, Bozakh, Bunib, Imeretian, Karabakh, Karachai, Kyasma, Lak, Lezgian, Mezekh, Shirvan, Tushin

Caucasian Mountain: breed group

Cauchois: (Caux, N Normandy, France)/m.lw/orig. from Oxford Down, Cotswold *et al.* × local/ extinct in 1950s

Caussenard: (Causses, S France)/ cw/♂ hd or pd, ♀ usu. pd/inc. Caussenard du Lot, Caussenard des Garrigues; formerly also Caussenard de la Lozère, Larzac, Ségala

Caussenard de la Lozère: (W Lozère, S France)/d.m/♂ often hd, ♀ pd/Roquefort breed/syn. *Lozère Causses*/extinct (absorbed by Blanc du Massif Central)

Caussenard des Garrigues: (Gard and E Hérault, S France)/m/face sometimes red; ♂ usu. hd, ♀ usu. pd/syn. *Caussenard du Gard et de l'Hérault*/rare

Caussenard du Lot: (Lot, S France)/ m.cw/black spectacles and ears; pd/HB 1955, BS/syn. *Race des Causses du Lot, Gascon à lunettes, Quercy, Quercynois*

Caux: *see* Cauchois

Central Anatolian Merino: (Turkey)/ fw.m/var. of Turkish Merino/orig. since 1952 at Konya state farm from German Mutton Merino (80%) × White Karaman/syn. *Anatolian Merino, Konya Merino*

Central Plateau: (N Massif Central, France)/inc. Ardes, Bizet, Limousin, Rava, Velay Black/Fr. *Races du Plateau central, Races du Massif Central Nord*/syn. *Auvergnat, Auvergne*

Central Pyrenean: (C and E Pyrenees, France)/m.mw/orig. from Merino × local Pyrenean/inc. Aure-Campan, Barégeois, Castillonnais, Lourdais, Tarasconnais; formerly only Castillonnais and Tarasconnais/BS/Fr. *Pyrénéen central, Races des Pyrénées centrales*

Central Rodopi: (S Bulgaria)/m.cw.d/
occ. speckled face; ♂ hd, ♀ usu.
pd/local unimproved var. of
Bulgarian Native/Bulg.
Srednorodopska, Rilo-Rodopska/
syn. *Rhodope*/rare
Central Stara Planina: (C Bulgaria)/
var. of Stara Planina/Bulg.
Srednostaroplaninska,
Srednoplaninska
Cerdagnole: *see* Catalan
Česká selská: *see* Šumava
Çeşme: *see* Sakiz
Cévennes, Cévenol: *see* Raqole
CFS: *see* Polish Mountain
Chacra: *see* Red Karaman
Chaffal: *see* Shafali
Chagra, Chakra: *see* Red Karaman
Chakri: *see* Magra
Chalkidiki: (Khalkidhiki peninsula,
Macedonia, Greece)/d.m.mw-
cw/white (with black or speckled
face), also black or grey; ♂ hd, ♀
pd/Ruda type/rare
Chall: *see* Shal
Chamar: (Mongolia)/♂ hd, ♀ hd or
pd; ft/superior var. of
Mongolian/Russ. *Chamarskaya*
Chamarita: (SE La Rioja and NE
Soria, Castille, Spain)/m.mw/
white, occ. black; usu. pd/syn.
Sampedrana (from San Pedro
Maurique)/[= small]
Champagne: (C France)/former var. of
Berrichon/extinct
Champagne Merino: (Aube, France)/
former var. of Précoce/HB 1925–
1929/Fr. *Mérinos champenois,*
or *de la Champagne*/extinct
Changthangi: (Ladakh, NE Kashmir,
India)/pa.cw/various colours/not
Chanthan
Chanothar: *see* Sonadi
Chanthan: *see* Changthangi
Chapan: (Kyrgyzstan)/cw/sft/sim. to
Darvas and Hetian
Chapper: *see* Chokla
Charmoise: (WC France)/m.sw/pd/
orig. 1844–1852 at La Charmoise
farm, Loir-et-Cher, from Romney
× (Berrichon, Merino, Solognot,
Touraine)/BS 1896, reorg. 1926,
HB 1927/not *Charmois*
Charollais: (Nièvre, France)/m.sw/

pd/orig. 1825 on from Leicester
Longwool × local/orig. of Swiss
Charollais/named 1963, HB 1963,
recog. 1974, BS; BS also GB 1977
**Charotar, Charotari, Charothar,
Charothri:** *see* Patanwadi
Chateng: *see* Bayinbuluke
Chatham Island: (New Zealand)/
feral/rare
Châtillonais: (Côte d'Or, France)/
former var. of Précoce/HB 1924–
1929/Fr. *Mérinos précoce du
Châtillonais*/syn. *Burgundy
Merino* (Fr. *Mérinos
bourguignon,* or *de la
Bourgogne*)/extinct
Chechen, Chechenskaya: *see*
Karachai
Chei-gai: *see* Tsigai
Cher Berrichon: *see* Berrichon du
Cher
Cherkassy: (Ukraine)/var. of Russian
Long-tailed (? with English or
Merino blood)/Russ.
Cherkasskaya/not *Cherkassian,
Cherkassky, Cherkasy,
Circassian*
Cherna Chervena: *see* Karnobat
Chernoglava plevenska: *see* Pleven
Blackhead
Chersolina: *see* Island Pramenka
(Cres)
Chevaldshay: (Orkney, Scotland)/
m/North Country Cheviot ×
North Ronaldsay, 1st cross
Chevali: *see* Shafali
Cheviot: (border of England and
Scotland)/m.mw/♂ usu. pd, ♀
pd/orig. at end of 18th c., ? with
Leicester Longwool blood/strain:
West Country Cheviot/vars:
Brecknock Hill Cheviot, North
Country Cheviot/orig. of Dala,
Rygja/named 1791; BS 1891, HB
1893; BS also USA 1891, New
Zealand 1949 (HB 1895); HB also
Australia, Canada/Nor.
Sjeviot/syn. *Border Cheviot,
South Country Cheviot*
Cheviot-Corriedale: (New Zealand)/
Cheviot × Corriedale, F$_1$
Chevlin: (New Zealand)/m.lw/orig.
(1950s) by P.G. Buckleton from
Lincoln × Cheviot/extinct 1960s

Chhanotar: *see* Chanothar

Chianina: (Val di Chiana, Tuscany, Italy)/local var. of Apennine with much Sopravissana blood

Chiapas: (S Mexico)/w.m.d/orig. from Spanish sheep in 16th c./ vars: Icsat, Mesha, Sacjol

Chietina: (Chieti, Abruzzo, Italy)/ local var. of Apennine

Chilhac: *see* Bizet

Chilludo: *see* Criollo

China (or Chinese) Fat-tailed: *see* Mongolian and its derivatives (Choubei, Han, Hu, Lanzhou Large-tail, Luan, Ningxia Black, Quanglin Large-tail, Shouyang, Taiku, Tan, Tong)

Chinese Finewool: *see* Chinese Merino

Chinese finewools: *see* Aohan Finewool, Chinese Merino, Erduos, Gadasu, Gansu Alpine Finewool, Inner Mongolian Finewool, Keerqin, North-East China Finewool, Shanxi Finewool, Xinjiang Finewool

Chinese Karakul: (W Xinjiang, China)/black; ♂ hd, ♀ pd/orig. since 1960s by grading Kuche and Kazakh to Karakul, also (Inner Mongolia) from Karakul × Mongolian

Chinese Merino: (Xinjiang and Inner Mongolia, China)/fw.m/orig. from Australian Merino × Xinjiang and Mongolian/syn. *Chinese Finewool*

Chios: (Khíos, Greece)/d.m.cw-mw/white with black or brown spots on face, belly and legs, sometimes black face; ♂ hd, ♀ usu. pd; lft; prolific/orig. of Sakiz

Chiraz: *see* Grey Shirazi

Chisqueta: *see* Pallaresa

Chistavina: *see* Pallaresa

Chita: (S Siberia, Russia)/mw/breed group/orig. at Voroshilov state farm from Buryat × finewool/ Russ. *Chitinskaya porodnaya gruppa*

Chiva: *see* Zel

Chleuh: *see* Berber

Chokla: (Churu and Sikar dists of Rajasthan, India)/cw.m/very

white shiny fleece, dark brown face; pd/sim. to Magra but smaller and with finer wool/orig. (with Rambouillet) of Avivastra/ syn. *Chapper*, *Rato Munda* (= sheep with dark brown or tan face), *Shekhawati*/not *Sherawati*/ declining through crossbreeding (Rambouillet, Russian Merino and Bharat Merino)

Choletais: (Maine, France)/orig. (with Leicester Longwool) of Bluefaced Maine (Bleu du Maine)/extinct

Cholistani: (Bahawalpur, Pakistan)/ m.cw/black head, neck and lop ears

Chotanagpuri: (Bihar, India)/cw/light grey or brown; pd; st/not *Chottanagpuri*

Cho-ten-jan: *see* Hetian

Choubei: (N Hubei, China)/cw.m/usu. white, sometimes black head; ♂ usu. hd, ♀ usu. pd; large ft/WG *Chou-pei*, Ger. *Dschau-bei*/not now recog.

Choufalié: *see* Shafali

Chou-pei: *see* Choubei

chubuku: *see* snow sheep

Chuisk Semifinewool: (Kazakhstan)/ m.w/orig. from Kazakh Meat-wool, Stavropol, Précoce and fat-rumped/Russ. *Chuiskaya polutonkorunnaya*/syn. *Chuisk meat-wool breed group*

Chundi: *see* Thalli

Chuntuk: (Ukraine)/m.cw/brown; pd; fr/sim. to Kalmyk/extinct

Churra alcarreña: *see* Ojalada

Churra algarvia: *see* Algarve Churra

Churra berciana: *see* Bierzo White

Churra campera: *see* Castilian Churro

Churra castellana: *see* Castilian Churro

Churra da Terra Quente: (Trás-os-Montes, NE Portugal)/ m.cw.d/ hd/Portuguese Churro type/HB 1990/syn. *Tarrincha*, *Terrincha*/ not *Badana*

Churra de Colmenar: *see* Colmenareña

Churra del Bierzo: *see* Bierzo White

Churra do Campo: (NE Castelo Branco, Beira Baixa, Portugal)/

m.cw.d/♂ hd, ♀ pd; small/
Portuguese Churro type

Churra entrefina: *see* Ojalada

Churra gallega: *see* Galician

Churra lebrijana: *see* Andalusian
Churro

Churra soriana: *see* Ojalada

Churra tensina: *see* Tensina

Churra turolense: *see* Ojalada

Churro: *see* Portuguese Churro,
Spanish Churro

Chusco: (Colombia) *see* Criolla Mora

Chusco: (Peru)/degenerate Criollo
type in Peru; orig. Criollo hd,
brown spots on face, from
Spanish (e.g. Churro, Lacho)/[=
funny, coarse]

Chushka: (W Ukraine and Moldova)/
fur, d/black or white; 50% ♂ hd
(small); lt/sim. to Reshetilovka
and Sokolki/orig. from Russian
Long-tailed/Russ. *Čuška*/ syn.
Tsushka

Ciavenasca: (NW Sondrio, Lombardy,
Italy)/m.cw/pd/nearly extinct

Cigaja: *see* Tsigai

Cikta: (Tardosbànya, Hungary)/m.cw/
♂ hd, ♀ pd/orig. from Zaupel in
18th c./HB 1974/rare

Çıldır: *see* Tuj

Cilician: *see* Anatolian red

Cine Capari: (Aydin prov., Turkey)/
cw/occ. with brown spots on feet
and belly/rare

Cinta: (Bergamo-Brescia, Italy)/
extinct 1976–1979

Circassian: *see* Cherkassy

Ciudad Rodrigo: (Salamanca, Spain)/
var. of Mestizo Entrefino-fino

Ciuta: (Val Masino, Sondrio,
Lombardy, Italy)/m/hd/sim. to
Ciavenasca/extinct

Cladore: (Connemara, Ireland)/occ.
coloured; pd/Northern Short-
tailed type but longer tail/syn.
Cladagh, Cottagh/[= shore]/
extinct (only crossbreds remain)

Closewool: *see* Devon Closewool

Cluj Merino: (Romania)/Romanian
Merino at Univ. of Agric. Sci.
and Vet. Med., Cluj-Napoca

Clun Forest: (SW Shropshire,
England)/sw.m/dark brown face
and legs; pd/? orig. from Hill

Radnor and Shropshire, with
Kerry Hill blood *c.* 1865/BS and
HB 1925; BS also France 1970,
USA 1973, HB also Canada/syn.
Clun

CMBLX: (Australia)/being bred to
Thai Longtail and Malin
(Malaysia)

Coastal Zackel: (Croatia) *see* Istrian
Milk

Cobb 101: (Suffolk, England)/m.sw/♂
hd, ♀ pd orig. 1960s from
Finnish Landrace *et al.* as ♂ line
for crossing on hill ewes/extinct

Coburger: (Coburg, N Bavaria,
Germany)/m.w/red head and
legs, born red; pd/recog. 1966
(HB) when nearly extinct/syn.
*Coburger Fuchsschaf, Eifel,
Fuchskopf* (= *Red Head*)/not
Foxhead, Koburg/rare

coda chiatta: *see* Campanian Barbary

Coete, Cohete: *see* Red Majorcan

Coimbatore: (WC Tamil Nadu, India)/
cw.m/usu. white with black or
brown head and neck; ♂ hd or
pd, ♀ pd; st/syn. *Coimbatore
White, Kuruba, Kurumba* or
Kurumbai (Karnataka), *Kurumbai
Adu*

Colbred: (Gloucestershire, England)/
m.mw/pd/orig. (1957 on) by
Oscar Colburn from East
Friesian, Clun Forest, Border
Leicester and Dorset Horn/BS
1962

Collines: *see* Houtland

Colmenareña: (Colmenar Viejo,
Madrid, Spain)/d.m.mw/white
with black points; pd or hd/syn.
*Churra de Colmenar, Oveja de
Colmenar*

Colombian African: *see* Red African

Colombian Criollo: (Colombia)/usu.
white; ♂ hd or pd, ♀ pd or
scurs/var. of Criollo

Colombian Woolless: *see* Red African

Columbia: (Idaho, USA)/mw.m/
pd/sim. to Panama/orig. (1912
on) from Lincoln × American
Rambouillet/BS 1942; HB also in
Canada/[? etym.]

Columbia-Southdale: (Middlebury,
Vermont, and Beltsville,

Maryland, USA)/m.mw/pd/orig. since 1943 from Columbia × Southdale

Combo-6: (S Illinois, USA)/synthetic orig. 1970s at Dixon Springs Agric. Centre from Border Leicester, Dorset Horn, Finnish Landrace, Rambouillet, Suffolk and Targhee/not *Comb-6*

Comeback: (Australia)/= Merino × (English Longwool × Merino)/BS 1976; eligible for Polwarth HB after 5 generations of *inter se* breeding

Comisana: (SE Sicily)/d.cw.m/ reddish-brown face; pd; semi-lop ears/orig. in late 19th and early 20th c. from Maltese × Sicilian/ HB and BSd 1942/syn. *Lentinese* (from Lentini), *Red Head* (It. *Testa rossa* or *Faccia rossa*/[from Comiso]

Commercial Corriedale: *see* Corriedale

Common Albanian: (S Albania)/ d.cw.m/usu. red or black spots on face and legs; ♂ hd, ♀ usu. pd/small plains type of Albanian Zackel/vars: Lushnja, South Albanian/Ger. *Kommunal*

Common Long-tailed: *see* Russian Long-tailed

Common Russian: *see* Russian Long-tailed

Commun des Alpes: *see* French Alpine

Congo Dwarf: *see* Bahu

Congo Long-legged: (Kibali-Ituri, NE DR Congo)/white or brown pied; hy; ♂ hd, ♀ pd; lop ears; st/African Long-legged group/ obs. syn. *Zaïre Long-legged/see also* Baluba

Constantinois: *see* Tunisian Barbary

Coolalee: (NSW, Australia)/m.sw/ pd/orig. 1968–1982 by Dr Ryan at Bugaldie from Suffolk, Wiltshire Horn, Hampshire, Lincoln, Leicester and Poll Dorset, as terminal sire line/ [Aboriginal word]

Coopworth: (S Island, New Zealand)/ m.mw/pd/orig. 1956–1968 from Border Leicester × New Zealand

Romney by I.E. *Coop* at Lincoln College/BS 1968; BS also Australia, USA 1985

Copper-Red: (E Bulgaria)/d.m/fawn, red or black; ♂ hd, ♀ pd/inc. Karnobat, Shumen/Bulg. *Mednochervena*/syn. *Karnobatoshumenska*/rare

Coquo: *see* Angola Maned

Corbières: (S Aude, S France)/mw/ extinct (absorbed by Lacaune in 1950s)

Córdoba de la Campaña: *see* Andalusian Merino

Corentyne White: (Guyana)/nucleus flock at Nat. Agric. Res. Inst., eventually for crossing with Virgin Islands White

Corino: (Patagonia, Argentina)/mw/ orig. since 1970 at "Monte Dinero", Rio Gallegos, Santa Cruz, from Mer*ino* × *Cor*riedale/ BS

corkscrew: *see* screwhorn

Cormo: (Tasmania, Australia)/fw.m/ pd/orig. since 1960 at "Dungrove" from Corriedale × Tasmanian Merino/BS; BS also USA 1976/[*Cor*riedale-Merin*o*]/ *see also* Argentine Cormo

Cormo argentino: *see* Argentine Cormo

Cornella White: (Bologna, Italy)/ d.m.(mw)/hd/Apennine group/It. *Cornella bianca*/nearly extinct

Cornetta: (Modena, Italy)/extinct 1976–1979

Cornigliese: (Corniglio, Emilia, Italy)/local var. of Apennine/ orig. from Vissana with Merino and Bergamasca blood/syn. *di Corniglio*/nearly extinct

Corriedale: (New Zealand)/mw.m/ pd/vars: Bond, Broomfield Corriedale/orig. 1880–1910 from Lincoln (or Leicester Longwool) × Merino/orig. of Canadian Corriedale/BS 1910, HB 1924; sim. orig. in Australia (BS); BS also USA 1915, S Africa 1953, GB, Argentina, Uruguay/*see also* Askanian Corriedale, Kazakh Corriedale, North Caucasus Mutton-Wool, Poznań Corriedale,

Soviet Corriedale, Soviet
Mutton-Wool, Tyan Shan

Corsican: (Corsica, France)/d.m.cw/
usu. white with black marks on
head and feet, also black, grey or
red; ♂ hd, ♀ pd or hd; often
tassels; small/HB 1975/Fr. *Corse*

Corsican mouflon: *see* mouflon

Corteno: (Brescia, Lombardy, Italy)/
m.(mw)/pd/Lop-eared Alpine
group, sim. to Bergamasca/It. *di
Corteno*/rare

Cotentin: (N Manche, France)/m.lw/
pd/sim. to Avranchin but larger/
orig. 1830–1900 from Leicester
Longwool × local (Roussin de la
Hague)/BS 1925; BS also GB/syn.
*Race du littoral nord de la
Manche*/rare

Cotswold: (England)/lw.m/pd/
English Longwool type/Leicester
Longwool blood/var.: Black
Cotswold (USA)/recog. 1862; BS
1892; BS also USA 1878; HB also
Canada/orig. (× Hampshire) of
Oxford Down and (× Marsh) of
German Whiteheaded Mutton/
rare in England

Cottagh: *see* Cladore

Crag, Cragg: *see* Limestone

Crau Merino: (France)/reddish/
formerly var. of Arles Merino,
now syn./Fr. *Mérinos de la Crau*

Creole: *see* Criollo

Cres: (Cres I, Croatia)/m.(w.d)/♂ hd,
♀ usu. pd/= local Island
Pramenka/orig. from local cw
improved with Merino/Cro.
Creska, It. *Chersolina*

Cretan Zackel: *see* Greek Zackel
(Sfakia, Sitia and Psiloris vars)

Crevant: (C France)/former var. of
Berrichon/not *Crevat*/extinct

Crickley Barrow: (Gloucestershire,
England)/m.sw/white with black
face and legs/orig. in 1970s by
Oscar Colburn from Clun Forest
and New Zealand Southdown

Criollo: (Spanish America, esp.
highlands of Bolivia, Colombia,
Ecuador, Guatemala, Mexico, Peru
and Venezuela)/m.cw/ white,
black or pied/cf. Crioulo, Gulf
Coast Native, Navajo-Churro/orig.

from Spanish Churro (? and
Spanish Merino) 1548–1812/vars:
Argentine, Colombian, Lucero,
Tarhumara, Uruguayan,
Venezuelan/syn. *Creole* (West
Indies), *Chilludo* (= hairy) or
Pampa (Argentina)/[= native]

Criolla Mora: (highlands of
Colombia)/m.cw/dark wool/orig.
from Spanish Churro and ?
Spanish Merino/syn. *Chusco*/
rare

Criollo negro: *see* Black Crioulo

Crioulo: (S Brazil)/cw.m/head and
legs sometimes red or black; ♂
hd, ♀ pd or hd (short)/orig. from
Portuguese Churro/var.: Black
Crioulo/syn. *Crioulo lanado* (=
wooled Criollo), *Curraleiro* (mid
São Francisco valley), *Fronteira*
(Rio Grande do Sul), *Serrana*
(Santa Catarina)/rare

Crioulo preto: *see* Black Crioulo

Crna vlaška: *see* Karakachan

Croatian Dairy: (Croatia)/d

Croatian Pramenka: *see* Dalmatian-
Karst, Island Pramenka, Istrian
Milk, Lika

Cross: (Scotland) *see* Greyface

Cruzado, Cruzo: *see* Entrefino

Cruzado Merino: (Hungary)/fw/
prolific/orig. since 1982 from
Hungarian Rambouillet graded to
Booroola Merino from New
Zealand

Cuban Hairy: *see* Pelibüey

Cubano Rojo: *see* Pelibüey

Cücch: *see* Tacòla

Cuorgné: *see* Savoiarda

Curkana: *see* Ţurcana

Curraleiro: *see* Crioulo

Čuška: *see* Chushka

Cutchi: (India) *see* Patanwadi

Cutchi: (Pakistan) *see* Kachhi

Cuu Phan Rang: (S Vietnam)

CVM: *see* Californian Variegated
Mutant

Cyprus Fat-tailed: (Cyprus)/d.cw.m/
usu. white, with black (or brown)
on face esp. eyes and nose, occ.
black, brown or white; ♂ usu.
hd, ♀ usu. pd/Near East Fat-
tailed type/sim. to Awassi

Cyprus mouflon: (Troödos mts,

Cyprus)/= *Ovis orientalis ophion*
Blyth/var. of red sheep/Gr.
agrinon/syn. *Cyprian mouflon,*
Cyprian red, Cyprian wild,
Cyprus urial, O. musimon var.
orientalis/rare

Cyrenaica: *see* Barki

Czarnogłowka: *see* German
Blackheaded Mutton, Polish
Blackheaded

Czarnogłowa owca mięsna: *see*
Polish Blackheaded Mutton

Czech: *see* Šumava

Czech Merino: (Czech Republic)/
fw.m/orig. from German Mutton
Merino, Caucasian and
Stavropol/HB 1942

Czech Mutton Merino: *see* Zirné
Merino

Czurkan: *see* Ţurcana

Dabene: *see* Duben

Daegwanryong Merino: (Korea)

Dagestan Mountain: (Russia)/m.mw/
orig. 1936–1950 from
Württemberg Merino (now
Merinolandschaf) × Gunib,
backcrossed to Württemberg and
then mated *inter se*/Russ.
Dagestanskaya gornaya

Daghma: *see* Beni Guil

Dağliç: (W Anatolia, Turkey)/
cw.m.d/black spots on head and
legs; ♂ usu. hd, ♀ pd; sft/? orig.
of Chios, Kamakuyruk/Turk.
Dağlıç/not *Daglich, Daglic*

Dahman: *see* D'man

Dakshini: (Orissa, India)/cw/local
var.

Dala: (Voss and Hordaland, Norway)/
m.w/pd/sim. to Cheviot/orig.
1860–1920 from Cheviot,
Leicester Longwool and Old
Norwegian/HB 1926/syn.
Voss/not *Dalas*/[= valley]

Daldale: (NSW, Australia)/m.mw/
pd/orig. in mid 1970s at
Wellington from Poll Dorset ×
(Border Leicester × Merino)

Dale-o'-Goyt: *see* Derbyshire
Gritstone

Dalesbred: (Upper Wharfdale, C
Pennines, England)/cw/hd/
Blackfaced Mountain type/orig.

from Swaledale with Scottish
Blackface blood/BS 1930, HB
1931/not *Dales-Bred*

Dall sheep: (Alaska and NW
Canada)/= *Ovis dalli*/white/var.
Stone or black sheep = *O.d.
stonei*; intermediate (grey) is
Fannin sheep/syn. *Alaskan
white, Dall's sheep*

Dalmatian-Karst: (Croatia and
Bosnia)/m.cw.d/usu. white with
white, black or spotted head,
also black or red-brown; ♂ hd, ♀
usu. pd/small Pramenka type/
orig. (with Merino) of
Dubrovnik/Serbo-cro.
Dalmatinska-Hercegovačka/syn.
Hum (Hercegovina), *Karst*
(Serbo-cro. *Krš*, It. *Carsolina*)

Damani: (Dera Ismail Khan,
Pakistan)/d.m.cw/tan or black
head or markings on head and
legs; pd; st; often tassels/syn.
Lama/not *Dhamani*

Damara: (N Namibia and S Africa)/
m/usu. brown, also pied, white
or black; hy-cw; ♂ usu. hd, ♀ pd;
lft/var. of Afrikander/improved
from late 1950s at Omatjenne
Res. Sta./BS Namibia 1966–
1973, revived 1986 (Southern
Africa); BS South Africa 1992/
not *Blinkhaar*

Damari: *see* Harnai

Damline: *see* ABRO Damline

Danadara: (Uzbekistan)/possible
orig. of Karakul/not *Danadar,
Donadar*/extinct

Dané Zaqla: (E Niger)/cw-hy/dwarf

Danish Finewool: (Denmark)/orig.
from Swedish Finewool and
Finnish Landrace/Dan. *Finuld
får*/rare

Danish Heath: (Denmark)/white face;
usu. pd; st/orig. from
Heidschnucke × Northern Short-
tailed/orig. of Danish Landrace/
Dan. *Hedefår*/extinct

Danish Landrace: (Jutland,
Denmark)/cw/grey head; pd/orig.
from Danish Heath crossed with
Merino in 18th c. and with
Leicester Longwool and Oxford
Down since 1900/Dan. *Dansk*

Landfår/syn. *Klitfår* (= dune sheep), *West Jutland Dune*/rare

Danish Whiteheaded Marsh: (Jutland, Denmark)/m.cw/pd/ imported from Germany and UK; same orig. as German Whiteheaded Mutton/Dan. *Hvidhovedet Marsk*/rare

Danube Finewool: (N Bulgaria)/ fw.m.d/orig. 1950–1967 from German Mutton Merino and Caucasian or Askanian, × Svishtov and Pleven Blackhead/HB/Bulg. *Dunavska t"nkorunna*/not *Dounavska*

Danube Merino: (S Romania)/fw.d.m/ usu. white, also pied or coloured/ orig. from Spancă (F$_1$ of Merino × Tsigai) selected F$_2$, F$_3$ or backcross to Merino/Rom. *Merinos dunărea*

Dargintsi: (Dagestan, Russia)/cw.m/ black or white; ♂ usu. hd, ♀ hd or pd/Caucasian Fat-tailed type/ Russ. *Darginskaya*/syn. *Gad*/not *Dargin, Darginci*

Darhad: (Hövsgül prov., Mongolia)/ head and neck may be black or brown; ♂ sometimes hd, ♀ pd; sft/var. of Mongolian/not *Darkhad, Darkhat*

Dartmoor: (Devon, England)/m.lw/ black or grey spots on nose; pd/English Longwool type/orig. (1820–1909) by selection from original Dartmoor, with Longwool blood/BS 1909, HB 1911/syn. *Greyface Dartmoor, Greyfaced Dartmoor, Improved Dartmoor*/rare/*see also* Whiteface Dartmoor

Darvaz: (Tajikistan)/cw.m/black, white, brown or pied; ♂ usu. hd, ♀ pd; sft/sim. to Chapan and Hotan/Russ. *Darvazskaya*/rare

Darvaz Mountain Mutton-Wool: (E Tajikistan)/m.mw-fw/breed group/orig. from Darvaz improved by Württemberg Merino (1941–1944 and 1948) and by Caucasian (since 1948)/ Russ. *Gornodarvazskaya myasosherstnaya porodnaya gruppa*/syn. *Gornodarvazskaya tonkorunnaya* (= finewool)

Dazdawi: (Iraq)/non-migrant var. of Iraqi Kurdi/not *Dazadi, Dizay*

Debouillet: (New Mexico, USA)/fw/ orig. 1920–1943 from Delaine Merino × Rambouillet/named 1947, BS 1954

Deccani: (C Maharashtra, NE Karnataka and W Andhra Pradesh, India)/m.cw/black or pied, also white or brown; ♂ usu. hd, ♀ pd; st/sim. to Bellary/var.: Sangamneri

Defaid Bal: *see* Balwen

Defaid Idloes: *see* Badger Faced Welsh Mountain

Degeres Mutton-Wool: (Kazakhstan)/ m/fr or sft/orig. 1931–1980 from Shropshire (and Précoce) × Kazakh Fat-rumped/Russ. *Degeresskaya myasosherstnaya*/ not *Degres*

Deiri: *see* Awassi

Delaimi: *see* Shafadi

Delaine Merino: (USA)/fw.(m)/var. of American Merino (C type)/BS 1882/[Fr. *delaine* wool is combing wool]/rare

Delimi: *see* Shafali

Del Sasso: *see* Steinschaf

Demane, Demmane: *see* D'man

Demonte, Demontina: *see* Sambucana

Derbyshire Gritstone: (Peak District, England)/mw.m/Blackfaced Mountain type, sim. to Lonk but pd/? orig. from Leicester Longwool × Lonk/BS 1892, HB 1893/syn. *Dale-o'-Goyt*

Derna, Dernawi: (Egypt) *see* Barki

Desert Sudanese: *see* Sudan Desert

deshi, desi: (Gujarat, India) *see* Patanwadi

desi: (India)/= local, indigenous/not *deshi*

Deslanado: (Brazil) *see* Brazilian Woolless

Deutsches Edelschaf: *see* Electoral-Negretti

Deutsches schwarzköpfiges Fleischschaf: *see* German Blackheaded Mutton

Deutsches veredeltes Landschaf: *see* Merinolandschaf

Deutsches Weisses Bergschaf: *see* German Mountain

Deutsches weissköpfiges
Fleischschaf: *see* German
Whiteheaded Mutton
Deux: *see* Belle-Ile
Devon: *see* Dartmoor, Devon
Closewool, Devon and Cornwall
Longwool, Devon Longwoolled,
Exmoor Horn, Whiteface Dartmoor
Devon and Cornwall Longwool:
(Devon and Cornwall, England)/
lw.m/pd/English Longwool type/
orig. (1977) by combining Devon
Longwoolled and South Devon
Devon Closewool: (N Devon,
England)/sw.m/pd/orig. in early
20th c. from Devon Longwoolled
× Exmoor Horn/BS, HB 1923/
syn. *Barnstaple*
Devon Longwoolled: (N Devon,
England)/lw.m/pd/English
Longwool type, sim. to South
Devon but smaller/orig. from
Leicester Longwool × Southam
Nott and Bampton Nott/ recog.
1870s; BS 1898–1977, HB
1900/syn. *Devon Longwool*/
extinct (combined with South
Devon to form Devon and
Cornwall Longwool 1977)
Dhamani: *see* Damani
Dhamari: (Dhamar, Yemen)/m.(d)/
white, usu. with fawn back and
legs; hy; pd; semi-lop ears;
ft/syn. *Burri, Jahrani*
Dhamda: (E Madhya Pradesh, India)/
m.cw/usu. black/not *Dhamada*
Dhitiki Thraki: *see* Thrace
Dhormendi: (S Maharashtra, India)/
black; hd/larger var. of Deccani
Dhormundi: *see* Godavari
Dilem: *see* Shafali
Dillène: *see* Shafali
Dinka: *see* Nilotic
Dishley, Dishley Leicester: *see*
Leicester Longwool
Dishley Merino: (France)/m.mw/orig.
1833–1900 from Leicester
Longwool × Merino/renamed
Île-de-France 1922/syn. *Alfort,
Grignon*/extinct
Dizay: *see* Dazdawi
Djallonké: *see* West African Dwarf
Djallonké: *see* Mossi (Burkina Faso),
Kirdi (Chad and Cameroon)

Djoydory: (Tajikistan)
dlinnosherstnaya: = longwool (Russ.)
DLS: (Quebec, Canada)/m.mw/pd/
orig. (1965 on) at Lennoxville
Res. Sta. from (Australian Dorset
× Border Leicester) ×
(Australian Dorset × Suffolk)
bred *inter se* since 1968/HB
1989/rare
Długowełnista owca polska: *see*
Polish Longwool
D'man: (oases of S Morocco and S
Algeria)/cw.m/black, brown,
white or pied; pd; lop ears;
prolific/syn. *Tafilalet* (Algeria)/
not *Dahman, Demane,
Demmane, Ndman*
Dobruja Finewool: *see* North-East
Bulgarian Finewool
Döhne Merino: (E Cape prov., S
Africa)/fw.m/pd/orig. 1939 at
Döhne Res. Sta. from German
Mutton Merino × South African
Merino/BS 1966
Domaći tip merina: *see* Vojvodina
Merino
domba ebor gomuk, domba ckor
gemuk: *see* Indonesian Fat-tailed
Donadar: *see* Danadara
Donggala, Donggola: (Indonesia) *see*
Indonesian Fat-tailed
Dongola: (Indonesia) *see* Indonesian
Fat-tailed
Dongola: (N Sudan)/cw/white or
black; pd/syn. *North Riverain* (or
Riverine) Wooled
Donji Vakuf: (Bosnia)/subvar. of
Privor var. of Bosnian Mountain/
Serbo-cro. *Donjevakufska*
Dora: (SE Andhra Pradesh, India)/
red-brown/var. of Nellore
Dorman: *see* Dormer (S Africa)
Dormer: (Australia)/mw.m/orig. from
Merino × (Merino × Dorset)
Dormer: (S Africa)/m.w/orig. (1941
on) at Elsenburg, W Cape prov.,
from Dorset Horn × German
Mutton Merino/recog. 1970; BS
1965/syn. *Dorman*/[*Dorset
Merino*]
Dorper: (S Africa)/m/usu. white with
black head, often black feet; hy-
cw; ♂ pd or small horns, ♀
pd/orig. *c.* 1942–1950, chiefly at

Grootfontein, from Dorset Horn ($\frac{1}{2}$) × Blackhead Persian ($\frac{1}{2}$)/var.: White Dorper/BS 1950/not *Dorpers*/[*Dorset Persian*]

Dorset: *see* Dorset Horn

Dorset Down: (S England)/sw.m/ brown face and legs; pd/Down type/orig. in early 19th c. from Southdown × Hampshire Down/BS 1906; HB also New Zealand, France 1966, Australia/ obs. syn. *Improved Hampshire Down, West Country Down*/rare in England

Dorset Horn: (S England)/sw.m/hd/? orig. from Portland (and Merino)/ vars: Pink-nosed Somerset, Poll Dorset, Polled Dorset/orig. of Dormer, Dorper/recog. 1862, BS 1891, HB; BS also USA 1890; HB Australia, Canada, S Africa, New Zealand/ syn. *Dorset* (USA)

Dorsian, Dorsie: *see* White Dorper

Doukkala: (SW Morocco)/cw.m/usu. white, head usu. black or brown; ♂ hd, ♀ pd/Atlantic Coast type/ Fr. *Doukkalide*/syn. *Doukkala-Abda*

Doulemi: *see* Shafali

Dounavska: *see* Danube

Doundoun: *see* Goundoun

Douro e Minho: *see* Entre Douro e Minho

Down: (S England)/coloured face; sw (5–15 cm); pd/inc. (all with Southdown blood) Dorset Down, Hampshire Down, Oxford Down, Shropshire, Southdown, Suffolk/ [Downs = chalk hills of S England]

Down-Cotswold: *see* Oxford Down

Drama Native: (Volax, E Macedonia, Greece)/m.d.cw/black, white or grey; hd or pd/var. of Greek Zackel/rare

Drasciani, Drashiani: *see* Sudan Desert

Drenthe Heath: (E Netherlands)/ white (usu. with reddish-brown or grey to black on face and legs; lamb spotted), black, brown or pied; ♂ hd, ♀ hd (small)/smaller var. of Dutch Heath/HB 1986, BS/Du. *Drentse Heideschaap*/ rare

Drysdale: (New Zealand)/cw.m/♂ hd, ♀ pd/orig. 1929–1967 by F.W. *Dry* of Massey Agric. College by selection of monogenic variants (N^d or *nrnr*) in New Zealand Romney/BS; BS also Australia/ syn. *Ennendale, N-type Romney*

Dschau-bei: *see* Choubei

Dub: (Bosnia)/subvar. of Vlašić var. of Bosnian Mountain/Serbo-cro. *Dubska, Dupska*

Dubasi: (Khartoum to Shendi, Gezira, Sudan)/pied/var. of Sudan Desert/syn. *Butana* (N of Khartoum), *Northern Riverain*/ not *Dubassi*/[Dubaseen tribe]

Duben: (Bulgaria)/m.cw.d/white or coloured; lt/Bulg. *D"benska*/not *Dabene*/rare

Dubrovnik: (Dalmatia, Croatia)/ d.m.mw/usu. white, occ. black or brown; ♂ usu. pd, ♀ pd/sim. to Pag Island/? orig. from Merino × Pramenka in late 18th and early 19th c., or relic of ancient finewool/Cro. *Dubrovačka, Dubrovačka ruda*, It. *Ragusa*/ syn. *Ragusa-Šipan* (It. *Giupanna*)/rare

Dugi Otok: *see* Island Pramenka

Dugli: *see* Red Karaman

Dulaimi: *see* Shafali

Dumari: *see* Harnai

dumba: (Iran, Afghanistan and Pakistan)/= fat-tailed/see Iran fat-tailed, Afghan, Pakistan fat-tailed

Dumbi: (E and N Sind, Pakistan)/ m.d.cw/black spots on face and ears; ♂ usu. hd, ♀ pd; ft

Dun, Dunface: *see* Tanface

Dunavska: *see* Danube

dune sheep: *see* Danish Landrace

Dupska: *see* Dub

Durmitor, Durmitorska: *see* Piva

Dusbay: *see* Duzbai

Dutch Black Blaze: (Netherlands)/ m/black-brown with white face blaze, feet and tail tip; pd/? orig. from Texel and Friesian with black colour gene from Drenthe and Schoonebeker/BS 1979, HB 1985/Du. *Zwartbles schaap*/syn. *Black Blazed, Six-point White*/ rare

Dutch Heath: (Netherlands)/cw/hd or pd; prim./vars: Drenthe, Kempen, Schoonebeker, Veluwe/ Du. *Heideschaap*

Duzbai: (Uzbekistan)/var. of Karakul/ not *Dusbay*

dwarf: *see* Aknoul, Bahu, Bangladeshi, Dané Zaqla, Garole, Kage, Southern Sudan, Ushant, West African Dwarf

Dymykh: (Azerbaijan)/ cw.m.d/ Caucasian Fat-tailed type/? Mazekh × Karabakh/not *Dymyh*, *Dymyk*/extinct

Dzhaidara: *see* Jaidara

Dzharo: *see* Jaro

East African Blackheaded: (W Uganda and Sukumaland, NW Tanzania)/m/pied (usu. white with black or brown head); hy; usu. pd; sft/East African Fat-tailed type/declining by crossing

EAST AFRICAN FAT-TAILED: hy; broad ft (usu. S-shaped)/inc. East African Blackheaded, Ethiopian Highland, Masai; *see also* African Long-fat-tailed

East Carpathian: (Poland)/var. of Polish Zackel

Eastern Butana: *see* Gash and Eastern Butana

Eastern Merino: (France) *see* Est à lain Mérinos

Eastern Serbian: *see* Pirot

East Friesian: (Germany)/d/pd; rat tail (woolless)/Marsh type/cf. Friesian Milk/var.: Black Milksheep/ part orig. of Pomeranian/HB 1890; BS also GB (as British Friesland), Switzerland, Argentina/Ger. *Ostfriesisches Milchschaf*, Swe. *Ostfriesiska mjölkfår*, Cz. *Východofríská ovce*/syn. *East Friesland Milch, German Milk Sheep, Milchschaf*

East Java Fat-tailed, East Javanese Fat-tailed: *see* Indonesian Fat-tailed

East Mongolian Semifinewool: (Mongolia)/w.m/orig. 1962–1982 from Altai, Transbaikal Finewool and Tsigia, × local coarsewool (Mongolian)

East Prussian: *see* Skudde

East Swiss: *see* Wildhaus

Ebeidi: *see* Ibeidi

Edilbaev: (N Kazakhstan)/m.cw.d/ black, red or brown; pd; fr/orig. from Kazakh Fat-rumped with Kalmyk blood/Russ. *Edil'baevskaya*/not *Edelbaev*

Eesti tumedapealine: *see* Estonian Darkheaded

Eesti valgepealine: *see* Estonian Whiteheaded

Egyptian: (Egypt)/inc. Near East Fat-tailed type (Barki, Fellahi, Ossimi, Rahmani), sft-lft (Ibeidi, Saidi), lt (Kurassi)

Eifel: *see* Coburger

El Awas: *see* Awassi

Elb, Elbfarbenes Schaff: *see* Frutigen

Elburz red: *see* red sheep

El'dari, El'darsk(aya): *see* Georgian Finewool Fat-tailed

Electoral Merino: *see* Saxony Merino

Electoral-Negretti: (Germany)/fw/ orig. from Saxony Merino and Austrian Negretti/Ger. *Deutsches Edelschaf* (= improved)/extinct

Elektoralschaf: *see* Saxony Merino

El Hammam: *see* Timahdit

Elliotdale: (Tasmania, Australia)/ cw.m/pd (or scurs)/orig. 1963–1977 at *Elliot* Res. Sta. from a Drys*dale* × (Border Leicester × Merino) ♂ × hairy Romney ♀♀; homozygous for mutant carpet wool gene/BS

Embrun, Embrunais: *see* French Alpine

Enderby Island: (New Zealand)/ feral/? extinct

Engadine Red: (Switzerland)/red-brown; pd/Lop-eared Alpine type/? orig. from Bergamasca × Steinschaf/HB 1985, BS 1992/ Ger. *Engadiner Fuchsschaf, Engadiner Landschaf, Fuchsfarbenes Engadinerschaf,* Romansch *Besch da Pader*/syn. *Paterschaf*/rare

English Halfbred: (Shropshire and Hereford, England)/m.mw/ mottled face and legs; pd/Border Leicester × Clun Forest, 1st cross/BS 1981

English Leicester: *see* Leicester Longwool

ENGLISH LONGWOOL: lw (15–40 cm); white face; pd/inc. (most with Leicester Longwool blood) Bluefaced Leicester, Border Leicester, Cotswold, Dartmoor, Devon and Cornwall Longwool, Leicester Longwool, Lincoln Longwool, Romney, Teeswater, Wensleydale, Whiteface Dartmoor/syn. *British Longwool*

English Merino: *see* Anglo-Merino

Ennendale: *see* Drysdale

Entre Douro e Minho: (NW Portugal)/ mw.m/usu. white (8% black); hd; small/Bordaleiro type/Port. *Bordaleiro de Entre Douro e Minho*

Entrefina pirenaica: *see* Ansotana, Roncalesa

ENTREFINO: (Spain)/m.mw/usu. white; pd/cf. Bordaleiro/? orig. from Churro and Merino/inc. Alcarreña, Aragonese, Castilian, Galician, Manchega, Segureña, Talaverana/syn. *Cruzado* or *Cruzo* (= crossbred)/[= medium fine]

Entrefino-fino: *see* Mestizo Entrefino-fino

Entre Minho e Douro: *see* Entre Douro e Minho

Entre-Sambre-et-Meuse: *see* Sambre-et-Meuse

Epirus: (Greece)/white with red, brown or black spots on head and legs/mt var. of Greek Zackel/Gr. *Ipiros*

Eppynt: *see* Beulah Speckled Face

Erbi: (Kuwait) *see* Arabi

Erdelyi Racka: *see* Ţurcana

Erduos: (Inner Mongolia, China)/fw/ ♂ hd or pd, ♀ pd/orig. from Soviet Merino and Caucasian, × Mongolian

Erek: (Turkmenistan)/cw/white, brown or black; st/Russ. *Érek*/ syn. *Geoclan*/extinct

Erik: (Amasiya, Armenia)/d.cw/ Caucasian Fat-tailed type/cf. Tuj (= *Herik*); sim. to Tushin but larger and with coarser fleece/ Russ. *Érik*

Erzurum: *see* Red Karaman

Escurial: (Spain)/former strain of Spanish Merino/Sp. *Escorial*/not *Eskurial*

Est à laine Mérinos: (Alsace-Lorraine, France)/fw.m/pd/orig. from Württemberg Merino (now Merinolandschaf) imported since 1870/named 1950; BS 1947; BS also GB/Fr. *Race de l'Est à laine Mérinos* (= breed in the east with Merino wool)/syn. *Mérino de l'Est* (= Eastern Merino)

Estonian Darkheaded: (C and N Estonia)/m.sw/pd/orig. (1940 on) from local graded to Shropshire/ Estonian *Eesti tumedapealine*, Russ. *Temnogolovaya éstonskaya*/syn. *Rakvereskaya* (from Rakvere)

Estonian Whiteheaded: (S Estonia)/ m.sw/pd/orig. from Cheviot × local coarsewooled/HB/Estonian *Eesti valgepealine*, Russ. *Belogolovaya éstonskaya*/rare

Estrela: *see* Serra da Estrela

Ethiopian Highland: (Ethiopia)/m.d/ often brown; hy, often with mane; ♂ usu. hd, ♀ usu. pd; ft/East African Fat-tailed type/vars with woolly undercoat: Akele-Guzai, Arsi, Rashaidi, Tucur; other local vars: Bonga, Wello; recog. breeds: Horro, Menz (cw)/syn. *Abyssinian* (It. *Pecora abissina*), *Begghié abesce*

European mouflon: *see* mouflon

Evdilon: (Ikaria, Greece)/d.m.cw/ white, often with small spots on head and legs; ♂ hd or pd, ♀ usu. pd; lft/? orig. from Anatolian fat-tailed × local thin-tailed/Gr. *Evdhílos*

Exmoor Horn: (N Devon and W Somerset, England)/mw.m/hd/ BS 1906, HB 1907/orig. (with Devon Longwoolled) of Devon Closewool/syn. *Porlock*

Fabrianese: (Fabriano, Ancona, Italy)/m.d.(cw)/pd; roman nose/orig. from local Apennine graded to Bergamasca/HB 1974

Faccia rossa: *see* Comisana

Faeroes: (Faroe Is)/Northern Short-tailed type/orig. from Icelandic × Old Norwegian/Dan. *Færøerne,* Swe. *Faerøsk får/*not *Faroe*

Fagas: (Poland)/Marsh type (? Friesian Milk) brought by Dutch settlers in 17th–18th c./part orig. of Pomeranian/extinct

Fannin sheep: (Alaska)/grey/intermediate between Dall sheep and Stone sheep/syn. *saddleback sheep*

Farahani: *see* Baluchi

Farahi: *see* Kandahari

Fardasca: *see* Ojalada

Farleton Knott: *see* Limestone

Faroe: *see* Faeroes

Fartass: (Morocco and Algeria)/pd var. of Beni Guil/not *Fartas*

Fasanese: (Fasano, Brindisi, Italy)/var. of Leccese with Pramenka blood

fat-rumped: (Africa, Aden; hy) *see* Somali

FAT-RUMPED: (Eurasia)/cw/inc.: (1) Chuntuk, Kalmyk, Karanogai; (2) Alai, Degeres, Edilbaev, Hissar, Jaidara, Kargalin, Kazakh, Kirgiz, Sary-Ja, Tajik, Turkmen; (3) Balkhi, Turki/Russ. *kurdyuk, kuiryuk*

Fat-rumped Merino: (Kyrgyzstan)/fw/fr/orig. from 3 crosses of Précoce on Kirgiz Fat-rumped/Russ. *Kurdyuchnyĭ Merinos/*syn. *Kurdyukos/*extinct

FAT-TAILED: (Africa)/inc. African Long-fat-tailed, Barbary, East African Fat-tailed, Egyptian

FAT-TAILED: (Asia)/inc. Afghan, Anatolian fat-tailed, Caucasian Fat-tailed, Chinese fat-tailed, Indonesian Fat-tailed, Iran fat-tailed, Near East Fat-tailed, Pakistan fat-tailed; (Arabia): Habsi, Hejazi, Najdi, Yemeni; (Central Asia): Darvaz, Karakul; (Siberia): Buryat, Kulunda, Siberian, Telengit

FAT-TAILED: (Europe)/inc. (Greece): Argos, Chios, Evdilon, Mytilene, Rhodes; (Italy): Campanian

Barbary, Sicilian Barbary; (Spain): Red Majorcan; (Ukraine): Malich, Voloshian

Fellahi: (Nile delta, Egypt)/cw.m/usu. brown; ♂ hd, ♀ usu. pd; ft/syn. *Baladi/*being replaced by Ossimi and Rahmani/[*fellah* = peasant]

Fellata: *see* Ballami, Uda

Feltrina: *see* Lamon

feral: *see* Arapawa Island, Boreray, Campbell Island, Chatham Island, Enderby Island, Hog Island, Hokonui, Mohaka, mouflon, Pitt Island, Raglan Peninsula, Santa Cruz Island, Soay

Fésüs: *see* Hungarian Combing Wool Merino

Finarda: (Po valley, N Italy)/ m.(mw)/pd/Lop-eared Alpine group/orig. from Bergamasca × Biellese

finewool: *see* Merino (esp. Tasmanian Merino)

Fingalway: (Ireland)/m/orig. in 1970s from Finnish Landrace × Galway

Finn-Dorset: (GB and Ireland)/Finnish Landrace × Dorset Horn

Finnish Landrace: (Finland)/w.m/white, occ. black, brown or grey; usu. pd; prolific/Northern Short-tailed type/BS 1918, HB 1923; BS also USA 1971, Canada, France 1966/Finn. *Suomenlammas, Suomalainen maatiais lammas,* Swe. *Finsk lantras,* Fr. *Finnois,* Cz. *Finskà ovce/*syn. *Finn, Finnish Native, Finnsheep, Improved Finnish Native*

Finncross: (China)/Finnish Landrace crossbred ewes (F_1) from either Rambouillet, Targhee or Columbia rams

Finnsheep: *see* Finnish Landrace

Finŭld får: *see* Danish Finewool

Flamand: *see* Flemish, Flemish Marsh

flämisch: *see* Leine

Flanders, Flandrin: *see* Flemish, Flemish Marsh

Flemish: (E and W Flanders, Belgium)/m.d.cw/pd; prolific/Marsh type/Du. *Vlaamse,* Fr.

Flamand/syn. *Flemish Landrace, Flemish Milk Sheep*/formerly also in Netherlands (Schouwen I. and Zeeland Flanders)/nearly extinct

Flemish Flock sheep: (Belgium)/ m.w/cream, occ. black or with black spots on face; pd/Flem. *Vlaamse Kuddeschaap*/rare

Flemish Marsh: (French Flanders)/ d.m.lw/st/Marsh type/vars: Artois, Cambrai, Picardy, St Quentin/Fr. *Flamand, Flandrin*/ extinct

Flevoland: (Netherlands)/m/prolific dam line/orig. 1975 on from Finnish Landrace × Île de France/BS

Flitsch: *see* Bovec

Florida Native: (USA)/var. of Gulf Coast Native/nearly extinct

Florina: (NW Macedonia, Greece)/ cw.d.m/usu. black rings around eyes, often spots on nose and ears; ♂ usu. hd, ♀ usu. pd/cross of mountain and lowland Zackels, with Ruda wool type/ syn. *Pelagonia*/rare

Fodata: *see* Vicentina

Fonte Bôa Merino: (Portugal)/strain of Portuguese Merino/orig. 1902–1926 at F.B. Zootechnical Sta. from Rambouillet × Spanish Merino

Fonthill Merino: (NSW, Australia)/ fw.m/♂ hd, ♀ pd/orig. in 1954 by Jim Maple-Brown of Goulburn from American Rambouillet × Australian Merino

Fornese, Forno: *see* Massese

Foulbé: *see* Fulani, Uda

Four-horned: *see* Hebridean, Jacob, Manx Loaghtan, Multihorned Shetland; also Icelandic, Karachai, Libyan Barbary, Multihorned Merino, Navajo-Churro, Uruguayan Criollo

Fouta, Fouta Djallon, Fouta Jallon: *see* West African Dwarf

Fouta Toro: *see* Toronké

Foxhead: (Belgium) *see* Brabant Foxhead

Foxhead: (Germany) *see* Coburger

Foza: *see* Vicentina

Frabosana: (Ligurian Alps, Italy)/ m.d.(cw)/usu. a few coloured spots, sometimes all-brown; hd; roman nose/sim. to Langhe/syn. *Rastela, Roaschina* or *Roascia* (Imperia)/[from Frabosa]/rare

Frafra: (Egypt)

Franconie: (E France)/black spectacles/extinct

French Alpine: (Hautes-Alpes, France)/m.mw/♂ hd, ♀ pd/cf. Lop-eared Alpine/orig. from local with Préalpes du Sud blood/HB 1952/Fr. *Commun des Alpes, Race des Alpes* or *Alpine*/obs. syn. *Barcelonnette* (E Basses-Alpes), *Embrun* (Fr. *Embrunais*), *Gap* (Fr. *Gapois*), *Trièves* (SE Isère)/nearly extinct (being absorbed by Préalpes du Sud)

French Blackheaded: (France, esp. Moselle)/m.sw/black head and legs; pd/orig. from Suffolk, Hampshire Down, Oxford Down, Southdown and (since 1945) German Blackheaded Mutton/BS and HB 1959 (Suffolk, Hampshire and Dorset Down now bred separately; BS is UPRA, Suffolk-Hampshire-Dorset)/Fr. *Moutons à tête noire*/ extinct

French Merino: (France)/inc. Arles Merino, Précoce, Rambouillet; formerly also Mérinos de Mauchamp, Mérinos du Naz, Roussillon Merino

French Southdown: (C and W France)/taller and heavier var. of Southdown from imports 1855–1962/HB 1947, BS

French Texel: (NE France)/more compact and less fat var. of Texel from imports 1933 on/BS 1935

Friesian Milk: (Netherlands)/d/pd; rat tail; prolific/Marsh type, sim. to East Friesian/HB 1908, BS 1950/Du. *Friese melkschaap*/ syn. *Friesian, West Friesian*/rare

Friesland: *see* East Friesian, Friesian Milk

Frisarta: (Arta, Greece)/d/pd/orig. since *c.* 1946 from East Friesian

× local already improved by Chios, Karagouniko and Zakynthos

Friserra: (Portugal)/d/orig. since 1962 from East Friesian × Serra da Estrela

Friulana: (Udine, Venetia, Italy)/ m.mw-cw.d/occ. dark spots on face/Lop-eared Alpine group/ orig. from Lamon and now absorbed by it/syn. *Furlana* [from Friuli]/extinct *c.* 1971

Frutigen: (Switzerland)/yellowish/ var. of Swiss Black-Brown Mountain/Ger. *Frutigschaf*/syn. *Elb, Elbfarbenes Schaf*

FSL: (Roquefort, S France)/d/orig. (1967 on) from $\frac{3}{8}$ East Friesian (EF), $\frac{3}{8}$ Sardinian (S), $\frac{1}{4}$ Lacaune (L), i.e. EF(EF × L) × S(S × L)

Fuchsfarbenes Engadinerschaf: *see* Engadine Red

Fuchskopf, Fuchsschaf: *see* Coburger

Fuhai Big-tail: (N Xinjiang, China)/ cw/usu. brown; ♂ usu. hd, ♀ hd or pd; ft/var. of Kazakh

FULANI: (Senegal to Cameroon)/Sahel type/inc. ara-ara, Balami, Peul-Peul, Toronké, Uda, Yankasa/Fr. *Peul, Peuhl, Peulh*/syn. *Foulbé, Fulbé, Pulfuli*

Fumex Black: (Megève, Haute Savoie, France)/ Fr. *OBS Mouton noir de Fumex*

Furlana: *see* Friulana

Futanké: (Mali)/black-spotted/var. of Toronké

Gacka: *see* Lika

gad: (N Baluchistan, Pakistan)/= *Ovis vignei cycloceros*/var. of (Afghan) urial in Kalat formerly called Blanford's urial (*O. v. blanfordi*)/ not *ghad*

Gad: *see* Dargintsi

Gadasu: (E Inner Mongolia, China)/ fw/orig. 1970s from Australian Merino × Polwarth

Gaddi: (S Jammu and Kashmir, C Himachal Pradesh and N Uttar Pradesh, India)/cw.pa.m/occ. coloured face or body; ♂ hd, ♀ usu. pd/syn. *Bhadarwah, Kanga Valley*/not *Guddi*

Gadertaler: *see* Pusterese

Gadik: (NE Afghanistan)/cw.m/♂ hd, ♀ hd or pd; ears short or long; sft/vars: Panjsher Gadik, Wakhan Gadik/not *Gadic*

Gaf: *see* Aljaf

Gala: (Apsheron peninsula, Azerbaijan)/var. of Shirvan

Galega: *see* Galician (Portugal)

Galega bragançana: *see* Bragança Galician

Galega mirandesa: *see* Miranda Galician

Galician: (Galicia, Spain)/m.(mw)/♂ usu. pd, ♀ pd; prolific/Entrefino type/Sp. *Gallega*; formerly called *Churra gallega* (= Galician *Churro*/syn. *Mariñana*

Galician: (Portugal) *see* Bragança Galician, Miranda Galician

Gallega: *see* Galician (Spain)

Galloway Blackface: (SW Scotland)/ strain of Scottish Blackface with finer wool/syn. *Newton Stewart*

Galway: (Clare, Galway, Roscommon and parts of adjacent counties, W Ireland)/m.lw/pd/Border(?) Leicester blood/split off from Roscommon in 1926, then absorbed by it/BS 1922, HB 1931

Gamde, Gamdi: *see* Taiz Red

Gammelnorsk: *see* Old Norwegian

Ganadi: *see* Taiz Red

Ganjam: (S Orissa, India)/m/brown or red; ♂ hd, ♀ pd/South India Hair type

Ganjia: (Xiahe, W Gansu, China)/ strain of Tibetan

Ganqin: (Gansu, Qinhai and Tibet, China)/strain of Tibetan

Gansu Alpine Finewool: (Qilan mts, C Gansu, China)/fw/orig. in mid-20th c. from Caucasian, Salsk and Stavropol × (Xinjiang Finewool × Mongolian and Tibetan)/syn. *Gansu Alpine Merino, Gansu Plateau Finewool, North-West China Merino, Northwestern Chinese Merino*

Gap, Gapois: *see* French Alpine

Garessina: (Tanaro, Inferno and Negrone valleys, Ligurian Alps, Cuneo, Italy)/m.d.(mw)/pd/sim. to Sambucana but smaller, with finer wool/It. *di Garessio*/syn. *Muma*/[from Garessio]/rare

Garfagnina White: (Garfagnana, NW Tuscany, Italy)/d.m.(cw)/♂ hd, ♀ usu. hd/Apennine group/sim. to Massese but white and larger/ former var.: Pavullese (with Bergamasca blood)/ It. *Garfagnina bianca*/rare

Garha, Gargha: (N Syria) *see* White Karaman

Garole: (Sunderbans, W Bengal, India)/m/light brown, occ. black, white or spotted; ♂ hd, ♀ pd; prolific; dwarf/cf. Bangladeshi

Garrigues Causses: *see* Caussenard des Garrigues

Gasc: *see* Gash and East Butana

Gascon: (Gascony, SW France)/local var. sim. to Landais and Lauraguais/extinct

Gascon à lunettes: *see* Caussenard du Lot

Gash and East Butana: (Sudan)/ many colours/var. of Sudan Desert/It. *Gasc*

Gatačka: *see* Lika

Gedek: (Azerbaijan, and Dagestan, Russia)/var. of Tabasaran/Russ. *Gëdek*/not *Gdek*/extinct

Gegica: *see* Gekika

Gekika: (Greece)/Ruda type/? = Luma (Albania)/not *Gegica*/ [*Gekides* are Albanian nomads]/extinct

Gentile a vello nero: *see* Carapellese

Gentile di Calabria: (toe of Italy)/ former var. of Gentile di Puglia with mw/orig. from Calabrian improved by Gentile di Puglia/ syn. *Improved Calabrian*/extinct

Gentile di Lucania: (S Italy)/former var. of Gentile di Puglia with mw/orig. from Gentile di Puglia × local/syn. *Basilicata*, *Improved Lucanian*

Gentile di Puglia: (S Italy)/fw-(mw).m.(d)/♂ hd, ♀ pd/orig. 15th c. on but chiefly 18th from Spanish Merino × local, also Saxony and Rambouillet blood in 19th c./former vars with less Merino blood: Gentile di Calabria, Gentile di Lucania,

Quadrella/HB 1942/syn. *Apulian Merino* (It. *Merino di Puglia*), *Improved Apulian*, *Italian Merino* (It. *Merino d'Italia*), *Merina Gentile*

Gentile moretta: *see* Carapellese

Geoclan: *see* Erek

Georgia Native: (USA) *see* Gulf Coast Native

Georgian: (Georgia) *see* Georgian Semifinewool Fat-tailed

Georgian: (Turkey) *see* Tuj

Georgian Fat-tailed Finewool: (Georgia)/m.fw/ft/orig. 1936–1958 at El'dri state farm from Tushin crossed with Soviet Merino and (since 1940) with Caucasian/Russ. *Gruzinskaya zhirnokhvostaya tonkorunnaya*/ syn. *Ėl'darskaya tonkorunnaya* (= *Eldari finewool*)/rare

Georgian Semifinewool Fat-tailed: (E Georgia)/m.mw/♂ hd, ♀ pd; ft/orig. 1931–1949 chiefly at Udabno farm from Rambouillet or Précoce × Tushin with 1 or 2 backcrosses to Merino ♂, 1 to Tushin ♀, and then *inter se* mating/Russ. *Gruzinskaya polutonkorunnaya zhirnokhvostaya*/ syn. *Georgian*/ rare

Georgian Thin-tailed Finewool: (E Georgia)/orig. 1950 on at Samgorsk state farm/Russ. *Tonkorunnaya toshchekhvostaya ovtsa gruzii*

German Blackheaded Mutton: (N and W Germany)/m.sw/pd/orig. 1870–1914 from Hampshire Down + Oxford Down (+ Shropshire + Suffolk)/orig. of Latvian Darkheaded, Lithuanian Blackheaded, Polish Blackheaded/HB 1920/Ger. *Deutsches schwarzköpfiges Fleischschaf*, Pol. *Czarnoglowka*/ syn. *German Hampshire*, *Improved Blackface Mutton*, *Teutoburg* (Westphalia)

German Hampshire: *see* German Blackheaded Mutton

German Heath: *see* Heidschnucke

German Improved Land: *see* Merinolandschaf

German Karakul: (E Germany)/
native graded to Karakul ($\frac{31}{32}$)/Ger.
Karakul Landschaf/ syn. *Land
Karakul*/rare

German Merino: (Germany)/formerly
inc. Electoral-Negretti and
Saxony; now see German Mutton
Merino/Russ. *Baldebukov*

German Milk Sheep: *see* East
Friesian

German Mountain: (S Bavaria,
Germany)/cw.m/pd/local mt
breeds (Carinthian, Goggel,
Steinschaf, Zaupel *et al.*) graded
to Bergamasca and Tyrol
Mountain since 1938/var.: Brown
Mountain/HB 1938/Ger.
Deutsches Weisses Bergschaf

German Mutton Merino: (Germany)/
m.fw/pd/orig. *c.* 1904 from
Précoce imported *c.* 1870/HB
1919; BS also Bulgaria, Spain/
Ger. *Merinofleischschaf*/ syn.
*German Merino, German
Précoce, Merino Meat, Merino
Mutton*

German Précoce: *see* German Mutton
Merino

German Whiteheaded Land: *see*
Merinolandschaf

German Whiteheaded Mutton: (NW
Germany)/m.lw/pd/cf. Danish
Whiteheaded Marsh/orig. in
mid-19th c. from Cotswold ×
Marsh (Butjadingen); recent
Texel blood/orig. of British
Oldenburg/HB 1885/Ger.
*Deutsches weissköpfiges
Fleischschaf*/syn. *Oldenburg
White Head, Whiteheaded
German, Whiteheaded Marsh,
Whiteheaded Oldenburg*

Gesel: *see* Red Karaman

Gessenay: *see* Saanen

Gezel: *see* Red Karaman

Gezira: (Sudan) *see* Watish

Gezirieh, Gezrawieh: (Iraq) *see*
Awassi

Ghachgai, Ghashgai: *see* Qashqai

Gharpala: *see* Kage

Ghezel: *see* Red Karaman

Ghiljai: (SE Afghanistan)/cw.m/white
with black or brown spots on
face and legs, or brown or black;

♂ hd or pd, ♀ pd; lop ears;
ft/sim. to Kandahari but larger
and with fleece of higher carpet-
wool quality/not *Gildjai*

Ghimi: (Fezzan, Libya)/cw-mw.m/ft/?
orig. from Algerian Arab ×
Libyan Barbary

Ghizel: *see* Red Karaman

Gıcık: (Turkey) *see* Herik or other
small breeds; also wrongly used
for Dağliç

Giganti di Bergamo: *see* Bergamasca

Gildjai: *see* Ghiljai

Gissar: *see* Hissar

Giupanna: *see* Dubrovnik

Glamorgan: *see* South Wales
Mountain

Glenara Improver: (Victoria,
Australia)/m/pd/orig. by George
Stewart of Langkoop from Dorset
Horn ($\frac{1}{2}$), Border Leicester ($\frac{1}{4}$),
Merino ($\frac{1}{4}$)

Glenvale: (NSW, Australia)/mw/var.
of Border Leicester/orig. 1993 by
Alan Luff/rare

Glossa: *see* Skopelos

Gmelin's sheep: *see* red sheep

Gobi-Altai: *see* Govĭ-Altay

Godavari: (NW Andhra Pradesh, and
Chanda, NE Maharashtra, India)/
m/white or reddish-brown; ?
pd/South India Hair type/syn.
Dhormundi/not *Godawari*

Goggel: (Bavaria, Germany)/cw/pd/
local term for mountain sheep
sim. to Steinschaf; probably a
cross/extinct

Gogo: *see* Tanzania Long-tailed

Goitred: *see* Zunu

Gojal: (Hunza valley, N Kashmir,
Pakistan)/d.m.cw/white and tan
with black patches and coloured
head and legs; ♂ usu. hd, ♀ pd;
sft

Gökçeada: (Turkey)/d.m.cw/black
spots around eyes, nose and ears;
♂ hd, ♀ pd or scurs/Island
Zackel type/syn. *Imroz*

Golemi: (Gijrokaster, S Albania)/
d.m.cw/white with black or
brown face; ♂ hd, ♀ pd

Gorki: (Russia)/m.sw/brown face;
pd/orig. 1936–1948 from
Hampshire Down × local

Northern Short-tailed/recog. 1950; HB/Russ. *Gor'kovskaya/* not *Gorky*

Gorno Altai: *see* Altai Mountain

Gornodarvazskaya: *see* Darvaz Mountain

Gornokarpatskie ovtsi: *see* Carpathian Mountain

Gornyĭ Korridel': *see* Soviet Mutton-Wool

Gornyĭ merinos: *see* Russian Mountain Merino

Gorodets: *see* Vyatka

Gorri Tippia: *see* Petite Manech

Goth: (Gotland, Sweden)/cw.m/usu. light to dark grey with black face, occ. pied; hd/var. of Swedish Landrace/orig. of Swedish Fur Sheep/named 1974, BS 1977; BS also GB/Swe. *Gutefår*, Ger. *Gotlandschaf*/syn. *Gotländska utegångsfår* (= Gotland outdoor sheep)/rare

Gotland: *see* Swedish Fur Sheep

Gotländisches Pelzschaf: *see* Swedish Fur Sheep

Gotlandschaf: *see* Goth

Gotländska utegångsfår: *see* Goth

Gotlandsk Pelsfår: *see* Swedish Fur Sheep

Goundoun: (Niger)/var. of Macina/ not *Doundoun, Koundoum/* [corruption of Goundam, town in Macina]

Gourane: (N Cameroon)/black; hy; ♂ hd, ♀ pd; long lop ears

Govĭ-Altay: (Mongolia)/head and hocks usu. brown, grey or black/ var. of Mongolian/not *Gobi-Altai, Gobi-Altay*

Grabs: (Switzerland)/orig. from Oxford Down × Wildhaus/orig. of Swiss Brownheaded Mutton/Ger. *Grabser*/extinct

Grammos: (NE Greece)/usu. white with red spot on head/mountain var. of Greek Zackel/Gr. *Gramoutsiano*

Gramoutsiano: *see* Grammos

Granada, Granadina: *see* Montesina

Graubünden: *see* Bündner Oberland

Graue gehörnte Heidschnucke: *see* Heidschnucke

Grazalema: (Sierra de Ronda,

Málaga, and Sierra de Grazalema, Cadiz, Spain)/d.m.w/hd/subvar. of Andalusian Merino selected for milk/Sp. *Merino de Grazalema*/rare

great Pamir sheep: *see* Marco Polo's sheep

great Tibetan sheep: *see* Marco Polo's sheep

Greek island breeds: *see* Chios, Evdilon, Kymi, Mytilene, Rhodes, Skopelos, Zakynthos; *see also* Island Zackel

Greek Zackel: (Greece)/d.m.cw/usu. white with black or red spots on head and legs, occ. coloured; ♂ hd, ♀ usu. pd/inc. Florina, Island Zackel (Lemnos, Levkimmi, Psiloris, Sfakia, Sitia), lowland vars (Karagouniko, Katsika), Vlakhiko (mountain vars) (Arvanitovlach, Boutsiko, Drama Native, Epirus, Grammos, Karakachan, Krapsa, Moraitiko)/ obs. syn. *Cretan Zackel, Macedonian, Parnassian*

Greenland: (Greenland)/m/black, grey, brown, white or mixed; hd/orig. from Faeroes and Icelandic/BS/Dan. *Grønlandsk Får*

Greyface: *see* Scottish Greyface

Greyface Dartmoor, Grey Faced Dartmoor: *see* Dartmoor

Greyfaced Lewis: *see* Island Blackface

Greyface Oldenbred: (England)/m/ British Oldenburg × Scottish Blackface or Swaledale, 1st cross/*see also* Welsh Oldenbred

Grey Horned Heidschnucke: *see* Heidschnucke

Grey Shirazi: (Fars, Iran)/m.cw.fur/♂ hd, ♀ pd; ft/sim. to Karakul but grey/var.: Zandi/Fr. *Chiraz*

Grignon: *see* Dishley Merino

Grisons: *see* Bündner Oberland

Gritstone: *see* Derbyshire Gritstone

Grivette: (Dauphiné, France)/ d.(m.w)/white, spotted at birth; pd/sim. to Rava/BS 1982

Gromark: (NSW, Australia)/m.mw-cw/pd/orig. 1965–1977 by Arthur Godlee at Tamworth from

Border Leicester × Corriedale selected for live weight, twinning and wool/BS 1979

Groningen Milk: (Netherlands)/d/ Marsh type, sim. to Friesian/Du. *Groningse melkschaap*/extinct

Grønlandsk Får: *see* Greenland

GrowBulk: (New Zealand)/being developed from New Zealand Romney, Dorset Horn and Texel

Grozny: (Dagestan, Russia)/fw/orig. 1929–1951 at Chervlennye Buruny stud farm from Australian Merino × Mazaev and Novocaucasian Merinos/HB/ Russ. *Groznenskaya*/not *Grozn*

Gruzinskaya: *see* Georgian

Guadalupe: (Spain)/former strain of Spanish Merino/not *Gouadaloupe*/extinct

Guangling Large-tail: (Shanxi, China)

Guddi: *see* Gaddi

Guerha: *see* White Karaman

Guide Black Fur: (Guide, Guian and Tengde counties, E Qinghai, China)/fur/black; hd; lop ears/orig. from Tibetan/rare

Guigou: (Morocco)/var. of Berber

Guillaumes: *see* Mourerous

Guinea: *see* West African

Guinea Long-legged: *see* Sahel type

Guirra: (coast of Valencia, S Castellón, and N Alicante, Spain)/m.(mw.d)/born dark brown, paling to red with age; pd; roman nose/syn. *Rocha, Rotxa* or *Roya* (= red), *Roja levantina* (= *Levant Red*), *Sudad* or *Sudat* (Alicante)

Guissar: *see* Hissar

Guizhou Mutton-Wool: (China)/ w.m/orig. from Corriedale and Romney, × (Xinjiang Finewool and Russian Merino, × local)

Gujarati, Gujerati, Gujrati: *see* Patanwadi

Gulf Coast Native: (SE USA)/cw-mw/ white, or tan to dark brown/cf. Criollo/local orig. from Spanish breeds *et al.* since 16th c./vars: Florida Native, Louisiana Native/ HB 1985, BS/syn. *Georgia Native, Pineywoods Sheep*/nearly extinct

Guligas: (Uzbekistan)/pink-roan/var. of Karakul

Gulijan, Gulijanska: *see* Svrljig

Güney Karaman: *see* Southern Karaman

Gunib: (Dagestan, Russia)/cw.m/ black; ♂ hd, 25% ♀ pd/ Caucasian Fat-tailed type, sim. to Lak/orig. (with Württemberg Merino) of Dagestan Mountain/ Russ. *Gunibskaya*

Guoerluosi Finewool: (China)

Gürcü: *see* Tuj

Gurez: (Kashmir North, Jammu and Kashmir, India)/cw-mw.d/usu. white, usu. pd; st

Gurktaler: (Carinthia, Austria)/ former local var. of Carinthian; ? with English Longwool blood/ extinct

Gutefår: *see* Goth

Gyzyl-Gojun: *see* Mazekh

Habashi, Habasi: *see* Habsi

Habsi: (Asir mts, Saudi Arabia)/m/ hy; ft/sim. to Hejazi but smaller/ syn. *Habashi, Habasi* (fem. *Habasiyah*), *Hagari, Hibsi*/[= Abyssinian]

Hadina: (E Niger)/cw-hy/black

Hadjazi: *see* Hejazi

Hagari: *see* Habsi

HAIR SHEEP: i.e. with fleece sim. to wild sheep/(Asia): see Habsi, Hejazi, South India hair, Yemeni hy breeds; (Africa): see African Long-legged, African Long-fat-tailed, East African Fat-tailed, Somali and Blackhead Persian, West African Dwarf; (America): see American Hair sheep

Hakasskaya: *see* Khakass

Halfbred: *see* English Halfbred, Kent Halfbred, New Zealand Halfbred, Romney Halfbred, Scottish Halfbred, Welsh Halfbred, Yorkshire Halfbred

Half-Finn: (Peru)/Finnish Landrace × Targhee, imported from USA and being crossed with Criollo and Junin

Halha: (Mongolia)/usu. black head/ var. of Mongolian/Russ. *Khalkhasskaya*/not *Halhas, Khalkha*

Halkali: (Turkey)/extinct

Hallenjoo: (Dera Ghazi Khan,
Pakistan)/grey with black
patches/larger var. of Khijloo
Hamadani, Hamadi: *see* Hamdani
Hamalé: *see* Hamari
Hamari: (SW Kordofan and SE
Darfur, Sudan)/brown/var. of
Sudan Desert/not *Amalé*,
Hamalé/[Hamar tribe]
Hamdani: (Arbil, Iraq)/cw.m.d/
lowland non-migrant var. of Iraq
Kurdi/not *Hamadani, Hamadi,
Hamdanya*
Hamitic Long-tailed: *see* Ancient
Egyptian
Hammam: (Morocco) *see* Timahdit
Hampshire Down: (England)/sw.m/
black-brown face and legs; pd/
Down type/orig. in early 19th c.
from Southdown × Wiltshire
Horn and Berkshire Knot/orig. of
Dorset Down, (× Cotswold) of
Oxford Down, (+ Oxford Down)
of German Blackheaded Mutton
and Polish Blackheaded, and (×
Northern Short-tailed) of Gorki/
recog. 1859; BS 1889, HB 1890;
BS also USA 1889, Australia; HB
also Canada, New Zealand, S
Africa, France 1957, Argentina/
syn. *Hampshire* (USA)
Hamra: (Algeria) *see* Beni-Guil
Hamra: (Syria) *see* Red Karaman
Hamyan: *see* Beni Guil
Han: (parts of Henan, Shanxi, Hebei,
Shandong and Jiangsu, China)/
mw-cw/prolific/vars: large ft (pd)
and small ft (hd)/orig. from
Mongolian/syn. *Han-yang* (=
Han sheep), *Shandong* (esp.
wool)/not *Hangyang, Hanjan,
Hanyan*
Hangay: (Tsagaan Nuur dist., Selenge
prov., Mongolia)/fw.m.d/♂ hd, ♀
pd/orig. from Mongolian graded
to Russian finewools (Altai *et
al.*)/not *Hangai, Khangai*
**Hangyang, Hanjan, Hanyan, Han-
yang:** *see* Han
Harcha: (Morocco)/main var. of Beni
Guil/not *Archa*
Hareki, Hargi, Harki: *see* Herki
Harnai: (NE Baluchistan, Pakistan)/
cw.m.d/coloured spots on head;

hd (short); ft/syn. *Damari,
Dumari*/not *Hasnai*
Harri: *see* Hejazi
Harrick: *see* Herki
Hasa, Hasah, Hasake: *see* Kazakh
Hashtnagri: (C of NWFP, Pakistan)/
m.cw/black face; pd; ft/not
Hastnagri
Hasnai: *see* Harnai
Hassan: (SC Karnataka, India)/
cw.m/white with black or light
brown markings; ♂ hd or pd, ♀
pd/syn. *Kolar*/not *Hassen*
Hastnagri: *see* Hashtnagri
Hausa: *see* Yankasa
Havasi: (Albania)/d/light brown; hd;
lop ears/nearly extinct
Hawaiian Black Buck: (Hawaii)/
black; hy; hd/? orig. from
American Hair Sheep
Hazake: *see* Kazakh
Hazaragie: (Hazarajat, C
Afghanistan)/cw.m/usu. reddish-
brown, occ. black or white with
brown belly; pd; ft/not *Hazara*
HEATH: (NW Europe)/cw/black face;
usu. hd; st/orig. of Danish
Landrace/see Danish Heath,
Dutch Heath, Heidschnucke,
Skudde, Wrzosówka
Hebridean: (Scotland)/black or dark
brown; usu. 4-horned/Northern
Short-tailed type/HB 1973, BS
1987/syn. (till 1977) *St Kilda*/
rare
Hebridean Blackface: *see* Boreray,
Island Blackface
Hedefår: *see* Danish Heath
Hedjazi: *see* Hejazi
hei: = black (Ch.)
Heideschaap: *see* Dutch Heath
Heidschnucke: (Lüneburg heath,
Hanover, Germany)/cw/grey with
black face, born black; hd; st or
rat tail/selected vars: White
Horned, White Polled/HB 1923,
BS/Ger. *Graue gehörnte
Heidschnucke* (= *Grey Horned
Heidschnucke*)/syn. *German
Heath, Lüneburg Heath*/not
*Heidschmucke, Heidschnukke,
Heidsnucke*
Hejari: (Saudi Arabia)/var. of Najdi
Hejazi: (W Saudi Arabia)/m/usu.

white; hy; ♂ pd or scurs, ♀ pd;
often earless; often tassels; sft/
syn. *Harri, Khazi, Mecca*/not
Hadjazi, Hedjazi

Hemşin: (NE Turkey)/cw.m/brown,
black or white; ♂ hd, ♀ usu. pd;
lt fat at base

Henan Fat-rumped: = ? Henan Large-
tailed

Henan Large-tailed: (S Hebei, E
Shandong and N Henan, China)/
lft/rare

Herati: *see* Kandahari

Hercegovina, Hercegovačka: *see*
Bosnian Mountain, Dalmatian-
Karst

Herdwick: (Lake District, NW
England)/cw.m/white, lamb
black face and legs and blue-roan
fleece; ♂ hd, ♀ pd/BA 1844, BS
1916, HB 1920/[? = let out in
herds; or = sheep pasture]

Hereford: *see* Ryeland

Herero: *see* Mondombes

Herigi: *see* Tuj

Herik: (N Anatolia, Turkey)/cw.m.d/
usu. white with dark spots on
head; ♂ hd, ♀ usu. pd; sft/sim.
to Dağlıç/syn. *Amasya Herik,
Heregi, Gıcık*/name also used for
Tuj

Herki: (N Iraq)/migrant var. of Iraq
Kurdi/syn. *Mosuli*/not *Hareki,
Hargi, Harki, Harrick, Herrik,
Hirik, Hirrick, Hirrik, Hurluck,
Mossul, Mousouli*/[Herki tribe
near Mosul]

Herrik: (Kurdistan) *see* Herki

Hetian: (W Xinjiang, China)/cw.m/
white with black or pied head; ♂
hd, ♀ hd or pd; sft/sim. to
Tibegolian/var.: Kargilik;
subvars: Kokyar, Saku-Bash/WG
Ho-t'ien/not *Cho-ten-jan, Hotan,
Khotan, Kotan*

Hexham Leicester: *see* Bluefaced
Leicester

Hibsi: *see* Habsi

High Fertility: (Ireland)/prolific/
population derived from ♂♂ and
♀♀ screened from commercial
flocks 1963–1965, with some
Finnish Landrace blood 1965–
1967: 22% Galway, 20% Cheviot,

15% Suffolk, 8% Border
Leicester, 5% Finnish Landrace,
30% unknown/extinct

High-Fertility Merino: *see* Booroola
Merino

Highland: *see* Scottish Blackface

Highland Half-bred: (Papua New
Guinea)/Corriedale × Priangan

Hill Radnor: (Black Mts, SE Wales)/
m.sw/tan or grey face and legs; ♂
usu. hd, ♀ pd/sim. to South
Wales Mountain/orig. from
Welsh Tanface/? orig. (with
Shropshire) of Clun Forest/BS
1926, HB 1955/syn. *Radnor,
Radnor Forest*/rare

Hinggan: (Inner Mongolia, China)/fw

Hirik, Hirrick, Hirrik: *see* Herki

Hissar: (Tajikistan)/m.cw/brown; pd;
fr/var.: Parkhar/Russ.
Gissarskaya/syn. *Uzbek*/not
Gissar, Guissar

Hissardale: (E Punjab, India)/fw/occ.
with brown or black patches; ♂
hd or pd, ♀ pd/orig. (? 1920s) at
Government livestock farm,
Hissar, from Australian Merino
($\frac{7}{8}$) × Bikaneri ($\frac{1}{8}$)/not *Hissar
Dale*/rare

Hog Island: (Virginia, USA)/white or
(10%) black; feral till 1974/orig.
from English breeds and
Merino/HB/ nearly extinct

Hokonui: (Southland, New Zealand)/
fw/white or coloured; hd or pd;
feral/? orig. from Tasmanian
Merino *c.* 1858/syn. *Hokonui
Hills, Hokonui Merino*/nearly
extinct

Holstein: *see* Wilstermarsch

Horned Cragg: *see* Limestone

Horro: (W Ethiopia)/m/tan, occ.
cream, brown, black or pied; pd;
hy; ft (triangular)

Hortobágy Racka: *see* Racka

Hotan, Ho-t'ien: *see* Hetian

Hottentot: (S Africa)/hy; ft/? orig.
from Near East Fat-tailed and
Ancient Egyptian (long-tailed)/
vars: Cape, Namaqua/orig. of
Afrikander/extinct

Houda: *see* Uda

Houtland: (Ronse, SE Flanders,
Belgium)/m.cw/brown and white
face and legs; occ. hd; roman

nose/orig. early 20th c. from Ardennes (i.e. Brabant Foxhead), Sambre-et-Meuse and Vlaamse (i.e. Flemish)/Fr. *Mouton des Collines*/syn. *Ardennais tacheté*/[= woodland]/rare

Hsin-Chiang: *see* Xinjiang Finewool

Hu: (S Jiangsu, China)/cw.pelt/usu. white; pd; sft; prolific/orig. from Mongolian/syn. *Huchow, Hu-yang* (= *Lake sheep*), *Ongti, Shanghai, Taihu, Wu, Wuxi*/not *Hujan, Huyan, Woozie, Wushing, Wusih*/rare

Hum: *see* Dalmatian-Karst

Hungarian Merino: (Hungary)/ fw.d.m/orig. from Racka (+ Bergamasca and Württemberg Merino) graded to Rambouillet, also Précoce, German Mutton Merino and Russian Merino blood/inc. Hungarian Mutton Merino (*Magyar Húsmerinó*)/ orig. of Transylvanian Merino/ Hung. *Magyar Fésüs Merinó* (= Hungarian Combing Wool Merino)

Hungarian Prolific Merino: (Hungary)/fw.m/pd; prolific/orig. 1980–1992 from Booroola Merino × Hungarian Merino/ recog. 1992/*see also* Prolific Merino, Szapora Merino

Hungarian Racka: *see* Racka

Hungarian Zackel: *see* Racka

Hurluck: *see* Herki

Huyan, Hu-yang: *see* Hu

Hvar: *see* Island Pramenka

Hvidhovedet Marsk: *see* Danish Whiteheaded Marsh

Hyfer: (NSW, Australia)/m/pd/prime lamb dam line selected for high fertility/orig. 1978–1991 from Booroola Merino ($\frac{1}{4}$), Trangie Fertility Merino ($\frac{1}{4}$) and Poll Dorset ($\frac{1}{2}$)

Ibeidi: (El Minya, Upper Egypt)/ white, usu. with brown head, occ. black; ♂ hd, ♀ usu. pd; sft/ not *Abidi, Ebeidi, Ibidi*/[= from Beni 'Ibeid]

IBERIAN: (mountain areas, Spain)/ m.mw/white with coloured

marks on face and legs; ♂ occ. hd, ♀ pd/inc. Montesina, Ojalada, Pallaresa/Sp. *Iberica*/ syn. *Serrana* (= mountain)

Ibicenca: *see* Ibiza

Ibidi: *see* Ibeidi

Ibiza: (Balearic Is, Spain)/d.m.cw/ pd/Sp. *Ibicenca*/rare

Icelandic: (Iceland)/w.d.m/usu. white with light brown head and legs, sometimes black, grey, brown or pied; usu. hd (sometimes 4 or more)/Northern Short-tailed type/orig. from Old Norwegian/var.: Kleifa/HB Canada 1986

Icsat: (Mexico)/white with black around eyes/var. of Chiapas

Ideal: *see* Polwarth

Ikzhaomen: (Inner Mongolia, China)/var. of Mongolian/Russ. *Ikzhaoménskaya*/not *Ikzhaomeng*

Île-de-France: (N France)/m.mw/ pd/orig. as Dishley Merino from Leicester Longwool × Merino, 1833–1900/BS 1922; BS also in GB 1982; HB also in Spain, S Africa

Ile Longue: (Kerguelen Archipelago)/ feral; lambing twice a year

Imeretian: (Abkhasia and Ajaria, Georgia)/♂ hd; lt-ft/Caucasian Fat-tailed type/Russ. *Imeretinskaya*/syn. *Imeritian*/ nearly extinct

Imi-n-Tanoute: (Morocco)/var. of Berber

Imperial: (Imperial Calcasieu parish, Louisiana, USA)/m.mw/face and legs white to tan; ♂ usu. hd, ♀ pd/orig. in late 18th c. from British breeds with blood of American Tunis/extinct

Improved Apulian: *see* Gentile di Puglia

Improved Ariano: *see* Quadrella

Improved Awassi: *see* Israeli Improved Awassi

Improved Black: (Italy) *see* Carapellese

Improved Blackface Mutton: (Germany) *see* German Blackheaded Mutton

Improved Blackheaded Mutton-Wool: *see* Swiss Brownheaded Mutton

Improved Border Leicester: (NSW, Australia)/m/pd/orig. from backcrossing Border Leicester × Merino to Border Leicester; 82% Border Leicester, 18% Merino/ syn. *Border Leicester Improved*, *BLI*

Improved Calabrian: *see* Gentile di Calabria

Improved Dartmoor: *see* Dartmoor

Improved Finnish Native: *see* Finnish Landrace

Improved Galway: (Ireland)/ Fingalway × Galway 1st cross

Improved German Country: *see* Merinolandschaf

Improved German Farm: *see* Merinolandschaf

Improved German Land: *see* Merinolandschaf

Improved Gorodets: *see* Vyatka

Improved Hampshire Down: *see* Dorset Down

Improved Haslingden: *see* Lonk

Improved Leicester: *see* Leicester Longwool

Improved Lucanian: *see* Gentile di Lucania

Improved Mongolian: (NE China)/ orig. from American Rambouillet × Mongolian in 1920s and 1930s/extinct (superseded by North-East China Finewool)

Improved Pirot: (Serbia)/improved local Pramenka/Serb. *Oplemenjena Pirotska*

Improved Sicilian: *see* Sicilian Barbary

Improved Šumava: *see* Šumava

Improved Valachian: (Moravia, Czech Republic)/m.cw/♂ hd, ♀ pd/orig. from Valachian improved by Texel, Cheviot and East Friesian/Cz. *Zušlechtěná valaška*/nearly extinct

Improved Valachian: (Slovakia)/ m.mw/orig. from Valachian improved by Texel, Lincoln and Border Leicester/Sl. *Zošl'achtená valaška*/rare

Improved Whiteheaded Land: *see* Merinolandschaf

Improved Whiteheaded Mountain: *see* Swiss White Mountain

Imroz: *see* Gökçeada

Indian fat-tailed: *see* Pakistan fat-tailed

Indian Merinos: *see* Bharat, Kashmir, Raymond

Indigenous Sheep of Malaysia: *see* Malin

Indonesian: *see* Indonesian Fat-tailed, Javanese Thin-tailed, Priangan, Semarang, Sumatran

Indonesian Fat-tailed: (E Java, Madura, Lombok and S Sulawesi)/m.(cw)/pd/ sft or lft/? orig. from fat-tailed breed of SW Asia (e.g. Baluchi from Iran) in 18th c./Indonesian *domba ebor gomuk*/syn. *Donggala* (Sulawesi), *East Java Fat-tailed Java(nese) Fat-tailed, Madurese* (Du. *Madoera*)/not *Dongala*, *Donggola*

Indre Berrichon: *see* Berrichon de l'Indre

Infantado: (Spain)/former strain of Spanish Merino/extinct

Ingessana: (S Sudan)/Sudan Desert × Nilotic intermediate

Ingush: *see* Karachai

Inner Mongolian Finewool: (China)/ orig. from Caucasian × Mongolian/Ch. *Neimenggu*/syn. *Inner Mongolian Merino*, *Neimonggol Merino*

INRA 401: (France)/m/pd/prolific strain orig. 1969 on at Bourges Exp. Sta. of Institut National de la Recherche Agronomique (INRA) from Romanov × Berrichon du Cher/recog. as breed 1980

Inverdale: (New Zealand)/prolific gene (FecX') in New Zealand Romney found in 1980 at Invermay Agric. Res. Centre, Mosgiel

Iomud: (Turkmenistan)/var. of Turkmen Fat-rumped/not *Iomut*/ extinct

Ipiros: *see* Epirus

Iran fat-tailed breeds: *see* Afshari,

Arabi, Bakhtiari-Luri, Baluchi,
Farahani, Grey Shirazi, Herki,
Kallakui, Karakul, Khorasan
Kurdi, Makui, Mehraban,
Moghani, Sangesari, Sanjabi,
Shal, Turki/syn. *dumba*, *Persian
fat-tailed*
Iran Thin-tailed: *see* Zel
Iraqi: *see* Awassi, Arabi, Iraq Kurdi/
not *Iraki*
Iraq Kurdi: (NE Iraq)/m.cw.d/black
head and legs; pd; ft/Kurdi type/
vars: Aljaf, Dazdawi, Hamdani,
Herki/not *Karadi*, *Karradi*, *Kordi*,
Kuradi
Irish Longwool: *see* Galway,
Roscommon
Irish Shortwool: *see* Wicklow
Cheviot
Island Blackface: (Hebrides; chiefly
Harris, Scotland)/face occ. brown
or grey/strain of Scottish
Blackface with finer wool/syn.
Greyfaced Lewis, *Lewis
Blackface*, *Stornoway Blackface*
Island Pramenka: (Dalmatia,
Croatia)/cw-mw.m.d/usu. white
with white or speckled head and
legs, occ. black or brown; ♂ hd,
♀ usu. pd/Pramenka type with
some Merino blood/inc. (with
increasing Merino blood): Brač,
Hvar and Kornat; Krk (It. *Veglia*)
and Rab; Cres (It. *Chersolina*),
Dugi Otok, Losinj and Olib; Silba
and Zlarin; culminating in Pag
Island/Cr. *Otočka Pramenka*/
syn. *Island Zackel*
Island Zackel: (Greece) *see* Lemnos,
Levkimmi, Psiloris, Sfakia, Sitia
ISM: *see* Malin
Israeli Improved Awassi: (Israel)/d/
var. of Awassi with Herki (Kurdi)
blood (1953–1957), selected for
milk/orig. (with East Friesian) of
Assaf/HB 1943
Issyk Kul argali: *see* arkhar
Istar Pramenka: *see* Istrian Milk
Istrian Milk: (E Istra and Croatian
coast, also Gorizia, Italy)/
d.m.cw/usu. white but also black
or brown; pd/Pramenka type, ?
influenced by Bergamasca or
Lamon/larger and smaller types/

Cro. *Istarska mlječna*, It.
Istriana, *Carsolina*/syn. *Istar
Pramenka* (Cro. *Istarska
Pramenka*), *Primorska* (=
coastal)/nearly extinct
Italian Merino: *see* Gentile di Puglia
İvessi: (Turkey) *see* Awassi
Ivrea: *see* Biellese
Iwessi: *see* Awassi

Jacob: (Great Britain)/mw.m/dark
brown patches; usu. 4-horned/?
orig. from Hebridean/BS 1969;
BS also USA 1988 (HB 1985); HB
Canada 1988/syn. *Jacob's sheep*,
Spanish Piebald, *Spotted*
Jaffna: (Sri Lanka)/m/often white
with tan or black patches, also
white, tan or black; hy; ♂ hd or
pd, ♀ usu. pd/orig. from South
India Hair type
Jahrani: *see* Dhamari
Jaidara: (Uzbekistan)/cw.m/white,
tan or brown/ ♂ hd or pd, ♀ pd;
fr/Russ. *Dzhaǐdara*
Jaisalmeri: (Jaisalmer, W Rajasthan,
India)/cw/brown or black face;
pd; long lop ears/not *Jaiselmeri*
J-AKI-1: (Hungary)/Swedish
Landrace × Hungarian Merino F₁
in early 1980s as prolific dam
line for use with Suffolk
♂/extinct
J-AKI-2: (Hungary)/Finnish Landrace
× J-AKI-1 F₁ in early 1980s as
prolific dam line for use with
Suffolk ♂/extinct
Jalauni: (SW Uttar Pradesh, India)/
cw.m.d/face may have coloured
markings; pd; long lop ears/syn.
Bundelkhandi or *Jubbulpuri* (N
Madhya Pradesh)/not *Jaluan*
Jaluan: *see* Jalauni
Jangli: *see* Magra
Jargalant: (Mongolia)
Jaro: (Azerbaijan)/d.cw.m/Caucasian
Fat-tailed type/Russ. *Dzharo*/
extinct
Javanese: *see* Indonesian Fat-tailed,
Javanese Thin-tailed, Priangan,
Semarang
Javanese Thin-tailed: (W and C Java,
Indonesia)/m.(cw)/usu. white,
also pied or coloured; ♂ hd, ♀

usu. pd; prolific/vars: Priangan, Semarang

Jędrzychowice Merino: (Poland)/fw/ orig. 1954–1964 at Osowa Sień State Breeding Centre from Polish Merino × (Polish Merino × Caucasian)/Pol. *Merynos wełnisty typu jędrzychowickiego*

Jezero-Piva, Jezero-pivska: *see* Piva

Jezerska, Jezersko-Solčavska: *see* Solčava

Jhalawani: (Pakistan)/var. of Rakhshani

Jhelum Valley: *see* Kashmir Valley

Jia Shike: (Qinghai, China)/w/hd or pd/orig. since 1970 from Polwarth × Tibetan

Jinzhong: (China)

Jodipi: (SE Andhra Pradesh, India)/ white with black spots on face/ var. of Nellore

Jomoor: *see* Samhoor

Joria: *see* Patanwadi

Jubbulpuri: *see* Jalauni

Jumli: (Jumla, NW Nepal)/m.cw/sim. to Baruwal but smaller, with finer wool than Kage/syn. (?) *Kiu* (or *Kew*)

Junin: (C Peru)/m.mw/white or black face; pd/orig. since 1940s at SAIS Tupac Amaru from Columbia, Corriedale, Panama, Romney and Warhill

Junken Merino: (China)

Jura: (Switzerland)/part orig. of Swiss Black-Brown/former vars: Berne, Solothurn/extinct

Kababish: *see* Kabashi

Kabarda: *see* Karachai

Kabarliavi: (SE Bulgaria)/dense-wool var. of Karnobat

Kabashi: (N Kordofan and N Darfur, Sudan)/many colours/var. of Sudan Desert/[Kababish tribe]

Kabyle: *see* Berber

Kachhi: (Great Rann of Cutch, Tharparkar, Sind, Pakistan)/ d.cw.m/black or tan face; pd; ears often vestigial; st/syn. *Cutchi, Kutchi*

Kae: (Laos)/m/composite of local and Mongolian/rare

Kage: (Pahar of Nepal)/m.(cw-hy)/

dirty white, often with light brown patches, occ. black; ♂ hd, ♀ hd or pd; often earless; st; small to dwarf/syn. *Gharpala* (= household-raised)/not *Cagi, Cago, Kagi*

Kaghani: (N of NWFP, Pakistan)/ m.cw/tan, grey or black; ♂ usu. hd, ♀ usu. pd; lop ears; st

Kail: (Neelam valley, Azad Kashmir, Pakistan)/m.cw/white, or white with black or brown head, or black or brown around eyes/ ♂ hd, ♀ usu. pd; lop ears

Kajli: (Sargodha, Gujurat and Mianwali, Pakistan)/m.cw/black nose, eyes and ear tips; ♂ hd or pd, ♀ pd; long lop ears; roman nose

Kalah: (Pakistan)/cw

Kalakou, Kala-Kouh, Kalaku, Kalakuh: *see* Kallakui

Kali: (Kotli, Azad Kashmir, Pakistan)/ m.cw/black; ♂ usu. hd, ♀ usu. pd

Kalinin: (Russia)/var. of Russian Longwool/orig. (1935 on) from Lincoln × Northern Short-tailed/Russ. *Kalininskaya*

Kallakui: (Varamin to Qom, NC Iran)/ cw.m.d/black marks on face and legs; ♂ usu. hd, ♀ pd/sim. to Baluchi/not *Calikui, Calokui, Kalakou, Kala-Kouh, Kalaku, Kalakuh, Kellakui*

Kalmyk: (Astrakhan, Russia)/cw.m/ fr/sim. to Chuntuk/Russ. *Kalmytskaya*/not *Kalmuck*/ extinct

Kamakuyruk: (NW Anatolia, Turkey)/ lt fat at base/Kivircik × Dağliç F$_1$, and breed derived from it/ syn. *Pirlak*/not *Kamakyuruk, Kamakuruk*/[= knife-blade tail]

Kambar: (Uzbekistan)/golden-brown/ var. of Karakul/not *Cambar, Kombar*

Kameng: *see* Tibetan

Kamieniec: (Olsztyn prov., Poland)/ m.lw/pd/Polish Longwool group/orig. 1954–1964 on Kamieniec farm of Susz State Sheep Imp. Centre, from Romney × (Texel or Leine, × Pomeranian)/ HB/Pol. *Kamieniecka*

Kanaltaler: (Austria)/d/black ear tips and spectacles/former var. of Carinthian/syn. *Canaltaler, Uggowitz*/extinct

Kandahari: (S and CW Afghanistan)/ cw/white with black spots on face and legs, or black or brown or mixed; ♂ hd, ♀ pd; ears long or very short; sft/sim. to Ghiljai but smaller/syn. *Farahi, Herati*

Kangal: (Sivas and Malatya, Turkey)/ local var. of White Karaman

Kangani: *see* Kenguri

Kangra Valley: *see* Biangi, Gaddi

Kapralin: *see* Kargalin

Kapstad: (Java, Indonesia)/m/rare

Karabakh: (Azerbaijan)/usu. reddish or greyish (dirty white or light brown), occ. black or red; ♂ hd or pd, ♀ pd/Caucasian Fat-tailed type/var.: Karadolakh/Russ. *Karabakh, Karabakhskaya*/not *Karabach, Karabagh*

Karacabey-Kivircik: (NW Anatolia, Turkey)/German Mutton Merino × Kivircik crosses at Karacabey stud farm/orig. of Karacabey Merino/syn. *Kirma* (= halfbred)/ extinct

Karacabey Merino: (NW Anatolia, Turkey)/mw.m.d/var. of Turkish Merino/orig. from Kivircik graded up (since 1928) with German Mutton Merino (95%) (via Karacabey-Kivircik)

Karachai: (N Caucasus, Russia)/usu. black, also grey, tan or white; 2–4 horns; lft/Caucasian Fat-tailed type/Russ. *Karachaevskaya*/syn. *Balkar, Chechen, Ingush, Kabarda, Ossetian*/not *Karachaev, Karatchayev*

Karachai Mountain Mutton-Wool: (N Caucasus, Russia)/orig. from North Caucasus Mutton-Wool × (Karachai × Cherkassy)

Karadi: *see* Kurdi

Karadolakh: (Azerbaijan)/var. of Karabakh/orig. 1964 on/Russ. *Karadolakhskaya*

Karagouniko: (Thessaly, Greece)/ d.m.cw/usu. black or white, also brown, pied or spotted; ♂ hd, ♀ pd/lowland breed of Greek Zackel/not *Karagunica*/ [Karagounides are farmers in plain of Thessaly]

karagöz: *see* Karayaka

Karakachan: (Balkans, esp. Macedonia)/d.m.cw/usu. black or brown, occ. white (*Bela vlaška*); ♂ hd, ♀ usu. pd/Zackel type/Bulg. *Karakachanska*/syn. *Karavlaška* or *Crna vlaška* (= black Vlach), *Kucovlaška* (Macedonia); *Macedonian Nomad, Romanian Nomad* (Bulgaria); *Karakatsan* or *Sarakatsan* (Greece)/not *Carakachanska, Karakachenski, Karakatschenski*/[Karakachan (= black shepherd) and Kutsovlach (= lame Vlach) are nomadic tribes speaking Greek and Romanian, respectively]

Karakaş: (Diyarbakir, Turkey)/local var. of White Karaman

Karakul: (Uzbekistan)/fur.d/various colour vars, also occ. white or pied; ♂ hd, ♀ pd; ft/vars: Arabi, Duzbai, Guligas, Kambar, Shirazi, Sur/orig. of American Karakul, Chinese Karakul, German Karakul, Large Karakul, Malich, Multifoetal Karakul, Pak Karakul, Sumber/HB; BS Namibia 1919, S Africa 1937, Argentina; HB also in Spain, Canada/Russ. *Karakul', Karakul'skaya*/syn. *AstKarakul', Astrakhan, Bukhara, Persian Lamb*/not *Bokara, Bokhara, Boukhara*

Karakul-Landschaf: *see* German Karakul

Karaman, Karamane: *see* Red Karaman, White Karaman

Karamaniko: (Crete) *see* Dağliç

Karamaniko: (Greece) *see* Argos (or other ft breeds)

Karamaniko Katsika: *see* Katsika

Karandhai: *see* Vembur

Karanogai: (Manych steppe, N Caucasus, Russia)/cw.m/fr/sim. to Kalmyk/Russ. *Karanogaĭskaya*/ syn. *Manych* (Russ. *Manychskaya*), *Nogai*/extinct

Karavlaška: *see* Karakachan

Karayaka: (N Anatolia, Turkey)/
cw.m.d/usu. white with black
eyes (*karagöz*) or black head and
legs (*çakrak*), occ. black or
brown; ♂ usu. hd, ♀ usu. pd; lt

Karbardinskaya: *see* Karachai

Karelin's argali: *see* arkhar

Kargalin Fat-rumped: (Aktyubinsk,
Kazakhstan)/m.cw/fr/breed
group/orig. since 1931 from
Degeres and Sary-Ja × (Edilbaev
× Fat-rumped)/Russ.
*Kargalinskie kurdyuchnye ovtsi,
Kargalinskaya porodnaya gruppa*

Kargilik: (W Xinjiang, China)/var. of
Hetian/subvars: Kokyar, Saku-
Bash/Russ. *Kargalyk*/[former
name of Yecheng]

Karha: *see* White Karaman

Karman: *see* Red Karaman, White
Karaman

Karnabat: *see* Karnobat

Karnah: (Muzaffarabad, Azad
Kashmir, Pakistan, and
neighbouring Indian Kashmir)/
sw/♂ hd, ♀ pd; st

Karnobat: (SE Bulgaria)/m.d.w/
yellow-grey to copper-red to
black; ♂ hd, ♀ pd/sim. to
Kivircik/? orig. from black
Tsigai/vars: Kabarliavi,
Rudavi/Bulg. *Karnobatska*, Rom.
Carnabat/syn. *Cherna Chervena*
(= black-red)/not *Karnabat*/rare

Karnobat Finewool: (SE Bulgaria)/
fw.m.d/orig. 1950–1967 from
Stavropol × Karnobat/Bulg.
Karnobatska t"nkorunna/syn.
South-East Bulgaria Finewool
(Bulg. *T"nkorunna Yugo-
istochna B"lgarska*)

Karnobatoshumenskaya: *see* Copper-
Red

Karnówka: (Poland)/white/var. of
Świniarka/[orig. from Karnów in
Lower Silesia]/? extinct

Kärntner: *see* Carinthian

Karradi: *see* Kurdi

Karrantzar: *see* Vasca Carranzana

Kars: *see* Tuj

Karst: *see* Dalmatian-Karst

Kärtner Brillenschaf: *see* Carinthian

Karuvai: *see* Kilakarsal

Kashmir Merino: (Kashmir, India)/
w.m/orig. from Delaine Merino
(1951–1952) × (Tasmanian
Merino × local Gaddi,
Bhakarwal and Poonchi, 1947
on), then Rambouillet and Soviet
Merino ♂♂ used; 50–75%
Merino blood

Kashmir Valley: (SW Kashmir,
India)/cw/usu. coloured; usu.
pd/mixed population/syn. *Valley*

Kashubian: (E coastal Poland)/orig.
var. of Pomeranian/Pol. *Kaszuby*/
syn. *Casubian, Kasubian*

Katafigion: (SE Macedonia, Greece)/
d.m.mw-cw/♂ hd, ♀ pd/Ruda
type/not *Katafiyion, Katafygion,
Kataphygion*/nearly extinct

Katahdin: (Maine, USA)/m/usu.
white, also tan or multicoloured;
hy; usu. pd; prolific/orig. (1957
on) by M. Piel, Abbot Village,
from Suffolk and Wiltshire Horn
× Virgin Islands White/BS 1986/
rare

Kataphygion: *see* Katafigion

Kathiawari: *see* Patanwadi

Katseno: *see* Krapsa

Katsika: (Ioánnina basin, Epirus,
Greece)/white with black around
eyes and on ears/lowland breed
of Greek Zackel/syn.
Karamaniko Katsika

Kavkazskaya tonkorunnaya: *see*
Caucasian

Kavkazskiĭ Rambulye: *see* Caucasian

Kazakh Arkhar-Merino: (SE
Kazakhstan)/m.fw/♂ hd, ♀ usu.
pd/orig. 1934–1949 from arkhar
× Merino followed by 2 Merino
top crosses/HB/Russ. *Kazakhskiĭ
arkharomerinos*

Kazakh Corriedale: (S Kazakhstan)/
orig. from Border Leicester,
Romney or Lincoln, × Kazakh
Finewool/Russ. *Kazakhskaya
Korridel*

Kazakh Fat-rumped: (Kazakhstan, E
Xinjiang, China, and Báyanölgiy
prov., Mongolia)/cw.m/reddish-
brown; ♂ hd, ♀ pd/vars: Altay,
Fuhai, Temir/orig. (with Kalmyk)
of Edilbaev, (with Shropshire) of
Degeres, (with Degeres and Sary-

Ja) of Kargalin/Russ.
Kazakhskaya kurdyuchnaya, Ch.
Hasahe or *Hazake*/syn. *West
Mongolian Fat-rumped* (Mong.
*Baruun mongolin uutsan
suult*)/not *Kazak, Hasa, Hasah*

Kazakh Finewool: (Alma-Ata,
Kazakhstan)/m.fw/pd/orig.
1931–1945 from Kazakh Fat-
rumped crossed first with
Précoce and then with American
Rambouillet/recog. 1946;
HB/Russ. *Kazakhskaya
tonkorunnaya*

Kazakh Semifinewool: (S
Kazakhstan)/m.mw/breed
group/orig. since 1945 from
Lincoln and North Caucasus
Mutton-Wool × (Précoce ×
Kazakh Fat-rumped) improved
by Hampshire *et al.*/Russ.
*Kazakhskaya polutonkorunnaya
porodnaya gruppa*

Kazanluk Semifinewool: (Bulgaria)/
m.mw/orig. from Romney and
(German Mutton and Merino and
Caucasian × Stara Zagora) ♂♂
(1964) × (German Mutton
Meriono × Panagyurishte)
♀♀/Bulg. *Mestna polut"nkorunna
Kazanl"shki raĭon*

Kazil: *see* Red Karaman

Kedabek Merino: (Kedabek,
Azerbaijan)/local Merino/orig.
late 19th c. from Mazaev,
Novocaucasian and other
Merinos/orig. (with Bozakh) of
Azerbaijan Mountain Merino/
extinct

Keelakaraisal: *see* Kilakarsal

Keerie: (Caithness, Scotland)/black/
Northern Short-tailed type/syn.
Rocky/not *Kerry*/[Gaelic *caora* =
sheep, cf. *keero* (Orkney) = feral
sheep]/extinct

Keerqin: (Inner Mongolia, China)/
strain of Chinese Merino/not
Kerqin, Keerquin

Keezha Karauvai, Keezhak(k)araisal:
see Kilakarsal

Kelakarisal: *see* Kilakarsal

Kelantan: *see* Malin

Keletfríz: (Hungary)/= East Friesian

Kellakui: *see* Kallakui

Kempen Heath: (S Limburg,
Netherlands)/light brown spotted
face; pd/var. of Dutch Heath/orig.
from local and Spanish Merino/
BS 1967/Du. *Kempische* (or
Kempense) *Heideschaap*, Fr.
Campinoise, Flem. *Kempens
Schaap*/rare; nearly extinct in
Belgium

Kendal Rough: *see* Rough Fell

Kenguri: (Raichur dist., EC
Karnataka, India)/m/usu. red,
also white, black or pied; ♂ hd,
♀ pd; st/South India Hair type/
syn. *Kangania, Keng, Kenga,
Teng-Seemai, Tenguri, Tonguri,
Yalag*

Kent Halfbred: (Kent, England)/sw/
pd/Southdown × Romney, 1st
cross/extinct

Kent Halfbreed: (S England)/Lleyn ×
Romney, 1st cross/BS 1988/syn.
KHB

Kent: *see* Romney

Kent or Romney Marsh: *see* Romney

Kentmere: *see* Rough Fell

Kermani: (SE Iran) *see* Baluchi

Kerqin: *see* Keerqin

Kerry: (SW Ireland) *see* Scottish
Blackface

Kerry Hill: (Montgomery Powys,
Wales)/sw.m/black spots on face
and legs; pd/Clun Forest blood
1840–1855/recog. 1809; BS 1893,
HB 1899/rare

Kesbır: *see* Çandır

Kesik: *see* Tuj

Kesme: *see* Çandır

Keustendil: *see* Kyustendil

Kew: ? *see* Jumli

Khakass: (Siberia, Russia)/smaller
var. of Krasnoyarsk at Moscow
and Askizskiĭ state farms and
Put k Kommunizma collective
farm/Russ. *Khakasskaya*/not
Hakasskaya

Khalkha, Khalkhas: *see* Halha

Khalkidhiki: *see* Chalkidiki

Khamseh: (E Farsistan, Iran)/cw/
often dark spots on face and legs;
♀ sometimes pd; ft/local var.

Khangai: *see* Hangay

Khazi: *see* Hejazi

KHB: *see* Kent Halfbreed

Kheri: (Rajasthan, India)

Khetrani: (Baluchistan, Pakistan)/var. of Bibrik

Khijloo: (Dera Ghazi Khan, Pakistan)/ cw.m/black marks on face and feet, or tan face and feet; sft/var.: Hallenjoo

Khios: *see* Chios

Khorasan Kurdi: (N Khorasan, Iran)/ m.d.cw/brown; pd; ft

Khorasani, Khorassan: (Iran) *see* Baluchi

Khotan: *see* Hetian

Khurasani: (Iran) *see* Baluchi

Kilakarsal: (Ramnad, Madurai and Thanjavur, Tamil Nadu, India)/ m/tan, usu. with black belly; ♂ hd, ♀ pd; usu. tassels; st/South India Hair type/syn. *Adikarasial, Karuvai, Keezha Karauvai, Keezhak(k)araisal, Ramnad Karuvi, Ramnad Red*/not *Keelakaraisal, Kelakarisal*

Kimi: *see* Kymi

Kipsigis: (W Kenya)/local var. of Masai/syn. *Lumbwa*/not *Kipsikis*

Kirdi: (N Cameroon and SW Chad)/ black/var. of West African Dwarf/ plains and mt vars/syn. *Djallonké, Kirdimi, Lakka, Massa, Poulfouli*

Kirgiz Fat-rumped: (Kyrgyzstan)/ cw.m/brown or black; hd (E) or pd (W)/orig. (with Précoce) of Alai and Fat-rumped Merino/Russ. *Kirgizskaya kurdyuchnaya*/not *Kirghiz*

Kirgiz Finewool: (Kyrgyzstan)/ fw.m/usu. white/orig. 1932–1956 at Juan Tyube state farm (*et al.*) by interbreeding of Württemberg Merino or Précoce, × [Rambouillet × (Novocaucasian or Siberian Merino × Kirgiz Fat-rumped)]/recog. 1956; HB/Russ. *Kirgizskaya tonkorunnaya*

Kirma: *see* Karacabey-Kivircik

Kirmani: (Turkey) *see* Red Karaman, White Karaman

Kisil-Karaman: *see* Red Karaman

Kiu: ? *see* Jumli

Kivircik: (NW Turkey) m.d.cw-mw/ white, with white or spotted face, also black or brown var.; ♂ hd, ♀ usu. pd/sim. to Karnobat and Tsigai; = Thrace (Greece)/ orig. of Karacabey Merino and (with Dağliç) of Kamakuyruk and Pirlak/Turk. *Kıvırcık*/not *Kivirçik, Kivirdjik*/[= curly]

Kizil, Kızıl-Karaman: *see* Red Karaman

Kleifa: (Iceland)/pd var. of Icelandic, ? with Cheviot and/or Border Leicester blood

Klein-Brabantse Voskop: *see* Brabant Foxhead

Klementina: *see* White Klementina

Klitfår: *see* Danish Landrace

Koburg: *see* Coburger

Kohai Ghizer: (Gilgit to Chatorkhand, N Kashmir, Pakistan)/cw.m/ white or tan with brown or black head and legs; pd; sft

Kokyar: (W Xinjiang, China)/sim. to Darvaz/subvar. of Kargilik var. of Hetian

Kolar: *see* Hassan

Kombar: *see* Kambar

Kommunal: (Albania) *see* Common Albanian

Konya Merino: *see* Central Anatolian Merino

Kooka: (Sind, Pakistan)/cw.m/head usu. black; ♂ hd, ♀ pd; long lop ears; st/not *Kuka*

Kopralin: *see* Kargalin

Koprivshten: (C Bulgaria)/Bulg. *Koprivshtenska*/rare

Kordi: *see* Kurdi

Koridel poznański: *see* Poznań Corriedale

Kornat: *see* Island Pramenka

Korridel': *see* Soviet Corriedale, Soviet Mutton-Wool

Kosovo: (S Serbia)/m.cw/black face and legs; ♂ usu. pd, ♀ pd/ Pramenka type/Serbo-cro. *Kosovska*

Kosse: *see* Skudde

Koszalin: (C coastal Poland)/var. of Pomeranian from local longwools improved by Kashubian, Texel and Leine

Kotan: *see* Hetian

Kotel: (EC Bulgaria)/m.cw.d/♂ hd, ♀ pd/Bulg. *Kotlenska*/syn.

aborigenna kotlenska (= native
Kotel)/rare
kouchari: *see* kushari
Koundoum: *see* Goundoun
Krainer Steinschaf: *see* Bovec
Krapsa: (Epirus, Greece)/var. of
Greek Zackel/syn. *Katseno*
Krasnoyarsk Finewool: (SC Siberia,
Russia)/fw/♂ hd or pd, ♀ pd/
orig. 1926 on by improving local
Mazaev and Novocaucasian
Merinos with Précoce, American
Rambouillet and Askanian/ vars:
Angara, Khakass, Uchum/recog.
1963; HB/Russ. *Krasnoyarskaya
tonkorunnaya*
krasnyĭ samukh: *see* red sheep
Krivovir: (E Serbia)/m.cw.d/white
with brownish-yellow head and
legs, occ. brown; ♂ hd, ♀ pd/
local Pramenka var./Serbo-cro.
Krivovirska/[from Krivi Vir]/
declining with spread of Svrljig
Krk: *see* Island Pramenka
Krš: *see* Dalmatian-Karst
Krukówka: (Poland)/black/var. of
Świniarka/not *Krukowska*/[Pol.
kruk = raven]/extinct
Krupnoplodnyĭ Karakul': *see* Large
Karakul
Krupnaya zadonskaya: *see* Transdon
Kucha, Kuchar, Kucharskaya: *see*
Kuche
Kuche: (W Xinjiang, China)/
pelt.m.cw/black, also pied, occ.
white or brown; ♂ hd or pd, ♀
usu. pd; st to sft/? orig. from
Karakul × Mongolian and
Kazakh Fat-rumped in late 19th
c./graded to Karakul since 1960s
to form Chinese Karakul/WG
K'u-ch'e, Russ. *Kucharskaya* or
Kuchėrskaya, Pol. *Kucze*/not
Kucha, Kuchar/extinct
Kuchugury: (Nizhnedevitsk,
Voronezh, Russia)/m.cw/usu.
black with white patch on head,
also white; usu. pd; lft/orig. in
late 19th c. from Voloshian ×
Russian Long-tailed/Russ.
Kuchugurovskaya/syn.
Voronezhskaya voloshskaya
Kuçi: (Kurveleshi-Gjirokaster, S

Albania)/d.m.cw/white with
black spots/not *Kugi*
Kucovlaška: *see* Karakachan
Kucze: *see* Kuche
Kuibyshev: (Russia)/m.lw/pd/orig.
1938–1948 from Romney ×
Cherkassy backcrossed to
Romney, or from Romney ×
Vagas (= Voloshian)/HB
kuirjuk, kuiryuk: *see* Fat-rumped
Kuka: *see* Kooka
Kulunda: (Altai, Siberia, Russia)/
cw.pelt/black; hd or pd; usu.
sft/? orig. from Siberian,
Mongolian and Kazakh Fat-
rumped/Russ. *Kulundinskaya*
Kumukh: *see* Lak
Kumyk: (N Dagestan, Russia/m.cw/
brown; ♂ hd/Caucasian Fat-
tailed type/Russ. *Kumykskaya*/
extinct
Kupres: (Bosnia)/usu. speckled face
and legs; usu. pd/var. of Bosnian
Mountain/Serbo-cro. *Kupreška*
Kuradi: *see* Kurdi
Kurassi: (Qena and Aswan, Upper
Egypt)/cw.m/usu. black or
brown, also fawn, pied or white;
pd; lt/cf. Dongola
Kurdi: (Kurdistan)/cw/black head;
pd; ft/inc. Khorasan Kurdi, Iraq
Kurdi, Sangesari, Sanjabi/not
Karadi, Karradi, Kordi, Kuradi
kurdyuk, kurdyuchnaya: *see* Fat-
rumped
Kurdyukos: *see* Fat-rumped Merino
Kuruba, Kurumba, Kurumbai Adu:
see Coimbatore
kushari: (Syria) = nomadic sheep/not
kouchari, kuschari
Kusman: (Dagestan, Russia)/var. of
Tabasaran/extinct
Kustendil: *see* Kyustendil
Kutchi: (India) *see* Patanwadi
Kutchi: (Pakistan) *see* Kachhi
**Kutsovlach, Kutsovlashka,
Kutzovlakh:** *see* Karakachan
Kyasma: (Amassi, Armenia)/
Caucasian Fat-tailed type/orig.
from Mazekh × Erik/Russ.
Kyas'ma, cf. Turk. *Kesme* (=
crossbred)
Kymi: (Euboea, Greece)/d.m.mw/
brown or black spots on face and

legs; ♂ hd, ♀ pd/orig. from
Skopelos/not *Kimi, Kyme*/rare
Kyustendil: (SW Bulgaria)/local
unimproved var. of Bulgarian
Native/Bulg. *Kyustendilska*/not
Keustendil, Kustendil/extinct
Kzyl-Karaman: *see* Red Karaman

Lacaune: (Tarn, S France)/d.m.(mw)/
pd/chief Roquefort breed/vars:
Lacaune Lait, Lacaune Viande/
orig. of Ardes, Blanc du Massif
Central/has absorbed Corbières,
Larzac, Lauraguais, Ruthenois,
Ségala/HB 1945, BS/syn.
Camarès (Aveyron), *Mazamet*
Lacho: (Vascongadas and Navarre,
Spain)/d.cw.m/coloured face and
feet; ♂ hd, ♀ hd or pd/sim. to
Churro but long wool; = Manech
(France)/vars: Cara Negra (dark
face), Cara Rubia (blond face)/
HB/Basque *Latsca* or *Latxa* (=
coarse)/syn. *Basque, Biscay*
Ladakh urial: *see* shapo
Ladoum: *see* Touabire
Laguna: (Philippines)/sw/pd/orig. from
Shropshire cross × local hd cw
Lak: (S Dagestan, Russia)/Caucasian
Fat-tailed type/sim. to Gunib but
smaller/Russ. *Lakskaya*/syn.
Kumukh
Lakens Kuddeschaap: (Brabant,
Belgium)/m.cw/beige head and
legs; pd/orig. (from 1890) by
selection of Domain Laken flock/
[*kudde* = flock]/nearly extinct
Lake sheep: (China) *see* Hu
Lakka: *see* Kirdi
Lama: *see* Damani
La Mancha: *see* Manchega
Lamkanni: (Bahawalpur, Pakistan)/
grey ears; no tassels/smaller var.
of Lohi/syn. *Lamochar*
Lamochar: *see* Lamkanni
Lamon: (Belluno, Venetia, Italy)/
m.(cw.d)/dark spots on face and
legs; pd/Lop-eared Alpine group/
BSd 1942/syn. *Feltrina*/rare
Lampuchhre: (S Tarai, Nepal)/m.cw/
occ. dark spots on body,
sometimes coloured head; ♂ hd,
♀ pd/syn. *Lohia, Tarai*/not
Lampuchera

Lanark Blackface: (Scotland)/strain
of Scottish Blackface with
coarser wool
Landais: (Landes, SW France)/
m.cw/coloured spots on head
and legs; ♂ usu. hd, ♀ pd/orig.
from Pyrenean/syn. *Landes de
Gascogne*/rare (by crossing with
Berrichon *et al.*)
Landcorp Lamb Supreme: (New
Zealand)/terminal sire flock orig.
1989–1990 by Landcorp
Farming, Hamilton/syn. *LLS*
Landes de Bretagne: (Brière, Brittany,
France)/usu. white with brown
spots on head and feet; ♂ usu.
pd, ♀ pd/sim. to Ushant but
larger/syn. *Briéron*/not *Biérois*/
rare
Landes de Gascogne: *see* Landais
Landim: (Mozambique, mainly S of
Limpopo)/pd; cone-shaped lft/
var. of Nguni
Land Karakul: *see* German Karakul
Land Merino: (Germany) *see*
Merinolandschaf
Langhe: (E Cuneo, Piedmont, Italy)/
d.m.cw/pd; semi-lop ears/sim. to
Frabosana/HB and BSd 1959/It.
delle Langhe, della Langa
Lannemezan: *see* Aure-Campan
Lantras: *see* Swedish Landrace
Lanzhou Large-tail: (Gansu, China)/
m.w/pd; ft/orig. 1962–1875 from
Tong × Mongolian/rare
Lara Polisi: *see* Pied Polisi
Large Karakul: (Ukraine)/m.fur/orig.
(1932 on) at Askania Nova from
Karakul × Hissar/Russ.
Krupnoplodnyǐ Karakul' (= Large
Progeny Karakul)
Large Polish: *see* Wielkopolska
Large-tailed Han: *see* Han
Larzac: (SE Aveyron, S France)/d.m/
var. of Caussenard/original
Roquefort breed/extinct
(absorbed by Lacaune)
La Spezia local: (La Spezia/Massa
Carrara, Italy)/m.d.cw/grey-white;
pd/Apennine group/It. *Locale*
Lasta: *see* Tucur
Lati: (Salt Range, NW Punjab,
Pakistan)/m.cw/tan or spotted
head; ft/syn. *Salt Range*/not *Latti*

laticauda: *see* Campanian Barbary
Latsca: *see* Lacho
Latti: *see* Lati
Latuka-Bari: *see* Mongalla
Latvian Darkheaded: (Latvia)/m.sw/
pd/orig. 1920–1940 from
Shropshire, Oxford Down and
German Blackheaded Mutton, ×
local Northern Short-tailed, bred
inter se since 1937/HB/Lat.
Latvijas tumšgalvas aitu, Russ.
Latviĭskaya temnogolovaya/syn.
*Latvian Blackfaced, Latvian
Blackheaded*
Latxa: *see* Lacho
Laughton: *see* Manx Loaghtan
Lauraguais: (Haute Garonne,
France)/m.d.sw/pd/syn.
Toulousain, Toulouse/not
Lauragais/extinct (absorbed by
Lacaune in 1940s)
Layda: *see* Marwari
Lebrija, Lebrijana: *see* Andalusian
Churro
Leccese: (Lecce, S Apulia, Italy)/
d.cw.m/white with black face
and legs, occ. black; ♂ usu. hd,
♀ pd/Moscia type/var.:
Fasanese/HB 1937/syn. *Moscia
leccese*/not *Lecca*
Lefkimi: *see* Levkimmi
Legagora: *see* Menz
Leicester Longwool: (England)/lw.m/
pd/English Longwool type/orig.
1755–1790 from Old Leicester/
var.: Black Leicester Longwool/
contributed to formation of most
English Longwool breeds/BS and
HB 1893; BS also Australia; HB
also USA 1888–*c*. 1930 (English
Leicester), Canada, New Zealand/
syn. *Bakewell Leicester, Dishley
Leicester, English Leicester,
Improved Leicester, Leicester,
New Leicester*/rare in England
Leine: (S Hanover, Germany)/w.m/
pd/improved by Leicester
Longwool, Cotswold and
Berrichon in 19th c. and by Texel
et al. in 20th/BSd and HB 1906/
obs. syn. (to 1886) *rheinisch,
flämisch*/old type rare
Lemnos: (Greece)/d/island var. of
Greek Zackel/Gr. *Limnos*

Lentinese, Lentini: *see* Comisana
Leonese Merino: (mt de Luna, León,
Spain)/var. of Spanish Merino
with less fine wool/syn. *Merino
entrefino, Merino trasterminante*
Lesbos: *see* Mytilene
Lessarkani: *see* Thalli
Lesvos: *see* Mytilene
Leszno: (Poznań, Poland)/Polish
Longwool group/orig. 1920–1939
from Berrichon and Polish
Strongwooled Merino *et al.* ×
local (white Świniarka)/Pol.
Owca leszczyńska/extinct
(absorbed by Wielkoplska)
Letelle Merino: (Orange Free State, S
Africa)/m.fw/orig. 1922–1938 by
J.P. van der Walt of Zastron from
Rambouillet/named 1945; BS
1951/not *Lettelle*/[name of
former Kaffir chieftain]
Levant: (France) *see* Préalpes du Sud
Levant Red: (Spain) *see* Guirra
Levézou: *see* Ségala
Levkimmi: (S Corfu, Greece)/m.d.cw/
black eyes, ears and nose; ♂ hd,
♀ usu. pd/island var. of Greek
Zackel/not *Lefkimi*/rare
Lewis Blackface: *see* Island Blackface
Lezgian: (SW Dagestan and NW
Azerbaijan)/usu. white or greyish
with coloured or spotted head
and feet, also black, red, grey or
pied; usu. hd; lft/Caucasian Fat-
tailed type/var.: Agul/Russ.
Lezginskaya
Libyan Barbary: (Libya)/m.cw.d/
white with brown or black (occ.
pied) face and legs, sometimes
pied or coloured; ♂ hd (occ. 4),
♀ usu. pd; ft/cf. Barki (Egypt)/
orig. of Barbary Halfbred,
Ghimi/syn. *Libyan Fat-tailed*
Lietuvos juvdgalvių: *see* Lithuanian
Blackheaded
Lika: (Croatia)/d.m.cw/white with
coloured or pied head and legs,
sometimes brown or black/
Pramenka type/HB/Cro. *Lička*/
syn. *Gacka* (Cro. *Gatačka*)
Limestone: (Lancashire-Cumbria,
England)/sim. to Whiteface
Woodland/syn. *Cragg, Farleton
Knott, Horned Cragg, Limestone*

Cragg, Silverdale, Warton Crag/extinct about 1900

Limousin: (Corrèze, C France)/ m.mw.(d)/pd/Central Plateau group/var.: Marchois/BS 1906, reorg. 1938, HB 1944

Linchuan: (Shanxi, China)/w.m/orig. from Corriedale and Romney, × local finewool

Lincoln Longwool: (E England)/lw.m/ pd/English Longwool type/orig. from Leicester Longwool × Old Lincoln/orig. of Kalinin, Liski, Tyan Shan/ BS and HB 1892; BS also USA 1891, Argentina; HB also New Zealand 1912, Canada/ syn. *Lincoln*/rare in GB

Linton: *see* Scottish Blackface

Lipe: (Morava valley, N Serbia)/d.m.cw/ black head and legs; ♂ hd, ♀ usu. pd/local Pramenka var./Serbo-cro. *Lipska*/declining by spread of Merino and crossing with Tsigai

Lipska: *see* Lipe

Liski: (Voronezh, Russia)/some dark spots on face and legs/var. of Russian Longwool/orig. (1936 on) from Lincoln × Mikhnov backcrossed to Lincoln ♂/ subvar.: Nizhnedevitsk/recog. 1978/Russ. *Liskinskaya*/syn. *Liskinskaya myasosherstnaya lyustrovaya poroda* (= mutton-wool lustre breed)/not *Luskin*

Lithuanian Blackheaded: (N Lithuania)/m.sw/orig. 1923–1934 from German Blackheaded Mutton and Shropshire, × local Northern Short-tailed/HB 1934/Lith. *Lietuvos juvdgalviu̧*, Russ. *Litovskaya chernogolovaya*

Lithuanian Coarsewooled: (SE Lithuania)/cw.m/grey, white, black or light brown; pd or hd/ Lith. *Vietines siurkšciavilnes*/ nearly extinct

Lithuanian Native: (Lithuania)/ Lith.*Vietines kiaules*

Litovskaya chernogolovaya: *see* Lithuanian Blackheaded

littoral nord de la Manche: *see* Cotentin

littoral sud de la Manche: *see* Avranchin

Livo: (Como, Italy)/m.cw-mw/straw-coloured; pd/Lop-eared Alpine group but straight profile/extinct

Llanwenog: (Cardigan, Wales)/m.sw/ black head and legs; pd/sim. to Clun Forest but smaller/orig. end of 19th c. from Shropshire and local blackfaced sheep/BS 1963/rare

Lleyn: (NW Wales)/m.mw/black nose; pd/? orig. in late 18th c. from Welsh Mountain with Roscommon and Leicester Longwool blood/BS 1970/not *Llŷn*

LLS: *see* Landcorp Lamb Supreme

Loaghtan, Loaghton, Loaghtyn: *see* Manx Loaghtan

Locale: *see* La Spezia local

Lockerbie Cheviot: *see* West Country Cheviot

Loghton: *see* Manx Loaghtan

Lohi: (S Punjab, Pakistan)/cw.m/head dark brown, tan (or black); pd; long lop ears usu. with tassel; st/var.: Lamkanni/syn. *Parkanni* (= tassel)

Lohia: *see* Lampuchre

Lojeña: (Sierra de Loja, SW Granada, Spain)/m.mw/black, white, red, pied or roan; usu. pd/syn. *Rabada* (or *Rabuda*) *de la Sierra de Loja*

Lomond Halfbred: (Scotland)

Long Island: *see* Bahama Native

Longmynd: (England)/black face; hd/orig. (with others) of Shropshire/extinct 1936

Longwool: *see* English Longwool, Polish Longwool, Russian Longwool

Lonk: (C and S Pennines, England)/ cw.m/Blackfaced Mountain type, sim. to Derbyshire Gritstone but hd/BS, HB 1905/syn. *Improved Haslingden*/[? = coarse herbage, or by corruption of Lancashire, or from *wlonk*, Old English for proud or haughty]

Lop-Eared Alpine: (Alps)/m.(cw-mw)/pd; pendent ears; roman nose/basic breed is Bergamasca; sim. breeds or derivatives: Alpagota, Bellunese, Biellese,

Brentegana, Brianzola, Brogne,
Cadorina, Carinthian, Corteno,
Finarda, Friulana, German
Mountain, Lamon, Livo, Paduan,
Pusterese, Saltasassi, Solčava,
Tyrol Mountain, Varesina,
Vicentina; cf. also French
Alpine, Swiss White Mountain/
It. *Razze alpine da carne* (=
Alpine meat breeds)/syn.
Sudanica (= *Sudanese*, from
incorrect association)
Lori, Lorri, Lory: *see* Luri
Lošinj: *see* Island Pramenka
Lot Causses: *see* Caussenard du Lot
Lötschental: (Valais, Switzerland)/
black/var. of Roux du Valais/rare
Louda: *see* Uda
Louisiana Native: (USA)/var. of Gulf
Coast Native/nearly extinct
Lourdais: (Lourdes, SW Hautes-
Pyrénées, France)/m.mw/white,
occ. brown pied; ♂ hd, ♀ usu.
hd/Central Pyrenean group/var.:
Barègeois/orig. from Béarnais
with Merino blood/HB 1975/rare
Louri: *see* Luri
Łowicz: (Łódź, Poland)/Polish
Lowland group/orig. 1924–1939
from Romney × local (white
Świniarka with Merino blood)/
part orig. of Żelazna/Pol. *Owca
łowicka*/extinct
Lowland: *see* Marsh
Lozère Causses: *see* Caussenard de la
Lozère
Lozère Mountain: *see* Blanc du
Massif Central
Luan: (SE Shanxi, China)/cw/white;
♂ hd, ♀ pd/sim. to Han but with
lft/local var. of Mongolian
orig./[former name of town of
Changchih]
Lublin: (SE Poland)/local population
of Polish Lowland group/orig.
late 1950s from Uhruska (i.e.
Leine and Romney, × Merino) in
Uhrusk, and Texel and Romney,
× Łowicz in Borowina, with
Corriedale blood since 1974/Pol.
*Polska owca nizinna w okręgu
lubelskim* (= Polish Lowland
sheep of the Lublin region)
Lucanian: *see* Gentile di Lucania

Lucero: (mts of S Mexico)/black with
white spot on head/var. of
Criollo
Lughdoan: *see* Manx Loaghtan
Lugnez: *see* Bündner Oberland
Luma: (NE Albania)/d.m.cw-mw/♂
hd, ♀ pd/Ruda type/? = *ruda* var.
of Šar Planina (Macedonia)/?
syn. *Gekika* (Greece)
Lumbwa: *see* Kipsigis
Lüneburg Heath: *see* Heidschnucke
Luo: (W Kenya)/local var. of Masai
Luri: (Iran)/var. of Bakhtiari-Luri/not
Lori, Lorri, Lory, Louri, Lury
Lushnja: (Albania)/var. of Common
Albanian/not *Lushnia*
Luskin: *see* Liski
Luzein: *see* Bündner Oberland

Maasai: *see* Masai
Macarthur Merino: *see* Camden Park
Macedonian: *see* Greek Zackel,
Karakachan, Ovče Polje, Šar
Planina
Macedonian Nomad: *see* Karakachan
Macheri: *see* Mecheri
machkaroa: *see* Speckled-face
Manech
Macina: (C delta of Niger, Mali)/
cw.m.d/usu. white with coloured
spots esp. around eyes and ears,
occ. pied or black; ♂ hd, ♀ pd or
occ. hd (small); lt/var.:
Goundoun/not *Massina*
Macou: *see* Makui
Madagascar: (Madagascar)/m/white
or light brown, often pied with
red, brown or black, often white
with black head; hy; ♂ hd or pd,
♀ pd; long fat sickle-shaped
tail/African Long-fat-tailed
group/syn. *Malagasy, Malgache*
Madoera: *see* Indonesian Fat-tailed
Madras Red: (NE Tamil Nadu, India)/
m/red or brown; ♂ hd, ♀ pd;
st/South India Hair type/sim. to
Nellore but smaller
Madurese: *see* Indonesian Fat-tailed
Maellana: (Maella, Zaragoza, Aragon,
Spain)/m.(mw)/reduced fleece
cover; pd/not var. of Aragonese
Maghreb thin-tailed breeds: *see*
Algerian Arab, Atlantic Coast
(Morocco), Beni Guil, Berber,

Boujaad, D'man, Raimbi, Sardi, Tadla, Timahdit

Magra: (E and S Bikaner, Rajasthan, India)/cw.m/very white shiny fleece, light brown around eyes; pd/syn. *Bikaneri* (obs.), *Bikaneri Chokla, Chakri, Jangli, Mogra* (Jodhpur)/declining (crossbreeding and low fertility)

Magyar Fésüs Merinó: *see* Hungarian Merino

Magyar Húsmerinó: *see* Hungarian Mutton Merino

Magyar juh: *see* Racka

Mähnerschaf: *see* aoudad

Maigaiti Large-tailed: (China)/ft

Maine: *see* Bleu du Maine, Maine à tête blanche, Rouge de l'Ouest

Maine à face bleu: *see* Bleu du Maine

Maine-Anjou: *see* Bleu du Maine

Maine à tête blanche: (NW France)/ m.lw/pd/sim. to Cotentin/orig. from Leicester Longwool (imported 1855–1890) × local and improved by Cotentin/syn. *Maine à face blanche, Mayenne White, Whitefaced Maine, Whiteheaded Maine*/extinct (disappeared 1950s with spread of Bleu du Maine)

Maine à tête bleue: *see* Bleu du Maine

Maiylambadi: *see* Mecheri

Majorcan: (Majorca, Balearic Is, Spain)/m.mw/occ. spots on head and legs; ♂ hd or pd, ♀ pd/larger lowland and smaller mountain vars/Sp. *Mallorquina*/not *Majorquina*/*see also* Red Majorcan

Makui: (Maku, NW Azerbaijan, Iran)/ cw.m/black spots on face and feet; ♂ usu. pd, ♀ pd or scurs; ft/sim. to White Karaman/Fr. *Macou*/not *Makoee*

Malagasy: *see* Madagascar

Malawi: (Malawi)/African Long-fat-tailed group/sim. to Sabi/obs. syn. *Nyasa*

Malaysian: *see* Malin

Malaysian Long-tail: (Malaysia)

Maleech: *see* Malich

Malgache: *see* Madagascar

Mali Samburu: *see* ? Banamba

Malich: (Crimea, Ukraine)/fur, d/ black, white or grey; ♂ hd, ♀ pd; ft/orig. from Karakul/not *Maleech, Malych*

Malin: (E of Malay peninsula)/m.cw-hy/usu. white, occ. shades of grey or brown, badger-face, black or pied; ♂ hd, ♀ usu. pd (or scurs); often earless; st/syn. *ISM* (= *Indigenous Sheep of Malaysia*), *Kelantan, Malaysian, Pahang, Thai*/[Malaysian indigenous]

Mallorquina: *see* Majorcan

Malpura: (E Rajasthan, India)/cw.m/ light brown face; pd; short ears/sim. to Sonadi

Maltese: (Malta)/d.m.cw/coloured markings on face; usu. pd/orig. of Comisana/rare

Malya: (Turkey)

Malych: *see* Malich

Manchada Paramuna: (highlands of Boyaca, Cundinamarca and Narino, Colombia)/m.w/brown or black speckled face; ♂ hd, ♀ hd or pd/orig. 1976–1986 (named) from Scottish Blackface × Criollo/[= speckled (face) of the *paramos* (highlands)]

Manche: *see* Avranchin, Cotentin

Manchega: (La Mancha, New Castille, Spain)/d.m.mw/usu. white, also black var; pd/ Entrefino type/milk and meat vars; also Black Manchega/orig. of Segureña, Talaverana/HB 1969, BS

Manchega pequeña: *see* Alcarreña

Mandya: (S Karnataka, India)/ m.hr/ pale red-brown patches on anterior; pd; st/South India Hair type/syn. *Bandur, Bannur*/not *Manday, Mandaya, Mandi*

Manech: (SW Basque country, France)/ d.m.cw/coloured face and legs; ♂ hd, ♀ hd or pd/ Pyrenean dairy group/= Lacho (Spain)/vars: Black-face Manech, Red-face Manech, Speckled-face Manech, Petite Manech/HB 1975/not *Manesch*/ [Basque = John; men of Soule call those of Basse Navarre and Labourd 'Manechac']

Manx Loaghtan: (Isle of Man)/brown,
formerly also white or black; 2–6
horns (4 preferred)/Northern
Short-tailed type, sim. to
Hebridean/HB, BS 1976; BS also
in GB 1988/syn. *Manx Loghtan*
(GB)/not *Laughtan, Loaghtyn,
Loghton, Lughdoan*/[Manx *lugh
dhoan* = mouse-brown]/rare
Manych, Manychskaya: *see*
Karanogai
Manych Stavropol: (N Caucasus,
Russia)/orig. from Australian
Merino × Stavropol/syn.
Manych type of Stavropol
Manze: *see* Menz
Marathwada: (S Maharashtra,
India)/m/black, red or pied; ♂
hd, ♀ pd/South India Hair type,
sim. to Nellore/not *Marhatwada*
Marchois: (France)/small var. of
Limousin (? with blood of
Berrichon de l'Indre)/extinct in
1930s
Marco Polo's sheep: (Pamirs, C Asia)/
= *Ovis ammon polii* Blyth/var. of
argali/syn. *great Tibetan sheep,
great Pamir sheep, nayan*
(Nepal), *Pamir argali*/rare
Mareb White: (Mareb, Yemen)/m/
white, face usu. black or pied;
hy; pd; ft/sim. to Tihami
Maremmana: (Maremma, Latium and
Tuscany, Italy)/former var. of
Sopravissana/syn. *Spanish
Mongrel* (It. *Bastarda spagnola*)
Bastarda maremmana/extinct
Marhatwada: *see* Marathwada
Marialva, Marialveira: *see* Badana
Mariñana: *see* Galician
Mariouti: *see* Barki
Marishka: *see* Maritsa
Marishka t'nkorunna: *see* Thrace
Finewool
Marismeña: *see* Andalusian Churro
Maritsa: (Maritsa valley,
Bulgaria)/m.w.d/pd/Bulg.
Marishka/not *Marshka*/rare
Maritsa Finewool: *see* Thrace
Finewool
Marmoucha: (Taza, Morocco)/vars of
Berber/Northern var.: small, white,
♂ usu. pd, ♀ pd/Southern var.:
larger, black-headed, ♂ hd, ♀ pd

Maroua: (N Cameroon)/West African
Dwarf × Fulani cross
Marquesado: (N Granada, Spain)/
larger var. of Segureña, with less
fleece cover/syn. *Marquesa,
Marqueseña*
Marrane: (NE Genoa, Italy)/m.cw/
straw-coloured or light brown;
pd/Apennine group/nearly
extinct
Marri: *see* Bibrik
Marschschaf: *see* Marsh
MARSH: (NW Europe)/pd; st or rat
tail/inc. Belgian Milk,
Butjadingen, East Friesian,
Fagas, Flemish, Flemish Marsh,
Friesian Milk, Groningen,
Pomeranian, Texel,
Wilstermarsch, Zeeland Milk/
orig. (with Cotswold) of Danish
Whiteheaded Marsh, German
Whiteheaded Mutton/Ger.
Marschschaf, Du. *Polderschaap*,
Pol. *Żuławy*/syn. *Lowland* (Ger.
Niederungsschaf)
Marshka: *see* Maritsa
Marthod: *see* Thônes-Marthod
Martunin: (Armenia)/var. of
Armenian Semicoarsewool/orig.
from Aragats × (Aragats ×
Balbas)/Russ. *Martuninskaya*
Marwari: (Jodhpur, Rajasthan, India)/
cw.m/black face; pd; small ears/
syn. *Layda, Marwadi* (N Gujarat)/
[Marwar = old name for Jodhpur]
Maryuti: *see* Barki
Masai: (N Tanzania and SC Kenya,
also Uganda)/m/red-brown, occ.
pied; hy; ♂ hd or pd, ♀ usu. pd;
sft-fr/East African Fat-tailed
type/inc. sheep of Bukusu,
Busia, Kipsigis, Luo, Nandi and
Samburu (all in Kenya)/syn. *Red
Maasai, Red Masai, Tanganyika
Short-tailed*/declining by
crossing
Mascherina: (C Italy)/m.d.mw/orig.
from Suffolk × Apennine
Masekh: *see* Mazekh
Masham: (N England)/mottled face
and legs; pd/Teeswater (or
Wensleydale) × Swaledale (or
Dalesbred), 1st cross/BS/syn.
Massam (wool), *Yorkshire cross*

(Scotland)/[town in N Yorkshire]/*see also* Scottish Masham, Welsh Masham

Mashona: *see* Sabi

Massa: (Cameroon) *see* Kirdi

Massese: (Massa, Tuscany, Italy)/ d.m.cw/grey or brown; hd; roman nose/Apennine group, sim. to Garfagnina but smaller and with darker head/HB 1971/syn. *Fornese* (from Forno)

Massif Central: *see* Central Plateau

Massina: *see* Macina

Matesina: (Dragoni, Caserta, Campania, Italy)/m.mw/dirty white or light hazel; ♂ hd, ♀ pd/orig. from Gentile di Puglia/ rare

Mati: (C Albania)/cw.d/red head and legs, and red hairs in fleece/var. of Albanian Zackel/syn. *Red Head*

Mauchamp: *see* Mérinos de Mauchamp

Maure: (W Africa)/Sahel type/inc. Black Maure, Touabire/syn. *Arab, Mauritanian, Moor, Moorish*/not *Mauretanian*

maxhar: *see* Speckled-face Manech

Mayenne Blue: *see* Bleu du Maine

Mayenne White: *see* Maine à tête blanche

Mayo Mountain: (Ireland)/extinct (absorbed by crossing with Scottish Blackface)

Mazaev Merino: (SE Ukraine)/fw/ orig. in mid-19th c. by P.D. Mazaev by improvement of (Russian) Infantado/orig. of Novocaucasian and other Russian Merinos/Russ. *Mazaevskiĭ Merinos*/extinct

Mazamet: *see* Lacaune

Mazandarani: *see* Zel

Mazekh: (Armenia)/m.d.cw/usu. red-brown, also black, red or grey; ♂ hd, ♀ pd/Caucasian Fat-tailed type, sim. to Balbas/cf. Çandır (= *Mazik*); ? = Red Karaman (Turkey)/Russ. *Mazekh, Mazekhskaya*/syn. *Gyzyl-Gojun* (= red sheep) (Azerbaijan)/not *Masekh, Mazech*

Mazik: *see* Çandır

Meatlinc: (Lincolnshire, England)/ m/pd/terminal meat sire/orig. (1963–1974) by H. Fell of Worlaby, Brigg, from Berrichon du Cher, Île-de-France and Charollais, × Suffolk and Dorset Down/HB/not *Meatline*

Meatmaster: (South Africa)/m/orig. by Univ. of Pretoria as mutton breed/extinct

Meat Merino: *see* German Mutton Merino

Mecca: *see* Hejazi

Mecheri: (Salem, W Tamil Nadu, India)/m/light brown, sometimes white patches; pd; st/South India Hair type/syn. *Maiylambadi, Mylambadi, Thuvaramchambali* (Coimbatore)/not *Macheri, Mechheri*

Medel, Medelser: *see* Tavetsch-Medels

Mednochervena: *see* Copper-Red

Mednochervena shumenska: *see* Shumen

Mehraban: (Hamadan, W Iran)/ m.d.cw/light brown, cream or grey, with dark face and neck; ft/sim. to Shal/not *Mehreban*

Meidob: (NW Darfur, Sudan)/? Zaghawa × Sudan Desert cross

Mejorado: *see* Aragonese (Improved)

Mele: (Neuenkirchen, Germany)/ mw.m/pd/orig. *c.* 1908 from Leicester Longwool × Merino/ [*Merino-Leicester*]/extinct (absorbed by German Mutton Merino 1934)

Melkschaap: *see* Belgian Milk Sheep

Menemen: (Izmir, Turkey)/orig. from Île de France × (Île de France × Tahirova), hence $\frac{3}{4}$ Île de France, $\frac{3}{16}$ East Friesian, $\frac{1}{16}$ Kivircik

Menemen Kivircik: (Izmir, Turkey)/ Sakiz × Kivircik cross at Ege Univ. Farm, Menemen

Mengali: *see* Baluchi

Menggu: *see* Mongolian

Menghi, Mengli: *see* Baluchi

Mennonite: (Saratov, Russia)/m.sw/ orig. from East Friesian/extinct

Menorquina: *see* Minorcan

Menz: (N Shoa, Ethiopia)/m.cw/ brown or black, usu. white spots

on head and legs; ♂ hd, ♀ pd;
ft/syn. *Legagora*/not *Manze,
Mens*
Meraisi: *see* Ossimi
Mergelland: (S Limburg,
Netherlands)/m.cw/beige spots
on neck; pd/sim. to Kempen/HB
1980, BS/rare
Meridale: (W Australia)/orig. by J.H.
Shepherd from AMS Merino
with blood from 23 British, New
Zealand and Australian breeds; ⅝
Merino, ⅜ other breeds
Merilín: (Uruguay)/mw.m/orig. since
1910 by J.M. Elorga from
Rambouillet (¾) and Lincoln
(¼)/BS/[*Meri*no-*Lin*coln]
Merina: *see* Spanish Merino
Merina andaluza entrefina: *see*
Andalusian Merino
Merina da Beira Baixa: *see* Beira
Baixa
Merina Gentile: *see* Gentile di Puglia
Merina serrana: *see* Spanish
Mountain Merino
MERINO: fw/♂ usu. hd, ♀ usu. pd/
orig. from Spanish Merino/see
American, Argentine, Askanian,
Australian, Azerbaijan
Mountain, Bulgarian, Chinese,
Czech, French, Gentile di Puglia,
Georgian, German, Hungarian,
Indian, Kazakh, Kirgiz, North
Kazakh, Polish, Portuguese,
Romanian, Russian,
Sopravissana, South African,
South Kazakh, Turkish,
Vojvodina /Fr. *Mérinos*, Pol.
Merynos, Russ. *Merinos*, Sp.
Merina
Merino Branco: *see* Portuguese
Merino (white vars)
Merino d'Arles: *see* Arles Merino
Merino de Barros: *see* Barros Merino
Merino de Grazalema: *see* Grazalema
Mérino de l'Est: *see* Est à laine
Mérinos
Merino del Pais: (Spain) *see* Spanish
Merino
Merino de montaña: *see* Spanish
Mountain Merino
Merino di Puglia: *see* Gentile di
Puglia
Merino d'Italia: *see* Gentile di Puglia

Merino entrefino: *see* Leonese
Merino
Merinofinn: (Poland) *see* prolific
Merinofinn Mf-40
Merinofleischschaf: *see* German
Mutton Merino
Merinolandschaf: (Germany)/mw.m/
pd/orig. in S Germany from
Merino (imported late 18th and
19th c.) × Württemberg Land/
orig. of Dagestan Mountain, Est à
laine Mérinos and Swiss White
Mountain/recog. 1887–1906 first
as *Württemberger* (syn.
*Württemberg Bastard,
Württemberg Improved Land,
Württemberg Merino*), then as
Deutsches veredeltes Landschaf
(*German Improved Land*) (also
*German Whiteheaded Land,
Improved German Country,
Improved German Farm,
Improved Whiteheaded Land*),
and finally named 1950/HB
1922; BS S Africa 1957, Spain/
syn. *Land Merino*/not
Wurtemburg
Merino Longwool: (Erfurt, E
Germany)/lw.m/pd/orig. (1971
on) from Merinolandschaf (½),
Lincoln (¼) and North Caucasus
Mutton-Wool (¼)/Ger.
Merinolangwollschaf, Cz.
Německá dlouhovlnná
Merino Meat, Merino Mutton: *see*
German Mutton Merino
Merino preto: (Portugal) *see* Black
Merino
Merinos a vello nero: *see* Carapellese
Mérinos bourguignon: *see*
Châtillonais
Mérinos champenois: *see*
Champagne Merino
Mérinos d'Arles: *see* Arles Merino
Mérinos de Mauchamp: (France)/
former var. of French Merino
with silky wool/extinct
Mérinos de Rambouillet: *see*
Rambouillet
Merinos de vest: (Romania) *see*
Transylvanian Merino
Merinos dunărea: *see* Danube
Merino
Mérinos du Naz: (France)/former var.

of French Merino with very fine wool/extinct

Mérinos précoce: *see* Précoce, Soissonais

Merino trasterminante: *see* Leonese Merino

Merynos cienkorunny: *see* Polish Merino

Merynos polski pogrubiony: *see* Polish Strongwooled Merino

Merynos wełnisto-mięsny: *see* Polish Merino

Merynos wełnisty typu jędrzychowickiego: *see* Jędrzychowice

Mesha: (Mexico)/brown var. of Chiapas/? orig. from Lacho

Mestizo Entrefino-fino: (CW Spain)/m.mw/pd/cf. Talaverana/orig. from Merino × Manchega or Castilian/vars: Campo de Calatrava, Ciudad Rodrigo, Villuercas

Mestna polut"nkorunna Kazanl"shki raĭon: *see* Kazanluk Semifinewool

Metaxomalicha: *see* Sitia

Metelini: *see* Mytilene

Metohija Whitehead: *see* Bardoka

meusse: (France)/= pd Précoce ♂

Michni: (Peshawar, Pakistan)/m.cw/tan or black ears, muzzle and around eyes; lft

Middle Atlas: *see* Timahdit

Mihnov, Mihnovskaya: *see* Mikhnov

Mikhnov: (Voronezh, Russia)/chestnut or black face; ♂ hd, ♀ pd; lt/orig. from Russian Long-tailed/crossed with Romney to form Ostrogozhsk and with Lincoln to form Liski/Russ. *Mikhnovskaya*/not *Mihnov*/nearly extinct

Milchschaf: (Germany) *see* East Friesian

Minho e Douro: *see* Entre Douro e Minho

Minnesota 100: (USA)/mw.m/orig. 1941–1944 at Minnesota Agric. Exp. Sta.; ½ Rambouillet, ¼ Border Leicester, ¼ Cheviot/extinct

Minnesota 101: (USA)/inbred Hampshire flock/extinct (absorbed by North Star Minnesota 103 in 1953)

Minnesota 102: (USA)/w.m/orig. since 1944 at Minnesota Agric. Exp. Sta. from Border Leicester × Shropshire with some Columbia and Targhee blood 1949/extinct

Minnesota 103: *see* North Star Minnesota 103

Minnesota 104: (USA)/inbred Hampshire flock/extinct (absorbed by North Star Minnesota 103 in 1953)

Minnesota 105: (USA)/orig. 1949–1954 from Columbia, Hampshire and Southdown/extinct

Minnesota 106: (USA)/Columbia flock/extinct

Minnesota 107: (USA)/inbred Shropshire flock closed since 1937/extinct

Minorcan: (Menorca, Balearic Is, Spain)/d.(cw)/usu. pd; lt/BS/Sp. *Menorquina*/rare

Minusinsk: (Siberia, Russia)/var. of Siberian/extinct

Minxian Black Fur: (S Gansu, China)/fur/black; ♂ hd, ♀ pd

Miranda Galician: (Terra Fria, NE Portugal)/m.cw/white with spots on face and legs, also black; hd/Portuguese Churro type/cf. Bragança Galician/HB 1989/Port. *Galega mirandesa*

Mirror: *see* Spiegel

Mitilíni: *see* Mytilene

Mnogoplodnyĭ Karakul: *see* Multifoetal Karakul

Modenese: *see* Pavullese, Zucca Modenese

Modern Romney: *see* New Zealand Romney

Modified Dorset: (Poland) *see* Polish Modified Dorset 59

Moghani: (Moghan steppe, NW Iran)/m.cw/occ. pale colour marks on head and feet; ♂ usu. pd, ♀ pd; ft/not *Mogani*, *Mugan*

Mogra: *see* Magra

Mohaka: (Hawke's Bay, New Zealand)/ fw; feral/rare

Mondegueira: (W Guarda, Beira Alta, Portugal)/m.d.cw/usu. small brown spots on head and legs;

hd/Portuguese Churro type/[from Mondego]

Mondombes: (Moçambes, SW Angola)/usu. black pied or brown pied; hy; ♂ hd, ♀ usu. pd/African Long-fat-tailed group/syn. *Herero, Mucubais*

Monegrina: (Los Monegros, Ebro valley, border of Huesca and Zaragoza, Aragon, Spain)/small var. of Aragonese with reduced fleece/rare

Mongalla: (Latuka-Bari, Sudan)/var. of Southern Sudanese

Mongolian: (Mongolia, and N China esp. Inner Mongolia)/cw.m.d/usu. white with black or brown head, also pied or self-coloured; ♂ hd, ♀ pd; ft/vars: Barga, Bayad, Baydrag, Chamar, Darhad, Govĭ-Altay, Halha, Ikzhaomen, Sutai, Ujumqin, Uzemchin/orig. of Choubei, Han, Hu, Lanzhou Large-tail, Luan, Quanglin Large-tail, Shouyang, Taiku, Tan, Tong and (with Karakul) of Kuche and Ningsia Black/Ch. *Menggu*/syn. *China Fat-tailed, Mongolian Fat-tailed*/not *Monggol*

Mongo-tibetan: *see* Tibegolian

Monselesana, Monselice: *see* Noventana

Monsetina: *see* Pallaresa

Montadale: (Missouri, USA)/mw.m/orig. (1933 on) from Cheviot (40%) × Columbia (60%)/BS 1945; HB also Canada/not *Montedale*/rare

Montafoner: (Montafon, Vorarlberg, Austria)/m/sim. to Zaupel, Steinschaf and Bündner Oberland/nearly extinct

Montagne Noire: (Ariège, France)/m/red spots on face/[= black mountain]/nearly extinct

Montagnes de la Lozère: *see* Blanc du Massif Central

Monta Khia: (USA)

Montañesa: *see* Ripollesa

Montedale: *see* Montadale

Montenegro: *see* Piva, Vasojević and Zeta Yellow vars of Pramenka

Montesina: (mts of Jaén and Granada, Andalucía, Spain)/m.mw/black ear tips, muzzle, around eyes and on lower legs; ♂ usu. pd, ♀ pd/Iberian type, sim. to Ojalada/not var. of Manchega/syn. *Granadina, Mora, Ojinegra, Sevillana*/[= mountain]/purebreds rare (being crossed with Segureña)

Montseny: *see* Pallaresa

Montsetina: *see* Pallaresa

Moor, Moorish: *see* Maure

Moor: *see* Red Karaman

moorit: (Scotland)/= brown/see Castlemilk Moorit, Manx Loaghtan, Shetland

Moorland, Moorschnucke: *see* Heidschnucke (White Polled var.)

Mora: (Spain) *see* Montesina

Morada Nova: (NE Brazil)/m/red (with white tail tip) or white; pd/American Hair Sheep group/orig. by selection from Brazilian Woolless/HB, BSd/[town in Ceará]

Moraitiko: (Thessaly, Greece)/black, or white with black spots on head, belly and legs/mt var. of Greek Zackel/? orig. from Karakachan × Karagouniko

Moretta: (Italy) *see* Carapellese

Morfe Common: (England)/coloured face; hd/sim. to Ryeland/part orig. of Shropshire/extinct

Mor-Karaman: *see* Red Karaman

Morlam: (Beltsville, Maryland, USA)/m.w/usu. pd, some hd/orig. (1961 on) from Rambouillet × Dorset Horn, Merino, Targhee, Columbia-Southdale, Hampshire and Suffolk, selected for frequent lambing

Moroccan: *see* (recog. breeds with BSd) Beni Guil, Boujaad, D'man, Sardi, Timahdit; *see also* Beni Ahsen, Berber, Doukkala, Siroua, South Moroccan, Tadla, Zemmour

Morvandelle: (Morvan, France)/extinct

Moscia: (S Italy)/d.cw.m/? Zackel type/inc. Altamurana, Leccese, Sciara/[= straight wool]

Moscia leccese: *see* Leccese

Mossi: (Yalenga, Burkina Faso)/m/ usu. black (or brown) forequarters and white hindquarters; ♂ hd, ♀ hd or pd; ♂ maned/? orig. from Toronké (or Peul Voltaïque) × West African Dwarf/syn. *Djallonké/* not *Moshi*

Mossul, Mosul, Mosuli: *see* Herki

motti: (France) = pd ♂ Arles Merino

mouflon: (Corsica and Sardinia)/= *Ovis musimon* (Pallas)/red-brown with dark back stripe, pale saddle patch and underparts; ♂ hd, ♀ hd or pd; st/? early feral/It. *muflone/*syn. *Corsican mouflon, European mouflon, musimon, musmon, Sardinian mouflon/*not *moufflon, muflon/*nearly extinct but successfully introduced into C Europe (Germany, Austria, Czech and Slovak Republics, Romania)

mouflon à manchettes: *see* aoudad

mouflon, oriental: *see* red sheep

Moulouya: *see* Tounsint, Zoulay

Mountain Corriedale: (Russia) *see* Soviet Mutton-Wool

Mountain Merino: *see* Azerbaijan, Russian, Spanish

Mountain Tsigai: (mts of N and S Bulgaria)/cw.m/orig. 1950–1967 from Tsigai × (Tsigai or German Mutton Merino × Replyan)/? = North Bulgarian Improved/Bulg. *Planinski tsigai*

Mountain Zackel: (Greece) *see* Vlakhiko

Mourerous: (upper Vésubie and Var valleys, French Alps)/m.mw/ white to yellowish with red to fawn head and legs; pd/BS 1983/ syn. *Péone, Rouge de Guillaumes/*[= red face]

Moussouli: *see* Herki

Mouton à tête noir: *see* French Blackheaded

Mouton à viande à tête brune: *see* Swiss Brownheaded Mutton

Mouton de montagne à laine Zoulai: *see* Zoulay

Mouton des Collines: *see* Houtland

Mouton laitier belge: *see* Belgian Milk Sheep

Mouton maure à poil long: *see* Black Maure

Mouton Mirroir: *see* Spiegel

Mouton nain d'Afrique occidentale: *see* West African Dwarf

Mouton Peul: (Cameroon) *see* Uda

Moyen Atlas: *see* Timahdit

Mucubais: *see* Mondombes

muflon, muflone: *see* mouflon

muflone berbero: *see* aoudad

Mug: *see* Wensleydale

Mugan: *see* Moghani

Muggs, Mugs: *see* Wensleydale

Mule: *see* North of England Mule, Scotch Mule, Welsh Mule

Multifoetal Karakul: (Ukraine)/fur/ black, also grey var./orig. (1935–1952) at Askania Nova from Karakul × (Karakul × Romanov), to increase twinning rate/Russ. *Mnogoplodnyĭ Karakul*

Multihorned: *see* Four-horned

Multihorned Merino: (Orange Free State, S Africa)/cw/? orig./rare

Multihorned Shetland: (Scotland)/♂ 4-hd, ♀ hd (2 or 4) or pd/var. of Shetland/syn. *Ronas Hill, Yell Shetland/*not *Rona's Hill, St Rona's Hill/*nearly extinct

Multinipple: (USA)/orig. 1923–1941 from Bell Multinippled × (Southdown × Rambouillet), + Suffolk × above/? extinct

Muma: *see* Garessina

Munjal: (Punjab, India)/m.cw/dark brown face; long lop ears/mixed population of Lohi and Nali

Murge: *see* Altamurana

Murle: (SE Sudan)/black pied or brown pied; ♂ usu. hd, ♀ usu. pd; sft/var. of Toposa with more Southern Sudanese blood

musimon, musmon: *see* mouflon

muthur belza: *see* Black-face Manech

muthur gorria: *see* Red-face Manech

muthur churria: (France) *see* Basque

Mutton Merino: *see* German, South African, Walrich

Mutton Synthetic: (India)

Muzaffarnagri: (Muzaffarnagar, Uttar Pradesh, India)/cw.m.(d)/occ. black or brown patches on face and legs; pd; lt/syn.

Bulandshari/not *Muzaffarnagari,
Muzzafarnagri, Muzzafarnargri*

Myanmar: *see* Burmese

myaso-sherstnaya: = meat-wool,
mutton-wool (Russ.)

Mylambadi: *see* Mecheri

Mytilene: (W Lésvos, Greece)/d.m.cw/
usu. white with spots on nose
and legs; sometimes coloured or
pied; ♂ hd, ♀ usu. pd; lft/sim. to
Kamakuyruk/? orig. from Turkish
breeds × local thin-tailed/Gr.
Mitilíni/syn. *Lesbos, Lesvos*/not
Metelini

**Na'ami, Na'amiyah, Naeemi,
Naeimi, Naimi:** *see* Ne'imi

Naeini: (Iran) *see* Baluchi

Najdi: (C Saudi Arabia)/hr/usu. black
with white head; ♂ pd or scurs,
♀ pd; lft/vars: Asali, Hejari/fem.
Najdiyah; syn. *Arabian Long-
tailed, Bedouin*/not *Nagdi,
Nedjed, Nejdi, Nidjy, Nigdi*

Nali: (N Rajasthan and S Haryana,
India)/cw.m/light brown face;
pd; long ears

Nalps: *see* Bündner Oberland

Nama: *see* Namaqua Afrikander

Namaqua: (S Africa and S Namibia)/
white; lft/var. of Hottentot/
extinct

Namaqua Afrikander: (NW Cape
prov., S Africa, and S Namibia)/
m/white, usu. with black or red-
brown head; hy-cw/ ♂ hd, ♀ usu.
pd; lft (corkscrew)/var. of
Afrikander/orig. from Namaqua
var. of Hottentot/Afrik.
Namakwa Afrikaner/syn. *Nama*
(Namibia)/rare (disappearing by
grading to Karakul; extinct in
Namibia)

Nami: *see* Ne'imi

Nandi: (W Kenya) *see* Masai

napo: *see* bharal

Nar: *see* Black Maure

Nasenspiegel: *see* Spiegel

Navajo-Churro: (Arizona/New
Mexico/Utah, USA)/cw/usu.
white with colour on face and
legs, also brown or pied; ♂ usu.
hd (occ. 4 or 0), ♀ usu. pd; small
to dwarf/cf. Criollo/orig. since

end 16th c. from Spanish
Churro/BS 1986/syn. *Navajo,
Navajo Four-horned, American
Four-horned*/not *Navaho*/old
type Navajo rare due to
upgrading to Rambouillet

Navargade: *see* Telingana

Navarra: (N Spain)/cf. Aragonese/Sp.
Rasa Navarra (*rasa* = smooth, i.e.
smooth fleece)

nayan: *see* Marco Polo's sheep

Naz: *see* Mérinos du Naz

Ndman: *see* D'man

Neahami: *see* Ne'imi

NEAR EAST FAT-TAILED: (NE Africa and
SW Asia)/cw.d.m/white, or
white with coloured face, or
black, brown or pied; ♂ usu. hd,
♀ usu. pd; ft (usu. large S-shaped
bilobed)/inc. Arabi (Iraq),
Awassi, Barbary, Barki, Cyprus
Fat-tailed, Fellahi, Kurdi,
Ossimi, Rahmani/syn. *Semitic
Fat-tailed*

Nedjed: *see* Najdi

Negretti: (Spain)/former strain of
Spanish Merino/Sp. *Negrete*/not
Negreo/extinct

Neimenggu, Neimonggol: *see* Inner
Mongolian

Ne'imi: (NW Iraq)/black or red face,
occ. red fleece/superior var. of
Awassi/not *Na'aimi, Na'amiyah,
Naeemi, Naeimi, Naimi, Nami,
Neahami, N'eimi, Niami,
Nu'ami, Nu'amieh, Nuamiyah*

Neini: (Iran) *see* Baluchi

Nejdi: *see* Najdi

Nellore: (SE Andhra Pradesh, India)/
m/♂ hd, ♀ pd; tassels; st/South
India Hair type/vars: Dora,
Jodipi, Palla

Nelson: *see* South Wales Mountain

Nemečka dlouhovlnná: *see* Merino
Longwool

Nepalese, Nepali: *see* Baruwal, Jumli,
Kage, Lampuchre, Tibetan

New Caucasian: *see* Novocaucasian

Newfoundland: (Canada)/m.mw/
sometimes dark face, or all black;
occ. hd/mixed orig. during 19th
and 20th c. from North Country
Cheviot and Dorset Horn with
some Scottish Blackface, Border

Leicester, Hampshire, Oxford
Down and Suffolk blood/nearly
extinct (declining by crossing)

New Kent: *see* Romney

New Leicester: *see* Leicester
Longwool

New Norfolk Horn: *see* Norfolk Horn

New South Wales Merino: (NSW,
Australia)/= Australian Merino
in NSW

Newton Stewart: *see* Galloway
Blackface

New Zealand Halfbred: (South
Island, New Zealand)/Longwool
(Lincoln, Leicester or Romney) ×
Merino, F_1 or F_2

New Zealand Merino: (New
Zealand)/= Australian Merino in
New Zealand/HB

New Zealand Romney: (New
Zealand)/orig. 1900–1910 from
Romney/orig. of Drysdale,
Tukidale/BS 1904, HB 1905/syn.
*Modern Romney, New Zealand
Romney Marsh, Romney*

New Zealand Wiltshire: (New
Zealand)/black nose and feet;
85% shed fleece; pd/orig. since
1974 from Australian Wiltshire
Horn with Poll Dorset ($\frac{1}{8}$–$\frac{1}{4}$) blood

Nguni: (Swaziland and KwaZulu)/m/
black, brown or pied; hy; ♂ hd or
pd, ♀ usu. pd; ears very small
(sometimes earless)/African
Long-fat-tailed group/vars:
Bapedi, Landim, Swazi, Zulu

Niami: *see* Ne'imi

Nidjy: *see* Najdi

Niederungsschaf: *see* Marsh

Nigdi: *see* Najdi

Nigerian Dwarf: *see* West African
Dwarf

Nilgiri: (W Tamil Nadu, India)/m.sw/
pd/orig. in early 19th c. from
Cheviot, Southdown and
Tasmanian Merino, × local hy
(Coimbatore) in Nilgiri hills/not
Nilagiri/rare

Nilotic: (Sudan)/var. of South
Sudanese/syn. *Dinka, Nuer,
Shilluk*

Ningxia Black: (Ningxia, China)/
cw.pelt/black or dark brown; ♂
hd, ♀ pd; ft/sim. to Kuche/orig.

from Karakul × Mongolian/WG
Ning-hsia hei/not *Ningsia*/not
now recog.

Nizhnedevitsk: (Russia)/subvar. of
Liski var. of Russian Longwool/
Russ. *Nizhnedevitskaya*

Nobile di Badia: *see* Pusterese

Nogai: *see* Karanogai

Noir de Bains: *see* Velay Black

Noir de Thibar: *see* Thibar

Noir du Velay: *see* Velay Black

Nolinsk: (Kirov, Russia)/var. of
Russian Northern Short-tailed/
Russ. *Nolinskaya*/extinct

Noord Hollander: *see* North Holland

Nord de la Manche: *see* Cotentin

Norfolk Horn: (Norfolk, Suffolk and
Cambridge, England)/m/black
face and legs; hd/nearly extinct
1973; revived as *New Norfolk
Horn* by grading up Suffolk,
Wiltshire Horn and Swaledale
(to $\frac{15}{16}$); name reverted to Norfolk
Horn 1984/orig. (× Southdown)
of Suffolk/HB 1978, BS
1995/syn. *Blackface Norfolk
Horned, Norfolk Horned, Old
Norfolk, Old Norfolk Horned*/
still nearly extinct

Norsk Pels-Sau: *see* Norwegian Fur
sheep

North: (England) *see* Scottish
Halfbred

North Bulgarian Improved:
(Bulgaria)/Tsigai type/? =
Mountain Tsigai

North Bulgarian Semifinewool:
(Bulgaria)/mw.m/orig. from
German Mutton Merino,
Caucasian, Merinolandschaf,
Romney, Lincoln and North
Caucasus

North Caucasus Merino: (Russia)/var.
of Soviet Merino

North Caucasus Mutton-Wool:
(Vorontsovo-Aleksandrovskoe
dist. of Stavropol territory,
Russia)/m.mw/usu. pd/orig.
1944–1960 at Stalin state farm
from Lincoln and Romney, ×
Stavropol (but Romney blood
culled since 1952)/Russ.
*Severokavkazskaya myaso-
sherstnaya*

North Caucasus Semifinewool:
(Karachaevo-Cherkessk and
Kabardino-Balkar, Russia)/
m.mw/orig. 1949–1957 from
Romney (or Soviet Corriedale) ×
(Précoce × Ossetian)/Russ.
*Severokavkazskaya gornaya
porodnaya gruppa
polutonkorunnykh ovets* (=
North Caucasus mountain breed
group of semifinewool sheep)

North Country: *see* Scottish Halfbred

North Country Cheviot: (N Scotland)/
var. of Cheviot (1792 on, esp.
1805–1820)/orig. of Wicklow
Cheviot/BS 1912, HB 1946; BS
also USA 1962; HB Canada/syn.
Caithness, Sutherland Cheviot

North-East Bulgarian Finewool: (NE
Bulgaria)/fw.m.d/orig. 1950–
1967 from German Mutton
Merino, Askanian and some
Caucasian, × local/var.: Shumen
Finewool/HB/syn. *Dobruja
Finewool* (Bulg. *Dobruzhanska
t"nkorunna*)

North-East China Finewool: (NE
China)/fw/orig. from Soviet
Merino and Stavropol ×
Mongolian/syn. *Northeast(ern)
Chinese Merino*

Northeast China Semifinewool: (NE
China)/mw.m/pd/orig. from
Corriedale × Mongolian

Northern Longwool: *see* Teeswater

Northern Riverain: (Sudan) *see*
Dubasi

NORTHERN SHORT-TAILED: (N Europe)/
often black, grey or brown/sim. to
Heath/inc. Faeroes, Finnish,
Gotland, Hebridean, Icelandic,
Manx Loaghtan, North Ronaldsay,
Old Norwegian, Romanov,
Russian, Shetland, Soay,
Spælsau, Swedish Landrace/syn.
Scandinavian Short-tail

Northern Sudanese: *see* Sudan
Desert

North Halfbred: *see* Scottish
Halfbred

North Holland: (Netherlands)/m/
prolific/orig. in 1970s from
Finnish Landrace × Texel/Du.
Noord Hollander

North Kazakh Merino: (N
Kazakhstan)/fw.m/orig. from
Beskaragai Merino combined
with Sulukol Merino/recog.
1976/Russ. *Severokazakhskiĭ
merinos*

North Nigerian Fulani: *see* Uda

North of England Mule: (England)/
m.lw/brown spots on head and
legs; pd/Bluefaced Leicester ×
Swaledale or Northumberland
Blackface, 1st cross/BS 1980/syn.
Mule

North Ossetian Semifinewool: (N
Caucasus, Russia)

**North Riveraine Wooled, North
Riverine Wooled:** *see* Dongola

North Ronaldsay: (Orkney,
Scotland)/white or grey, occ.
black or brown; ♂ hd, ♀ pd;
small/Northern Short-tailed
type/HB 1974/pronounced
Ronaldshay/orig. seaweed-eating
flock on North Ronaldsay, also
flock on Linga Holm 1975/rare

North Star Minnesota 103: (USA)/
mw.m/orig. since 1889 by W.W.
Bell of Beaver Creek from
Rambouillet, Oxford Down and
Lincoln; to Minnesota Agric.
Exp. Sta. 1947, Minnesota 101
and 104 blood in 1953/extinct

North Sudanese: *see* Sudan Desert

North Sumatran: *see* Sumatran

North Ukrainian Semifinewool: (N
Ukraine)/mw.w/orig. from
Argentine Romney ($\frac{1}{2}$), Azov
Tsigai ($\frac{1}{4}$), English Romney ($\frac{1}{8}$) and
Précoce ($\frac{1}{8}$)

Northumberland Blackface: (NE
England)/strain of Scottish
Blackface

North-West China Merino: *see* Gansu
Alpine Finewool

Norwegian: *see* Dala, Norwegian Fur
Sheep, Old Norwegian, Rygja,
Spælsau, Steigar

Norwegian Fur Sheep: (Norway)/
m.fur/grey or white; pd/orig.
from Swedish Fur Sheep × Old
Norwegian/named 1968; HB/
Nor. *Norsk Pels-Sau*/syn.
Norwegian Pelt

Nostrale: (France) *see* Brigasque

nostrale: (Italy) *see* Biellese, Brigasco, Pinzirita

nostrana: (Campania, Italy) *see* Campanian Barbary

Nostrana: (Passo della Cisa, Parma/ Massa Carrara, Italy)/m.cw- mw/usu. pd/Apennine group/? orig. from Garfagnina/[= local]/ rare

No-tail: (S Dakota, USA)/w.m/pd/ orig. from two Kazakh Fat- rumped ♂♂ (imported in 1913) and European breeds, with Rambouillet blood 1926, Southdown 1935, Columbia 1940s/? extinct

Noventana: (Italy)/larger var. of Paduan, with finer wool/syn. *Monselesana* (from Monselice)/ [from Noventa]/extinct

Novocaucasian Merino: (N Caucasus, Russia)/fw/orig. in late 19th c. from Mazaev Merino, improved by German Merino and Rambouillet/orig. of most Russian Merinos/Russ. *Novokavkazskiĭ Merinos*/syn. *New Caucasian, New Caucasian Mazaev*/extinct

N-type Romney: *see* Drysdale

Nu'ami, Nu'amieh, Nuamiyah: *see* Ne'imi

Nuba Maned: (Nuba mts, Sudan)/all colours except black; ♂ hd, ♀ pd/var. of South Sudanese/syn. *Nuba Mountain Dwarf*

Nuer: *see* Nilotic

Nungua Blackhead: (Legon, Ghana)/ m/hy/orig. at Agric. Res. Sta. of Univ. of Ghana from Blackhead Persian × West African Dwarf/ purebred nearly extinct

Nyasa: *see* Malawi

Oberhasli-Brienz: (Switzerland)/part orig. of Swiss White Mountain/ extinct

OBS Mouton noir de Fumex: *see* Fumex Black

Occidental: (Spain) *see* Andalusian Churro

October breed group: (Orenburg, Russia)/fw/orig. from Grozny or Stavropol (1947–1953) ×

(Précoce × coarsewooled) (1928–1946)/Russ. *Oktyabrskaya porodnaya gruppa*/extinct (joined with Orenburg Finewool to form South Ural)

Ödemiş: (Küçük Menderes valley, W Turkey)/m.d.cw/black or brown face; ♂ usu. pd, ♀ pd; ft

Oetztaler: *see* Ötztaler

Ofche Hulmski, Oftshe Hulmski: *see* Panagyurishte

Ogaden: *see* Somali

Ogiç: (Albania)/= unshorn Shkodra ♂; usu. black head; horns trained in upward spiral

Ojalada: (mts of Guadalajara, Soria, Teruel, Zaragoza, Castellón and Tarragona, Spain)/m.mw/white with black around eyes and on ear tips, muzzle and feet; ♂ usu. pd, ♀ pd/Iberian type, sim. to Montesina/syn. *Churra alcarreña, Churra entrefina, Churra soriana, Churra turolense, Fardasca* (Teruel), *Ojinegra* (= black-eyed), *Serrana* (= mt), *Serranet* (Tarragona)/[= button-holed]

Ojinegra: *see* Montesina, Ojalada

Oktyabrskaya porodnaya gruppa: *see* October breed group

Old Berkshire: *see* Berkshire Knot

Oldenbred: *see* Greyface Oldenbred, Welsh Oldenbred

Oldenburg: *see* British Oldenburg

Oldenburg White Head: *see* German Whiteheaded Mutton

Old Norfolk, Old Norfolk Horned: *see* Norfolk Horn

Old Norwegian: (Sunnhordland, W Norway)/m.w/many colours; hd or pd/Northern Short-tailed type/orig. of Icelandic, Faeroes, Spælsau/Nor. *Gammelnorsk*/syn. *Sunnhordland, Utegangarsau, Vildsau* (= wild sheep)/not *Sundhordland*/rare

Old Scottish Shortwool: *see* Tanface

Old Southdown: *see* Sussex

Old Wiltshire Horned: *see* Wiltshire Horn

Olib: *see* Island Pramenka

Olkusz: (Kraków, Poland)/m.w/pd; prolific/local var. of Polish

Longwool group/orig. from
Friesian Milk and Wilstermarsch
or Pomeranian/Pol. *Olkuska*/not
Olkcucyska/nearly extinct

Omani: (Oman)/m.cw/white, black,
brown or pied; ♂ hd or pd, ♀ pd

Omsk Semifinewool: (Siberia,
Russia)/m.mw/breed group/orig.
from Lincoln × local/Russ.
Omskaya myaso-sherstnaya (=
meat-wool) *polutonkorunnaya
porodnaya gruppa*

Ongti: *see* Hu

oorial: *see* urial

Oparino: (NW Kirov, Russia)/m.sw/
usu. white with coloured spots
on head, neck and legs, occ.
black or brown; ♂ hd, ♀ usu. pd/
orig. from English mutton breeds
× local Russian Long-tailed/
Russ. *Oparinskaya*/not *Oparina*

Oplemenjena Pirotska: *see* Improved
Pirot

Orecchiuta: *see* Sicilian Barbary

Orenburg Finewool: (Orenburg,
Russia)/orig. at Karl Marx state
farm from Caucasian and Grozny
(1952 on) × (Précoce *et al.* ×
local coarsewooled)/recog. as
breed group 1962/Russ.
*Orenburgskaya tonkorunnaya
porodnaya gruppa*/extinct
(joined with October breed group
to form South Ural)

Orhon: (N Mongolia)/mw.m/♂ occ.
hd, ♀ pd/orig. 1943–1961
(recog.) from Altai × [Tsigai ×
(Précoce × Mongolian)]/Russ.
Orkhonskaya/not *Orkhon*

oriental mouflon: *see* red sheep

Orkhon, Orkhonskaya: *see* Orhon

Orkney: *see* North Ronaldsay

Osemi: *see* Ossimi

Osetinskaya: *see* Karachai

Ossetian: *see* Karachai

Ossimi: (Lower Egypt)/cw.m/white
with brown head, often brown
neck, and occ. brown spots; ♂
usu. hd, ♀ usu. pd; ft/syn.
Meraisi/not *Auseymy*, *Ausimi*
(correct), *Awsemy*, *Osemi*,
Ousimi/[from Ausim]

Ostfriesisches Milchschaf: *see* East
Friesian

Ostfriesiska mjölkfår: *see* East
Friesian

Ostrogozhsk: (NW Voronezh,
Russia)/lw.m/breed group/orig.
before 1963 from Romney ×
Mikhnov, bred *inter se* or
backcrossed to Romney/Russ.
*Ostrogozhskaya porodnaya
gruppa*

Otočka Pramenka: *see* Island
Pramenka

Otter: *see* Ancon

Ötztaler: (Tyrol, Austria)/local var.
sim. to Tyrol Mountain/syn.
Oetztaler/extinct

Ouda, Oudah: *see* Uda

Ouessant, Ouessantin: *see* Ushant

Ouled Djellal, Ouled Jellal: *see*
Algerian Arab

Ousemi: *see* Ossimi

Ousi, Oussi: *see* Awassi

Outaouais Arcott: (Canada)/m/face
white or mottled; pd; prolific/
orig. 1966–1988 (recog.) at
Animal Res. Centre, Ottawa,
from Finnish Landrace (49%),
Shropshire (26%), Suffolk (21%)
and 6 other breeds (4%), crossed
1968–1974, selected for litter
size and early lambing 1977–
1988; for crossing with Canadian
Arcott/HB 1988/obs. syn. *ARC
(dam) Strain 2*

Ovca pogórza: *see* Polish Hill

Ovče Polje: (E Macedonia)/m.cw.d/
white, with head and legs partly
or wholly black or brown, occ.
black, brown or grey; ♂ hd, ♀
usu. pd/Pramenka type/Serbo-
cro. *Ovčepoljska*/not *Ovce
Polian*/[= sheep plain]

Oveja de Colmenar: *see* Colmenareña

Ovis ammon: *see* argali

Ovis ammon karelini: *see* arkhar

Ovis ammon polii: *see* Marco Polo's
sheep

Ovis canadensis: *see* bighorn

Ovis dalli: *see* Dall sheep

Ovis dalli stonei: *see* Stone sheep

Ovis musimon: *see* mouflon

Ovis nivicola: *see* snow sheep

Ovis orientalis: *see* red sheep, urial

Ovis orientalis armeniana: *see*
Armenia red sheep

Ovis polii: *see* argali
Ovis vignei: *see* urial
Owca bocheńska: *see* Bochnia
Owca leszczyńska: *see* Leszno
Owca łowicka: *see* Łowicz
Owca pomorska: *see* Pomeranian
Owca śląska: *see* Silesian
Oxford Down: (England)/sw.m/dark
brown face and legs; pd/Down
type/orig. *c.* 1830 from Cotswold
× Hampshire Down (and
Southdown)/orig. (+ Hampshire
Down) of German Blackheaded
Mutton, and (+ Grabs) of Swiss
Brownheaded Mutton/recog.
1851; BS 1888, HB 1889; BS also
USA 1882; HB also Canada,
Switzerland, Czech Rep./syn.
Oxford (USA)/obs. syn. *Down-
Cotswold, Oxfordshire Down*

Padova, Padovana: *see* Paduan
Paduan: (Venetia, Italy)/d.m.mw/pd/
Lop-eared Alpine group/var.:
Noventana/orig. of Carinthian,
Solčava/It. *Padovana*/extinct
1970s (crossed with Lamon)
Pag Island: (Croatia)/d.m.mw/white,
occ. black; ♂ usu. hd, ♀ usu. pd;
st/sim. to Dubrovnik/? orig. from
Merino × Pramenka in early 19th
c./Cro. *Paška*, It. *Pago-Selve*/syn.
Silba/not *Paka, Pashka, Pog*/*see
also* Island Pramenka
Pagliarola: (Abruzzo and Molise,
Italy)/m.(cw-mw)/yellowish-
white, also reddish-black; pd/
Apennine group/[= straw-eater]
Pagota: *see* Alpagota
Pago-Selve: *see* Pag Island
Pahang: *see* Malin
Pahari: (Muzaffarabad, Azad
Kashmir, Pakistan)/m.cw.d/white
with tan, brown or black head, or
pied; usu. pd; sometimes sft
Paka: *see* Pag Island
Pak Awassi: (Punjab, Pakistan)/
d.cw.m/orig. from Awassi ×
Kachhi/rare
Pakistan fat-tailed breeds: *see*
Baluchi, Bibrik, Dumbi, Gojal,
Harnai, Hashtnagri, Khijloo,
Kohai Ghizer, Latti, Michni,
Rakhshani, Tirahi, Waziri

Pak Karakul: (Punjab, Pakistan)/fur/
orig. since 1965 at Rakh
Kairewala farm, Muzaffargarh,
from Karakul × Kachhi/rare
Palas Merino: (Constanta region, SE
Romania)/fw.d.m/♂ usu. hd, ♀
usu. pd/orig. 1920–1950 at Palas
Animal Breeding Sta., Constanta,
from Tsigai, Ţurcana and Stogoşă,
graded to Rambouillet (1926–
1934) and to German Mutton
Merino (1928–1942), with
further Rambouillet and
Stavropol blood/HB 1926/Rom.
Merinos de Palas
Palla: (SE Andhra Pradesh, India)/
white/var. of Nellore
Pallaresa: (Pallars, Lérida, Catalonia,
Spain)/m.(mw)/white, usu. with
black or red points; ♂ occ. hd, ♀
pd/Iberian type/syn.
*Benasquina, Chisqueta,
Chistavina, Sisqueta, Tisqueta* or
Xisqueta (from Catalan
Txisqueta), *Monsetina* or
Montsetina (from Montseny)/not
Catalan
Palmera: (La Palma, Canary Is,
Spain)/m.cw/occ. black, ♂ hd, ♀
pd/? orig. from Andalusian
Churro/nearly extinct
Paloma: *see* Segureña; also
(confusingly) Alcarreña,
Ansotana
Pålsfår: *see* Swedish Fur Sheep
Pamir: (Tajikistan)/cw.m/sim. to
Kirgiz Fat-rumped/Russ.
Pamirskaya/extinct
Pamir argali: *see* Marco Polo's sheep
Pamir Finewool: (Tajikistan)/orig.
from Merino × Darvaz
Pampa: *see* Criollo
Pampara: (Georgia)/var. of Tushin,
with longer wool of lower
density
Pampinta: (La Pampa, Argentina)/
m.d.w/pd/orig. 1980–1990 at
Anguil Res. Sta. from Corriedale
($\frac{1}{2}$) and East Friesian ($\frac{2}{4}$)
Panagiurište, Panaguirishte: *see*
Sredna Gora
Panagyurishte, Panagyurska: *see*
Sredna Gora
Panama: (Idaho, USA)/mw.m/pd/

sim. to Columbia/orig. (1912 on) from Rambouillet × Lincoln/HB 1951/[exhibited at *Panama-Pacific Exposition*, San Francisco, 1915]/extinct 1990–1993

Panjsher Gadik: (Afghanistan)/usu. black or brown/var. of Gadik

Pantanwadi: *see* Patanwadi

Parai: *see* Tirahi

Parasi: *see* White Karaman

Pardina: *see* Ripollesa

Parkanni: *see* Lohi

Parkent Mutton-Wool: (Uzbekistan)/ m.w/Russ. *Parkentskaya myaso-sherstnaya*

Parkhar: (Tajikistan)/improved var. of Hissar

Parnassian: *see* Greek Zackel

Pashka, Paška: *see* Pag Island

Patanwadi: (Mahsana, Kutch and Saurashtra, N Gujarat, India)/ cw.d.m/brown face and legs; pd; st; occ. tassels/syn. *Buti* (= tasselled), *Charotari, Cutchi* or *Kutchi, desi* or *deshi, Gujarati, Joria* (wool type), *Kathiawari, Vadhiyari*/not *Charothar, Charothri, Gujerati, Gujrati, Pantanwadi, Pattanwadi*/[from Patan]

Paterschaf: *see* Engadine Red

Pattanwadi: *see* Patanwadi

Paular: (Spain)/former strain of Spanish Merino/extinct

Pavullese: (Pavullo, Modena, Emilia, Italy)/m.w.d/var. of Garfagnina White with Bergamasca blood/ syn. *Appennino-Modenese, Balestra* (= crossbow, i.e. horn shape), *Modenese*/extinct 1976–1979

Peat: (Switzerland)/orig. of Bündner Oberland, Steinschaf/syn. *Turbary* (Ger. *Torfschaf*)/[remains found in neolithic lake dwellings]/extinct

Pechora: (Komi, Russia)/lw.m/breed group/orig. 1937–1950 from Romney × local Russian Northern Short-tailed, backcrossed twice to Romney and then bred *inter se*/Russ. *Pechorskaya porodnaya gruppa,*

Pechorskaya polutonkorunnaya (= semifinewooled)/rare

Pecora abissina: *see* Ethiopian Highland

Pecora bianca delle Montagne: *see* Tyrol Mountain

Pecora della Roccia, Pecora delle Rocce: *see* Steinschaf

Pecora del Sasso: *see* Steinschaf

Pecora somala a testa nera: *see* Somali

Pedi: *see* Bapedi

Pelagonia: *see* Florina

Pelibüey: (Cuba and Mexico, chiefly Gulf coast)/m/usu. tan, red, white or pied; pd; ♂ usu. mane and throat ruff/American Hair Sheep group/syn. *Carnero de pelo de buey, Cuban Hairy, Cubano Rojo* (= *Red Cuban*), *Peligüey, Tabasco* (Mexico)/[= ox-haired]

Pellagonia: *see* Florina

Pelo de Boi: *see* Brazilian Woolless

Pelona: *see* Red African

Pembroke Hill: (Wales)/hd/local var. of Welsh Mountain, with browner face/syn. *Prescelly Mountain*/ not *Precelly*/extinct in 1930s

Penistone: *see* Whitefaced Woodland

Péone: *see* Mourerous

Peppin: (Australia)/mw/mw strain of Australian Merino/developed 1840–1870 at Messrs Peppin's stud, Wanganella, NSW (and elsewhere), with English Longwool blood/syn. *Wanganella*

Perales: (Spain)/former strain of Spanish Merino/extinct

Perendale: (New Zealand)/m.mw/ pd/orig. 1938–1960 (chiefly since 1947) from Cheviot × New Zealand Romney/BS 1960, HB 1961; BS also Australia/[Prof. G. *Peren* of Massey Agric. College]

Permer: (Nigeria)/m/usu. pied, esp. white with black head; hy-w/ orig. in Göttingen, Germany, from Blackhead Persian × German Merino and transported to Ibadan/[*Persian-Merino*]

Persian: (S Africa)/smooth haired; fr/vars: Blackhead Persian, Red and Roan Persian, Redhead Persian, Skilder, Wooled Persian/Somalian or Saudi Arabian orig. from group of hy fr sheep shipwrecked 1869 off Cape coast with black heads, some progeny with red heads/ [sheep boarded in vicinity of Aden in *Persian* Gulf, hence name]

Persian fat-tailed: *see* Iran fat-tailed

Persian Lamb, Persian Lambrakhan: *see* Karakul

Persian Red: (S Africa) *see* Redhead Persian, Wooled Persian

Persian Thin-tailed: *see* Zel

Perthshire Blackface: (Scotland)/ strain of Scottish Blackface with coarser wool/syn. *Perth Blackface*

Perugian Lowland: (Italy)/Apennine type with much Bergamasca blood

Pešter, Pešterska: *see* Sjenica

Peštersko-sjenička: *see* Sjenica

Petite Manech: (France)/black, red or white with red and black spots, coloured head/var. of Manech/ syn. *Gorri tippia, Xaxi Ardia*/rare

Petit Oranais: *see* Beni Guil

Petrokhan Tsigai: (Bulgaria)/orig. from Mutton Merino and Tsigai, × Sofia White

Pettadale: (Shetland, Scotland)/m.w/ pd/orig. (1959 on) from Romney ($\frac{1}{4}$) × Shetland ($\frac{3}{4}$) on Kergood estate

Petzen: (Austria)/former var. of Carinthian/extinct

Peuhl, Peul, Peulh: *see* Fulani

Peul-Peul: (C Senegal and S Mauritania)/usu. white with black or red spots, or whole red; hd/Fulani group of Sahel type/ syn. *Peul, Poulfouli* (Mauritania)

Peul Voltaïque: (N Burkina Faso)/ white with some spotting/var. of Toronké

Phan Rang: (Ninh Thuan, Vietnam)/ m/rare

Pialdo: (Alentejo, Portugal)/black/ name coined for dominant black Merino/[*P*igmented *A*lentejo *d*ominant]

Picardy: (French Flanders)/var. of Flemish Marsh/Fr. *Picard*/extinct

Piebald: *see* Jacob

Pied: (W Africa) *see* Uda

Piedmont, Piedmont Alpine: *see* Biellese, Savoiarda

Pied Polisi: (C Albania)/d.cw.m/black or dark red spectacles; ♂ hd, ♀ usu. hd/var. of Albanian Zackel/ Alb. *Lara Polisi*

Piemontese alpina: *see* Biellese, Savoiarda

Pineywoods Sheep: *see* Gulf Coast Native

Pink-nosed Somerset: (England)/var. of Dorset Horn/syn. *Somerset Horn*/extinct

Pinzirita: (Sicily, Italy)/d.m.cw/black or brown marks on face and legs; ♂ hd, ♀ pd/orig. (with Barbary) of Sicilian Barbary, and (with Maltese) of Comisana/HB/syn. *Siciliana locale, Comune siciliana, nostrale* (= local)/not *Pinzerita, Pinzonita, Piperina, Piperita*/[= speckled]

Piperina, Piperita: *see* Pinzirita

Pirenaica: *see* Ripollesa

Pirlak: *see* Kamakuyruk

Pirot: (SE Serbia)/m.cw/white, black or grey with black spots on face and legs; ♂ hd, ♀ usu. pd/local Pramenka/Serbo-cro. *Pirotska*/ syn. *Eastern Serbian*/nearly extinct (being absorbed by Svrljig)

Pitt Island: (Chatham Is, New Zealand)/fw/usu. coloured; hd/ feral/orig. in 1840s from Saxony Merino from Hutt Valley/rare

Piva: (N Montenegro)/m.cw.d/usu. white with spotted head and legs, occ. black or grey/ Pramenka type/Serbo-cro. *Pivska*/syn. *Durmitor* (Serbo-cro. *Durmitorska*), *Jezero-Piva* (Serbo-cro. *Jezero-pivska*) (= *Lake Piva*)

plain sheep: *see* Ovče Polje

plain type sheep: *see* Shahabadi

Planinski tsigai: *see* Mountain Tsigai

Plateau central: *see* Central Plateau

plateaux de l'Est: *see* Beni Guil

Plateaux de l'Ouest: *see* Tadla

plenno-mięsna: *see* prolific meat

plenno-wełnista: *see* prolific wool

Pleven Blackhead: (N Bulgaria)/ d.cw/born black; ♂ hd, ♀

pd/Bulgarian Native improved by Tsigai/Bulg. *Chernoglava plevenska*, Ger. *Plewener*/syn. *Pleven Black-Face*/not *Plevin*, *Plevna*

Plezzana, Plezzo: *see* Bovec

plickus: *see* Xinjiang Finewool

Plovdiv Merino: *see* Thrace Finewool

Plovdiv-Purvomai: (Maritsa valley, Bulgaria)/lw.d.m/var. of White South Bulgarian/orig. from Tsigai × local Bulgarian with Merino blood/orig. (with Merino) of Thrace Finewool/Bulg. *Plovdivsko-P"rvomaĭska*/syn. *Plovdiv, Purvomai*/extinct

Podhale: (S Poland)/var. of Polish Zackel/Pol. *Podhalánska*

Pog: *see* Pag Island

Pogórza: *see* Polish Hill

Polderschaap: *see* Marsh

Polish Anglo-Merino: *see* Polish Corriedale

Polish Blackheaded: (NE Poland)/= German Blackheaded Mutton/ Pol. *Czarnogłowa*

Polish Blackheaded Mutton: (Poland)/m/pd/synthetic sire line orig. 1976–1990 at Poznań Agric. Univ. from Suffolk (90%), East Friesian (5%), and Wielkopolska and Polish Merino (5%)/Pol. *Czarnogłowa owca mięsna*/syn. *Blackheaded Synthetic*/rare

Polish Corriedale: (C Poland)/w.m/ pd/Polish Lowland group/orig. (1962 on) from Lincoln Longwool (35%) × Polish Merino/HB 1977/Pol. *Polski koridel*/obs. syn. *Polish Anglo-Merino* (Pol. *Anglomerynos*) (which also had Romney and Leine blood)/rare

Polish Heath: *see* Wrzosówka

Polish Highland: *see* Polish Mountain

Polish Hill: (Poland)/Polish Longwool group/Pol. *Ovca pogórza*

Polish Improved Mountain: *see* Polish Mountain

POLISH LONGWOOL: (Poland)/m.lw/ pd/orig. from East Friesian,

Leine, Texel and esp. Romney, × local (e.g. Świniarka)/inc. Bochnia, Kamieniec, Leszno, Olkusz, Polish Hill, Pomeranian, Silesian/Pol. *Długowełnista owca polska*

POLISH LOWLAND: (Poland)/m.mw/ pd/orig. from English Longwool, Polish Merino and local/inc. Polish Corriedale, Wielkopolska; also (extinct) Łowicz, Poznań, Poznań Corriedale; also Białystok and Lublin populations and Żelazna flock/Pol. *Polska nizinna*

Polish Merino: (Poland)/fw.m/pd/ orig. chiefly in late 19th and 20th c. from Précoce, named 1946, improved by Caucasian Merino 1952–1957/vars: Jędrzychowice Merino, Polish Strongwooled Merino/HB/Pol. *Merynos polska*/ syn. *Merynos cienkorunny* (= finewool), *Merynos wełnisto-mięsny* (= wool-mutton)

Polish Modified Dorset 59: (Poland)/ synthetic sire line/orig. at Poznań Agric. Univ. from Dorset Horn (31%), Texel (25%), Polish Merino (19%), Île-de-France (9%), Wielkopolska (6%), Berrichon du Cher (5%), East Friesian (5%)

Polish Mountain: (S Poland)/m.d.cw/ usu. white, occ. black, grey or pied; ♂ hd, ♀ pd/orig. from Polish Zackel improved by Transylvanian Zackel (imported 1911–1913 and 1935–1937) and East Friesian (imported 1925–1927 and after 1946)/HB/Pol. *Polska owca górska*/syn. *Cakiel-Fryz-Siedmiogrodzka* (= *Zackel-Friesian-Transylvanian*), *CFS*, *Polish Highland, Polish Improved Mountain*

Polish prolific lines: *see* prolific 05, prolific 09, prolific meat 08, prolific meat 10, prolific Merinofinn Mf-40, prolific Wielkopolska 10, prolific wool 04

Polish Strongwooled Merino: (Poland)/m.mw/var. of Polish Merino with Berrichon du Cher

and Île-de-France blood/Pol.
Merynos polski pogrubiony

Polish synthetics: *see* Polish
Blackheaded Mutton, Polish
Modified Dorset 59, Polish
prolific lines, Polish
Whiteheaded Meat 06, Polish
Whiteheaded Mutton

Polish Whiteheaded Meat 06:
(Poland)/synthetic sire line/orig.
at Poznań Agric. Univ. from
Texel (50%), Île-de-France
(18%), East Friesian (11%),
Berrichon du Cher (9%),
Wielkopolska (6%), Polish
Merino (6%)

Polish Whiteheaded Mutton:
(Poland)/m/pd/synthetic line for
crossing/orig. 1976–1990 at
Poznań Agric. Univ. from Texel
(53%), Île de France (19%), East
Friesian (11%), Berrichon du
Cher (6%), Wielkopolska (6%)
and Polish Merino (5%)/Pol.
Białogłowa owca mięsna/rare

Polish Zackel: (S Poland)/cw.d.pelt/
white, also black or brown; hd or
pd/vars: Podhale, Tatra, East
Carpathian/ orig. of Polish
Mountain/Pol. *Cakiel*

Polisi: *see* Pied Polisi

Polje: *see* Ovče Polje

Poll Dorset: (SE Australia)/pd/orig.
(1937 on) from Dorset Horn and
Ryeland or Corriedale/BS 1954;
also in England/syn. *Polled
Dorset* (USA)

Polled Rambouillet: (USA)/pd var. of
American Rambouillet

Poll Merino: (Australia)/pd var. of
Australian Merino/BS also in
Uruguay; HB also in New
Zealand

Polska nizinna: *see* Polish Lowland

Polska owca górska: *see* Polish
Mountain

**Polska owca nizinna w okręgu
białystockim:** *see* Białystok

**Polska owca nizinna w okręgu
lubelskim:** *see* Lublin

Polski koridel: *see* Polish Corriedale

Poltava Fur-Milch: (Ukraine)/sim. to
Chushka/inc. Reshetilovka,
Sokolki

polugrubosherstnaya: =
semicoarsewool (Russ.)

polutonkorunnaya: = semifinewooled
(Russ.)

Polwarth: (Colac, W Victoria,
Australia)/mw.m/orig. *c.* 1880
from Merino × (English
Longwool, usu. Lincoln, ×
Merino), i.e. Comeback with at
least 5 generations of *inter se*
breeding/hd and pd vars/BS; BS
also GB, Argentina (Ideal),
Uruguay/syn. *Ideal* (S America)

Polypay: (Idaho, USA)/m.w/pd/orig.
(1969 on) at US Sheep Exp. Sta.,
Dubois from (Targhee × Dorset)
× (Rambouillet × Finnish), bred
for 2 lambs twice a year/BS 1979;
HB also Canada

Pomarancina: (Pomarance, Pisa,
Italy)/m.d.cw/usu. pd/Apennine
group/rare

Pomeranian: (Pomorze, Poland)/
lw.m.(d)/white (earlier also
brown or black); pd/orig. Marsh
type, now Polish Longwool
group/orig. from Fagas (? =
Friesian Milk) in 18th c., East
Friesian and Wilstermarsch
blood 1929–1939, later Texel
blood/vars: Kashubian, Koszalin/
BS 1984, HB/Pol. *Owca
pomorska*

Pomeranian Coarsewool: (NE
Germany, chiefly Rügen I)/cw/
grey with black head, born black;
pd/HB/Ger. *Rauwolliges
Pommersches Landschaf*/syn.
Pomeranian Roughwoolled/
nearly extinct

Poogal: *see* Pugal

Poonchi: (Punch, SW Kashmir,
Pakistan and India)/cw.m/
sometimes black or brown head
and legs, or pied; ♂ hd, ♀ pd; st

Porakani: *see* Thalli

Porlock: *see* Exmoor Horn

Portland: (Dorset, England)/tan face,
lamb has red fleece; hd/HB 1975,
BS 1992/orig. (with Merino) of
Dorset Horn/rare

PORTUGUESE CHURRO: (NE Portugal)/
white or coloured, ♂ hd, ♀ hd or
pd/inc. Algarve Churro,

Bragança Galacian, Churra da Terra Quente, Churra do Campo, Miranda Galician, Mondegueira/ syn. *Bordaleiro Churro*/[= coarsewooled]

Portuguese Merino: (C Portugal)/ fw.d.m/orig. from Bordaleiro crossed with Spanish Merino (since 15th c.), Rambouillet (since 1903) and, predominantly, Précoce (since 1929)/vars: Beira Baixa, Black Merino, Fonte Bôa

Poulfouli: (Cameroon) *see* Kirdi

Poulfouli: (Mauritania) *see* Peul-Peul

Poznań: (W Poland)/Polish Lowland group/orig. (1948 on) from Romney ($\frac{9}{16}$), Merino ($\frac{1}{4}$), Leszno ($\frac{1}{8}$), Świniarka ($\frac{1}{16}$)/Pol. *Poznańska*/extinct

Poznań Corriedale: (Poland)/orig. (1962 on) from Poznań × [Poznań × (Romney × Polish Merino)]/orig. of Wielkopolska/ Pol. *Koridel poznański*/extinct

Poznań synthetics: *see* Polish prolific lines, Polish synthetics

PRAMENKA: (Bosnia, Croatia, Macedonia, Montenegro and Serbia)/m.cw.d/usu. white, also black, or white with black on head and legs; ♂ usu. hd, ♀ usu. pd/Zackel type/inc. Baljuša, Bardoka, Bosnian Mountain, Dalmatian-Karst, Island Pramenka, Istrian Milk, Karakachan, Kosovo, Krivovir, Lika, Lipe, Ovče Polje, Pirot, Piva, Šar Planina, Sjenica, Stogoš, Svrljig, Zeta Yellow/orig. (with Merino) of Dubrovnik, Pag Island/syn. *Yugoslav(ian) Zackel*/[Serbo-cro. *pramen* = staple or lock, hence *pramenka* = with open fleece]

Préalpes du Sud: (Drôme, SW Hautes-Alpes and NE Vaucluse, SE France)/d.(m.cw)/pd/named 1947, HB 1948, BS/obs. syn. *Levant, Quint, Sahune, St Nazaire, Savournon, Valdrôme*

Preanger: *see* Priangan

Precelly: *see* Pembroke Hill

Précoce: (France)/fw.m/♂ hd or pd (= Meusse), ♀ pd/orig. from Spanish Merino imported 1799–1811, selected for early maturity/HB 1929 by fusion of those of Champagne, Châtillonais and Soissonais Merinos; HB also in Portugal, Spain/orig. of German Mutton Merino, Polish Merino, Portuguese Merino/Fr. *Mérinos précoce*/rare in France

Prescelly Mountain: *see* Pembroke Hill

Priangan: (W Java, Indonesia)/ram fighting, m.(cw)/usu. black or pied, occ. grey or tan; ♂ hd, ♀ pd; often earless; st (fat at base)/ var. of Javanese Thin-tailed (? with Afrikander and Merino blood in 19th c.)/syn. *Garut*/not *Preanger, Prianger*

Priangarskiĭ Merinos: *see* Angara Merino

Priaral'skaya: *see* South Kazakh Merino

Priazov: *see* Azov Tsigai

Prijevorska: *see* Privor

Primorska: (Croatia) *see* Istrian Milk

Privor: (Bosnia)/usu. white muzzle and black face; ♂ usu. hd, ♀ usu. pd/var. of Bosnian Mountain/ subvar.: Donji Vakuf/Serbo-cro. *Privorska*/syn. *Prijevorska*

"Prolific": (N England)/m/orig. by G.L.H. Alderson, Haltwhistle, Northumberland, from Bluefaced Leicester, Poll Dorset and Lleyn/ extinct

prolific 05: (Poland)/orig. at Poznań Agric. Univ. from East Friesian ($\frac{5}{8}$) and Polish Merino ($\frac{3}{8}$)

prolific 09: (Poland)/mw/pd/ synthetic dam line, orig. 1976–1992 at Poznań Agric. Univ. from Polish Merino (44%), East Friesian (31%) and Finnish Landrace (25%)

Prolific Bábolna: (Bábolna, Hungary)/ orig. 1970s from Romanov (47%), Finnish Landrace (44%) and Hungarian Merino (9%)/obs. syn. *Tetra line*

Prolific Finn: (Poland) *see* Finnish Landrace

prolific meat 08: (Poland)/m.mw/ pd/synthetic dam line, orig. 1976–1992 at Poznań Agric.

Univ. from East Friesian (37%),
Texel (26%), Polish Merino
(24%), Île de France (9%) and
Berrichon du Cher (5%)/Pol.
plenno-mięsna 08

prolific meat 10: (Poland)/m.mw/pd/
synthetic dam line, orig.
1976–1992 at Poznań Agric.
Univ. from Poll Dorset and Dorset
Horn (31%), Texel (26%), Polish
Merino and Wielkopolska (24%),
Île de France (9%), East Friesian
(6%) and Berrichon du Cher
(4%)/Pol. *plenno-mięsna 10*

Prolific Merino: (Hungary)/orig.
1975–1979 from Romanov ×
Hungarian Merino ♀
backcrossed to Hungarian
Merino ♂/extinct/*see also*
Hungarian Prolific Merino,
Szapora Merino

prolific Merinofinn Mf-40: (Poland)/
synthetic dam line/orig. from
Polish Merino and Finnish
Landrace

prolific Wielkopolska 10: (Poland)/
synthetic dam line/orig. at
Poznań Agric. Univ. from
Wielkopolska (½), Polish Merino
(¼) and Romanov (¼)

prolific wool 04: (Poland)/mw/pd/
synthetic dam line, orig. 1976–
1990 at Poznań Agric. Univ. from
East Friesian (⅝) and Wielkopolska
(⅜)/Pol. *plenno-wełnista 04*

Prong Horn: *see* Zackel

**Prostaya derevenskaya dlinno-
toshchekhvostaya:** *see* Russian
Long-tailed

prosti: (Bulgaria) *see* Țurcana

Provence Merino: *see* Arles Merino

Pseudois nayaur: *see* bharal

Psiloris: (C Crete, Greece)/d.m.cw/
white with black spots on face; ♂
hd, ♀ usu. pd/island var. of
Greek Zackel/Gr. *Psiloritiana*/
syn. *Anogia*

puchia: *see* Pakistan fat-tailed

Pugal: (W Bikaner, Rajasthan, India)/
cw.m/black face; pd/not *Poogal*

Puglia, Pugliese: *see* Apulian

Pulfuli: *see* Fulani

Punch: *see* Poonchi

Purvomai: *see* Plovdiv-Purvomai

Pusterese: (Val Pusteria, Bolzano,
Italy)/m.cw/yellowish-white;
pd/Lop-eared Alpine group/orig.
since 1750 from Bergmasca ×
Steinschaf, with Lamon blood/
syn. *Badiota, Badertaler* (Ger.) or
Nobile di Badia (It.) (Val Badia),
Pustera gigante, Tedesca (=
German) *di Pusteria* or *Val di
Pusteria, Sextner, Tauferer* (Valle
Aurina)/rare

Pyrenean: (France) *see* Pyrenean
dairy breeds, Central Pyrenean

Pyrenean dairy breeds: (W Pyrenees,
France)/inc. Basco-Béarnais,
Manech/BS/Fr. *Races ovines
laitières des Pyrénées*

Pyrenean Semifinewool: *see*
Ansotana, Roncalesa

**Pyrénéen central, Pyrénées
centrales:** *see* Central Pyrenean

**Pyrénéen central à extrémités
charbonées:** *see* Tarasconnais

**Pyrénéen central à extrémités
rousses:** *see* Castillonnais

**Pyrénéen central à extrémités
tachetées:** *see* Tarasconnais

Pyrny: (W Ukraine)/♂ hd, ♀ hd
(short) or pd/var. of Voloshian
with shorter tail/not *Pyrnai*

Qashqai: (Fars, Iran)/cw.m/coloured
spots on head and legs; ♀ pd;
ft/local var./Fr. *Ghachgai*/not
Ghashgai

Qinghai Black Tibetan: (Qinghai,
China)/cw.m/black; pd/orig.
from Tibetan

Qinghai Semifinewool: (Qinghai,
China)/lw.m/♂ hd, ♀ pd/orig.
from Tsigai × (Xinjiang
Finewool × Tibetan) + Romney
× above

Qezel, Qizil: *see* Red Karaman

Quadrella: (Campania, Italy)/m.d/hd
or pd/formerly var. of Gentile di
Puglia/obs. syn. *Improved
Ariano, Spanish Ariano* (It.
Spagnola Arianese/rare

Quanglin Large-tail: (Shanxi,
China)/m.cw/♂ hd, ♀ pd;
sft/orig. from Mongolian

Quercy, Quercynois: *see* Caussenard
du Lot

Queue fine de l'Ouest: (Tunisia) *see* Algerian Arab

Quint: *see* Préalpes du Sud

Rab: *see* Island Pramenka

Rabada de la sierra de Loja: *see* Lojeña

Rabo gordo: *see* Brazilian Somali

Rabo Largo: (Bahia, NE Brazil)/m.cw-hy/white, red or red pied, or white with red head; hd; lft/? orig. from imported hy ft breed (Asian origin but shipped from S Africa mid-19th c.), × Crioulo/HB/[= broad tail]

Rabuda de la sierra de Loja: *see* Lojeña

Race côtière atlantique: *see* Atlantic Coast

Race d'Aure et de Campan: *see* Aure-Campan

Race des plateaux de l'Est: *see* Beni Guil

Race des Plateaux de l'Ouest: *see* Tadla

Race de Thônes et de Marthod: *see* Thônes-Marthod

Race du littoral nord de la Manche: *see* Cotentin

Race du littoral sud de la Manche: *see* Avranchin

Races des Pyrénées centrales: *see* Central Pyrenean

Races du Massif Central Nord: *see* Central Plateau

Races du Plateau central: *see* Central Plateau

Racka: (Hungary)/d.m.cw/white (with brown face and legs), or black; vertical corkscrew horns/Zackel type/BS 1983/Rom. *Raţca*/syn. *Hortobágy Racka, Hungarian Racka, Hungarian Zackel, Magyar juh* (= Hungarian sheep)/not *Radska, Rasko, Ratska, Ratzka*/[= Zackel; *Racz* = Serbian]/rare/*see also* Ţurcana

Rackulja: *see* Stogoš

Radmani: (S Yemen)/hy/white; pd; ft; often earless/? = Tihami/syn. *Sha'ra* (= hairy)

Radnor, Radnor Forest: *see* Hill Radnor

Radomir: (SW Bulgaria)/mottled head and feet/var. of Breznik

Radska: (Romania) *see* Racka

Raglan Peninsula: (New Zealand)/feral/orig. from New Zealand Romney

Ragusa, Ragusa-Šipan: *see* Dubrovnik

Rahmani: (Beheira, Lower Egypt)/cw.m/brown, fading with age; ♂ hd, ♀ usu. pd; often earless; ft/[from Rahmaniya]

Raimbi: (Algeria)/red-brown/sim. to Algerian Arab/not *Rembi, Rumbi*/[Arabic = red]

Raqole: *see* Rayole

Rajasthani: *see* Chokla, Jaisalmeri, Magra, Malpura, Marwari, Nali, Pugal, Sonadi

Rakhshani: (W Baluchistan, Pakistan)/m.cw.d/black or brown head or muzzle; ♂ hd, ♀ pd; long lop ears; ft/vars: Jhalawani, Sarawani/not *Rakshini*

Rakvere, Rakvereskaya: *see* Estonian Darkheaded

Rambouillet: (France)/fw.m/♂ hd, ♀ pd/orig. from 367 Spanish Merinos of many strains imported from Segovia in 1786 and bred only at Bergerie Nationale de Rambouillet near Paris (40 more imported 1801 had little influence)/orig. of American Rambouillet/HB/Fr. *Mérinos de Rambouillet*/rare

Ramliç: (Turkey)/m.fw/orig. since 1969 at Istanbul Univ. from American Rambouillet (65%) × Dağliç (35%)

Ramnad Karuvi: *see* Kilakarsal

Ramnad Red: *see* Kilakarsal

Ramnad White: (Ramanathapuram, Tamil Nadu, India)/m/usu. white, occ. pied; ♂ hd, ♀ pd; st/South India Hair type

Rampur Bushair: (Himachal Pradesh and N Uttar Pradesh, India)/cw.pa/white, often with tan face, or occ. pied or brown; ♂ hd, ♀ usu. pd; long lop ears; st/not *Rampur Bushahr, Rampur Bushier*

Randile: (E of L Turkana, Kenya)/various colours; hy/ft/? = Samburu var. of Masai

Rasa aragonese: *see* Aragonese

Rasa navarra: *see* Navarra

Rashaidi: (NE Eritrea)/cw/brown, white, red or pied; pd/? var. of Ethiopian Highland, or orig. from Yemen/It. *Rasciaida*

Rasko: *see* Racka

Raso: *see* Aragonese

Rastela: *see* Frabosana

Raţca, Ratska, Ratzka: *see* Racka

Rato Munda: *see* Chokla

Rauwolliges Pommersches Landschaf: *see* Pomeranian Coarsewool

Rava: (W Puy-de-Dôme, C France)/ m.cw/white with black spots on extremities, or black; pd/Central Plateau group/HB 1973/not *Ravas, Ravat*

Raymond Merino: (Dhule, Maharashtra, India)/fw/orig. 1973 on/inc. Australian Merino × Polwarth, Merino × Chokla, Merino-Polwarth × Chokla, and Merino-Chokla × Deccani

Rayole: (Cévennes, C France)/m.cw/ face and legs spotted; sometimes brown or grey; ♂ hd, ♀ hd or pd/Causses type, ? with Barbary blood/HB 1980, BS/syn. *Cévenol* (from Cévennes)/not *Raqole*/rare

Razlog: (SW Bulgaria)/mw/orig. 1955–1967 from Romney and Lincoln, × Merino

Recka: (Albania)/d.m.cw/usu. white; ♂ hd, ♀ pd

Red African: (N and C Colombia and W Venezuela)/m/yellow-red to dark red; pd; ♂ sometimes maned/American Hair Sheep group/Sp. *Roja africana*/syn. *Africana, Camura, Colombian Woolless, Pelona, West African*

Red and Roan Persian: (S Africa)/ var. of Persian/nearly extinct

Red Cuban: *see* Pelibüey

Red Engadine: *see* Engadine Red

Red-face Manech: (SW Basque country, France)/d.m.cw/red face and legs; ♂ hd, ♀ pd/var. of Manech/Basque *muthur gorria*

Red Head: (Albania) *see* Mati

Red Head: (France) *see* Rouge de l'Ouest

Red Head: (Germany) *see* Coburger

Red Head: (Sicily) *see* Comisana

Redheaded Maine: *see* Rouge de l'Ouest

Redhead Persian: (S Africa)/white with red head/cf. Blackhead Persian/var. of Persian

Red Karaman: (NE Turkey)/cw.m.d/ red or brown; ♂ usu. pd, ♀ pd/var.: Hemşin/Turk. *Kızıl-Karaman* (*kızıl* = red-brown), *Mor-Karaman* (*mor* = maroon)/ syn. *Dugli, Erzurum, Hamra* (Arab., = red) or *Shagra* (Fr. *Chacra, Chagra, Chakra*) (Syria); *Gesel, Gezel, Ghezel, Kazil, Khezel, Khizel, Kizil, Qezel* or *Qizil* (Iran); *Turkish Brown*/not *Erzerum, Moor, Morekaraman*/ *see also* White Karaman

Red Majorcan: (S Majorca, Balearic Is, Spain)/m.(d.mw)/white with red head and feet, born red; ♂ hd or pd, ♀ usu. pd; sft/? orig. from Barbary/Sp. *Roja mallorquina (de cola ancha)* (= with broad tail)/syn. *Borde* (= crossbred), *Coete* or *Cohete* (*coe* = tail), *Pigmentada mallorquina*/rare

Red Masai: *see* Masai

red sheep: (Middle East)/= *Ovis orientalis* Gmelin, *orientalis, gmelini* and *ophion* sections/ reddish with paler underparts; ♂ hd, ♀ pd; st/intermediate between urial and mouflon/type sp. is Elburz red; vars inc. Cyprus mouflon, Anatolian red, Armenian red, also subspp. in various parts of Iran/? part origin of domestic sheep

red sheep: (Azerbaijan) *see* Gyzyl-Gojun (= Mazekh in Armenia)

Red Woolless: (Brazil) *see* Brazilian Woolless

Rehamna-Sraghna: (Morocco)/var. of South Moroccan/not *Rehamma-Srarhna, Sghrana, Srarna*

Rembi: *see* Raimbi

Replenska, Replianska, Repljaner: *see* Replyan

Replyan: (NE Bulgaria)/m.cw.d/ mottled head and feet; ♂ usu. hd, ♀ pd/local unimproved var.

of Bulgarian Native/orig. (with Tsigai) of Mountain Tsigai/Bulg. *Replenska, Replyanska*, Ger. *Repljaner*/syn. *Belogradchik*/not *Replianska*

Reshetilovka: (Ukraine)/black; ♂ hd, ♀ hd or pd (20–25%)/orig. from Russian Long-tailed/Russ. *Reshetilovskaya*/extinct

rheinisch: *see* Leine

Rhiw Hill: (S Caernarvon, Wales)/ black or mottled face; pd/local var. sim. to Welsh Mountain/not *Rhuy*/extinct

Rhodes: (Dodecanese, Greece)/ d.m.cw/usu. white with black spots on face and legs; ♂ hd, ♀ pd; sft/? orig. from thin-tailed × fat-tailed cross/Gr. *Ródhos*, It. *Rodi*/syn. *Caramanitica*

Rhodesian: *see* Sabi

Rhodope: *see* Central Rodopi

Rhodope Tsigai: (C Rhodopes, Bulgaria)/Tsigai type endangered by recent land reforms; proposal to use purebred Tsigai nucleus flocks to create selected lines (milk yield and fertility) for pyramid breeding programme/ rare

Rhön: (C Germany)/w.m/black head; pd/recog. 1844, HB 1890/rare but increasing

Rhuy: *see* Rhiw Hill

Rideau Arcott: (Canada)/m/face white, tan or mottled; pd; prolific/orig. 1968–1988 (recog.) at Animal Res. Centre, Ottawa, from Finnish Landrace (40%), Suffolk (20%), East Friesian (14%), Shropshire (9%), Dorset (8%) and 4 other breeds (9%), crossed 1968–1974, selected for litter size and early lambing 1977–1988; for crossing with Canadian Arcott/HB 1988/obs. syn *ARC (dam) Strain 3*/rare

Rijnlam: (Netherlands)/breeding programme of Cofok Co./♀ line A (for use with Texel ♂): German Whiteheaded Mutton × Romanov; ♀ line B (for use with Texel or Suffolk ♂): German Mutton Merino × (Barbados Black Belly × Friesian Milk)

Rila Monastery: (SW Bulgaria)/d.w/ ♂ hd, ♀ pd/local Bulgarian Native improved by Tsigai/Bulg. *Rilomonastirska* or *Rilskiĭ Monastir*, Ger. *Rilokloster*/syn. *Rila*/extinct

Rilokloster: *see* Rila Monastery

Rilo-Rodopska: *see* Central Rodopi

Ripollesa: (Gerona, Catalonia, Spain)/m.(mw)/white with pigmented face and legs; ♂ hd, ♀ usu. hd; lt/? orig. from Merino × Tarasconnais/syn. *Bergerá, Bergueda, Caralpina* (from Caralp), *Montañesa, Pardina, Pirenaica, Sardana, Solsonenca, Vicatana* or *Vigatana* (from Vich)

Roaschia, Roaschina: *see* Frabosana

Rocce, Roccia: *see* Steinschaf

Rocha: *see* Guirra

Rocky: *see* Keerie

Rocky Mountain: *see* bighorn

Ródhos: *see* Rhodes

Rodi: *see* Rhodes

Rodopa, Rodopi: *see* Central Rodopi

Rodrigaise: (Rodrigues I, Mauritius)/ hy/white to light brown; small

Rogaland: *see* Rygja

Roja africana: *see* Red African

Roja levantina: *see* Guirra

Roja mallorquina: *see* Red Majorcan

Romanian Karakul: (Romania)/ Karakul imported in 19th c.

Romanian Merino: *see* Danube Merino, Palas Merino, Transylvanian Merino

Romanian Nomad: *see* Karakachan

Romanian Zackel: *see* Țurcana

Romanov: (Yaroslavl, Russia)/ pelt.cw/grey with black head and legs and usu. white face stripe and feet; hd or pd; prolific/orig. in late 17th c. from Russian Northern Short-tailed/HB; BS France (HB 1963), Canada 1986, USA/Russ. *Romanovskaya*

Romashkov, Romashkovski: *see* Volgograd

Romeldale: (California, USA)/mw.m/ orig. (1915 on) by A.T. Spencer from New Zealand Romney × American Rambouillet/coloured var.: Californian Variegated Mutant/HB

Romnelet: (Alberta, Canada)/mw.m/ orig. 1935–1947 from Romney × Rambouillet

Romney: (Kent, S England)/lw.m/pd/ English Longwool type/orig. from Old Romney Marsh/var.: New Zealand Romney/part orig. of Kuibyshev, Ostrogozhsk, Pechora/BS 1895, HB; BS also USA 1911, Australia, Argentina, Uruguay; HB also Canada, Poland, South Africa, Falkland Is, Czech Rep./syn. *Kent, Kent or Romney Marsh, New Kent, Romney Marsh*

Romney-Corriedale: (New Zealand)/ New Zealand Romney × Corriedale, F$_1$

Romney Halfbred: (England)/North Country Cheviot × Romney, 1st cross

Romsdown: (Tasmania, Australia)/ orig. from Southdown × Romney/extinct

Romshire: (Victoria, Australia)/m/ Wiltshire Horn × Romney, F$_1$

Ronaldsay, Ronaldshay: *see* North Ronaldsay

Ronas Hill, Rona's Hill: *see* Multihorned Shetland

Roncalesa: (Roncal and Salazar valleys, NE Navarre, Spain)/var. of Aragonese with longer wool and semi-open fleece/syn. *Churra navarra, Entrefina pirenaica* (with Ansotana), *Salazenca* (in Salazar valley)/not *Salacenca*

Ronderib Afrikander: (NC Cape prov., S Africa)/m/white; hy-cw; ♂ usu. hd, ♀ hd or pd; sft/var. of Afrikander/orig. from Cape var. of Hottentot/former vars: Blinkhaar, Steekhaar/BS 1937–?/ Afrik. *Ronderib Afrikaner*/[= round rib, i.e. oval in cross-section, not flat like Merino]/rare (being displaced by Dorper and Merino)

Ronderib Merino: (S Africa)

Roquefort breeds: (S France)/i.e. milk used for Roquefort cheese/inc. Larzac (orig. type), Lacaune, Caussenard de la Lozère; also used: Basco-Béarnais, Corsican, Manech

Roscommon: (W Ireland)/lw.m/ pd/Leicester Longwool blood/ absorbed by Galway/BS 1895–1926/ syn. *Irish Longwool, Roscommon Longwool*/extinct *c.* 1977

Rosset: (Grisanche, Rhêmes and Savaranche valleys, Aosta, Italy)/ m.cw/yellowish-white with red-brown spots around eyes and on legs; ♂ usu. pd, ♀ pd/sim. to Savoiarda/nearly extinct

Rotxa: *see* Guirra

Rouge de Guillaumes: *see* Mourerous

Rouge de la Hague: *see* Roussin de la Hague

Rouge de l'Ouest: (Maine-et-Loire, NW France)/m.sw/wine-red face; pd/sim. to Bleu du Maine (same orig.)/named 1963; HB 1968; BS also GB 1986/syn. *Tête rouge du Maine* (= *Redheaded Maine*)/[= *Western Red*]

Rouge de Roussillon: *see* Roussillon Red

Rouge du Littoral: *see* Roussillon Red

Rough Fell: (NW England)/cw/hd/ Blackfaced Mountain type/BS 1926, HB 1927/syn. *Kendal Rough, Kentmere, Rough*/very local

Roumloukion: (C Macedonia, Greece)/m.d.cw/usu. white, occ. with speckled or black head, or all black or brown, or pied; ♂ hd, ♀ pd/Ruda type

Roundrib: *see* Ronderib

Roussillon Merino: (E Pyrenees, France)/orig. in late 18th c. from Spanish Merino/extinct *c.* 1940 (by crossing with Central Pyrenean)

Roussillon Red: (coast of Pyrénées-Orientales, France)/m.mw/ yellowish-white with red or pied head and legs and red kemp under neck; pd/orig. from local with Merino and Barbary blood/ also in Germany/Fr. *Rouge du Roussillon*/syn. *Rouge du Littoral*/nearly extinct

Roussin de la Hague: (N Manche, France)/brown face and legs,

born brown; pd/orig. early 20th
c. from Leicester Longwool ×
Southdown, with Suffolk and
Avranchin blood in 1960/orig.
(with Leicester Longwool) of
Cotentin/HB 1983, BS 1978; BS
also GB/syn. *Berca, Rouge de la
Hague, Roussin*/rare

Roux-de-Bagnes: (Val de Bagnes,
Valais, Switzerland)/red-brown;
usu. pd/part orig. of Swiss Black-
Brown Mountain/syn. *Bagnes,
Bagnerschaf, Roux-du-Pays*/?
extinct

Roux du Valais: (Upper Valais,
Switzerland)/m.(cw)/red-brown;
hd; semi-lop ears/var.:
Lötschental/HB/Ger. *Walliser
Landschaf*/syn. *Valais Red*/
nearly extinct

Roya: *see* Bilbilitana, Guirra

Roya bilbilitana: *see* Bilbilitana

Rubia de El Molar: *see* Somosierra
Blond

Rubia de Somosierra: *see* Somosierra
Blond

Rubia Serrana: *see* Somosierra Blond

RUDA: (Balkans)/wool type, more
uniform and slightly less coarse
than Zackel or Pramenka (due to
Tsigai blood)/inc. Ruda types in
Albania and Greece (Chalkidiki,
Katafigion, Luman,
Roumloukion, Serrai, Thrace),
Kivircik (Turkey), Ruda
Pramenka vars in former
Yugoslavia (Ovće Polje, Šar
Planina, Sjenica, Svrljig); ? also
some Bulgarian breeds
(Karnobat, Rila Monastery,
Shumen, White Klementina,
White South Bulgarian)/[= curly,
wavy]

Ruda: (Montenegro) *see* Dubrovnik

Rudavi: (Bulgaria)/soft-wool var. of
Karnobat/cf. Ruda

Rumbi: *see* Raimbi

Ruşetu 1: (Romania)/mw.m/orig. at
Ruşetu Exp. Sta. from (Romney
× Tsigai) × (Corriedale × Tsigai)

Russian Long-tailed: (S European
Russia)/m.cw/black or white; ♂
hd, ♀ pd/vars: Bityug, Bokino,
Cherkassy/orig. of Chushka,

Kuchugury, Mikhnov,
Reshetilovka, Sokolki/Russ.
*Prostaya derevenskaya dlinno-
toshchekhvostaya* (= Common
village long-thin-tailed)/syn.
*Common Russian, Common
Long-tailed*/extinct

Russian Longwool: (Voronezh and
Kalinin, Russia)/m.lw/pd/orig.
from Lincoln × local
coarsewools/vars: Kalinin, Liski
(inc. Nizhnedevitsk)/recog. 1978/
Russ. *Russkaya dlinnosherstnaya*

Russian Merinos: *see* Altai,
Askanian, Caucasian, Grozny,
Krasnoyarsk Finewool, Russian
Mountain Merino, Salsk, South
Ural, Soviet Merino, Stavropol,
Transbaikal Finewool, Volgograd,
Vyatka

Russian Mountain Merino: (N
Caucasus, Russia)/fw/orig. from
mouflon × Merino backcrossed
to Merino ♀ ♀/Russ. *Gornyĭ
merinos*

Russian Northern Short-tailed: (N
European Russia)/pelt.cw/black,
grey, white or pied; ♂ usu. hd, ♀
usu. pd/var.: Nolinsk/orig. of
Romanov/Russ. *Severnaya
korotkokhvostaya*/extinct

Russian Perseair: *see* Wooled Persian

Russian Red Wooled Persian: *see*
Wooled Persian

Russkaya dlinnosherstnaya: *see*
Russian Longwool

Ruthenian Zackel: *see* Carpathian
Mountain

Ruthenois: (S France)/extinct
(absorbed by Lacaune)

Rwanda-Burundi: (Burundi, Rwanda
and E Kivu, DR Congo)/m/usu.
black pied/African Long-fat-
tailed group, sim. to Tanzania
Longtailed/*see also* Burundi

Rya: (C Sweden)/m.cw/usu. white,
occ. black, grey or brown; long
wavy shiny 'rya' wool; pd/var. of
Swedish Landrace/BS 1979/Swe.
Ryafår/syn. *Swedish Carpet
Wool sheep*/[= carpet]/rare

Ryeland: (England)/sw.m/pd/recog.
18th c./BS 1909, HB; HB also
New Zealand 1925, Australia/

syn. *Hereford*/[= sandy rye-growing land in S Hereford]

Rygja: (Rogaland, Norway)/m.sw/face and legs sometimes coloured; pd/orig. (1850 on) from Cheviot × Old Norwegian with Leicester Longwool or Oxford Down blood/named 1924, HB 1926/[= from Rogaland]

Saane: (Denmark)/rare

Saanen: (Switzerland)/sim. to Simmental/part orig. of Swiss Black-Brown Mountain/Fr. *Gessenay*/syn. *Saanerland*/extinct

Sabi: (S Zimbabwe)/m/usu. fawn, brown or red, also black or pied; hy; ♂ hd or pd, ♀ usu. pd/ African Long-fat-tailed group/ inc. Mashona/obs. syn. *Rhodesian*/rare (indiscrimate crossing with Dorper and Wiltshire)

Sächsisches Elektoralschaf: *see* Saxony Merino

Sacjol: (Mexico)/black var. of Chiapas with white spot on head/? orig. from black Manchega, or possibly from black Castilian or Canary Island

Sackni: (Saudi Arabia)

saddleback sheep: *see* Fannin sheep

Sahabadi: *see* Shahabadi

SAHEL type: (N of W Africa)/m/usu. white or pied; hy; ♂ hd (long twisted), ♀ usu. pd; lop ears; tassels common/inc. Fulani breeds, Maure breeds, Tuareg/ syn. *Guinea Long-legged, Sahelian, West African Long-legged*

Sahune: *see* Préalpes du Sud

Saidi: (Asyut, Upper Egypt)/cw.m/ black or brown; pd; lft/var.: Sanabawi/syn. *Sohagi*/[= valley, i.e. of Nile]

St Croix: (USA and Canada)/m/ white, occ. brown; hy; pd/orig. from Virgin Islands White, imported to mainland USA (Utah State Univ.) 1975/BS 1988/rare

St Croix Blackbelly-Barbados Cross: (N Sumatra, Indonesia)/m/cross of St Croix with Barbados Black

Belly/imported from USA 1994/rare

St Elizabeth: (Jamaica)/pd/? orig. from imported African Hair and European breeds/rare

St Gallen: *see* Wildhaus

St Gironnais, St Girons: *see* Castillonnais

St Jean de Maurienne: *see* Thônes-Marthod

St Kilda: *see* Hebridean

St Nazaire: *see* Préalpes du Sud

St Quentin: (French Flanders)/var. of Flemish Marsh/extinct

St Rona's Hill: *see* Multihorned Shetland

Sakar: (Bulgaria)/m.cw.d/occ. colour around eyes; ♂ hd; lt/Bulg. *Sakarska*

Sakiz: (Izmir, Turkey)/d.cw.m/white with black spots around mouth and eyes and on ears and legs; ♂ hd, ♀ usu. pd; lt with fat at base/ orig. from Chios/Turk. *Sakız*/syn. *Çeşme*/not *Sakis, Sakkes*

Saku-Bash: (Xinjiang, China)/sim. to Chapan/subvar. of Kargilik var. of Hetian

Salacenca: *see* Roncalesa

Salamali: *see* White Karaman

Salazenca: *see* Roncalesa

Saloia: (Lisbon, Portugal)/d.mw.m/ usu. white with pale brown head and legs, occ. brown; ♂ hd, ♀ pd or hd/Bordaleiro type/HB 1988

Salsk: (Rostov, N Caucasus, Russia)/ fw/orig. 1932–1949 from American Rambouillet × Novocaucasian Merino/HB/Russ. *Sal'skaya, Sal'skaya poroda tonkorunnykh ovets* (= Salsk breed of finewool sheep)/not *Salsky*

Saltasassi: (N Novara, Piedmont, Italy)/m.(cw)/pd/Lop-eared Alpine group/rare

Salt range: *see* Lati

Salz: (Ebro valley, Spain)/d.m.mw/ pd/orig. since late 1970s from Romanov × Aragonese

Salzburg Steinschaf: *see* Steinschaf

Sambre-et-Meuse: (S Belgium)/m.w/ grey-speckled head and legs; pd/Flem. *Samber en Maas*/syn. *Entre-Sambre-et-Meuse*/extinct *c*.

1950; name now used for nearly extinct local breed with English blood in Melsbroek dist.

Sambucana: (Sambuco to Demonte, SW Cuneo, Italy)/m.mw/♂ usu. pd, ♀ pd/sim. to Garessina but larger/HB/syn. *Demontina*/ nearly extinct (from crossbreeding with Biellese)

Samburu: (Mali) ? *see* Banamba

Samburu: (N Kenya)/variable type and colour/var. of Masai

Samhoor: (Iran)/var. of Sanjabi/syn. *Jomoor*

Sampedrana: *see* Chamarita

Sampeirina: (Sampeyre, Piedmont, Italy)/rare

Samur: *see* Tabasaran

SA Mutton Merino: *see* South African Mutton Merino

Sana'a White: (NE and N of Ṣanʻā, Yemen)/m.cw/pd; ft

Sanabawi: (Sanbo, Upper Egypt)/red, with red, black or white head; sometimes hd/var. of Saidi with smaller tail

Sandjabi: *see* Sanjabi

Sandyno: (Nilgiris, W Tamil Nadu, India)/fw.m/orig. since 1973 at Sheep Breeding Res. Sta., Sandynallah, from Merino (⅝) × Nilgiri

Såne: (Denmark)/m.w/born black, turning brown or grey-brown/ orig. from Shropshire, Karakul and Rygja/recog. 1991/Dan. *Sånefår*

Sangamneri: (Ahmedhagar, Maharashtra, India)/strain of Deccani

Sangesari: (N and E of Tehran, Iran)/ m.cw/brown, also black; pd; ft; small/Kurdi type/not *Sangsar*, *Sangsari*

Sanjabi: (Kermanshah, Iranian Kurdistan)/cw.m/brown face, feet and, sometimes, tail; pd; lft/Kurdi type/vars: Calhoor, Samhoor/not *Sandjabi*, *Sinjabi*

Sanjiang: (Qinghai-Tibet plateau, China)/forest var. of Tibetan/[= 3 rivers]

Santa Cruz Island: (California, USA)/cw/feral on Santa Cruz I.

since *c.* 1920/? orig. from Merino, Rambouillet and Navajo-Churro/HB for mainland flocks/ nearly extinct

Santa Inês: (Bahia, NE Brazil)/m/ white, red, black or pied; hy; pd; lop ears/American Hair Sheep group/orig. from Bergamasca × Brazilian Woolless since late 1940s/BSd/syn. (white var.) *Pelo de Boi de Bahia* (= *Bahia Ox-haired*)

Saracatsanica: *see* Karakachan

Saradgy, Saradzhinskaya, Saraja: *see* Sary-Ja

Sarakatzan: *see* Karakachan

Sarawani: (Pakistan)/var. of Rakhshani

Sarda: *see* Sardinian

Sardana: *see* Ripollesa

Sardarsamand: (Rajasthan, India)/ orig. (1935 on) from Australian Merino × Marwari

Sardi: (C Plateau, Morocco)/m.cw/ black around eyes and on nose; hd/orig. from Beni Meskine

Sardinian: (Sardinia, Italy; also in France)/d.(m.cw)/♂ occ. hd, ♀ pd/former vars: large lowland (white, pd, with Merino and Barbary blood, syn. *Cagliari*, *Campidano*), small mountain (occ. black, ♂ hd, ♀ often hd)/ orig. of Tunisian milk sheep/HB 1927/It. *Sarda*

Sardinian mouflon: *see* mouflon

Šar Planina: (W Macedonia)/ m.cw.d/♂ hd, ♀ pd/Pramenka type/cf. Luma/Serbo-cro. *Šarplaninska*/syn. *Šar Mountain*/not *Schar Planina*, *Sharplanian*

Sartuul: (Dzavhan prov., W Mongolia)/var. of Mongolian

Sary-Ja: (SE Turkmenistan)/cw.m/ grey; pd; fr/improved by Degeres since 1950/var.: Ashkhabad/part orig. of Alai, Kargalin, Tajik/ HB/Russ. *Saradzhinskaya*/not *Saradgy*, *Saraja*

Sasi Ardi: (Guipúzcoa, Basque provs and Navarre, Spain)/cw/blond; pd; small

Sasso: *see* Steinschaf

Sataya: (Mongolia)/? = Sutai

Savoiarda: (W Turin, Piedmont, Italy)/d.m.cw/black spots on face and legs; ♂ hd, ♀ hd or pd; semi-lop ears/sim. to Rosset; cf. Thônes-Marthod/syn. *Cuorgné, di Torino, Piemontese alpina, Razza delle Alpi*/[from Savoy]/ nearly extinct (being crossed with Biellese)

Savournon: *see* Préalpes du Sud

Savoyard: *see* Thônes-Marthod

Saxon Merino: (Australia) *see* Tasmanian Merino

Saxony Merino: (Germany)/fw/orig. from Escurial strain (with some Negretti and Infantado strains) of Spanish Merino imported 1765–1815/part orig. of Australian Merino/Ger. *Sächsisches Elektoralschaf*/syn. *Electoral Merino*/extinct

Sayaguesa: (Sayago, Zamora, Spain)/ m.cw/small var. of Spanish Churro

Scandinavian Short-tail: *see* Northern Short-tailed

Schar Planina: *see* Šar Planina

Schnalser: *see* Tyrol Mountain, Val d'Ultimo

Schoonebeker: (Schoonebeek, SE Drenthe, Netherlands)/usu. white with red or grey on face and legs, also black, brown or pied; pd/var. of Dutch Heath/ orig. from Drenthe and Veluwe/ HB 1990, BS/syn. *Schoonebeker Heideschapp*/not *Schoonebecker*/ rare

Schumen, Schumener: *see* Shumen

Schwarzbraunes Bergschaf: *see* Swiss Black-Brown Mountain

Schwarzes Milchschaf: *see* Black Milk sheep

Schwyz: (Switzerland)/part orig. of Swiss White Mountain/syn. *Uri* (Ger. *Urner Landschaf*)/extinct

Sciara: (Calabria, Italy)/d.(m.cw)/ usu. pd/Moscia type/var.: Urbascia/syn. *Calabrese* (*Calabrian*), *Moscia calabrese*/? extinct

Scimenzana: *see* Akele Guzai

Sciucria: *see* Baraka

Scopelos: *see* Skopelos

Scotch: *see* also Scottish

Scotch Halfbred: (Scotland, England)/m.mw/pd/Border Leicester × Cheviot, 1st cross/ BS/syn. *Baumshire, North, North Country, North Halfbred, Scottish Halfbred*

Scotch Horn: *see* Scottish Blackface

Scotch Mule: (Scotland)/mottled brown face; pd/Bluefaced Leicester × Scottish Blackface, F₁/BS 1986

Scottish Blackface: (Scotland and N England)/cw.m/black or pied face; hd/Blackfaced Mountain type/strains inc. Galloway and Island (finer wool), Lanark and Perthshire (coarser wool), Northumberland; var.: Boreray/ BS *c.* 1890; BS also USA 1907, Argentina 1925–1950, 1989 on/syn. *Blackface, Blackfaced Highland, Kerry* (Ireland), *Linton, Scottish Mountain, Scotch Blackface, Scotch Horn*

Scottish Greyface: (S Scotland and N England)/cw.m/speckled face/ Border Leicester × Scottish Blackface (or Swaledale), 1st cross/syn. *Cross, Greyface*

Scottish Halfbred: *see* Scotch Halfbred

Scottish Masham: (Scotland)/ Teeswater (or Wensleydale) × Scottish Blackface, 1st cross

Scottish Soft-wool: *see* Tanface

Screwhorn: (SE Europe)/type with vertical corkscrew (spiral) horns (♂), now seen in Zackel/orig. named *Walachian* (1780, 1865), altered to *Zackel* (1888)

Scutari: *see* Shkodra

Seeboden, Seebodner: *see* Solčava

Seeland: *see* Solčava

Seeländer: (Austria)/pre-1918 local var. (or syn.) of Carinthian/ extinct

Seeland-Sulzbach: *see* Solčava

Ségala: (Causses de Rodez, Aveyron, S France)/d/Caussenard type; Roquefort breed/syn. *Ségala-Levézou*/extinct (absorbed by Lacaune)

Segezia Triple Cross: (Foggia, Apulia, Italy)/m.d.mw/pd/recent orig. from Merinolandschaf × (Île-de-France × Gentile di Puglia)/It. *Trimeticcia de Segezia*/rare

Segureña: (Segura mts and R valley, SE Spain)/m.d.mw/white, occ. blond spots on face and legs (*rubisca*) or all-brown (*mora*); pd/Entrefino type/orig. from Manchega/var.: Marquesado/BS, HB 1982/syn. (white only) *Cabreña* (Almería and Castellón), *Paloma* (Granada)

Selay ardiga: *see* Vasca Carranzana

Semarang: (C Java, Indonesia)/ m.(cw)/white, ♂ hd, ♀ pd/var. of Javanese Thin-tailed

Semitic Fat-tailed: *see* Near East Fat-tailed

Senales: *see* Tyrol Mountain, Val d'Ultimo

Senese: (Siena, Tuscany, Italy)/var. of Apennine/syn. *Senese delle Crete*

Sennybridge Cheviot: *see* Brecknock Hill Cheviot

Serbian (Zackel): *see* Baljuša, Bardoka, Kosovo, Krivovir, Lipe, Pirot, Sjenica, Stogoš, Svrljig

Serena Merino: (Badajoz, Spain)/var. of non-migratory Spanish Merino/Sp. *Merino de la Serena*/ not *Serrena*

Seres, Seris: *see* Serrai

Serra da Estrela: (NC Portugal)/ d.mw.m/white usu. with brown spots on head and legs, or black; hd/Bordaleiro type/HB 1986/not *Serra da Estrêla*

Serrai: (NE Macedonia, Greece)/ d.m.cw/usu. white with black marks on head and legs; ♂ hd, ♀ usu. pd/Ruda type/not *Seres, Seris, Serres*

Serrana: *see* Iberian (esp. Ojalada), Spanish Mountain Merino

Serranet: *see* Ojalada

Serrena: *see* Serena Merino

Serres: *see* Serrai

Setswana: *see* Tswana

Severnaya korotkokhvostaya: *see* Russian Northern Short-tailed

Severokavkazskaya: *see* North Caucasus

Severokazakhskiĭ merinos: *see* North Kazakh Merino

Severtzov wild: (Nuratau mts, Russia)/= arkhar

Sevillana: *see* Montesina

Sextner: *see* Pusterese

Sfakia: (W Crete, Greece)/d.m.cw/ usu. white with black spectacles; ♂ hd or pd, ♀ usu. pd/island var. of Greek Zackel/Gr. *Sfakiana*/ syn. *Adromalicha*/not *Sphakia*

Sghrana: *see* Rehamna-Sraghna

sha: *see* shapo

Shaanbei: (China)/fw

Shafali: (C Iraq and Syria)/red or black/arable var. of Awassi/cf. Arabi (Iraq)/Fr. *Chaffal, Chevali, Choufalié*/syn. *Delaimi, Delimi, Dilem, Dillène, Douleimi, Dulaimi*/not *Ashfal, Shaffal, Shevali*

Shaffal: *see* Shafali

shagra: (Syria) *see* redfaced Awassi, Red Karaman

Shahabadi: (Bihar, India)/cw/white or grey, sometimes black spots (e.g. on face); pd/syn. *'plain type sheep'*/not *Sahabadi, Shahbadi*

Shal: (Qazvin, Iran)/m.cw.d/black, grey or brown/sim. to Mehraban/ not *Chall, Shahl, Shall*

sham: *see* shapo

Shambali: (E Kordofan, Sudan)/many colours/var. of Sudan Desert

Shami: (Syria) *see* Awassi

Shandong: *see* Han

Shanghai: *see* Hu

Shangu: (Qinghai-Tibet plateau, China)/var. of Tibetan/[= valley]

Shantung: *see* Han

Shanxi Finewool: (China)/fw/orig. since 1920s from Merino × Mongolian/WG *Shansi*/syn. *Shanxi Merino*

shapo: (Ladakh and Astor, Kashmir)/= *Ovis vignei vignei* Blyth/var. of urial/syn. *O. orientalis vignei, Astor urial, Ladakh urial, urin*; ♂ is *sha*, ♀ is *sham*/not *shapu*/[Tibetan *sha-pho* = wild sheep]/rare

Sha'ra: *see* Radmani

Sharka: (Czech Republic)/fw.m/orig. from German Mutton Merino ×

Romney with some Askanian, Caucasian and Stavropol blood

Sharplanian: *see* Šar Planina

Shekhawati, Sherawati: *see* Chokla

Shetland: (Scotland)/mw.m/white, brown, occ. black, grey or piebald/Northern Short-tailed type/vars: improved (usu. white; mw; ♂ hd, ♀ pd); moorit (brown; hy-w; hd); Multihorned Shetland/orig. (with English Longwool) of Pettadale and (with Soay) of Castlemilk Moorit/BS 1927, reformed 1985; HB USA 1981, BS 1991/declining by crossing

Shetland-Cheviot: (N Scotland)/face white or spotted/North Country Cheviot × Shetland, 1st cross/BS

Shevali: *see* Shafali

Shilluk: *see* Nilotic

Shimenzana: *see* Akeke Guzai

Shinwari: *see* Baluchi

Shirazi: (Iran) *see* Grey Shirazi

Shirazi: (Russia)/grey (lethal when homozygous)/var. of Karakul/not *Shiraz*

Shirvan: (E and C Azerbaijan)/usu. off-white, also brown, black or pied/Caucasian Fat-tailed type, sim. to Karabakh/var.: Gala/Russ. *Shirvanskaya*/purebreds rare (being crossed with Azerbaijan Mountain Merino)

Shkodra: (NW Albania)/cw.d.m/ yellow, with light brown face and legs; ♂ hd, ♀ pd/var. of Albanian Zackel/= Zeta Yellow (Montenegro)/Alb. *Shkodrane*, It. *Scutari*/unshorn ♂ is *Ogiç*/not *Skutari*

Shoa: *see* Menz

short-tailed: (Europe) *see* Heath (some), Marsh, Northern Short-tailed

short-tailed: (Asia) *see* Erek, Tibetan, Nepalese; also many Pakistan breeds (e.g. Baltistani, Bhadarwah, Buchi, Damani, Kachhi, Kaghani, Kooka, Lohi, Poonchi, Thalli), Indian breeds (e.g. Bellary, Chotanagpuri, Coimbatore, Deccani, Gurez, Karnah, Patanwadi, Rampur

Bushair) and most South India Hair breeds

Shouyang: (Shanxi, China)/cw/♂ hd, ♀ pd; ft/local var. of Mongolian orig., sim. to Taiku but smaller and with heavier ft/not *Showyang*/extinct

Shropshire: (England)/sw.m/black-brown face and legs; pd/Down type/orig. early 19th c. from Southdown × heath sheep of Cannock Chase, Longmynd, Morfe Common *et al.*/orig. of Estonian Darkheaded/named 1848, BS 1882, HB 1883; BS also USA 1884 (HB 1889); HB also Australia, New Zealand, Canada/ rare in GB

Shugur: (N Sudan)/sandy to red/ commonest var. of Sudan Desert/ syn. *Ashgur*/not *Shugor*/[Arabic = fawn]

Shukria: *see* Baraka

Shumen: (NE Bulgaria)/m.w.d/ coloured; ♂ hd, ♀ pd/? orig. from Tsigai/Bulg. *Mednochervena shumenska* (= copper-red Shumen), Ger. *Schumener*/not *Schumen*, *Shuman*/rare

Shumen Finewool: (Bulgaria)/type of North-East Bulgarian Finewool/ Bulg. *Shumenska t"nkorunna*

Siamese Longtail: *see* Thai Longtail

Siberian: (S Siberia, Russia)/cw/ black; sft/local vars: Minusinsk, Tuva *et al.*; *see also* Buryat, Kulunda, Telengit/Russ. *Sibirskaya*/extinct

Siberian Merino: (SW Siberia)/var. of Soviet Merino/orig. in early 20th c. from Mazaev and Novocaucasian Merinos × local coarsewooled (Siberian and Kulunda) and Kazakh Fat-rumped/Russ. *Sibirskiĭ Merinos*/syn. *Siberian Soviet Merino*

Siberian Rambouillet: *see* Altai

Siberian type of Soviet Mutton-Wool: (Siberia, Russia)/m.lw/orig. from Lincoln and Romney, × local

Sibiriskaya: *see* Siberian

Sichuan Semifinewool: (Sichuan, China)/w.m/orig. since 1970s

from Border Leicester and
Romney, × [Border Leicester ×
(Xinjiang Finewool × local
Tibetan)]

Sicilian, Siciliana locale: *see*
Pinzirita

Sicilian Barbary: (C Sicily, Italy)/cw-
mw.m.d/usu. dark spots on face
and legs; pd; lop ears; fat at base
of tail/orig. from Tunisian
Barbary × Pinzirita/BSd 1942,
HB 1977/It. *Barbaresca della
Sicilia, Barbaresca siciliana*/syn.
Orecchiuta (= long-eared),
Siciliana migliorata (= *Improved
Sicilian*)

Sicilo-Sarde: *see* Tunisian milk
sheep

Sidi Tabet cross: (Tunisia)/mw/
black/orig. from Portuguese
Black Merino × Thibar

Siena, Sienese: *see* Senese

Sikkim: *see* Bonpala

Silba: *see* Pag Island

Silesian: (Lower Silesia, Poland)/
d.cw/pd/Polish Longwool group,
sim. to Bochnia/orig. 1932–1954
from East Friesian × local/Pol.
Owca śląska

Silverdale: *see* Limestone

Simmental: (Switzerland)/sim. to
Saanen/part orig. of Swiss Black-
Brown Mountain/extinct

Sinjabi: *see* Sanjabi

Sinkiang: *see* Xinjiang

Sin'tszyanskaya: *see* Xinjiang
Finewool

Šipan: *see* Dubrovnik

Sipli: (Bahawalpur, Pakistan)/cw.m/
white with brown or white face
and ears; pd

Siromeat: (NSW, Australia)/m.cw/
orig. (to 1989) by CSIRO at
Armidale from Dorset Horn ($\frac{3}{8}$),
Cheviot ($\frac{3}{8}$), Corriedale ($\frac{1}{4}$)/?
extinct

Siroua: (SW High Atlas, Morocco)/
d.cw/white or black; ♂ hd, ♀ pd;
♀ often earless/var. of Berber

Sisqueta: *see* Pallaresa

Sitia: (E Crete, Greece)/m.d.cw/white
with black spots on head, belly
and legs, often coloured or pied;
♂ hd, ♀ pd or hd/island var. of

Greek Zackel/syn.
Metaxomalicha/rare (by crossing
with Sfakia and Psiloris)

Six-point White: *see* Dutch Black
Blaze

Sjenica: (SW Serbia)/m.d.cw/usu.
black around mouth, ears and
eyes; ♂ usu. hd, ♀ usu. pd/
Pramenka type/being improved
by Corriedale and Précoce/var.:
Vasojević/Serbo-cro. *Sjenička,
Peštersko-sjenička, Sjeničko-
pešterska*/syn. *Pešter*/not
Sjenichka

Sjeviot: (Norway) *see* Cheviot

Skilder: (S Africa)/black with white
spots or red speckled/var. of
Persian/syn. *Speckled Persian*

Skopelos: (N Sporades, Greece)/
d.m.mw/usu. white with black or
brown spots on face and legs; ♂
usu. hd, ♀ usu. pd/? orig. from
Ayios Evstratios or from
Chalkidiki/orig. of Kymi/syn.
Glossa/not *Scopelos*

Skudde: (orig. E Prussia, now
Germany)/cw/grey-white, occ.
black, rarely brown; ♂ hd, ♀
scurs or pd/Heath type/BS 1984;
BS also Switzerland 1991/syn.
Kosse/[? from Skuodas in SW
Lithuania]/rare

Skutari: *see* Shkodra

Skye Farm Romney: (Hawkes Bay,
New Zealand)/m.lw/pd/orig. late
1960s from New Zealand
Romney × (South Suffolk ×
Coopworth)

Śląska: *see* Silesian

Slovakian Merino: (Slovakia)/fw.m/
orig. from (Romney × North
Caucasus Mutton-Wool) × Czech
Merino

Small-tailed Han: *see* Han

snow sheep: (N and E Siberia,
Russia)/= *Ovis nivicola*/grey-
brown with dark brown stripe
across muzzle, and white
underparts, rump and posterior
face of legs/syn. *Asiatic bighorn,
chubuku* (Yakutsk)

Soay: (St Kilda, Scotland)/w-hy/dark
brown with pale belly, or light
brown; ♂ hd, ♀ hd or pd/

Northern Short-tailed type/ feral on St Kilda, also HB flocks GB mainland/HB 1974, BS 1994/ rare

Socotra: (Socotra I, Indian Ocean)/ earless

Sofia-Breznik: *see* Breznik

Sofia White: (SW Bulgaria)/black head and feet/var. of Breznik

Sohagi: *see* Saidi

Soissonais: (Aisne, France)/former var. of Précoce/HB 1925–1929/ Fr. *Mérinos précoce du Soissonais*/extinct

Sokolki: (Ukraine)/fur, d/grey (lethal when homozygous) or occ. black; ♂ hd, ♀ 20–25% hd/orig. from Russian Long-tailed/Russ. *Sokol'skaya*/not *Sokol, Sokolka, Sokolov, Sokolsky*

Solčava: (N Slovenia)/m.cw-mw.[d]/ white or black-and-white; pd/ Lop-eared Alpine group/orig. in late 18th and early 19th c. from Bergamasca and Paduan × local, via Seeländer var. of Carinthian/ HB 1983/Sn. *Solčavsko-Jezerska, Jezersko-Solčavska,* Ger. *Sulzbach-Seeland, Seeland-Sulzbach*/syn. *Seeboden* (Ger. *Seebodner*)/not *Solca, Solcava, Soltschava*

Solognot: (E Loir-et-Cher, C France)/ m.sw/greyish with red-brown head and legs; pd/sim. to Berrichon/HB 1948, BS/[from Sologne]/rare

Solothurn: (Switzerland)/brown/var. of Jura/extinct

Solsonenca: *see* Ripollesa

Somali: (Somalia, also E Ethiopia and N Kenya)/m/white with black head; hy; pd; fr/var.: Toposa/orig. of Blackhead Persian (S Africa)/ syn. *Berbera Blackhead, Blackhead Persian, Blackheaded Somali* (It. *Pecora somala a testa nera*), *Ogaden*

Somali Arab: (coast of Somalia)/cw-hy/white; pd; ft/? orig. from Radmani (Aden)

Somali Brasileiro: *see* Brazilian Somali

Sombourou: *see* Banamba

Somerset, Somerset Horn: *see* Pink-nosed Somerset

Somosierra Blond: (NE Madrid, Spain)/d.m.(cw)/white with pale brown face and legs/ hd or pd/ Sp. *Rubia di Somosierra*/syn. *Churra de El Molar, Rubia de El Molar, Rubia serrana*

Sonadi: (S Rajasthan and N Gujarat, India)/cw.d.m/light brown face, neck and legs; pd; very long ears/ sim. to Malpura but smaller/ syn. *Chanothar*/not *Chhanotar*

Sonmez: (Turkey)

Sopravissana: (C Apennines, esp. Latium and Umbria, Italy)/fw-mw.d.m/♂ hd, ♀ pd/orig. from Vissana crossed with Spanish Merino and Rambouillet in 18th and early 19th c. and improved by American and Australian Merinos in 20th c./former var.: Maremmana/HB 1942/syn. *Upper Visso*

Sous: (Morocco)

South African Merino: (S Africa)/fw/ orig. (1789 on) from Spanish, Saxony, Rambouillet and American Merinos but chiefly from Australian Merino/HB 1906/*see also* Döhne Merino, Letelle Merino, Multihorned Merino, South African Mutton Merino

South African Mutton Merino: (S Africa)/fw.m/pd/German Mutton Merino imported since 1932/ recog. 1971 (name change); BS 1947; BS also Australia/Afrik. *Suid-Afrikaanse Vleismerino*/ syn. *SA Mutton Merino*/*see also* Walrich Mutton Merino

South Albanian: (Albania)/var. of Common Albanian

Southam Nott: (Devon, England)/ orig. (+ Bampton Nott and × Leicester Longwool) of Devon Longwoolled/not *Southern Notts*/[nott = pd]/extinct

South Australian Merino: (S Australia)/ strongwool strain of Australian Merino, with English Longwool blood 1840–1870/inc. Bugaree Merino

South Bulgarian Corriedale:
(Bulgaria)
South Bulgarian Finewool: (SC
Bulgaria)/orig. 1943–1967 from
Merino × Stara Zagora/syn.
Stara Zagora Finewool/extinct
(now inc. in Thrace Finewool)
South Bulgarian Semifinewool: (S
Bulgaria)/mw.m/orig. from
Romney or North Caucasus ×
(German Mutton Merino and
Caucasian, × local)
South Country Cheviot: *see* Cheviot
Southdale: (Middlebury, Vermont,
USA)/m.w/pd/orig. 1930–1943
from Southdown × Corriedale/
extinct (crossed with Columbia
to form Columbia-Southdale)
South Devon: (S Devon and
Cornwall, England)/lw.m/pd/
English Longwool type, sim. to
Devon Longwoolled but larger/
BS and HB 1904/syn. *South Dum*
(= soft wool)/extinct (combined
with Devon Longwoolled to form
Devon and Cornwall Longwool
1977)
South Dorset Down: (New Zealand)/
m.sw/orig. from Dorset Down ×
Southdown/BS
Southdown: (Sussex, England)/sw.m/
grey-brown face and legs; pd/
Down type/orig. from Sussex by
selection 1780–1829/var.: French
Southdown/basis of Down
breeds/BS and HB 1892; BS also
USA 1882, Australia, New
Zealand 1923 (HB 1893); HB also
Canada/rare in GB
Southdown Norfolk: *see* Suffolk
South Dum: *see* South Devon
South-East Bulgaria Finewool: *see*
Karnobat Finewool
Southern: (Chile) *see* Austral
Southern: (W Africa) *see* West
African Dwarf
Southern Goat: (Algeria) *see* Tuareg
Southern Karaman: (S Anatolia,
Turkey)/black/var. of White
Karaman/Turk. *Güney Karaman*
Southern Notts: *see* Southam Nott
South Hampshire: (New Zealand)/
orig. 1970s from Southdown ×
Hampshire/HB

SOUTH INDIA HAIR: (S India)/m/inc.
Ganjam, Godavari, Kenguri,
Kilakarsal, Madras Red, Mandya,
Marathwada, Mecheri, Nellore,
Ramnad White, Tiruchy Black,
Vembur; orig. of Jaffna
South Kazakh Merino: (Jambul to
Kazalinsk, S Kazakhstan)/fw.m/
orig. 1944–1964 from Caucasian,
Stavropol and Grozny 1942 ×
(Novocaucasian and Soviet
Merino 1932 × Kazakh Fat-
rumped)/recog. 1966/HB/Russ.
Yuzhnokazakhskiı̆ Merinos/syn.
Aral (Russ. *Priaral'skaya*)
South Madras (red): *see* Kilakarsal,
Madras Red, Mecheri
South Moroccan: (Morocco)/hd/orig.
from Tadla × Berber/vars:
Rehamna-Sraghna, Zemrane
South-Steppe Caucasian: (Russia)/fw
South Sudanese: (Sudan S of lat. 11°
N)/white, usu. with black or tan
patches; hy; hd or pd; ♂ usu.
with ruff; dwarf/vars: Mongalla,
Nilotic, Nuba Maned/syn.
*Southern Sudan, Southern
Sudan Dwarf, Sudanese Maned*
South Suffolk: (Canterbury, New
Zealand; also in Australia)/
m.sw/black face/orig. (1938 on)
from Suffolk × Southdown,
crossed both ways and F$_1$ bred
inter se/HB/syn. *South Suffolk
Halfbred*
South Ural: (Orenburg, Russia)/fw.m/
♂ hd, ♀ pd/formed by
combining Orenburg Finewool
and October breed groups/recog.
1968/Russ. *Yuzhnoural'skaya*/
syn. *South Ural Merino*
South Wales Mountain: (South
Wales)/tan face, bare belly,
kempy fleece (often red)/larger
var. of Welsh Mountain/BS 1948,
HB/syn. *Glamorgan, Nelson*
South-West Longwool: (England)/see
Dartmoor, Devon and Cornwall
Longwool, Whiteface Dartmoor
Sovetskaya myaso-sherstnaya: *see*
Soviet Mutton–Wool
Sovetskiı̆ Korridel': *see* Soviet
Corriedale
Soviet Corriedale: (Russia)/mw.m/

orig. 1926–1936 from Lincoln × Rambouillet/Russ. *Sovetskiĭ Korridel'*/extinct

Soviet Merino: (Russia)/fw.m/orig. in 20th c. (esp. 1925–1946) from Mazaev and Novocaucasian Merinos improved by American Rambouillet, Askanian and Caucasian and other improved Merino breeds/vars: North Caucasus, Siberian/named 1938/Russ. *Sovetskiĭ Merinos*

Soviet Mutton-Wool: (N Caucasus, Russia)/m.lw/pd/orig. from Karachai × finewool ♀♀ mated since 1950 to North Caucasus Mutton-Wool, Lincoln (and Liski) ♂♂/var.: Siberian type/ Russ. *Sovetskaya myaso-sherstnaya*/syn. (till 1985) *Mountain Corriedale* (Russ. *Gornyĭ Korridel'*)

Soviet Rambouillet: *see* Altai, Askanian, Caucasian

Spælsau: (W Norway)/m.w/usu. white, sometimes coloured; hd or pd/Northern Short-tailed type/orig. from Old Norwegian with Icelandic and Faeroes blood/HB 1947/syn. *Old Norwegian Short Tail Landrace*/[= bobtail sheep]

Spagnola arianese: *see* Quadrella

Spancă: (S Romania)/mw.d.m/Palas Merino × Tsigai, F_1/selected F_2, F_3 and backcrosses called Danube Merino/not *Spanga*/[? = Spanish]

Spanish Ariano: *see* Quadrella

Spanish Churro: (Duero valley, NW Spain)/d.(m.cw)/black eyes, ears, nose and feet; ♂ usu. hd, ♀ usu. pd/sim. to Lacho/vars: Andalusian, Castilian, Sayaguesa, Tensina/orig. of Algarve Churro/BS/[Sp. *churro* = rustic, coarsewooled]

Spanish Merino: (Spain, esp. Extremadura, W Andalucía and parts of Castille)/fw.m/white, also black var.; ♂ occ. pd, ♀ pd/former strains: Escurial, Guadalupe, Infantado, Negretti, Paular, Perales; migratory

(*trashumante*) and non-migratory types; vars: Andalusian, Barros, Leonese, Serena, Serrana (Spanish Mountain Merino)/orig. of Merino type; orig. (with Churro) of Entrefino/HB 1978, BS 1982/rare

Spanish Mongrel: (Italy) *see* Maremmana

Spanish Mountain: *see* Iberian

Spanish Mountain Merino: (mts of Castille, C Spain)/var. of migratory Merino which has become non-migratory in summer range/Sp. *Merino de montaña, Merina serrana*

Spanish Piebald: *see* Jacob

Speckled-face, Specklefaced Mountain: (Wales) *see* Beulah Speckled Face, Welsh Hill Speckled Face

Speckled-face Manech: (SW Basque country, France)/d.m.cw/ speckled face and legs/var. of Manech/Basque *machkaroa, maxharo*

Speckled Halfbred: (Wales)/Border Leicester × Speckled Face, 1st cross

Speckled Persian: (S Africa) *see* Skilder

Spelsau: *see* Spælsau

Sphakia: *see* Sfakia

Spiegel: (Austria)/black spectacles (*Nasenspiegel*)/former var. of Carinthian/orig. (with Bergamasca) of Tyrol Mountain and of Pusterese/[= mirror]/ nearly extinct

Spiegel: (Prättigau, NE Graubunden, Switzerland)/cw/black around eyes and ear tips/pd/HB/related to Austrian breed/syn. *Mirror sheep* (Fr. *mouton miroir*/rare

Spotted: *see* Jacob

Sraghna, Srarhna, Srarna: *see* Rehamna-Sraghna

Sredna Gora: (C Bulgaria)/m.cw.d/ white or coloured, with black spots around eyes; ♂ hd or pd, ♀ pd/local unimproved var. of Bulgarian Native/Bulg. *Srednogorska* (= Central Mountain)/syn. *Panagyurska* (=

from Panagyurishte),
Panagiurište, Panaguirishte; obs.
syn. *Ofche Hulmski, Talschaf*
(Ger.) (= valley sheep)/rare
Sredna Gora Semifinewool: (C
Bulgaria)
Srednorodopska: *see* Central Rodopi
Srednostaraplaninska: *see* Central
Stara Planina
Srem, Sremska: *see* Birka
Stara Planina: (Balkan Mts, C
Bulgaria)/m.cw.d/black spots on
head and legs; ♂ pd/inc. Central
Stara Planina, West Stara
Planina/Bulg. *Staroplaninska*
Stara Planina Tsigai: (Balkan Mts, C
Bulgaria)/recently approved (*c.*
1997) new breed/Bulg.
Staroplaninski Tsigai
Stara Zagora: (C Bulgaria)/m.d.w/♂
hd/local Bulgarian Native
improved by Tsigai/Bulg.
Starozagorska
Stara Zagora Finewool: (Bulgaria)/
type of Thrace Finewool/Bulg.
Starozagorska t"nkorunna
Stavropol: (N Caucasus, Russia)/fw/
orig. on Sovetskoe Runo state
farm from Novocaucasian and
Mazaev Merinos (1923),
improved by American
Rambouillet (1928) and
Australian Merino (1936)/part
orig. (with Australian Merino) of
Manych Stavropol/recog. 1950;
HB/Russ. *Stavropol'skaya,
Stavropol'skaya poroda
tonkorunnykh ovets* (= Stavropol
breed of finewool sheep)/syn.
Stavropol Merino
Steekhaar: (S Africa)/cw/kempy var.
of Ronderib Afrikander (not
recog. by BS)/[= rough hair]/
nearly extinct
Steigar: (Steigen, N Norway) m.cw/
grey; pd/orig. 1940s from North
Country Cheviot × local/recog.
1954; HB
Steiner: (Steiner alps, Austria)/
former var. of Carinthian/extinct
Steinschaf: (Tyrol, Austria)/m.cw/
white, brown-black, or grey with
black head and legs; ♂ usu. hd,
♀ usu. pd; small; prim./sim. to

Bündner Oberland and Zaupe/
orig. (with Bergamasca) of
Carinthian and Tyrol Mountain,
Pusterese and German
Mountain/BS 1974/syn. *Pecora
della Roccia* or *del Sasso* (= rock
sheep) (Val Venosta, Italy),
Tiroler Steinschaf/[= stone
sheep, i.e. living on rocky
slopes]/rare
Steppe Screwhorn: *see* Stogoš
Steppe Voloshian: (N Caucasus and
SW Siberia, Russia)/m.w/white,
rarely black; ♂ usu. hd, ♀ pd;
lft/typical var. of Voloshian/orig.
from Zackel with some ft blood/
nearly extinct
Stepska vitaroga: *see* Stogoš
Stogoman: *see* Stogoşă
Stogoš: (S Banat, Serbia)/m.cw.d/
white, usu. with brown or yellow
face and legs, occ. brown or
black; ♂ vertical screw horns, ♀
pd/Pramenka type/syn. *Rackulja,
Stepska vitaroga* (= *Steppe
screwhorn*), *Vlaška vitaroga* (=
Vlach screwhorn)
Stogoşă: (Romania)/Tsigai ×
Ţurcana, F₁/syn. *Stogoman*/not
*Stogoš, Stogosa, Stogosch,
Stogoso*/[Rom. *stog* = with fleece
in shape of haycocks]
Stone sheep: (NW British Columbia,
Canada)/= *Ovis dalli stonei*/black
or brown with white nose, rump
and posterior face of legs/var. of
Dall sheep/syn. *Stone's sheep*
stone sheep: (Austria) *see* Steinschaf
Stornoway Blackface: *see* Island
Blackface
Stranja: (SE Bulgaria)/cw.m/tan,
black or speckled face, ears and
legs, born black; ♂ hd, ♀ pd/
Bulgarian Native/Bulg.
Strandzhanska/not
Strandjanska/rare
Sudad: *see* Guirra
Sudan Desert: (Sudan N of lat. 12°
N)/d.m/hy; ♂ usu. pd, ♀ pd; lop
ears; long fleshy tail/African
Long-legged group/cf. Baraka/
chief vars: Dubasi, Shugur,
Watish; other vars: Baggara, Bija,
Gash and East Butana, Hamari,

Kabashi, Shambali/syn. *Desert Sudanese, Drashiani* (It. *Drasciani*), *North Sudanese, Sudanese Desert*

Sudanese: *see* Dongola, South Sudanese, Sudan Desert, Zaghawa

Sudanese Maned: *see* South Sudanese

Sudanica: *see* Lop-eared Alpine

Sudat: *see* Guirra

Sud de la Manche: *see* Avranchin

Suffolk: (England)/sw.m/black face and legs; pd/Down type/orig. from Southdown × Norfolk Horn in early 19th c./orig. of South Suffolk, White Suffolk/recog. 1810, named 1859, BS 1886, HB 1887; BS also USA 1892, Australia, South Africa 1959 (HB 1906), Ireland, Netherlands; HB also Canada, France 1957, Germany 1973, New Zealand, Czech Rep./syn. *Blackface, Southdown Norfolk*

Suffolk Whiteface: *see* White Suffolk

Suid-Afrikaanse Vleismerino: *see* South African Mutton Merino

Sulukol Merino: (Kustanai, NW Kazakhstan)/fw/orig. on Sulukol state farm from finewool × Kazakh Fat-rumped, selected since 1958 towards type with $\frac{5}{8}$ Askanian blood/Russ. *Sulukil'skiĭ Merinos*/extinct (combined with Beskaragai Merino to form North Kazakh Merino)

Sulzbach, Sulzbach-Seeland: *see* Solčava

Sumatran: (N Sumatra, Indonesia)/ m.(cw)/white, light brown or black-bellied/sim. to Javanese Thin-tailed but smaller and with shorter tail/syn. *Sumatra Thin-tail*

Šumava: (Bohemia, Czech Republic)/ w.m/white; ♂ hd, ♀ pd/orig. colour on head and legs, also black or pied, Zackel type sim. to Valachian but heavier and less coarse wool; improved (*zušlechtěná*) by selection and limited use of Texel, Tsigai and Merinolandschaf/HB 1954/Cz. *Šumavska*/syn. *Bohemian Land* (Cz. *Česká selská*, Ger. *Böhmerwald*)/not *Shumava*

Sumber: (Dornogovi prov., SE Mongolia)/fur/black, grey or brown/orig. since 1950 at Sumber state farm from Karakul × Mongolian/recog. 1991/syn. *Sumber Karakul*

Sungai Putih: (Indonesia)/orig. from local Sumatran ($\frac{1}{2}$), St Croix ($\frac{1}{4}$), Barbados Black Belly ($\frac{1}{4}$)

Sunnhordland: *see* Old Norwegian

Suomenlammas: *see* Finnish Landrace

Sur: (Uzbekistan)/agouti (or golden)/ var. of Karakul

Surkhandarin: (Kazakhstan)/breed type/Russ. *Surkhandarinskiĭ porodnyi typ*

Sussex: (England)/orig. of Southdown/syn. *Old Southdown*/extinct

Sutai: (Darib dist., Khobot prov., Mongolia)/var. of Mongolian

Sutherland Cheviot: *see* North Country Cheviot

Svanka: (Svanetski mts, Georgia)/ cw/dark; ft; prolific/extinct

Svea: (Sweden)/synthetic

Svensk Finullsfår: *see* Swedish Finewool

Svensk Lantras: *see* Swedish Landrace

Svishtov: (N Bulgaria)/cw.m.d/face and legs often black or spotted; pd/Bulgarian Native, improved/ Bulg. *Svishtovska*, Ger. *Swistower*/not *Svistov*/rare

Svrljig: (E Serbia)/m.d.cw/black on head and legs, occ. all black or grey; ♂ usu. hd, ♀ pd/Pramenka type/being improved by Corriedale/Serbo-cro. *Svrljiśka*/syn. *Gulijanska* (from Gulijan)/rare

Swaddle: *see* Swaledale

Swaledale: (Pennines, England)/ cw.m/black face with grey-white muzzle; hd/ Blackfaced Mountain type/orig. of Dalesbred/BS, HB 1919/syn. *Swaddle*

Swazi: (Swaziland)/black, brown or reddish, also pied; ♂ usu. hd, ♀ pd; carrot-shaped tail/var. of Nguni

Swedish Carpet Wool sheep: *see* Rya

Swedish Finewool: (Sweden)/orig. 1920 from old native Swedish breeds with some influence from Finnish fw sheep/Swe. *Svensk Finullsfår*/syn. *Swedish Landrace* (Swe. *Svensk lantras*)/ extinct

Swedish Fur Sheep: (Sweden)/pelt/ grey; pd/var. of Swedish Landrace/orig. from Goth selected for curl and colour since 1920/Swe. *Pålsfår*/syn. *Gotland, Swedish Pelt, Gotländisches Pelzschaf* (Ger.)/BS in GB 1992 (HB 1983)

Swedish Landrace: (Sweden)/m.w/ orig. from Northern Short-tailed/ inc. white Landrace breeds (Swedish Finewool and Rya) and grey breeds (Goth and Swedish Fur); also used as syn. for Swedish Finewool Landrace/ Swe. *Svensk lantras*/syn. *Swedish Native*

Swedish Pelt: *see* Swedish Fur Sheep

Swifter: (Netherlands)/m/prolific/ orig. 1967–1971 at Swifter farm, Univ. Wageningen, from Texel × Flemish/BS/not *Swift*

Świniarka: (Kielce, Poland)/w.d/♂ hd, ♀ hd or pd; kempy fleece; st; prim./vars: Karnówka, Krukówka/orig. (with Romney) of Łowicz/not *Swinarka*/[Pol. *świnia* = pig]/nearly extinct

Swiss Black-Brown Mountain: (W Switzerland)/m.sw/pd/orig. 1938 by union of Frutigen, Jura, Roux-de-Bagnes, Saanen and Simmental/HB/Ger. *Schwarzbraunes Bergschaf, Gebirgsschaf*, Fr. *Brun noir du payes, Brun noir des alpes, Brun noir des montagnes*

Swiss Blackheaded Mutton: *see* Swiss Brownheaded Mutton

Swiss Brownheaded Mutton: (N Switzerland)/m.sw/pd/orig. 1938 from Grabs + Oxford Down/HB

1900/Ger. *Braunköpfiges Fleischschaf*, Fr. *Mouton à viande à tête brune*/syn. *Improved Blackheaded Mutton-Wool* (Ger. *Veredeltes schwarzköpfiges Fleischwollschaf*), *Swiss Blackheaded Mutton*

Swiss Charollais: (W Switzerland)/ m/pd/orig. from Swiss White Alpine graded to Charollais/ recog. 1990, HB 1991/Fr. *Charollais suisse*

Swiss White Alpine: (Switzerland)/ m.sw/orig. (1936 on) from Swiss White Mountain with 50–75% Île-de-France blood/HB/Fr. *Blanc des Alpes*, Ger. *Weisses Alpenschaf*/syn. *West Swiss White* (Fr. *Blanc de la Suisse Occidentale*), *White Improved* (Ger. *Weisses Edelschaf*, Fr. *Blanc des Alpes amelioré*)

Swiss White Mountain: (E Switzerland)/m.sw/pd/Lop-eared Alpine type/orig. 1929–1938 from Württemberg Merino (Merinolandschaf) × native (Appenzell, Bündner Oberland, Oberhasli-Brienz, Schwyz, Wildhaus)/orig. (with Île-de-France) of Swiss White Alpine and absorbed by it/Fr. *Blanc des Montagnes*, Ger. *Weisses Gebirgsschaf, Weisses Bergschaf*/syn. *Improved Whiteheaded Mountain* (Ger. *Veredeltes weissköpfiges Gebirgsschaf*)/extinct

Swistower: *see* Svishtov

Synthetic Hairsheep: (Italy)/new synthetic orig. from crossing Malaysian longtail with Cameroon, F_1

Syrian: *see* Awassi

Syrmian: *see* Birka

Szapora Merino: (Hungary)/mw.m/♂ hd, ♀ pd/syn. *Prolific Merino*/ rare

Szendrő New: (Hungary)/m/new breed being selected for reproductive performance and meat production at farm in Szendrő since 1997

Tabasaran: (Dagestan, Russia)/
m.d.cw/brown or red, also black;
♂ hd or scurs, ♀ pd; often
earless/Caucasian Fat-tailed
type/vars: Gedek, Kusman/Russ.
Tabasaranskaya/syn. *Samur*/
extinct

Tabasco: *see* Pelibüey

Tacòla: (N Vercelli, Piedmont, Italy)/
m/pd; reduced ears/smaller var.
of Biellese/syn. *Bertuna*,
Cücch/rare

Tadla: (W Morocco plateaux)/m.cw/
white with coloured legs; hd/
var.: Beni Meskine/orig. (with
Berber) of South Moroccan,
Timahdit, Zaian; ? orig. of
Boujaad/Fr. *Race des Plateaux
de l'Ouest*

Tadmit: (Algeria and W Tunisia)/
m.mw/♂ hd, ♀ pd/orig. *c.* 1925
from Algerian Arab, ? with
Merino blood/syn. *Queue fine de
l'Ouest* (= thin-tailed of the west)
(Tunisia)

Tadzhikskaya: *see* Tajik

Tafilalet: *see* D'man

Tahirova: (Ege region, Turkey)/
d.m.w/orig. 1964 on from East
Friesian (¾) × Kivircik

Taihu: *see* Hu

Taiku: (Shanxi, Hebei and Shaanxi,
China)/cw/white face; ♂ hd, ♀
pd; sft/sim. to Mongolian but
smaller/WG *T'ai-ku*

Taiz Red: (Ta'izz, Yemen)/m/brown
or red; hy; pd; ft/syn. *Ganada*
(from Al Ganad)/not *Gainde*,
Gamdi

Tajik: (Tajikistan)/cw.m/tan face and
legs, born tan; ♂ hd, ♀ pd; fr/
orig. (1947–1963) from Sary-Ja
(and Lincoln) × Hissar/recog.
1963/Russ. *Tadzhikskaya*/syn.
Tajik Semicoarsewooled

Talaverana: (W Toledo, Spain)/
m.mw.(d)/pd/Entrefino type,
sim. to Mestizo Entrefino-fino/
orig. since late 19th c. from
Merino and Manchega, with
some Merinolandschaf blood
since 1960/[from Talavera de la
Reina]

Talin Tsagaan: (Mongolia)

Talschaf: *see* Sredna Gora

Talybont Welsh: (SE Breconshire,
Wales)/tan face/larger strain of
Welsh Mountain with longer
kemp-free fleece

Tan: (N Ningxia and neighbouring
areas, China)/cw.m.pelt/white,
usu. with black or brown head
and legs; ♂ hd, ♀ pd or scurs;
sft/orig. from Mongolian/syn.
Tan-yang (= Tan sheep)/not
Tang, Tanjan, Tanyan, Tanyang

Tanface: (Scotland and N England)/
cf. Welsh Tanface/syn. *Dun*,
*Dunface, Old Scottish Shortwool,
Scottish Soft-wool*/extinct
(displaced by Blackfaced
Mountain, late 18th c. in N
Scotland)

Tang: *see* Tan

Tanganyika Long-tailed: *see*
Tanzania Long-tailed

Tanganyika Short-tailed: *see* Masai

Tanjan: *see* Tan

Tanyan, Tan-yang: *see* Tan

Tanzania Long-tailed: (Tanzania)/m/
various colours; hy; ♂ often hd,
♀ usu. pd; often earless; lft, sft or
lt; sometimes tassels/African
Long-fat-tailed group, sim. to
Rwanda-Burundi/syn. *Gogo*,
Tanganyika Long-tailed, Ugogo

Tarai: *see* Lampuchre

Taraki: *see* Baluchi

Tarasconnais: (C and E Pyrenees,
France)/m.mw/white, coloured
spots on heads and legs now
rare, occ. red, brown or black; ♂
hd, ♀ hd or pd/Central Pyrenean
group/= Aranesa (Spain)/BS
1975/syn. *Ariègeois* (from
Ariège), *Pyrénéen central à
extremités tachetées*, or
charbonées (= with spotted
extremities)/[from Tarascon]

Targhee: (Idaho, USA)/mw.m/pd/
orig. (1926 on) from Rambouillet
× (Lincoln × Rambouillet) +
Rambouillet × [Corriedale ×
(Lincoln × Rambouillet)]/BS
1951; HB Canada/not *Targee*/
[Targhee National Forest]

Targhi, Targi, Targui: *see* Tuareg

Tarhumara: (Mexico)/var. of Criollo

Tarina: (Taro, Emilia, Italy)/local var. of Apennine

Tarrincha: *see* Churra da Terra Quente

Tarset: (Mexico)/m/Tabasco (Pelibüey) × Dorset

Tasmanian Merino: (Australia)/ finewool strain of Australian Merino/orig. from Saxony Merino imported 1830/syn. *finewool, Saxon Merino*

Tatra: (Poland)/hd/var. of Polish Zackel

Tauferer: *see* Pusterese

Tauter: (Tautra I, Trondheim fjord, Norway)/sw/pd/orig. from British breeds (? Ryeland) imported 1770–1788, ? Merino blood/nearly extinct

Tavetsch: *see* Bündner Oberland

Tavetsch-Medels: (Graubunden, Switzerland)/cw/hd/being used to breed back to Tavetsch (i.e. Bündner Oberland)/Ger. *Tavetsch-ähnliches Medelserschaf* (= Tavetsch-like Medels sheep), Romansch *Tujetsch-Medel*/rare

Tchabale, Tchapera: *see* Brigasque

Tedesca di Pusteria: *see* Pusterese

Teeswater: (Teesdale, Durham, England)/m.lw/white or grey face; pd/English Longwool type, sim. to Lincoln/orig. of Wensleydale/BS and HB 1949/ syn. *Northern Longwool*

Tekin: (Turkmenistan)/var. of Turkmen Fat-rumped/extinct

Telangana: *see* Telingana

Telengit: (Altai, Siberia, Russia)/ m.cw/usu. white with black or red head and neck, also black pied or red; sft/sim. to Mongolian/Russ. *Telengitskaya*/ syn. *Altai, Altaïskaya*

Telingana: (Hyderabad, Andhra Pradesh, India)/cw.m/usu. black/ local var./syn. *Navargade*/not *Telangana*

Temir: (Kazakhstan)/var. of Kazakh Fat-rumped/Russ. *Temirskaya*

Temnogolovaya èstonskaya: *see* Estonian Darkheaded

Tendasque: *see* Brigasque

Tengchong: (W Yunnan, China)/var. of Tibetan

Teng-Seemai: *see* Kenguri

Tenguri: *see* Kenguri

Tensina: (Tena, Huesca, Spain)/d.m/ hd or pd/var. of Spanish Churro/Sp. *Churra tensina*

Terra Quente: *see* Churra da Terra Quente

Terrincha: *see* Churra da Terra Quente

Testa rossa: *see* Comisana

Tête noire: *see* French Blackheaded

Tête rouge du Maine: *see* Rouge de l'Ouest

Teteven: (C Bulgaria)/Bulg. *Tetevenska*/rare

Tetra line: *see* Prolific Bábolna

Teutoburg: *see* German Blackheaded Mutton

Texas Barbado: *see* Barbado

Texel: (Netherlands)/m.lw/pd/Marsh type/orig. in late 19th and early 20th c. from Leicester Longwool, Lincoln *et al.*, × local (Old Texel)/vars: Blue Texel, French Texel/orig. (with Flemish) of Swifter/HB 1909, BS; BS also Ireland 1976, UK 1973, Argentina, USA 1991; HB also Germany 1962, Finland, Luxembourg 1968/ Du. *Texelaar, Texelse*/syn. *Improved Texel* (Du. *Verbeterde Texelse*)/[pronounced *Tessel*]

Thai Longtail: (Thailand)/m.(cw)/ syn. *Siamese Longtail*

Thalli: (Thal desert, Punjab, Pakistan)/cw.d.m/black, brown or pied head; pd; long or short ears; st/syn. *Buti, Chundi, Lessarkani, Porakani, Tilari*/not *Thali, Thall*

Thibar: (Tunisia)/mw/black, occ. with white spot on head or tail/orig. (1911 on) from Arles Merino × Tadmit/orig. of Sidi Tabet cross/HB 1945/Fr. *Noir de Thibar*/syn. *Black Merino*/rare (severe inbreeding problems)

Thoka: (Iceland)/prolific gene in Iceland sheep/[name of 1st known carrier, ♀ born 1950]

Thônes-Marthod: (Arly and Arc valleys, Savoy, France)/

m.cw.(d)/black spectacles, nose tip, ears and feet; hd/cf. Savoiarda (Italy)/combined and named 1955; BS 1980, revived 1993/Fr. *Race de Thônes et de Marthod*/syn. *Mauriennais, St Jean de Maurienne* (Italy), *Savoyard*

Thrace: (NE Greece)/d.m.cw-mw/ spotted head and legs; hd/= Kivircik (Turkey)/Gr. *Thraki*/syn. *Western Thrace* (Gr. *Dhitiki Thraki*)

Thrace Finewool: (Maritsa valley, S Bulgaria)/fw.m.d/orig. 1943– 1967 from Caucasian, Rambouillet and German Mutton Merino, × South Bulgarian and Plovdiv-Purvomai, and inc. South Bulgarian Finewool/var.: Stara Zagora Finewool/Bulg. *Trakiiska t"nkorunna*/syn. *Maritsa Finewool* (Bulg. *Marishka t"nkorunna*), *Plovdiv Merino, Thracian*

Three-rivers: *see* Sanjiang

Thribble Cross: (California, USA)/ mw/orig. (1903 on) from Delaine Merino × (Cotswold × American Merino)/[= 3-way cross]/? extinct

Thuvaramchambali: *see* Mecheri

Tian Shan argali: *see* arkhar

Tibegolian: (E Qinghai, SW Gansu and S Ningxia, China)/cw/ sometimes black head; hd; st or sft/sim. to Hetian/orig. from Mongolian × Tibet/syn. *Mongo-tibetan*/not now recog.

Tibetan: (Qinghai-Tibet plateau, China, also N Nepal, also N Sikkim and Kameng, Arunachal Pradesh, India)/m.cw.(pa)/usu. black or brown head and legs; hd; st/vars: Sanjiang, Shangu, Tengchong, Yaluzangbu; strains: Ganjia, Ganqin/orig. of Qinghai Black Tibetan/Ch. *Xizang, Zang, Zangxi*/syn. (Nepal) *Bhanglung, Bhote, Bhotia, Bhyanglung*

Tien Shan: *see* Tyan Shan

Tiflis-Herik: *see* Tuj

Ṭigae, Ṭigai, Ṭigaia, Ṭigaie: *see* Tsigai

Tihami: (E coast of Yemen)/m/white; hy; pd; ft/[from Tihama]

Tilari: *see* Thalli

Timahdit: (Middle Atlas, Morocco)/ m.cw/brown face; hd/orig. from Tadla, Berber and Beni Guil/HB/ syn. *Azrou, Bekrit, El Hammam, El Hammam-Azrou, Middle Atlas*/not *Timadhit, Timadit, Timahdet, Timhadite*

Tirahi: (Kohat, Bannu and Peshawar, Pakistan)/m.d.cw/light tan, brown or black; ft/syn. *Afridi, Parai*

Tiroler, Tirolese: *see* Tyrol Mountain

Tiroler Steinschaf: *see* Steinschaf

Tiruchy Black: (Tiruchirapalli, Tamil Nadu, India)/m.hr/♂ hd, ♀ pd; st/South India Hair type/syn. *Tiruchy Karungurumbai*/not *Tiruch, Trichi, Trichy*

Tisqueta: *see* Pallaresa

Tjan Shan: *see* Tyan Shan

Tlyarota: (Dagestan, Russia)/var. of Avar, with Lezgian and Tushin blood/Russ. *Tlyarotinskaya*/ extinct

T"nkorunna Yugo-istochna B"lgarska: *see* Karnobat Finewool

Tom-Tom: (Bahrain)

Tong: (N Shaanxi, China)/m.cw-mw/white; pd; large ft; tassels/ orig. from Mongolian/orig. (with Mongolian) of Lanzhou Large-tail/WG *T'ung*/syn. *Tongyang* (= Tong sheep), *Tongzhou, Tungchow*/not *Tung, Tungyang, Tunjan, Tunjang, Tunyang*

Tonguri: *see* Kenguri

tonkorunnaya, tonkorunnykh: = finewool (Russ.)

Tonkorunnaya toshchekhvostaya ovtsa gruzii: *see* Georgian Thin-tailed Finewool

Toposa: (SE Sudan)/white, or white with black or brown on head; ♂ usu. hd, ♀ sometimes hd; fr/var. of Somali with some Nilotic blood/*see also* Murle

Torddu: *see* Badger Faced Welsh Mountain

Torfschaf: *see* Peat

Torguud: (Bulgan, Altay and Üyonch dists, Hovd prov., Mongolia)/ w.m/light grey or white with

brown, tan or black head and
legs; ♂ hd, ♀ hd or pd; ft/orig.
since 1962 from Sary-Ja ✕
Mongolian/Russ. *Torgudskaya*/
not *Torgud*

Torino: *see* Savoiarda

Torki: *see* Turki

Toronké: (WC Mali and N Burkina
Faso)/m.(d)/red or black pied, or
white or spotted, also red or
black/Fulani group of Sahel
type/vars: Banamba, Futanké,
Peul Voltaïque/syn. *Fouta Toro,
Large Peul*

Torwen: (Wales)/reversed badger-face
colour var. of Badger Faced
Welsh Mountain/[= white belly]

Touabire: (S Mauritania, N Senegal
and N Mali)/m.d/usu. white (also
pied); ♂ hd, ♀ pd; short-haired;
lop ears/Sahel type/orig. (with
Peul-Peul) of Waralé/syn.
Ladoum (Mauritania), *Tuabir,
White Arab, White Maure*

Touareg: *see* Tuareg

Toulousain, Toulouse: *see* Lauraguais

Tounfite: (between Rich and
Moulouya valley, Morocco)/
white, occ. with black marks on
head and body; hd/var. of Berber

Tounsint: (Moulouya valley,
Morocco)/var. of Beni Guil/not
Tousimet, Tousint

Trakiiska t"nkorunna: *see* Thrace
Finewool

Trangie Fertility: (NSW, Australia)/♂
usu. hd, ♀ hd or pd/strain of
Australian Merino by selection of
Peppin flock for multiple births,
with some Booroola blood in 1965

Transbaikal Finewool: (Tsitsin,
Siberia, Russia)/fw.m/orig.
1927–1956 from local
coarsewool (Buryat) graded to
Précoce, Novocaucasian and
Siberian Merinos (1927–1930)
and later to Altai and Grozny/
HB/Russ. *Zabaĭkal'skaya
tonkorunnaya*

Transcarpathian Finewool:
(Transcarpathian Ukraine)/fw/
orig. from (Australian Merino, ✕
Askanian or Altai) ✕ (Précoce ✕
local finewool)

Transcaspian urial: *see* arkal

Transdon: (Russia)/usu. pd; lft/var. of
Voloshian/Russ. *Krupnaya
zadonskaya* (= large Transdon)

Transvaal Kaffir: *see* Bapedi

Transylvanian Black: *see* Ţurcana

Transylvanian Merino: (NW
Romania)/fw.m.d/orig. from
Hungarian Combing Wool
Merino/HB 1936/Rom. *Merinos
transilvănean*/syn. *Merinos de
vest* (= *Western Merino*)

Transylvanian White: *see* Ţurcana

Transylvanian Zackel: *see* Ţurcana

Travnik, Travnicka: *see* Vlašić

Trentarka: *see* Bovec

Trichi, Trichy: *see* Tiruchy

Trièves: *see* French Alpine

Trimeticcia di Segezia: *see* Segezia

Trun: (C Calvados, N France)/m.w/
red head and legs/? orig. from
Cauchois ✕ Solognot/Fr. *Trunier,
Trunois*/extinct in 1960s

Tsigai: (SE Europe)/mw-cw.d.m/ dirty-
white and black vars, white var.
may have white face (Rom. *Ţigaie
bela*), red face (*Ţigaie ruginie*) or
black face (*Ţigaie bucălae*); ♂ hd,
♀ pd/var.: Azov/orig. of Ruda
type; orig. (with Replyan) of
Mountain Tsigai/HB (Hungary)
1974/ Rom. *Ţigaie*, Cz., Pol.,
Hung. and Serbo-cro. *Cigaja*,
Russ. *Tsigaiskaya*, Ger. *Zigaja*, Fr.
Tzigaqa, Ch. *Chei-gai*/not *Tsigaia,
Tsigay*/rare in Czech Rep.,
Hungary, Serbia

Tsurcana: *see* Ţurcana

Tsushka: *see* Chushka

Tswana: (Botswana and SW
Zimbabwe)/m/usu. white or
black-and-white/African Long
fat-tailed group/syn. *Setswana*

Tuabir: *see* Touabire

Tuareg: (N Mali)/m.d/white (usu.
with spots), also pied or red; ♂
usu. hd, ♀ usu. pd/Sahel type/
sing. *Targ(u)i*; Fr. *Touareg*/syn.
Southern Goat (Algeria; pd)

Tucur: (Lasta, Amhara and Welo,
Ethiopia)/cw-hy.m/white, brown
or pied/var. of Ethiopian
Highland with woolly
undercoat/not *Tucu*

Tudelana: (Navarre, Spain)/var. of
Castilian/extinct

Tuj: (Çildir, NE Turkey)/m.cw.d/
sometimes dark marks around
eyes and on feet; ♂ hd, ♀ pd; sft
or fr/orig. from Tushin/syn.
Çıldır, Georgian (Turk. *Gürcü*),
Kars, Kesik, Tiflis-Herik (Turk.
Herigi), *Tunç*

Tujetsch-Medel: *see* Tavetsch-Medels

Tukidale: (Hawkes Bay, New
Zealand)/cw.m/♂ hd, ♀ pd/hairy
strain of New Zealand Romney
with dominant gene N^t/cf.
Drysdale/orig. 1966 on property
of M. Coop, Tuki Tuki/BS
Australia

Tunç: *see* Tuj

Tung, T'ung, Tungchow, Tungyang:
see Tong

Tunis: *see* American Tunis

Tunisian Barbary: (Tunisia)/m.cw/
white with black or red-brown
head, occ. black or white; ♂ hd,
♀ usu. pd; ft/orig. of Campanian
Barbary, Sicilian Barbary/HB and
BSd 1947/Fr. *Barbarin*/syn.
(Algeria) *Constantinois, Moutons
de l'Oued Souf, Tunisien*

Tunisian milk sheep: (Tunisia)/
d.m.cw/white, black, brown or
grey; ♂ hd, ♀ usu. pd/orig. from
Sardinian (chiefly) and Sicilian
breeds/syn. *Sardinian, Sicilo-
Sarde*/rare (severe inbreeding
problems)

Tunjan, Tunyan, Tunyang: *see* Tong

Turbary: *see* Peat

Ţurcana: (Bihor Mts and
Transylvanian Alps, Romania)/
d.cw.m/black (*curkan*), grey
(*brumării*) or white (*purza*)/
Zackel type/grey var. (*brumării*)
used for crossing with Karakul in
NE/Hung. *Erdelyi Racka*/syn.
*Romanian Zackel, Transylvanian
Zackel, Tsurcana*/not *Curkana,
Czurkan, Turkana, Tzourcana,
Zurkana*/[= Zackel]

turchessa: *see* Campanian Barbary

Turkana: (Romania) *see* Ţurcana

Turkana: (Kenya)/nomadic rangeland
type

Türkgeldi: (Thrace, Turkey)/d.m.w/

orig. from Tahirova × (Tahirova
× Kivircik), hence $\frac{9}{16}$ East
Friesian, $\frac{7}{16}$ Kivircik

Turki: (Gorgan, NE Iran)/m.cw/black
around eyes; pd; ft/not *Torki,
Turkey, Turky*

Turki: (NE Afghanistan)/hy.m/usu.
brown (yellowish to black); ♂
usu. pd, ♀ pd; fr; tassels
common; very large

Turkish: (Italy) *see* Campanian
Barbary (turchessa)

Turkish Brown: *see* Red Karaman

Turkish Merino: (Turkey)/vars:
Central Anatolian Merino,
Karacabey

Turkmen Fat-rumped:
(Turkmenistan)/cw/usu. grey/
vars: Iomud, Tekin/Russ.
Turkmenskaya kurdyuchnaya/
extinct

Turolense: (Teruel, Aragon, Spain)/
small var. of Aragonese with less
wool

Tushin: (Georgia)/hd/lft or sft/
Caucasian Fat-tailed type/vars:
Budiani, Pampara/orig. (with
finewool) of Georgian Finewool,
Georgian Semifinewool; orig. of
Tuj/Russ. *Tushinskaya*/not
Tushetian, Tushino

Tuva: (Siberia, Russia)/white with
black head, or pied/var. of
Siberian/extinct

Twisted Horn: *see* Zackel

Txisquet: *see* Pallaresa

Tyan Shan: (Kyrgyzstan)/lw.m/orig.
1938–1950 from Précoce,
Novocaucasian Merino and
Württemberg Merino × Kirgiz
Fat-rumped, crossed with
Lincoln since 1950/recog. 1955/
Russ. *Tyanshanskaya*/syn. *Tien
Shan*

Tyrol Mountain: (Tyrol, Austria, and
Bolzano, Italy)/cw.m/white face,
occ. pied or black; ♂ hd, ♀
pd/cf. German Mountain/Lop-
eared Alpine group, sim. to
Carinthian but better wool and
longer ears/orig. from
Bergamasca × Steinschaf and
Spiegel var. of Carinthian/Ger.
Tiroler Bergschaf, It. *Tirolese*

della Montagna or *delle Rocce*/syn. *Bergschaf, Carinthian Mountain* (Carinthia), *Tiroler, White Mountain* (Ger. *Weisses Bergschaf* or *Gebirgsschaf*, It. *Pecora bianca delle Montagne*); obs. syn. (Italy) *Val Senales* (It.) or *Schnalser* (Ger.), *Val d'Ultimo* (It.) or *Ultnerschaf* (Ger.)

Tzigae, Tzigai, Tzigaia, Tzigaya: *see* Tsigai

Tzourcana, Tzurckana, Tzurkana: *see* Ţurcana

UAS: (Karnataka, India)/strain at *U*niv. Agric. Sciences, Bangalore, from (Southdown × Mandya) × Deccani

Uchum: (Siberia, Russia)/larger var. of Krasnoyarsk Finewool at Uchum stud farm/Russ. *Uchumskaya*

Uda: (N Nigeria, S Niger, C Chad, N Cameroon and W Sudan)/m/ front half black (or brown), back half white; ♂ hd, ♀ usu. pd/ Fulani group of Sahel type/Fr. *Oudah bicolore*/syn. *Bali-Bali* (Niger), *Bororo* or *Fellata* (Chad and W Sudan)/*Foulbé* (Chad and N Cameroon), *Houda, Louda, North Nigerian Fulani, Ouda, Peul* (Cameroon), *Pied*/not *Auda*

udad: *see* aoudad

Uggowitz: *see* Kanaltaler

Ugogo: *see* Tanzania Long-tailed

Uhruska: (Lublin, CE Poland)/pd/ subvar. of Lublin var. of Polish Lowland/orig. late 1950s in Uhrusk from Leine and Romney ♂♂, × Merino ♀♀/rare

Ujumqin: (Inner Mongolia, China)/ larger var. of Mongolian/= Uzemchin (Mongolia)/syn. *Wuzhumuqin*

Ukrainian Mountain: (W Ukraine)/ cw.m/breed group/orig. 1950 from Tsigai × local Carpathian Mountain/Russ. *Ukrainskaya gornaya porodnaya gruppa*

Ultimo: *see* Tyrol Mountain, Val d'Ultimo

Uluchshennaya gorodetskaya: *see* Vyatka

Ultnerschaf: *see* Tyrol Mountain, Val d'Ultimo

Upper Visso: *see* Sopravissana

Urbascia: (Calabria, Italy)/dark brown or black/var. of Calabrian (now Sciara)/extinct

Uri: *see* Schwyz

urial: (C Asia)/= *Ovis vignei* Blyth/ shades of red-brown with pale underparts; ♂ and ♀ hd; st (up to 10 cm)/intermediate between argali and red sheep (sometimes used to inc. latter)/vars in Kashmir, Afghanistan, Punjab, N Baluchistan and S Central Asia, inc. arkal, gad, shapo/orig. of domestic sheep *Ovis 'aries'* Linnaeus/syn. *Ovis orientalis* Gmelin, *vignei* section/not *oorial*

urin: *see* shapo

Urner Landschaf: *see* Schwyz

Uruguayan Criollo: (Uruguay)/cw/ white; hd (4 in ♂)/selected by A.G. Gallinal, San Pedro, Cerro Colorado/rare

Uruguayan Rambouillet: (Uruguay)/ fw/orig. from Rambouillet imported *c.* 1830–1840/rare

Ushant: (Brittany, France)/m.cw/ often black, also white; ♂ hd, ♀ pd; dwarf/BS 1977; BS also Netherlands/Fr. *Ouessant, Ouessantin*/syn. *Breton Dwarf*/ rare; now on mainland only (Pays de Loire)

Ussy: *see* Awassi

Utegangarsau: *see* Old Norwegian

Uzbek: (Tajikistan) *see* Hissar

Uzbek Mutton-Wool: (Uzbekistan)/ m.w/black spots on head; pd/ orig. since 1955 from Lincoln × (Caucasian and Lincoln, × Jaidara)

Uzemchin: (Erdeni and Tsagaan dists, Suhbaatar prov., Mongolia)/ cw.m/♂ hd, ♀ pd/var. of Mongolian/= Ujumqin (China)

Vadhiyari: *see* Patanwadi

Vagas: (Poland) *see* Fagas

Vagas, Vagaskaya: (Russia) *see* Voloshian

Valachian: (E Moravia, Czech Republic, and Slovakia)/d.m.cw/ white or

black; ♂ hd, ♀ usu. pd/ Zackel
type/orig. of Improved Valachian/
Cz. *Valaška*/syn. *Original
Valoshian, Original Walachian,
Wallachian*/nearly extinct

**Valachian, Valahian, Valahskaya,
Valakhian:** (Russia) *see*
Voloshian

Valais Blacknose: (Switzerland)/
cw.m/hd/HB/Fr. *Valais nez noir*,
Ger. *Walliser Schwarznasenschaf*/
syn. *Blacknosed Swiss, Visp,
Visperschaf*

Valais Red: *see* Roux du Valais

Valakhskaya: *see* Voloshian

Valašská: *see* Valachian

Val Badia: *see* Pusterese

Valdichiana: *see* Chianina

Val di Pusteria: *see* Pusterese

Valdrôme: *see* Préalpes du Sud

Val d'Ultimo: (Bolzano, Italy)/former
syn. of Tyrol Mountain now
restricted to red-brown var., 90%
in Val d'Ultimo and 10% in Val
Senales (Ger. *Schnalserschaf*)/cf.
Brown Mountain (Germany and
Austria), Engadine Red/HB, BS/
Ger. *Ultnerschaf*/syn. *Uttererschaf*/
rare

Valle Aurina: *see* Pusterese

Valle del Belice: (Sicily, Italy)/d/orig.
from Pinzirita, Comisana and
Sardinian

Valley: *see* Kashmir Valley

Val Senales: *see* Tyrol Mountain

Vandor: (S Africa)/orig. 1944 by C.J.
van Vuuren of Zingfontein, near
Philipstown, from Dorset Horn ×
Van Rooy; + German Merino
1957–8/recog. 1987; BS 1968/
[*Van* Rooy, *Dor*set]/nearly extinct

Van Rooy: (Orange Free State, S
Africa)/m/all white; hy; pd;
small fr/orig. (from 1906) at
Koppieskraal farm, Bethulie, by
J.C. van Rooy from one white
Blinkhaar Ronderib Afrikander
♂ × Rambouillet ♀♀, inbreeding
and selection, later also pd
Wensleydale ♂ and subsequently
Blackhead Persian and Blinkhaar
to upgrade/BS 1948/Afrik. *Van
Rooy-Persie*/syn. *Van Rooy White
Persian*

Varesina: (Varese, Lombardy, Italy)/
m.cw/pd/Lop-eared Alpine
group/sim. to or var. of
Bergamasca/rare (disappearing
by crossing with Bergamasca and
Biellese)

Var Merino: *see* Arles Merino

Varzese: (Varzi, Emilia, Italy)/local
var. of Apennine

Vasca Carranzana: (Basque
provinces, Spain)/d.m.cw/head
and legs shades of reddish-
yellow, fleece white; ♂ hd or pd,
♀ pd/= Basco-Béarnais
(France)/Basque *Selay ardiga* (=
meadow sheep)/syn. *Carranzana*
(from Carranza) (Basque
Karrantzar), *Vasca* (= *Basque*)/
rare

Vasojević: (Lim valley, NE
Montenegro)/now var. of Sjenica
(by grading)/Serbo-cro.
Vasojevička

Veglia: *see* Krk

Velay Black: (Haute-Loire, C France)/
m.mw/black with white spot on
forehead and white tail tip; pd;
lt/Central Plateau group/BS 1931,
HB 1970/Fr. *Noir du Velay*/ syn.
Bourbonnais, Noir de Bains

Veluwe Heath: (C and E
Netherlands)/♂ scurred or pd, ♀
pd; lt/var. of Dutch Heath/Du.
Veluwse Heideschaap/rare

Vembur: (Tirunelveli, S Tamil Nadu,
India)/m/white with red or fawn
spots; ♂ hd, ♀ pd; st/South India
Hair type/syn. *Karandhai*/not
Bembur

Venda: (Limpopo Valley, S Africa)/
black, white or reddish brown; ♂
usu. pd, ♀ pd; ft/sim. to Bapedi

Vendéen: (Vendée, W France)/m.sw/
grey to dark brown head and legs;
pd/BS 1967; BS also GB 1984

Venezuelan Criollo: (W Venezuela
mts)/sometimes coloured or
spotted face; ♂ hd or pd, ♀
pd/var. of Criollo

Verbeterde Texelse: *see* Texel

Veredeltes: *see* Improved

**Veredeltes schwarzköpfiges
Fleischwollschaf:** *see* Swiss
Brownheaded Mutton

Veredeltes weissköpfiges Gebirgsschaf: *see* Swiss White Mountain

Vermont Merino: (USA)/var. of American Merino (A type)/HB 1879/extinct (but backbreeding in progress to develop phenotypic equivalent)

Versilia: *see* Massese

Viat: *see* Vyatka

Vicanere: *see* Bikaneri

Vicatana: *see* Ripollesa

Vicentina: (Vicenza, Venetia, Italy)/ var. of Lamon with finer wool/ syn. *di Foza, Fodata*/nearly extinct

Victoria Merino: (Australia)/= Australian Merino in Victoria

Vietines kiaules: *see* Native Lithuanian

Vietines siurkšciavilnes: *see* Lithuanian Coarsewooled

Vigatana: *see* Ripollese

Villnösser: (Val d'Isarco, Bolzano, Italy)/orig. from Carinthian/ nearly extinct

Virgin Islands White: (British and US Virgin Is)/m/occ. tan or pied; pd; mane in ♂; prolific/American Hair Sheep group/? orig. from West African stock with possible influence of Wiltshire Horn/orig. of Katahdin, St Croix/syn. *White Virgin Islander*/obs. syn. *African Hair sheep*/not *Virgin Island White*

Visp, Visperschaf: *see* Valais Blacknose

Vissana: (Visso, C Apennines, Italy)/ m.d.cw/occ. black or pied; ♂ usu. pd, ♀ pd/Apennine group/ orig. (with Merino) of Sopravissana/rare

Vitoroga: *see* Stogoš

Vlaamse: *see* Flemish

Vlach: *see* Karakachan

Vlach screwhorn: *see* Stogoš

Vlakhiko: (Balkans)/mt vars of Greek Zackel/inc. Arvanitovlach, Boutsiko, Drama Native, Epirus, Grammos, Karakachan, Krapsa, Moraitiko/syn. *Vlach, Vlahiko*

Vlashka: *see* Karakachan

Vlašić: (Bosnia)/var. of Bosnian Mountain/subvar.: Dub/Serbo-cro. *Vlašićka*/syn. *Travnik* (Serbo-cro. *Travnicka*)

Vlaška: *see* Karakachan

Vlaška vitoroga: *see* Stogoš

Vogan: (S Togo)/m.d/red pied, black pied or brown-and-black; hy; ♂ usu. hd, ♀ usu. pd/orig. from West African Dwarf × Sahelian

Vojvodina Merino: (NE Serbia)/fw.m/ ♂ usu. hd, ♀ pd/orig. in 19th c. from Hungarian and other Merinos/Serbo-cro. *Vojvodanski Merino*/syn. *Domaći tip merina*; obs. syn. *Yugoslav Merino*

Volgograd: (NE Volgograd, Russia)/ m.fw/♂ usu. pd, ♀ pd/orig. (1932–1978) from Novocaucasian Merino and Précoce, × Kazakh and Astrakhan Fat-rumped, transferred in 1945 to Romashkovski state farm and mated to Caucasian and Grozny ♂♂/recog. as breed group 1963, as breed 1978/Russ. *Volgogradskaya*/ syn. *Romashkovski* (Russ. *Romashkovskaya*)

Volokolamsk: (Moscow, Russia)/ m.sw/crossbred group (1936 on) from (Hampshire Down × Tsigai) × Northern Short-tailed/extinct

Voloshian: (Slovakia) *see* Valachian

Voloshian: (Ukraine, to Urals and Caucasus, Russia)/m.cw/white or black; ♂ hd or pd, ♀ pd; lt-lft/? orig. from Zackel with ft blood/ vars: Carpathian Mountain, Pyrny, Steppe (Caucasus and Siberia), Transdon/orig. of Kuchugury/Russ. *Valakhskaya, Voloshskaya*/syn. *Vagas*(?), *Valachian, Valahian, Valakhian, Volosh, Walachian, Wallachian Zackel, Woloschian*/nearly extinct

Voronezh, Voronezhskaya: *see* Kuchugury

Voskop: *see* Brabant Foxhead

Voss: *see* Dala

Vrin: *see* Bündner Oberland

Vyatka: (Nolinsk, Kirov, and Gorodets, Gorki, Russia)/m.fw/ orig. 1936–1956 from 2 or 3 crosses of Rambouillet or Précoce on Northern Short-tailed

(Nolinsk)/inc. Improved
Gorodets (Russ. *Uluchshennaya
gorodetskaya*)/Russ. *Vyatskaya,
Vyatskaya myaso-sherstnaya* (=
mutton-wool), *Vyatskaya
tonkorunnaya* (= finewool)/not
Viat

Východofríská ovce: *see* East Friesian

Waila: *see* Balami

Wakhan Gadik: (Afghanistan)/usu.
white/var. of Gadik

Walachenschaf: (Germany)/mw/
spotted legs and face/nearly
extinct

Walachian: (Czech and Slovak
Republics) *see* Valachian

Walachian: (SE Europe) *see* Zackel

Walachian: (Ukraine and Russia) *see*
Voloshian

Walachian Screwhorn: (Serbia) *see*
Ţurcana

Walcheren Milk: *see* Zeeland Milk

Walcherse melkschaap: *see* Zeeland
Milk

Waldschaf: *see* Bavarian Forest

Wallachian: (Czech and Slovak
Republics) *see* Valachian

Wallachian: (Ukraine and Russia) *see*
Voloshian

Walliser Landschaf: *see* Roux du
Valais

Walliser Schwarznasen: *see* Valais

Walrich Mutton Merino: (S Africa)/
fw.m/orig. since 1930 by *Wal*ter
A. Higgs of *Rich*mond, Zastron,
from pd Précoce × South African
Merino/BS 1960, recog. 1965/
Afrik. *Walrich Vleis Merino*/
extinct (merged with Döhne
Merino)

Wanganella: *see* Peppin

Waralé: (Senegal)/m/hy/orig. (1975
on) at Agric. Res. Inst., Dahra,
from Touabire × Peul-Peul

Warhill: (USA)/orig. from Merino,
Columbia, Corriedale, Panama
and Rambouillet

Waridale: (NSW, Australia)/fw.m/
pd/orig. in 1970s at New
England Univ., Armidale, from
medium-wool Peppin Merino ($\frac{1}{2}$)
× Border Leicester ($\frac{1}{4}$) and Poll
Dorset ($\frac{1}{4}$)

Warton Crag: *see* Limestone

Watish: (Gezira, Sudan)/white/var. of
Sudan Desert (? with Nilotic
blood)

Waziri: (Waziristan, NW Pakistan)/
cw.m/black, brown or spotted
head, occ. also body; ♂ usu. pd,
♀ pd; lop ears; ft

Wealden Four-quarter: (Kent,
England)/d/orig. since 1971 from
Romney ♀ with 4 teats, bred to
Clun Forest ♂ and selected for 4
teats; also Friesian, Finnish,
Booroola Merino and Suffolk
blood

Weiladjo: *see* Balami

Weisses Alpenschaf: *see* Swiss White
Alpine

Weisses Bergschaf: *see* Tyrol
Mountain

Weisses Edelschaf: *see* Swiss White
Alpine

Weisses Gebirgsschaf: *see* Swiss
White Mountain

Weisse gehörnte Heidschnucke: *see*
White Horned Heath

Welo: (Ethiopia)/local var. of
Ethiopian Highland

Welsh Badger-faced: *see* Badger Face
Welsh Mountain

Welsh Bleu: (Wales)/white, blue or
mottled face/Bleu du Maine ×
Beulah Speckled Face, Lleyn,
Welsh Hill Speckled Face or
Welsh Mountain, 1st cross/BS
1990

Welsh Halfbred: (Wales)/m.mw/
pd/Border Leicester × Welsh
Mountain, 1st cross/BS 1955

Welsh Hill Speckled Face: (mid-
Wales)/black markings on nose,
eyes, ears, knees and feet; ♂ hd
or pd, ♀ pd/derivative of Welsh
Mountain (? with Kerry Hill
blood) of larger size and finer
fleece/BS 1969/not *Welsh Hill
Speckled*

Welsh Masham: (Wales)/Teeswater
(or Wensleydale) × Welsh
Mountain (or Speckled Face), 1st
cross

Welsh Mountain: (Wales)/w.m/white
or light tan face; ♂ hd, ♀ pd/orig.
from Welsh Tanface/ colour vars:

Badger Face Welsh Mountain, Balwen; local strains: Cardy, Talybont/derived breeds: Black Welsh Mountain, South Wales Mountain, Welsh Hill Speckled Face, *see also* Beulah Speckled Face; part orig. of Lleyn /BS 1905, HB 1906 (inc. Pedigree (improved) and Hill Flock sections)/syn. *Whiteface Welsh*

Welsh Mountain Badger Faced: *see* Badger Face Welsh Mountain

Welsh Mule: (Wales)/m.lw/white or mottled face; pd/Bluefaced Leicester × Welsh Mountain, Welsh Hill Speckled Face or Beulah Speckled Face, 1st cross, developed 1970s/BS 1979

Welsh Oldenbred: (England)/m/ British Oldenburg × Welsh Mountain, 1st cross/*see also* Greyface Oldenbred

Welsh Speckled Face: *see* Beulah Speckled Face, Welsh Hill Speckled Face

Welsh Tanface: (Wales)/cf. Tanface (Scotland)/orig. of Welsh Mountain/extinct

Wensleydale: (N Yorkshire, England)/ lw.m/blue face and legs; pd/ English Longwool type/orig. *c.* 1839–1860 from Leicester Longwool ram "Bluecap" × Teeswater/named 1876, BS and HB 1890; HB also Canada/syn. *Mugs, Wensleydale Blueface, Wensleydale Longwool, Yorkshire-Leicester* (Scotland)/rare

Wera: *see* Bangladeshi

West African: *see* Macina (wooled), Sahel type and West African Dwarf (hairy); syn. *Guinea*

West African Dwarf: (coast of W and C Africa)/m/white usu. with black patches or occ. black (S Senegal to Nigeria), or black, pied, tan with black belly and tricolour also common (in W of C Africa, Cameroon to Angola); ♂ usu. hd, ♀ usu. pd; hy; ♂ throat ruff and mane; trypanotolerant/ Dwarf Forest type in S, larger Savanna type in N of W Africa/ var.: Kirdi/Fr. *Mouton nain*

d'Afrique occidentale/syn. *Djallonké, Fouta Djallon, Futa Jallon, Guinean, Pagan, Southern, West African Maned*/local names: *Cameroon Dwarf, Nigerian Dwarf, et al.*

West African Long-legged: *see* Sahel type

West African Maned: *see* West African Dwarf

West Country Cheviot: (SW Scotland)/ ♂ often hd/strain of Cheviot/syn. *Lockerbie Cheviot*

West Country Down: *see* Dorset Down

Western: (USA) *see* Whiteface Western

Western, Western Horn: (England) *see* Wiltshire Horn

Western Red: (France) *see* Rouge de l'Ouest

Western Merino: (Romania) *see* Transylvanian Merino

Western Thin-tailed: (Tunisia) *see* Algerian Arab

Western Thrace: *see* Thrace

West Friesian: *see* Friesian Milk

West Indian: *see* Bahama Native, Barbados Black Belly, Virgin Islands White

West Jutland Dune: *see* Danish Landrace

West Kanem: (N and W of Mao, Chad)/cw-hy/rare

West Kazakhstan Mutton-Wool: (W Kazakhstan)/m.mw/orig. since 1948 from (Stavropol × Lincoln) × (Tsigai × Précoce × local cw), bred *inter se* since 1952/Russ. *Zapadnokazakhstanskaya myaso-sherstnaya*

West Mongolian Fat-rumped: *see* Kazakh

West Pyrenean: (France) *see* Pyrenean Dairy

West Stara Planina: (Bulgaria)/var. of Stara Planina/Bulg. *Zapadnostaroplaninska*/not *Weststaroplaninska*/rare

West Swiss White: *see* Swiss White Alpine

White Arab: (W Africa) *see* Touabire

White Bororo: *see* Balami

White Dorper: (Transvaal, S Africa)/ m/all-white; hy; usu. pd/var. of

Dorper/orig. (1946 on) at Pretoria from Dorset Horn × Blackhead Persian/BS 1960/syn. *Dorsian* (= *Dorset Persian*), *Dorsie*

Whiteface: (USA) *see* Whiteface Western

White Face Dartmoor: (Devon, England)/m.lw/occ. speckled face; ♂ hd or pd, ♀ pd/English Longwool type/orig. *c.* 1900 from Leicester Longwool × original Dartmoor/revived by BS 1950, HB 1951/syn. *Widecombe Dartmoor*, *Whiteface Dartmoor*, *Whitefaced Dartmoor*

Whitefaced Maine: *see* Maine à tête blanche

Whitefaced Woodland: (S Pennines, England)/cw/♂ hd, ♀ pd/sim. to Limestone/HB 1974, BS 1986/syn. *Penistone*, *Woodland Whiteface*, *Woodlands Horned*/not *Whiteface Woodlands*/rare

Whiteface Western: (USA)/whitefaced ewes of wool breeds from western states, usu. grade Rambouillet, or various mixtures of Rambouillet with Merino, Columbia, Panama or Corriedale/syn. *Western*, *Whiteface*

White Fulani: *see* Balami, Yankasa

Whiteheaded German: *see* German Whiteheaded Mutton

Whiteheaded Maine: *see* Maine à tête blanche

Whiteheaded Marsh: *see* Danish Whiteheaded Marsh, German Whiteheaded Mutton

Whiteheaded Mutton: *see* German Whiteheaded Mutton, Polish Whiteheaded Mutton

Whiteheaded Oldenburg: *see* German Whiteheaded Mutton

White Horned Heath: (S Oldenburg, Germany)/var. of Heidschnucke/HB 1936/Ger. *Weisse gehörnte Heidschnucke*/rare

White Hornless, White Hornless Heidschnucke: *see* White Polled Heidschnucke

White Hornless Moorland: *see* White Polled Heidschnucke

White Improved: *see* Swiss White Alpine

White Karaman: (C Anatolia, Turkey)/m.d.cw/black on nose and occ. around eyes; ♂ usu. pd, ♀ pd/sim. to Makui/vars: Kangal, Karakaś, Southern/Turk. *Ak-Karaman*/syn. (Syria) *Barazi* (*Barasi*, *Brazi*, *Brézi*, *Parasi*), *Garha* (*Agrah*, *Akrah*, *Gargha*, *Guerha*, *Karha*), *Salamali*/not *Caraman*, *Karamane*, *Karman*, *Kirmani*, *Salami*/see also Red Karaman

White Klementina: (Bulgaria)/w.d/orig. at state farm Klementina (now G. Dimitrov) near Plovdiv from White Karnobat (1910) and Romanian Tsigai (1916) with some Merino blood/Bulg. *Byala* (or *Bela*) *Klementinska*

White Kranjina: *see* Belakranjina

White Maure: *see* Touabire

White Methonian: *see* Bardoka

White Mountain: *see* German Mountain, Tyrol Mountain

White Persian: *see* Van Rooy

White Polled Heidschnucke: (Lower Saxony, Germany)/var. of Heidschnucke/Ger. *Weisse hornlöse Heidschnucke*, or *Moorschnucke* (*White Hornless Moorland*)/syn. *White Hornless*

White Slatina: *see* Belaslatina

White South Bulgarian: (Bulgaria)/name used to inc. Plovdiv-Purvomai and Stara Zagora/Bulg. *Byala yuzhnob"lgarska*

White Suffolk: (NSW, Australia)/m.w/orig. (1977 on) from Poll Dorset × Suffolk, and Border Leicester × Suffolk, F_2 selected for white face/BS/syn. *Suffolk Whiteface*, *White-faced Suffolk*

White Swiss: *see* Swiss White

White Uda: *see* Balami

White Wooled Mountain: (Orange Free State, S Africa)/m.cw/♂ usu. hd, ♀ usu. pd/orig. by A.D. Wentworth, Trompsburg, from German Mutton Merino × (Dorset Horn × Blackhead Persian) 1942 but never formally recog./? extinct

White Woolless: (Brazil) *see* Brazilian Woolless

Wicklow Cheviot: (Ireland)/strain of
North Country Cheviot by
grading of Wicklow Mountain/
BS and HB 1943/syn. *Irish
Shortwool, Wicklow*
Widecombe Dartmoor: *see* Whiteface
Dartmoor
Wielkopolska: (NW Poland)/m.w/
pd/Polish Lowland group/orig.
1948–1977 at Poznań Agric.
Univ. from Poznań × Poznań
Corriedale/HB 1977/[= Large
Polish]
wild: *see* aoudad, argali, bharal,
bighorn, Dall sheep, feral, mouflon,
red sheep, snow sheep, urial
Wildhaus: (Switzerland)/orig. (with
Oxford Down) of Grabs and (with
Württemberg Merino) of Swiss
White Mountain/syn. *East Swiss,
St Gallen*/extinct
Willamette: (Oregon, USA)/m.mw/
usu. pd/orig. (1952 on) in
Willamette valley by Oregon
State Univ. from Cheviot ×
Columbia, and Dorset Horn ×
Columbia, crossed reciprocally
and selected for weight and score
Wilstermarsch: (Holstein, Germany)/
d.m.w/pd; st/Marsh type/syn.
Wilster-Dithmarscher/extinct
Wiltiper: (Hartley-Gatooma,
Zimbabwe)/m/usu. black, also
white or brown; hy; hd or
pd/orig. (1946 on) from Wiltshire
Horn × Blackhead Persian/
[*Wiltshire Persian*]
Wiltshire Horn: (England)/m/hd;
woolless, i.e. very short shedding
fleece/orig. (+ Berkshire Knot
and × Southdown) of Hampshire
Down; orig. of New Zealand
Wiltshire/BS and HB 1923; BS
also Australia; HB also USA/obs.
syn. *Old Wiltshire Horned,
Western, Western Horn*/not
Wiltshire Horned/rare in GB
Woila: (Cameroon) *see* Balami
Woloschian: *see* Voloshian
**Woodlands Horned, Woodland
Whiteface:** *see* Whitefaced
Woodland
Wooled Afrikander: *see*
Bezuidenhout Afrikander

Wooled Persian: (S Africa)/m.cw/
usu. brown; pd; ft/var. of
Persian/orig. from Arabi
imported from Iran by Moss and
Wardrop *c.* 1915/syn. *Persian
Red, Russian Perseair, Russian
Red Wooled Persian*/extinct
Woolless: (Brazil) *see* Brazilian
Woolless
Woozie: *see* Hu
Wrzosówka: (Białystok, Poland)/
cw.m.pelt/grey with black head,
born black; ♂ hd, ♀ pd; st/Heath
type; cf. Heidschnucke/being
improved by Romanov/HB 1975,
BSd 1982/syn. *Polish Heath*/not
Wrzozowka/rare
Wu: *see* Hu
Württemberg, Württemberger: *see*
Merinolandschaf
**Württemberg Bastard, Württemberg
Improved Land:** *see*
Merinolandschaf
Württemberg Land: (Germany)/orig.
from Zaupel/orig. of
Merinolandschaf/extinct
Württemberg Merino: *see*
Merinolandschaf
Wushing, Wusih, Wuxi: *see* Hu
Wuzhumuqin: *see* Ujumqin

Xaxi Ardia: *see* Petite Manech
Xinjiang Finewool: (W Xinjiang,
China)/fw.m/orig. since 1935
from Kazakh Fat-rumped and
Mongolian ♀♀ with
Novocaucasian Merino and
Précoce ("*plickus*") ♂♂/WG
Hsin-Chiang, Russ.
Sin'tszyanskaya/syn. *Sinkiang
Fine-wool, Sinkiang Merino,
Xinjiang Merino*/not *Sintsiang*
Xisquet: *see* Pallaresa
Xizang: *see* Tibetan

Yalag: *see* Kenguri
Yaluzangbu: (Qinghai-Tibet plateau,
China)/valley var. of Tibetan/not
Yarlung Zangbo
Yambali: (Bulgaria)
Yankasa: (N and NC Nigeria)/m/ white
with black eyes and nose; small
horns or pd; short semi-lop ears;
♂ usu. maned/Fulani group of

Sahel type, ? with blood of West African dwarf/syn. *Hausa, White Fulani, U'ankasa*/[Hausa, = local]

Yarlung Zangbo: *see* Yaluzangbu

Yazdi: (Iran) *see* Baluchi

Yell Shetland: *see* Multihorned Shetland

Yemeni: (Yemen)/m/pd; often earless; ft/see (hy) Dhamari, Mareb White, Taix Red, Tihami, and (cw) Aansi, Amran Black, Amran Grey, Sana'a White and Yemen White

Yemen White: (E and NE of N Yemen)/m.cw/pd; ft

Yeroo: *see* Yoroo

Yiecheng: (S Xinjiang, China)/cw/♂ usu. hd, ♀ usu. pd; sft

Yorkshire Cross: *see* Masham

Yorkshire Halfbred: (Yorkshire Wolds, England)/Suffolk × Leicester Longwool, 1st cross/ extinct

Yorkshire-Leicester: *see* Wensleydale

Yoroo: (Tavin, Selenge prov., Mongolia)/mw.m.d/pd; lt/orig. from North Caucasus (Mutton-Wool), Kuibyshev and Romney, × local Mongolian/recog. 1981/ not *Yeroo*

Yugoslav Merino: *see* Vojvodina Merino

Yugoslav(ian) Zackel: *see* Pramenka

Yunnan Semifinewool: (Yunnan, China)/w.m/orig. since 1970 from Romney × local/not *Yungui, Yunnai*

Yuzhnokazakhskiĭ Merinos: *see* South Kazakh Merino

Yuzhnoural'skaya: *see* South Ural

Zabaĭkal'skaya: *see* Transbaikal

ZACKEL: (SE Europe)/cw.d.m/usu. white, also black, brown or pied; ♂ hd (long spiral corkscrew, screwhorn), ♀ hd or pd; usu. lt/inc. Albanian, Bulgarian, Greek, Karakachan, Moscia, Polish, Pramenka, Racka, Ruda, Šumava, Ţurcana, Valachian, Voloshian/Pol. *Cakiel*/syn. *Prong Horn, Screw Horn, Twisted Horn*/not *Zakel*/[Ger. *Zacke* = prong, referring to straight vertical horns of Racka]

Zackel-Friesian-Transylvanian: *see* Polish Mountain

Zadonskaya: *see* Transdon

Zaghawa: (NW Darfur, Sudan and E Chad)/usu. black; hy; ♂ hd, ♀ pd/= Black Maure (W Africa)

Zagoria: (Tepelene, S Albania)/ d.m.cw

Zaian: (Khenifra, Morocco)/orig. from Tadla with Berber blood

Zaïre Long-legged: *see* Congo Long-legged

Zakynthos: (Ionian Sea, Greece)/ m.d.(cw)/white, occ. black spots on head; ♂ hd or pd, ♀ pd/? orig. from Bergamasca/It. *Zante*/rare

Zandi: (Qom, Iran)/black/var. of Grey Shirazi

Zandir: *see* Çandır

Zangxi: *see* Tibetan

Zante: *see* Zakynthos

Zaobei Large Tail: (China)/extinct

Zapadnostaroplaninska: *see* West Stara Planina

Zapadnokazakhstanskaya myaso-sherstnaya: *see* West Kazakhstan Mutton-Wool

Zaupel: (S and C Germany)/cw/sim. to Steinschaf/orig. of Bavarian Forest, Cikta, German Mountain, Württemberg Land/extinct

Zeeland Milk: (Walcheren, Zeeland, Netherlands)/d/pd; rat tail; prolific/Marsh type/BS/Du. *Zeeuwse Melkschaap*/syn. *Walcherse Melkschaap* (= *Walcheren Milk sheep*)/not *Zealand*/rare

Zel: (Mazandaran, N Iran)/d.cw.m/ white, sometimes with colour on head and legs, sometimes all black or brown, or pied; ♂ hd, ♀ pd/syn. *Chiva, Iran(ian) Thin-tailed, Mazandarani, Persian Thin-tailed*/not *Mazenderani*

Żelazna: (C Poland)/Polish Lowland group/pd/orig. 1955 from Polish Merino × (Leicester Longwool × Łowicz)/Pol. *Żeleżnieńska*/ [Żelazna exp. farm of Warsaw Agric. Univ.]/rare

Zembrane: *see* Zemrane

Zemmour: (NW Morocco)/m.cw/ white with pale brown face; ♂ hd, ♀ pd/Atlantic Coast type

Zemrane: (Morocco)/var. of South Moroccan with more Berber blood than Rehamna-Sraghna/ not *Zembrane*

Zenith: (Victoria, Australia)/mw.m/ pd/orig. 1947 by L.L. Bassett of Donald from Merino ($\frac{7}{8}$) and Lincoln ($\frac{1}{8}$)/BS 1955/rare

Zenu: *see* Zunu

Zeta Yellow: (S Montenegro)/d.m.cw/ brownish-yellow head and legs; small/Pramenka type; cf Shkodra/Serbo-cro. *Zetska Žuja*

Zigai, Zigaia, Zigaja, Zigaya: *see* Tsigai

Zimbabwe: *see* Sabi

Zillertal: (Tyrol, Austria)/local var. of Steinschaf/extinct

Zirné Merino: (Czech Rep.)/m.fw/♀ pd/imported from Germany 1960/Cz. *Žírné merino* (*ZM*)/syn. *Fleisch Merino, Nemecká výkrmová merino* (Cz. *Nemecká* = German), *Merinofleischschaf* (Ger.), *Merino Mutton*/rare

Zlarin: *see* Island Pramenka

Zlatusha: (N and SW Bulgaria)/ mw/orig. 1965–1967 from German Mutton Merino and Merinolandschaf × Sofia White/Fr. *Zlatoucha*

Zošl'achtená valaška: *see* Improved Valachian

Zoulay: (Upper Moulouya valley, Morocco)/orig. from Tounsint × Berber/syn. *Berber à laine Zoulai, Mouton de montagne à laine Zoulai*

Zucca Modenese: (Emilia, Italy)/ d.m.cw/pd; roman nose/ Apennine group/extinct

Žuja: *see* Zeta Yellow

Zuławy: *see* Marsh

Zulu: (Zululand, S Africa)/var. of Nguni sim. to Swazi

Zunu: (Angola)/hy; lt/syn. *Goitred, Zenu*

Zurkana: *see* Ţurcana

Zušlechtěná valašká: *see* Improved Valachian

Zwartbles schaap: *see* Dutch Black Blaze

Glossary

Some of the more common words used in breed names in countries of origin are defined below. The list includes geographical and descriptive terms; it is far from comprehensive but will give guidance where no English translation or foreign-language dictionary is to hand. It should be remembered, of course, that in many languages word endings vary according to gender, or in plurality, and not all forms of such words are given here.

Afrikaans

angorahaar:	mohair
berg:	mountain
bok:	goat
buffel:	buffalo
dal:	valley
esel:	ass
groot:	large
haar:	hair
klein:	small
koei:	cow
kol:	spot
korthoring:	shorthorned
mannetjie:	bull
melk:	milk
noord:	north
oos(te):	east
perd:	horse
poenskop:	polled
ponie:	pony
rooi:	red
skaap:	sheep
suide:	south
vark:	pig
vee:	cattle
vleis:	meat
weste:	west

Arabic

a'lā:	upper
barqā:	hill
djebel, jabal:	mountains
hamra:	red
saghir:	small
shugur:	fawn-coloured

Bulgarian

byala:	white
cherno:	black
chernoglava:	blackheaded
chernosharena:	black spotted
cherveno:	red
dolna:	lower
govedo:	cattle
kafyava:	brown

kon:	horse
koza:	goat
k"soroga:	shorthorned
mestna:	native, local
mlechna:	milk, dairy
planina:	mountain
p"stra:	pied
sivo:	grey
sredna:	middle, central
t"nkorunna:	finewool

Burmese

myinn:	horse
nam:	river
nwar:	cattle
shweni:	red
shwewar:	yellow
taung:	mountain
wet:	pig
ye:	island

Chinese

bai:	white
bei:	northern
da:	large
daer:	big-eared
dong:	eastern
er:	ear
er:	middle
hai:	sea
hei:	black
hu:	lake
hua:	spotted, pied
huang:	yellow
jiang:	river
ling:	mountain range
mao:	hairy
Menggu:	Mongolian
nan:	southern
nei:	inner
niu:	cattle
pingyan:	plain
qi:	an administrative division
qiuling:	hills
shamo:	desert

shan:	mountain(s)
shandi:	mountain area
shang:	upper
shui niu:	water cattle (= buffalo)
tao:	hill(s)
xi:	west
xia:	lower
xian:	county
xiao:	small
yang:	sheep
Zhang:	Tibet
zhong:	middle
zhu:	pig

Czech

(Note that Slovakian terms have not been listed but are broadly similar.)

bezrohá:	polled
bílé:	white
černostrakatý:	black pied
červené:	red
červenostrakatý:	red pied
Český:	Czech
chladnokrevník:	coldblood
dlouhovlnná:	longwool
fríské:	Friesian
hnědý:	brown
hora:	mountain
klusák:	trotter
kůň:	horse
koza:	goat (pl. **kozy**)
kraj:	region
krátkosrstá:	short-haired
malé:	small
masné:	meat
Německé:	German
nížinný:	lowland
ovce:	sheep
prase:	pig
skot:	cattle
stará:	old
strakatý:	spotted, pied
Švýcarské:	Swiss
teplokrevník:	warmblood
ušlechtilé:	improved
velka:	large

východo:	east
zušlechtěná:	improved

Danish

blåh:	blue
dal:	valley
hede:	heath
hest:	horse
hvidhovedet:	whiteheaded
korthorn:	shorthorn
kvæg:	cattle
lille:	small
malkekvæg:	milk (dairy) cattle
marsk:	marsh
nordre:	northern
øster:	eastern
rødbroget:	red pied
rødt:	red
sønder:	southern
sortbroget:	black pied
traver:	trotter
varmblod:	warmblood
vester:	western
vidt:	white

Dutch

aalstreep:	eelstripe
blaarkop:	whiteheaded
blauw:	blue
bont:	pied
draver:	trotter
dubbeldoel:	dual-purpose
geit:	goat
heide:	heath
hoog:	high
melkras:	dairy type/breed
oost:	east
oud:	old
paard:	horse
rood:	red
roodbont:	red pied
schaap:	sheep
trek:	(heavy) draught
verbeterd:	improved
Vlaamse:	Flemish
vleesras:	beef type/breed

wit:	white
witrug:	whiteback
zuid:	south
zwart:	black
zwartbont:	black pied

Finnish

hevonen:	horse
karja:	cattle

German

alt-:	old
Berg:	mountain
blässiger:	white-marked
braun:	brown
braunscheck:	brown-and-white
Büffel:	buffalo
bunt:	coloured
doppelnutzung:	dual-purpose
edel:	noble (= improved, thoroughbred)
Esel:	ass
Fleckvieh:	spotted cattle
Fleisch:	meat
Fleischrind:	beef cattle
Fleischwoll:	mutton-wool
Gebirge:	mountains
gebirgsvieh:	mountain cattle
gehörnte:	horned
gelbes:	yellow
gelbscheck:	yellow pied
Gelbvieh:	yellow cattle
gemsfarbige:	chamois-coloured
grau:	grey
Halbblut:	halfblood
Heide:	heath
hoch:	high
Höhenvieh:	highland cattle
hornlöse:	hornless
hummel:	polled
Kaltblut:	coldblood
klein:	small
lichter:	pale, light
Niederungs:	lowlands

ost:	eastern
Pferd:	horse
rot:	red
rotbunt:	red pied
Rotfleckvieh:	red spotted (pied) cattle
rückenblessen, rückenscheck:	whiteback
Schaf:	sheep
schecken, scheckig:	spotted, pied
schlappohr:	lop-eared
schwarz:	black
schwarzbunt:	black pied
Schwein:	pig
stehohr:	prick-eared
strahlen:	striped
süd:	south
traber:	trotter
veredelte:	improved
Vieh:	cattle
Vollblut:	fullblood (= blood horse)
Wald:	forest
Warmblut:	warmblood
weiss:	white
weissköpfiges:	whiteheaded
Wolle:	wool
Ziege:	goat
zwerg:	dwarf

Hungarian

barna:	brown
birka:	sheep
félvér:	halfbred
hidegvérű:	coldblood
hus:	meat
juh:	sheep
kis:	small
Magyar:	Hungarian
nagy:	large
tarka:	spotted, pied
tejelő:	dairy
tincse:	curly
uj:	new

Indonesian

babi:	hog
baharu:	new
barat:	western
besar:	large
kambing:	goat
karbo:	buffalo
kecil:	small
perahan:	dairy
sapi:	cattle
sawah:	swamp
selatan:	southern
tengah:	middle
timur:	east(ern)
utara:	northern

Italian

asino:	ass
bastarda:	crossbred
bestiame:	cattle
bianca:	white
bruno:	brown
bufalo:	buffalo
camosciata:	chamois-coloured
capra:	goat
castana:	chestnut-coloured
cavallino:	pony
cavallo:	horse
cinta:	belted
fulva:	fawn
gentile:	improved
grigia:	grey
laticauda:	broadtail
migliorata:	improved
montanaro:	mountain
nera:	black
nostrale, nostrana:	local
orecchiuta:	long-eared
pecora:	sheep
pelatella:	plucked (i.e. hairless)
pezzata:	pied
porco:	pig
Pugliese:	Apulian
rossa:	red

Sarda: Sardinian
screziata: streaked, speckled
strisciata: striped
toro: bull
vacca: cow
vecchio: old
vello: fleece

Japanese

aka: red
baffarō: buffalo
buta: pig (domestic)
Chosen: Korean
hitsuji: sheep
ino: wild pig
kairyo: improved
kuro: black
mukaku: polled
roba: ass
tankatu: shorthorned
uma: horse
ushi: cattle
yagi: goat

Korean

nam: southern
puk: northern
san: mountain

Latvian

brūnū: brown
jaun: new
latvijas: Latvian
liel: large
maz: small

Lithuanian

baltnugariai: whiteback
baltųjų: white
juvdgalvių: blackheaded

lietuvos: Lithuanian
semieji: light grey
siurkščiavilnes: coarsewool
vietines: local

Mongolian

baruun: west(ern)
chuluu, nuruu: mountains
dund: middle
dzüün: eastern
Neimenggu: Inner Mongolia
omno: southern
tchicki: ear
uul: mountains

Polish

biała: white
białogłowa: whiteheaded
białogrzbietka: whitebacked
bor: forest
brązowy: brown
ciemny: dark
cienkorunny: finewool
czarno-biała: black-and-white
czarnogłowa: blackheaded
czarny: black
czerwona: red
długoucha: long-eared
długowełnista: longwool
dolny: lower
duży: large
głowa: head
góra, gory: mountain
gorka: hill
gruby: fat
kón: horse
las: forest
Małopolska: Little Poland
mały: small
mięsna: meat
mleko: milk
niebieski: blue
niemiecki: German
nizinna: lowland
nowa: new

ostroucha:	short-eared, prick-eared	lanada:	wooled
owca:	sheep	leiteiro:	dairy, milking
plenno:	prolific	malhado:	spotted, pied
północ:	north	meísta:	halved
południe:	south	mestiço:	crossbred
pstra:	spotted	mocho:	polled
róg:	horn	mouro:	Moorish, i.e. dark
rosyjski:	Russian		
stary:	old	nanico:	dwarf
świnia:	swine	negra:	black
szary:	grey	nordestino:	northeastern
szlachteny:	thoroughbred, purebred	ocidental:	western
		ovelha:	sheep
ucho:	ear	pântano:	swamp
uzlachetniona:	improved	parda:	brown, grey, fawn-coloured
wełna:	wool		
wełnisto-mięsny:	wool-mutton	pequeño:	small
wielko:	large	pintado:	pied
Wielkopolska:	Great Poland	pónei:	pony
wschód:	east	porco:	pig
zachód:	west	preto:	black, dark
zwisłoucha:	lop-eared	rabo gordo:	fat rump
		repartida:	divided
		rosilho:	roan
		serra:	mountain range
Portuguese (including Brazil)		sertanejo:	inhabitant of interior
		sertão:	interior of country, backwoods, upcountry
alemão:	German		
amarelo:	yellow		
azul:	blue	sul:	south
baio:	bay-coloured	touro:	bull
bordo:	curly	vermelho:	red
branco:	white		
burro:	ass		
búfalo:	buffalo	**Romanian**	
cabra:	goat		
camurça:	chamois	alb:	white
carne:	meat	balta:	marsh
cavalo:	horse	bălta:	white
charneco:	heath	băltată:	spotted, pied
chato:	flat (e.g. short-nosed)	brună:	brown
		cal, calul:	horse, pony
crioulo:	native	carne:	meat
deslanada:	woolless	crukan:	black
escuro:	dark	mare:	large
galega:	Galician	mic:	small
garrano:	pony	mocanița:	mountain peasant, shepherd
hipismo:	horse-racing		
jumento:	ass		

munte:	mountain	nemets:	German
negru:	black	nizinna:	lowland
nou:	new	novaya:	new
românească:	Romanian	ovtsi:	sheep
roşie:	red	pestraya:	pied
semigreu:	light draught	polugrubo-	
surǎ:	grey	sherstnaya:	semicoarsewool
trapaš:	trotter	poluton-	
vechi:	old	korunnaya:	semifinewool
		pomes':	cross
		poroda:	breed

Russian

		porodnaya	
		gruppa:	breed group
bekonnaya:	bacon-type	pri-:	near
belaya:	white	pukh:	down or
belogolovaya:	white-headed		undercoat (true
bol'shaya:	large		wool, cashmere)
buraya:	brown		
cherno:	black	ryabaya:	spotted
cherno-pestraya:	black pied	rysak,	
dlinno-		rysistaya:	trotter
sherstnaya:	longwool	salnaya:	lard
dlinno-		seraya:	grey
toshchekhvostaya:	long-thin-tailed	severna,	
dlinnoukhaya:	long-eared	severo:	north(ern)
dolina:	valley	shuba:	pelt
gibridnaya:	hybrid	skot:	cattle
gora:	mountain	smushka:	lambskin
gornaya:	of the mountain	sredne:	middle, central
grubosherstnye:	coarse-haired	temnogolovaya:	dark-headed
gruzins:	Georgian	tonkorunnaya:	finewool
karpats:	Carpathian	toshchekhvostaya:	thin-tailed
kavkazs:	Caucasian	tyazhelovoz:	heavy draught
kon':	horse	uluchshennaya:	improved
korotkokhvostaya:	short-tailed	upryazhnaya:	draught (harness,
korotkoukhaya:	short-eared		carriage)
koza:	goat		
krasnaya:	red	velik:	large
krupnaya:	large	verkhovaya:	saddle horse
kurdyuchnaya,		visloukhaya:	lop-eared
kurdyuk:	fat-rumped	vostochny:	eastern
les(o):	forest	yugo, yuzhno:	southern
malaya:	small	Za-:	Trans-
mestnaya,		zapadny:	western
mestnye:	native, local	zavodskaya	
molochnye:	dairy type	poroda:	improved breed
myasnyĭ typ:	meat type	zhirnokhvostaya:	fat-tailed
myaso-			
sherstnaya:	mutton-wool		

Scandinavian (Swedish, Norwegian)

allmoge:	peasantry
åsno:	ass
aust:	east(ern)
boskap:	cattle
buffel:	buffalo
dal:	valley
får:	sheep
fjäll:	mountain
förädlad:	improved
get, geit:	goat
gris:	pig
häst, hest:	horse
hede, het:	heath
hornet:	horned
hvitt:	white
ko:	cow
kollet, kullig:	polled
låglands:	lowland
lille:	small
ny:	new
øster, østra:	eastern
ponny:	pony
raukolle,	
rautte kollet:	red polled
rautt:	red
röd, rødt:	red
rödbrokig:	red pied
rødkolle:	red polled
skog:	woods
slette:	lowland
sør:	southern
stora:	large
svart:	black
svin:	swine
tjur:	bull
travare, traver:	trotter
väst, vest:	western
vit:	white

Serbo-Croatian

beloglave:	whitehead
bijel:	white
brdski:	mountain
crna:	black
crven:	red
gorni:	upper

goveče:	cattle
hladnokrvan:	coldblood
Hrvatski:	Croatian
konj:	horse
koza:	goat
magarac:	ass
mala:	small
meso:	meat
mlječna:	milk
mrko-smeda:	dark brown
Njemački:	German
oplemenjena:	improved
ovca:	sheep
planina:	mountains
plava:	blue
polukrvnji:	halfblood
prase:	pig
šarena:	pied, spotted
siv:	grey
smedje:	brown
stara:	old
stoka:	cattle
svinja:	swine
velika:	large
vitaroga:	screwhorn
žut:	yellow

Slovenian

belo:	white
crno-belo:	black-and-white
crnopasasta:	black-belted (i.e. saddleback)
domača:	local
govedo:	cattle
kasač:	trotter
konj:	horse
koze:	goat
križana:	crossbred
ovca:	sheep
pšenična:	wheaten-coloured
rjavo:	brown
žláhtna:	improved

Spanish

amarillo:	yellow
asno:	ass

austral:	south(ern)
bahía:	bay
berrenda:	pied
blanco:	white
búfalo:	buffalo
burdo:	coarse
burro:	ass
caballo:	horse
cabra, cabrío:	goat
campiñesa:	of the countryside
cárdena:	grey
casta:	breed
castaño:	chestnut brown
cebú:	zebu
cerdo:	pig
cervat:	fawn-coloured
chino:	hairless
churro:	rustic, coarse-wooled
cola ancha:	broad tail
con cuernos:	horned
criollo:	native
cruzo, cruzado:	crossbred
de vega:	lowland
entrefina:	semifine(wool)
entrepelada:	with hair
gallega:	Galician
ganado:	cattle
jaca:	pony
lechero:	dairy
manchada:	spotted
mejorada:	improved
mestizo:	mixed, crossbred
moreno:	dark brown
morucha:	dark, black
negro:	black
oreja:	ear
oscura:	dark
oveja:	sheep
pardo:	brown
pelón:	hairless, bald
pío:	piebald
puerco:	pig
rasa:	smooth
rojo:	red
romo:	polled
rubia:	blond

serrano:	highland
sur:	south
toro:	bull
vaquero:	cattle
vasco:	Basque

Thai

ban:	village
dong:	mountain
kao, khao:	white
kwai:	buffalo
nam:	river

Turkish

adi:	ordinary
ak:	white
Anadolu:	Anatolian
ati:	horse
bati:	west
boz:	grey
büyük:	large
çandır:	crossbred
cenubî:	south
dag:	mountain
doğu:	east
eski:	old
güney:	south
Gürcü:	Georgian
Kafkas:	Caucasian
kamakuyruk:	knifeblade tail
kara:	black
keçi:	goat
kesme:	crossbred
kıl-keçi:	hair goat
kırmızı:	red
kızıl:	red-brown
kıvırcık:	curly
küçük:	small
mor:	maroon (colour)
sarkî:	east
tiftik:	mohair
yeni:	new
yerli:	native

Vietnamese

ba, bonom:	mountain	**heo**:	pig
da:	river	**hon**:	island
deo:	hill	**nam**:	south(ern)
		phu:	mountain

Selected Bibliography of Publications on Livestock Breeds

General

Alderson, L. (ed.) (1990) *Genetic Conservation of Domestic Livestock.* CAB International, Wallingford, UK (for the Rare Breeds Survival Trust), 242 pp.

Alderson, L. and Bodó, I. (eds) (1992) *Genetic Conservation of Domestic Livestock*, Vol. 2. CAB International, Wallingford, UK, 304 pp.

Henson, E.L. (1992) *In situ Conservation of Livestock and Poultry.* FAO Animal Production and Health Paper No. 99. FAO, Rome.

International Committee for World Congresses on Genetics Applied to Livestock Production (1994) *Proceedings 5th World Congress, Genetics Applied to Livestock Production*, University of Guelph, Guelph, Ontario, Canada, 7–12 August 1994

International Committee for World Congresses on Genetics Applied to Livestock Production (1998) *Proceedings 6th World Congress, Genetics Applied to Livestock Production*, Armidale, Australia, January 11–16 1998.

Kukovics, S. (ed.) (1998) *Sheep and Goat Production in Central and Eastern European Countries. Proceedings of the Workshop, Budapest, Hungary, 29 November–2 December 1997.* REU Technical Series (1998) No. 50, Food and Agriculture Organization (FAO), Rome, vii + 357 pp.

Maijala, K. (ed.) (1991) *Genetic Resources of Pig, Sheep and Goat.* Elsevier Science Publishers, Amsterdam, xvii + 556 pp.

National Research Council (BOSTID) (1991) *Microlivestock: Little-known Small Animals with a Promising Economic Future.* National Academy Press, Washington, DC.

Scherf, B.D. (ed.) (2000) *World Watch List for Domestic Animal Diversity*, 3rd edn. FAO, Rome, 726 pp.

Cattle

Cai Li and Wiener, G. (1995) *The Yak.* FAO, Bangkok, Thailand.

Felius, M. (1995) *Cattle Breeds: an Encyclopedia.* Elsevier Business Information, Doetinchem, The Netherlands, 799 pp.

Hickman, C.G. (ed.) (1991) *Cattle Genetic Resources*. Elsevier Science
    Publishers, Amsterdam, xiv + 313 pp.
Maule, J.P. (1990) *The Cattle of the Tropics*. Centre for Tropical Veterinary
    Medicine, Edinburgh, 200 pp.
Millar, P., Lauvergne, J.J. and Dolling, C. (eds) (2000) *Mendelian Inheritance in
    Cattle 2000*. EAAP Publication No. 101. Wageningen Pers, Wageningen,
    The Netherlands, 590 pp.
Porter, V. (1991) *Cattle – a Handbook to the Breeds of the World*. Christopher
    Helm (A. and C. Black), London, 400 pp.
Payne, W.J.A. and Hodges, J. (eds) (1997) *Tropical Cattle – Origins, Breeds and
    Breeding Policies*. Blackwell Science, Oxford.

Goat

Gall, C. (1996) *Goat Breeds of the World*. CTA, Wageningen, 186 pp.
Porter, V. (1996) *Goats of the World*. Farming Press, Ipswich, UK, 174 pp.

Horse

Hendricks, B. (1995) *International Encyclopedia of Horse Breeds*. Oklahoma
    University Press, Norman, Oklahoma, 486 pp.

Pig

Porter, V. (1993) *Pigs – a Handbook to the Breeds of the World*. Helm
    Information, Mountfield, East Sussex, UK, 256 pp.

Sheep

Enzlin, G.J. (ed.) (1995) *Colour and Island Sheep of the World*. Ram Press,
    Haarlem, The Netherlands.
Fahmy, M.H. (ed.) (1996) *Prolific Sheep*. CAB International, Wallingford, UK, 542 pp.

Africa

FAO (1999) L'état des ressources génétiques dans l'espèce bovine dans
    l'Afrique subsaharienne. *Bulletin d'Information sur les ressources
    génétiques animaux* 25.
Planchenault, D. and Boutonnet, J.P. (1997) Conservation de la diversité des
    ressources génétiques animales dans les pays d'Afrique francophone sub-
    saharienne. *Animal Genetic Resources Information* 21, 1–22.
Rege, J.E.O. (1999) The state of African cattle genetic resources. I.
    Classification framework and identification of threatened and extinct
    breeds. *Animal Genetic Resources Information* 25, 1–25.
Rege, J.E.O. and Tawah, C.L. (1999) The state of African cattle genetic resources.

II. Geographical distribution, characteristics and uses of present-day breeds and strains. *Animal Genetic Resources Information* 26, 1–25.

Rege, J.E.O., Aboagye, G.S. and Tawah, C.L. (1994) Shorthorn cattle of West and Central Africa. *World Animal Review* 78, 2–48.

Wilson, R.T. (1991) *Small Ruminant Production and the Small Ruminant Genetic Resource in Tropical Africa.* FAO Animal Production and Health Paper No. 88, 231 pp.

Algeria

Khemici, E., Mamou, M., Lounis, A., Bounihi, D., Ouachem, D., Merad, T. and Boukhetala, K. (1996) Etude des ressources génétiques caprines de l'Algérie du Nord a l'aide des indices de primarité. *Animal Genetic Resources Information* 17, 69–80

Botswana

Senyatso, E.K. and Masilo, B.S. (1996) Animal genetic resources in Botswana. *Animal Genetic Resources Information* 17, 57–68.

Burkina Faso

Nianogo, A.J., Saufo, R., Kondombo, S.C. and Neya, S.B. (1996) Le point sur les ressources génétiques en matière d'élevage au Burkina Faso. *Animal Genetic Resources Information* 17, 13–31.

Cameroon

Messine, O., Tanya, V.N., Mbah, D.A. and Tawah, C.L. (1995) Ressources génétiques animales du Cameroun. Passé, présent et avenir: le cas des ruminants. *Animal Genetic Resources Information* 16, 51–69.

Seignobos, C. and Thys, E. (eds) (1998) [Bos taurus *cattle and man: Cameroon, Nigeria.*] Édition de l'ORSTOM, Paris, 399 pp. (In French.)

Côte d'Ivoire

Yapi-Gnaore, C.V., Oya, B.A. and Ouattara Zana (1996) Revue de la situation des races d'animaux domestiques de Côte d'Ivoire. *Animal Genetic Resources Information* 19, 99–118.

Ethiopia

FARM-Africa (1996) *Goat Types of Ethiopia and Eritrea. Physical Description and*

Management Systems. FARM-Africa, London, UK, and ILRI (International Livestock Research Institute), Nairobi, Kenya. 76 pp. inc. 23 colour plates.

Morocco

Boujenane, I. (1999) *Les Ressources Génétiqus Ovines au Maroc.* Actes Editions, Rabat, Morocco.

South Africa

Campbell, Q. (1995) *The Indigenous Sheep and Goat Breeds of South Africa.* Quentin Peter Campbell, Bloemfontein, South Africa.
Camper, J.P., Hunlun, C. and van Zyl, G.J. (1998) *South African Livestock Breeding.* South African Stud Book and Livestock Improvement Association, Bloemfontein, South Africa.
Ramsay, K., Harris, E. and Kotzé, A. (2000) *Landrace Breeds: South Africa's Indigenous and Locally Developed Farm Animals.* Farm Animal Conservation Trust (FACT), Pretoria, South Africa.

Sudan

Mufarrih, M.E. (1991) Sudan Desert sheep: their origin, ecology and production potential. *World Animal Review* 66, 23–31

Togo

Hadzi, Y.N. (1996) Les populations de betail presentes au Togo. *Animal Genetic Resources Information* 17, 39–56.

America

Brazil

Associaça de Criadores de Ovinos e Caprinos do Estada do Ceará (ACOCECE) (undated) *O Berro, No. 11: Edição comemorativa da IV Conferéncia internacional sobre caprinos.* EMBRAPA, Brazil.
EMBRAPA (1990) *Suinos Nacionalis.* Centro Nacional de Pesquisa de Recursos Genéticos e Biotecnologia, Brasilia, 22 pp.
Machado, T.M. (1995) Le peuplement des animaux de ferme et l'élevage de la chèvre au Brésil avec une étude du polymorphisme visible de la chèvre du Ceará. DSc Thesis, Université de Paris-sud, Paris, 257 pp. + tables 112 pp.
Mariante, A. da S. (2001) Conservação de recursos genéticos animals no Brasil. [Conservation of animal genetic resources in Brazil.] *Ação Ambiental*, Vol. III, No. 15 (Dec/Jan 2001). (Journal published by Federal University of Viçosa, Minas Gerais, Brazil.) (In Portuguese, with English summary.)

Mariante, A. da S. and de Bem, A.R. (1993) Animal genetics resources conservation programme in Brazil. *Animal Genetic Resources Information* 10, 9–32

Mariante, A. da S., Albuquerque, M. do S.M., Egito, A.A. do and McManus, C. (1999) Advances in Brazilian animal genetic resources conservation programme. *Animal Genetic Resources Information* 25, 109–123.

USA

Bixby, D.E., Christman, C.J., Erhman, C.J. and Sponenberg, D.P. (1994) *Taking Stock: The North American Livestock Census*. McDonald and Woodward Publishing Co., Blacksburg, Virginia, 182 pp.

Christman, C.J., Sponenberg, D.P. and Bixby, D.E. (1997) *A Rare Breeds Album of American Livestock*. American Livestock Breeds Conservancy, Pittsboro, North Carolina.

Mayer, J.J. and Brisbin, I.L. Jr (1991) *Wild Pigs of the United States: Their History, Morphology, and Current Status*. University of Georgia Press, Athens, Georgia.

Vorwald Dohner, J. (2001) *The Encyclopedia of Historic and Endangered Livestock and Poultry Breeds*. Yale University Press, New Haven and London, 496 pp.

Asia

Devendra, C. (ed.) (1989) *Goat Meat Production in Asia*. Proceedings of Workshop, Tando Jam, Pakistan, 13–18 March 1988. International Development Research Centre, Ottawa.

Dmitriev, N.G. and Ernst, L.K. (eds) (1989) *Animal Genetic Resources of the USSR*. FAO Animal Production and Health Paper No. 65. FAO, Rome.

China

Chen Youchun (ed.) (1990) *Characteristics of Chinese Yellow Cattle Ecospecies and their Course of Utilization*. Agricultural Publishing House, Beijing, 278 pp. (Partly in Chinese, partly in English.)

Editorial Committee (1992) [*Swine Breeding in China*]. Sichuan Publishing House of Science and Technology, Chengdu, 279 pp. + 35 plates. (In Chinese.)

Wu Ming-Che, Chang Hsiu-Luan and Tai Chein (eds) (1992) *Genetic Resources Information of Native Livestock and Poultry Breeds in Taiwan*. Taiwan Livestock Research Institute, Hsinhua, Tainan, 134 pp.

You-Chun Chen and Tiequan Wang (1996) Mini-horses in China. *Animal Genetic Resources Information* 18, 25–29.

Japan

Obata, T., Takeda, H. and Oishi, T. (1994) Japanese native livestock breeds. *Animal Genetic Resources Information* 13, 13–24.

Mongolia

Batsukh, G. and Zagdsuren, E. (1991) Sheep breeds of Mongolia. *World Animal Review* 68, 11–25

Nepal

Wilson, R.T. (1997) Animal genetic resources and domestic animal diversity in Nepal. *Biodiversity and Conservation* 6, 233–251.

Vietnam

Molenat, M. and Tran The Thong (1990) Le production porcine au Viet Nam et son amélioration. *World Animal Review* 68, 26–37.

Yemen

Hasnain, H.U., Al Hokhief, A.A. and Al Iryani, A.R.F. (1991) Goats in Yemen. *Animal Genetic Resources Information* 8, 38–45.
Hasnain, H.U., Al Hokhief, A.A. and Al Iryani, A.R.F. (1994) Sheep and cattle in Yemen. *Animal Genetic Resources Information* 13, 65–71.

Europe

Árnason, Th., Jensen, P., Klemetsdal, G., Ojala, M. and Philipsson, J. (1994) Experience from application of animal breeding theory in Nordic horse breeding. *Livestock Production Science* 40, 9–19.
Dmitriev, N.G. and Ernst, L.K. (eds) (1989) *Animal Genetic Resources of the USSR.* FAO Animal Production and Health Paper No. 65. FAO, Rome.
Flamant, J.C., Boyazoglu, J. and Nardone, A. (eds) (1996) *Cattle in the Mediterranean Area.* Proceedings special session Cattle Commission, Madrid, 1992. EAAP Publication No. 86. Wageningen Pers, Wageningen, 123 pp.
Frahm, K. (1990) *Rinderrassen in der Ländern der Europäischen Gemeinschaft*, 2nd edn. Ferdinand Enke Verlag, Stuttgart, 241 pp.
Matassino, D., Boyazoglu, J. and Cappuccio, A. (eds) (1997) *International Symposium on Mediterranean Animal Germplasm and Future Human Challenges. A Joint EAAP/FAO/CIHEAM Symposium*, Benevento, Italy, November 1995. Wageningen Pers, Wageningen.
Ollivier, L., Labroue, F., Glodek, P., Gandini, G. and Delgado, J.V. (2001) *Pig Genetic Resources in Europe.* EAAP Publication No. 104. Wageningen Pers, Wageningen, The Netherlands, 150 pp.
Pro Specie Rara (1995) *Landwirtschaftliche Genressourcen der Alpen.* Fondation Pro Specie Rara, St Gall, Switzerland, 544 pp. (In German, French, Italian and Slovenian.)
Rieux-Laucat, B. and Rossier, E. (1994) *Les Races d'Équidés Ménacées dans la C.E.E.* Centre d'Etude et de Recherche sur l'Economie et l'Organisation

des Productions Animales, Direction Générale de l'Environnnement, Commission des Communautés Européennes, 30 pp. + 5 annexes. Annexe 6: Fiches descriptives des races présentes dans la C.E.E., 240 pp.

Sambraus, H.H. (1992) *A Colour Atlas of Livestock Breeds*. Wolfe Publishing, London (English translation), 272 pp.

Sambraus, H.H. (1994) *Gefährdete Nutztierrassen. Ihre Zuchtgeschichte, Nutzung un Bewahrung*. Verlag Eugen Ulmer, Stuttgart, 384 pp.

Simon, D.L. and Buchenauer, D. (1993) *Genetic Diversity of European Livestock Breeds*. EAAP Publication No. 66, Wageningen Pers, Wageningen, 581 pp.

Belgium

Anon. (1992) Themanummer schapen. Rasbeschrijvingen. *De Ark* Oct.–Dec. 1992, 4–28.

Croatia

Wilson, R.T. (1995) Livestock production and animal genetic resources in Croatia. *Animal Genetic Resources Information* 16, 41–50.

Czech Republic and Slovakia

Spacek, F. (1987) *Atlas Plemen Hospodářských Zvířat*. Státní Zemědělské Nakladatelstvı, Prague, 258 pp.

France

Anon. (1990) The French pig breeds and their performance. *Bulletin d'Elevage Français* 23, 13–20.

Audiot, A. (1995) *Races d'Hier pour l'Élevage de Demain*. Institut National de la Recherche Agronomique, Paris, 229 pp.

Durand-Tardif, M. and Planchenault, D. (2000) *Base de donées nationale France: situation des ressources génétiques. Bovins–ovins–caprins–porcins, 2000*. Bureau des Ressources Génétiques, Paris.

Institut de l'Elevage (1995) *Cryoconservation. Espèces bovine, caprine et ovine. Races à petits et trés petits effectifs. (Etat des stocks de semence) 1995*. Compte-Rendu No. 2397, Département Génétique et Contrôle des Performances, Institut de l'Elevage, Paris, 34 pp. + 8 tables.

Pujol, R. (ed.) (1995) *L'Ane*. Ethnozootechnie No. 56, Sociéte d'Ethnozootechnie, Clermont-Ferrand, France, 140 pp.

Pujol, R., Blanc, H. and Lizet, B. (eds) (1999) *Poneys*. Ethnozootechnie No. 64, Sociéte d'Ethnozootechnie, Paris, France, 140 pp.

Roque, M., Soison, P. and Adnet-Goffinet, L. *Vaches de Montagne – Montagnes à Vaches*. Editions Quelqe part sur terre, Montsalvy, France

Sopexa (2000) *Livestock from France*. Special issue, *Bulletin de l'Élevage français*. Société pour l'Expansion des Ventes des Produits Agricoles et Alimentaries, Paris. (In English; also available in French or Spanish.)

Germany

Guintard, C. and Denis, B. (1996) Pour un standard de l'Aurochs de Heck. _Varia/Ethnozootechnie_ 57, 25–30.

Italy

Gandini, G. and Rognoni, G. (eds) (1997) _Atlante Etnografico delle Popolazioni Equine e Asinine Italiane per la Salvaguardia delle Risorse Genetiche._ [_An Ethnographic Atlas of Italian Horse and Ass Populations for the Conservation of Genetic Resources._] CittàStudiEdizioni di UTET Libreria srl, Milan. (In Italian.)
Rubino, R. (1993) Goat breeds in Italy. _Animal Genetic Resources Information_ 11, 47–62.

Spain

Delgado, J.V., Capote, J.F., Fresno, M. del R. and Camacho, M.E. (1990) _Agrocanarias 90. Exposición de Animales Domésticos Autóctonos Canarios._ Consejería de Agricultura y Pesca, Gobierno de Canarias, Tenerife, 16 pp.
Delgado, J.V., Rodero, E., Camacho, M.E. and Rodero, A. (1992) _Razas Autóctonas Andaluzas en Peligro de Extinción._ Consejería de Agricultura y Pesca, Junta de Andalucía, 40 pp.

Sweden

Hallander, H. (1989) _Svenska Lantraser, deras Betydelser förr och nu._ Bockförlaget Blå Ankan AB, Veberöd.

United Kingdom

CAB International (2002) _Animal Health and Production Compendium_, 2002 edn. CAB International, Wallingford, UK.
Henson, E. (1994) _British Sheep Breeds._ Shire Publications, Princes Risborough, UK.
Porter, V. (2002) _British Cattle._ Shire Publications, Princes Risborough, UK.
Porter, V. (1999) _British Pigs._ Shire Publications, Princes Risborough, UK.

Databases

DAD-IS of FAO: http://www.fao-org/dad-is/
EAAP Animal Genetic Data Bank: http://www.tiho-hannover.de/